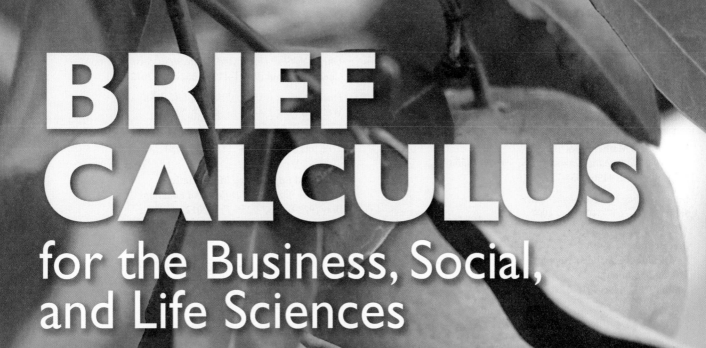

BRIEF
CALCULUS
for the Business, Social, and Life Sciences

The Jones & Bartlett Learning Series in Mathematics

The Jones & Bartlett Learning International Series in Mathematics

THIRD EDITION

BRIEF CALCULUS

for the Business, Social, and Life Sciences

Bill Armstrong
Lakeland Community College

Don Davis
Lakeland Community College

JONES & BARTLETT
LEARNING

World Headquarters
Jones & Bartlett Learning
5 Wall Street
Burlington, MA 01803
978-443-5000
info@jblearning.com
www.jblearning.com

Jones & Bartlett Learning books and products are available through most bookstores and online booksellers. To contact Jones & Bartlett Learning directly, call 800-832-0034, fax 978-443-8000, or visit our website, www.jblearning.com.

Production Credits

Chief Executive Officer: Ty Field
President: James Homer
SVP, Editor-in-Chief: Michael Johnson
SVP, Chief Marketing Officer: Alison M. Pendergast
Executive Publisher: Kevin Sullivan
Senior Developmental Editor: Amy Bloom
Director of Production: Amy Rose
Production Editor: Tiffany Sliter

Production Assistant: Eileen Worthley
Senior Marketing Manager: Andrea DeFronzo
V.P., Manufacturing and Inventory Control: Therese Connell
Composition: Aptara®, Inc.
Cover & Title Page Design: Scott Moden
Director of Photo Research and Permissions: Amy Wrynn
Cover & Title Page Image: © inacio pires/ShutterStock, Inc.
Printing and Binding: Courier Companies
Cover Printing: Courier Companies

To order this product use ISBN: 978-1-4496-9516-3

Library of Congress Cataloging-in-Publication Data
Armstrong, Bill (William A.)
    [Brief calculus]
    Brief calculus for the business, social, and life sciences / Bill Armstrong and Don Davis.—Third edition.
                pages cm
    Revised edition of: Brief calculus: solving problems in business, economics, and the social and behavioral sciences
    Includes index.
    ISBN 978-0-7637-9328-9 (casebound)—ISBN 0-7637-9328-0 (casebound) 1.   Calculus.   I. Davis, Don, 1958–   II. Title.
    QA303.2.A75 2014
    515—dc23
2012029910

6048

Printed in the United States of America
17   16   15   14   13   10 9 8 7 6 5 4 3 2

Dedication

I dedicate this book to my father, Thad Davis. His humility, his thoughtfulness, and his insight have served to be a lifelong style guide.

—Don Davis

I dedicate this book to my wife, Lisa, and sons, Austin and Dylan. Your love, support, and understanding in everything I do are amazing.

—Bill Armstrong

Contents

Preface

As with other brief or applied calculus textbooks, *Brief Calculus for the Business, Social, and Life Sciences, Third Edition* may be used in either a one- or two-term course for students majoring in economics, business, or social or behavioral sciences. The topics are organized for maximum flexibility so that the text may be adapted to any college or university's curriculum. We have crafted this book around five key principles designed to address students' needs:

1. Present the mathematics in a language that students can read and understand
2. Teach good problem-solving techniques and provide ample practice
3. Use real data applications to keep the text interesting
4. Provide timely reinforcement of algebra and other essential skills
5. Let instructors decide whether to incorporate technology in the classroom

1. Present the Mathematics in Language that Students Can Read and Understand

In the third edition, we continue to use a conversational, easy-to-read style that evokes the one-on-one communication of a tutorial session. When students find that they can understand the clear presentation and follow the interesting, real-world examples, we believe that they will get into the habit of reading the text. Although we have written a text that is accessible, we have been careful not to sacrifice the proper depth of coverage and necessary rigor required of applied calculus. We are confident that both objectives have been met.

2. Teach Robust Problem-Solving Techniques and Provide Ample Practice

Problem-Solving Method

New to the third edition, our clearly developed problem-solving method is the single most distinctive and user-friendly feature of the book, and it has been enhanced through the structure of highlighted examples. Frequently, applied mathematics instructors hear students comment, "I don't even know how to begin the problem." Or, "If the problem was just set up for me, I could solve it." Because skills such as setting up problems and writing solutions in their proper contexts can be a major challenge for brief and applied calculus students, we have integrated and highlighted a *problem-solving method* throughout the text. We use the phrases *Understand the Situation*, *Perform the Mathematics*, *Interpret the Results*, and *Try It Yourself* to identify four critical steps in problem solving.

- When solving an application example, we first need to *Understand the Situation*. In these clearly labeled paragraphs we state the quantity we are trying to determine and identify the method used to determine the solution. Frequently, we reveal the thought processes necessary to attack the problem.
- After we understand the situation, we *Perform the Mathematics*. This is the typical solution step in which acquired mathematical skills are used to determine a numerical result.

- After determining a numerical result, we *Interpret the Result*. In the interpretation step, we write a concluding sentence that conveys the meaning of the numerical answer within the context of the application.
- Finally, to reinforce the techniques mentioned above, we offer students a *Try It Yourself* option. Here, students are directed to related odd-numbered exercises at the end of the section in order to reinforce the skills gained through solving the example.

Exercises

The comprehensive exercise sets continue to be the heart of our text. *New to the third edition* is the expanded structure of the exercise sets, including two new components: Vocabulary Exercises and Concept and Writing Exercises. We feel that mathematics is a language and Vocabulary Exercises reinforce that notion. We also believe that there is more to collegiate-level mathematics than simply giving a numerical answer, and therefore the Concept and Writing Exercises give the student an opportunity to explore concepts in theoretical form. A typical exercise set contains Vocabulary Exercises, numerous skill exercises with varying levels of difficulty, a generous selection of Application Exercises that, new to the third edition, are titled by subject and discipline, some Concept and Writing Exercises, and a Section Project. Prior market research indicated that many applied calculus textbooks fail to provide a sufficient number of exercises for the student to be able to grasp the course content. With greater than 3500 exercises, we are confident that *Brief Calculus for the Business, Social, and Life Sciences, Third Edition* has more than enough exercises to meet students' needs.

3. Use Real Data Applications to Keep It Interesting

Students in this course tend to be very pragmatic; they want to know why they must learn the mathematical content in this course. Including many *real data-modeling* applications in examples and exercises helps to answer their unstated question and provides motivation, interest, and currency. Many of the models, parameters, and scenarios in these examples and exercises are based on data gathered from the U.S. Statistical Abstract, the Census Bureau, and other reliable sources, which are always clearly cited. In the third edition, we have updated hundreds of mathematical models based on the published results of the 2010 census and other current sources. For example, in the integral calculus chapters, we include real data-modeling applications that examine the rate of change in the unemployment rate of selected European countries from 2000–2010 (*Source*: Google Public Data). Other examples of real data-modeling applications include

- The rate of change in total paper and paperboard waste generated in the United States (*Source*: U.S. Environmental Protection Agency).
- The rate of change of the amount of carbon dioxide emissions by the United States (*Source*: U.S. Census Bureau).
- The rate of change in the total number of inmates in the U.S. Federal Bureau of Prisons (*Source*: U.S. Federal Bureau of Prisons).

The quantity, quality, and variation of these types of applications are simply not found in most applied calculus textbooks. We believe that real data-modeling applications not only keep the course content relevant and fresh, but compel students to interpret the numerical solution within the context of the problem they have solved.

4. Provide Timely Reinforcement of Algebra and Other Essential Skills

One of the major challenges faced by students—and frustrations encountered by instructors—is weak preparation in algebra and other essential skills. Even students who have proficient algebra skills are often rusty and unsure of which algebraic tool to apply. Further, the content demands of an applied calculus course do not allow for extensive time to be spent on review. In an effort to address this pervasive problem, we developed the From Your Toolbox feature.

From Your Toolbox

When appropriate, the From Your Toolbox feature directs students to read background material in the Algebra Review or other appendices. In addition, this feature is used to review previously introduced definitions, theorems, or properties as needed. By providing a brief review when it is needed, students stay on task and do not need to flip back to hunt through previous sections for key information. Each Toolbox is titled, clearly stating what timely skill or concept is being reviewed.

5. Let Instructors Decide When and Whether to Incorporate Technology

As graduates and instructors of The Ohio State University, we began using graphing-calculator technology in the classroom long before it was fashionable to do so. Based on our years of experience in this area, we have seen the strengths of using a graphing calculator and the drawbacks as well. Our philosophy is to let the instructor, rather than the text, determine how much or how little graphing calculators, spreadsheets, or other desktop or handheld applications are used in the classroom. Consequently, we developed the Technology Option in our text, which allows each instructor to decide whether or not to use technology in the curriculum.

 Technology Option

These optional, highlighted sections are easy to find, or to skip, and typically follow selected examples. The content of these parts mirrors the traditional presentation but shows how the answer to a particular example may be found using a graphing calculator. Although keystroke commands are not given, we provide answers to general questions students may have. All screenshots included in the text are from the Texas Instruments TI-84 calculator.

 Exercises that Assume Technology

Exercises that assume the use of a graphing calculator are clearly marked so that they can be assigned or skipped as desired by the instructor. We recognize that the graphing calculator is simply a tool to be used in the understanding of mathematics. We have been very careful to introduce the technology only where it is appropriate and not to let its use overshadow the mathematics.

Content Features and Highlights

Rate-of-Change Theme

Because we believe that it is important for students to understand that calculus is the study of rates of change, we have highlighted this theme throughout the text. Beginning in Chapter 1, the basic algebraic functions are reviewed in a concise and comprehensive manner. As each type of function is introduced, the average rate of change of the function on a closed interval is presented in examples and exercises. From the onset, appropriate units and an emphasis on interpreting, rather than merely finding a numerical answer, are stressed. For example, the *Interpret the Result* to Example 5 in Section 1.3 reads,

> This means that during the period from 2002 to 2007, U.S. production of fuel ethanol and biodiesel increased at an average rate of 24.34 million barrels per year.

This example illustrates for the student how using the difference quotient determined the average rate of change of a nonlinear function on a closed interval. We have found that by introducing the difference quotient in this manner early and often in Chapter 1, we create a smooth transition to computing the instantaneous rate of change and the derivative in Chapter 2.

The remaining Chapters continue to highlight the rate-of-change theme along with the importance of proper units and a sound interpretation of the solution. If a student who uses our text is asked. "What is calculus?" we are confident that the student will proclaim, "It is the study of rates of change!"

Use of the Differential

To supplement the rate-of-change theme, we have paid particular attention to the use of the differential in applied calculus topics. The differential is introduced in Chapter 3, right after the basic derivative rules. The differential is then used as a mathematical tool to introduce new topics in later sections.

- In Section 3.5 marginal analysis is introduced by means of the differential, and then we show that the marginal business functions are a type of differential.
- In Chapter 5 the differential is used to motivate the formula for elasticity of demand.
- In Chapter 6 we revisit the differential when we introduce integration by u-substitution.

In contrast to many applied calculus texts, which teach the differential as an isolated topic, we present it as a powerful tool that is useful in both differential and integral calculus.

The Definite Integral as a Continuous Sum

We have found that many of the applications of the definite integral in applied calculus textbooks tend to be contrived, esoteric, and difficult to interpret. To address this shortcoming, we have focused on the use of the definite integral to compute a total accumulation. In Chapter 6, students learn that integrating a rate function (that is, a derivative), on an interval gives the total accumulation of the dependent variable values on that interval. For example, the total difference in cost $C(b) - C(a)$ can be determined by integrating the marginal cost function $MC(x)$ over the interval $[a, b]$. This central idea is used throughout the integral calculus unit, resulting in applications that are richer and more interesting than those found in most texts.

Chapter Features

Chapter Openers

The first page of each chapter lists the sections included in that chapter along with a photo and representative graphs or figures that foreshadow the fundamental ideas presented in the chapter within the context of an application. What We Know reiterates what information has been learned in previous chapters, and Where Do We Go explains what topics will be covered. The chapter opener creates a roadmap to guide the student through the book and underscores the connections between topics.

Section Objectives

New to the third edition, each section begins with a numbered list of Objectives that are presented in that section. Within the section text, the objective name and number appears with its related topic. This unique feature is another distinction between our text and other applied calculus texts. This feature was developed with the student in mind; it provides the student with a clear understanding of what should be learned in the section.

Flashbacks

Selected examples used earlier in the textbook are revisited in the Flashback feature. The Flashback carries over an applied example from a previous section and then extends the content of the example by considering new questions. In this manner, new topics are introduced in a more natural way within a familiar context. Moreover, the Flashback often reviews necessary skills and concepts from previous chapters. We believe that this pedagogical technique of using applications previously discussed allows students to concentrate on new topics that use familiar applications.

Try It Yourself

At the end of each example employing the problem-solving method, as well as at other strategic points in each section, an example is followed by a Try It Yourself option. Each Try It Yourself directs the student to complete some odd-numbered exercises in that section's exercise set. Guiding the student to an exercise that parallels the example problem helps to reinforce the topic at hand and ensures that the recently introduced skill or concept is practiced immediately. This pedagogical tool promotes interaction between the text and the student and helps students to develop good study habits, Students who use the Try It Yourself feature will quickly learn to take ownership of the course material.

Notes

Notes appear after many definitions, theorems, and properties. These are used to clarify a mathematical idea verbally and to provide students with additional insight into the material. Many times these notes echo what a professor might state in the classroom to help the students understand the definition, theorem, or property.

Section Projects

At the end of each exercise set, a Section Project presents a series of questions that ask students to explore the idea that is presented. Some of the projects are based on real data, others give a step-by-step procedure that can be used to solve classic problems in applied calculus. Instructors can use Section Projects as standard hand-in assignments, collaborative activities, or for demos in class.

Section Summary

At the end of each section, a short Summary highlights the important concepts of the section. Section summaries are a convenient resource that students may use while completing exercises or as a springboard for test review.

Chapter Review Exercises

An extensive set of Chapter Review Exercises is designed to augment the exercises in each section of the chapter. These exercises are presented by section. By identifying the section to which the review exercise aligns, students will not have to guess which section to look to for more practice on the given topic.

Additional Resources

Instructor Resources

The following online resources are provided for instructors. To request access, please contact your sales representative or visit www.jblearning.com.

> *Complete Solutions Manual*—Written by Mike McCraith and checked for accuracy by the authors, this resource contains complete solutions to all of the exercises and section projects.
> *Test Bank*
> *PowerPoint Lecture Outlines*

Student Resources

Student Solution Manual—Written by Mike McCraith and checked for accuracy by the authors, this electronic resource contains solutions to the odd-numbered section and review exercises from the text.

Companion Website—Access to the student companion website, available at go.jblearning.com /BriefCalculus, is included with each new copy of the text. The site provides the resources to enhance student learning including practice quizzes, an interactive glossary, crossword puzzles, and interactive flashcards.

WebAssign

WebAssign, the leading provider of powerful online instructional tools for faculty and students, allows instructors to create assignments online and electronically transmit them to their classes. Students provide their answers online and WebAssign automatically grades the assignment, giving students instant feedback on their performance.

Much more than just a homework grading system, WebAssign delivers secure online testing, customizable precoded questions extracted from numerous exercises in this textbook, and unparalleled customer service.

Instructors who adopt this program for use in their classrooms will have access to a digital version of this textbook. Students who purchase the access code for the WebAssign program set up by the instructor will also have access to the digital version of the text.

With WebAssign, instructors can:

- Create and distribute algorithmic assignments using questions specific to this textbook,
- Grade, record, and analyze student responses and performance instantly,
- Offer more practice exercises, quizzes, and homework, and
- Upload resources to share and communicate with students seamlessly.

For more detailed information and to sign up for free faculty access, please visit www.webassign.net. For information on how students can purchase access to WebAssign bundled with this textbook, please contact your Jones & Bartlett Learning account representative.

Acknowledgments

We owe a debt of gratitude to many individuals who helped us shape and refine this and earlier editions of *Brief Calculus*. Reviewers have included Martin Bonsangue, California State University at Fullerton; Fred Bakenhus, St. Philips College; Biswa Datta, Northern Illinois University; Matthew Hudock, St. Philips College; Anthony Macula, SUNY Genesco; Thomas Ordayne, University of South Carolina at Spartanburg; Rene Barrientos, Miami-Dade Community College; Mark Burtch, Arizona State University; Adrienne Goldstein, Miami-Dade Community College; John Grima, Glendale Community College; Michael Kirby, Tidewater Community College; Zhuangyi Lui, University of Minnesota at Duluth; Martha Pratt, Mississippi State University; Michele Clement, Louisiana State University; Karabi Datta, Northern Illinois University; and Richard Witt, University of Wisconsin-Eau Claire; Biswa Nath Datta, Northern Illinois University.

We offer a very special thank you to an extremely valuable team member, Mike McCraith. Mike served as an accuracy checker for all of the new and updated exercises. His professional, high-quality work on creating both the Instructor and Student Solutions Manuals was valued for its precision and careful proofreading. Mike has also brought his expertise to creating the Test Bank and PowerPoint lecture outlines for this text.

Many people involved in the management and production of the text deserve recognition. Amy Bloom has provided high-quality, professional guidance as our developmental editor. Amy Rose and Tiffany Sliter provided tremendous leadership as the production editors. They were always available at a moment's notice to handle any crisis we had, no matter how big or small. Mark Bergeron worked through several iterations of the design and produced a truly modern-looking text. Many thanks to WebAssign for giving some of our exercises their special treatment. We appreciate the quality services provided by Aptara in preparing the graphs, tables, and art in the text. Amy Wrynn did a spectacular job in researching and finding the perfect photos that we requested. Thanks to Eileen Worthley for her abilities to wrestle the printer into submission and get page proofs out to us in a timely fashion. A huge thanks to the Jones & Bartlett Learning sales and marketing staff for their sales efforts and enthusiasm. A special thanks goes out to Tim Anderson for believing in this project.

Finally, we owe personal thanks to our families. Our wives, Lisa and Melissa, continue to be supportive and patient through the entire process. To our children, Austin and Dylan, and Randy, Rusty, and Ronnie, who continue to understand that sometimes their father needs to focus on meeting a deadline and is temporarily unavailable.

Functions, Modeling, and Average Rate of Change

© Yuri Arcurs/ShutterStock, Inc. **(a)**

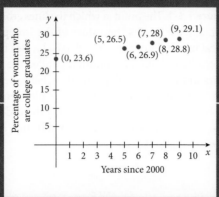

(b)

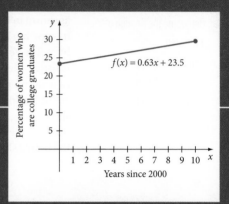

(c)

For the past couple of decades, more and more women have been entering college and earning degrees. Figure (b) plots data gathered by the U.S. Census Bureau showing the percentage of females who have earned a four-year college degree. Figure (c) gives the function that models the data. The data plotted in Figure (b) and the function modeling the data in Figure (c) show that the percentage of females who earned a four-year college degree rose steadily during the period 2000–2009.

What We Know

We begin our study of calculus having learned the basics of algebra. These include equation solving, factoring, and the graphing of functions.

Where Do We Go

In this chapter, we will review the function concept, the properties of various type of functions, and the average rate of change of functions over a closed interval. The average rate of change will review the process of finding the slope of a line and will lay the groundwork for a future calculus concept.

Chapter Sections

1. Create a scatterplot.
2. Evaluate a function.
3. Determine if a graph represents a function.
4. Determine the domain of a function.
5. Determine the domain and range of a price-demand function.

1.1   The Coordinate System and Functions

Plotting Points

We start our study of functions by examining how they appear visually. The graph on which functions are plotted is called the **Cartesian plane**.

The Cartesian plane can be thought of as two number lines that are perpendicular to one another, as shown in **Figure 1.1.1**. The point at which the lines cross is called the **origin**. The horizontal number line, or horizontal axis, locates the values of the **independent variable**, which is usually denoted by x. The vertical number line, or vertical axis, locates the values of the **dependent variable**, which is usually denoted by y. These are commonly called the x and y axes, respectively. Points are plotted as **ordered pairs** in the form (independent variable, dependent variable), which, most of the time, are of the form (x, y). Before we continue, let's plot some points on a Cartesian plane. **Figure 1.1.2** shows the graph of some ordered pairs. To plot the ordered pair $(3, -1)$, we start at the origin and move to the left three units and then up one unit. In a similar fashion, we plotted the remainder of the points shown in Figure 1.1.2.

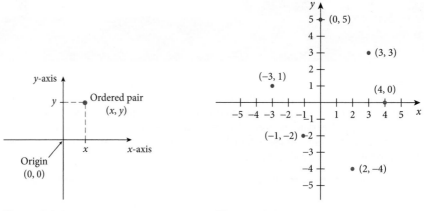

Figure 1.1.1	Figure 1.1.2

In the study of applied calculus, we often use tabular data as a basis for forming a mathematical model. When we plot data, we are producing what is called a **scatterplot**. Frequently, the values of the independent variable are a measure of time, primarily in years. The dependent variable then represents some phenomenon in business, life, or social science that is related to time.

Example 1: Creating a Scatterplot

The data in **Table 1.1.1** represent the percentage of women in the United States who are college graduates (four years of college or more) for the years 2000 to 2009. Let x represent the number of years since 2000, and let y represent the percentage of women who are college graduates (four years of college or more). Construct ordered pairs to represent the data, and plot the data.

Perform the Mathematics

To avoid working with large values of the independent variable x, it is convenient to let x represent the number of years since 2000 (see **Table 1.1.2**).

The year 2000 corresponds with $x = 0$, 2005 corresponds with $x = 5$, and so on. Notice that we determine these values by taking the year and subtracting 2000. This process is called **standardizing the values**. From Table 1.1.2 we construct the ordered pairs $(0, 23.6)$, $(5, 26.5)$, $(6, 26.9)$, $(7, 28.0)$, $(8, 28.8)$ and $(9, 29.1)$. Using these ordered pairs we get the scatterplot shown in **Figure 1.1.3**.

Table 1.1.1

Year	Percentage of women who are college graduates
2000	23.6
2005	26.5
2006	26.9
2007	28.0
2008	28.8
2009	29.1

Source: U.S. Census Bureau

Table 1.1.2

Years since 2000	Percentage of women who are college graduates
0	23.6
5	26.5
6	26.9
7	28.0
8	28.8
9	29.1

© Supri Suharjoto/ShutterStock, inc.

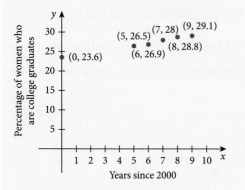

Figure 1.1.3

▶ **Try It Yourself**

Some related Exercises are 69 and 71.

Notice in Example 1 that the possible values of x came from the set $\{0, 5, 6, 7, 8, 9\}$ and the possible values of y came from the set $\{23.6, 26.5, 26.9, 28.0, 28.8, 29.1\}$. These sets of numbers have special names in mathematics. The set of all possible values of the independent variable, in this case x, is called the **domain**; the set of all possible values of the dependent variable, in this case y, is called the **range**. These two sets of numbers are critical in the study of functions.

Function Notation and Evaluating Functions

Many times, a dependent relationship exists between two phenomena. The price of a concert ticket may **depend** on the popularity of the band, the cost of manufacturing may depend on the quantity manufactured, and the life expectancy of a person may depend on the year in which she or he was born. These examples exhibit a direct relationship, or correspondence, between **independent variable** values (price, quantity, year) and the **dependent variable** values (popularity, cost, life length). We write these relationships between the independent and dependent variables using **functions**.

A function f can be thought of as a process in which an independent value x in the domain is mapped to a dependent value y in the range. This is illustrated in **Figure 1.1.4**.

Functions are usually expressed by first naming the function with a letter such as f or g, and then writing the independent variable letter in parentheses. For example, in the expression $f(x)$ (read "f of x"), f represents the function and $f(x)$ represents the value of f at x, where x is the independent variable. It is often convenient to represent $f(x)$ with the dependent variable y.

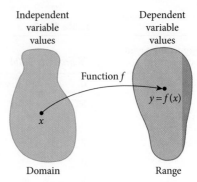

Figure 1.1.4

Functions are not only expressed as tables and graphs, but also as mathematical expressions. For example, the function $f(x) = \dfrac{x+2}{5}$ means: "Take the independent value and add 2; then divide that sum by 5."

The process of "plugging in" various x-values into a function is called **evaluating** the function. To evaluate the function $f(x) = \dfrac{x+2}{5}$ at $x = 13$, denoted by writing $f(13)$, we substitute 13 for every x. This gives

$$f(x) = \frac{x+2}{5}$$

$$f(13) = \frac{13+2}{5} = \frac{15}{5} = 3$$

OBJECTIVE 2

Evaluate a function.

Example 2: Evaluating Functions

For the function $g(x) = -2x + 6$, evaluate $g(-2), g(0)$, and $g(3)$, and write the results as ordered pairs.

Perform the Mathematics

Evaluating the function, we get $g(-2) = -2(-2) + 6 = 4 + 6 = 10$, which produces the ordered pair $(-2, 10)$.

Now $g(0) = -2(0) + 6 = 0 + 6 = 6$. This gives the ordered pair $(0, 6)$.

Finally, $g(3) = -2(3) + 6 = -6 + 6 = 0$, which produces the ordered pair $(3, 0)$. ∎

Now let's take a look at a relationship that is not a function. **Figure 1.1.5** shows the graph of relationship R. Because $(11, 3)$ and $(11, -3)$ are on the graph, R assigns both 3 and -3 to the value 11.

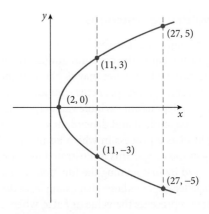

Figure 1.1.5

Applying our definition of a function, we see that R is not a function. A visual way to tell whether a graph represents a function is to use the **vertical line test** (see marginal definition).

Example 3: Determining If a Graph Represents a Function

Use the vertical line test to determine which of the graphs in **Figures 1.1.6(a), (b),** and **(c)** do not represent functions.

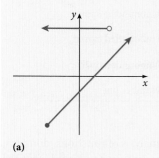

(a)

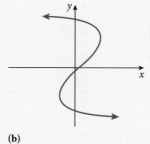

(b)

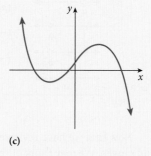

(c)

Figure 1.1.6

Perform the Mathematics

We see in Figure 1.1.6(c) that vertical lines will intersect the graph at only one place, but in Figures 1.1.6(a) and (b) they will intersect at more than one point. So the first two graphs do not represent functions, whereas the third graph is a function. ∎

▶ *Try It Yourself*

Some related Exercises are 21 and 23.

OBJECTIVE 3

Determine if a graph represents a function.

DEFINITION

Vertical Line Test

If every vertical line drawn through a graph intersects the graph at only one point, then the graph represents a function.

NOTE: An alternative way to state the Vertical Line Test is as follows: If one vertical line can be drawn that intersects the graph in more than one point, then the graph does not represent a function.

Many times, we need to express the domain and range of a function as some portion of the real number line. To write this in a convenient form, we use **interval notation**. Let's say that we have the inequality notation $2 \leq x < 5$: that is, all the x-values from 2 to 5, including 2. We can conveniently write this portion of the real number line using interval notation as $[\,2, 5\,)$. Intervals with two endpoints are called **finite intervals**. If the endpoints are included, the interval is **closed**. If the endpoints are not included, the interval is **open**.

Finite Intervals

Number Line	Interval Notation	Inequality Notation	Interval Type
a b	$[a, b]$	$a \leq x \leq b$	Closed
a b	(a, b)	$a < x < b$	Open
a b	$(a, b]$	$a < x \leq b$	Half-open
a b	$[a, b)$	$a \leq x < b$	Half-open

Intervals that extend indefinitely in at least one direction are called **infinite intervals**. We use the **infinity** symbol ∞ to indicate that the numbers extend positively (to the right on the number line) without bound. We use $-\infty$ to represent **negative infinity** to show that the numbers extend negatively (or to the left on the number line) without bound. Since ∞ and $-\infty$ are only **concepts** and not real numbers themselves, we write these endpoints as open, using parentheses.

Infinite Intervals

Number Line	Interval Notation	Inequality Notation	Interval Type
a	$[a, \infty)$	$x \geq a$	Unbounded closed
a	(a, ∞)	$x > a$	Unbounded open
 *b*	$(-\infty, b]$	$x \leq b$	Unbounded closed
 *b*	$(-\infty, b)$	$x < b$	Unbounded open
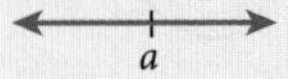 *a*	$(-\infty, \infty)$	All x in $\mathbb{R}$	Number line

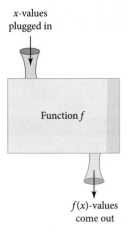

x-values plugged in

Function *f*

f(*x*)-values come out

Figure 1.1.7

Now that we have reviewed interval notation, let's return to the analysis of functions, domains, and ranges. We may think of the domain as the set of all possible values that can be *plugged in* for the independent variable of the function. The range is the set of numbers that *come out* of the function. An illustration of this *plug-in* and *come-out* process is given in **Figure 1.1.7**.

To find numbers that cannot be in the domain, think of the number properties learned in algebra. We cannot divide by zero; for example, $\frac{3}{0}$ is not defined. We cannot take the square root of a negative number; that is, $\sqrt{-9}$ does not have a **real** number solution. (You might have heard of complex numbers and the imaginary unit i, but in applied calculus we consider only real numbers.) So, in general, when we look for the domain values of functions, we exclude values that can make the denominator of a fraction zero or make the expression under a square root (or any even index root) function negative. (Another restriction will arise in Section 4.1 when we study logarithmic functions.)

OBJECTIVE 4

Determine the domain of a function.

Example 4: Finding the Domain of a Function Algebraically

Determine the domain of the following functions. Show the domain on a real number line and by writing in interval notation.

a. $f(x) = \sqrt{5x - 2}$ **b.** $g(x) = \dfrac{\sqrt{x - 2}}{x^2 - 4x}$

Perform the Mathematics

a. To find the domain for f, we need the x-values so that the radicand (the expression under the square root symbol) is greater than or equal to zero. As an inequality, this means that we need

$$5x - 2 \geq 0$$
$$5x \geq 2$$
$$x \geq \frac{2}{5}$$

This domain is shown on the number line in **Figure 1.1.8**. In interval notation, the domain is given as $[\frac{2}{5}, \infty)$.

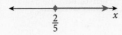

$\frac{2}{5}$ *x*

Figure 1.1.8

b. This function requires us to examine the restrictions on both the numerator and denominator. In the numerator, we see that we need values of x such that

$$x - 2 \geq 0$$
$$x \geq 2$$

For the denominator, we can factor x from each term to get

$$x^2 - 4x = x(x - 4)$$

If we set the factored form equal to zero and solve, we get

$$x(x - 4) = 0$$
$$x = 0 \text{ or } x = 4$$

So we must also exclude the values of 0 and 4 from the domain, since they make the denominator of g zero. The resulting domain is shown on the number line in **Figure 1.1.9**. Because the domain of g is the x-values for which both the numerator and denominator are defined, the domain of g is $[2, 4) \cup (4, \infty)$.

Figure 1.1.9

■

> ▶ *Try It Yourself*
> Some related Exercises are 49 and 53.

Before continuing, let's summarize how to **determine the domain of a function** algebraically.

Using purely algebraic techniques, finding the range of a function can be difficult. Many functions require techniques learned in calculus to determine the range, but the range is relatively easy to determine when we see the graph of the function. We illustrate how in Example 5.

Example 5: Determining Domains and Ranges from a Graph

Determine visually the domain and range for the function in **Figure 1.1.10** and write the answer using interval notation.

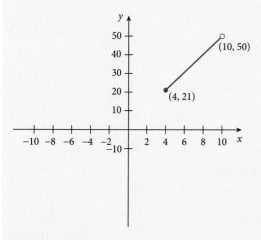

Figure 1.1.10

Determining the Domain of a Function

To determine the domain of a function, we exclude independent variable values that

1. Make the denominator of a fraction zero.
2. Produce a negative result under an even-indexed radical.

OBJECTIVE 5

Determine domains and ranges from a graph.

Perform the Mathematics

First, notice that the left endpoint of the function is a solid dot, and the right endpoint is a hollow dot. This means that the domain is the half-open interval $[4, 10)$. Notice that the lowest point on the graph is at the ordered pair $(4, 21)$ and the highest point is near $(10, 50)$. The y-coordinates give us the range of $[21, 50)$.   ■

DEFINITION

Price-Demand Function

The **price-demand function** p gives us the price at which people buy exactly x units of product.

Now let's introduce a function that we will see frequently in this text, the **price-demand function**. Generally, as the price of a product decreases, more and more people will buy the product. So for the price-demand function, as the x-values increase, the p-values decrease. This is shown in **Figure 1.1.11**.

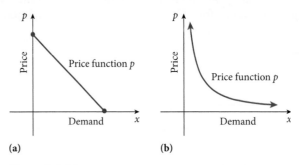

(a)   (b)

Figure 1.1.11

OBJECTIVE 6

Determine the domain and range of a price-demand function.

Example 6: Determining the Domain and Range of a Price-Demand Function

The price-demand function for electronic organs sold at Red River Mall is given by

$$p(x) = 16{,}000 - 320x$$

The graph of p is given in **Figure 1.1.12**. Here, x represents the number of electronic organs that people buy at Red River Mall weekly, and $p(x)$ represents the price per organ in dollars.

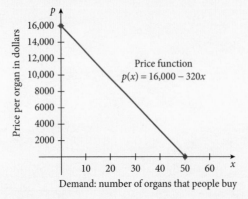

Figure 1.1.12

   a.   Determine the domain and range of the function.

   b.   Evaluate $p(15)$ and interpret.

Perform the Mathematics

a. Notice that the function applies only for x-values up to 50 units. For other values, $p(x)$ is not defined. We may express this restriction on x using interval notation by writing

$$p(x) = 16{,}000 - 320x \quad [0, 50]$$

We can see that the possible values for x are in the interval $[0, 50]$, which gives us our domain. Since the range is the set of all possible function values, we can see from the graph that the range is $[0, 16{,}000]$.

b. Evaluating the price-demand function at $x = 15$, we get

$$p(15) = 16{,}000 - 320(15) = 16{,}000 - 4800 = 11{,}200$$

This means that when the weekly demand for organs is 15, the price of an organ is $11,200. ∎

Summary

In this section we have reviewed the fundamental properties of functions.

- A **function** is a rule that assigns to each element in the domain one and only one element in the range.

- The **domain** of a function is the set of all possible independent variable values.

- The **range** of a function is the set of all possible dependent variable values.

- The **price-demand function** p gives us the price $p(x)$ at which people buy exactly x units of product.

Section 1.1 Exercises

Vocabulary Exercises

1. The _____ of a function is the set of all possible dependent variable values.

2. The domain of a function is the set of all possible _____ variable values.

3. A function is a rule, or set of ordered pairs, that assigns each element in the _____ one and only one element in the range.

4. We call a graph of plotted ordered pairs a _____.

5. The process of "plugging in" x-values to a function to get the function values is called _____.

6. The interval $[a, b]$ is an example of a _____ interval.

Skill Exercises

In Exercises 7–10, make a scatterplot of the tabular data.

7.

x	$f(x)$
-2	3
1	5
4	4

8.

x	$g(x)$
1	2.1
2	2.3
3	2.9

9.

x	-4	-3.5	-2	2
$f(x)$	-5	-6	-8	-11

10.

x	0	-1	-2
$f(x)$	0	1.1	1.5

In Exercises 11–16, use the scatterplots to make a table of ordered pairs.

11.

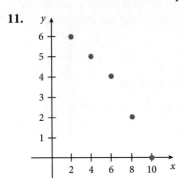

12.

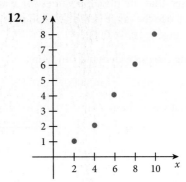

13.

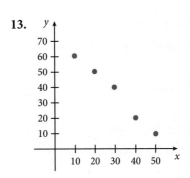

14.

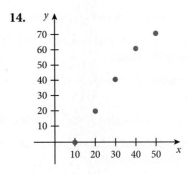

15.

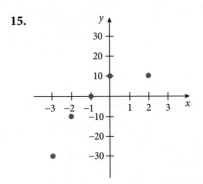

16.

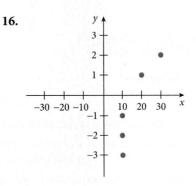

For Exercises 17–20, determine if the given table represents a function.

17.

Domain	Range
0	1
2	1
3	2
4	5

18.

Domain	Range
-5	-2
-3	1
-2	5
-1	7
0	10
1	6

19.

Domain	Range
−3	1
−2	0
−2	3
−1	4
0	5
1	6

20.

Domain	Range
2	1
−2	3
0	0
−1	2
−3	7
2	4

For Exercises 21–26, determine if the graph represents a function.

21.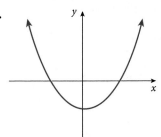

22.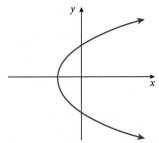

23.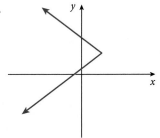

24.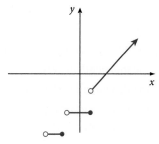

25.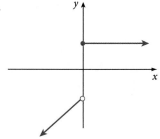

26.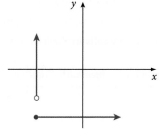

For Exercises 27–37, let $f(x) = 2x - 1$ and let $g(x) = x^2$. Evaluate each expression and write the solutions as an ordered pair.

27. $f(2)$

28. $f(-1)$

29. $f(1)$

30. $f(-2)$

31. $g(3)$

32. $g(1)$

33. $g(0)$

34. $f\left(\dfrac{1}{2}\right)$

35. $g\left(\dfrac{1}{2}\right)$

36. $f(-0.5)$

37. $g(-0.25)$

38. Use the graph of the function f to answer parts (a) through (f):

   (a) What is the independent variable?

   (b) What is the dependent variable?

   (c) What is the value of $f(x)$ when $x = 0$?

   (d) What is the value of $f(3)$?

   (e Write the domain of f using interval notation.

   (f) Write the range of f using interval notation.

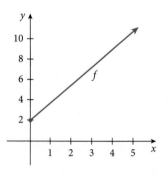

39. Use the graph of the function f to answer parts (a) through (f):

   (a) What is the independent variable?

   (b) What is the dependent variable?

   (c) What is the value of $f(t)$ when $t = 10$?

   (d) What is the value of $f(30)$?

   (e) Write the domain of f using interval notation.

   (f) Write the range of f using interval notation.

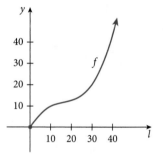

40. Use the graph of the function g to answer parts (a) through (g):

   (a) What is the independent variable?

   (b) What is the dependent variable?

   (c) What is the value of $g(x)$ when $x = 20$?

   (d) What is the value of $g(10)$?

   (e) What are the values of x when $g(x) = 50$?

   (f) Write the domain of g using interval notation.

   (g) Write the range of g using interval notation.

41. Complete the empty cells in the following table.

Interval Notation	Inequality Notation	Number Line
$[-\frac{1}{2}, 5)$		
	$x \geq -2$	
		-3
	$-1 < x < 10$	
$(-\infty, 3) \cup (3, \infty)$		

For Exercises 42–53, determine the domain of the function algebraically and write the domain using interval notation.

42. $f(x) = 2x$

43. $f(x) = -10x$

44. $f(x) = \dfrac{x^2 - 36}{x - 6}$

45. $f(x) = 3x^2 - 10$

46. $f(x) = x^3 - 6x^2$

47. $f(x) = \dfrac{x - 10}{5 - x}$

48. $f(x) = \sqrt{x - 6}$

49. $f(x) = \sqrt{6 - x}$

50. $f(x) = \dfrac{\sqrt{x - 1}}{x - 4}$

51. $f(x) = \sqrt{x^2 + 3}$

52. $f(x) = \dfrac{10x}{x^2 - 25}$

53. $f(x) = \dfrac{x - 1}{2x^2 - 4x}$

For Exercises 54–57, graph the given function on your calculator and use the graph to estimate the domain and range of the function. Write the domain and range using interval notation.

 54. $f(x) = x^2 - 4$

55. $f(x) = \dfrac{x^2 + 5}{10}$

56. $f(x) = \sqrt{9 - x}$

57. $f(x) = \dfrac{x - 2}{x^2 + 6}$

For Exercises 58–63, use the graph to write the domain and range of the function using interval notation.

58.

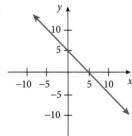

59.

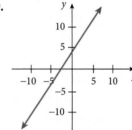

60.

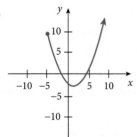

61.

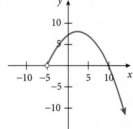

62.

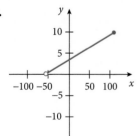

63.

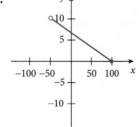

Application Exercises

64. Macroeconomics—Price-Demand: The price-demand function for the new Hanford hand-held gaming device can be modeled by

$$p(x) = 120 - 0.1x \quad 0 \le x \le 80$$

where x represents the number of units sold per day, and $p(x)$ represents the price per unit.

(a) Complete the table of values. That is, determine the price when the demand is $x = 0, 10, 20, 30, \ldots, 80$ units per day.

x	0	10	20	30	40	50	60	70	80
$p(x)$									

(b) Make a graph of the function p on the interval $[0, 80]$.

(c) If $[0, 80]$ is the domain, what is the corresponding range of the function?

(d) Evaluate $p(45)$ and interpret.

65. **Macroeconomics—Price-Demand:** The price-demand function for the new Teddy Bear line of lingerie is given by

$$p(x) = \frac{165 - x}{4} \quad 0 \le x \le 35$$

where x represents the number of units sold per day, and $p(x)$ represents the price of the unit.

(a) Complete the following table of values for p.

x	0	5	10	15	20	25	30	35
$p(x)$								

(b) Make a graph of the function p on the interval $[0, 35]$.

(c) If $[0, 35]$ is the domain, what is the corresponding range of the function?

(d) Evaluate $p(22)$ and interpret.

66. **Macroeconomics—Price-Demand:** For the price-demand function in Exercise 64, determine the desired price for each unit if the number of units sold is 65 units per day.

67. **Macroeconomics—Price-Demand:** For the price-demand function in Exercise 65, determine the desired price for each unit if the number of units sold is 18 units per day.

In Exercises 68–73, standardize the value of the independent variable based on its definition and make a scatterplot of the data.

 68. **Construction—Housing Starts:** Let x represent the number of years since 2000.

Year	Number of Housing Starts in the Northeastern United States (in thousands)
2000	154
2001	149
2002	158
2003	163
2004	175
2005	190
2006	167
2007	143
2008	121
2009	62

Source: U.S. Census Bureau

69. Construction—Housing Starts: Let x represent the number of years since 2000.

Year	Number of Housing Starts in the Midwestern United States (in thousands)
2000	318
2001	330
2002	350
2003	374
2004	356
2005	357
2006	280
2007	210
2008	135
2009	97

Source: U.S. Census Bureau

 70. Nutrition—Milk Consumption: Let x represent the number of years since 1980.

Year	American Per Person Consumption of Whole Milk (in gallons)
1980	17.0
1990	10.5
1995	8.6
2000	8.1
2004	7.3
2005	7.0
2006	6.7
2007	6.4
2008	6.1

Source: U.S. Department of Agriculture

 71. Nutrition—Tea Consumption: Let x represent the number of years since 1980.

Year	American Per Person Consumption of Tea (in gallons)
1980	7.3
1990	6.9
1995	7.9
2000	7.8
2004	8.0
2005	8.0
2006	8.4
2007	8.4
2008	8.0

Source: U.S. Department of Agriculture

72. Law—Supreme Court Opinions: Let x represent the number of years since 1990.

Year	Number of Signed U.S. Supreme Court Opinions
1990	112
1995	75
2000	77
2005	69
2006	67
2007	67
2008	74
2009	73

Source: Office of the Clerk, U.S. Supreme Court

 73. Law—Supreme Court Cases: Let x represent the number of years since 1990.

Year	Total Cases on U.S. Supreme Court Docket (in thousands)
1990	6.3
1995	7.6
2000	9.0
2005	9.6
2006	10.3
2007	9.6
2008	9.0
2009	9.3

Source: Office of the Clerk, U.S. Supreme Court

Concept and Writing Exercises

74. Explain in a sentence why the number zero cannot be part of the domain of the function $f(x) = \dfrac{3}{x}$.

75. Is zero a value in the domain of the function $f(x) = \dfrac{x}{3}$? Explain why or why not.

76. Explain in a sentence why -25 cannot be a value in the range of the function $f(x) = \sqrt{x}$.

77. Given the function $f(x) = \dfrac{x - 3}{x - 5}$, how many real number values are excluded from the domain? Explain your answer in a brief sentence.

78. Given the function $f(x) = \dfrac{x}{x^2 - 4}$, how many real number values are excluded from the domain? Explain your answer in a brief sentence.

79. Fill in the blanks to complete a method for finding the domain of a function. *When determining the domain of a function, we exclude real number values that make the denominator of a fraction equal to* _____, *and we exclude values that make the radicand of a square root function* _____.

Section Project

Determine if the following situations can be represented by functions. If they can be represented by a function, describe its domain and range.

(a) A person's astrological sign is a function of his or her birth date.

(b) A person's income is a function of how long he or she works each day.

(c) The voter turnout for an election is a function of the weather.

(d) The distance a delivery person is from a pizzeria is a function of the number of deliveries made.

(e) The number of U.S. Congresspersons a state has is a function of its population.

1.2   Linear Functions and Average Rate of Change

The first specific type of function that we will study is called a **linear function**. When we say things like "The cost of a phone call is a function of the length of the call" or "The price of a cab ride is a function of the distance to the destination," we are referring to linear functions.

Another critical concept that we address in this section is **average rate of change**. This kind of rate is a common component of linear functions. If we know the average rate of change of a linear function (also known as the **slope** of the line), it is usually a short process to write the function itself. Statements such as "On average, the price of used cars is going up \$100 per year" or "My computer equipment is depreciating at about \$80 each month" are using the average rate of change concept.

Linear Functions

We begin our investigation of linear functions with an example of how it is used in an everyday application. Suppose that The Fashion Mystique wants to print promotional flyers for its annual sidewalk sale. The store manager goes to the local copy center and finds that there is a \$5 setup charge and a 10-cent copying fee for each flyer produced. The manager is handed the pricing schedule shown in **Table 1.2.1**.

Table 1.2.1

Number of copies	Cost
100	\$ 15
500	\$ 55
1000	\$105

There must be a functional relationship between the values in the two columns, since the cost of the copies is a **function** of the number of copies made. Because the cost of the job **depends** on the number of copies produced, we believe that the number of copies is the independent variable, and the cost is the dependent variable.

This function consists of two parts. The first is the setup charge, which is a flat fee that is assessed no matter how many copies are made. It is what economists refer to as a **fixed cost**. The other part is the copying fee, which changes as the number of copies changes. Economists call this type of cost **variable costs**. We can use this combination of fixed and variable costs to verify the cost of making 500 flyers. For example, the \$5 setup fee plus 10 cents times 500 for the copying fee gives

$$\text{Cost for 500} = (\text{price per copy}) \cdot (\text{number of copies}) + (\text{setup fee})$$
$$= (0.10) \cdot 500 + 5$$
$$= 50 + 5 = 55 \text{ or } \$55$$

This is an exact numerical representation of what a linear cost model looks like. We now define the general form of a **linear cost function**.

SECTION OBJECTIVES

1. Determine a cost function.
2. Compute an average rate of change.
3. Write a linear function for a line through two points.
4. Determine *x*- and *y*-intercepts of the graph of a linear function.
5. Write a depreciation model as a linear function.
6. Identify increasing and decreasing functions.
7. Determine marginal cost.
8. Evaluate a piecewise-defined function.
9. Graph an absolute value function.

DEFINITION

Linear Cost Function

The cost $C(x)$ of producing x units of a product is given by the linear **cost function**

$C(x) = (\text{variable costs}) \cdot (\text{units produced}) + (\text{fixed costs})$

NOTE: Some cost functions are not linear because the variable cost is a function itself. However, every cost function is made up of a variable cost expression and a fixed cost.

Example 1: Determining a Cost Function

Determine a cost function, C, for the printing job just described.

Perform the Mathematics

The variable costs are 10 cents per copy, and the fixed costs are 5 dollars, which means that the cost function for the printing job can be written as

$$C(x) = 0.10x + 5$$

where x represents the number of copies made and $C(x)$ represents the cost, in dollars, of making the copies. **Figure 1.2.1** shows a graph of the cost function and the points from the pricing schedule given in Table 1.2.1.

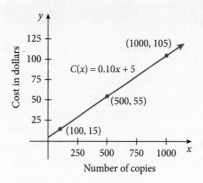

Figure 1.2.1

DEFINITION

Linear Function

A **linear function** has the form

$$f(x) = mx + b$$

where m is the **average rate of change** or **slope** of the line and b is a **constant**, where m and b are real numbers.

The cost function in Example 1 is one example of a linear function. Notice that the three points of the graph all *line up*. This is precisely why we call functions of this type **linear**. A **linear function** is made up of two key parts: a fixed number called a constant, and a variable part that is the product of a number and a variable.

The number given by m has a very important meaning. Let's return to the printing job described in Example 1 and use the graph to compute the cost per copy. To do this, we calculate the difference between two of the points (500, 55) and (1000, 105) as follows:

$$\text{Cost/copy} = (\text{difference in cost})/(\text{difference in copies})$$

$$= \frac{105 - 55}{1000 - 500} = \frac{50}{500} = \frac{1}{10} = 0.1$$

We say that the cost of this printing job is increasing at an average rate of $0.10 per copy. If we think of the cost as the variable y and the copies as x, then this scenario serves to illustrate the **average rate of change**.

When referring to a linear function, the average rate of change is also called the **slope** of the line. See **Figure 1.2.2**.

DEFINITION

Average Rate of Change

The **average rate of change** of y with respect to x is

$$\frac{\text{difference in } y}{\text{difference in } x} = \frac{y_2 - y_1}{x_2 - x_1}$$

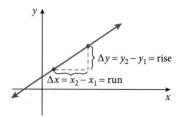

Figure 1.2.2

$$\text{slope} = \frac{\text{rise}}{\text{run}} = \frac{\Delta y}{\Delta x} = \frac{y_2 - y_1}{x_2 - x_1}$$

DEFINITION

Slope

The **slope** of a line, denoted by m, is a measurement of the steepness of the line. Given two points on a line, (x_1, y_1) and (x_2, y_2), the slope of the line is computed by

$$m = \frac{\Delta y}{\Delta x} = \frac{y_2 - y_1}{x_2 - x_1}$$

The slope of the line also gives the average rate of change of y with respect to x.

Example 2: Computing an Average Rate of Change   **OBJECTIVE 2**

For the points $P = (2, -3)$ and $Q = (4, 2)$:   Compute an average rate of
                                                                             change.
a.   Compute the average rate of change between the points.

b.   If the independent variable value increases by 1 unit, how will this affect the dependent
         variable value?

Perform the Mathematics

a.   Here we will let P be the first point and Q be the second point. This means that (x_1, y_1) is the
         ordered pair $(2, -3)$, and (x_2, y_2) is the ordered pair $(4, 2)$. Using the definition of average
         rate of change we get

$$m = \frac{y_2 - y_1}{x_2 - x_1} = \frac{2 - (-3)}{4 - 2} = \frac{5}{2} = 2.5$$

b.   Since the average rate of change is $m = 2.5$, this means that if x, the independent variable,
         increases by 1 unit, then the dependent variable value will increase by 2.5 units. This change
         is illustrated in **Figure 1.2.3**.

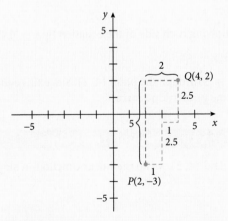

Figure 1.2.3

▶ **Try It Yourself**

Some related Exercises are 9 and 11.

If we switch P and Q in Example 2, we still get the average rate of change as being 2.5. This illustrates that the average rate of change between two points has only one numerical value. In other words, the **slope of a line is unique**.

Equations of Lines

There are several ways to write equations of lines. The form in the definition of a linear function, $f(x) = mx + b$, is called the **slope-intercept form**. Since linear equations are critical in our study of graphing and calculus, let's explore other ways to write a linear function.

We said at the beginning of this section that if the average rate of change of a line is known, it is a short process to write its linear function. To show how easy it is, let's say that we know that the slope of a line is $m = 3$ and a point on the line is $(4, 2)$. Now we call any other point on the line (x, y). See **Figure 1.2.4**.

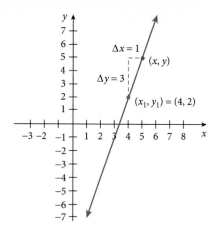

Figure 1.2.4

Using the definition of average rate of change, $m = \dfrac{y_2 - y_1}{x_2 - x_1}$, gives us

$$3 = \frac{y - 2}{x - 4}$$

where $(x_1, y_1) = (4, 2)$ and $(x_2, y_2) = (x, y)$. Multiplying each side of the equation by $x - 4$ yields

$$y - 2 = 3(x - 4)$$

We now have an equation for the line with slope $m = 3$ through the point $(4, 2)$. Since this equation uses a known slope and a known point, it is called the point-slope form.

OBJECTIVE 3

Write a linear function for a line through two points.

Example 3: Writing a Linear Function for a Line Through Two Points

For the line passing through the points $(2, -3)$ and $(4, 2)$, write the linear function in slope-intercept form $f(x) = mx + b$.

Perform the Mathematics

In Example 2, we found the slope between these two points to be $m = 2.5$. Using the point-slope form and replacing (x_1, y_1) with $(2, -3)$, we get

$$y - (-3) = 2.5(x - 2)$$
$$y + 3 = 2.5x - 5$$
$$y = 2.5x - 8$$

Since y is another name for $f(x)$, we have that the slope-intercept form is $f(x) = 2.5x - 8$. ∎

Technology Option

Figure 1.2.5 has a graph of the line found in Example 3.

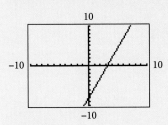

Figure 1.2.5

Throughout the textbook, the viewing window for calculator-generated graphs will appear on the perimeter of the graph. In text, we denote the viewing window by writing $[X_{min}, X_{max}]$ by $[Y_{min}, Y_{max}]$.

Intercepts

Two important points on the graph of a linear function are its **intercepts**. These are the points where the graph crosses the x (independent) and y (dependent) axes, as shown in **Figure 1.2.6**.

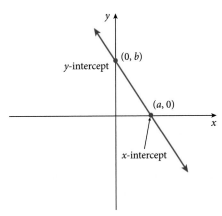

Figure 1.2.6

The figure also suggests how to **determine the intercepts**.

Determining Intercepts

1. To determine the y-intercept, we replace x with zero and solve the resulting equation for y.

2. To determine the x-intercept, we replace y with zero and solve the resulting equation for x.

Example 4: Finding the Intercepts of a Function

Determine the x- and y-intercepts of the graph of the linear function

$$y = \frac{2}{3}(x + 1) + 3$$

Perform the Mathematics

To find the y-intercept, we replace x with zero and solve for y. This gives

$$y = \frac{2}{3}(0 + 1) + 3 = \frac{2}{3} + 3 = \frac{11}{3}$$

OBJECTIVE 4

Determine x- and y-intercepts of the graph of a linear function.

So the y-intercept is $(0, \frac{11}{3})$. To get the x-intercept, we replace y with zero and solve for x to get

$$0 = \frac{2}{3}(x + 1) + 3$$

$$-3 = \frac{2}{3}(x + 1)$$

$$\frac{-9}{2} = x + 1$$

$$\frac{-11}{2} = x$$

Thus the x-intercept is $(\frac{-11}{2}, 0)$. These points are highlighted on the graph in **Figure 1.2.7**.

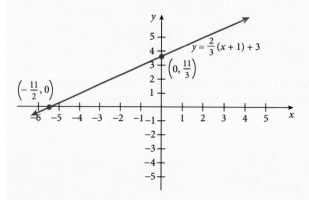

Figure 1.2.7

Notice that the y-intercept can be written in the form $(0, b)$. If we evaluate the linear function $f(x) = mx + b$ when $x = 0$, we get $f(0) = m(0) + b = b$. This means that the y-intercept of the graph of a linear function is the same as the constant b. This is why the form $f(x) = mx + b$ is called the **slope-intercept form** of a line. Now let's look at how intercepts can be used in an application.

OBJECTIVE 5

Write a depreciation model as a linear function.

Example 5: Writing a Depreciation Model and Interpreting Intercepts

The value of many products is said to **depreciate**, meaning that they decrease in value as time goes on. The amount that the value of the product loses each year is called the **depreciation rate**. Many businesses incorporate depreciation models into their fiscal plans and for scheduling upgrades of equipment. For example, the Clark Clipboard Company recently purchased a metal press for $15,000. The press is estimated to depreciate at a rate of $2500 per year.

a.   Write a linear function in the form $f(x) = mx + b$ for the value of the press after x years.

b.   Determine the intercepts, and interpret each.

Perform the Mathematics

a. We know that when the press is new, meaning that its age is $x = 0$, its value is $15,000. Thus, $f(0) = 15,000$, which means that the point $(0, 15,000)$ is on the graph. Notice that this point is the y-intercept. Now, because the rate at which the value of the press **decreases** is $2500 per year, the average rate of change in the press's value is -2500 dollars per year. Since the average rate of change is the same as the slope of a line, we have $m = -2500$. So the value of the press after x years is

$$f(x) = mx + b$$
$$f(x) = -2500x + 15,000$$

b. From part (a) we know that the y-intercept is $(0, 15,000)$. To find the x-intercept, we set $f(x)$ equal to zero and solve the resulting equation for x.

$$0 = -2500x + 15,000$$
$$-15,000 = -2500x$$
$$\frac{-15,000}{-2500} = x$$
$$6 = x$$

The y-intercept, at the point $(0, 15,000)$, means that when the press has just been purchased, its value is $15,000. The x-intercept at $(6, 0)$ means that after 6 years of use, the press has a value of $0. Notice that the reasonable domain of $f(x)$ is $[0, 6]$. A graph of this depreciation function is shown in **Figure 1.2.8**.

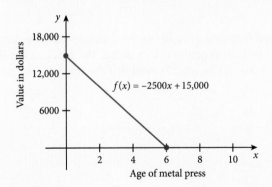

Figure 1.2.8

∎

▶ **Try It Yourself**

Some related Exercises are 67 and 69.

We determined in Example 5 that, as time went on, the values of the metal press decreased. On the other hand, as the x-values increased in Example 4, so did the function values. These examples serve as motivation for the definition of **increasing and decreasing functions**.

Identify increasing and decreasing functions.

Example 6: Identifying Increasing and Decreasing Functions

Classify the function as increasing, decreasing, or neither. State the slope of the line.

a. $h(x) = 2x - 1$

b. $g(x) = -3x + 5$

c. $f(x) = 4$

Perform the Mathematics

a. The graph of $h(x) = 2x - 1$ is shown in **Figure 1.2.9**.

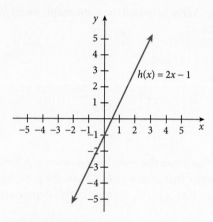

$h(x) = 2x - 1$

Figure 1.2.9

The graph "runs uphill" when viewed from left to right, so the function values increase as the x-values increase. This means that the function is increasing. The function has the slope-intercept form $h(x) = mx + b$. So we can directly read that the slope of the line is $m = 2$.

b. **Figure 1.2.10** has a graph of $g(x) = -3x + 5$.

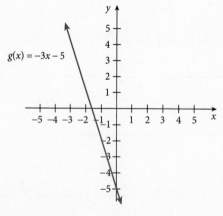

$g(x) = -3x - 5$

Figure 1.2.10

The graph "runs downhill" when viewed from left to right, so the function values decrease as the x-values increase. This means that the function is decreasing. The function has the slope-intercept form $g(x) = mx + b$, so the slope of the line is $m = -3$.

c. The graph of $f(x) = 4$ is in **Figure 1.2.11**.

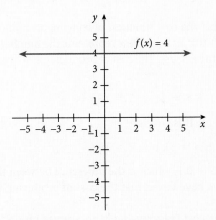

Figure 1.2.11

For this function, the function values stay the same as the x-values increase. We call this function a **constant function**, because the function is made up of only a constant. Notice that if we write the function in the form $f(x) = mx + b$, it would look like $f(x) = 0 \cdot x + b$. This means that the slope of the line is $m = 0$.   ■

A natural question to ask is "What is the slope of a vertical line?" A line of this type has the form $x = a$, where a is a constant. Since the change in x is zero, the formula for slope, $m = \dfrac{y_2 - y_1}{x_2 - x_1}$, would have zero in the denominator. Since division by zero is not defined, we say that **vertical lines have undefined slope**. We summarize lines and their graphs in **Table 1.2.2**.

Table 1.2.2

Type of Line	Value of Slope, m	Diagram
Increasing	Positive	
Decreasing	Negative	
Horizontal	Zero	
Vertical	Undefined	

OBJECTIVE 7

Determine marginal cost.

Marginal Cost

Often in economics it is desirable to know the cost of producing the $(x + 1)$st item of a product. This is commonly called the **marginal cost**.

At the beginning of the section, we found that the cost function of copying promotional flyers for The Fashion Mystique was given by $C(x) = 0.10x + 5$, where x represents the number of copies produced and $C(x)$ represents the cost of producing x flyers. The cost of producing 110 flyers is given by $C(110)$. This cost is

$$C(110) = 0.10(110) + 5 = 11 + 5 = 16$$

The cost of producing 111 flyers is given by $C(111)$. This cost is

$$C(111) = 0.10(111) + 5 = 11.1 + 5 = 16.1$$

DEFINITION

Linear Cost Function

For linear cost functions, variable costs and marginal costs are equal.

The value of $C(111) - C(110)$ is then $16.1 - 16 = 0.1$, which is the difference between the cost of producing 110 and 111 flyers. In other words, we have found that the cost of producing the 111th flyer is $0.10 or 10 cents.

Earlier in this section we also stated that the cost is increasing at an average rate of $0.10 per copy. Thus, the marginal cost for the 111th copy is the same as the average rate of change for the cost of the copies. The fact that, for a **linear cost function**, the marginal cost and the slope (given by the variable costs) are the same is no coincidence.

We will study marginal functions in more depth in Chapter 3.

Example 7: Determining Marginal Costs

The ProAudio Company manufactures DVD disks. It determines that the weekly fixed costs are $14,000, and variable costs are $2.60 per disk.

a. Determine the linear cost function C, and interpret $C(1500)$.

b. Identify the marginal cost. At a 1500-per-week production level, what is the cost of manufacturing the 1501st disk?

Perform the Mathematics

a. The manufacturing process has fixed costs of $14,000 and variable costs of $2.60, so the linear cost function is

$$C(x) = 2.60x + 14,000$$

Evaluating at $x = 1500$ gives us

$$C(1500) = 2.60(1500) + 14,000 = 3900 + 14,000 = 17,900$$

So the weekly cost of producing 1500 DVD disks is $17,900.

b. Since the marginal cost and the variable cost are the same for a linear function, the marginal cost for the manufacturing process is $2.60. Thus, at a 1500-disk-per-week production level, the cost of manufacturing the 1501st disk is $2.60. ∎

Piecewise-Defined Functions

Often one simple function alone cannot represent everyday applications. For example, a taxi may charge a flat fee of $3 up to the first mile driven and an additional $0.25 for 1 mile and for each mile afterward. To express a function for the cost of the taxi, we must write the function in two parts. Up to the first mile, the x-values go from 0 to 1, where x is the number of miles, and the cost is given by the constant function, which we will call $y_1 = 3$. For x-values greater than or equal to 1, we have to add 25 cents for each mile ridden to the \$3. So for the x-values greater than or equal to 1, the cost is given by a second piece that we will call $y_2 = 0.25x + 3$. We can join the two pieces y_1 and y_2 to get the cost of riding x miles as

$$f(x) = \begin{cases} 3, & 0 < x < 1 \\ 0.25x + 3, & x \geq 1 \end{cases}$$

A graph of f is given in **Figure 1.2.12**. This kind of function, which is defined for specific intervals of the domain, is called a **piecewise-defined function**. Let's try an example of graphing this new type of function.

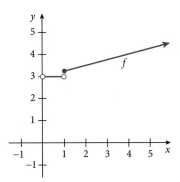

Figure 1.2.12
Graph of
$$f(x) = \begin{cases} 3, & 0 < x < 1 \\ 0.25x + 3, & x \geq 1 \end{cases}$$

Example 8: Graphing a Piecewise-Defined Function

For the piecewise-defined function $f(x) = \begin{cases} x + 2, & x < 1 \\ 3 - x, & x \geq 1 \end{cases}$

a.   Evaluate $f(0), f(1)$, and $f(3)$.

b.   Make an accurate graph of the function.

OBJECTIVE 8

Evaluate a piecewise-defined function.

Perform the Mathematics

a.   Since $x = 0$ is in the interval $(-\infty, 1)$, we evaluate using the "top" piece to get $f(0) = 0 + 2 = 2$. To get $f(1)$, we see that $x = 1$ is the left endpoint of our interval $[1, \infty)$, so we use the "bottom" piece to evaluate and get $f(1) = 3 - 1 = 2$. Finally, using the bottom piece again to evaluate $f(3)$, we get $f(3) = 3 - 3 = 0$.

b.   On the interval of x-values $(-\infty, 1)$, we plot the linear equation $y = x + 2$ shown in **Figure 1.2.13(a)**. Then, on the interval $[1, \infty)$, we plot the equation $y = 3 - x$ as in Figure 1.2.13(b). Finally, the graph of f is found by plotting the two graphs shown in Figures 1.2.13(a) and (b) on the same Cartesian plane, as shown in Figure 1.2.13(c).

We use a solid dot to show which endpoint is included when $x = 1$ and a hollow dot to indicate that the endpoint is not included when $x = 1$.

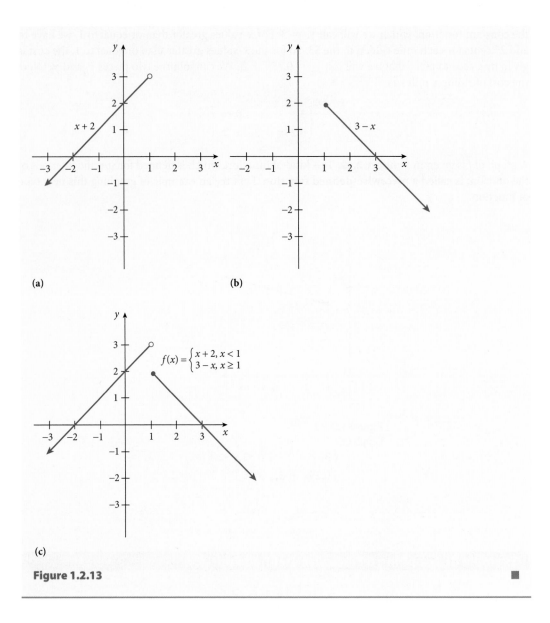

(a)

(b)

$$f(x) = \begin{cases} x + 2, x < 1 \\ 3 - x, x \geq 1 \end{cases}$$

(c)

Figure 1.2.13

Absolute Value Function

A piecewise-defined function that is often used is the **absolute value function**. Recall from algebra that the absolute value of any real number x, denoted by $|x|$, is defined by

$$|x| = \begin{cases} -x, & x < 0 \\ x, & x \geq 0 \end{cases}$$

In words, when taking the absolute value of any number, just write that number if the value is positive or zero, but if the number is negative, take its opposite. For example, $|8| = 8$ since $8 > 0$, and $|-5| = -(-5) = 5$, since $-5 < 0$. Notice how the definition of absolute value looks like a piecewise-defined function. It seems natural to define the **absolute value function** as a piecewise-defined function.

A graph of $f(x) = |x|$ is shown in **Figure 1.2.14**. Notice that the domain of the function is $(-\infty, \infty)$, the set of real numbers. Since an absolute value cannot be negative, the range of the absolute value function is $[0, \infty)$.

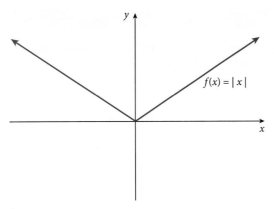

Figure 1.2.14

There are two key points to recognize about this type of function. First, the graph of the function $f(x) = |x|$ has a sharp "corner" at the origin. Second, the variable x inside the absolute value bars can be replaced by **any** algebraic expression.

Example 9: Rewriting and Graphing Absolute Value Functions

For the function $f(x) = |x - 5|$, graph the function and rewrite $f(x) = |x - 5|$ as a piecewise-defined function.

Perform the Mathematics

A graph of $f(x) = |x - 5|$ is given in **Figure 1.2.15**. We see from the graph that the corner has shifted to the right 5 units to the point $(5, 0)$. Also from the graph, the piecewise-defined function will have one piece for x-values less than 5 and another for x-values greater than or equal to 5. For $x \geq 5$, we see that $|x - 5| = (x - 5) = x - 5$. For $x < 5$, we have to take the opposite of $x - 5$ to get $|x - 5| = -(x - 5) = -x + 5$. This application of the definition of absolute value allows us to write the function in piecewise-defined form as

$$f(x) = |x - 5| = \begin{cases} -x + 5, & x < 5 \\ x - 5, & x \geq 5 \end{cases}$$

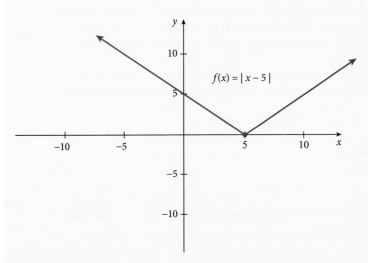

Figure 1.2.15   ■

◀▶ Technology Option

Most graphing calculators use abs(to denote absolute value. For the function in Example 9, **Figure 1.2.16(a)** shows what is entered in the $y =$ editor to produce the graph shown in Figure 1.2.16(b).

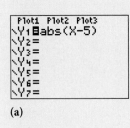

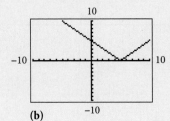

(a)   (b)

Figure 1.2.16

Summary

In this section, we learned that the slope of a line is also its **average rate of change** and is computed by $m = \dfrac{y_2 - y_1}{x_2 - x_1}$. We also found that the value of the slope indicated whether the linear function was increasing or decreasing:

- If $m > 0$, the function is increasing.
- If $m < 0$, the function is decreasing.
- If $m = 0$, the function is the constant function.
- If m is undefined, the line is vertical.

The slope, along with a point on the line, can be used to write linear equations in two main forms.

- **Slope-intercept form**: $f(x) = mx + b$
- **Point-slope form**: $y - y_1 = m(x - x_1)$

The linear cost function has the form $f(x) = mx + b$, where m is the variable costs and b represents the fixed costs. This same model, $f(x) = mx + b$, can be used as a depreciation model, where b represents the cost of a new item, and m is the depreciation rate.

Functions that have different rules based on the x-values are called **piecewise-defined functions**, and a special type of piecewise-defined function is the **absolute value function**.

Section 1.2 Exercises

Vocabulary Exercises

1. A function of the form $f(x) = mx + b$ is called a _____ function.

2. Another term used to describe the average rate of change of a linear function is

   _____.

3. The equation in the form $y - y_1 = m(x - x_1)$ said to be in _____-_____ form.

4. The points where a line crosses the x-axis and the y-axis are called the _____.

5. A function of the form $f(x) = b$, where b is a real number, is called a _____ function.

6. A function that has different algebraic rules for different parts of its domain is called a _____ - _____ function.

Skill Exercises

For Exercises 7–14, determine the average rate of change between the two given points.

7. $(4, 8)$ and $(5, 3)$ **8.** $(4, 8)$ and $(3, 5)$ **9.** $(2, 3)$ and $(-4, 8)$

10. $(2, 2)$ and $(-4, 4)$ **11.** $(5, 6.1)$ and $(7, 8.3)$ **12.** $(0, 1.25)$ and $(4, 5)$

13. $(a - 1, b - 1)$ and (a, b), where a and b are constants.

14. $(1 - a, 1 - b)$ and $(a - 1, b)$, where a and b are constants.

For Exercises 15–20, determine the average rate of change for the given graphs.

15.

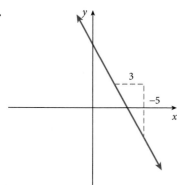

16.

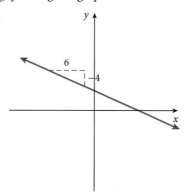

17.

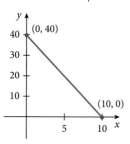

18.

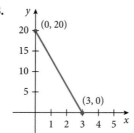

19.

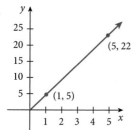

20.

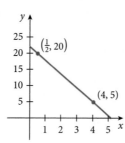

For Exercises 21–28, write an equation of the line in point-slope form if possible. Then graph the function.

21. The line contains the points $(2, 6)$ and $(5, -3)$.

22. The line contains the points $(3, 2)$ and $(-1, -2)$.

23. The line given by the values in the table:

x	f(x)
3	5
5	13

24. The line given by the values in the table:

x	f(x)
2	4
0	4

25. The line has a y-intercept at -3 and passes through the origin.

26. The line has an x-intercept at 2 and a y-intercept at 4.

27. The line is vertical and passes through the point $(-2, 9)$.

28. The line is horizontal and passes through the point $(-3, 1)$.

For Exercises 29–39, write the equation of the line in slope-intercept form.

29. A line has the slope of 3 and passes through the point $(8, 4)$.

30. A line has a slope $-\frac{5}{8}$ and $f(0) = 3$.

31. A line has a slope of 5 and a y-intercept at -13.

32. A line has an average rate of change of $-\frac{2}{3}$ and an x-intercept at 3.

33. A line has an average rate of change of 0.3 and an x-intercept at -4.

34. The graph contain the data points $(1, 3)$ and $(-10, 1)$.

35. The graph contains the data points in the table:

x	$f(x)$
-1	2
3	1

36. The graph contains the data points in the table:

x	$f(x)$
0	2
-3	4

37. A line has an x-intercept at $(12, 0)$ and a y-intercept at $(0, \frac{3}{2})$.

38. A linear function f has function values $f(5) = 1$ and $f(-7) = -1$.

39. A linear function f has function values $f(0) = 6$ and $f(4) = 1.3$.

In Exercises 40–46, determine the x-intercept and the y-intercept for each graph.

40. The given function and graph:

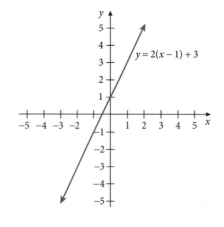

$y = 2(x - 1) + 3$

41. The given function and graph:

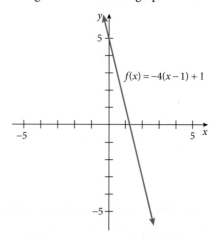

$f(x) = -4(x - 1) + 1$

42. The graph of the function $f(x) = \dfrac{x}{2} - 4$.

43. The graph of the function $f(x) = 120 - 0.3x$.

44. The graph of the function $f(x) = \dfrac{750 - 35x}{10}$.

45. The graph of the function $f(x) = 0.2(x - 5.1) + 10$.

46. The graph of the point-slope equation $y - 3 = -2(x + 4)$.

For Exercises 47–55, classify each of the linear functions as increasing or decreasing, and then identify the slope of a line.

47. $f(x) = -4.5(x - 2)$

48. $f(x) = \dfrac{100 - 5x}{2}$

49. $f(x) = \dfrac{500 + 60x}{15}$

50. $f(x) = 4(x - 6) + 8$

51. $f(x) = 5 + 4x$

52. The graph of the function whose points are in the table:

x	f(x)
1	3
7	5

53. The graph of the function whose points are in the table:

x	f(x)
10	5
1	2

54. The function whose graph is shown:

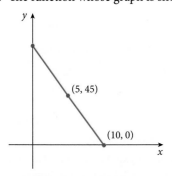

55. The function whose graph is shown:

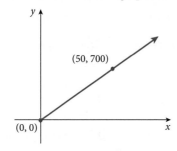

For Exercises 56–61, make an accurate graph of the given piecewise-defined function.

56. $f(x) = \begin{cases} 4 - 0.2x, & x < 5 \\ 5, & x \geq 5 \end{cases}$

57. $f(x) = \begin{cases} 2x, & x \leq 0 \\ -3x, & x > 0 \end{cases}$

58. $f(x) = \begin{cases} -3, & x < 4 \\ 3, & x > 4 \end{cases}$

59. $f(x) = \begin{cases} \dfrac{x + 6}{2}, & x \leq 1 \\ \dfrac{7}{2}, & x > 1 \end{cases}$

60. $f(x) = \begin{cases} -4, & x < 5 \\ x - 2, & x \geq 5 \end{cases}$

61. $f(x) = \begin{cases} -2, & x < 0 \\ 2x + 4, & 0 \leq x \leq 5 \\ -x, & x > 5 \end{cases}$

For Exercises 62–67, complete the following.

(a) *Make an accurate graph of the absolute value function.*

(b) *Rewrite the absolute value function as a piecewise-defined function.*

62. $f(x) = |x + 1|$

63. $f(x) = |x + 3|$

64. $f(x) = |6 - 2x|$

65. $f(x) = |10 - x|$

66. $f(x) = |3x - 15|$

67. $f(x) = |4x + 12|$

Application Exercises

Business—Depreciation: Answer Exercises 68–71 using the depreciation model.

68. A new photocopy machine initially costs $25,000 and depreciates at a rate of $1250 per year.

    (a) Write the depreciation model in the form $f(x) = mx + b$.

    (b) Determine the number of years that it takes for the machine to depreciate to $10,000.

69. A new multimedia computer costs $800 and depreciates at a rate of $150 per year.

    (a) Write the depreciation model in the form $f(x) = mx + b$.

    (b) How long will it take for the computer to depreciate to the point that it has no worth?

70. A factory invests $90,000 in a new machine press that depreciates at a rate of $6000 per year.

    (a) Write the depreciation model for the problem situation.

    (b) How long will it take for the press to have half of its original value?

71. The R-383 sports car costs $40,000 and depreciates $3000 per year.

    (a) Write the depreciation model for the problem situation.

    (b) How much will the car be worth five years after its purchase?

Economics—Cost Function: In Exercises 72–77, write the linear cost function for the problem situation in the form $C(x) = mx + b$.

72. The variable costs are $70 per unit, and the fixed costs are $10,000.

73. The variable costs are $147.75 per unit, and the fixed costs are $2700.

74. Use the information given in the table.

Number of Units	Cost (in Dollars)
5	1250
23	5580

75. Use the information given in the table.

Number of Units	Cost (in Dollars)
10	7000
50	30,000

76. The fixed costs are $14,500 per unit, and it costs $100,100 to produce 75 units of the product.

77. The variable costs are $80 per unit, and the cost of producing 40 units is $3850.

78. Economics—Cost Function: The Dirt Rocket Company produces spokes for off-road motorcycles. It determines that the daily cost in dollars of producing x quick-replace spokes is

$$C(x) = 1250 + 3x$$

    (a) What are the fixed and variable costs for producing the spokes?

    (b) Evaluate $C(400)$, and interpret.

    (c) How many spokes can be produced for $1820?

    (d) What is the marginal cost for the spokes?

    (e) If the current daily production level is 100 spokes, what is the cost in dollars of producing the 101st spoke?

79. Economics—Cost Function: The Fontana Vitamin Company determines that the cost in dollars of producing x bottles of a new supplement is given by the linear cost function

$$C(x) = 1.25x + 550$$

    (a) What are the fixed and variable costs for producing the supplement?

    (b) Evaluate $C(70)$, and interpret.

    (c) How many bottles of the supplement can be produced for $693.75?

    (d) What is the marginal cost for the supplement?

    (e) If the current production level is 70 bottles, what is the cost of producing the 71st bottle of the supplement?

80. Economics—Cost Function: The Get-It-Now florist determines that the cost in dollars of arranging and delivering x get-well bouquets is given by the cost function $C(x) = 10 + 32x$.

(a) What are the fixed and variable costs for producing the bouquets?

(b) Evaluate $C(15)$, and interpret.

(c) How many bouquets can be produced and delivered for $8010?

(d) What is the marginal cost for the bouquets?

(e) If the current production level is 40 bouquets, what is the cost of producing the 41st bouquet?

81. Economics—Cost Function: The Lesky Truck Rental Company charges a flat fee of $20 plus 15 cents per mile driven for renting a moving truck.

(a) What are the fixed costs? What are the variable costs?

(b) Write a linear function C, where x represents the number of miles driven and $C(x)$ represents the rental costs.

(c) What would be a realistic domain for this function?

(d) Evaluate $C(120)$, and interpret.

(e) What is the cost of the 121st mile driven? What is this cost called?

82. Economics—Cost Function: Consider the daily cost schedule for the cost of producing x Never Die light bulbs shown in the table.

Bulbs Produced	Cost (in Dollars)
0	1050
100	1250
215	1480

(a) What are the fixed costs? Explain how you determined this from the table.

(b) How much does the cost increase for each additional bulb produced? How do you classify this kind of cost?

(c) Write a linear cost function C, where x represents the number of bulbs produced, and $C(x)$ represents the cost in dollars.

(d) What would the cost be if 260 bulbs were produced?

83. Economics—Cost Function: Consider the price schedule for the repair cost of a Ripen microwave oven shown in the table.

Hours Labor	Repair Cost (in Dollars)
0	50
1	65
2	80

(a) What are the fixed costs? Explain how you determined this from the table.

(b) How much does the repair cost increase for each additional hour of labor? How do you classify this kind of cost?

(c) Write a linear cost function C, where x represents hours of labor, and $C(x)$ represents the repair cost in dollars.

(d) What would the repair cost be if the microwave oven took 4.5 hours to fix?

84. Banking—Branch Offices: In order to increase consumer access to personal banking, many banking institutions have added branch offices in grocery stores, supermarkets, and department stores. The number of branch offices in the United States from 2000 to 2011 can be modeled by the linear function

$$f(x) = 3.31x + 60.3 \quad 0 \le x \le 11$$

© andipantz/iStockphoto

where x represents the number of years since 2000 and $f(x)$ represents the number of branch offices in thousands. (*Source:* U.S. Federal Deposit Insurance Corporation.)

(a) By inspecting the model, fill in the blanks to complete the following sentence:

From 2000 to 2011, the number of branch offices _____ by an average of _____ thousand branches per year.

(b) Evaluate $f(9)$ and interpret.

(c) According to the model, during what year were there 73.54 thousand branch offices?

 85. Banking—Main Offices: While banks were increasing their numbers of branch offices, they consolidated, and consequently decreased, their numbers of main bank offices. The number of main offices in the United States from 2000 to 2011 can be modeled by the linear function

$$f(x) = -0.15x + 8.28 \quad 0 \le x \le 11$$

where x represents the number of years since 2000, and $f(x)$ represents the number of main banking offices in thousands. (*Source:* U.S. Federal Deposit Insurance Corporation.)

(a) By inspecting the model, fill in the blanks to complete the following sentence:

From 2000 to 2011, the number of main offices _____ by an average of _____ thousand main banks per year.

(b) Evaluate $f(5)$ and interpret.

(c) According to the model, during what year were there 7.08 thousand main bank offices?

Concept and Writing Exercises

86. Sometimes the slope is referred to as "rise over run." Explain the meaning of that phrase using a complete sentence.

87. In a linear depreciation model $f(x) = mx + b$, what can we say about the value of m? Explain your answer in a complete sentence.

88. The slope between the points $(x_1, 3)$ and $(2, 6)$ is $m = 0.3$. Determine the value of x_1.

89. The average rate of change between the ordered pairs $(8, 3)$ and $(6, y_2)$ is $m = \frac{5}{2}$. Determine the value of y_2.

90. Suppose a line contains the points $(1, 5)$ and $(6, y_2)$. What values of y_2 are necessary so that the line is increasing? Write your answer using interval notation.

91. Suppose a line contains the points $(-6, y_1)$ and $(7, 6)$. What values of y_1 are needed so that the line is decreasing? Write your answer using interval notation.

 ## Section Project

"China has passed the U.S. to become the world's biggest energy consumer, according to new data from the International Energy Agency, a milestone that reflects both China's decades-long burst of economic growth and its rapidly expanding clout as an industrial giant." (*Source:* "China Tops US in Energy," *Wall Street Journal.*) The amount of energy consumed in the United States from 2000 to 2011 decreased slightly and can be modeled by

$$f(x) = -2.1x + 2216 \quad 0 \le x \le 11$$

where x represents the number of years since 2000, and $f(x)$ represents the annual energy use in millions of tons of oil. In 2000, China's energy consumption was 1275 million tons of oil and increased at an average rate of 93 millions of tons of oil per year through 2011.

(a) Write a linear model for China's energy consumption in the form

$$g(x) = mx + b \quad 0 \le x \le 11$$

where x represents the number of years since 2000, and $g(x)$ represents the annual energy use in millions of tons of oil.

© pdtnc/ShutterStock, Inc.

(b) Evaluate $f(5)$ and $g(5)$, and interpret each value.

(c) Evaluate $f(11)$ and $g(11)$, and interpret each value.

(d) Inspect the results from parts (b) and (c). How do these results show that China has passed the United States to become the world's biggest energy consumer?

 (e) Graph the functions f and g in the viewing window [0, 11] by [1000, 2500] and use the `Intersect` command to determine the point of intersection of the two functions. Round the x-value to the nearest whole number. Interpret this result.

(f) Verify the solution to part (e) algebraically by setting the functions f and g equal to each other and solving for x.

1.3 Quadratic Functions and Secant Line Slope

SECTION OBJECTIVES

1. Evaluate a quadratic function.

2. Determine the zeros of a quadratic function.

3. Determine the zeros of a quadratic function by using the Quadratic Formula.

4. Use the slope of a secant line to give the average rate of change of a quadratic function on a closed interval.

In this section, we turn our attention to quadratic functions. This family of functions has many applications in the physical laws of motion, economics, and the social sciences. We then extend our knowledge of the average rate of change by discussing the secant line slope of a function on an interval.

Properties of Quadratic Functions

After the linear function, the next most common type of function we study is the **quadratic function**. The quadratic function may be called a **second-degree function** since the largest exponent for the independent variable is 2.

DEFINITION

Quadratic Function

A function of the form $f(x) = ax^2 + bx + c$ is called a **quadratic function**, where a, b, and c are real numbers and $a \neq 0$.

Since there are no x-values that can make a denominator zero or give us the square root of a negative number, the domain of a quadratic function is the set of real numbers $(-\infty, \infty)$. The shape of the graph of a quadratic function is called a **parabola**.

Example 1: Evaluating a Quadratic Function

OBJECTIVE 1

Evaluate a quadratic function.

The first Earth Day was April 22, 1970, and in the time since then, Americans have become more aware of recycling programs to help save natural resources. The total paper and paperboard waste that is generated in the United States can be modeled by

$$f(x) = -0.15x^2 + 3.38x + 69.38 \quad 1 \leq x \leq 19$$

where x represents the number of years since 1989, and $f(x)$ represents the total paper and paperboard waste generated in millions of tons. Evaluate $f(13)$ and interpret the result. (*Source*: U.S. Statistical Abstract).

Perform the Mathematics

Since x represents the number of years since 1989, $x = 13$ corresponds to the year 2002. To evaluate $f(13)$ requires us to substitute 13 for every occurrence of x in the model. Evaluating the model at $x = 13$ gives us

$$f(13) = -0.15(13)^2 + 3.38(13) + 69.38$$
$$= -25.35 + 43.94 + 69.38$$
$$= 87.97$$

© Geoffrey Kuchera/Dreamstime.com

This means that in 2002 about 87.97 million tons of paper and paperboard waste was generated in the United States. See **Figure 1.3.1**.

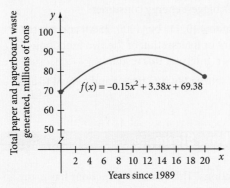

Figure 1.3.1 Graph of $f(x) = -0.15x^2 + 3.38x + 69.38$ on $[1, 19]$.

DEFINITION

Vertex of a Parabola

The x-coordinate of the **vertex** of a parabola given by $f(x) = ax^2 + bx + c$ is

$$x = \frac{-b}{2a}$$

The y-coordinate can be found by evaluating $f\left(\frac{-b}{2a}\right)$.

Figure 1.3.1 shows that the graph appears to have a peak or maximum value when x has a value of about 11. This peak is called the **vertex** of the parabola. To the left of the vertex, the graph appears to be increasing. To the right, the graph appears to be decreasing.

The vertex of a parabola is one of its most important features, since it gives us the maximum or minimum value of the function. Usually, finding a maximum or minimum value requires tools from calculus, but in the case of quadratic functions, we can use a simple formula to find the x-coordinate of the **vertex**. (You will derive this property in Chapter 5.)

The way that the parabola given by $f(x) = ax^2 + bx + c$ opens is determined by the value of a, which is called the **leading coefficient**. If a is a positive number, the graph opens upward, and if a is negative, the graph opens downward. If the graph of a parabola opens up, the vertex is a **minimum**, and if the graph opens down, the vertex is a **maximum**. See **Figures 1.3.2(a)** and **(b)**.

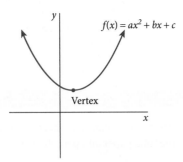

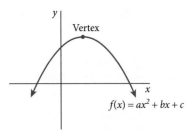

(a) If $a > 0$, then the parabola opens up, and the vertex is a minimum.

(b) If $a < 0$, then the parabola opens down, and the vertex is a maximum.

Figure 1.3.2

Consequently, if we know the y-coordinate of the vertex and the direction in which the graph opens, we can determine the range of any quadratic function. For example, consider the function $f(x) = -x^2 + x + 6$. Since the leading coefficient of the quadratic function is $a = -1$, we know that the graph opens down. This means that the vertex is a maximum. The x-coordinate of the vertex is given by $x = \frac{-b}{2a} = \frac{-1}{2(-1)} = \frac{1}{2}$. The y-coordinate of the vertex is

$$f\left(\frac{1}{2}\right) = -\left(\frac{1}{2}\right)^2 + \left(\frac{1}{2}\right) + 6$$

$$= -\frac{1}{4} + \frac{1}{2} + 6 = \frac{25}{4}$$

Since the maximum function value is $y = \frac{25}{4}$, this gives the range of the function as $(-\infty, \frac{25}{4}]$. This is shown in **Figure 1.3.3**.

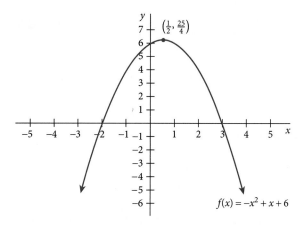

Figure 1.3.3

Determining Zeros of Quadratic Functions

The real **zeros** of a function f are given by the x-intercepts of its graph. They can also be called the **roots** of the function, because they are the solutions to the equation $f(x) = 0$. Many times, the zeros, or roots, of a function can be found algebraically by the process of **factoring**. We just set $f(x)$ equal to 0 and factor, as shown in Example 2.

Example 2: Determining the Zeros of a Quadratic Function by Factoring

OBJECTIVE 2

Determine the zeros of a quadratic function by factoring.

Determine the zeros of the function $f(x) = 4x^2 + 6x - 4$. Check the result numerically.

Perform the Mathematics

To find the zeros for this function, we need to solve the quadratic equation $4x^2 + 6x - 4 = 0$. So, by factoring, we get

Given equation	$4x^2 + 6x - 4 = 0$
Factor out the constant 2	$2(2x^2 + 3x - 2) = 0$
Divide both sides by 2	$2x^2 + 3x - 2 = 0$
Factor the expression $2x^2 + 3x - 2$	$(x + 2)(2x - 1) = 0$
Set each factor equal to zero	$(x + 2) = 0$ or $(2x - 1) = 0$
Solve for x	$x = -2$ or $x = \dfrac{1}{2}$

We can check our solutions by evaluating $f(x) = 4x^2 + 6x - 4$ at both solutions to see if the function value is zero. Checking at $x = -2$, we get

$$f(-2) = 4(-2)^2 + 6(-2) - 4$$
$$= 16 - 12 - 4 = 0$$

At the second zero $x = \frac{1}{2}$, we get

$$f\left(\frac{1}{2}\right) = 4\left(\frac{1}{2}\right)^2 + 6\left(\frac{1}{2}\right) - 4$$
$$= 1 + 3 - 4 = 0$$

So the zeros of the function $f(x) = 4x^2 + 6x - 4$ are $x = -2$ and $x = \frac{1}{2}$. ∎

Many times, the quadratic that we wish to solve is not easily factorable. To algebraically solve these types, we use the **quadratic formula**. The derivation of this formula appears in Appendix C.

DEFINITION

Quadratic Formula

The zeros of the quadratic function $f(x) = ax^2 + bx + c$ are given by the **quadratic formula**

$$x = \frac{-b \pm \sqrt{b^2 - 4ac}}{2a}$$

$b^2 - 4ac$ is called the **discriminant**.

If $b^2 - 4ac > 0$, the graph looks like

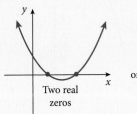 or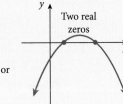

If $b^2 - 4ac = 0$, the graph looks like

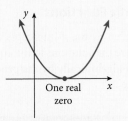 or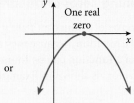

If $b^2 - 4ac < 0$, the graph looks like

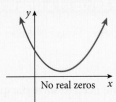 or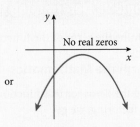

OBJECTIVE 3

Determine the zeros of a quadratic function by using the Quadratic Formula.

Example 3: Determining Zeros with the Quadratic Formula

Determine the zeros of the quadratic function $f(x) = 2x^2 - 6x + 3$.

Perform the Mathematics

To find the zeros of the function, we need to solve the quadratic equation $2x^2 - 6x + 3 = 0$. Because the expression $2x^2 - 6x + 3$ is not easily factorable, we determine the zeros by evaluating the Quadratic Formula using $a = 2$, $b = -6$, and $c = 3$. This gives us

$$x = \frac{-(-6) \pm \sqrt{(-6)^2 - 4(2)(3)}}{2(2)}$$

$$= \frac{6 \pm \sqrt{36 - 24}}{4} = \frac{6 \pm \sqrt{12}}{4}$$

$$= \frac{6 \pm 2\sqrt{3}}{4} = \frac{2(3 \pm \sqrt{3})}{4} = \frac{3 \pm \sqrt{3}}{2}$$

So the two real zeros of the function $f(x) = 2x^2 - 6x + 3$ are $x = \dfrac{3 - \sqrt{3}}{2}$ and $x = \dfrac{3 + \sqrt{3}}{2}$. ∎

▶ *Try It Yourself*

Some related Exercises are 21 and 27.

Technology Option

Sometimes even the Quadratic Formula can be unwieldy when locating zeros of a quadratic function. In some situations, we use a calculator to approximate the solutions to quadratic equations. To approximate the zeros of $f(x) = 2x^2 - 6x + 3$, we can use the `Zero` or `Root` command on our calculator. **Figure 1.3.4(a)** shows the approximation of the zero of the leftmost x-intercept, and the rightmost is shown in Figure 1.3.4(b). So the approximations are $x \approx 0.63$ and $x \approx 2.37$, rounded to the nearest hundredth.

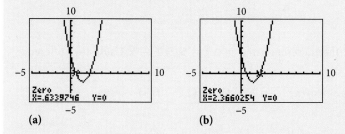

(a)   (b)

Figure 1.3.4

Secant Line Slope and the Difference Quotient

In Section 1.2, we introduced the average rate of change of a line. At times, we want to know what the average rate of change would be for a **nonlinear function** over a specified closed interval. We can find this rate by computing the slope of a **secant line**, which is a line that passes through two points on a curve.

To determine the slope of a secant line, let's suppose that a curve defined by a nonlinear function f has two distinct points, $P = (x_1, y_1)$ and $Q = (x_2, y_2)$. The average rate of change, or slope, of a line passing through P and Q is

$$m = \frac{y_2 - y_1}{x_2 - x_1} = \frac{f(x_2) - f(x_1)}{x_2 - x_1}$$

We call the change from P to Q in the x-coordinate Δx (read "delta x"). This means that $\Delta x = x_2 - x_1$. If we let $x_1 = x$, then $x_2 = x + \Delta x$. See **Figure 1.3.5**.

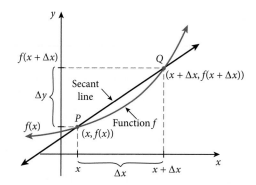

Figure 1.3.5

OBJECTIVE 4

Use the slope of a secant line to give the average rate of change of a quadratic function on a closed interval.

DEFINITION

Secant Line Slope and the Difference Quotient

The **average rate of change** of a function f on an interval $[x, x + \Delta x]$ is equivalent to the **slope of the secant line** through the points $(x, f(x))$ and $(x + \Delta x, f(x + \Delta x))$, denoted m_{sec}, and is given by the **difference quotient**

$$m_{sec} = \frac{f(x + \Delta x) - f(x)}{\Delta x}$$

where $\Delta x \neq 0$. The units of this average rate of change are $\dfrac{\text{units of } f}{\text{units of } x}$.

NOTE: The denominator of the difference quotient, Δx, is also called the **increment in x**. The numerator $f(x + \Delta x) - f(x)$ is the same as Δy and is called the **increment in y**.

Observe in Figure 1.3.5 that the coordinates of the points P and Q are $(x, f(x))$ and $(x + \Delta x, f(x + \Delta x))$, respectively. Now, determining the $\dfrac{\text{change in } y}{\text{change in } x}$, we get

$$\text{Secant line slope} = \frac{\text{change in } y}{\text{change in } x} = \frac{f(x + \Delta x) - f(x)}{(x + \Delta x) - x} = \frac{f(x + \Delta x) - f(x)}{\Delta x}$$

This fraction represents the **average rate of change** between the points P and Q. The fraction $\dfrac{f(x + \Delta x) - f(x)}{\Delta x}$ is also called the **difference quotient**.

Example 4: Determining the Secant Line Slope

Find the average rate of change of $f(x) = 2x^2 + 1$ on the interval $[2, 5]$.

Perform the Mathematics

The first x-value is $x = 2$, and the increment in x is $\Delta x = 5 - 2 = 3$. Using the difference quotient gives us

$$m_{sec} = \frac{f(x + \Delta x) - f(x)}{\Delta x} = \frac{f(2 + 3) - f(2)}{3} = \frac{f(5) - f(2)}{3}$$

$$= \frac{(2(5)^2 + 1) - (2(2)^2 + 1)}{3} = \frac{51 - 9}{3} = \frac{42}{3} = 14$$

Since the slope of the secant line on $[2, 5]$ is 14, we conclude that the average rate of change of f on $[2, 5]$ is 14, or $14 \dfrac{\text{units of } f}{\text{units of } x}$. The graph of f and the secant line is shown in **Figure 1.3.6**.

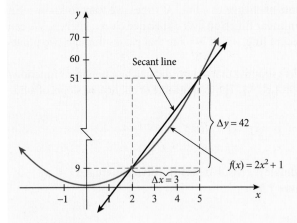

Figure 1.3.6

▶ *Try It Yourself*

Some related Exercises are 39 and 41.

In applications, the secant line slope can be an indicator of the general trend in the rate of change of some phenomena on some interval. This is illustrated in Example 5.

Example 5: Interpreting a Secant Line Slope

During the first decade of the 21st century, the United States has been searching for alternative fuel sources as the demand for oil has increased. The production of fuel ethanol from corn as well as biodiesel has increased since 2000 in the United States. The amount of fuel ethanol and biodiesel produced in the United States can be modeled by

$$f(x) = 3.36x^2 - 12.62x + 49.54 \quad 1 \le x \le 10$$

where x is the number of years since 1999, and $f(x)$ is the U.S. production of fuel ethanol and biodiesel in millions of barrels. (*Source*: U.S. Statistical Abstract.) Determine the average rate of change in U.S. production of fuel ethanol and biodiesel from 2002 to 2007, and interpret. Include appropriate units.

Understand the Situation

For the years 2002 to 2007, the corresponding x-values are $x = 3$ and $x = 8$, respectively. Thus, $\Delta x = 8 - 3 = 5$.

Perform the Mathematics

Now we compute the difference quotient as

$$m_{\text{sec}} = \frac{f(x + \Delta x) - f(x)}{\Delta x} = \frac{f(3 + 5) - f(3)}{5}$$

$$= \frac{f(8) - f(3)}{5} = \frac{163.62 - 41.92}{5} = \frac{121.7}{5}$$

$$= 24.34$$

Interpret the Result

This means that during the period from 2002 to 2007, U.S. production of fuel ethanol and biodiesel increased at an average rate of 24.34 million barrels per year.

 Try It Yourself

Some related Exercises are 49 and 51. ■

© David R. Frazier Photolibrary, Inc./Alamy

Summary

This section featured the introduction of the **quadratic function**, $f(x) = ax^2 + bx + c$, which is a second-degree function with $a \ne 0$. We found that every graph of a quadratic function is a parabola, and the **vertex** of the parabola was found at the coordinates $\left(\dfrac{-b}{2a}, f\left(\dfrac{-b}{2a} \right) \right)$. Zeros, also known as roots of the quadratic function, or x-intercepts of the graph, can be found in three ways:

- Algebraic methods such as factoring

- The **Quadratic Formula**, $x = \dfrac{-b \pm \sqrt{b^2 - 4ac}}{2a}$

- The capabilities of the graphing calculator

Finally, we introduced the **slope of the secant line** and found that this type of average rate of change could be computed by the difference quotient

$$m_{\text{sec}} = \frac{f(x + \Delta x) - f(x)}{\Delta x}, \Delta x \ne 0$$

We found that the **difference quotient** could determine the average rate of change over an interval for a nonlinear function. As you will see in upcoming sections, the difference quotient will be used for more than just quadratic functions.

Section 1.3 Exercises

Vocabulary Exercises

1. The shape of a quadratic function is that of a _____.

2. A quadratic function is sometimes referred to as a _____ - _____ function.

3. The minimum or maximum ordered pair associated with a quadratic function is called its

   _____.

4. The values found by solving the equation $ax^2 + bx + c = 0$ are called its roots. The associated graphical values of $f(x) = ax^2 + bx + c$ are called its _____.

5. The radicand of the quadratic formula is called the _____.

6. The average rate of change of a nonlinear function on a closed interval $[x, x + \Delta x]$ is called the _____ line slope.

Skill Exercises

For Exercises 7–16, complete the following.

(a) *Determine the coordinates of the vertex algebraically.*

(b) *Use the solution from part (a) to determine the interval where the function is increasing and where it is decreasing.*

7. $f(x) = x^2 + 6x + 5$
8. $f(x) = x^2 - 4x + 2$
9. $f(x) = -x^2 + 6x + 6$
10. $f(x) = -2x^2 - 4x + 5$
11. $f(x) = 5x^2 + 6x - 3$
12. $f(x) = 5x^2 - 7x + 2$
13. $f(x) = 0.14x^2 + 0.5x - 0.3$
14. $f(x) = -0.81x^2 + 3.24x - 0.4$
15. $f(x) = -0.9x^2 - 1.8x + 0.5$
16. $f(x) = -0.35x^2 + 2.8x - 0.3$

For Exercises 17–24, determine the zeros of the quadratic function by factoring.

17. $f(x) = x^2 - 16$
18. $f(x) = x^2 - 49$
19. $f(x) = -2x^2 - 4x$
20. $f(x) = -13x^2 - 39x$
21. $f(x) = x^2 - 5x + 6$
22. $f(x) = x^2 + 2x - 8$
23. $f(x) = 6x^2 - 5x - 50$
24. $f(x) = 8x^2 + 14x + 3$

For Exercises 25–30, determine the real number zeros of each quadratic function using the quadratic formula. If the function does not have real number roots, write "no real zeros."

25. $f(x) = x^2 - x - 1$
26. $f(x) = x^2 - 3x - 2$
27. $f(x) = 11x^2 - 7x + 1$
28. $f(x) = 4x^2 - 12x + 11$
29. $f(x) = 2x^2 + 2x + 1$
30. $f(x) = 9x^2 - 12x + 8$

In Exercises 31–36, use the `Zero` *command on your graphing calculator to approximate the zeros of each quadratic function. Round the zeros to the nearest hundredth. If the function does not have real number roots, write "no real zeros."*

 31. $f(x) = x^2 - 2x - 24$ 32. $f(x) = x^2 - 2x - 15$

 33. $f(x) = 4x^2 - 12x + 9$ 34. $f(x) = 9x^2 + 24x + 16$

 35. $f(x) = 1.25x^2 - 5.1x + 8.3$ 36. $f(x) = 1.53x^2 - 3x + 2.67$

For Exercises 37–42, compute the slope of the secant line over the indicated interval.

37. $f(x) = -5x + 1$

   (a) x changes from $x = 0$ to $x = 6$.

   (b) x changes from $x = 0$ to $x = 3$.

   (c) x changes from $x = 3$ to $x = 4$.

38. $f(x) = 3x - 5$

   (a) x changes from $x = 0$ to $x = 8$.

   (b) x changes from $x = 0$ to $x = 4$.

   (c) x changes from $x = 4$ to $x = 8$.

39. $f(x) = x^2 + x$

   (a) x changes from $x = 0$ to $x = 4$.

   (b) x changes from $x = 0$ to $x = 2$.

   (c) x changes from $x = 2$ to $x = 4$.

40. $f(x) = x^2 - 2x$

   (a) x changes from $x = 0$ to $x = 6$.

   (b) x changes from $x = 0$ to $x = 3$.

   (c) x changes from $x = 3$ to $x = 6$.

41. $f(x) = x^2 - x + 3$

   (a) If $x = 1$ and $\Delta x = 2$.

   (b) If $x = 1$ and $\Delta x = 1$.

   (c) If $x = 1$ and $\Delta x = 0.5$.

42. $f(x) = 2x^2 - 3$

   (a) If $x = 2$ and $\Delta x = 3$.

   (b) If $x = 2$ and $\Delta x = 1$.

   (c) If $x = 2$ and $\Delta x = 0.5$.

Application Exercises

43. **Law Enforcement—Citations:** The number of parking citations that the Sampsonburg Police Department gave over a five-year period can be modeled by

$$f(x) = \frac{1}{5}x^2 + 100x + 30 \quad 0 \le x \le 5$$

where x represents the number of years since records of the citations began to be taken, and $f(x)$ represents the number of citations given. Determine the secant line slope $m_{sec} = \dfrac{f(5) - f(0)}{5}$ and interpret the result.

44. **Manufacturing—Sunglasses:** The number of unbreakable sunglasses sold at the U-C-Me specialty store during a long-term sales promotion can be modeled by

$$f(x) = \frac{1}{10}x^2 + 50x + 10 \quad 0 \le x \le 10$$

where x represents the number of months since the promotion began, and $f(x)$ represents the number of sunglasses sold. Determine the secant line slope $m_{sec} = \dfrac{f(10) - f(0)}{10}$ and interpret the result.

45. **Physics—Ballistics:** If a rock is thrown from the ground with an initial velocity of 80 feet per second, then its height can be modeled by

$$s(t) = -16t^2 + 80t \quad 0 \le t \le 5$$

where t represents the number of seconds since the rock was thrown, and $s(t)$ represents the rock's height in feet.

   (a) Evaluate $s(3)$, and interpret.

   (b) Determine the average rate of change in the rock's height for $t = 1$ to $t = 3$ and interpret.

46. **Physics—Ballistics:** If a model rocket is launched from a 3-foot platform with an initial velocity of 150 feet per second, then its height can be modeled by

$$s(t) = -16t^2 + 150t + 3$$

where t represents the number of seconds since launch, and $s(t)$ represents the height in feet.

   (a) Evaluate $s(8)$, and interpret.

   (b) Determine the average rate of change in the rocket's height for $t = 5$ to $t = 8$ and interpret.

47. Biology—Bacterial Growth: The number of bacteria in a colony after t hours is given by

$$g(t) = t^2 + 8t + 2000 \quad 0 \le t \le 24$$

where t is the number of hours since the colony was established, and $g(t)$ represents the number of bacteria.

(a) Evaluate $g(3)$ and interpret.

(b) Determine the average rate of change in the increase of the colony's population for $t = 3$ to $t = 6$, and interpret.

48. Botany—Plant Growth: The number of seeds dispersed by a plot of dandelions each day in April can be modeled by

$$f(x) = 40x - x^2 \quad 1 \le x \le 30$$

where x represents the number of days in April, and $f(x)$ represents the number of seeds dispersed.

(a) Evaluate $f(7)$ and interpret. Does the graph of the model open up or down? How do you know this?

(b) On which day(s) of the month is the dispersion of seeds the greatest?

(c) Determine the average rate of change in the dispersion of the seeds between $x = 24$ and $x = 29$, and interpret.

 49. Law—Supreme Course Docket: The total number of cases on the docket of the U.S. Supreme Court has increased over the past few decades. The number of cases on the Supreme Court docket from 1980 to 2011 can be modeled by the quadratic function

$$f(x) = -1.79x^2 + 219.65x + 4891 \quad 0 \le x \le 31$$

where x represents the number of years since 1980, and $f(x)$ represents the number of cases on the U.S. Supreme Court annually. (*Source:* Office of the Clerk, U.S. Supreme Court.)

(a) Use the model to determine the number of U.S. Supreme court cases in 1985.

(b) In 1972, then–Chief Justice Warren Burger warned of the increasing Court workload, stating, "something must be done to arrest the increase in docketed cases." (*Source: Los Angeles Times.*) Determine m_{sec} when $x = 5$ and $\Delta x = 25$, and interpret the result.

© Gary Blakeley/ShutterStock, Inc.

 50. Pathology—Botulism Cases: Botulism is a rare but serious illness caused by *Clostridium botulinum* bacteria. The bacteria may enter the body through wounds, or they may live in improperly canned or preserved food. (*Source:* National Center for Biotechnology Information.) The number of cases of botulism in the United States from 1980 to 2011 can be modeled by the quadratic function

$$f(x) = -0.11x^2 - 0.48x + 87.77 \quad 0 \le x \le 31$$

where x represents the number of years since 1980, and $f(x)$ represents the number of cases of botulism detected annually. (*Source:* Centers for Disease Control and Prevention.)

(a) According to the model, how many botulism cases were there in 2000?

(b) Recent regulations, including increased inspections of canned goods and banning Botox injections for minors, have resulted in a steady decrease in the number of cases of the deadly disease. Determine m_{sec} when $x = 20$ and $\Delta x = 5$, and interpret the result.

 51. Medicine—Nonprofit Groups: Groups such as the American Cancer Society and the American Heart Association are part of a growing number of nonprofit health and medical associations. The number of these associations from 1980 to 2011 can be modeled by the function

$$f(x) = 1.17x^2 + 33.46x + 1506 \quad 0 \le x \le 31$$

© Getty Images

where x represents the number of years since 1980, and $f(x)$ represents the number of non-profit health and medical associations in the United States each year. (*Source: The Encyclopedia of Associations.*)

(a) Compute m_{sec} for $x = 0$ and $\Delta x = 10$. Interpret this value.

(b) Now compute m_{sec} for $x = 10$ and $\Delta x = 10$, and m_{sec} for $x = 20$ and $\Delta x = 10$.

(c) According to the results of parts (a) and (b), during which decade did the number of non-profit health and medical associations grow the least?

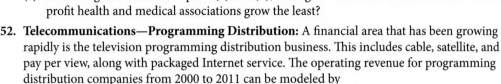 **52. Telecommunications—Programming Distribution:** A financial area that has been growing rapidly is the television programming distribution business. This includes cable, satellite, and pay per view, along with packaged Internet service. The operating revenue for programming distribution companies from 2000 to 2011 can be modeled by

$$f(x) = x^2 - 1.6x + 63.9 \quad 0 \le x \le 11$$

where x represents the number of years since 2000, and $f(x)$ represents the operating revenue in billions of dollars annually. (*Source:* U.S. Census Bureau.)

(a) Evaluate $f(10)$ and $f(0)$, then subtract the two function values. Interpret the meaning of this value.

(b) Divide the result from part (a) by 10. What does this number represent?

Concept and Writing Exercises

53. In this section, we found how to compute the slope of a secant line. Describe what a secant line is relative to a quadratic function.

54. Explain how we can use the coefficient of a quadratic function to determine if a vertex is a minimum value or a maximum value.

Use the following information for Exercises 55 and 56.

Another way to write a quadratic function is the form $f(x) = a(x - h)^2 + k$, where (h, k) are the coordinates of the vertex, and a is a real number constant. This means we can write a quadratic function if we know the coordinates of the vertex and another point on the parabola so that we can solve for the value of a. We can then use algebra to rewrite the function in the standard form $f(x) = ax^2 + bx + c$.

55. Write a quadratic function in standard form for a parabola whose vertex is at $(-3, 1)$, knowing that $f(1) = 33$.

56. Write a quadratic function in standard form for a parabola whose vertex is at $(-4, -6)$ and whose y-intercept is at $(0, -14)$.

57. For the function $f(x) = x^2 - 6x$, we have $x = 0$ and know that $m_{sec} = -3$. Determine the value of Δx.

58. For the function $f(x) = 2x^2 - 4x + 6$, we have $x = 1$ and know that $m_{sec} = 8$. Determine the value of Δx.

Section Project

Many adults have found that a simple way to reduce their caloric intake is to switch from whole milk to reduced-fat or skim milk. The per capita consumption of whole milk from 1980 to 2011 can be modeled by the function

$$f(x) = 0.01x^2 - 0.71x + 16.8 \quad 0 \le x \le 31$$

Courtesy of Keith Weller/USDA ARS

where x represents the number of years since 1980, and $f(x)$ represents the per capita consumption of whole milk in gallons annually. (*Source:* U.S. Department of Agriculture.)

(a) Does the graph of f open up or open down? How can you tell this by inspecting the function?

(b) Determine the slope of the secant line of f from 1980 to 1990. Interpret the answer.

The per capita consumption of reduced-fat and skim milk from 1980 to 2011 can be modeled by the function

$$g(x) = -0.01x^2 + 0.48x + 10.83 \quad 0 \le x \le 31$$

where x represents the number of years since 1980, and $f(x)$ represents the per capita consumption of reduced-fat and skim milk in gallons annually.

(c) Does the graph of g open up or open down? How can you tell this by inspecting the function?

(d) Determine the slope of the secant line of g from 1980 to 1990. Interpret the answer.

(e) Compare the results from parts (b) and (d). What do these values mean in terms of the consumption of whole milk versus reduced-fat and skim milk?

 (f) Graph the functions f and g in the viewing window [0, 31] by [0, 20]. Use the `Intersect` command to determine the point of intersection of the functions. Round to the nearest whole number the x-value of the point of intersection of these functions. What is the meaning of this value?

SECTION OBJECTIVES

1. Evaluate operations of functions.

2. Determine a revenue function using operations of functions.

3. Determine a profit function using operations of functions.

4. Determine break-even points.

5. Compute an average rate of change of a revenue function on a closed interval.

6. Compute an average rate of change of a polynomial function on a closed interval.

7. Determine the end behavior of a graph of a polynomial function.

8. Evaluate composition of functions.

1.4   Operations on Functions and Composite Functions

In this section we extend our discussion to functions of higher degree. In general, we can call this family of functions **polynomial functions**. While continuing to build up our tools to prepare for the study of calculus, we will also discuss **operations on functions**, which are commonly used to determine the **business functions**. Finally, we will revisit the difference quotient and examine more applications of the secant line as well as look at **composite functions**.

Operations on Functions

The basics of arithmetic with real numbers—addition, subtraction, multiplication, and division—are called mathematical **operations**. These operations can also be performed on functions.

DEFINITION

Operations on Functions

Let f and g be functions. Then, for all x-values for which both f and g exist, we have

1. The **sum** of f and g is $(f + g)(x) = f(x) + g(x)$.

2. The **difference** of f and g is $(f - g)(x) = f(x) - g(x)$.

3. The **product** of f and g is $(f \cdot g)(x) = f(x) \cdot g(x)$.

4. The **quotient** of f and g is $\left(\dfrac{f}{g}\right)(x) = \dfrac{f(x)}{g(x)}$, provided that $g(x) \ne 0$.

Example 1: Evaluating Operations of Functions

Let $f(x) = x^2 + x - 3$ and $g(x) = 4x + 2$. Evaluate the following.

a. $(f + g)(1)$

b. $(f - g)(1)$

c. $(f \cdot g)(1)$

d. $\left(\dfrac{f}{g}\right)(1)$

Perform the Mathematics

We first evaluate the two functions at $x = 1$ and get

$$f(1) = (1)^2 + 1 - 3 = -1 \quad \text{and} \quad g(1) = 4(1) + 2 = 6$$

a. From the definition of sum of functions, we get

$$(f + g)(1) = f(1) + g(1) = -1 + 6 = 5$$

b. From the definition of difference of functions, we get

$$(f - g)(1) = f(1) - g(1) = -1 - 6 = -7$$

c. From the definition of product of functions, we get

$$(f \cdot g)(1) = f(1) \cdot g(1) = (-1)(6) = -6$$

d. From the definition of the quotient of functions, we get

$$\left(\frac{f}{g}\right)(1) = \frac{f(1)}{g(1)} = \frac{-1}{6} \qquad \blacksquare$$

▶ *Try It Yourself*

Some related Exercises are 13 and 15.

The critical restriction when using operations with functions occurs when computing the quotient $\left(\dfrac{f}{g}\right)(x)$. We must make sure to exclude all values from the domain that can make the denominator zero.

Functions of Business

The operations on functions are commonly used in the functions of business, such as price-demand and cost functions. Other types of business functions are illustrated in everyday life. For example, let's say that a youngster has a lemonade stand in her front yard and that she charges 50 cents for each cup of lemonade. If 35 cups are sold in an afternoon, the amount that the youngster makes is $\$0.50(35) = \$17.50$. This amount earned represents the **revenue**. We computed this revenue by multiplying the price per cup, called the **unit price**, times the quantity sold. We can generalize this notion to form the **revenue function**.

Determine a revenue function using operation of functions.

DEFINITION

Revenue Function

The revenue generated by selling a certain quantity of a product at a certain price is

$$(\text{Total revenue}) = (\text{quantity sold}) \cdot (\text{unit price})$$

For x units sold at a price given by the price function $p(x)$, the **revenue function** R as a function of x is given by

$$R(x) = x \cdot p(x)$$

Example 2: Determining a Revenue Function

Table 1.4.1

Demand x	Price $p(x)$
10	$125
30	$75

The Professional Bookstore knows from past sales records that the weekly number of units demanded, x, of a certain pack of computer tutorial software at price p is shown in **Table 1.4.1**.

a. Use Table 1.4.1 to find a linear price function, p, written in slope-intercept form for the tutorial software.

b. Use the result from part (a) to determine the revenue function R.

c. Determine the vertex of the graph of R and interpret each coordinate.

Perform the Mathematics

a. To find the slope-intercept form, we must compute the slope m from the data given in Table 1.4.1. This gives us

$$m = \frac{75 - 125}{30 - 10} = \frac{-50}{20} = -\frac{5}{2}$$

Using the point-slope form with $(x_1, y_1) = (10, 125)$, we find that the price function is

$$y - y_1 = m(x - x_1)$$

$$y - 125 = -\frac{5}{2}(x - 10)$$

$$y - 125 = -\frac{5}{2}x + 25$$

$$y = -\frac{5}{2} + 150$$

Substituting $p(x)$ for y gives us

$$p(x) = -\frac{5}{2}x + 150$$

b. To get the revenue function R, we need to multiply the price function by the quantity x. This gives us

$$R(x) = x \cdot p(x) = x \cdot \left(-\frac{5}{2}x + 150\right) = -\frac{5}{2}x^2 + 150x$$

Notice that this revenue function gives a graph that is a parabola that opens down. See **Figure 1.4.1**. This is a common shape for revenue functions.

c. The point where revenue is a maximum is at the vertex. Recall that the x-coordinate of the vertex is given by $x = \dfrac{-b}{2a}$. For this revenue function, we know that $a = -\frac{5}{2}$ and $b = 150$, so we get

$$x = \frac{-150}{2\left(-\dfrac{5}{2}\right)} = \frac{-150}{-5} = 30$$

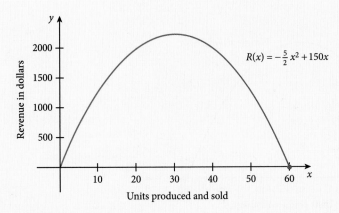

$$R(x) = -\frac{5}{2}x^2 + 150x$$

Figure 1.4.1

Evaluating the revenue function at $x = 30$, we get

$$R(30) = -\frac{5}{2}(30)^2 + 150(30) = -2250 + 4500 = 2250$$

So the weekly revenue is at its maximum of $2250 when 30 packs of tutorial software are sold.　■

Let's return to our young entrepreneur and her lemonade stand. If the cost of making the lemonade was $4.25, then the **profit** that is made during the afternoon when she sold 35 cups is $17.50 − $4.25 = $13.25. This same model applies no matter how small or large the business venture may be. The profit made is simply revenue minus cost. This relationship between these two quantities allows us to form the **profit function**.

Example 3: Determining a Profit Function

The Flashbroc Company makes designer ties. It determines that the weekly cost and revenue functions, in dollars, for producing and selling x designer ties are

$$C(x) = 30x + 50 \quad \text{and} \quad R(x) = 90x - x^2$$

respectively.

a. Determine the profit function.

b. Determine the vertex of the graph of P, and interpret each coordinate.

Perform the Mathematics

a. Since the definition of the profit function is $P(x) = R(x) - C(x)$, we use it and get

$$P(x) = R(x) - C(x)$$
$$= (90x - x^2) - (30x + 50)$$
$$= -x^2 + 60x - 50$$

So the profit function for producing and selling x units is $P(x) = -x^2 + 60x - 50$. A graph of P is given in **Figure 1.4.2**.

b. The x-coordinate of the vertex is $x = \dfrac{-b}{2a} = \dfrac{-60}{2(-1)} = 30$. The y-coordinate is given by

$$P(30) = -(30)^2 + 60(30) - 50 = -900 + 1800 - 50 = 850$$

DEFINITION

Profit Function

The profit generated after producing and selling a certain quantity of a product is given by

Profit = Revenue − Cost

For x units of a product, the **profit function** P is

$$P(x) = R(x) - C(x)$$

where R and C are the revenue and cost functions, respectively.

NOTE: Notice that the last two definitions for revenue and profit use operations on functions. The revenue function is a product of functions, and the profit function is the difference of functions.

OBJECTIVE 3

Determine a profit function using operations of functions.

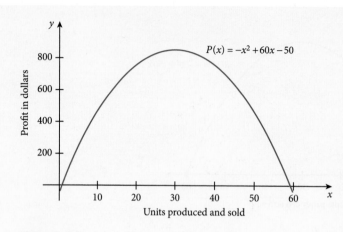

Figure 1.4.2

This means that a maximum profit of $850 will be realized when 30 designer ties are produced and sold each week. ■

Intuitively, when the revenue is equal to the cost (that is, when money coming in is equal to money going out), we have reached the **break-even point**. Mathematically, this occurs at the point where the graphs of the cost and revenue functions intersect. See **Figure 1.4.3**.

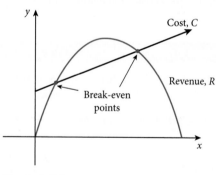

Figure 1.4.3

Algebraically, this means that the break-even point occurs where $R(x) = C(x)$. If the cost is more than the revenue, the result is a **loss**, and when the revenue is greater than the cost, a **profit** is the result.

OBJECTIVE 4

Determine the break-even points.

Example 4: Determining the Break-Even Points

In Example 2, we showed that the revenue function for the tutorial software packs was $R(x) = -\frac{5}{2}x^2 + 150x$. Now assume that the cost function associated with the tutorial software is given by $C(x) = 7.5x + 1505$.

a. Determine the break-even point(s) and interpret.

b. Determine the profit function P.

Perform the Mathematics

a. To determine the break-even point(s), we need the x-values such that $R(x) = C(x)$. Setting $R(x) = C(x)$ and solving for x yields

$$R(x) = C(x)$$

$$-\frac{5}{2}x^2 + 150x = 7.5x + 1505$$

$$-\frac{5}{2}x^2 + 142.5x - 1505 = 0$$

$$-5x^2 + 285x - 3010 = 0$$

$$x^2 - 57x + 602 = 0$$

Because this quadratic is not easily factorable, we use the quadratic formula to get

$$x = \frac{57 \pm \sqrt{(-57)^2 - 4(1)(602)}}{2(1)}$$

$$= \frac{57 \pm \sqrt{841}}{2} = \frac{57 \pm 29}{2}$$

So $x = \dfrac{57 + 29}{2}$, or $x = \dfrac{57 - 29}{2}$. These solutions simplify to $x = 43$ and $x = 14$, which gives us the x-coordinates of the break-even points. The y-coordinates are determined by evaluating either the cost function C or the revenue function R at these two x-values. (Recall from algebra that a point of intersection satisfies both equations.) Using the cost function C, we have

$$C(x) = 7.5x + 1505$$

$$C(14) = 7.5(14) + 1505 = 1610$$

and

$$C(43) = 7.5(43) + 1505 = 1827.5$$

So the two break-even points are (14, 1610) and (43, 1827.5). The break-even points tell us that when 14 packs of tutorial software are sold, the cost and revenue are equal at $1610. Likewise, when 43 packs of tutorial software are sold, the cost and revenue are equal at $1827.50.

b. Using the definition of the profit function, we have

$$P(x) = R(x) - C(x)$$

$$= \left(-\frac{5}{2}x^2 + 150x\right) - (7.5x + 1505)$$

$$= -\frac{5}{2}x^2 + 142.5x - 1505$$

So the profit function for the tutorial software is $P(x) = -\frac{5}{2}x^2 + 142.5x - 1505$. ∎

Using the cost, revenue, and profit functions in Example 4, we observe that the break-even points on the cost and revenue functions lie directly above the zeros of the profit function P. See **Figure 1.4.4**.

Relationship Between Profit, Revenue, and Cost

1. Break-even occurs when $R(x) = C(x)$. This is equivalent to $P(x) = 0$. Graphically, $P(x) = 0$ at the x-intercepts.

2. Profit occurs when $R(x) > C(x)$. This is equivalent to $P(x) > 0$, when the graph of P is above the x-axis.

3. A loss occurs when $R(x) < C(x)$. This is equivalent to $P(x) < 0$, when the graph of P is below the x-axis.

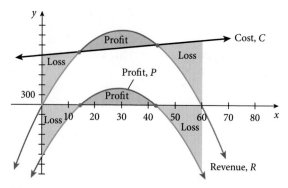

Figure 1.4.4

Thus, another way to find the break-even points is to find the zeros of the profit function. A profit is realized in the interval where the graph of P lies above the x-axis (that is, $P(x) > 0$). Let's summarize the **relationship between P, R, and C** before continuing.

We can also apply the difference quotient to business functions to approximate the average rate of change on specific intervals.

Example 5: Determining the Secant Line Slope of a Business Function

OBJECTIVE 5

Compute an average rate of change of a revenue function on a closed interval.

The Midwest Manufacturing Company manufactures titanium lock washers. Its revenue function is given by

$$R(x) = -0.02x^2 + 8x$$

where x represents the number of titanium lock washers sold, and $R(x)$ represents the amount of revenue generated in dollars. Find the average rate of change in revenue as the number of washers sold increases from 50 to 150 washers and interpret.

Understand the Situation

To find the desired average rate of change, we need to compute the difference quotient

$$m_{sec} = \frac{R(x + \Delta x) - R(x)}{\Delta x}$$

Perform the Mathematics

With $x = 50$ and $\Delta x = 150 - 50 = 100$, we get

$$m_{sec} = \frac{R(50 + 100) - R(50)}{100} = \frac{R(150) - R(50)}{100}$$

$$= \frac{-0.02(150)^2 + 8(150) - (-0.02(50)^2 + 8(50))}{100}$$

$$= \frac{750 - 350}{100}$$

$$= \frac{400}{100} = 4$$

Interpret the Result

This means that as the number of titanium lock washers sold increases from 50 to 150, the revenue increased at an average rate of $4 per washer. See **Figure 1.4.5**.

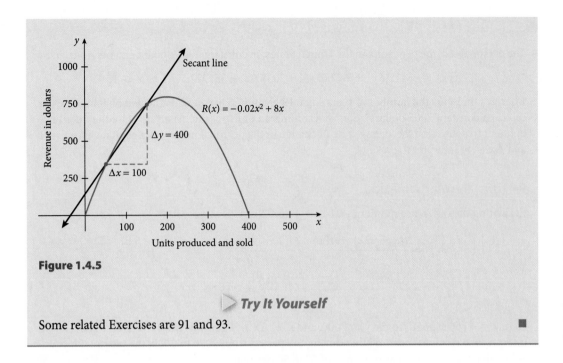

Figure 1.4.5

▷ *Try It Yourself*

Some related Exercises are 91 and 93.

Polynomial Functions

So far we have studied linear, or first-degree, functions and quadratic, or second-degree, functions. Now let's take a look at functions of higher degree. We call this general family of functions **polynomial functions**.

DEFINITION

Polynomial Function

A **polynomial function** of degree n has the form

$$f(x) = a_n x^n + a_{n-1} x^{n-1} + \cdots + a_1 x + a_0$$

where $a_0, a_1, a_2, \ldots, a_n$ are real number constants where $a_n \neq 0$ and n is a whole number. The **degree** of the polynomial function is given by n.

Using this definition, we see that $f(x) = 2x^3 - 4x^2 - x + 5$ is a third-degree polynomial function, also called a **cubic** function. The functions $f(x) = \sqrt{x} + 3$ and $g(x) = \frac{4}{x} + 8$ are not polynomial functions, because neither can be written with whole number exponents.

A few **properties of polynomial functions** make them very convenient to use mathematically.

DEFINITION

Properties of Polynomial Functions

1. The domain of every polynomial function is the set of real numbers, $(-\infty, \infty)$. This means that there are no restrictions on the numbers evaluated in polynomial functions.
2. There are no holes or breaks in the graphs of polynomial functions.
3. We say that polynomial functions are **continuous**. This means that their graphs are smooth and unbroken.

Compute an average rate of change of a polynomial function on a closed interval.

© Taewoon Lee/ShutterStock, Inc.

Example 6: Using a Polynomial Model

The personal income per capita in the United States, in constant 2005 dollars, can be modeled by

$$f(x) = -41.25x^3 + 677.65x^2 - 2638.9x + 35,772.5 \quad 1 \le x \le 10$$

where x represents the number of years since 1999, and $f(x)$ represents the personal income per capita in constant 2005 dollars. (*Source*: U.S. Bureau of Economic Analysis.) Use the difference quotient to determine the average rate of change in the personal income per capita where $x = 1$ and $\Delta x = 9$. Interpret the results.

Perform the Mathematics

Evaluating the difference quotient with $x = 1$ and $\Delta x = 9$, we get

$$m_{sec} = \frac{f(x + \Delta x) - f(x)}{\Delta x} = \frac{f(1 + 9) - f(1)}{9} = \frac{f(10) - f(1)}{9}$$

$$= \frac{35,898.5 - 33,770}{9} = \frac{2128.5}{9} = 236.50$$

Here $x = 1$ corresponds to the year 2000, and $x + \Delta x = 10$ corresponds to the year 2009. So we interpret the result as meaning that during the period from 2000 to 2009, the personal income per capita increased at an average rate of $236.50 per year. ∎

End Behavior of Functions

At times we want to know the values of a function for extremely large positive and extremely large negative values of the independent variable x. This determines what is called the **end behavior** of a function. To begin to see what happens, let's take a look at the quadratic function $f(x) = 2x^2 - 6x + 3$. **Table 1.4.2** shows the function values for extremely large positive and negative values of x. From this table, it appears that as the x-values grow both negatively and positively large, the function values grow positively large. Table 1.4.2 illustrates what end behavior is all about. We commonly describe the end behavior symbolically. We write $x \to -\infty, f(x) \to \infty$ to denote that, as we move from right to left on the x-axis, the $f(x)$-values grow positively large. (This is read "as x approaches negative infinity, $f(x)$ approaches infinity.") We write $x \to \infty, f(x) \to \infty$ to denote that, as we move from left to right on the x-axis, the $f(x)$-values grow large positively. (This is read "as x approaches infinity, $f(x)$ approaches infinity.")

Table 1.4.2

X	$f(x) = 2x^2 - 6x + 3$
$-100,000$	20,000,600,003
$-10,000$	200,060,003
-1000	2,006,003
-100	20,603
0	3
100	19,403
1000	1,994,003
10,000	199,940,003
100,000	19,999,400,003

To determine the end behavior of a polynomial function, all we need to do is examine the leading term, $a_n x^n$, of the polynomial. We can summarize the results with the **leading coefficient test**.

DEFINITION

Leading Coefficient Test

1. Assume that $f(x)$ is an **even-degree** polynominal function with leading term $a_n x^n$ and leading coefficient a_n.

If the leading coefficient is	as $x \to -\infty$	as $x \to \infty$	The typical graph is
Positive ($a_n > 0$)	$f(x) \to \infty$	$f(x) \to \infty$	
Negative ($a_n < 0$)	$f(x) \to -\infty$	$f(x) \to -\infty$	

2. Assume that $f(x)$ is an **odd-degree** polynominal function with leading term $a_n x^n$ and leading coefficient a_n.

If the leading coefficient is	as $x \to -\infty$	as $x \to \infty$	The typical graph is
Positive ($a_n > 0$)	$f(x) \to -\infty$	$f(x) \to \infty$	
Negative ($a_n < 0$)	$f(x) \to \infty$	$f(x) \to -\infty$	

Example 7: Determining the End Behavior of a Polynomial Function

OBJECTIVE 7

For the cubic function $f(x) = x^3 - 3x^2 - 6x + 8$, determine the end behavior of the graph of the function.

Determine the end behavior of a graph of a polynomial function.

Perform the Mathematics

From case 2 of the Leading Coefficient Test, because f is a third, or odd, degree polynomial function and the leading coefficient, $a_n = 1$, is positive, we know that as $x \to -\infty$, $f(x) \to -\infty$ and as $x \to \infty$, $f(x) \to \infty$. This is shown graphically in **Figure 1.4.6**.

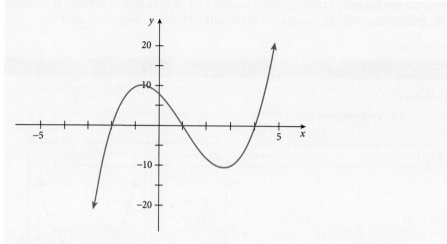

Figure 1.4.6 Graph of $f(x) = x^3 - 3x^2 - 6x + 8$.

Composite Functions

Our final topic discussed in this section is the composition of functions. We will see that **composite functions** occur frequently. In Section 1.1, we learned that when we apply a function to an independent variable value, a dependent variable value is returned. Now let's suppose that the dependent value of the one function becomes the independent value of another. To illustrate, suppose that a refinery's underwater supply line ruptures, resulting in an oil spill that is fairly circular. The petroleum company, the local fisheries, and those who are concerned about the environment would like to know how much area the spill will cover as time passes.

Assume that past experiments show that the radius of the oily, circular slick increases at a rate of about 0.7 feet per second. As a function of time t, the radius of the slick is given by

$$r(t) = 0.7t$$

Because the area of the oil spill depends on its radius, the area is given by the function

$$A(r) = \pi r^2$$

Because the radius is given by $r(t) = 0.7t$, to find the area of the slick as a function of time, we can substitute $r(t) = 0.7t$ into $A(r) = \pi r^2$. This yields

$$A(t) = \pi(0.7t)^2 = \pi \cdot 0.49t^2 = 0.49\pi t^2.$$

Figure 1.4.7 suggests that we are nesting the functions, meaning that we could write $A(t) = A(r(t))$. This is exactly what a **composite function** is—a *nesting* of one function inside another.

DEFINITION

Composite Function

A function h is a **composition** of the functions f and g if

$$h(x) = f(g(x))$$

NOTE: The composition of functions $f(g(x))$ could also be denoted as $(f \circ g)(x)$.

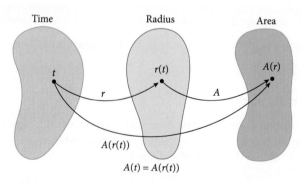

Figure 1.4.7

Example 8: Evaluating Composite Functions

Consider the functions $f(x) = \dfrac{3}{x}$ and $g(x) = 2x - 1$. Evaluate $f(g(6))$ and $g(f(6))$.

OBJECTIVE 8

Evaluate composition of functions.

Perform the Mathematics

Starting from the inside of the expression $f(g(6))$ and working our way out, we first evaluate

$$g(6) = 2(6) - 1 = 11$$

To evaluate $f(g(6))$, we substitute 11 for $g(6)$ to get

$$f(g(6)) = f(11) = \frac{3}{11}$$

In the case of $g(f(6))$, we first start by evaluating $f(6)$ to get

$$f(6) = \frac{3}{6} = \frac{1}{2}$$

So, to evaluate $g(f(6))$, we get

$$g(f(6)) = g\left(\frac{1}{2}\right) = 2 \cdot \frac{1}{2} - 1 = 0 \qquad \blacksquare$$

▶ *Try It Yourself*

Some related Exercises are 65 and 67.

Notice in Example 8 that $f(g(6))$ and $g(f(6))$ gave different results. This shows that generally $f(g(x)) \neq g(f(x))$. In Chapter 4, we will see a special case when these compositions are the same.

When determining the domain of a composition of functions, we have to take care that the dependent values of the *inside* function are all members of the domain of the second. Knowing the range values of the inside function takes on added importance.

Building up functions through composition is a very exacting process. However, **decomposing** a function into simpler parts offers several options. This is illustrated in Example 9.

Example 9: Decomposing Functions

Determine the functions f and g for each composition so that $f(g(x)) = h(x)$.

a. $h(x) = \sqrt{(x + 5)^3}$ **b.** $h(x) = (5x^3 + 6)^2$

Perform the Mathematics

a. Here we could let the *inside function* be $g(x) = (x + 5)^3$ and the *outside function* $f(x) = \sqrt{x}$. Checking this, we have

$$f(g(x)) = f[(x + 5)^3] = \sqrt{(x + 5)^3} = h(x)$$

b. For this function, we could let $g(x) = 5x^3 + 6$ and $f(x) = x^2$. Checking this, we get

$$f(g(x)) = f[(5x^3 + 6)] = (5x^3 + 6)^2 = h(x) \qquad \blacksquare$$

Summary

This section began with operations on functions. We learned that we can add, subtract, multiply, and divide functions, much as we do with real numbers. Then we discussed the functions of business:

- The **revenue function** $R(x) = x \cdot p(x)$

- The **profit function** $P(x) = R(x) - C(x)$

- The **break-even point(s)** occur when $R(x) = C(x)$, or equivalently, where $P(x) = 0$

Then we turned our attention to the properties of polynomial functions. These functions have the form

$$f(x) = a_n x^n + a_{n-1} x^{n-1} + \cdots + a_1 x + a_0$$

where $a_0, a_1, a_2, \ldots, a_n$ are real number constants where $a_n \neq 0$ and n is a whole number. The end behavior can be found by using the **Leading Coefficient Test**.

We concluded the section with a look at **composite functions**.

Section 1.4 Exercises

Vocabulary Exercises

1. Addition, subtraction, multiplication, and division are collectively called the mathematical _____.

2. The quotient of functions $\left(\dfrac{f}{g}\right)$ is defined for all values except those that make the value of $g(x)$ _____.

3. The total revenue of a product is determined by the quality sold times the _____.

4. The difference between the revenue and cost functions gives us the _____ function.

5. The point at which the graphs of the cost and revenue functions intersect is the _____ - _____ point.

6. A function whose graph is smooth and unbroken is called a _____ function.

Skill Exercises

For Exercises 7–10, simplify the functions for the following operations. Then write the domain for the resulting function.

(a) $(f + g)(x)$ **(b)** $(f - g)(x)$

(c) $(f \cdot g)(x)$ **(d)** $\left(\dfrac{f}{g}\right)(x)$

7. $f(x) = 6x + 3; \quad g(x) = 4x - 1$ 8. $f(x) = -2x + 9; \quad g(x) = -5x - 2$

9. $f(x) = \sqrt{x + 5}; \quad g(x) = x^2 + 5$ 10. $f(x) = \sqrt{x - 3}; \quad g(x) = x - 5$

For Exercises 11–18, let $f(x) = 2x^2 - 4x$ and $g(x) = 5x + 1$. Evaluate each of the following.

11. $(f + g)(0)$ 12. $(f + g)(-3)$

13. $(f \cdot g)(4)$ 14. $(f \cdot g)(-2)$

15. $(f - g)(2)$

16. $(f - g)(-1)$

17. $\left(\dfrac{f}{g}\right)(5)$

18. $\left(\dfrac{f}{g}\right)(-1)$

For Exercises 19–26, the price-demand function p is given. Determine the revenue function R.

19. $p(x) = 2.55$

20. $p(x) = 87.10$

21. $p(x) = -3.1x$

22. $p(x) = -0.91x$

23. $p(x) = -0.3x + 20$

24. $p(x) = -\dfrac{1}{2000}x + 3$

25. $p(x) = 50 - 0.1x$

26. $p(x) = 30 - 0.12x$

For Exercises 27–30, the revenue and cost function are given.

(a) Determine the break-even point(s).

(b) Determine how much revenue must be generated to reach the break-even point.

27. $R(x) = 50x;\quad C(x) = 20x + 170$

28. $R(x) = 8x;\quad C(x) = 5000$

29. $R(x) = 25x - 0.25x^2;\quad C(x) = 2x + 5$

30. $R(x) = 100x - x^2;\quad C(x) = 20x + 4$

For Exercises 31–34, the revenue and cost functions are given.

(a) Graph R and C in the same viewing window.

(b) Use the `Intersect` command on your graphing calculator to determine the break-even point(s).

(c) Determine how much revenue must be generated to determine the break-even point.

 31. $R(x) = 25x - 0.25x^2;\quad C(x) = 2x + 5$

 32. $R(x) = 100x - x^2;\quad C(x) = 20x + 4$

 33. $R(x) = -2.1x^2 + 500x;\quad C(x) = 80x + 9500$

 34. $R(x) = 95.2x - 5x^2;\quad C(x) = 155 + 20x$

For Exercises 35–38, the revenue and cost function are given.

(a) Write the profit function $P(x) = R(x) - C(x)$.

(b) Determine the zeros of the profit function and interpret.

(c) Determine the vertex of the graph of the profit function and interpret each coordinate.

35. $R(x) = 120x - 6x^2;\quad C(x) = 240 + 2x$

36. $R(x) = 1200x - 37x^2;\quad C(x) = 4300 + 148x$

37. $R(x) = 95.2x - 5x^2;\quad C(x) = 155 + 20x$

38. $R(x) = -2.1x^2 + 500x;\quad C(x) = 80x + 9500$

For Exercises 39–44, determine if the given function is a polynomial function. Classify each polynomial function as linear, quadratic, or cubic.

39. $f(x) = 7$

40. $f(x) = -30$

41. $f(x) = 3x - \dfrac{5}{x}$

42. $f(x) = \dfrac{4}{x^2} + 3x + \dfrac{1}{2}$

43. $f(x) = x^2 + 5x + 2$

44. $f(x) = x^3 + x + 1$

In Exercises 45–54, determine the end behavior of the function.

45. $f(x) = 3x$

46. $f(x) = -17x$

47. $f(x) = -2.1x$

48. $f(x) = 3.8x$

49. $f(x) = -3x^2 + 10x$

50. $f(x) = 5.5x^2 + 4$

51. $f(x) = -x^3 + 2x - 1$

52. $f(x) = 11x + 0.1x^3$

53. $f(x) = -x^6 - x$

54. $f(x) = x^6 - 2x^3 - x^2$

In Exercises 55–60, the price-demand p is given; along with the fixed costs and variable costs.

(a) Determine the profit function $P(x) = R(x) - C(x)$.

(b) Determine the coordinates of the vertex of the profit function P. Then interpret each coordinate of the vertex.

55. $p(x) = 105.70 - 0.89x$
Variable costs are $80 per unit
Fixed costs are $61.80

56. $p(x) = 37.80 - x$
Variable costs are $29 per unit
Fixed costs are $10.15

57. $p(x) = 50 - x$
Variable costs are $20 per unit
Fixed costs are $200

58. $p(x) = -10x + 1040$
Variable costs are $500 per unit
Fixed costs are $6650

59. $p(x) = 40 - x$
Variable costs are $20 per unit
Fixed costs are $80

60. $p(x) = -x + 30$
Variable costs are $5 per unit
Fixed costs are $60

For Exercises 61–66, use your graphing calculator to complete the following.

(a) Use the Zero command on your calculator to determine the values where $P(x) = 0$.

(b) Write an interpretation for the zeros found in part (a).

 61. The profit function determined in Exercise 55.

 62. The profit function determined in Exercise 56.

 63. The profit function determined in Exercise 59.

 64. The profit function determined in Exercise 60.

For Exercises 65–72, complete the following.

(a) *Find the composition of functions $g(f(2))$ and $f(g(2))$.*

(b) *Find the composition of functions $g(f(x))$ and $f(g(x))$.*

65. $f(x) = 3x - 1$; $g(x) = 8x + 2$

66. $f(x) = 6x - 9$; $g(x) = 7x + 5$

67. $f(x) = 5x - 3$; $g(x) = x^2 + 3x + 4$

68. $f(x) = 3x^2 + x + 5$; $g(x) = x - 1$

69. $f(x) = x^3$; $g(x) = \dfrac{1}{x}$

70. $f(x) = \dfrac{2}{x}$; $g(x) = x + 3$

71. $f(x) = \sqrt{x + 1}$; $g(x) = x^2 + 2$

72. $f(x) = \sqrt{x + 2}$; $g(x) = 4x - 3$

For Exercises 73–80, determine the functions f and g so that $f(g(x)) = h(x)$ for each composite function h, then check the answer. Note that there is more than one correct answer.

73. $h(x) = (x + 3)^2$

74. $h(x) = (5x + 6)^4$

75. $h(x) = \left(\dfrac{1}{x + 3}\right)^2$

76. $h(x) = \left(\dfrac{5}{x - 1}\right)^4$

77. $h(x) = \sqrt[4]{x - 2}$

78. $h(x) = \sqrt[3]{2x + 3}$

79. $h(x) = 2 - 3\sqrt{x}$

80. $h(x) = 9 + 2\sqrt{x}$

For the functions f and g in Exercises 81–88, show that $f(g(x)) = x$.

81. $f(x) = 8x$; $g(x) = \dfrac{1}{8}x$

82. $f(x) = \dfrac{3}{4}x$; $g(x) = \dfrac{4}{3}x$

83. $f(x) = 7x - 10$; $g(x) = \dfrac{x + 10}{7}$

84. $f(x) = \dfrac{x - 3}{4}$; $g(x) = 4x + 3$

85. $f(x) = x^3 + 6$; $g(x) = \sqrt[3]{x - 6}$ **86.** $f(x) = x^5 - 9$; $g(x) = \sqrt[5]{x + 9}$

87. $f(x) = \sqrt{x - 1}$; $g(x) = x^2 + 1$, for $x \geq 0$

88. $f(x) = \sqrt[4]{x - 1}$; $g(x) = x^4 + 1$, for $x \geq 0$

Application Exercises

89. **Finance—Weekly Sales:** The Handi-Neighbor hardware store sells PowerDriver hammers and finds that the amount made in dollars for the weekly sales is given by

$$f(x) = 9.75x \quad 0 \leq x \leq 50$$

The competing U-Do-It hardware store finds that the weekly sales for selling PowerDriver hammers is

$$g(x) = \frac{4}{5}x^2 + 6x \quad 0 \leq x \leq 50$$

In both functions, x represents the number of PowerDriver hammers sold.

  (a) Evaluate $(f - g)(20)$ and interpret.

  (b) Evaluate $(f + g)(20)$ and interpret.

90. **Finance—Operation Expenses:** The Nuthin'-But-Socks specialty store finds that the monthly expenses for operation during its first year of business is given by

$$f(x) = \frac{1}{5}x^2 + 300 \quad 1 \leq x \leq 12$$

where $x = 1$ corresponds to January, $x = 2$ corresponds to February, and so on, and $f(x)$ is the monthly expenses in hundreds of dollars. The monthly revenue during the first year of business is given by

$$g(x) = 27.5x \quad 1 \leq x \leq 12$$

where $x = 1$ corresponds to January, $x = 2$ corresponds to February, and so on, and $g(x)$ is the monthly revenue in hundreds of dollars.

  (a) Evaluate $(g - f)(6)$ and interpret.

  (b) Did the store break even before the end of the first year?

91. **Business—Revenue:** ActivLife Incorporated determines that the revenue generated by producing and selling x of their LifeMax ellipticals can be modeled by the function

$$R(x) = -\frac{x^2}{15} + 150x \quad x \geq 0$$

where x represents the number of ellipticals produced and sold, and $R(x)$ represents the revenue in dollars. Determine m_{sec} on the closed interval $[5, 20]$ and interpret the result.

92. **Business—Revenue:** The accounting department at BackStreet LLC has found that their daily revenue function for producing their Boomer-Blocker noise-canceling headphones is given by

$$R(x) = 110x - \frac{x^2}{73} \quad [0, 50]$$

where x represents the number of headphones produced and sold daily, and $R(x)$ represents the daily revenue in dollars. Determine m_{sec} on the closed interval $[10, 20]$ and interpret the result.

93. **Business—Profit:** The Brimfield Company computes their profit function for making and selling x touch-screen ski gloves can be modeled by the function

$$P(x) = -\frac{x^2}{4} + 23x - 5 \quad [0, 70]$$

where x represents the number of gloves sold, and $P(x)$ represents the profit in dollars. Determine the average rate of change in profit as the number of gloves produced and sold increases from 35 to 45.

94. Business—Profit: The Casitas Fashion Warehouse finds that the profit generated by selling x pairs of their Cut-n-Run blue jeans can be expressed by the profit function

$$P(x) = -4 + 80x - x^2 \quad [0, 110]$$

where x represents the number of Cut-n-Run blue jeans sold, and $P(x)$ represents the profit in dollars. Determine the average rate of change in profit as the number of jeans sold increases from 90 to 105.

For Exercises 95 and 96, the models f represent per capita consumption of food items. Per capita is a Latin phrase meaning "for each person." The population of the United States can be modeled by various functions, but for these exercises, we will use the linear model

$$g(x) = 2.85x + 224.36 \quad 0 \le x \le 31$$

where x represents the number of years since 1980, and g(x) represents the U.S. population from 1980 to 2011 in millions of people. (Source: U.S. Census Bureau.)

 95. Nutrition—Ice Cream Consumption: One of Americans' favorite treats is ice cream. The per capita consumption of ice cream in the United States from 1980 to 2011 can be modeled by

$$f(x) = -0.11x + 17.44 \quad 0 \le x \le 31$$

where x represents the number of years since 1980, and $f(x)$ represents the per capita consumption of ice cream, measured in pounds per person annually. (*Source:* U.S. Department of Agriculture.)

(a) Evaluate $f(28)$ and $g(28)$, and interpret each value.

(b) Evaluate $(f \cdot g)(28)$, and interpret the product. Keep in mind the units for the function g.

(c) Simplify the product of functions $(f \cdot g)$. What does this function represent?

© Thomas M. Perkins/ShutterStock, Inc.

 96. Nutrition—Orange Consumption: Many people who have adjusted their diets to eat more fresh fruit have been challenged with steadily increasing prices for fresh fruit. "Orange juice fortified with calcium may help fortify bones and fight osteoporosis, and the vitamin C naturally found in oranges may inhibit cancer," said Deepak Chopra, MD. (*Source:* The Chopra Center.) The per capita consumption of oranges in the United States from 1980 to 2011 can be modeled by

$$f(x) = -0.18x + 14.43 \quad 0 \le x \le 31$$

where x represents the number of years since 1980, and $f(x)$ represents the per capita consumption of oranges, measured in pounds per person annually. (*Source:* U.S. Department of Agriculture.)

(a) Examine the coefficient of x in the model f. Interpret the meaning of this value.

(b) Simplify the product of functions $(f \cdot g)$. What does this function represent?

(c) Evaluate $(f \cdot g)(28)$, and interpret the product. Keep in mind the units for the function g.

© Anupan Aupanusorn/Dreamstime.com

97. Transportation—Highway Construction: Federal highway spending is funded mainly through gas and other fuel taxes that are paid into the Highway Trust Fund. The annual *receipts* into the Highway Trust Fund from 1990 to 2011 can be modeled by the function

$$f(x) = 0.24x^2 + 1.92x + 77.86 \quad 0 \le x \le 21$$

where x represents the number of years since 1990, and $f(x)$ represents the annual receipts into the Highway Trust Fund in billions of dollars. Conversely, the annual *outlays* from the Highway Trust Fund from 1990 to 2011 can be modeled by the function

$$g(x) = 0.15x^2 + 3.06x + 75.29 \quad 0 \leq x \leq 21$$

where x represents the number of years since 1990, and $f(x)$ represents the annual outlays from the Highway Trust Fund in billions of dollars. (*Source:* U.S. Federal Highway Administration.)

© AbleStock

(a) Simplify the difference of functions $(f - g)$. Interpret the meaning of this operation on functions.

(b) Evaluate $(f - g)(11)$ and interpret the result.

 98. Defense—Veteran Spending: A portion of the U.S. defense budget includes veterans' outlays. These outlays include budgets for veterans' hospitals, educational scholarships, and disability payments. The annual outlays for U.S. veterans from 1990 to 2011 can be modeled by

$$f(x) = 0.3x^2 - 2.22x + 35.4 \quad 0 \leq x \leq 21$$

where x represents the number of years since 1990, and $f(x)$ represents the veterans' outlays in billions of dollars. The total annual Department of Defense budget during the same time period can be modeled by the function

$$g(x) = 2.23x^2 - 1.97x + 329.6 \quad 0 \leq x \leq 21$$

where x represents the number of years since 1990, and $g(x)$ represents the annual U.S. defense budget in billions of dollars. (*Source:* U.S. Office of Management and Budget.)

Courtesy of Master Sgt. Ken Hammond, U.S. Air Force/U.S. Department of Defense

(a) Evaluate $f(11)$ and $g(11)$, and interpret each of these values.

(b) Use the results of part (a) to compute $\left(\dfrac{f}{g}\right)(11) \cdot 100\%$. What does this percentage represent?

(c) Use your calculator to graph $y_3 = \dfrac{y_1}{y_2} \cdot 100 = \left(\dfrac{f}{g}\right)(x) \cdot 100$, and graph y_3 in the viewing window $[0, 21]$ by $[0, 12]$. Use the `Minimum` command on your graphing calculator to determine when the percentage of the U.S. defense budget allocated to veterans was at its lowest. What is the percentage?

99. **Physics—Volume Analysis:** Suppose a spherical balloon is being inflated with the radius increasing at an average rate of 1.3 inches per second.

(a) Write the radius $r(t)$ as a function of time t. Assume that when $t = 0, r = 0$.

(b) The volume of a sphere as a function of its radius is given by $V(r) = \frac{4}{3}\pi r^3$. Simplify $V(t) = V(r(t))$, and interpret the meaning of this composition.

(c) Evaluate $V(t)$ when $t = 6$ and interpret the result.

100. **Medicine—Pill Size:** The Top Drug Company has developed a new type of pill in the shape of a ball. The spherical pill is dropped into a glass of water and dissolves. Tests show that when the 2-centimeter pill is dropped into the water, its radius decreases at an average rate of 0.003 centimeters per second.

(a) Write the radius $r(t)$ as a function of time t. In this case, $r(t) = 2$ when $t = 0$.

(b) Knowing that the surface area of a sphere as a function of its radius is $S(r) = 4\pi r^2$, simplify $S(t) = S(r(t))$, and interpret the meaning of this composition.

(c) Evaluate $S(t)$ when $t = 20$ and interpret.

Concept and Writing Exercises

101. We know that polynomial functions are continuous for all real numbers. Suppose that we have two polynomial functions f and g. Is the difference of the functions $(f - g)(x)$ also continuous on $(-\infty, \infty)$? Explain your answer in a complete sentence.

102. We call addition, subtraction, multiplication, and division the four basic function operations. Explain why you think exponentiation is not considered a basic function operation.

For Exercises 103–106, consider a fourth-degree (quartic) polynomial function
$f(x) = a_4x^4 + a_3x^3 + a_2x^2 + a_1x + a_0$ *and a fifth-degree (quintic) polynomial function*
$g(x) = b_5x^5 + b_4x^4 + b_3x^3 + b_2x^2 + b_1x + b_0.$

103. Determine the coefficient of the quadratic term of the sum $(f + g)(x)$.

104. What is the leading (that is, fifth-degree) term of the difference $(f - g)(x)$?

105. Without computing it, what is the degree of the product $(f \cdot g)(x)$? Explain how you determined your answer.

106. Without computing it, what is the degree of the quotient $\left(\dfrac{g}{f}\right)(x)$? Explain how you determined your answer.

Section Project

Medicaid is an assistance program that serves low-income people of every age in which medical bills are paid from federal, state, and local tax funds. Recent cuts have caused concern among senior citizens. "People are calling us every day who are being denied, with very serious health problems," said Anne Ronan, of the Arizona Center for Law in the Public Interest. (*Source: The Arizona Republic.*) The annual payments from the Medicaid assistance program from 2000 to 2011 can be modeled by

$$f(x) = 15{,}766x + 1{,}751{,}232 \quad 0 \le x \le 11$$

where x represents the number of years since 2000, and $f(x)$ represents the annual payments from the Medicaid assistance program in millions of dollars. The number of Medicaid beneficiaries during the same years can be modeled by

$$g(x) = -0.34x^2 + 4.55x + 42.96 \quad 0 \le x \le 11$$

where x represents the number of years since 2000, and $g(x)$ represents the number of Medicaid beneficiaries, in millions of people. (*Source: U.S. Centers for Medicare & Medicaid Services.*)

(a) Evaluate $\left(\dfrac{f}{g}\right)(9)$, and interpret the result.

(b) Graph the function $y_3 = \left(\dfrac{y_1}{y_2}\right)(x) = \left(\dfrac{f}{g}\right)(x)$ in the viewing window [0, 11] by [30,000, 50,000]. Explain the meaning of this quotient using a brief sentence.

(c) Use the `Minimum` command to determine the minimum of the quotient of the functions $\left(\dfrac{f}{g}\right)$ for $0 \le x \le 11$.

(d) Round the x-coordinate of the ordered pair found in part (c) to the nearest whole number. How can we interpret this value?

(e) Use the result from part (d) to determine the interval in which $\left(\dfrac{f}{g}\right)$ is decreasing and the years in which $\left(\dfrac{f}{g}\right)$ is increasing on [0, 11], and interpret.

(f) Compute the slope of the secant line of $\left(\dfrac{f}{g}\right)$ for $x = 7$ and $\Delta x = 4$ and interpret.

1.5   Rational, Radical, and Power Functions

SECTION OBJECTIVES

1. Analyze rational functions near an excluded domain value.

2. Determine a vertical asymptote.

3. Determine a horizontal asymptote.

4. Determine x- and y-intercepts of a graph of a rational function.

5. Sketch a graph of a rational function.

6. Use a radical function in an application.

7. Apply a power function.

So far, the functions that we have studied are polynomial functions. However, many of the functions that we study in calculus are not polynomials. In this section, we study **rational functions**. These functions are made up of polynomials, but their graphs do not resemble those that we have seen. Then we will examine **radical functions**, which are functions involving square roots, cube roots, and the like and having restricted domains and ranges. Finally, we will investigate **power functions**, which have fractions, not whole numbers, in their exponents. Once again we will see how these functions are used in applications, and we will calculate the slope of a secant line over an interval and interpret it as an average rate of change.

Rational Functions

Expressions for rational functions have fractions whose numerator and denominator are polynomials. Here are some examples of rational functions:

$$f(x) = \frac{x}{x+1}, \quad g(x) = \frac{2x^2 - 1}{x - 2}, \quad y = \frac{x - 3}{x^2 + x + 1}$$

A function such as $f(x) = \dfrac{\sqrt{x+2}}{x+5}$ is not a **rational function**, because the numerator is not a polynomial.

DEFINITION

Rational Function

A **rational function** has the form

$$y = \frac{f(x)}{g(x)}$$

where f and g are polynomials and $g(x) \neq 0$.

Notice that the domain of a rational function is determined by the set of all real numbers that make the denominator not equal to zero. Since there are usually x-values that make the denominator zero, rational functions frequently have breaks in their graphs.

Example 1: Finding Vertical Asymptotes Using Tables

OBJECTIVE 1

Analyze rational functions near an excluded domain value.

Consider the rational function $f(x) = \dfrac{2x}{x - 3}$.

a. Determine the domain.

b. Complete **Table 1.5.1** and describe what happens to the function values for x-values close to $x = 3$.

Table 1.5.1

x	2	2.9	2.99	2.999	2.9999	3.0001	3.001	3.01	3.1	4
$f(x)$										

Perform the Mathematics

a. The denominator, $x - 3$, is zero when $x = 3$. So the domain of f is the set of all real numbers except 3 or, using interval notation, $(-\infty, 3) \cup (3, \infty)$.

b.   Table 1.5.2

x	2	2.9	2.99	2.999	2.9999	3.0001	3.001	3.01	3.1	4
$f(x)$	−4	−58	−598	−5998	−59,998	60,002	6002	602	62	8

Table 1.5.2 shows that for x-values very close to but less than 3, the function values get negatively large or, symbolically, $f(x) \to -\infty$. Also, for x-values very close to but greater than 3, the function values grow positively large, or $f(x) \to \infty$. Since x cannot equal 3, the graph of $f(x) = \dfrac{2x}{x-3}$ will never cross the vertical line $x = 3$. This line is called a **vertical asymptote**. A graph of f is given in **Figure 1.5.1**.

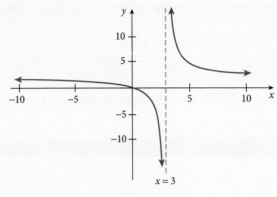

Figure 1.5.1 Graph of $f(x) = \dfrac{2x}{x-3}$.

Finding Vertical Asymptotes

Consider the rational function $h(x) = \dfrac{f(x)}{g(x)}$, where f and g are polynomials. If there is a value c that makes the denominator zero, yet the numerator is not zero, then the vertical line $x = c$ is a vertical asymptote for the graph of h.

Making numerical tables is a sound, yet time-consuming, way to determine vertical asymptotes. Algebraically, we see that for the function $f(x) = \dfrac{2x}{x-3}$, the value $x = 3$ makes the denominator equal to zero and yet makes the numerator a nonzero real number (in this case, 6). This is the common procedure for **finding vertical asymptotes**.

The graphs of some rational functions have more than one vertical asymptote, as illustrated in Example 2.

OBJECTIVE 2

Determine a vertical asymptote.

Example 2: Determining Vertical Asymptotes

Find the vertical asymptotes of the graph of the function $f(x) = \dfrac{x-4}{x^2+x-2}$.

Perform the Mathematics

We start by factoring the denominator. This gives

$$f(x) = \frac{x-4}{x^2+x-2} = \frac{x-4}{(x+2)(x-1)}$$

The denominator of the rational function is zero when $x = -2$ and $x = 1$. Because neither of these values makes the numerator equal to zero, the graph of $f(x) = \dfrac{x-4}{x^2+x-2}$ has vertical asymptotes at the lines $x = -2$ and $x = 1$. See **Figure 1.5.2**.

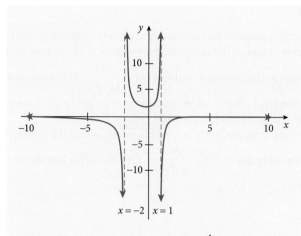

Figure 1.5.2 Graph of $f(x) = \dfrac{x - 4}{x^2 + x - 2}$.

We stated that the vertical asymptotes are identified by values that make the denominator zero, yet the numerator is not zero. But what if a value makes both the numerator and denominator zero? Let's say we have the function $g(x) = \dfrac{x^2 - 4}{x - 2}$. The value $x = 2$ is excluded from the domain because this number makes the denominator zero. Since the factored form of the function is

$$g(x) = \frac{x^2 - 4}{x - 2} = \frac{(x + 2)(x - 2)}{x - 2}$$

we can algebraically reduce the numerator and denominator by $(x - 2)$ and rewrite the function as

$$f(x) = x + 2 \quad x \neq 2$$

This means that $g(x) = \dfrac{x^2 - 4}{x - 2}$ and $f(x) = x + 2$ are identical for all values of x except $x = 2$. Hence, the graphs of $g(x) = \dfrac{x^2 - 4}{x - 2}$ and $f(x) = x + 2$ are the same except for when $x = 2$. The graph of $f(x) = x + 2$ is in **Figure 1.5.3(a)**, and the graph of $g(x) = \dfrac{x^2 - 4}{x - 2}$ is shown in Figure 1.5.3(b). Notice that the graph of $g(x) = \dfrac{x^2 - 4}{x - 2}$ is the same as the graph of $f(x) = x + 2$, except there is a "hole" in the graph when $x = 2$.

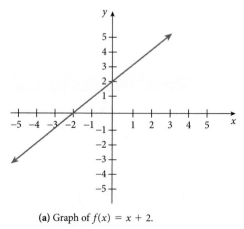

(a) Graph of $f(x) = x + 2$.

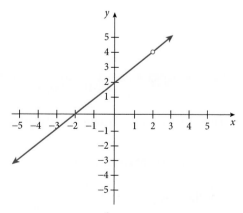

(b) Graph of $g(x) = \dfrac{x^2 - 4}{x - 2}$ with a hole when $x = 2$.

Figure 1.5.3

End Behavior of Rational Functions

In Section 1.4, we said that the end behavior of a polynomial function describes the type of numerical values that a function takes on for extreme values of the independent variable x. To see how end behavior works for rational functions, we return to the rational function $f(x) = \dfrac{2x}{x - 3}$. The numerical output for extreme x-values is shown in **Table 1.5.3**. The . . . at the end of the number in the second column means that the decimal values continue on, but are not written.

From Table 1.5.3, it appears that as x gets very large positively and very large negatively, the function values are getting close to 2. **Figure 1.5.4** shows a graph of $f(x) = \dfrac{2x}{x - 3}$, and we notice that the line $y = 2$ is a **horizontal asymptote** for the graph.

Table 1.5.3

x	$f(x)$
$-100{,}000$	$1.99994\ldots$
$-10{,}000$	$1.9994\ldots$
-1000	$1.994\ldots$
-100	$1.94\ldots$
0	0
100	$2.061855\ldots$
1000	$2.006018\ldots$
$10{,}000$	$2.000600\ldots$
$100{,}000$	$2.000060\ldots$

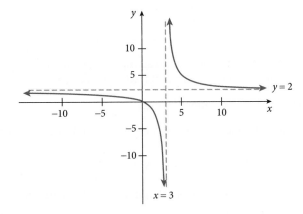

Figure 1.5.4 Graph of $f(x) = \dfrac{2x}{x - 3}$ has a vertical asymptote at $x = 3$ and a horizontal asymptote at $y = 2$.

Horizontal asymptotes can be determined quickly by inspecting the degree of the numerator and the degree of the denominator. Notice that both the numerator and denominator of f have a degree of 1, and the ratio of the leading coefficients is $\frac{2}{1} = 2$. It is not a coincidence that this ratio is the horizontal asymptote, as this is one procedure for **finding horizontal asymptotes**. (A more rigorous analysis of horizontal asymptotes is presented in Chapter 2.)

OBJECTIVE 3

Determine a horizontal asymptote.

Example 3: Determining Horizontal Asymptotes

Find the horizontal asymptote of the graph of $f(x) = \dfrac{2x^2 - 5x - 3}{x^2 - 16}$.

Perform the Mathematics

For the function $f(x) = \dfrac{2x^2 - 5x - 3}{x^2 - 16}$, the numerator and denominator are both quadratic (degree of 2), so the horizontal asymptote is given by the coefficients of the highest degree terms, $y = \frac{2}{1} = 2$. ∎

► *Try It Yourself*
Some related Exercises are 11 and 13.

Another key feature of the graphs of rational functions are the intercepts. The process for determining intercepts is the same as we have shown for other functions, as illustrated in Example 4.

Example 4: Determining the Intercepts of the Graph of a Rational Function

> **OBJECTIVE 4**
>
> Determine x- and y-intercepts of a graph of a rational function.

Determine the x- and y-intercepts of the graph of $f(x) = \dfrac{2x^2 - 5x - 3}{x^2 - 16}$.

Perform the Mathematics

To find the y-intercept, we substitute 0 for x. This yields

$$f(0) = \frac{2(0)^2 - 5(0) - 3}{(0)^2 - 16} = \frac{-3}{-16} = \frac{3}{16}$$

So the y-intercept is at the point $(0, \frac{3}{16})$. To find the x-intercept, we substitute 0 for y (or in this case, $f(x)$) and solve for x. This gives

$$0 = \frac{2x^2 - 5x - 3}{x^2 - 16}$$

Factoring gives us

$$0 = \frac{(2x + 1)(x - 3)}{(x + 4)(x - 4)}$$

The only time a fraction equals 0 is when the numerator equals 0. Thus,

$$0 = (2x + 1)(x - 3)$$

when $x = -\frac{1}{2}$ and $x = 3$. (It is important to check that neither of these x-values makes the denominator equal 0.) We conclude that the x-intercepts occur at $(-\frac{1}{2}, 0)$ and $(3, 0)$. ∎

Now that we know how to find vertical asymptotes, horizontal asymptotes, and intercepts, let's put it all together and sketch a graph of a rational function by hand.

Example 5: Sketching a Rational Function

> **OBJECTIVE 5**
>
> Sketch a graph of a rational function.

Sketch a graph of $f(x) = \dfrac{2x^2 - 5x - 3}{x^2 - 16}$. Label all asymptotes and intercepts.

Perform the Mathematics

The factored form of $f(x)$ is

$$f(x) = \frac{2x^2 - 5x - 3}{x^2 - 16} = \frac{(2x + 1)(x - 3)}{(x + 4)(x - 4)}$$

Vertical asymptotes are at $x = -4$ and $x = 4$, whereas the horizontal asymptote is at $y = 2$. In Example 4, we determined that the y-intercept is at $(0, \frac{3}{16})$ and the x-intercepts are at $(-\frac{1}{2}, 0)$ and $(3, 0)$. By plotting a couple of other points, $f(-5) = 8$ and $f(5) = \frac{22}{9}$, we get the graph in **Figure 1.5.5**.

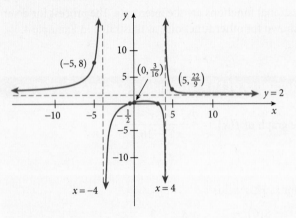

Figure 1.5.5 Graph of $f(x) = \dfrac{2x^2 - 5x - 3}{x^2 - 16}$. Vertical asymptotes at $x = -4$ and $x = 4$. Horizontal asymptote at $y = 2$. ∎

Example 6: Analyzing a Flu Epidemic

The number of people affected by the flu in Spring Point, an isolated town with a population of 4000, can be modeled by

$$f(t) = \frac{8000t + 6000}{2t + 5}$$

where $f(t)$ represents the number of people affected by the flu after t weeks. See **Figure 1.5.6**.

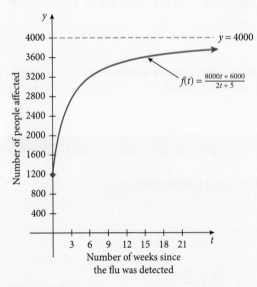

Figure 1.5.6

a. Evaluate $f(3)$, and interpret.

b. Find the average rate of change for $t = 3$ and $\Delta t = 2$, and interpret.

Perform the Mathematics

a. Evaluating the function at $t = 3$, we get

$$f(3) = \frac{8000(3) + 6000}{2(3) + 5} = \frac{30,000}{11} \approx 2727.27$$

This means that after 3 weeks, about 2727 people had been affected by the flu.

b. Because an average rate of change over an interval is given by the slope of a secant line over the interval, we use the difference quotient to determine the slope of the secant line.

$$m_{\text{sec}} = \frac{f(t + \Delta t) - f(t)}{\Delta t} = \frac{f(3 + 2) - f(3)}{2} = \frac{f(5) - f(3)}{2}$$

$$\approx \frac{3066.67 - 2727.27}{2} \approx 169.7$$

This means that from the end of the third week to the end of the fifth week, the number of people affected by the flu had increased at an average rate of about $170 \, \frac{\text{people}}{\text{week}}$.

Radical Functions

To begin to understand radical functions, let's recall a definition from algebra as shown in the Toolbox.

From Your Toolbox: Rational Exponents

A **rational exponent** has the form a/b, where a and b are integers, and has the property

$$x^{a/b} = \sqrt[b]{x^a} = (\sqrt[b]{x})^a$$

where $\sqrt{}$ is the **radical sign**, x is the **radicand**, and b is the **root index**.

Notice that the Toolbox implies that any function written in radical form can be written as a function with a rational exponent, and vice versa. Knowing how to rewrite a function between radical and rational exponent form is essential for success in calculus. The **Radical/Rational Exponent Functions** definition relates these two families of functions.

Many times, $g(x)$ is a polynomial expression such as $5x - 1$ or $8 - x$. For example, for the function $g(x) = x^{2/3}$, the root index is 3, so we can write the function in radical form as $g(x) = \sqrt[3]{x^2}$. Similarly, for the function $f(x) = (2x + 1)^{1/2}$, the function has a rational exponent of $\frac{1}{2}$, so we can rewrite f as $f(x) = \sqrt{2x + 1}$. This kind of function is called a **square root function**.

The **domains for radical/rational exponent functions** are determined by the value of the root index. Let's say that we have the function $f(x) = \sqrt[3]{x - 1}$. Since we can determine the cube root of any real number, the domain of this function is the set of real numbers $(-\infty, \infty)$. This can be verified visually in **Figure 1.5.7**.

If the function has an even index, such as a square root function, we have to be more cautious. Recall from Section 1.1 that we found that the domain of $g(x) = \sqrt{2x - 5}$ is $[\frac{5}{2}, \infty)$ by determining the x-values that make $2x - 5$ greater than or equal to zero. Thus, the domain of a radical function depends on whether the root index is even or odd.

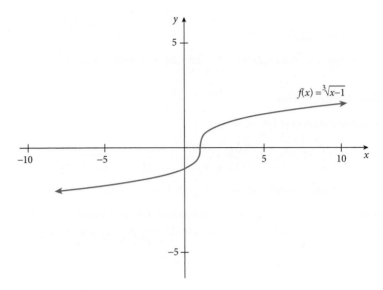

Figure 1.5.7

In practice, many of the radical functions that we study are square root functions. Let's take a look at an application of this class of radical functions.

OBJECTIVE 6

Use a radical function in an application.

Example 7: Using a Square Root Function

A sapling of a certain kind of tree grows according to the mathematical model

$$g(x) = 10\sqrt{x} + 0.75$$

where x represents the number of years since the sapling was planted, and $g(x)$ represents the height of the tree in feet after x years.

a. How tall will the sapling be 20 years after being planted? Round to the nearest hundredth of a foot.

b. How long will it take for the tree to be 30 feet tall?

Perform the Mathematics

a. To determine the height of the tree after 20 years, we need to evaluate $g(20)$. This gives us

$$g(20) = 10\sqrt{20} + 0.75 \approx 45.47$$

So, according to the model, the tree will be about 45.47 feet tall in 20 years.

b. In this case, we have to find the x-value(s) so that the $g(x)$-value is 30. That is, we need to find x such that $g(x) = 30$. Substituting 30 for $g(x)$ and solving for x gives us

$$10\sqrt{x} + 0.75 = 30$$
$$10\sqrt{x} = 29.25$$
$$\sqrt{x} = 2.925$$
$$x = (2.925)^2 \approx 8.56$$

So, according to the model, the tree will be 30 feet tall in about 8.56 years.

Power Functions

Some rational exponent functions are easier to study mathematically when the exponents are written as decimals, such as $f(x) = x^{0.145}$. We will call this class of functions the **power functions**.

The power function has various properties based on the values of a and b. We will study these properties in two cases. If a is a positive number, then the power function appears much like the polynomial function that we studied in Section 1.4. Much of the shape of the graph of these functions is determined by whether b is between zero and 1 or b is greater than 1.

For example, consider the table of values and graphs of the power functions $f(x) = x^{1.5}$ and $g(x) = x^{0.7}$, where f and g are defined for $x \geq 0$. See **Table 1.5.4**. The graph in **Figure 1.5.8** supports the numerical notion that f increases more rapidly than g. Finally, we observe that both functions are increasing and that both functions are **positive exponent power functions**.

DEFINITION

Power Functions
A function of the form

$$f(x) = a \cdot x^b$$

Is called a **power function**, where a and b are real numbers.

Table 1.5.4

x	$f(x)$	$g(x)$
0	0	0
1	1	1
2	2.83	1.62
3	5.20	2.16
4	8	2.64
5	11.18	3.09
6	14.70	3.51
7	18.52	3.90
8	22.63	4.29
9	27	4.66
10	31.62	5.01

Some values are rounded to the nearest hundredth.

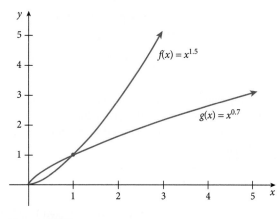

Figure 1.5.8

DEFINITION

Positive Exponent Power Functions

For the power function of the form $f(x) = a \cdot x^b$ with $a > 0$, we have

If the value of b is	The graph	Diagram
$0 < b < 1$	Opens down	
$b > 1$	Opens up	

OBJECTIVE 7

Apply a power function.

Example 8: Using a Positive Exponent Power Function

Many Americans enjoy watching sports either on television or in person. The amount of money spent on admission to spectator sporting events in the United States can be modeled by

$$f(x) = 11.27x^{0.25} \quad 1 \leq x \leq 9$$

where x represents the number of years since 1999, and $f(x)$ represents the amount of money spent on admission to spectator sports in billions of dollars. (*Source*: U.S. Statistical Abstract.) Determine the slope of the secant line of the graph of f for $x = 1$ and $\Delta x = 8$, and interpret.

Perform the Mathematics

Evaluating the secant line slope gives us

$$m_{sec} = \frac{f(x + \Delta x) - f(x)}{\Delta x} = \frac{f(9) - f(1)}{8}$$

$$= \frac{11.27(9)^{0.25} - 11.27(1)^{0.25}}{8} \approx \frac{19.52 - 11.27}{8} = 1.03125$$

Noting that $x = 9$ corresponds to 2008 and $x = 1$ corresponds to 2000, we interpret the result as saying that between the years 2000 and 2008, the amount of money spent on admission to spectator sports in the United States increased at an average rate of about $1.03125 billion per year. ■

In Example 8, note that when we evaluated $11.27(9)^{0.25}$, we raised 9 to the 0.25 power first and *then* multiplied by 11.27. Remember from order of operations in algebra: exponentiation first, then multiplication.

Now let's examine power functions for which the exponent is a negative number. The domain used for these functions is $(0, \infty)$ and a is a positive number. Consider the power functions $f(x) = x^{-1.8}$ and $g(x) = x^{-0.2}$, where f and g are defined for $x > 0$. Rewriting with positive exponents, we get $f(x) = x^{-1.8} = \dfrac{1}{x^{1.8}}$ and $g(x) = x^{-0.2} = \dfrac{1}{x^{0.2}}$. Because $0.2 = \frac{1}{5}$, we could write g as $g(x) = \dfrac{1}{\sqrt[5]{x}}$. Notice that when $x = 0$, the denominators of f and g are zero. This is why zero is not part of the domain of these functions. The graphs of $f(x) = x^{-1.8}$ and $g(x) = x^{-0.2}$ are shown in **Figure 1.5.9**. Notice that since $x = 0$ is not in the domain, the graphs of f and g have the y-axis as a vertical asymptote. We can see from the graph that both functions are decreasing.

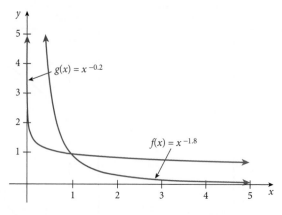

Figure 1.5.9

So when a power function has a negative exponent, the domain is the set of all positive real numbers. The functions are decreasing, and their graphs have vertical asymptotes. This kind of power function is frequently used as a model for a decreasing phenomenon.

Summary

In this section, we discussed the rational functions, which included finding the intercepts and asymptotes of their graphs. Then we examined radical functions and saw the relationship between radical and rational exponent functions. Finally, we examined the power functions. The investigation of all these functions is key in our understanding of calculus.

The types of functions we discussed in this section include:

- **Rational function**: $h(x) = \dfrac{f(x)}{g(x)}$, f and g are polynomials, $g(x) \neq 0$
- **Radical/rational exponent function**: $f(x) = \sqrt[b]{[g(x)]^a}$ or $f(x) = [g(x)]^{a/b}$
- **Square root function**: $f(x) = a\sqrt{x} + b$
- **Power function**: $f(x) = a \cdot x^b$

Vertical Asymptote Rule for the Rational Function $h(x) = \dfrac{f(x)}{g(x)}$

For $h(x) = \dfrac{f(x)}{g(x)}$, if there is a value c so that $g(c) = 0$, yet $f(c) \neq 0$, then $x = c$ is a vertical asymptote.

Horizontal Asymptote Rule for the Rational Function $h(x) = \dfrac{f(x)}{g(x)}$

- If the degree of f is largest, then there is no horizontal asymptote.
- If the degree of g is largest, then the x-axis is the horizontal asymptote.
- If the degrees of f and g are the same, then the horizontal asymptote is the reduced form of the leading coefficients of f and g.

Section 1.5 Exercises

Vocabulary Exercises

1. A line that continually approaches a given curve but does not meet it at any finite distance is called an _____.

2. A function made from the quotient of two polynomial functions is called a _____ function.

3. A function of the form $f(x) = \sqrt[b]{[g(x)]^a}$ is called a _____ function.

4. A function of the form $f(x) = [g(x)]^{a/b}$ is called a _____ _____ function.

5. For real numbers a and b, a function of the form $f(x) = a \cdot x^b$ is called a _____ function.

6. The expression under a square root symbol is called the _____.

Skill Exercises

For Exercises 7–16, complete the following.

    (a) Write the domain for f using interval notation.

    (b) Identify any holes or vertical asymptotes for the graph of f.

7. $f(x) = \dfrac{12x^3}{3x}$

8. $f(x) = \dfrac{15x^2}{25x^3}$

9. $f(x) = \dfrac{3x - 12}{x - 2}$

10. $f(x) = \dfrac{2x - 16}{x - 5}$

11. $f(x) = \dfrac{3x + 6}{x + 2}$

12. $f(x) = \dfrac{x^2 - 1}{x + 1}$

13. $f(x) = \dfrac{x^2 + 2x + 1}{x^2 + 4x + 3}$

14. $f(x) = \dfrac{6x^2 + x - 2}{8x^2 + 2x - 3}$

15. $f(x) = \dfrac{4x^2 + 8x + 3}{2x^2 - x - 6}$

16. $f(x) = \dfrac{x^2 + x - 30}{x^2 - x - 20}$

For Exercises 17–26, identify any horizontal asymptotes for the graph of f.

17. The function in Exercise 7.

18. The function in Exercise 8.

19. The function in Exercise 9.

20. The function in Exercise 10.

21. The function in Exercise 11.

22. The function in Exercise 12.

23. The function in Exercise 13.

24. The function in Exercise 14.

25. The function in Exercise 15.

26. The function in Exercise 16.

For Exercises 27–36, determine the x-intercepts and y-intercept for the graph of f.

27. $f(x) = \dfrac{x}{2x + 6}$

28. $f(x) = \dfrac{x - 1}{3x + 9}$

29. $f(x) = \dfrac{6x - 3}{2x + 4}$

30. $f(x) = \dfrac{7x + 4}{3x - 1}$

31. $f(x) = \dfrac{x^2 - x - 6}{x^2 - 7x + 12}$

32. $f(x) = \dfrac{x^2 - 5x + 4}{x^2 + 2x - 3}$

33. $f(x) = \dfrac{x - 2}{x^2 + 6x + 9}$

34. $f(x) = \dfrac{x^2 + 2x}{3x^2 + 18x + 24}$

35. $f(x) = \dfrac{x^2 - 4}{x^3 + 2x^2 - 3x}$

36. $f(x) = \dfrac{4x^2 - 9}{x^3 + 3x^2 + 4x}$

For Exercises 37–42, determine the x-intercepts and y-intercept graphically.

37. $f(x) = \dfrac{x^2 - 2x + 5}{x + 1}$

38. $f(x) = \dfrac{x^2 - 2x + 1}{x - 2}$

39. $f(x) = \dfrac{2x^3 - x^2 + 3x - 2}{x + 1}$

40. $f(x) = \dfrac{x^3 - 2x + 1}{x - 1}$

41. $f(x) = \dfrac{2x^3 - x^2 + 1}{2 - x}$

42. $f(x) = \dfrac{3x^2 - x + 5}{x^2 - 4}$

For Exercises 43–48, sketch a graph of f by hand on the interval $1 \le x \le 10$. Label all asymptotes and intercepts.

43. $f(x) = \dfrac{x}{x + 2}$

44. $f(x) = \dfrac{2x}{x - 4}$

45. $f(x) = \dfrac{x + 4}{x - 4}$

46. $f(x) = \dfrac{x - 2}{x + 2}$

47. $f(x) = \dfrac{6 - 3x}{x - 6}$

48. $f(x) = \dfrac{4 - 4x}{x - 2}$

For Exercises 49–56, complete the following.

(a) *Rewrite the given radical function as a rational exponent function.*

(b) *Write the domain of f using interval notation.*

49. $f(x) = \sqrt[4]{x - 1}$

50. $f(x) = \sqrt{2x + 3}$

51. $f(x) = \sqrt[3]{4 - x}$

52. $f(x) = \sqrt[3]{5x - 8}$

53. $f(x) = \sqrt{(x - 4)^3}$

54. $f(x) = \sqrt[4]{(6x + 1)^5}$

55. $f(x) = \dfrac{1}{\sqrt{3x + 4}}$

56. $f(x) = \dfrac{3}{\sqrt{7x - 2}}$

For Exercises 57–64, complete the following.

 (a) *Rewrite the rational exponent function as a radical function.*

 (b) *Write the domain of f using interval notation.*

57. $f(x) = (4x + 3)^{1/2}$

58. $f(x) = (8x - 9)^{1/2}$

59. $f(x) = (3x + 1)^{1/3}$

60. $f(x) = (7x + 9)^{1/3}$

61. $f(x) = (6x - 1)^{3/2}$

62. $f(x) = (4x - 5)^{3/2}$

63. $f(x) = (6x + 3)^{-1/2}$

64. $f(x) = (7x - 2)^{-1/2}$

Application Exercises

65. Ecology—Cost-Benefit Analysis: Environmental scientists and municipal planners often are guided by **cost-benefit models**. These mathematical models estimate the cost of removing a pollutant from the atmosphere as a function of the percentage of pollutant removed. Let's suppose that a cost-benefit function is given by

$$f(x) = \frac{30x}{100 - x} \quad 0 \le x < 100$$

where x represents the percentage of the pollutant removed and $f(x)$ represents the associated cost in millions of dollars.

 (a) Evaluate $f(85)$ and interpret.

 (b) Complete the following table:

x	5	50	70	90	95
$f(x)$					

 (c) Many cost-benefit functions exhibit the very high cost of removing the final percentage of a pollutant. To calculate this behavior, evaluate $f(99.9) - f(95)$. This difference represents the approximate cost of removing the final 5% of the pollutant.

 (d) Why can we not compute $f(100) - f(95)$ to get the *actual* cost of removing the final 5% of the pollutant?

66. Ecology—Cost-Benefit Analysis: The cost-benefit function for removing a certain pollutant from the atmosphere is given by

$$f(x) = \frac{20x}{100 - x} \quad 0 \le x < 100$$

where x represents the percentage of the pollutant removed and $f(x)$ represents the associated cost in millions of dollars.

 (a) Evaluate $f(70)$ and interpret.

 (b) Complete the following table:

x	5	50	70	90	95
$f(x)$					

 (c) Evaluate $f(99.9) - f(95)$. This represents the approximate cost of removing the final 5% of the pollutant.

 (d) Why can we not compute $f(100) - f(95)$ to get the *actual* cost of removing the final 5% of the pollutant?

67. Psychology—Memory Studies: The students in an anatomy class were asked to memorize a list of 20 parts of the human body. After each class, a student was chosen and asked to write as many of these anatomical parts as she could. The average number of parts remembered is given by the mathematical model

$$f(x) = \frac{20x - 18}{x} \quad x \geq 4$$

where x represents the number of days since the list was distributed and $f(x)$ represents the average number of items that were remembered.

  (a) Evaluate $f(10)$ and interpret.

  (b) Evaluate $f(100)$ and interpret.

68. Psychology—Memory Studies: The technicians at the Arp Brothers auto parts factory are given a checklist of 40 items to inspect for quality control. Over the next three months, a technician is selected and asked to write the checklist from memory. The mathematical model for the technician's performance is given by

$$f(x) = \frac{80x - 36}{2x} \quad 1 \leq x \leq 120$$

where x represents the number of days since the checklist was distributed and $f(x)$ represents the average number of inspection items that was remembered.

  (a) Evaluate $f(2)$ and $f(30)$ and interpret each.

  (b) Find the horizontal asymptote and interpret its meaning.

  (c) The State Regulatory Agency requires technicians to remember a minimum of 25 items after three months. According to the model, will the technicians meet this requirement?

69. Economics—Sales Analysis: The Slaybaugh Satellite Company is manufacturing a new low-cost satellite dish and promotes its sales through an aggressive sales campaign. The income from sales is given by the sales function

$$S(x) = \frac{120x^2 - 600x + 3}{2x^2 - 10x + 1} \quad x \geq 5$$

where x represents the amount spent on advertising in thousands of dollars, and $S(x)$ represents the income from sales measured in hundreds of thousands of dollars.

  (a) Evaluate $S(10)$ and interpret.

  (b) Find the horizontal asymptote and interpret its meaning.

70. Physics—Highway Speed: A slightly banked highway corner will safely accommodate traffic at a speed given by the model

$$f(x) = \frac{29}{20}x^{1/2} \quad x \geq 0$$

where x represents the radius of the corner in feet, and $f(x)$ represents the speed at which the traffic can travel safely in miles per hour.

  (a) Evaluate $f(20)$ and interpret.

  (b) If the highway planners expect the traffic to travel at a speed of 64 miles per hour, what radius should the corner be?

71. Physics—Sight Lines: In a forest fire tower, the distance that an observer is able to see into the forest is related to the height of the fire tower via the function

$$f(x) = \frac{7}{5}x^{1/2} \quad x \geq 0$$

where x represents the height of the tower in feet, and $f(x)$ represents the distance that the observer can see in miles.

(a) Evaluate $g(70)$ and interpret.

(b) If the observer is required to see 29 miles into the forest, how high must the tower be?

72. **Physics—Sight Lines:** The distance that a person can see from an airplane to the horizon on a cloudless day is given by

$$f(x) = \frac{61}{50}\sqrt{x} \quad x \geq 0$$

where x represents the altitude of the plane in feet, and $f(x)$ is the distance that a person can see in miles.

(a) Evaluate $f(32{,}000)$ and interpret.

(b) Find the slope of the secant line for $x = 36{,}000$ and $\Delta x = 4000$ and interpret.

73. **Botany—Plant Species:** A biologist has shown that the number of plant species in the South American rain forest is related to the area of the land studied via the model

$$f(x) = 28.1\sqrt[3]{x} \quad x \geq 0$$

where x represents the area studied in square miles, and $f(x)$ represents the number of plant species.

(a) Evaluate $f(300)$ and interpret.

(b) Find the slope of the secant line for $x = 1728$ and $\Delta x = 469$ and interpret.

74. **Civil Engineering—City Growth:** A city planner has projected that the population of a newly developed suburb will grow for the next four years according to the mathematical model

$$f(x) = 10{,}000 + 20x^{3/2} + 30x \quad 0 \leq x \leq 48$$

where x is the number of months from the present, and $f(x)$ is the suburb's population.

(a) Evaluate $f(12)$ and interpret.

(b) Find the slope of the secant line for $x = 12$ and $\Delta x = 24$ and interpret.

 75. **Government Finance—Food Assistance:** The U.S. federal government has many supplemental nutrition assistance programs, such as the National School Lunch Program, food stamps, and the WIC program. The total cost of these programs from 1990 to 2011 can be modeled by the power function

$$f(x) = x^{1.17} + 16.75 \quad 0 \leq x \leq 21$$

where x represents the number of years since 1990 and $f(x)$ represents the total food program assistance cost in billions of dollars annually. (*Source:* U.S. Department of Agriculture.)

(a) Determine the y-intercept and interpret its meaning.

(b) Compute m_{sec} for $x = 0$ and $\Delta x = 10$ and interpret its meaning.

(c) Now compute m_{sec} for $x = 10$ and $\Delta x = 10$. During which decade did the amount of federal food assistance grow more rapidly?

76. **Pathology—Diabetes Proliferation:** A chronic illness that has proliferated during the past few decades is diabetes. The total amount spent in the United States annually for diabetes medicines from 1990 to 2011 can be modeled by the power function

$$f(x) = x^{1.19} + 0.76 \quad 0 \leq x \leq 21$$

where x represents the number of years since 1990 and $f(x)$ represents the amount spent on diabetes and metabolic medicines in billions of dollars annually. (*Source: Pharmaceutical Preparations* Industrial Report.)

(a) Determine the y-intercept and interpret its meaning.

(b) Compute m_{sec} for $x = 0$ and $\Delta x = 10$ and interpret its meaning.

(c) Now compute m_{sec} for $x = 10$ and $\Delta x = 10$. During which decade did the amount spent on diabetes medicines grow more rapidly?

 77. International Economics—Unemployment in Ireland: Suppose a social worker wants to explore the number of unemployed in Ireland to see how rapidly the unemployment rate has been increasing. The number unemployed in Ireland annually from 2000 to 2010 can be modeled by

$$f(x) = \frac{2.03x - 26.34}{0.03x - 0.35} \quad 0 \le x \le 10$$

where x represents the number of years since 2000 and $f(x)$ represents the number of unemployed in Ireland, measured in thousands. (*Source:* Google public data.)

(a) Determine the vertical asymptote of f. Is the vertical asymptote in the reasonable domain of the model?

(b) Assume that the social worker wants to determine, on average, how rapidly the number of unemployed has been increasing from 2005 to 2010. Determine this value, and interpret.

 78. Economics—Consumer Spending: Social networking and Internet commerce have contributed to the amount of spending on video and audio products during the past two decades. The amount spent on audio and video products in the United States from 1990 to 2010 can be modeled by

$$f(x) = \frac{3.39x + 44.41}{-0.01x + 0.82} \quad 0 \le x \le 20$$

where x represents the number of years since 1990 and $f(x)$ represents the amount spent on audio and video products, measured in billions of dollars. (*Source:* U.S. Bureau of Economic Analysis.)

(a) Determine the vertical asymptote of f. Is the vertical asymptote in the reasonable domain of the model?

(b) Suppose an industry analyst wants to know, on average, how rapidly the amount spent on these electronic products increased from 2000 to 2005. Determine this value and interpret the result.

© Yuri Arcurs/ShutterStock, Inc.

Concept and Writing Exercises

For Exercises 79–82, consider the general form of a power function $f(x) = ax^b + c$, where a, b, and c represent real numbers and $b > 0$.

79. Determine the coordinates of the y-intercept of f.

80. Find the x-intercept of f by solving $f(x) = 0$ for x.

81. Compute the general form for the secant line slope by simplifying m_{sec} for $x = x_1$ and an increment Δx.

82. Does f have any asymptotes? Explain your answer.

For Exercises 83 and 84, consider a rational function of the form $f(x) = \dfrac{ax + b}{cx + d}$, where a, b, c, and d represent real numbers.

83. Determine the coordinates of the x-intercept and y-intercept of f.

84. Determine the equations for the vertical asymptote and horizontal asymptote of f.

 ### Section Project

In a July 22, 2010, Associated Press interview, Treasury Secretary Tim Geithner said, "We are likely to have to take a broader look at corporate tax reform next year," adding it was likely to be one of the areas the fiscal commission appointed by President Obama would examine to make recommendations on deficit reduction. (*Source:* Associated Press.) The amount collected in U.S. corporate taxes annually from 1990 to 2010 can be modeled by

$$f(x) = 1.65x + 13.13 \quad 0 \leq x \leq 20$$

where x represents the number of years since 1990 and $f(x)$ represents the amount in corporate taxes collected in billions of dollars. The amount of corporate profits of U.S. corporations from 1990 to 2010 can be modeled by

$$g(x) = 2.45x^2 + 26.52x + 419.01 \quad 0 \leq x \leq 20$$

where x represents the number of years since 1990 and $g(x)$ represents the corporate profits in billions of dollars. (*Sources:* Associated Press, U.S. Bureau of Economic Analysis).

© David L. Lewis/Wishing Well Productions/ShutterStock, Inc.

(a) Write the function $h(x) = \dfrac{f(x)}{g(x)} \cdot 100, \quad 0 \leq x \leq 20$, to determine the ratio of taxes to profit as a percentage.

(b) Graph h in the viewing window [0, 20] by [0, 4], then evaluate and interpret $h(3)$.

(c) Determine the horizontal asymptote of the function h. How could this be interpreted?

(d) Is it possible to determine the equations for the vertical asymptotes of h? By what means could these equations be computed?

(e) On average, during which decade did the ratio of taxes to profit increase the least, during the 1990s or the 2000s? Compute secant line slopes to support your conclusion.

Section 1.1 Review Exercises

1. Use the table to construct ordered pairs, and plot them in the Cartesian plane.

x	-1	2	2.8	4
$f(x)$	3.5	1	-1.7	2

2. Make a table of values for x and y based on the scatterplot.

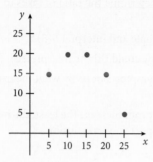

3. For the function $A(r) = \pi r^2$, identify the independent variable and the dependent variable.

 4. **Government Economics—Military Construction:** Make a scatterplot of the data in the **Table R.1**, where x represents the number of years since 2000 and y represents the amount spent on military construction annually, measured in billions of dollars. (*Source:* U.S. Office of Management and Budget.)

Table R.1

Year	Amount spent on military construction (in billions of dollars)
2000	5.1
2003	6.7
2004	6.1
2005	7.3
2006	9.5
2007	14.0
2008	22.1
2009	26.8
2010	22.9

Source: U.S. Office of Management and Budget

For Exercises 5 and 6, determine if what is given represents a function. If it is not a function, explain why it is not.

5. The values given in the table:

Domain	1	2	3	4	5
Range	5	3	1	3	5

6. The given graph:

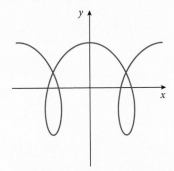

Section 1.2 Review Exercises

For Exercises 7 and 8, let $f(x) = x^2 - 4$, and let $g(x) = 5x + 6$. Evaluate each expression and write the solution as an ordered pair.

7. $f(3)$

8. $g(-3)$

9. Use the graph of the function $y = f(x)$ to answer the following.

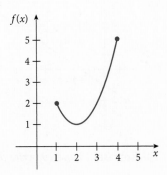

   (a) Name the independent variable.

   (b) Name the dependent variable.

   (c) Determine the value of $f(x)$ when $x = 3$.

   (d) Evaluate $f(1)$.

   (e) Write the domain of f using interval notation.

   (f) Write the range of f using interval notation.

10. Solve the following inequality and write the solution using interval notation:

$$3x - 7 \geq 20$$

For Exercises 11–13, determine the domain of the function algebraically and write the domain using interval notation.

11. $f(x) = \sqrt{-4 + 2x}$

12. $f(x) = \dfrac{x + 3}{x^2 + 3x - 10}$

13. $f(x) = \dfrac{x + 3}{x^2 - 4}$

14. Economics—Price-Demand: The price-demand function for the new SwipeType keyboard is given by

$$p(x) = 35 - 0.2x$$

where x represents the number of units sold each day and $p(x)$ represents the price per unit, measured in dollars.

   (a) Complete the table of value for the various production levels:

x	0	10	20	30	40	50
$p(x)$						

   (b) Make a graph of p for $0 \leq x \leq 50$.

   (c) Using the domain given in part (b), write the corresponding range.

   (d) Evaluate $p(35)$ and interpret the result.

15. Inspect the graph, and write the domain and range of *f* using interval notation.

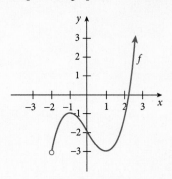

For Exercises 16 and 17, determine the average rate of change.

16. Between the points $(3, 7)$ and $(-2, 14)$.

17. For the given graph:

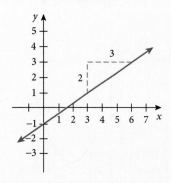

For Exercises 18 and 19, write the equation of the line in point-slope form.

18. A line that contains the points $(2, 1)$ and $(8, 19)$.

19. A line that has an *x*-intercept at -4 and *y*-intercept at 5.

For Exercises 20 and 21, write the equation of the line in slope-intercept form.

20. A line that has a slope of $\frac{3}{7}$ and passes through the point $(-3, 2)$.

21. The points in the given table:

x	y
-4	-10
12	2

22. Determine the *x*-intercept and the *y*-intercept of the function $f(x) = \dfrac{60 - 9x}{4}$.

23. **Microeconomics—Depreciation:** A Crossroads comfort-style bicycle costs \$800 and depreciates at a rate of \$40 per year.

   (a) Write a depreciation function for the bicycle of the form $f(x) = mx + b$.

   (b) In how many years will the bicycle be worth \$600?

24. Classify the function as increasing or decreasing, and identify the slope of the line given by the function

$$f(x) = -3(2 - 5x)$$

25. **Economics—Cost Function:** Managers who oversee the production of the Heart-Hopper heart rate monitor know that the fixed costs are \$2600 and know that it costs \$3950 to make 18 heart rate monitors.

(a) Write a linear cost function of the form $C(x) = mx + b$.

(b) Use the model in part (a) to determine the cost of producing 20 heart rate monitors.

(c) How much does it cost to produce the 21st heart rate monitor?

26. **Economics—Cost Analysis:** The manager of Pizza Pizzazz determines that the cost function for their specialty pizzas is given by

$$C(x) = 6.5x + 300$$

where x represents the number of pizzas prepared and delivered and $C(x)$ represents the cost in dollars.

(a) Identify the fixed and variable costs for the specialty pizzas.

(b) Evaluate $C(36)$ and interpret the result.

(c) Determine how many pizzas can be prepared and delivered for $703.

(d) What is the marginal cost of the specialty pizzas?

27. Make a graph of the piecewise-defined function f and accurately label the endpoints:

$$f(x) = \begin{cases} 1 - 0.5x, & x \le 1 \\ 3, & x > 1 \end{cases}$$

28. Make a graph of the function $f(x) = |12 - 4x|$, then rewrite f as a piecewise-defined function.

29. **Economics—Cost Analysis:** Consider the daily cost schedule for producing CasaVerde ceiling fans:

Fans Produced	Cost (in Dollars)
0	2500
50	5500
220	15,700

(a) By inspecting the table of values, determine the fixed costs.

(b) How much does it cost for each additional ceiling fan that is produced? What do we call this cost?

(c) Write a cost function C, where x represents the number of ceiling fans produced, and $C(x)$ represents the cost.

(d) What would the cost be if 84 ceiling fans were produced?

Section 1.3 Review Exercises

For Exercises 30 and 31, complete the following.

(a) Determine the coordinates of the vertex algebraically.

(b) Write the range of the function using interval notation.

30. $f(x) = x^2 - 4x + 3$

31. $f(x) = -2x^2 - 6x + 12$

For Exercises 32 and 33, complete the following.

(a) Determine the vertex.

(b) Use the result from part (a) to determine the intervals where the function is increasing and where it is decreasing.

32. $f(x) = 5x^2 + 20x$

33. $f(x) = -x^2 + 3.2x - 7.8$

For Exercises 34 and 35, find the zeros of the quadratic function by factoring.

34. $f(x) = x^2 - 144$

35. $f(x) = 2x^2 + 3x - 20$

36. Find the real number roots of the quadratic function $f(x) = 3x^2 - 6x + 2$ by using the quadratic formula. If the discriminant is a negative number, write *no real roots*.

37. Use the `Zero` or `Root` capabilities of your calculator to approximate the zeros of the quadratic function $g(x) = 2.4x^2 + 1.7x - 5.3$. Round to the nearest hundredth. If the graph has no *x*-intercepts, write *no real roots*.

38. Consider the function $f(x) = -1.8x + 5.3$. Use the difference quotient to determine the slope of the line over the following intervals:

(a) As *x* changes from $x = 0$ to $x = 2$

(b) As *x* changes from $x = 0$ to $x = 5$

(c) As *x* changes from $x = 4$ to $x = 7$

39. Consider the function $g(x) = 2x^2 - 3x$. Use the difference quotient to determine the slope of the secant line over the following intervals:

(a) If $x = 0$ and $\Delta x = 1$

(b) If $x = 0$ and $\Delta x = 2$

(c) If $x = 0$ and $\Delta x = 4$

40. Public Finance—Property Tax: The average homeowner's property tax in a certain community can be modeled by

$$f(x) = -15x^2 + 300x + 1200 \quad 0 \le x \le 9$$

where *x* represents the number of years since 2002, and $f(x)$ is the average property tax.

(a) Does the graph of the model open up or down? How do you know this?

(b) The domain of the model consists of *x*-values in the interval $[0, 9]$. What is the corresponding range?

(c) Determine the average rate of change in the property tax from 2004 to 2009.

41. Career Planning—Computer Engineers: A profession that has seen a rise and fall recently is computer systems engineering. Some see the change resulting in reduced quality of computer software design. "For example, rarely is the processor the bottleneck, but so many people look to 'tight loops' to speed up performance when it's often an I/O issue," said project management professional Bernard Hayes. (*Source:* Computerworld.) The number of computer systems engineers employed from 2004 to 2008 can be modeled by the quadratic function

$$f(x) = -7.85x^2 + 97.05x - 206.6 \quad 4 \le x \le 8$$

where *x* represents the number of years since 2000 and $f(x)$ represents the number of computer systems engineers employed, measured in thousands. (*Source:* National Science Foundation.)

(a) By inspecting the function, is the graph of *f* concave up or concave down? Explain how you determine your answer.

(b) Complete **Table R.2**.

Table R.2

x	f(x)
4	
5	
6	
7	
8	

(c) Evaluate and interpret $f(5)$.

(d) According to the table, during what year was the greatest number of computer systems engineers employed?

(e) Determine the coordinates of the vertex and round the x-coordinate to the nearest whole number. Is the result the same as in part (d)?

(f) Determine m_{sec} for $x = 4$ and $\Delta x = 3$. Interpret the result.

Section 1.4 Review Exercises

For Exercises 42 and 43, functions f and g are given. Use the functions to algebraically simplify and write the domain of the following.

(a) $(f + g)(x)$

(b) $(f - g)(x)$

(c) $(f \cdot g)(x)$

(d) $\left(\dfrac{f}{g}\right)(x)$

42. $f(x) = x^2 + 4$; $g(x) = 2x + 3$

43. $f(x) = \sqrt{x + 2}$; $g(x) = x - 2$

For Exercises 44–47, let $f(x) = 3x - 5$ and $g(x) = (x + 2)^2$. Evaluate each of the following:

44. $(f + g)(2)$

45. $(f - g)(4)$

46. $(f \cdot g)(-2)$

47. $\left(\dfrac{f}{g}\right)(3)$

48. Demographics—Population Growth: The population of Eastbrook from 2000 to 2010 can be modeled by

$$f(x) = 80{,}300 + 800x \quad 0 \le x \le 10$$

where x represents the number of years since 2000, and $f(x)$ represents the population. The population of Westbrook during the same time period is given by

$$g(x) = 62{,}600 - 200x \quad 0 \le x \le 10$$

(a) Evaluate $(f + g)(4)$ and interpret.

(b) Evaluate $(f - g)(8)$ and interpret.

49. For the price function $p(x) = 145 - 0.15x$, determine the revenue function R.

50. Economics—Profit Analysis: Suppose that the revenue and cost functions for a certain product are given by

$$R(x) = 18x - 0.08x^2 \quad \text{and} \quad C(x) = 5x + 275$$

(a) Graph R and C in the same viewing window.

(b) Use algebra or the `Intersect` command on your calculator to determine the break-even point.

(c) Determine how much revenue must be generated to reach the break-even point.

51. Economics—Profit Analysis: For the revenue and cost functions $R(x) = 32x - 0.08x^2$ and $C(x) = 16x + 400$, determine the profit function P, find the coordinates of the vertex, and interpret.

52. For a certain product, the price-demand function is $p(x) = 400 - 0.1x$. The variable costs are $225 per unit, and the fixed costs are $3000.

(a) Determine the linear cost function C.

(b) Determine the revenue function R.

(c) Determine the profit function $P(x) = R(x) - C(x)$.

(d) Use the `Zero` command on your calculator to find the zeros of the profit function P. These are the break-even points.

(e) Find the vertex of the graph of the profit function P.

(f) Determine the demand level that yields the maximum profit and find the maximum profit.

53. Determine if each function is a polynomial function. Identify each polynomial as linear, quadratic, cubic, or quartic.

(a) $f(x) = x^2 + 5x - 3$

(b) $f(x) = 17$

(c) $f(x) = \dfrac{1}{3}x - 4$

(d) $f(x) = \dfrac{3}{4}x^4 - \dfrac{2}{3}x^3 + \dfrac{1}{2}x^2 + x$

For Exercises 54 and 55, determine the end behavior of the function.

54. $f(x) = 5x - 2x^3$

55. $f(x) = 3x^2 - 5x$

56. Economics—Revenue Analysis: The Disma Department Store finds that the revenue generated by selling x dresses is given by

$$R(x) = 68x - 0.3x^2 \quad x \geq 0$$

(a) Evaluate $R(42)$ and interpret.

(b) Find the average rate of change for $x = 40$ and $\Delta x = 10$, and interpret.

 57. Consider the polynomial function

$$f(x) = x^3 - 6x^2 + 6x + 4.$$

(a) Determine the end behavior of the function.

(b) Use the `Maximum` and `Minimum` commands on the calculator to approximate the peaks and valleys of the function. Estimate the points to the nearest hundredth.

(c) Use the solution from part (b) to determine the intervals where f increases and where it decreases.

Section 1.5 Review Exercises

For Exercises 58–61, complete the following.

(a) Write the domain using interval notation.

(b) Identify any holes or vertical asymptotes in the graph.

(c) Identify any horizontal asymptotes.

58. $f(x) = \dfrac{x + 3}{x - 1}$

59. $f(x) = \dfrac{x^2}{2x + 3}$

60. $f(x) = \dfrac{x^2 - x - 6}{x^2 + 3x + 2}$

61. $f(x) = \dfrac{2x^2 - x - 15}{x^2 - 9}$

For Exercises 62–65, determine the x- and y-intercepts for the graph of the given function.

62. $f(x) = \dfrac{2x}{x - 5}$

63. $f(x) = \dfrac{x^2 - 8x + 16}{x^2 - 9}$

64. $f(x) = \dfrac{x^2 + 5x + 6}{x^2 + x - 2}$

65. $f(x) = \dfrac{12}{x^2 + 16}$

66. Determine the x-intercepts and the y-intercept for the function $f(x) = \dfrac{3x^2 - 5x + 2}{x + 1}$.

67. Sketch a graph by hand for the function $f(x) = \dfrac{x + 3}{x - 2}$ on the interval $1 \leq x \leq 10$. Label the asymptotes and the intercepts.

68. Economics—Cost-Benefit Analysis: Consider the cost-benefit function for removing a certain pollutant from the atmosphere.

$$f(x) = \dfrac{26x}{100 - x} \quad 0 \leq x < 100$$

where x represents the percentage of the pollutant removed, and $f(x)$ represents the associated cost in millions of dollars.

(a) Evaluate $f(60)$ and interpret.

(b) Evaluate $f(95)$ and interpret.

69. For the function $f(x) = \sqrt[4]{(x - 2)^3}$, complete the following.

(a) Rewrite the radical function as a rational exponent function.

(b) Write the domain using interval notation.

70. For the function $f(x) = (2x - 7)^{3/5}$, complete the following.

(a) Rewrite the rational exponent function as a radical function.

(b) Write the domain using interval notation.

71. **Physics—Free Fall:** The time required for an object to fall a distance $f(x)$ is given by the model

$$f(x) = \frac{1}{4}\sqrt{x} \quad x \geq 0$$

where x represents the distance in feet, and $f(x)$ represents the time in seconds.

(a) Evaluate $f(64)$ and interpret.

(b) How long does it take an object to fall a distance of 144 feet?

72. **Biology—Entomology:** A biologist has shown that the number of known insect species in a certain desert is related to the area of the land studied by the model

$$f(x) = 16.3\sqrt[3]{x} \quad x \geq 0$$

where x represents the area studied in square miles, and $f(x)$ represents the number of insect species.

(a) Evaluate $f(343)$ and interpret.

(b) Find the slope of the secant line for $x = 1000$ and $\Delta x = 331$ and interpret.

Limits and the Derivative

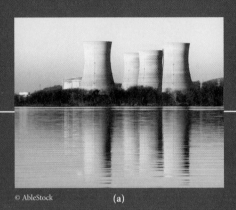

© AbleStock

(a)

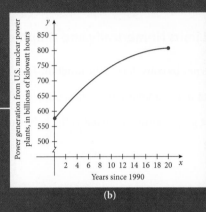

(b)

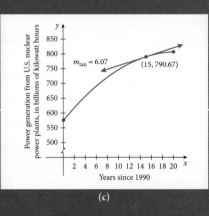

(c)

One of the serious challenges facing the world is that of energy. Alternative energy sources, such as solar arrays and wind turbine farms, are becoming more popular. The United States recently has considered building more nuclear power plants in an attempt to meet the energy demand of the country. Figure (b) shows a model of the power generated by U.S. nuclear power plants, in billions of kilowatt hours from 1990 to 2010. From Figure (c), we see that in 2005, the amount of power generated by U.S. nuclear power plants was increasing at a rate of 6.07 billion kilowatt hours per year.

What We Know

In Chapter 1, we reviewed algebra and also reviewed an important rate of change called an average rate of change. We saw how an average rate of change is equivalent to the slope of a secant line over an interval.

Where Do We Go

In this chapter, we will see how the limit concept is used to introduce a new type of rate of change called an instantaneous rate of change. We will see how an instantaneous rate of change is equivalent to the slope of a tangent line at a specific point.

Chapter Sections

2.1 Limits and Continuity

2.2 Limits and Asymptotes

2.3 Approximating Instantaneous Rates of Change

2.4 The Derivative of a Function
Chapter Review Exercises

2.1   Limits and Continuity

The two main branches of calculus, differential calculus and integral calculus, depend on the limit concept. To help grasp the ideas that calculus is based on, we first provide a practical introduction to limits. We use a combined numerical, graphical, and algebraic approach to examine the concept of a limit.

Determining Limits Numerically and Graphically

We begin our journey by considering the function $f(x) = \dfrac{x^2 - 4}{x - 2}$ and its graph shown in **Figure 2.1.1**. Since $x = 2$ is not in the domain of the function, in other words $f(2)$ is undefined, there appears to be a "hole" in the graph. However, what is the behavior of $f(x) = \dfrac{x^2 - 4}{x - 2}$ as x gets very, very close to the value of 2?

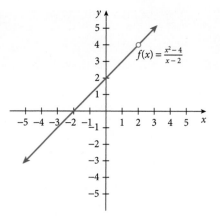

Figure 2.1.1

By *behavior*, we mean what is happening to the function values (or the y-values if you prefer) as x approaches 2. One way to answer this question is by constructing a table to numerically analyze the behavior of f as x gets closer and closer to 2. Because we could approach 2 from the **left side** of 2 or from the **right side** of 2, we must include values of x less than 2 and values greater than 2. See **Table 2.1.1**.

Table 2.1.1

	x approaches 2 from the left →						← x approaches 2 from the right				
x	1	1.9	1.99	1.999	1.9999	2	2.0001	2.001	2.01	2.1	3
$C(x)$	3	3.9	3.99	3.999	3.9999		4.0001	4.001	4.01	4.1	5

NOTE: We intentionally left a blank below $x = 2$ for two reasons. First, the function is not defined at $x = 2$. Second, we wish to emphasize that in the limit process, we do not care what is happening at $x = 2$, but only in the behavior of the function as x gets close to 2.

It appears from **Table 2.1.1** that if we start to the left of $x = 2$ or to the right of $x = 2$, as we allow x to approach 2, our functional values are approaching 4. We can say that,

$$\text{"The limit of } \frac{x^2 - 4}{x - 2}\text{, as } x \text{ approaches 2, is 4."}$$

Using an arrow for the word "approaches" and *lim* as shorthand for the word "limit," the mathematical notation for this English sentence is

$$\lim_{x \to 2} \frac{x^2 - 4}{x - 2} = 4$$

You should interpret this limit notation to mean that as x gets closer and closer to 2, from both sides of 2, $\frac{x^2 - 4}{x - 2}$ gets closer and closer to 4. Notice that Figure 2.1.1 graphically supports our numerical work in Table 2.1.1. At this time, numerically and graphically, we believe that $\lim_{x \to 2} \frac{x^2 - 4}{x - 2} = 4$.

Technology Option

Once we enter the function $f(x) = \frac{x^2 - 4}{x - 2}$ into the $y =$ editor of the graphing calculator, we can numerically and graphically analyze a limit. **Table 2.1.2** shows the result of using the `Table` command on the graphing calculator to numerically analyze $\lim_{x \to 2} \frac{x^2 - 4}{x - 2}$.

Table 2.1.2

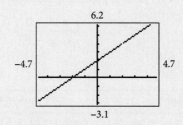

Figure 2.1.2

    Figure 2.1.2 shows a graph of $f(x) = \frac{x^2 - 4}{x - 2}$, and **Figure 2.1.3** is the result of utilizing the `Zoom In` command. Notice in Figure 2.1.3(b) and (c) that we have used the `Trace` command to get as close to 2 as possible from the left side of 2 and from the right side of 2, respectively. Because of the limitations of a graphing calculator, you may not see a hole in the graph at $x = 2$. Using `Zdecimal` or selecting the x-axis window so that $x = 2$ is the midpoint of the graphing interval should show the hole.

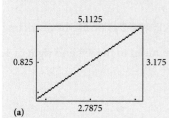
(a)

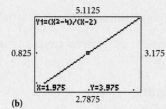
(b)

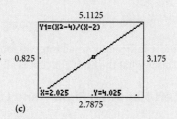
(c)

Figure 2.1.3

Example 1: Analyzing a Limit

For $f(x) = \frac{x^2 - 1}{x - 1}$, construct a table of values around $x = 1$ and guess the value of $f(x)$ as $x \to 1$. Use the graph of f in **Figure 2.1.4** to graphically support your numerical work.

OBJECTIVE 1

Estimate limits numerically and graphically.

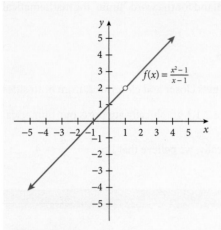

Figure 2.1.4

Perform the Mathematics

Using a calculator, we can quickly construct **Table 2.1.3**. We notice that as $x \to 1$ from both sides (left and right) of 1, $f(x)$ is approaching 2. Hence we conjecture that the limit is 2 and write

$$\lim_{x \to 1} \frac{x^2 - 1}{x - 1} = 2$$

Table 2.1.3

	$x \to 1$ *from left*					$x \to 1$ *from right*			
x	0	0.9	0.99	0.999	1	1.001	1.01	1.1	2
$f(x)$	1	1.9	1.99	1.999		2.001	2.01	2.1	3

The graph of f in Figure 2.1.4 appears to support our numerical work. That is, from the graph it looks like the closer we get to $x = 1$, from both sides of $x = 1$, the closer the function values get to 2. So numerically and graphically we believe that

$$\lim_{x \to 1} \frac{x^2 - 1}{x - 1} = 2 \qquad \blacksquare$$

By table building and graphing, we can make $f(x) = \dfrac{x^2 - 1}{x - 1}$ as close to 2 as we like by restricting x to a sufficiently small interval *around* 1. By "interval around 1," we mean to the left and right of 1. This is what the limit concept is all about!

Left-Hand and Right-Hand Limits

In Example 1, we constructed Table 2.1.3 so that we approached $x = 1$ both from the left side and from the right side. We also made sure that the graph of f in Figure 2.1.4 showed x-values to the left and right of $x = 1$. Because this **left-side** and **right-side** analysis is so important in the limit process, we introduce notation for it.

OBJECTIVE 2

Estimate left-hand and right-hand limits.

DEFINITION

Left-Hand and Right-Hand Limits

1. $\displaystyle\lim_{x \to a^-} f(x)$ means "the limit as x approaches a from the left side of a" and is called the **left-hand limit**.
2. $\displaystyle\lim_{x \to a^+} f(x)$ means "the limit as x approaches a from the right side of a" and is called the **right-hand limit**.

Using this notation, we are ready for the following important **definition of the limit**.

DEFINITION

Limit

For any function f, $\lim\limits_{x \to a} f(x) = L$ means that, as x approaches a, $f(x)$ approaches L. Alternatively, if

$\lim\limits_{x \to a^-} f(x) = L$ and $\lim\limits_{x \to a^+} f(x) = L$, then $\lim\limits_{x \to a} f(x) = L$.

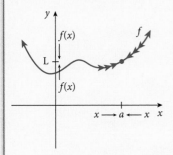

Figure 2.1.5 Illustrating the definition of a limit.

OBJECTIVE 3

Define limit.

NOTES:

1. If the left-hand limit does not equal the right-hand limit, then there is no limit. In this case, we say that the limit **does not exist**. See **Figure 2.1.6(a)**.

2. The existence of $\lim\limits_{x \to a} f(x)$ does **not** depend on whether $f(a)$ is defined. See Figure 2.1.6(b).

3. The existence of $\lim\limits_{x \to a} f(x)$ does not depend on the value of $f(a)$ is $f(a)$ is defined. See Figure 2.1.6(c).

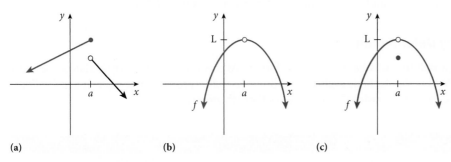

(a)   **(b)**   **(c)**

Figure 2.1.6 (a) $\lim\limits_{x \to a} f(x)$ does not exist because $\lim\limits_{x \to a^-} f(x) \neq \lim\limits_{x \to a^+} f(x)$. (b) $\lim\limits_{x \to a} f(x)$ exists even though $f(a)$ is undefined. (c) $f(a)$ is defined, but does *not* equal $\lim\limits_{x \to a} f(x)$.

Example 2 illustrates that the existence of $\lim\limits_{x \to a} f(x)$ does not depend on the value of $f(a)$ if $f(a)$ is defined.

Example 2: Analyzing a Limit

Estimate $\lim\limits_{x \to 2} f(x)$ for $f(x) = \begin{cases} 1 - x, & x \leq 2 \\ 4, & x > 2 \end{cases}$

Perform the Mathematics

From **Table 2.1.4** and from the graph of f in **Figure 2.1.7**, we conclude that

$$\lim\limits_{x \to 2^-} f(x) = -1, \text{ whereas } \lim\limits_{x \to 2^+} f(x) = 4$$

Since the left-hand limit **does not equal** the right-hand limit, we conclude that $\lim\limits_{x \to 2} f(x)$ does not exist. Notice that graphically this corresponds to a jump in the graph. See Figure 2.1.7.

Table 2.1.4

	$x \rightarrow 2^-$					$x \rightarrow 2^+$		
x	1	1.9	1.99	1.999	2	2.001	2.01	2.1	3
$f(x)$	0	-0.9	-0.99	-0.999		4	4	4	4

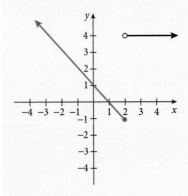

Figure 2.1.7   ■

Again, we want to point out that in Example 2, $f(2) = -1$, yet $\lim\limits_{x \to 2} f(x)$ does not exist. Remember that the existence of the limit $\lim\limits_{x \to a} f(x)$ does *not* depend on the value of $f(a)$.

Algebraically Determining Limits

So far, our approach to limits has been very informal. At this time, we will briefly state some **limit theorems** to aid us in the algebraic evaluation of limits, as well as provide some mathematical validity. Later in the text, we will occasionally use some of these theorems.

OBJECTIVE 4

Examine Limit Theorems.

Limit Theorems

If a, b, and n are real numbers, then

1. $\lim\limits_{x \to a} c = c$
2. $\lim\limits_{x \to a} x = a$
3. $\lim\limits_{x \to a} [c \cdot f(x)] = c \cdot \lim\limits_{x \to a} f(x)$
4. $\lim\limits_{x \to a} [f(x) \pm g(x)] = \lim\limits_{x \to a} f(x) \pm \lim\limits_{x \to a} g(x)$
5. $\lim\limits_{x \to a} [f(x) \cdot g(x)] = \lim\limits_{x \to a} f(x) \cdot \lim\limits_{x \to a} g(x)$
6. $\lim\limits_{x \to a} \dfrac{f(x)}{g(x)} = \dfrac{\lim\limits_{x \to a} f(x)}{\lim\limits_{x \to a} g(x)}$, provided $\lim\limits_{x \to a} g(x) \neq 0$
7. $\lim\limits_{x \to a} [f(x)]^n = [\lim\limits_{x \to a} f(x)]^n$

NOTES:

1. The first limit theorem simply states that the limit of a constant is that constant.

2. Instead of memorizing these theorems, we suggest the following alternative, which we call the **Substitution Principle**.

Substitution Principle

When attempting to algebraically determine a limit, first try direct substitution. In other words, when attempting to find $\lim\limits_{x \to a} f(x)$, first try to compute $f(a)$.

Example 3: Analyzing Limits Algebraically

OBJECTIVE 5

Determine limits algebraically.

Determine the following limits.

a. $\lim\limits_{x \to 2} 7$ **b.** $\lim\limits_{x \to 1}(2x^2 - 3x + 5)$ **c.** $\lim\limits_{x \to 3} \sqrt{2x - 1}$

Perform the Mathematics

a. Because 7 is a constant, we use the first Limit Theorem and have

$$\lim\limits_{x \to 2} 7 = 7$$

b. Here, we simply substitute and get

$$\lim\limits_{x \to 1}(2x^2 - 3x + 5) = 2(1)^2 - 3(1) + 5 = 4$$

c. Again, we simply substitute and have

$$\lim\limits_{x \to 3} \sqrt{2x - 1} = \sqrt{2(3) - 1} = \sqrt{5} \qquad \blacksquare$$

▶ *Try It Yourself*

Some related Exercises are 27 and 29.

At this time you may be asking yourself, "When does the Substitution Principle fail?" The answer to this question can be found in Examples 1 and 2. In Example 2, the Substitution Principle fails because we have a piecewise-defined function. In Example 1, the Substitution Principle fails because substituting produces a fraction of the form $\frac{0}{0}$. We call the fraction $\frac{0}{0}$ an **indeterminate form**. When we try substitution and get the indeterminate form $\frac{0}{0}$, we have to use other techniques to determine the limit.

Recall that in Example 1, we determined numerically and graphically the limit

$$\lim\limits_{x \to 1} \frac{x^2 - 1}{x - 1} = 2$$

To determine this limit algebraically, we can do the following:

$$\lim\limits_{x \to 1} \frac{x^2 - 1}{x - 1}$$

Factor $= \lim\limits_{x \to 1} \dfrac{(x - 1)(x + 1)}{x - 1}$

Cancel, provided $x \neq 1$ $= \lim\limits_{x \to 1}(x + 1)$

Substitution Principle $= 1 + 1 = 2$

What we have really done using algebra is to determine another function, $g(x) = x + 1$, that is equal to $f(x) = \dfrac{x^2 - 1}{x - 1}$, for all values of x, except $x = 1$. In other words,

$$f(x) = \frac{x^2 - 1}{x - 1} = \frac{(x - 1)(x + 1)}{x - 1} = x + 1 = g(x), \text{ provided that } x \neq 1$$

This allows us to write

$$\lim_{x \to 1} \frac{x^2 - 1}{x - 1} = \lim_{x \to 1} \frac{(x - 1)(x + 1)}{x - 1} = \lim_{x \to 1}(x + 1) = 2$$

OBJECTIVE 6

Analyze indeterminate form.

Example 4: Analyzing a Limit Involving $\frac{0}{0}$

Determine $\lim\limits_{x \to 3} g(x)$, where $g(x) = \dfrac{x^2 - 9}{x - 3}$.

Perform the Mathematics

We try substituting, which gives

$$\lim_{x \to 3} \frac{x^2 - 9}{x - 3} = \frac{(3)^2 - 9}{3 - 3} = \frac{0}{0}$$

Since we have the indeterminate form $\frac{0}{0}$, we decide to try a little algebra. We will factor the numerator and cancel as follows:

$$\lim_{x \to 3} \frac{x^2 - 9}{x - 3}$$

Factor
$$= \lim_{x \to 3} \frac{(x - 3)(x + 3)}{x - 3}$$

Cancel, provided that $x \neq 3$
$$= \lim_{x \to 3}(x + 3)$$

Substitute
$$= 3 + 3 = 6$$

Once again, notice that even though $g(3)$ is undefined, we determined that $\lim\limits_{x \to 3} g(x)$ exists. This reinforces the fact that the limit as x approaches 3 is *not* dependent on $g(3)$. See **Figure 2.1.8**.

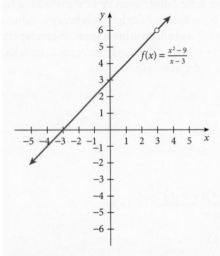

Figure 2.1.8   ∎

▶ *Try It Yourself*

Some related Exercises are 37 and 49.

Example 5: Analyzing a Limit Involving $\frac{0}{0}$

Determine $\lim\limits_{x \to 0} \dfrac{|x|}{x}$.

Perform the Mathematics

If we try substituting, we get

$$\lim_{x \to 0} \frac{|x|}{x} = \frac{0}{0}$$

By letting $f(x) = \dfrac{|x|}{x}$, we can use a table and a graph to determine this limit. See **Table 2.1.5** and **Figure 2.1.9**.

Table 2.1.5

	$x \to 0^-$					$x \to 0^+$			
x	-1	-0.1	-0.01	-0.001	0	0.001	0.01	0.1	1
$f(x)$	-1	-1	-1	-1		1	1	1	1

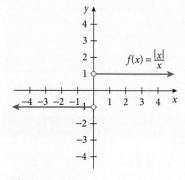

Figure 2.1.9

As we can see numerically and graphically, since the left-hand limit does not equal the right-hand limit, we conclude that $\lim\limits_{x \to 0} \dfrac{|x|}{x}$ does not exist. The algebraic verification of this is left for you in Exercise 100. ∎

For Example 6, we need to recall the first Limit Theorem, which states that the limit of a constant is that constant.

Example 6: Analyzing Limits Involving Two Variables

Determine the following limits.

a. $\lim\limits_{h \to 0} (3x + 2h)$ **b.** $\lim\limits_{h \to 0} \dfrac{5xh + 2h^2}{h}$

Perform the Mathematics

a. We notice that two variables, x and h, are involved. Remember that as $h \to 0$, the variable x acts as a **constant**; only the value of h changes. With this in mind, we try the Substitution Principle, which yields

$$\lim_{h \to 0} (3x + 2h) = 3x + 2(0)$$
$$= 3x$$

b. Again, two variables are involved, x and h. Proceeding as in part (a) gives

$$\lim_{h \to 0} \frac{5xh + 2h^2}{h} = \frac{5x(0) + 2(0)^2}{0}$$

$$= \frac{0}{0}$$

Because this is an indeterminate form, we try some algebra:

$$\lim_{h \to 0} \frac{5xh + 2h^2}{h}$$

Factor $\qquad\qquad = \lim_{h \to 0} \frac{h(5x + 2h)}{h}$

Cancel, provided $h \neq 0$ $\qquad = \lim_{h \to 0} (5x + 2h)$

Substitution $\qquad\qquad = 5x + 2(0) = 5x$ ∎

▶ *Try It Yourself*

Some related Exercises are 55 and 57.

In Example 7, we look at another limit that will be quite important in Section 2.4.

OBJECTIVE 7

Analyze the limit of a difference quotient.

Example 7: Analyzing a Limit of a Difference Quotient

For $f(x) = x^2$, compute $\lim\limits_{h \to 0} \dfrac{f(2 + h) - f(2)}{h}$.

Perform the Mathematics

Here we have

$$\lim_{h \to 0} \frac{f(2 + h) - f(2)}{h} = \lim_{h \to 0} \frac{(2 + h)^2 - (2)^2}{h}$$

Substituting 0 for h produces the indeterminate form $\frac{0}{0}$. But, if we try some algebra, we get

$$\lim_{h \to 0} \frac{f(2 + h) - f(2)}{h}$$

Evaluate f at $(2 + h)$ and 2 $\qquad = \lim_{h \to 0} \dfrac{(2 + h)^2 - (2)^2}{h}$

$(2 + h)^2 = 4 + 4h + h^2$ $\qquad = \lim_{h \to 0} \dfrac{(4 + 4h + h^2) - 4}{h}$

Simplify $\qquad\qquad\qquad = \lim_{h \to 0} \dfrac{4h + h^2}{h}$

Factor $\qquad\qquad\qquad\quad = \lim_{h \to 0} \dfrac{h(4 + h)}{h}$

Cancel, provided $h \neq 0$ $\qquad = \lim_{h \to 0} (4 + h)$

Substitute $\qquad\qquad\qquad 4 + 0 = 4$ ∎

Continuity

We conclude this section by addressing what it means for a function to be **continuous** at a point (and over an interval). A function is said to be **continuous** if its graph has no breaks in it such as holes, gaps, or jumps. This means that the graph of the function can be drawn without lifting a pencil off the paper. If the graph of a function has a hole, gap, or jump at $x = a$, we say that the function is **discontinuous** at $x = a$. Rather than list every possibility, we supply a definition for **continuity** to determine whether a function is continuous at $x = a$.

Figure 2.1.10 shows a graph of three different functions, each discontinuous at $x = a$. Notice that the function in Figure 2.1.10(a) violates condition 1 of our continuity definition, the function in Figure 2.1.10(b) violates condition 2, and the function in Figure 2.1.10(c) violates condition 3.

> ### DEFINITION
>
> **Continuity**
>
> A function f is said to be continuous at the point $x = a$ if all of the following are true:
> 1. $f(a)$ is defined
> 2. $\lim\limits_{x \to a} f(x)$ exists
> 3. $\lim\limits_{x \to a} f(x) = f(a)$

> **NOTE:** A function is continuous on open interval (b, c) if it is continuous for all x in the interval.

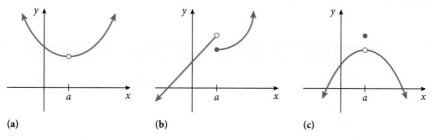

(a) **(b)** **(c)**

Figure 2.1.10 **(a)** $f(a)$ is not defined. **(b)** $\lim\limits_{x \to a} f(x)$ does not exist. **(c)** $\lim\limits_{x \to a} f(x) \neq f(a)$.

Example 8: Determining Continuity

> ### OBJECTIVE 8
>
> Determine continuity.

Determine the continuity of $g(x) = \dfrac{x^2 - 4}{x - 2}$ at $x = 1$ and $x = 2$.

Perform the Mathematics

According to the definition, three items must be true for continuity to exist. For $x = 1$,

1. Is $g(1)$ defined? Yes; $g(1) = \dfrac{(1)^2 - 4}{1 - 2} = 3$.

2. Does $\lim\limits_{x \to 1} \dfrac{x^2 - 4}{x - 2}$ exist? Yes; the substitution principle shows that

$$\lim_{x \to 1} \frac{x^2 - 4}{x - 2} = \frac{(1)^2 - 4}{1 - 2} = 3$$

3. Does $\lim\limits_{x \to 1} \dfrac{x^2 - 4}{x - 2} = g(1)$? Yes; $3 = 3$.

So we conclude that $g(x) = \dfrac{x^2 - 4}{x - 2}$ is continuous at $x = 1$.

For $x = 2$,

1. Is $g(2)$ defined? No!

Because $g(2)$ is not defined, hence violating condition 1 of the definition of continuity, we conclude that $g(x) = \dfrac{x^2 - 4}{x - 2}$ is not continuous at $x = 2$. ∎

▶ *Try It Yourself*

Some related Exercises are 77 and 79.

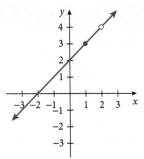

Figure 2.1.11 Graph of $g(x) = \dfrac{x^2 - 4}{x - 2}$.

Notice in Example 8 that $\displaystyle\lim_{x \to 2} \frac{x^2 - 4}{x - 2}$ exists, as seen in **Figure 2.1.11**. In fact, at the beginning of this section we determined that $\displaystyle\lim_{x \to 2} \frac{x^2 - 4}{x - 2} = 4$. But since $g(2)$ is not defined, there is a hole in the graph at the point $(2, 4)$. This type of discontinuity is said to be **removable**.

Continuity for Specific Functions

From Chapter 1 and our work thus far in Chapter 2, you may have a hunch that continuity can be determined for certain families of functions just by recognition. In fact, we can determine continuity for certain functions as indicated in Theorem 2.1.

Theorem 2.1

1. A polynomial function is continuous for all real x.
2. A rational function is continuous for all real x except those values of x for which the denominator is 0.
3. A radical function, or rational exponent function, is continuous for all real x in its domain.

Example 9: Determining Intervals of Continuity

Use Theorem 2.1 to determine intervals where the following functions are continuous.

a. $f(x) = \dfrac{2x + 1}{(x - 1)(2x + 1)}$

b. $g(x) = \sqrt{x - 7}$

Perform the Mathematics

a. Since this is a rational function, we conclude that it is continuous everywhere except at $x = 1$ and at $x = -\frac{1}{2}$, the values of x that make the denominator 0. The function is continuous on $(-\infty, -\frac{1}{2}) \cup (-\frac{1}{2}, 1) \cup (1, \infty)$. See **Figure 2.1.12**.

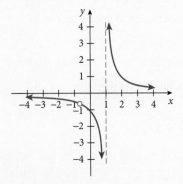

Figure 2.1.12

b. The domain of this radical function is $[7, \infty)$, so we conclude that the function is continuous on $[7, \infty)$. See **Figure 2.1.13**.

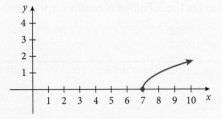

Figure 2.1.13

▶ **Try It Yourself**

Some related Exercises are 85 and 89.

Summary

In this introductory section on limits, we have presented the limit concept numerically, graphically, and algebraically. When analyzing a limit of a function, it is important to remember that in order for $\lim\limits_{x \to a} f(x) = L$, we must have the **left-hand limit**, $\lim\limits_{x \to a^-} f(x)$, and the **right-hand limit**, $\lim\limits_{x \to a^+} f(x)$, both equal to L. We used **limit theorems** and the substitution principle to determine limits. In addition, we used algebra to find the limits of an indeterminate form $\frac{0}{0}$.

- The limit $\lim\limits_{x \to a} f(x) = L$ means that as x approaches a, $f(x)$ approaches L. Alternatively, if, $\lim\limits_{x \to a^-} f(x) = L$ and $\lim\limits_{x \to a^+} f(x) = L$, then $\lim\limits_{x \to a} f(x) = L$.
- **Substitution Principle**: When trying to find $\lim\limits_{x \to a} f(x)$, first try computing $f(a)$.

We concluded this section with a brief look at continuity.

Section 2.1 Exercises

Vocabulary Exercises

1. The abbreviation lim represents the word _____.

2. The notation $x \to a^-$ means "as x approaches a from the _____." In this case, we would be computing a _____ - _____ limit.

3. The notation $x \to a^+$ means "as x approaches a from the _____." In this case, we would be computing a _____ - _____ limit.

4. When we use a table of values to estimate a limit, we are estimating the limit _____.

5. When we look or make a picture of a function in order to estimate a limit, we are estimating the limit _____.

6. When we use the limit theorems to find the limit of a function, we are determining the limit _____.

7. When attempting to determine the limit $\lim\limits_{x \to a} f(x)$ by evaluating $f(a)$, we are using the _____ Principle.

8. If we evaluate the limit of a rational function and get a reduced fraction in the form $\frac{0}{0}$, we call this form _____.

Skill Exercises

For Exercises 9–18, complete the tables to numerically estimate the limits.

9. (a) $\lim\limits_{x \to 1^-} (2 - 3x)$ (b) $\lim\limits_{x \to 1^+} (2 - 3x)$ (c) $\lim\limits_{x \to 1} (2 - 3x)$

x	0	0.9	0.99	0.999	1	1.001	1.01	1.1	2
$f(x) = 2 - 3x$					?				

10. (a) $\lim\limits_{x \to -2^-} (2x - 3)$ **(b)** $\lim\limits_{x \to -2^+} (2x - 3)$ **(c)** $\lim\limits_{x \to -2} (2x - 3)$

x		-3	-2.1	-2.01	-2.001	-2	-1.999	-1.99	-1.9	-1
$f(x) = 2x - 3$						?				

11. (a) $\lim\limits_{x \to -2^-} (2x^4 - 3x^3 + 2x - 4)$ **(b)** $\lim\limits_{x \to -2^+} (2x^4 - 3x^3 + 2x - 4)$

(c) $\lim\limits_{x \to -2} (2x^4 - 3x^3 + 2x - 4)$

x		-3	-2.1	-2.01	-2.001	-2	-1.999	-1.99	-1.9	-1
$f(x) = 2x^4 - 3x^3 + 2x - 4$						?				

12. (a) $\lim\limits_{x \to 1^-} (-2x^3 + 3x - 2)$ **(b)** $\lim\limits_{x \to 1^+} (-2x^3 + 3x - 2)$ **(c)** $\lim\limits_{x \to 1} (-2x^3 + 3x - 2)$

x		0	0.9	0.99	0.999	1	1.001	1.01	1.1	2
$f(x) = -2x^3 + 3x - 2$						?				

13. (a) $\lim\limits_{x \to 1^-} \dfrac{x^2 - 1}{x - 1}$ **(b)** $\lim\limits_{x \to 1^+} \dfrac{x^2 - 1}{x - 1}$ **(c)** $\lim\limits_{x \to 1} \dfrac{x^2 - 1}{x - 1}$

x	0	0.9	0.99	0.999	1	1.001	1.01	1.1	2
$f(x) = \dfrac{x^2 - 1}{x - 1}$					?				

14. (a) $\lim\limits_{x \to 3^-} \dfrac{x^2 - 3x}{x^2 - 9}$ **(b)** $\lim\limits_{x \to 3^+} \dfrac{x^2 - 3x}{x^2 - 9}$ **(c)** $\lim\limits_{x \to 3} \dfrac{x^2 - 3x}{x^2 - 9}$

x	2	2.9	2.99	2.999	3	3.001	3.01	3.1	4
$f(x) = \dfrac{x^2 - 3x}{x^2 - 9}$					?				

15. (a) $\lim\limits_{x \to -2^-} \dfrac{x^3 + 8}{x + 2}$ **(b)** $\lim\limits_{x \to -2^+} \dfrac{x^3 + 8}{x + 2}$ **(c)** $\lim\limits_{x \to -2} \dfrac{x^3 + 8}{x + 2}$

x		-3	-2.1	-2.01	-2.001	-2	-1.999	-1.99	-1.9	-1
$f(x) = \dfrac{x^3 + 8}{x + 2}$						?				

16. (a) $\lim\limits_{x \to 1^-} \dfrac{x^3 - 1}{x - 1}$ **(b)** $\lim\limits_{x \to 1^+} \dfrac{x^3 - 1}{x - 1}$ **(c)** $\lim\limits_{x \to 1} \dfrac{x^3 - 1}{x - 1}$

x	0	0.9	0.99	0.999	1	1.001	1.01	1.1	2
$f(x) = \dfrac{x^3 - 1}{x - 1}$					?				

17. $f(x) = \begin{cases} 3x - 1, & x > 0 \\ x^2 + 1, & x \le 0 \end{cases}$

(a) $\lim\limits_{x \to 0^-} f(x)$ **(b)** $\lim\limits_{x \to 0^+} f(x)$ **(c)** $\lim\limits_{x \to 0} f(x)$

x		-1	-0.1	-0.01	-0.001	0	0.001	0.01	0.1	1
$f(x) = \begin{cases} 3x - 1, & x > 0 \\ x^2 + 1, & x \le 0 \end{cases}$						?				

18. $f(x) = \begin{cases} x^2 + 1, & x \geq 0 \\ 2x^2 - 1, & x < 0 \end{cases}$

(a) $\lim\limits_{x \to 0^-} f(x)$

(b) $\lim\limits_{x \to 0^+} f(x)$

(c) $\lim\limits_{x \to 0} f(x)$

x		-1	-0.1	-0.01	-0.001	0	0.001	0.01	0.1	1
$f(x) = \begin{cases} x^2 + 1, & x \geq 0 \\ 2x^2 - 1, & x < 0 \end{cases}$						?				

For Exercises 19–26, use your calculator to graph the given function f. Use the `Zoom In` and `Trace` commands to graphically estimate the indicated limits. Verify your estimate numerically.

19. $f(x) = x^2 - 5x - 2; \lim\limits_{x \to 1} f(x)$

20. $f(x) = x^2 - 8x + 15; \lim\limits_{x \to 3} f(x)$

21. $g(x) = \dfrac{x^3 + 1}{x + 1}; \lim\limits_{x \to -1} g(x)$

22. $g(x) = \dfrac{x + 1}{x^3 + 1}; \lim\limits_{x \to -1} g(x)$

23. $f(x) = (x - 2)(x + 1); \lim\limits_{x \to 3} f(x)$

24. $f(x) = \dfrac{x - 2}{x + 1}; \lim\limits_{x \to 3} f(x)$

25. $f(x) = \dfrac{\sqrt{x} - 2}{x - 4}; \lim\limits_{x \to 4} f(x)$

26. $f(x) = \dfrac{\sqrt{x} - 3}{x - 9}; \lim\limits_{x \to 9} f(x)$

For Exercises 27–52, determine the indicated limit algebraically by using the Substitution Principle and the Limit Theorems.

27. $\lim\limits_{x \to -2} (3x + 1)$

28. $\lim\limits_{x \to 2} (-2x^2 + 50x)$

29. $\lim\limits_{x \to 10} \sqrt{x - 5}$

30. $\lim\limits_{x \to -4} \sqrt{x^2 + 9}$

31. $\lim\limits_{x \to 3} \dfrac{x^2 - 9}{x - 3}$

32. $\lim\limits_{x \to 5} \dfrac{x^2 - 25}{x - 5}$

33. $\lim\limits_{x \to 0} |x - 2|$

34. $\lim\limits_{x \to 1} |x + 1|$

35. $\lim\limits_{x \to 1} \dfrac{|x - 1|}{|x|}$

36. $\lim\limits_{x \to -1} \dfrac{x + 1}{|x + 1|}$

37. $\lim\limits_{x \to -1} \dfrac{x^2 - 1}{x + 1}$

38. $\lim\limits_{x \to -1} \dfrac{x + 1}{x^2 - 1}$

39. $\lim\limits_{x \to 0} \sqrt{2x + 3}$

40. $\lim\limits_{x \to 1} \sqrt{5x - 1}$

41. $\lim\limits_{x \to -5} \dfrac{x^2 - 25}{x - 5}$

42. $\lim\limits_{x \to 5} \dfrac{x - 5}{x^2 - 25}$

43. $\lim\limits_{x \to 1} \dfrac{x^2 - 1}{x + 1}$

44. $\lim\limits_{x \to 1} \dfrac{x - 1}{x^2 - 1}$

45. $\lim\limits_{x \to 2} (x + 1)^2 (3x - 1)^3$

46. $\lim\limits_{x \to -1} (x + 2)^2 (3x + 2)$

47. $\lim\limits_{x \to 0} \dfrac{x^2 - x - 6}{x - 3}$

48. $\lim\limits_{x \to 3} \dfrac{x - 3}{x^2 - x - 6}$

49. $\lim\limits_{x \to -2} \dfrac{x + 2}{x^2 + 5x + 6}$

50. $\lim\limits_{x \to -2} \dfrac{x^2 + 5x + 6}{x + 2}$

51. $\lim\limits_{x \to 4} \dfrac{\sqrt{x} - 2}{x - 4}$ (Hint: Multiply the numerator and denominator by $\sqrt{x} + 2$.)

52. $\lim\limits_{x \to 9} \dfrac{\sqrt{x} - 3}{x - 9}$ (Hint: Multiply the numerator and denominator by $\sqrt{x} + 3$.)

53. Use the graph of f in **Figure 2.1.14** to determine the following limits.

  (a) $\lim\limits_{x \to -1} f(x)$

  (b) $f(-1)$

  (c) $\lim\limits_{x \to 3^-} f(x)$

  (d) $\lim\limits_{x \to 3^+} f(x)$

  (e) $\lim\limits_{x \to 3} f(x)$

  (f) $f(1)$

  (g) $\lim\limits_{x \to 1} f(x)$

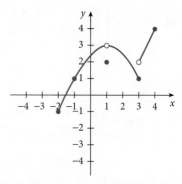

Figure 2.1.14

54. Use the graph of g in **Figure 2.1.15** to determine the following limits.

  (a) $\lim\limits_{x \to 2^-} g(x)$

  (b) $\lim\limits_{x \to 2^+} g(x)$

  (c) $\lim\limits_{x \to 2} g(x)$

  (d) $g(2)$

  (e) $\lim\limits_{x \to -2^+} g(x)$

  (f) $\lim\limits_{x \to -2} g(x)$

  (g) $g(-2)$

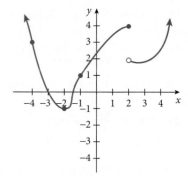

Figure 2.1.15

Find the indicated limit for Exercises 55–62.

55. $\lim\limits_{h \to 0} (2x + h)$

56. $\lim\limits_{h \to 0} (3x^2 + 3h)$

57. $\lim\limits_{h \to 0} \dfrac{3xh + h^2}{h}$

58. $\lim\limits_{h \to 0} \dfrac{2xh + h^2}{h}$

59. $\lim\limits_{h \to 0} \dfrac{4x^2h + 2h^2}{h}$

60. $\lim\limits_{h \to 0} \dfrac{4x^3h^2 + 3xh^3 + 5h^4}{h}$

61. $\lim\limits_{h \to 0} \dfrac{\sqrt{x + h} - \sqrt{x}}{h}$ (Hint: Multiply the numerator and denominator by $\sqrt{x + h} + \sqrt{x}$.)

62. $\lim\limits_{h \to 0} \dfrac{\sqrt{2x + 2h} - \sqrt{2x}}{h}$ (Hint: Multiply the numerator and denominator by $\sqrt{2x + 2h} + \sqrt{2x}$.)

Complete the following for each function in Exercises 63–72.

(a) *Evaluate and simplify $f(1 + h)$.*

(b) *Evaluate the limit* $\lim\limits_{h \to 0} \dfrac{f(1 + h) - f(1)}{h}$.

63. $f(x) = 2x + 3$

64. $f(x) = 3x - 1$

65. $f(x) = x^2 - 1$

66. $f(x) = x^2 + 2$

67. $f(x) = x^2 - 2x + 3$

68. $f(x) = x^2 + 3x - 1$

69. $f(x) = |x|$

70. $f(x) = \dfrac{1}{x}$

71. $f(x) = \sqrt{x}$

72. $f(x) = \sqrt{5x}$

In Exercises 73–82, determine the continuity of the given function at the indicated point.

73. $f(x) = 3x - 2; \quad x = -2$

74. $f(x) = 2x + 6; \quad x = -3$

75. $f(x) = x^2 - x - 6; \quad x = 3$

76. $f(x) = 2x^3 + 3x^2 - 5; \quad x = 2$

77. $f(x) = \dfrac{x + 1}{x - 3}$; $x = 3$

78. $f(x) = \dfrac{x - 5}{x + 2}$; $x = -2$

79. $f(x) = \dfrac{x^2 - 25}{x - 5}$; $x = 5$

80. $f(x) = \dfrac{x^2 - 25}{x - 5}$; $x = 0$

81. $f(x) = \begin{cases} x + 2, & x \le 1 \\ x^2 + 3, & x > 1 \end{cases}$; $x = 1$

82. $f(x) = \begin{cases} x - 3, & x \le 0 \\ x^2 + x - 3, & x > 0 \end{cases}$; $x = 0$

In Exercises 83–90, determine intervals where the following functions are continuous.

83. $f(x) = 7x^2 - 3.2x + 10.5$

84. $f(x) = 4x^3 - 2x^2 + 1.3x - 5$

85. $f(x) = \dfrac{2x - 3}{x + 5}$

86. $f(x) = \dfrac{3x - 7}{2x + 1}$

87. $f(x) = \dfrac{x + 1}{(x + 1)(x - 3)}$

88. $f(x) = \dfrac{2x - 3}{(2x - 3)(x + 4)}$

89. $f(x) = \sqrt{2x + 3}$

90. $f(x) = \sqrt{3x - 2}$

Application Exercises

91. Business Management—Advertising: The LectroScoot Company determines that the number of E21 model electric scooters it sells is related to the amount of advertising dollars spent and can be modeled by the function

$$N(x) = 2000 - \frac{520}{x} \quad 1 \le x \le 20$$

where x represents the amount spent in advertising, in thousands of dollars and $N(x)$ represents the number of scooters sold annually.

(a) Evaluate $N(10)$ and interpret.

(b) Evaluate $\lim\limits_{x \to 10} N(x)$ and interpret.

92. Business Management—Advertising: The LectroScoot Company also finds that the number of E25 model electric scooters it sells is related to the amount of advertising dollars spent and can be modeled by the function

$$N(x) = 2200 - \frac{750}{x} \quad 1 \le x \le 20$$

where x represents the amount spent in advertising, in thousands of dollars and $N(x)$ represents the number of scooters sold annually.

(a) Evaluate $N(10)$ and interpret.

(b) Evaluate $\lim\limits_{x \to 10} N(x)$ and interpret.

93. Microeconomics—Per Unit Cost: Suppose that the total cost of producing x of the E21 model electric scooters is given by the cost function $C(x) = 22{,}500 + 7.35x$. The cost to produce each scooter is given by the average cost function. For the E21 model scooters, the average cost is given by the function

$$AC(x) = \frac{C(x)}{x} = \frac{22{,}500 + 7.35x}{x} \quad 1 \le x \le 20$$

where x represents the number of E21 scooters produced and $AC(x)$ represents the average cost.

(a) Evaluate $AC(10)$ and interpret.

(b) Evaluate $\lim\limits_{x \to 10} AC(x)$ and interpret.

© Unni Bente Knag Langedal/ShutterStock, Inc.

94. **Microeconomics—Per Unit Cost:** Suppose that the total cost of producing x of the E25 model electric scooters is given by the cost function $C(x) = 33{,}125 + 6.38x$. The cost to produce each scooter is given by the average cost function. For the E25 model scooters, the average cost is given by the function

$$AC(x) = \frac{C(x)}{x} = \frac{33{,}125 + 6.38x}{x} \qquad 1 \le x \le 20$$

where x represents the number of E25 scooters produced and $AC(x)$ represents the average cost.

   (a) Evaluate $AC(10)$ and interpret.

   (b) Evaluate $\lim\limits_{x \to 10} AC(x)$ and interpret.

95. **Business Management—Pricing:** Lisa's Lease-a-Car charges $25 per day plus $0.05 per mile to rent one of its mid-sized cars. If c represents the cost to drive the rental car m miles, the cost is represented in **Figure 2.1.16**. Use the graph to answer the following.

   (a) Evaluate $c(100)$ and interpret.

   (b) Evaluate $\lim\limits_{m \to 100} c(m)$ and interpret.

   (c) Evaluate $c(200)$ and interpret.

   (d) Evaluate $\lim\limits_{m \to 200} c(m)$ and interpret.

   (e) Give a possible explanation for the jump in the graph at $m = 200$.

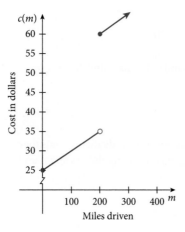

Figure 2.1.16

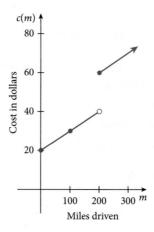

Figure 2.1.17

96. **Business Management—Pricing:** Dylan's Truck Rental Company charges $20 per day plus $0.10 per mile to rent a moving truck. If c represents the cost to drive the truck m miles, the cost is represented in **Figure 2.1.17**. Use the graph to answer the following.

   (a) Evaluate $c(50)$ and interpret.

   (b) Evaluate $\lim\limits_{m \to 50} c(m)$ and interpret.

   (c) Evaluate $c(100)$ and interpret.

   (d) Evaluate $\lim\limits_{m \to 100} c(m)$ and interpret.

   (e) Give a possible explanation for the jump in the graph at $m = 200$.

Concept and Writing Exercises

97. Suppose for a given function f that as $x \to a$, the limit exists, and $\lim\limits_{x \to a} f(x) = b$. Does that mean that $f(a) = b$? Either explain your answer or give a counterexample.

98. Suppose for a function f that $\lim\limits_{x\to a} f(x) = k$, where k represents a real number and $\lim\limits_{x\to a} g(x) = 0$. If $h(x) = f(x)g(x)$, does $\lim\limits_{x\to a} h(x) = 0$? Either explain your answer or give a counterexample.

99. Consider the functions $f(x) = x + 1$ and $g(x) = \dfrac{x^2 - 1}{x - 1}$.

   (a) Complete the tables for f and g.

x	0	0.9	0.99	0.999	1	1.001	1.01	1.1	2
$f(x) = x + 1$					?				
$g(x) = \dfrac{x^2 - 1}{x - 1}$					?				

   (b) Compare $f(1)$ to $g(1)$. Is $x = 1$ in the domain of f and of g?

   (c) Compare $\lim\limits_{x\to 1} f(x)$ to $\lim\limits_{x\to 1} g(x)$. Does the limit of f and g exist as $x \to 1$?

100. Recall that the absolute value function is defined as $f(x) = |x| = \begin{cases} x, & x \geq 0 \\ -x, & x < 0 \end{cases}$. Use this definition to algebraically verify that $\lim\limits_{x\to 0} \dfrac{|x|}{x}$ does not exist.

101. Suppose that for a function f, $\lim\limits_{x\to 3} f(x) = 5$, and that for a function g, $\lim\limits_{x\to 3} g(x) = 0$. Let the function h be defined by $h(x) = \dfrac{f(x)}{g(x)}$. What is $\lim\limits_{x\to 3} h(x)$? Explain your answer.

102. Consider the function $f(x) = \sqrt[3]{x^2 + x}$. If we wish to determine $\lim\limits_{x\to 9} f(x)$, can we first evaluate $\lim\limits_{x\to 9}(x^2 + x)$, then take the cube root of that result? Explain why or why not.

103. Explain in a sentence the meaning of the equation $\lim\limits_{x\to 3} g(x) = 4$.

104. Explain in a sentence the meaning of the equations $\lim\limits_{x\to 1^-} f(x) = 5$ and $\lim\limits_{x\to 1^+} f(x) = 5$.

105. Consider the piecewise-defined function $f(x) = \begin{cases} 2 - x, & x < -1 \\ x + 4, & x > -1 \end{cases}$. Does $f(-1)$ exist? Does $\lim\limits_{x\to 1} f(x)$ exist? Explain your answers.

106. Consider the piecewise-defined function $g(x) = \begin{cases} 3 - x, & x < 0 \\ x + a, & x \geq 0 \end{cases}$. Determine the value of a so that $\lim\limits_{x\to 0} g(x)$ exists.

107. Let $f(x) = x^2 - 3x$ and $g(x) = \sqrt{x - 10}$. If $h(x) = f(g(x))$, determine $\lim\limits_{x\to 15} h(x)$.

108. Consider the absolute value function $f(x) = \dfrac{x^2 - 2x - |x - 2|}{x - 2}$. Determine the one-sided limits $\lim\limits_{x\to 2^-} f(x)$ and $\lim\limits_{x\to 2^+} f(x)$.

 Section Project

During the 1990s, officials at the World Track and Field Association (known as the IAAF) realized that the world-class men's javelin throwers had reached the point where they could throw from one end of a stadium to the other, possibly endangering spectators and other athletes. So in the name of safety, the center of gravity of the javelin was moved forward so that the implement would not fly as far. The world season's best in the javelin from 1986 to 2006 can be modeled by the piecewise-defined function

$$f(x) = \begin{cases} 0.12x^2 - 0.11x + 86.48 & 0 \leq x \leq 10 \\ -0.31x + 96.46 & 10 < x \leq 20 \end{cases}$$

© Gregory Kendall/ShutterStock, Inc.

where x represents the number of years since 1986, and $f(x)$ represents the world's season best javelin throw, measured in meters. (*Source:* IAAF.)

(a) Determine $\lim\limits_{x \to 10^{-}} f(x)$ and interpret.

(b) Determine $\lim\limits_{x \to 10^{+}} f(x)$ and interpret.

(c) For international play, a soccer field is 110 meters in length. If the IAAF had not changed the rules, during what year could it be projected that a thrower could throw a javelin from one end of the field to the other?

SECTION OBJECTIVES

1. Analyze infinite limits and vertical asymptotes.

2. Apply infinite limits.

3. Analyze limits at infinity and horizontal asymptotes.

4. Apply limits at infinity.

2.2   Limits and Asymptotes

In Section 2.1, we introduced the limit concept and analyzed limits numerically, graphically, and algebraically. In this section, we look at the behavior of functions via limits when the independent variable increases without bound or decreases without bound. Graphically, this is important to the understanding of **horizontal asymptotes**. We also analyze situations when the dependent variable increases and decreases without bound. This situation will be closely related to **vertical asymptotes**.

Infinite Limits and Vertical Asymptotes

Example 1 begins our discussion of the relationship between infinite limits and vertical asymptotes.

Example 1: Analyzing a Limit

Consider $f(x) = \frac{1}{x}$. Numerically and graphically estimate $\lim\limits_{x \to 0} \frac{1}{x}$.

Perform the Mathematics

We begin our analysis by constructing a table of values as shown in **Table 2.2.1**.

Table 2.2.1

	$x \to 0^{-}$					$x \to 0^{+}$			
x	-1	-0.1	-0.01	-0.001	0	0.001	0.01	0.1	1
$f(x)$	-1	-10	-100	-1000		1000	100	10	1

Numerically, it appears that the left-hand limit and the right-hand limit are not equal, which means that the limit does not exist. As $x \to 0^{+}$, $\frac{1}{x}$ is positive and appears to be increasing without bound. We will indicate this "increasing without bound" behavior by writing

$$\lim_{x \to 0^{+}} \frac{1}{x} = \infty$$

Likewise, as $x \to 0^{-}$, $\frac{1}{x}$ is negative and appears to be decreasing without bound. We will indicate this "decreasing without bound" behavior by writing

$$\lim_{x \to 0^{-}} \frac{1}{x} = -\infty$$

These behaviors are shown graphically in **Figure 2.2.1**. The graph seems to support our numerical work.

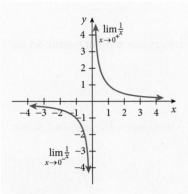

Figure 2.2.1

Since both the table and the graph suggest $\lim\limits_{x \to 0^-} \dfrac{1}{x} \neq \lim\limits_{x \to 0^+} \dfrac{1}{x}$, we conclude that $\lim\limits_{x \to 0} \dfrac{1}{x}$ does not exist. ■

NOTE: Technically speaking, in neither case does the limit $\lim\limits_{x \to 0^+} \dfrac{1}{x}$ or $\lim\limits_{x \to 0^-} \dfrac{1}{x}$ exist, because ∞ and $-\infty$ are not real numbers; they are merely concepts. To ensure that these concepts are understood, let's consider another example.

Example 2: Analyzing an Infinite Limit

Consider $f(x) = \frac{1}{x^2}$. Use **Table 2.2.2** and **Figure 2.2.2** to determine the following:

a. $\lim\limits_{x \to 0^-} \dfrac{1}{x^2}$ **b.** $\lim\limits_{x \to 0^+} \dfrac{1}{x^2}$ **c.** $\lim\limits_{x \to 0} \dfrac{1}{x^2}$

Table 2.2.2

	$x \to 0^-$					$x \to 0^+$			
x	-1	-0.1	-0.01	-0.001	0	0.001	0.01	0.1	1
$f(x)$	1	100	10,000	1,000,000		1,000,000	10,000	100	1

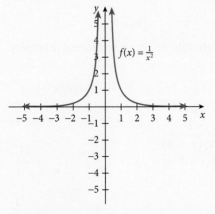

Figure 2.2.2

Infinite Limits and Vertical Asymptotes

The line $x = a$ is a **vertical asymptote** of the graph of f if any of the following are true.

- If as $x \to a^-$ the function values $f(x)$ increase or decrease without bound, that is, $\lim\limits_{x \to a^-} f(x) = \infty$ or $\lim\limits_{x \to a^-} f(x) = -\infty$, then $x = a$ is a vertical asymptote.

- If as $x \to a^+$ the function values $f(x)$ increase or decrease without bound, that is, $\lim\limits_{x \to a^+} f(x) = \infty$ or $\lim\limits_{x \to a^+} f(x) = -\infty$, then $x = a$ is a vertical asymptote.

NOTE: If both the left-hand and right-hand limits exhibit the same behavior, we say that $\lim\limits_{x \to a} f(x) = \infty$ (or $-\infty$).

Perform the Mathematics

a.   Table 2.2.2 and Figure 2.2.2 both suggest that, as $x \to 0^-$, $\frac{1}{x^2}$ increases without bound. So we write

$$\lim_{x \to 0^-} \frac{1}{x^2} = \infty$$

b.   Table 2.2.2 and Figure 2.2.2 both suggest that, as $x \to 0^+$, $\frac{1}{x^2}$ increases without bound. So we write

$$\lim_{x \to 0^+} \frac{1}{x^2} = \infty$$

c.   From our knowledge of left-hand and right-hand limits, the results from parts (a) and (b) suggest that

$$\lim_{x \to 0} \frac{1}{x^2} = \infty \qquad \blacksquare$$

It is worth mentioning one more time that in Example 2, none of the limits really exist, because $-\infty$ and ∞ are not real numbers. We are using the symbols $-\infty$ and ∞ to describe the behavior of the function near $x = 0$.

Notice in Examples 1 and 2 that the line $x = 0$, also known as the y-axis, serves as a **vertical asymptote** for the graphs. We now summarize the connection between **infinite limits and vertical asymptotes**.

OBJECTIVE 1

Analyze infinite limits and vertical asymptotes.

Example 3: Analyzing Limits at Excluded Domain Values

Consider $f(x) = \dfrac{x - 1}{x^2 - 1}$. The domain for f is all real numbers except -1 and 1.

a.   Determine $\lim\limits_{x \to 1} \dfrac{x - 1}{x^2 - 1}$.

b.   Determine $\lim\limits_{x \to -1} \dfrac{x - 1}{x^2 - 1}$.

Perform the Mathematics

a.   Utilizing the substitution principle produces the indeterminate form $\frac{0}{0}$, so we need to employ some algebraic manipulation:

$$\lim_{x \to -1} \frac{x - 1}{x^2 - 1}$$

Factor
$$= \lim_{x \to -1} \frac{x - 1}{(x - 1)(x + 1)}$$

Cancel, provided $x \neq 1$
$$= \lim_{x \to -1} \frac{1}{x + 1}$$

Substitute
$$= \frac{1}{1 + 1} = \frac{1}{2}$$

b. Again, substitution yields $\frac{0}{0}$, so proceeding with the same algebra as in part (a) gives

$$\lim_{x\to-1}\frac{x-1}{x^2-1}$$

Factor

$$=\lim_{x\to-1}\frac{x-1}{(x-1)(x+1)}$$

Cancel, provided $x\neq1$

$$=\lim_{x\to-1}\frac{1}{x+1}$$

We attempt to substitute at this time, but this gives $\frac{1}{0}$, which we know is undefined. We decide to make a table and see if, numerically, some light can be shed on what is happening here. See **Table 2.2.3**.

Table 2.2.3

	$x\to-1^-$					$x\to-1^+$			
x	-2	-1.1	-1.01	-1.001	-1	-0.999	-0.99	-0.9	0
$f(x)$	-1	-10	-100	-1000		1000	100	10	1

Since the table suggests that

$$\lim_{x\to-1^-}\frac{x-1}{x^2-1}\neq\lim_{x\to-1^+}\frac{x-1}{x^2-1}$$

we conclude that $\lim_{x\to-1}\dfrac{x-1}{x^2-1}$ does not exist. **Figure 2.2.3** supports our conjecture.

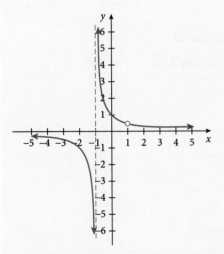

Figure 2.2.3 Graph of $f(x)=\dfrac{x-1}{x^2-1}$ has a vertical asymptote at $x=-1$. ∎

▶ **Try It Yourself**

Some related Exercises are 17 and 19.

Example 4: Analyzing Limits in an Applied Setting

The cost, $C(x)$, in thousands of dollars of removing x% of a city's pollutants discharged into a lake is given by

$$C(x) = \frac{113x}{100 - x}$$

A graph of C is given in **Figure 2.2.4**. Determine $\lim\limits_{x \to 100^-} C(x)$, and interpret.

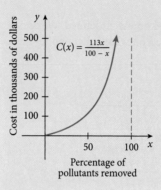

Figure 2.2.4

Understand the Situation

Our independent variable, x, represents a percentage of pollutants removed. It is not reasonable to remove a negative percentage, nor is it reasonable to remove more than 100%. From this analysis and Figure 2.2.4, we determine that the reasonable domain for C is [0, 100). For this application, $\lim\limits_{x \to 100^+} C(x)$ does not exist, and $\lim\limits_{x \to 100^-} C(x)$ should tell us the cost as the city tries to remove all (100%) of the pollutants.

Perform the Mathematics

Figure 2.2.4 suggests that $\lim\limits_{x \to 100^-} C(x) = \infty$. We construct **Table 2.2.4** to support our graphical belief that $\lim\limits_{x \to 100^-} C(x) = \infty$.

Table 2.2.4

$x \to 100^-$				
x	90	99	99.9	99.99	99.999
$C(x)$	1017	11,187	112,887	1,129,887	11,229,887

Interpret the Results

This means that as the city tries to remove all of the pollutants (100%), the cost to do so is "increasing without bound." In other words, it is cost prohibitive to remove 100% of the pollutants.

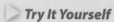 **Try It Yourself**

Some related Exercises are 43 and 44.

Limits at Infinity and Horizontal Asymptotes

We now turn our attention to analyzing the behavior of a function as the independent variable values increase without bound (heads to ∞) and as the independent variable value decreases without bound (heads to $-\infty$). To introduce this concept, we return to a problem first encountered in Section 1.5.

Flashback: Spring Point Epidemic Revisited

In Section 1.5 we modeled a flu epidemic in Spring Point, an isolated town with a population of 4000, with the function

$$f(t) = \frac{8000t + 6000}{2t + 5}$$

where $f(t)$ represented the number of people affected by the flu after t weeks. See **Figure 2.2.5**. How many people were affected after 10 weeks?

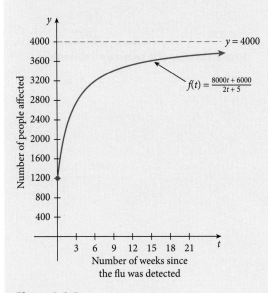

Figure 2.2.5

Perform the Mathematics

To determine the number of people affected by the flu after 10 weeks, we simply substitute 10 for t and get

$$f(10) = \frac{8000(10) + 6000}{2(10) + 5} = 3440$$

So after 10 weeks, 3440 people had been affected by the flu. ∎

Notice in the Flashback that after 10 weeks, 3440 of 4000 people had been affected by the flu. Intuitively, for any value of t the largest value that $f(t)$ can have is 4000, the entire population of the town! Also, it seems reasonable that as t increases (that is, as the number of weeks since the flu outbreak increases) the number of people affected by the flu, $f(t)$, should increase. Putting this information together, namely that as t increases $f(t)$ increases, yet $f(t)$ cannot exceed Spring Point's population of 4000, leads us to claim that

$$\lim_{t \to \infty} \frac{8000t + 6000}{2t + 5} = 4000$$

Notice how this is supported by the **horizontal asymptote** in Figure 2.2.5.

Example 5: Analyzing Limits at Infinity

Consider $f(x) = \dfrac{6x - 3}{2x + 4}$.

a. Use **Table 2.2.5** and **Figure 2.2.6** to estimate $\displaystyle\lim_{x \to \infty} \dfrac{6x - 3}{2x + 4}$.

b. Use **Table 2.2.6** and Figure 2.2.6 to estimate $\displaystyle\lim_{x \to -\infty} \dfrac{6x - 3}{2x + 4}$.

Table 2.2.5 $x \to \infty$

x	$f(x)$
1	0.5
10	2.375
100	2.9265
1000	2.9925

Some values rounded to nearest ten thousandth.

Table 2.2.6 $x \to -\infty$

x	$f(x)$
-1	-4.5
-10	3.9375
-100	3.0765
-1000	3.0075

Some values rounded to nearest ten thousandth.

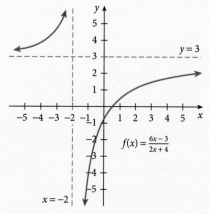

Figure 2.2.6 Graph of $f(x) = \dfrac{6x - 3}{2x + 4}$.

Perform the Mathematics

a. Table 2.2.5 and Figure 2.2.6 suggest that $\displaystyle\lim_{x \to \infty} \dfrac{6x - 3}{2x + 4} = 3$. In other words, as x increases without bound, $\dfrac{6x - 3}{2x + 4}$ gets close to 3.

b. Table 2.2.6 and Figure 2.2.6 suggest that $\displaystyle\lim_{x \to -\infty} \dfrac{6x - 3}{2x + 4} = 3$. In other words, as x decreases without bound, $\dfrac{6x - 3}{2x + 4}$ gets close to 3.   ∎

As seen in the Flashback and in Example 5, for some functions there is a relationship between limits at infinity and horizontal asymptotes.

Consider $f(x) = \dfrac{1}{x}$, $g(x) = \dfrac{1}{x^{5/3}}$, and $h(x) = \dfrac{1}{x^{1/2}}$ and their graphs in **Figures 2.2.7, 2.2.8**, and **2.2.9**, respectively.

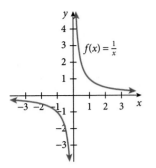

Figure 2.2.7

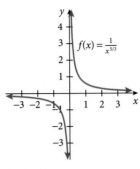

Figure 2.2.8

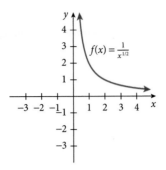

Figure 2.2.9

> **Special Limits at Infinity**
>
> 1. For n, a positive real number, $\lim\limits_{x \to \infty} \dfrac{1}{x^n} = 0$.
> 2. For n, a positive real number, $\lim\limits_{x \to -\infty} \dfrac{1}{x^n} = 0$, provided that x^n is a real number for negative values of x.

NOTE: All Limit Theorems from Section 2.1 are true for limits at infinity.

Figure 2.2.7 suggests that $\lim\limits_{x \to \infty} \dfrac{1}{x} = 0$ and $\lim\limits_{x \to -\infty} \dfrac{1}{x} = 0$, and the line $y = 0$ is a horizontal asymptote for the graph. Figure 2.2.8 suggests that $\lim\limits_{x \to \infty} \dfrac{1}{x^{5/3}} = 0$ and $\lim\limits_{x \to -\infty} \dfrac{1}{x^{5/3}} = 0$, and the line $y = 0$ is a horizontal asymptote. Figure 2.2.9 suggests that $\lim\limits_{x \to \infty} \dfrac{1}{x^{1/2}} = 0$, whereas $\lim\limits_{x \to -\infty} \dfrac{1}{x^{1/2}}$ does not exist, because negative numbers are not in the domain. Also, the line $y = 0$ is a horizontal asymptote here as well. These three functions and their respective graphs demonstrate some **special limits at infinity**.

Example 6 illustrates the usefulness of these special limits at infinity while confirming a graphing technique first presented in Section 1.5.

Example 6: Determining Horizontal Asymptotes

OBJECTIVE 3

Analyze limits at infinity and horizontal asymptotes.

For $f(x) = \dfrac{x^2 + 1}{2x^2 - 1}$, determine $\lim\limits_{x \to \infty} f(x)$ and $\lim\limits_{x \to -\infty} f(x)$, and determine the horizontal asymptote for the graph of f.

Perform the Mathematics

Because we have a rational function, the algebraic manipulation for problems of this type is rather straightforward:

> Divide every term in the numerator and denominator by x^b, where b is the larger degree of the numerator and the denominator.

For this problem, since both degrees are 2, we have $b = 2$. We divide every term in the numerator and denominator by x^2. This gives

$$f(x) = \frac{x^2 + 1}{2x^2 - 1} = \frac{(x^2 + 1) \div x^2}{(2x^2 - 1) \div x^2}$$

$$= \frac{\dfrac{x^2}{x^2} + \dfrac{1}{x^2}}{\dfrac{2x^2}{x^2} - \dfrac{1}{x^2}} = \frac{1 + \dfrac{1}{x^2}}{2 - \dfrac{1}{x^2}}$$

So, to determine the limit algebraically, we have

$$\lim_{x \to \infty} \frac{x^2 + 1}{2x^2 - 1} = \lim_{x \to \infty} \frac{1 + \dfrac{1}{x^2}}{2 - \dfrac{1}{x^2}} = \frac{\lim\limits_{x \to \infty}\left(1 + \dfrac{1}{x^2}\right)}{\lim\limits_{x \to \infty}\left(2 - \dfrac{1}{x^2}\right)}$$

$$= \frac{\lim\limits_{x \to \infty} 1 + \lim\limits_{x \to \infty} \dfrac{1}{x^2}}{\lim\limits_{x \to \infty} 2 - \lim\limits_{x \to \infty} \dfrac{1}{x^2}} = \frac{1 + 0}{2 - 0} = \frac{1}{2}$$

The algebraic determination for $\lim\limits_{x \to -\infty} \dfrac{x^2 + 1}{2x^2 - 1}$ is identical. We immediately conclude that the horizontal asymptote is the line $y = \frac{1}{2}$. ■

Try It Yourself

Some related Exercises are 23 and 25.

Example 7: Applying Limits at Infinity

Pharmacological studies have determined that the amount of medication present in the body is a function of the amount given and how much time has elapsed since the medication was administered. For a certain medication, the amount present in milliliters, $A(t)$, can be approximated by

$$A(t) = \frac{2t + 5}{10t^2 + 1} \qquad t > 0$$

where t is the number of hours since the medication was administered. Determine $\lim\limits_{t \to \infty} A(t)$ and interpret.

Understand the Situation

In determining $\lim\limits_{t \to \infty} \dfrac{2t + 5}{10t^2 + 1}$, we can employ the algebraic techniques illustrated in Example 6 and use the special limits at infinity.

Perform the Mathematics

We have a rational function where the degree of the numerator is 1 and the degree of the denominator is 2. Our algebraic manipulation is to divide every term in the numerator and denominator by t^2. This gives

$$\lim_{t \to \infty} \frac{2t + 5}{10t^2 + 1}$$

Divide numerator and denominator by t^2 $\qquad = \lim\limits_{t \to \infty} \dfrac{\dfrac{2t}{t^2} + \dfrac{5}{t^2}}{\dfrac{10t^2}{t^2} + \dfrac{1}{t^2}}$

Simplify

$$= \lim_{t \to \infty} \frac{\dfrac{2}{t} + \dfrac{5}{t^2}}{10 + \dfrac{1}{t^2}}$$

Limit Theorems

$$= \frac{\lim\limits_{t \to \infty}\left(\dfrac{2}{t} + \dfrac{5}{t^2}\right)}{\lim\limits_{t \to \infty}\left(10 + \dfrac{1}{t^2}\right)}$$

Limit Theorems

$$= \frac{\lim\limits_{t \to \infty} \dfrac{2}{t} + \lim\limits_{t \to \infty} \dfrac{5}{t^2}}{\lim\limits_{t \to \infty} 10 + \lim\limits_{t \to \infty} \dfrac{1}{t^2}}$$

Special Limits at Infinity

$$= \frac{0 + 0}{10 + 0} = 0$$

Table 2.2.7 and **Figure 2.2.10** support our work that $\lim\limits_{t \to \infty} \dfrac{2t + 5}{10t^2 + 1} = 0$.

Table 2.2.7

	$t \to \infty$			
t	0	1	10	100	1000
$A(t)$	5	0.63646	0.02498	0.00205	2×10^{-4}

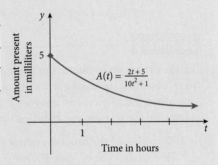

Figure 2.2.10

Interpret the Result

This means that as the number of hours since administering the medication increases without bound, the amount present in the body approaches 0 milliliters.

▷ **Try It Yourself**

Some related Exercises are 51 and 53. ■

Our final example in this section applies limits at infinity to a business function.

Example 8: Analyzing Average Cost

The total cost, in dollars, to produce x units of a certain product is given by $C(x) = 22{,}500 + 7.35x$. The **average cost**, AC, is given by

$$AC(x) = \frac{C(x)}{x} = \frac{22{,}500 + 7.35x}{x}$$

Determine $\lim\limits_{x \to \infty} AC(x)$ and interpret.

Perform the Mathematics

Since $AC(x) = \dfrac{22{,}500 + 7.35x}{x}$ is already written as a fraction with a single term in the denominator, we may break it apart as follows:

$$\lim_{x\to\infty} \frac{22{,}500 + 7.35x}{x} = \lim_{x\to\infty}\left(\frac{22{,}500}{x} + \frac{7.35x}{x}\right) = \lim_{x\to\infty}\left(\frac{22{,}500}{x} + 7.35\right)$$

$$= 0 + 7.35 = 7.35$$

This means that as the number of units produced increases without bound, the average total cost is approaching \$7.35. In other words, the cost per unit is approaching \$7.35 as the number of units produced increases without bound. ■

▶ *Try It Yourself*

Some related Exercises are 49 and 50.

Summary

In this section, we continued our study of limits. We introduced the symbols ∞ (and $-\infty$) to represent an increasing without bound (decreasing without bound) behavior. We examined the relationship between **infinite limits** and **vertical asymptotes**. We also examined the relationship between **limits at infinity** and **horizontal asymptotes**.

- If as $x \to a^-$ the function values $f(x)$ increase or decrease without bound, that is, $\lim\limits_{x\to a^-} f(x) = \infty$ or $\lim\limits_{x\to a^-} f(x) = -\infty$, then $x = a$ is a vertical asymptote.
- If as $x \to a^+$ the function values $f(x)$ increase or decrease without bound, that is, $\lim\limits_{x\to a^+} f(x) = \infty$ or $\lim\limits_{x\to a^+} f(x) = -\infty$, then $x = a$ is a vertical asymptote.
- For any function f, if $\lim\limits_{x\to \pm\infty} f(x) = L$, then the line $y = L$ is a **horizontal asymptote** for the graph of f.
- For n, a positive real number, $\lim\limits_{x\to\infty} \dfrac{1}{x^n} = 0$.
- For n, a positive real number, $\lim\limits_{x\to -\infty} \dfrac{1}{x^n} = 0$, provided that x^n is a real number for negative values of x.

Section 2.2 Exercises

Vocabulary Exercises

1. If the dependent variable values of a function get closer and closer to a line, without touching, we call that line a(n) _____.

2. The symbol _____ is used to represent the idea that values, whether independent or dependent, grow negatively without bound.

3. If as the function values f increase or decrease as the independent values approach a real number a from the left or right side, we say the graph of f has a _____ asymptote at $x = a$.

4. Technically speaking, if $\lim\limits_{x \to a} f(x) = \infty$, then as x approaches a, we say the limit
_____.

5. We say that a function f has a _____ _____ if the dependent
variable values of f approach a single real number as $x \to \pm\infty$.

6. For a function f, if $\lim\limits_{x \to \infty} f(x) = 0$, then f has a horizontal asymptote at the
_____ - _____.

Skill Exercises

For Exercises 7–14, complete the given table to numerically estimate the following limits. Use the symbols
∞ *and* $-\infty$ *where applicable.*

7. (a) $\lim\limits_{x \to 0^-} \dfrac{1}{x^3}$ 　　(b) $\lim\limits_{x \to 0^+} \dfrac{1}{x^3}$ 　　(c) $\lim\limits_{x \to 0} \dfrac{1}{x^3}$

x	-0.1	-0.01	-0.001	-0.00001	0	0.00001	0.001	0.01	0.1
$f(x) = \dfrac{1}{x^3}$									

8. (a) $\lim\limits_{x \to 0^-} \dfrac{2}{x^3}$ 　　(b) $\lim\limits_{x \to 0^+} \dfrac{2}{x^3}$ 　　(c) $\lim\limits_{x \to 0} \dfrac{2}{x^3}$

x	-0.1	-0.01	-0.001	-0.00001	0	0.00001	0.001	0.01	0.1
$f(x) = \dfrac{2}{x^3}$									

9. (a) $\lim\limits_{x \to 1^-} \dfrac{1}{(x-1)^2}$ 　　(b) $\lim\limits_{x \to 1^+} \dfrac{1}{(x-1)^2}$ 　　(c) $\lim\limits_{x \to 1} \dfrac{1}{(x-1)^2}$

x	0.9	0.99	0.999	0.99999	1	1.00001	1.001	1.01	1.1
$f(x) = \dfrac{1}{(x-1)^2}$									

10. (a) $\lim\limits_{x \to 1^-} \dfrac{3}{(x-1)^2}$ 　　(b) $\lim\limits_{x \to 1^+} \dfrac{3}{(x-1)^2}$ 　　(c) $\lim\limits_{x \to 1} \dfrac{3}{(x-1)^2}$

x	0.9	0.99	0.999	0.99999	1	1.00001	1.001	1.01	1.1
$f(x) = \dfrac{3}{(x-1)^2}$									

11. (a) $\lim\limits_{x \to -2^-} \dfrac{x+2}{x^2-x-6}$ 　　(b) $\lim\limits_{x \to -2^+} \dfrac{x+2}{x^2-x-6}$ 　　(c) $\lim\limits_{x \to -2} \dfrac{x+2}{x^2-x-6}$

x	-2.1	-2.001	-2.0001	-2	-1.9999	-1.999	-1.9
$f(x) = \dfrac{x+2}{x^2-x-6}$							

12. (a) $\lim\limits_{x \to 3^-} \dfrac{x+2}{x^2-x-6}$ 　　(b) $\lim\limits_{x \to 3^+} \dfrac{x+2}{x^2-x-6}$ 　　(c) $\lim\limits_{x \to 3} \dfrac{x+2}{x^2-x-6}$

x	2.9	2.99	2.999	2.99999	3	3.00001	3.001	3.01	3.1
$f(x) = \dfrac{x+2}{x^2-x-6}$									

13. (a) $\displaystyle\lim_{x\to 0^-}\frac{13{,}250 + 2.35x}{x}$ (b) $\displaystyle\lim_{x\to 0^+}\frac{13{,}250 + 2.35x}{x}$ (c) $\displaystyle\lim_{x\to 0}\frac{13{,}250 + 2.35x}{x}$

x	−0.1	−0.01	−0.001	−0.00001	0	0.00001	0.001	0.01	0.1
$f(x) = \dfrac{13{,}250 + 2.35x}{x}$									

14. (a) $\displaystyle\lim_{x\to 0^-}\frac{4.25x - 23{,}350}{x}$ (b) $\displaystyle\lim_{x\to 0^+}\frac{4.25x - 23{,}350}{x}$ (c) $\displaystyle\lim_{x\to 0}\frac{4.25x - 23{,}350}{x}$

x	−0.1	−0.01	−0.001	−0.00001	0	0.00001	0.001	0.01	0.1
$f(x) = \dfrac{4.25x - 25{,}350}{x}$									

In Exercises 15–22, determine the limit algebraically.

15. $\displaystyle\lim_{x\to 0}\frac{2}{x^3}$

16. $\displaystyle\lim_{x\to 0}\frac{13{,}250 + 2.35x}{x}$

17. $\displaystyle\lim_{x\to -2}\frac{x + 2}{x^2 - x - 6}$

18. $\displaystyle\lim_{x\to 3}\frac{x + 2}{x^2 - x - 6}$

19. $\displaystyle\lim_{x\to 3}\frac{x - 3}{x^2 - 9}$

20. $\displaystyle\lim_{x\to -3}\frac{x - 3}{x^2 - 9}$

21. $\displaystyle\lim_{x\to -1}\frac{x^2 - x + 1}{x^3 + 1}$

22. $\displaystyle\lim_{x\to 3}\frac{x^2 - 9}{x - 3}$

In Exercises 23–30, determine the indicated limit algebraically using the method in Example 6. Verify your result numerically.

23. $\displaystyle\lim_{x\to -\infty}\frac{2x + 5}{x - 1}$

24. $\displaystyle\lim_{x\to \infty}\frac{2x + 5}{x - 1}$

25. $\displaystyle\lim_{x\to \infty}\frac{3x^2 - x + 2}{2x^2 + x - 5}$

26. $\displaystyle\lim_{x\to -\infty}\frac{3x^2 - x + 2}{2x^2 + x - 5}$

27. $\displaystyle\lim_{x\to \infty}\frac{2x^2 + 2x + 1}{5x^3 + 3x - 5}$

28. $\displaystyle\lim_{x\to \infty}\frac{3x^2 - 2x + 5}{2x^3 + x^2 - 2x + 3}$

29. $\displaystyle\lim_{x\to -\infty}\frac{2x^2 + 2x + 1}{5x^3 + 3x - 5}$

30. $\displaystyle\lim_{x\to -\infty}\frac{3x^2 - 2x + 5}{2x^3 + x^2 - 2x + 3}$

For Exercises 31–34, determine an equation for the horizontal asymptote for the graph of f. Consult your work performed in Exercises 23–30.

31. $f(x) = \dfrac{2x + 5}{x - 1}$

32. $f(x) = \dfrac{3x^2 - x + 2}{2x^2 + x - 5}$

33. $f(x) = \dfrac{2x^2 + 2x + 1}{5x^3 + 3x - 5}$

34. $f(x) = \dfrac{3x^2 - 2x + 5}{2x^3 + x^2 - 2x + 3}$

In Exercises 35–42, use the graph of f in Figure 2.2.11 to determine the indicated limit.

35. $\displaystyle\lim_{x\to 2^+} f(x)$

36. $\displaystyle\lim_{x\to 2^-} f(x)$

37. $\displaystyle\lim_{x\to 2} f(x)$

38. $\displaystyle\lim_{x\to -1^-} f(x)$

39. $\displaystyle\lim_{x\to -1} f(x)$

40. $\displaystyle\lim_{x\to -3} f(x)$

41. $\displaystyle\lim_{x\to \infty} f(x)$

42. $\displaystyle\lim_{x\to -\infty} f(x)$

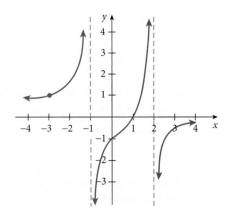

Figure 2.2.11

Application Exercises

43. Ecology—Removing Pollution: The city of Gilbertsburg is interested in analyzing the cost involved in removing local water pollution. The cost of removing pollutants in Lake Sutrane near the city of Gilbertsburg is a function of the amount of money invested in its removal and can be modeled by

$$C(x) = \frac{233x}{100 - x}$$

where x represents the percentage of pollutants removed, and $C(x)$ represents the cost of removing the pollutants, measured in thousands of dollars.

(a) Determine the reasonable domain of C.

(b) Evaluate $C(40)$ and interpret.

(c) Determine $\lim\limits_{x \to 100^-} C(x)$ and interpret.

44. Ecology—Removing Pollution: The city of Deverville is interested in analyzing the cost involved in removing local water pollution. The cost of removing pollutants in Lake Kirkstan near the city of Deverville is a function of the amount of money invested in its removal and can be modeled by

$$C(x) = \frac{93x}{100 - x}$$

where x represents the percentage of pollutants removed, and $C(x)$ represents the cost of removing the pollutants, measured in thousands of dollars.

(a) Determine the reasonable domain of C.

(b) Evaluate $C(40)$ and interpret.

(c) Determine $\lim\limits_{x \to 100^-} C(x)$ and interpret.

45. Marketing—Advertising Budget: The number of dollars spent on advertising for a product influences the number of items of the product that will be purchased by customers. The number of Krypto-Cases sold for a new smartphone is a function of the amount of advertising spent and can be modeled by

$$N(x) = 2000 - \frac{520}{x} \qquad x > 0$$

where x represents the amount spent on advertising, measured in thousands of dollars, and $N(x)$ represents the number of cases purchased. Determine $\lim\limits_{x \to \infty} N(x)$ and interpret.

46. Marketing—Advertising Budget: The number of dollars spent on advertising for a product influences the number of items of the product that will be purchased by customers. The

number of Find-it-Now GPS devices is a function of the amount of advertising spent and can be modeled by

$$N(x) = 3200 - \frac{750}{x} \qquad x > 0$$

where x represents the amount spent on advertising, measured in thousands of dollars, and $N(x)$ represents the number of devices purchased. Determine $\lim_{x \to \infty} N(x)$ and interpret.

47. **Marketing—Advertising Budget** (continuation of Exercise 45): Graph the model

    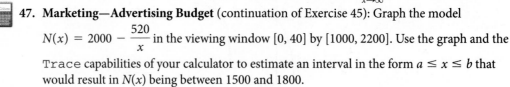

    $N(x) = 2000 - \frac{520}{x}$ in the viewing window [0, 40] by [1000, 2200]. Use the graph and the Trace capabilities of your calculator to estimate an interval in the form $a \le x \le b$ that would result in $N(x)$ being between 1500 and 1800.

48. **Marketing—Advertising Budget** (continuation of Exercise 46): Graph the model

    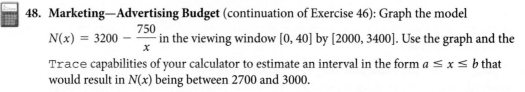

    $N(x) = 3200 - \frac{750}{x}$ in the viewing window [0, 40] by [2000, 3400]. Use the graph and the Trace capabilities of your calculator to estimate an interval in the form $a \le x \le b$ that would result in $N(x)$ being between 2700 and 3000.

49. **Manufacturing—Backpacks:** The cost of producing WallyWorld children's backpacks can be modeled by the cost function $C(x) = 13{,}700 + 6.85x$, where x represents the number of backpacks produced, and $C(x)$ represents the cost in dollars. The average, or per-unit, cost of producing x backpacks is given by the average cost function

$$AC(x) = \frac{C(x)}{x} = \frac{13{,}700 + 6.85x}{x} \qquad x > 0$$

Determine $\lim_{x \to \infty} AC(x)$ and interpret.

50. **Manufacturing—Key Fobs:** The cost of producing Rally Racing key fobs can be modeled by the cost function $C(x) = 15{,}200 + 4.85x$, where x represents the number of fobs produced, and $C(x)$ represents the cost in dollars. The average, or per-unit, cost of producing x key fobs is given by the average cost function

$$AC(x) = \frac{C(x)}{x} = \frac{15{,}200 + 4.85x}{x} \qquad x > 0$$

Determine $\lim_{x \to \infty} AC(x)$ and interpret.

51. **Conservation—Bass Population:** The local game commission decided to stock Lake Lindy with 100 bass. The subsequent population of bass in the lake can be modeled by

$$P(x) = \frac{10(10 + 7x)}{1 + 0.02x} \qquad x \ge 0$$

where x represents the number of months since the lake was stocked, and $P(x)$ represents the bass population.

   (a) Use the model to determine the bass population one year after the lake was stocked.

   (b) Determine $\lim_{x \to \infty} P(x)$ and interpret. This is called the *limiting size* of the population.

52. **Conservation—Trout Population:** The local game commission decided to stock Lake Clouster with 200 trout. The subsequent population of trout in the lake can be modeled by

$$P(x) = \frac{20(10 + 7x)}{1 + 0.02x} \qquad x \ge 0$$

where x represents the number of months since the lake was stocked, and $P(x)$ represents the trout population.

   (a) Use the model to determine the trout population one year after the lake was stocked.

   (b) Determine $\lim_{x \to \infty} P(x)$ and interpret. This is called the *limiting size* of the population.

53. International Economics—Unemployment in Ireland: Unemployment has been acute in Ireland in the past few years. "Unemployment compensation is not a good measure for unemployment. A lot of people who lost their jobs in the last three years are not entitled to be on it," said senior policy analyst John Sweeney. (*Source: The Belfast Times.*) The number unemployed in Ireland annually from 2000 to 2010 can be modeled by

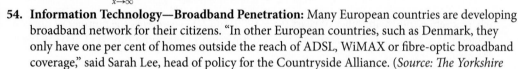

$$f(x) = \frac{-2.03x + 26.34}{-0.03x + 0.35} \qquad 0 \le x \le 10$$

where x represents the number of years since 2000, and $f(x)$ represents the number of unemployed in Ireland, measured in thousands. (*Source:* Google public data.)

(a) Determine the equation for the vertical asymptote of f. Is the vertical asymptote in the reasonable domain of f?

(b) Determine $\lim\limits_{x \to \infty} f(x)$ and interpret.

54. Information Technology—Broadband Penetration: Many European countries are developing broadband network for their citizens. "In other European countries, such as Denmark, they only have one per cent of homes outside the reach of ADSL, WiMAX or fibre-optic broadband coverage," said Sarah Lee, head of policy for the Countryside Alliance. (*Source: The Yorkshire Times.*) The broadband penetration rate in Denmark from 2003 to 2010 can be modeled by a linear, a quadratic, and a cubic function. The models are:

$$f(x) = 4.65x - 7.47$$
$$g(x) = -0.55x^2 + 11.26x - 25.12$$
$$h(x) = -0.23x^3 + 3.5x^2 - 11.46x + 14$$

where x represents the number of years since 2003, and $f(x), g(x)$, and $h(x)$ represent the percentage of the Danish population with access to broadband. (*Source:* Google public data.)

(a) Determine $\lim\limits_{x \to \infty} f(x)$ and interpret.

(b) Determine $\lim\limits_{x \to \infty} g(x)$ and interpret.

(c) Find $\lim\limits_{x \to \infty} h(x)$ and interpret.

(d) Which of the three models, if any, gives the most reasonable estimate for the upper limit of the broadband rate in Denmark? Explain your answer.

55. Economics—Telecommunications: Many historians have called the period from 1995 to 2010 the Communications Age because of the proliferation of handheld phones throughout the world. "We live in the communication age, although some people mistakenly call it the Information Age," said Andrew Kantor of CyberSpeak. (*Source: USA Today.*) The number of mobile cellular subscribers globally can be modeled by

$$f(x) = \frac{2089x - 9113}{-x + 25.89} \qquad 0 \le x \le 20$$

where x represents the number of years since 1990, and $f(x)$ represents the number of mobile cellular subscribers internationally, measured in millions. (*Source:* International Telecommunication Union.)

(a) Evaluate $f(10)$ and interpret.

© Hemera/Thinkstock

(b) Determine $\lim\limits_{x \to 25.89^+} f(x)$ and $\lim\limits_{x \to 25.89^-} f(x)$. Are these values in the reasonable domain of f?

(c) Determine $\lim\limits_{x \to \infty} f(x)$. Is this value in the "reasonable range" of f?

56. Economics—Consumer Spending: Social networking and Internet commerce have contributed to the amount of spending on video and audio products during the past two decades. The amount spent on audio and video products in the United States from 1990 to 2010 can be modeled by

$$f(x) = \frac{3.39x + 44.41}{-0.01x + 0.82} \qquad 0 \le x \le 20$$

where x represents the number of years since 1990, and $f(x)$ represents the amount spent on audio and video products, measured in billions of dollars. (*Source:* U.S. Bureau of Economic Analysis.)

(a) Evaluate $f(15)$ and interpret.

(b) Determine $\lim_{x \to 82^-} f(x)$ and $\lim_{x \to 82^+} f(x)$. Are these values in the reasonable domain of f?

(c) Determine $\lim_{x \to \infty} f(x)$. Is this value in the "reasonable range" of f?

 57. Political Science—Political Action Committees: A political action committee, or PAC, is a committee formed by a special-interest group to raise money for their favorite political candidates. Nonconnected PACs are made up of individuals or groups of U.S. citizens not connected to a corporation, a labor party, or a political party, such as the National Rifle Association and Emily's List. The number of nonconnected PACs formed from 1980 to 2010 can be modeled by

$$f(x) = \frac{5.54x + 27.49}{0.002x + 0.068} \qquad 0 \le x \le 30$$

where x represents the number of years since 1980, and $f(x)$ represents the number of nonconnected PACs. (*Source:* U.S. Federal Election Commission.) Determine $\lim_{x \to \infty} f(x)$ and interpret the result.

 58. Agriculture—Fuel Ethanol Production: One of the renewable fuels that have been promoted over the past few decades is corn-based ethanol. "Without the renewable-fuels standard, ethanol really wouldn't be a viable business," said Bill Day of ethanol producer Valero. (*Source: The National Journal.*) The annual amount of corn-based ethanol produced from 1980 to 2010 can be modeled by

$$f(x) = \frac{0.41x + 8.63}{-0.035x + 1.12} \qquad 0 \le x \le 30$$

where x represents the number of years since 1980, and $f(x)$ represents the amount of corn-based ethanol produced in millions of barrels. (*Source:* U.S. Energy Information Administration.)

(a) Determine the equation of the vertical asymptote for the graph of f, and write it in the form $x = a$.

(b) Use the value of a from part (a) to determine $\lim_{x \to a^-} f(x)$, and interpret the result.

Concept and Writing Exercises

59. In your own words, explain why we say that $\lim_{x \to 0} \dfrac{1}{x}$ does not exist but we say that $\lim_{x \to 0} \dfrac{1}{x^2} = \infty$.

60. Consider the function $f(x) = \dfrac{5x^n + 6}{cx^4 + 10}$, where c is a real number and n is a whole number. For what values of n does $\lim_{x \to \infty} f(x) = 0$?

61. Consider the function $f(x) = \dfrac{5x^n + 6}{cx^4 + 10}$, where c is a real number and n is a whole number. If $n = 4$, find $\lim_{x \to \infty} f(x)$.

62. Suppose that $f(x) = \dfrac{k}{x^4}$, where k represents a real number. If $k < 0$, determine $\lim_{x \to 0} f(x)$.

63. Let $f(x) = \dfrac{k}{x^5}$. If $k < 0$, find $\lim_{x \to 0^-} f(x)$ and $\lim_{x \to 0^+} f(x)$.

64. For the function $f(x) = \dfrac{2x^3}{5x^3 + x^2 - 6x + 7}$, explain why $f(x) \to \dfrac{2}{5}$ as $x \to \infty$.

65. Explain in a sentence the meaning of the equation $\lim_{x \to 0} g(x) = \infty$.

66. Explain in a sentence the meaning of the equation $\lim_{x \to \infty} f(x) = 5$.

67. Consider the function $g(x) = \dfrac{x^2 - 12}{x - a}$. Find the value(s) of a so that $\lim\limits_{x \to a} g(x)$ does exist.

68. For the function $f(x) = \dfrac{ax^2 - 3x}{3x^2 - 9}$, find the value of a so the graph of f has a horizontal asymptote.

For Exercises 69–71, consider the function $f(x) = \dfrac{2x^4 + 7x^2 - 7x + 2}{x^n + 5x - 6}$, *where n represents a whole number.*

69. For what value(s) of n does $\lim\limits_{x \to \infty} f(x) = \infty$?

70. For what value(s) of n does $\lim\limits_{x \to \infty} f(x) = 2$?

71. For what value(s) of n does $\lim\limits_{x \to \infty} f(x) = 0$?

*In addition to horizontal and vertical asymptotes, a rational function can also have **oblique** or **slant** asymptotes. For Exercises 72 and 73 complete the following:*

(a) *Graph the functions f and g.*

(b) *Use long division to divide f.*

(c) *Complete the table:*

x	f(x)	g(x)
10		
100		
1000		

(d) Show that $\lim\limits_{x \to \infty} f(x) = \lim\limits_{x \to \infty} g(x)$.

72. $f(x) = \dfrac{3x^2 + 1}{x}, g(x) = 3x$

73. $f(x) = \dfrac{5x^2 - 6x + 3}{x - 1}, g(x) = 5x - 1$

74. Based on the results of Exercises 72 and 73, fill in the blank to complete the following statement. The end behavior of the rational function $h(x) = \dfrac{f(x)}{g(x)}$ is the same as the _____ when the function f is divided by the function g.

 Section Project

The Special Supplemental Nutrition Program for Women, Infants and Children (known as WIC) is a federal assistance program for healthcare and nutrition of low-income pregnant women, breastfeeding women, and infants and children under the age of five. The annual federal cost of the WIC program from 1990 to 2010 can be modeled by the rational function

$$f(x) = \frac{21x + b}{-0.01x + 0.49} \qquad 0 \le x \le 20$$

where x represents the number of years since 1990, and $f(x)$ represents the annual federal cost of the WIC program, measured in millions of dollars. (*Source:* U.S. Department of Agriculture.)

© Congressional Quarterly/Getty Images

(a) Knowing that \$3503 million was spent on the WIC program in 2006, determine the value of b. Round your answer to the nearest whole number.

(b) Using the model along with the information found in part (a), evaluate and interpret $f(20)$.

(c) Determine $\lim\limits_{x \to \infty} f(x)$. Why is this value not appropriate for the problem situation?

(d) Determine the equation of the vertical asymptote for the model f, and write the equation in the form $x = a$. What year does this vertical asymptote correspond to?

(e) Using the value of a from part (d), we see that $\lim\limits_{x \to a^-} f(x) = \infty$. Why is this sort of answer unreasonable for the problem situation?

SECTION OBJECTIVES

1. Determine an average rate of change.

2. Compare tangent line to secant line.

3. Use tangent line slope approximation to approximate an instantaneous rate of change.

4. Apply an instantaneous rate of change.

2.3   Approximating Instantaneous Rates of Change

The study of calculus is the study of rates of change. In Chapter 1, we reviewed the **average rate of change** concept. Now we wish to study the **instantaneous rate of change**. This type of rate of change is one of the central topics of differential calculus. In this section, we will see how these rates of change are different from one another. We will also see how we can approximate the instantaneous rate of change for a given function.

Thinking About Rate of Change

In Chapter 1, we saw that the average rate of change of a function f on a closed interval was given by the **slope of a secant line** for a given point $(x, f(x))$ and a given increment in x denoted by Δx. To **simplify computations in the remainder of this chapter, we will use h in the place of** Δx. This gives us the secant line slope formula shown in the Toolbox. We first saw this formula in Section 1.2.

⚒ From Your Toolbox: Slope of a Secant Line

The **slope of a secant line** between the points $(x, f(x))$ and $(x + h, f(x + h))$, denoted by m_{sec}, is given by the difference quotient

$$m_{sec} = \frac{f(x + h) - f(x)}{h}$$

where $h = \Delta x$ is the increment in x.

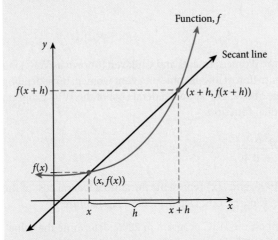

Figure 2.3.1

One of our main uses of the secant line slope was to use it to give us the average rate of change of a function on a closed interval. In Example 1 we review how to do this.

OBJECTIVE 1

Example 1: Determining an Average Rate of Change

Determine an average rate of change.

Compute the average rate of change of the function $f(x) = x^2 + 3$ over the interval $[2, 5]$.

Understand the Situation

Since the average rate of change is given by the slope of the secant line over an interval, we use the formula $m_{sec} = \dfrac{f(x + h) - f(x)}{h}$ to compute the slope of the secant line. We are given the interval $[2, 5]$, which means that $x = 2$ and $h = 5 - 2 = 3$.

Perform the Mathematics

So we get

$$m_{sec} = \frac{f(x + h) - f(x)}{h} = \frac{f(2 + 3) - f(2)}{3}$$

$$= \frac{f(5) - f(2)}{3} = \frac{28 - 7}{3} = \frac{21}{3} = 7$$

Interpret the Result

The slope of the secant line on the interval $[2, 5]$ is $m_{sec} = 7$. This means that the average rate of change of $f(x) = x^2 + 3$ over the interval $[2, 5]$ is $7\dfrac{\text{dependent units}}{\text{independent units}}$. See **Figure 2.3.2**.

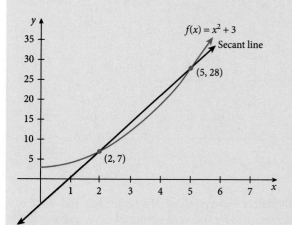

Figure 2.3.2

▷ *Try It Yourself*

Some related Exercises are 19 and 21.

Thinking About Instantaneous Rate of Change

We now pose the following question: At what rate is the function in Example 1, $f(x) = x^2 + 3$ changing when $x = 2$? We are not asking for the average rate of change over some interval as was done in

Example 1. Instead the question asks for the rate of change at a particular *point*. This type of rate of change is called an **instantaneous rate of change** and is equivalent to the slope of the line **tangent** to the curve at a specific point. What do we mean by **tangent line**? Consider the curve in **Figure 2.3.3**.

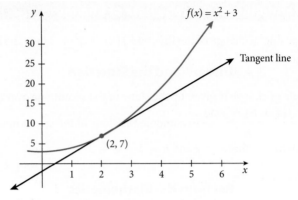

Figure 2.3.3

The steepness of the curve varies from point to point. The measure of the steepness of the curve at the point $(2, 7)$ is equivalent to finding the slope of the tangent line through the point. (We call the point through which the tangent line passes the **fixed point** or the **point of tangency**.) With this in mind, the following serves as a working definition of **tangent line**.

DEFINITION

Tangent Line

The **tangent line** to a curve at a point A is the line through A whose slope matches the steepness of the curve at point A. We can say that the tangent line slope is equal to the slope of the curve at A.

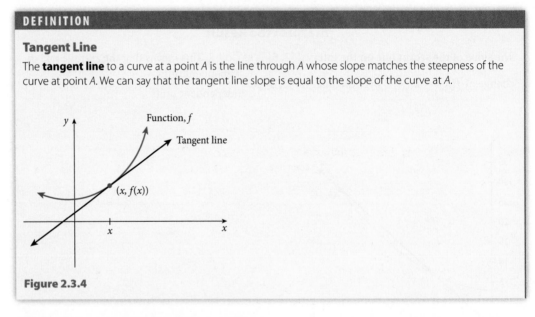

Figure 2.3.4

Figures 2.3.1 and 2.3.4 illustrate the main differences between slope of a secant line and slope of a tangent line. The slope of the **secant line** is determined by two points on the curve, and it gives an **average rate of change**. The slope of the **tangent line** is associated with a single point on the curve, and it gives an **instantaneous rate of change**. Before we begin the task of approximating tangent line slope, Example 2 contrasts these two rates of change.

OBJECTIVE 2

Compare tangent line to secant line.

Example 2: Classifying Rates of Change

Classify each of the following as exhibiting average rate of change or instantaneous rate of change.

a.   The line l_1 that is shown in **Figure 2.3.5**.

b.   The line l_2 that is shown in **Figure 2.3.6**.

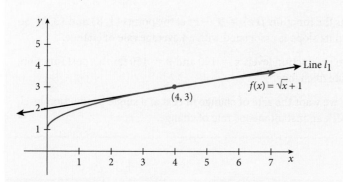

Figure 2.3.5

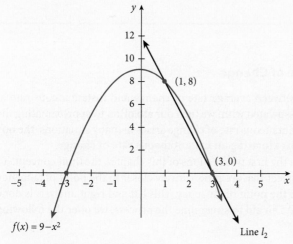

Figure 2.3.6

c. The Country Day Company determines that the daily cost of producing a lawn tractor is given by

$$C(x) = 100 + 40x - 0.001x^2 \qquad 0 \le x \le 300$$

where x represents the number of tires produced each day, and $C(x)$ represents the daily cost in dollars. Suppose we want the rate of change in cost when the production changes from $x = 100$ to $x = 150$.

d. The Kelomata Company determines that the daily cost of producing patio swings can be modeled by

$$C(x) = 15,000 + 100x - 0.001x^2 \qquad 0 \le x \le 200$$

where x represents the number of patio swings produced each day, and $C(x)$ represents the daily cost in dollars. Suppose we want the rate of change in cost at a production level of $x = 75$.

Perform the Mathematics

In order to distinguish between the two types of rates of change, we can examine the number of points on the curve that a line would pass through. If two points are given, we have a secant line and its slope is associated with an average rate of change. If one point is given, we have a tangent line and its slope is associated with an instantaneous rate of change.

a. Here we have a line touching the graph of the function $f(x) = \sqrt{x} + 1$ at the point $(4, 3)$. So the line l_1 is a tangent line and its slope is associated with an instantaneous rate of change.

Approximating an Instantaneous Rate of Change

To approximate the slope of a tangent line and use it to approximate an instantaneous rate of change, we do the following:

1. Determine the point of tangency.

2. Select points on the curve that are progressively closer to the point of tangency. This is achieved by letting h, the increment in x reviewed at the beginning of the section, approach 0. That is, we are going to let $h \to 0$.

3. Calculate the slope of the secant line through each of these points and the point of tangency. The slope of the tangent line is the limit of the slopes of these secant lines.

b. In this case the line intersects the function $f(x) = 9 - x^2$ at the points $(1, 8)$ and $(3, 0)$. So the line l_2 is a secant line and its slope is associated with an average rate of change.

c. Since we are considering two production levels $x = 100$ and $x = 150$ for this cost function, we are seeking an average rate of change.

d. For the given cost function, we want the rate of change in cost at a single production level, $x = 75$. So in this case we seek an instantaneous rate of change. ■

▶ *Try It Yourself*

Some related Exercises are 15 and 17.

Approximating Instantaneous Rate of Change

Now that we understand the difference between average rate of change and instantaneous rate of change and the types of lines that each is associated with, we turn our attention to approximating the instantaneous rate of change. Because instantaneous rate of change arises in many situations, the box to the left shows the process we use in **approximating an instantaneous rate of change**.

Item 3 revisits a concept introduced in the first two sections of this chapter, the limit concept. To ensure that the limit exists, the points we select that are progressively closer to the point of tangency will be points to the left and to the right of the point of tangency. This left and right analysis is done by letting $h \to 0^-$ and by letting $h \to 0^+$. As an aid to streamline the process, we offer the following as a way to organize our work.

Approximating an Instantaneous Rate of Change

To approximate the instantaneous rate of change of a function f at a given point $(x, f(x))$, we use the following steps.

1. Complete a table with the following columns for $h \to 0^+$:

h	Interval, $[x, x + h]$	$m_{sec} = \dfrac{f(x + h) - f(x)}{h}$
1		
0.1		
0.01		
0.001		

2. Complete a table with the following columns for $h \to 0^-$:

h	Interval, $[x, x + h]$	$m_{sec} = \dfrac{f(x + h) - f(x)}{h}$
-1		
-0.1		
-0.01		
-0.001		

Example 3: Approximating an Instantaneous Rate of Change

OBJECTIVE 3

Use tangent line slope approximation to approximate an instantaneous rate of change.

Approximate the instantaneous rate of change of $f(x) = x^2 + 3$ at $x = 2$.

Perform the Mathematics

We need to compute two tables of values: one for $h = 1, 0.1, 0.01, 0.001$ (that is, $h \to 0^+$), and one for $h = -1, -0.1, -0.01, -0.001$ (that is, $h \to 0^-$). The result of letting $h \to 0^+$ is shown in **Table 2.3.1**. The secant lines over the intervals in Table 2.3.1 are shown in **Figure 2.3.7**.

Table 2.3.1

h	Interval, $[2, 2 + h]$	$m_{sec} = \dfrac{f(2 + h) - f(2)}{h}$
1	$[2, 3]$	$\dfrac{f(2 + h) - f(2)}{h} = \dfrac{f(3) - f(2)}{1} = \dfrac{12 - 7}{1} = 5$
0.1	$[2, 2.1]$	$\dfrac{f(2 + h) - f(2)}{h} = \dfrac{f(2.1) - f(2)}{0.1} = \dfrac{7.41 - 7}{0.1} = 4.1$
0.01	$[2, 2.01]$	$\dfrac{f(2 + h) - f(2)}{h} = \dfrac{f(2.01) - f(2)}{0.01} = \dfrac{7.0401 - 7}{0.01} = 4.01$
0.001	$[2, 2.001]$	$\dfrac{f(2 + h) - f(2)}{h} = \dfrac{f(2.001) - f(2)}{0.001} = \dfrac{7.004001 - 7}{0.001} = 4.001$

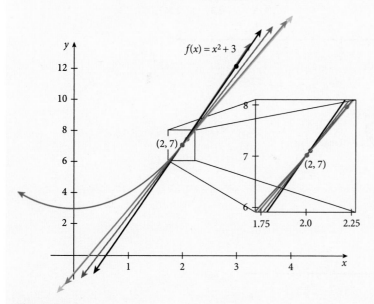

Figure 2.3.7

The result of letting $h \to 0^-$ is shown in **Table 2.3.2**. The secant lines over the intervals in Table 2.3.2 are shown in **Figure 2.3.8**.

From inspecting the values of the two tables, as h becomes smaller and smaller, the slopes of the secant lines get closer and closer to 4. So we approximate the instantaneous rate of change of $f(x) = x^2 + 3$ at $x = 2$ as 4. This rate of change, which is the slope of the tangent line of $f(x) = x^2 + 3$ at the point $(2, f(2))$, is shown graphically in **Figure 2.3.9**.

Table 2.3.2

h	Interval, $[2, 2 + h]$	$m_{sec} = \dfrac{f(2 + h) - f(2)}{h}$
-1	$[1, 2]$	$\dfrac{f(2 + h) - f(2)}{h} = \dfrac{f(1) - f(2)}{-1} = \dfrac{4 - 7}{-1} = 3$
-0.1	$[1.9, 2]$	$\dfrac{f(2 + h) - f(2)}{h} = \dfrac{f(1.9) - f(2)}{-0.1} = \dfrac{6.61 - 7}{-0.1} = 3.9$
-0.01	$[1.99, 2]$	$\dfrac{f(2 + h) - f(2)}{h} = \dfrac{f(1.99) - f(2)}{-0.01} = \dfrac{6.9601 - 7}{-0.01} = 3.99$
-0.001	$[1.999, 2]$	$\dfrac{f(2 + h) - f(2)}{h} = \dfrac{f(1.999) - f(2)}{-0.001} = \dfrac{6.996001 - 7}{-0.001} = 3.999$

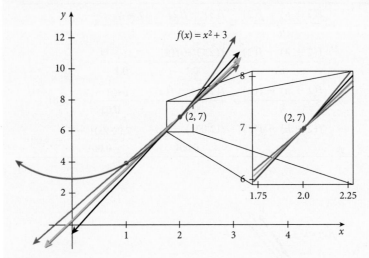

Figure 2.3.8

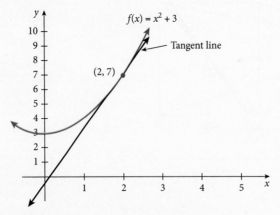

Figure 2.3.9 Slope of tangent line at $(2, 7)$ is approximated as 4. This also gives the instantaneous rate of change of $f(x) = x^2 + 3$ at $x = 2$.

▶ **Try It Yourself**

Some related Exercises are 33 and 35.

▶◀ **Technology Option**

The computations for approximating the instantaneous rate of change of a function can be tedious, so we can use the program MSEC in Appendix B to help complete the two tables. For the program, the function f is entered in Y_1 and the x-value is entered when the program is executed. The results of Example 3 using the MSEC program are shown in **Figure 2.3.10(a)** and in Figure 2.3.10(b).

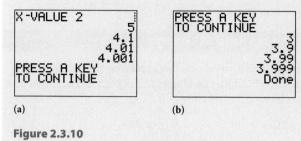

(a)   (b)

Figure 2.3.10

Thinking About Applying Instantaneous Rates of Change

To see how we can approximate the instantaneous rate of change in an application, let's return to a scenario first encountered in Section 1.3. At that time we saw that U.S. production of fuel ethanol and biodiesel for the years 2000 to 2009 can be modeled by

$$f(x) = 3.36x^2 - 12.62x + 49.54 \qquad 1 \le x \le 10$$

where x is the number of years since 1999, and $f(x)$ is the U.S. production of fuel ethanol and biodiesel in millions of barrels. (*Source:* U.S. Statistical Abstract.) To determine the average rate of change for the U.S. production of fuel ethanol and biodiesel from 2003 to 2006, we need to compute the slope of the secant line. For this scenario we have $x = 2003 - 1999 = 4$ and $h = 2006 - 2003 = 3$, so the slope of the secant line is

$$m_{sec} = \frac{f(x+h) - f(x)}{h} = \frac{f(4+3) - f(4)}{3}$$

$$= \frac{f(7) - f(4)}{3} = \frac{125.84 - 52.82}{3} = 24.34$$

This means that from 2003 to 2006, the U.S. production of fuel ethanol and biodiesel increased at an average rate of 24.34 million barrels per year. See **Figure 2.3.11**.

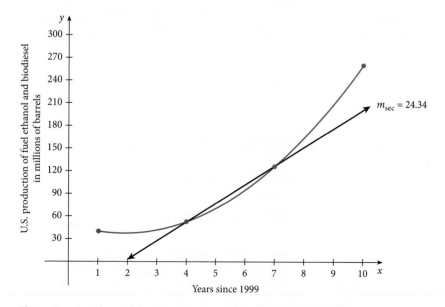

Figure 2.3.11 Slope of the secant line, m_{sec}, gives the average rate of change.

Now suppose we want to determine at what rate the U.S. production of fuel ethanol and biodiesel was changing in 2006. We are not asking for an average rate of change over some time interval. Instead, the question being posed is asking for a rate of change at a particular **point**, namely $(7, f(7))$. Here we need to determine the **instantaneous rate of change**. We now know that this is given by the slope of the line **tangent** to the curve at the given point. We approximate the tangent line slope the same way we did in Example 3.

OBJECTIVE 4

Apply an instantaneous rate of change.

© Jim Parkin/ShutterStock, Inc.

Example 4: Applying an Instantaneous Rate of Change

The production of fuel ethanol from corn as well as biodiesel has increased since 2000 in the United States. The amount of fuel ethanol and biodiesel produced in the United States can be modeled by

$$f(x) = 3.36x^2 - 12.62x + 49.54 \qquad 1 \le x \le 10$$

where x is the number of years since 1999, and $f(x)$ is the U.S. production of fuel ethanol and biodiesel in millions of barrels. (*Source:* U.S. Statistical Abstract.) Approximate the instantaneous rate of change of f at the value $x = 7$, and interpret the approximation.

Perform the Mathematics

We first note that $x = 7$ corresponds to the year 2006. Here we will complete tables just as we did in Example 3. Completing the values of m_{sec} as $h \to 0^+$ gives us the results shown in **Table 2.3.3**, and completing the values of m_{sec} as $h \to 0^-$ gives us the results shown in **Table 2.3.4**.

Table 2.3.3

h	Interval, $[7, 7 + h]$	$m_{\text{sec}} = \dfrac{f(7 + h) - f(7)}{h}$
1	$[7, 8]$	$\dfrac{f(8) - f(1)}{1} = \dfrac{163.62 - 125.84}{1} = 37.78$
0.1	$[7, 7.1]$	$\dfrac{f(7.1) - f(7)}{0.1} = \dfrac{129.3156 - 125.84}{0.1} = 34.765$
0.01	$[7, 7.01]$	$\dfrac{f(7.01) - f(7)}{0.01} = \dfrac{126.184536 - 125.84}{0.01} = 34.4536$
0.001	$[7, 7.001]$	$\dfrac{f(7.001) - f(7)}{0.001} = \dfrac{125.8744234 - 125.84}{0.001} = 34.42336$

Table 2.3.4

h	Interval, $[7 + h, 7]$	$m_{\text{sec}} = \dfrac{f(7 + h) - f(7)}{h}$
-1	$[6, 7]$	$\dfrac{f(6) - f(7)}{-1} = \dfrac{94.78 - 125.84}{-1} = 31.06$
-0.1	$[6.9, 7]$	$\dfrac{f(6.9) - f(7)}{-0.1} = \dfrac{122.4316 - 125.84}{-0.1} = 34.084$
-0.01	$[6.99, 7]$	$\dfrac{f(6.99) - f(7)}{-0.01} = \dfrac{125.496136 - 125.84}{-0.01} = 34.3864$
-0.001	$[6.999, 7]$	$\dfrac{f(6.999) - f(7)}{-0.001} = \dfrac{125.8055834 - 125.84}{-0.001} = 34.41664$

Tables 2.3.3 and 2.3.4 suggest that the approximate instantaneous rate of change is about 34.42. This means that in 2006, the U.S. production of fuel ethanol and biodiesel was increasing at a rate of about 34.42 million barrels per year.   ■

Summary

- The **slope of the secant line** between the points $(x, f(x))$ and $(x + h, f(x + h))$, denoted by m_{sec}, is given by the difference quotient

$$m_{sec} = \frac{f(x + h) - f(x)}{h}$$

  where $h = \Delta x$ is the increment in x.

- For a curve given by the function f, the **tangent line** to the curve at the point $(x, f(x))$ is the line whose slope is the same as the slope of the curve at $(x, f(x))$.

- To **approximate the instantaneous rate of change** of a function f at a given point $(x, f(x))$, we use the following steps:

  1. Complete a table for $h \to 0^+$.
  2. Complete a table for $h \to 0^-$.
  3. Examine the values of m_{sec} as $h \to 0^+$ and as $h \to 0^-$ to approximate the instantaneous rate of change.

Section 2.3 Exercises

Vocabulary Exercises

1. The _____ line to a curve at a point is a line whose slope matches the steepness of the curve at that point.

2. The notation m_{tan} is used to denote the _____ of a tangent line.

3. The increment in x, denoted by Δx, has an alternative notation of _____.

4. The slope of a secant line, denoted by m_{sec}, gives the _____ rate of change between two points on a curve.

5. The slope of a tangent line is also called the _____ rate of change.

6. The expression $\dfrac{f(x + h) - f(x)}{h}$ is called the _____

   _____.

Skill Exercises

In Exercises 7–18, classify each of the following as exhibiting average rate of change or instantaneous rate of change.

7.
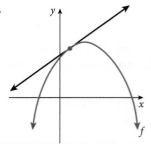

8.
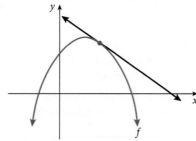

9.

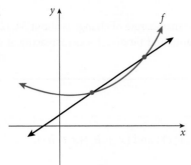

10.

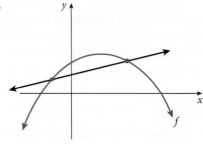

11.

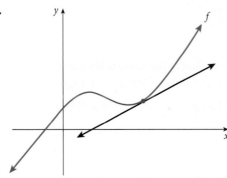

12.

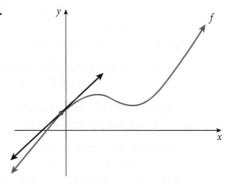

13. **Car Security—Auto Theft:** The StopCop Company has determined that the cost to produce auto theft devices can be modeled by

$$C(x) = (3x + 6)^{1.5} + 30 \qquad 0 \le x \le 50$$

where x represents the number of auto theft devices produced in hundreds, and $C(x)$ represents the production costs in thousands of dollars. The rate of change in cost when production changes from $x = 10$ to $x = 15$.

14. **Cost Analysis—Cameras:** The SnapPic Company determines that the cost of producing disposable cameras can be modeled by

$$C(x) = 60 + \sqrt{3x + 5} \qquad 0 \le x \le 40$$

where x represents the number of disposable cameras produced each shift, and $C(x)$ represents the cost of production in hundreds of dollars. The rate of change in cost when the production changes from $x = 20$ to $x = 35$.

15. **Transportation—Ethanol Consumption:** The annual consumption of ethanol in the United States from 1996 to 2010 can be modeled by the quadratic function

$$f(x) = 1.99x^2 + 6.26x + 29.76 \qquad -4 \le x \le 10$$

where x represents the number of years since 2000, and $f(x)$ represents the annual consumption of ethanol in millions of barrels. (*Source:* U.S. Department of Energy.) The rate of change in cost from 1996 to 2000.

16. **Personal Finance—Median Income:** The median family income for freshmen entering college from 1980 to 2010 can be modeled by

$$f(x) = -0.02x^2 + 2.42x + 24.89 \qquad 0 \le x \le 20$$

where x represents the number of years since 1980, and $f(x)$ represents the median family income in thousands of dollars. (*Source:* U.S. Census Bureau.) The rate of change in the median family income for freshmen when x changes from 0 to 12.

17. **Medicine—Pancreatic Transplants:** The number of pancreatic transplants performed from 1990 to 2010 can be modeled by the function

$$f(x) = -0.49x^3 + 11.88x^2 - 35.53x + 56.94 \qquad 0 \le x \le 20$$

where x represents the number of years since 1990, and $f(x)$ represents the number of pancreatic transplants performed. (*Source:* U.S. Department of Health and Human Services.) The rate of change in the number of transplants performed when $x = 10$.

18. **Medicine—Intestinal Transplants:** The number of intestinal transplants performed from 1990 to 2010 can be modeled by the function

$$f(x) = 0.73x^2 - 1.58x + 1.04 \qquad 0 \le x \le 20$$

where x represents the number of years since 1990, and $f(x)$ represents the number of intestinal transplants performed. (*Source:* U.S. Department of Health and Human Services.) The rate of change in the number of transplants performed when $x = 14$.

In Exercises 19–32, compute the average rate of change of the function over the indicated closed interval.

19. $f(x) = x^2 + x \quad [1, 4]$

20. $f(x) = x^2 - x \quad [1, 4]$

21. $f(x) = 2x^3 + 3x + 1 \quad [2, 4]$

22. $f(x) = x^3 - 2x - 3 \quad [2, 5]$

23. $f(x) = 2x^2 + x - 3 \quad [-2, 1]$

24. $f(x) = 3x^2 - 2x + 2 \quad [-3, 0]$

25. $f(x) = 3\sqrt{x} \quad [1, 9]$

26. $f(x) = 2\sqrt{x} \quad [1, 16]$

27. $f(x) = 2x^{0.2} \quad [0, 20]$

28. $f(x) = 1.3x^{0.3} \quad [0, 25]$

29. $f(x) = 2.1x^{-0.2} \quad [0, 15]$

30. $f(x) = 3.2x^{-0.4} \quad [0, 5]$

31. $f(x) = \dfrac{2x - 3}{x + 1} \quad [0, 4]$

32. $f(x) = \dfrac{4x + 1}{x - 2} \quad [0, 5]$

33. Between which pairs of consecutive points on the curve is the average rate of change positive? Negative? Zero?

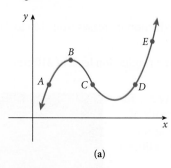

(a)

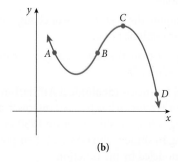

(b)

34. At which points on the given curve is the tangent line slope positive? Negative? Zero?

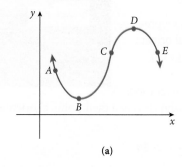

(a)

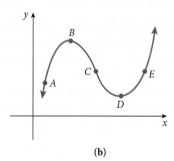

(b)

In Exercises 35–46, estimate the instantaneous rate of change of the given function at the indicated x-value by completing the table for given values of h.

h	-1	-0.1	-0.01	-0.001	0.001	0.01	0.1	1
m_{sec}								

35. $f(x) = 3x; x = 2$

36. $f(x) = 4x; x = 2$

37. $f(x) = x^2 + x; x = 3$

38. $f(x) = x^2 - x; x = -3$

39. $f(x) = 2x^3 + 3x + 1; x = 4$

40. $f(x) = x^3 - 2x - 3; x = -1$

41. $f(x) = 3\sqrt{x}; x = 4$

42. $f(x) = 2\sqrt{x}; x = 9$

43. $f(x) = 2.1x^{0.2}; x = 2$

44. $f(x) = 1.3x^{0.3}; x = 4$

45. $f(x) = \dfrac{2.1}{x^{0.2}}; x = 3$

46. $f(x) = \dfrac{1.3}{x^{0.3}}; x = 5$

Application Exercises

47. Business Management—Cost Analysis: The Seas Beginning Company determines that the cost of producing their Deluge Buster rain coat is given by the cost function

$$C(x) = 1000 + 35x - 0.01x^2 \quad 0 \le x \le 300$$

where x represents the number of raincoats produced each day, and $C(x)$ represents the cost in dollars.

   (a) Determine the average rate of change in cost as daily production increases from 100 to 150 coats per day.

   (b) Approximate the instantaneous rate of change in cost at a production level of 200 coats a day.

48. Business Management—Cost Analysis: The Kanky-Krank Company determines that its weekly cost of producing automatic transmissions for cars is given by the cost function

$$C(x) = 55,000 + 600x - 1.25x^2 \quad 0 \le x \le 230$$

where x represents the number of transmissions produced each week, and $C(x)$ represents the cost in dollars.

   (a) Compute the average rate of change in cost as daily production increases from 180 to 210 transmissions per day.

   (b) Approximate the instantaneous rate of change in cost at a production level of 210 transmissions a day.

 49. Political Science—Ideological Affiliation: For the past 40 years, *The American Freshman* has tracked interests, affiliations, and trends of incoming college freshmen. From 1980 to 2010, the percentage of incoming freshmen who classifed their political orientation as liberal can be modeled by the function

$$f(x) = 0.01x^2 - 0.03x + 21.83 \quad 0 \le x \le 30$$

where x represents the number of years since 1980, and $f(x)$ represents the percentage of freshmen who classify their political orientation as liberal. Approximate the instantaneous rate of change of f when $x = 26$ and interpret.

Welcome Freshmen!

© fotohorst/ShutterStock, Inc.

 50. Political Science—Ideological Affiliation (Continuation of Exercise 49): From 1980 to 2010, the percentage of incoming freshmen who classified their political orientation as conservative can be modeled by the function

$$f(x) = 0.09x + 19.69 \quad 0 \le x \le 30$$

where x represents the number of years since 1980, and $f(x)$ represents the percentage of freshmen who classify their political orientation as conservative. Approximate the instantaneous rate of change of f when $x = 26$ and interpret.

 51. Public Management—NEH Funding: The National Endowment for the Humanities (NEH) is an independent grant-making agency of the U.S. federal government used to support research, education, and public programs in the humanities. The annual amount in federal research grants to the NEH from 1990 to 2010 can be modeled by the rational function

$$f(x) = \frac{-193.53x + 6637}{12.36x + 264} \quad 0 \le x \le 20$$

where x represents the number of years since 1990, and $f(x)$ represents the amount in federal research grants to the NEH, measured in millions of dollars. (*Source:* U.S. National Endowment for the Humanities). Approximate the instantaneous rate of change of f when $x = 7$ and interpret.

52. **Public Management—NEA Funding:** Established by the U.S. Congress in 1965, the National Endowment for the Arts (NEA) is a public agency that provides support for excellence in the arts as well as leadership in art education. The number of federal grants awarded by the NEA from 1990 to 2010 can be modeled by the rational function

© Getty Images

$$f(x) = \frac{10{,}460x + 1{,}775{,}015}{35x + 408} \qquad 0 \le x \le 20$$

where x represents the number of years since 1990, and $f(x)$ represents the number of grants awarded. (*Source:* U.S. Endowment for the Arts Annual Report). Approximate the instantaneous rate of change of f when $x = 6$ and interpret.

Concept and Writing Exercises

53. Explain in a few sentences the difference between the average rate of change of a function and the instantaneous rate of change of a function.

54. For a given function f, explain in a few sentences the difference between the tangent line of a function and the secant line of a function.

55. Let $f(x) = 4x - 5$. Determine the average rate of change of f between any two values $x = a$ and $x = b$.

56. Let $f(x) = x^2 - 4x + 1$. Determine the average rate of change of f between any two values $x = a$ and $x = b$.

57. Consider the function $f(x) = \sqrt{2x - 4}$. Knowing that the slope of the secant line from $x = 4$ to $x = b$ is $m_{\text{sec}} = \frac{1}{4}$, determine the value of b.

58. Consider the function $f(x) = \dfrac{x + 3}{x - 1}$. Knowing that the slope of the secant line from $x = a$ to $x = 5$ is $m_{\text{sec}} = \frac{1}{2}$, determine the value of a.

In Exercises 59 and 60, a piecewise-defined function is given. Estimate the instantaneous rate of change of f at the indicated x-value, if possible. Do this by completing the table.

h	-1	-0.1	-0.01	-0.001	0.001	0.01	0.1	1
m_{sec}								

If the function does not have an instantaneous rate of change at the indicated value, explain why.

59. $f(x) = \begin{cases} x + 2, & x \le 1 \\ 3x, & x > 1 \end{cases}; \quad x = 1$

60. $f(x) = \begin{cases} 3, & x < 0 \\ 4x, & x \ge 0 \end{cases}; \quad x = 0$

In Exercises 61 and 62, the polynomial form of a function is given where a, b, and c represent real numbers. Estimate the instantaneous rate of change of the given function f at $x = 2$ by completing the table.

h	-1	-0.1	-0.01	-0.001	0.001	0.01	0.1	1
m_{sec}								

61. $f(x) = ax + b$

62. $f(x) = ax^2 + bx + c$

Section Project

There is another method for approximating the instantaneous rate of change of a function. We start with an *alternative formula* for the average rate of change of f between the points $(x, f(x))$ and $(c, f(c))$ given by

$$m_{\text{sec}} = \frac{f(x) - f(c)}{x - c} \qquad x \ne c$$

This formula allows us to compute m_{sec} for values of c that are close to x in value. To illustrate, consider the function $f(x) = \sqrt{x} + 5$, and suppose we wish to approximate the instantaneous rate of change of f at $x = 4$. The calculations for $c = 5$ and $c = 4.1$ are shown in the table.

c	$x - c$	$m_{sec} = \dfrac{f(x) - f(c)}{x - c}$
5	$4 - 5 = -1$	$\dfrac{f(4) - f(5)}{4 - 5} = \dfrac{(\sqrt{4} + 5) - (\sqrt{5} + 5)}{4 - 5} = \dfrac{2 - \sqrt{5}}{-1} \approx 0.236$
4.1	$4 - 4.1 = -0.1$	$\dfrac{f(4) - f(4.1)}{4 - 4.1} = \dfrac{(\sqrt{4} + 5) - (\sqrt{4.1} + 5)}{-0.1} \approx 0.248$
4.01		
4.001		

(a) Complete the missing entries in the table.

(b) Explain in a sentence what the values in the right-hand column of the table represent.

(c) Complete the next table.

c	$x - c$	$m_{sec} = \dfrac{f(x) - f(c)}{x - c}$
3	$4 - 3 = 1$	$\dfrac{f(4) - f(3)}{4 - 3} = \dfrac{(\sqrt{4} + 5) - (\sqrt{3} + 5)}{4 - 3} = \dfrac{2 - \sqrt{3}}{1} \approx 0.268$
3.9	$4 - 3.9 = 0.1$	$\dfrac{f(4) - f(3.9)}{4 - 3.9} = \dfrac{(\sqrt{4} + 5) - (\sqrt{3.9} + 5)}{0.1} \approx 0.252$
3.99		
3.999		

(d) Explain in a sentence what the values in the right-hand column of the table represent.

(e) Compare the values in the two tables, and approximate the instantaneous rate of change $f(x) = \sqrt{x} + 5$ at $x = 4$.

SECTION OBJECTIVES

1. Compute an instantaneous rate of change.

2. Compute a derivative.

3. Apply the derivative.

4. Compute the derivative of a function.

5. Analyze nondifferentiable functions.

2.4 The Derivative of a Function

In Section 2.3, we analyzed the difference between an average rate of change and an instantaneous rate of change. We saw that an instantaneous rate of change is given by the slope of a tangent line. The process we outlined to approximate this type of rate of change employed table building and computing secant line slope over smaller and smaller intervals. This process actually used the limit concept presented in Section 2.1 and 2.2. In this section, we formalize and generalize the process of determining an instantaneous rate of change. This generalization will allow us to compute an instantaneous rate of change for **any** value of x, and it will also give us the **exact** value of the instantaneous rate of change.

Difference Quotient and the Derivative

Suppose that we wish to determine the instantaneous rate of change at x for the function graphed in **Figure 2.4.1**. In Section 2.3, we had a specific value for x. Here, we consider *any* value for x.

Using the process outlined in Section 2.3, we apply the limit concept by letting h get smaller and smaller. In other words, we take the limit of the difference quotient as $h \to 0$, and this limit gives the **instantaneous rate of change**. For any value of x, we can formalize the process as follows.

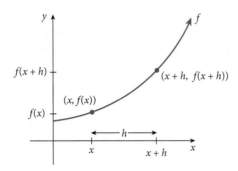

Figure 2.4.1

DEFINITION

Instantaneous Rate of Change/Slope of a Tangent Line

The **instantaneous rate of change** of a function, f, at x is equal to the **slope of the line tangent** to the graph of f at x and is given by

$$\text{Instantaneous rate of change} = m_{\tan} = \lim_{h \to 0} \frac{f(x + h) - f(x)}{h}$$

provided that the limit exists. Its units of measurement are units of f per units of x.

NOTE: $m_{\tan}$ means "slope of the tangent line."

Example 1: Determining an Instantaneous Rate of Change

For $f(x) = x^2 + 3$, determine the instantaneous rate of change at $x = 2$.

Perform the Mathematics

From the preceding definition, we know that

$$\text{Instantaneous rate of change} = m_{\tan} = \lim_{h \to 0} \frac{f(x + h) - f(x)}{h}$$

Substituting $x = 2$ gives us

$$
\begin{aligned}
\text{Instantaneous rate of change} = m_{\tan} &= \lim_{h \to 0} \frac{f(2 + h) - f(2)}{h} \\
&= \lim_{h \to 0} \frac{[(2 + h)^2 + 3] - [2^2 + 3]}{h} \\
&= \lim_{h \to 0} \frac{[4 + 4h + h^2 + 3] - [7]}{h} \\
&= \lim_{h \to 0} \frac{7 + 4h + h^2 - 7}{h} \\
&= \lim_{h \to 0} \frac{4h + h^2}{h} \\
&= \lim_{h \to 0} \frac{h(4 + h)}{h} \\
&= \lim_{h \to 0} (4 + h) \\
&= 4
\end{aligned}
$$

OBJECTIVE 1

Compute an instantaneous rate of change.

So the instantaneous rate of change of $f(x) = x^2 + 3$ at $x = 2$ is 4. This is the exact slope of the line tangent to the graph of $f(x) = x^2 + 3$ at $x = 2$. See **Figure 2.4.2**.

Derivative

For a function f, the **derivative of f at x,** denoted $f'(x)$ (this is read "f prime of x"), is defined to be

$$f'(x) = \lim_{h \to 0} \frac{f(x + h) - f(x)}{h}$$

provided that the limit exists. The units of $f'(x)$ are units of f per unit of x.

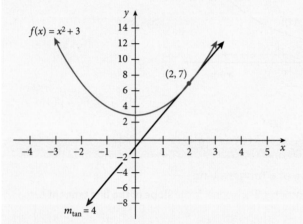

Figure 2.4.2 Slope of the line tangent to the graph of $f(x) = x^2 + 3$ at $x = 2$ is 4.

NOTE: The derivative of f at x gives an instantaneous rate of change of f at x.

The definition given before Example 1 is so important that it has been given a special name, the **derivative**. We now present one of the most important definitions in mathematics, the **definition of the derivative.**

Compute a derivative.

Compute $f'(2)$ for $f(x) = x^2 + x$. Interpret the result.

Perform the Mathematics

To compute $f'(2)$, we use the definition of the derivative

$$f'(x) = \lim_{h \to 0} \frac{f(x + h) - f(x)}{h}$$

where x has a value of 2. Proceeding, this gives us

$$f'(2) = \lim_{h \to 0} \frac{f(2 + h) - f(2)}{h}$$

$$= \lim_{h \to 0} \frac{[(2 + h)^2 + (2 + h)] - [(2)^2 + 2]}{h}$$

$$= \lim_{h \to 0} \frac{[4 + 4h + h^2 + 2 + h] - 6}{h}$$

$$= \lim_{h \to 0} \frac{h^2 + 5h}{h}$$

Now we rely on the algebraic techniques learned in Section 2.1 to evaluate this limit. Employing the Substitution Principle yields the indeterminate form $\frac{0}{0}$, so we factor the numerator and cancel:

$$f'(2) = \lim_{h \to 0} \frac{h(h + 5)}{h}$$

$$= \lim_{h \to 0} (h + 5) = 5$$

We have now determined that $f'(2) = 5$, which means that the instantaneous rate of change of $f(x) = x^2 + x$ at $x = 2$ is 5. Alternatively, we could say that the slope of the curve given by $f(x) = x^2 + x$ at $x = 2$ is 5, or we could even say that the slope of the line tangent to the graph of $f(x) = x^2 + x$ at $x = 2$ is 5. See **Figure 2.4.3**.

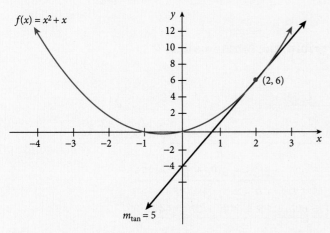

Figure 2.4.3 The notation $m_{\tan}$ represents "slope of tangent line."

▶ *Try It Yourself*

Some related Exercises are 7 and 9.

◀ **Technology Option**

Many graphing calculators have a Draw Tangent command. **Figure 2.4.4** was generated using such a command. Notice that the lower left-hand corner of the screen displays the equation of the tangent line in slope-intercept form. From this, we can see that the slope of the tangent line is 5, which supports our work in Example 2.

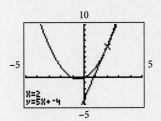

Figure 2.4.4

Example 3: Computing a Derivative in an Applied Setting

OBJECTIVE 3

Apply the derivative.

The total cost of producing q recreational vehicles can be modeled by

$$C(q) = 100 + 60q + 3q^2$$

where q is the number of vehicles produced, and $C(q)$ is in hundreds of dollars. Compute $C'(5)$ and interpret.

© Jim Parkin/ShutterStock, Inc.

Understand the Situation

To compute $C'(5)$, we will use the definition of the derivative,

$$C'(q) = \lim_{h \to 0} \frac{C(q + h) - C(q)}{h}$$

where q has the value of 5.

Perform the Mathematics

Proceeding this gives us

$$
\begin{aligned}
C'(5) &= \lim_{h \to 0} \frac{C(5 + h) - C(5)}{h} \\[2mm]
&= \lim_{h \to 0} \frac{[100 + 60(5 + h) + 3(5 + h)^2] - [100 + 60(5) + 3(5)^2]}{h} \\[2mm]
&= \lim_{h \to 0} \frac{100 + 300 + 60h + 3(25 + 10h + h^2) - 475}{h} \\[2mm]
&= \lim_{h \to 0} \frac{100 + 300 + 60h + 75 + 30h + 3h^2 - 475}{h} \\[2mm]
&= \lim_{h \to 0} \frac{90h + 3h^2}{h}
\end{aligned}
$$

As in Example 2, employing the substitution principle yields the indeterminate form $\frac{0}{0}$, so we factor and cancel:

$$
\begin{aligned}
C'(5) &= \lim_{h \to 0} \frac{h(90 + 3h)}{h} \\[2mm]
&= \lim_{h \to 0} (90 + 3h) = 90
\end{aligned}
$$

Interpret the Result

So we have determined that $C'(5) = 90$ which means that the instantaneous rate of change is $9000 per vehicle when the production level is 5 vehicles. Alternatively, we could say that when the production level is 5 vehicles, the total cost is increasing at a rate of $9000 per vehicle. **Figure 2.4.5** shows a graph of C along with the line tangent at the point $(5, 475)$.

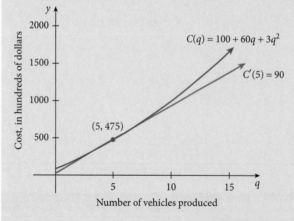

Figure 2.4.5

▷ **Try It Yourself**

Some related Exercises are 71 and 73.

Derivatives in General

Before we continue with how to compute a derivative in general, we need to present some **facts about the derivative**.

> **Facts About the Derivative**
>
> 1. Recall that the derivative of a function f can be found for any value x where the derivative exists. Thus, the derivative of a function, f', is itself a function. This means that it has a domain for which it is defined.
> 2. For any c in the domain of f', $f'(c)$ is a number that gives the slope of a line tangent to the graph of f at $x = c$. Also, $f'(c)$ gives the instantaneous rate of change of f at $x = c$.
> 3. The process of determining $f'(x)$ is called **differentiation**.

By taking full advantage of the definition of the derivative coupled with the fact that the derivative is a function, we can compute a formula for the derivative. In other words, recall the total cost function in Example 3, which was given by $C(q) = 100 + 60q + 3q^2$. Suppose that we were asked to compute $C'(3)$, $C'(7)$, $C'(13)$, and $C'(15)$. At first glance, it appears that we would have the long and tedious task of duplicating our work in Example 3 for these four different values of q. However, the definition of the derivative tells us that the derivative is a *function*, which means that we should be able to determine a rule for this function. In Example 4, we illustrate how to use the definition of the derivative to do this.

Example 4: Computing a Derivative Formula

OBJECTIVE 4

Compute the derivative of a function.

Compute $C'(q)$ for $C(q) = 100 + 60q + 3q^2$.

Perform the Mathematics

The process is the same as in Example 3, except here we have our independent variable q in place of 5:

$$
\begin{aligned}
C'(q) &= \lim_{h \to 0} \frac{C(q + h) - C(q)}{h} \\
&= \lim_{h \to 0} \frac{[100 + 60(q + h) + 3(q + h)^2] - [100 + 60q + 3q^2]}{h} \\
&= \lim_{h \to 0} \frac{[100 + 60q + 60h + 3(q^2 + 2qh + h^2)] - 100 - 60q - 3q^2}{h} \\
&= \lim_{h \to 0} \frac{100 + 60q + 60h + 3q^2 + 6qh + 3h - 100 - 60q - 3q^2}{h} \\
&= \lim_{h \to 0} \frac{60h + 60qh + 3h^2}{h} \\
&= \lim_{h \to 0} \frac{h(60 + 6q + 3h)}{h} \\
&= \lim_{h \to 0}(60 + 6q + 3h) = 60 + 6q
\end{aligned}
$$

So $C'(q) = 60 + 6q$. A huge advantage in having this formula for the derivative is that we can now evaluate it for different values of our independent variable very quickly. For example,

$$C'(3) = 60 + 6(3) = 78$$
$$C'(5) = 60 + 6(5) = 90$$
$$C'(7) = 60 + 6(7) = 102$$

Example 5: Computing and Using a Derivative Formula

Determine $f'(x)$ for $f(x) = x^2 + 3x$. Also, determine the slope of the line tangent to the graph of f at $x = -2$.

Perform the Mathematics

Utilizing the definition of the derivative gives us

$$f'(x) = \lim_{h \to 0} \frac{f(x+h) - f(x)}{h}$$

$$= \lim_{h \to 0} \frac{[(x+h)^2 + 3(x+h)] - (x^2 + 3x)}{h}$$

$$= \lim_{h \to 0} \frac{x^2 + 2xh + h^2 + 3x + 3h - x^2 - 3x}{h}$$

$$= \lim_{h \to 0} \frac{2xh + h^2 + 3h}{h}$$

$$= \lim_{h \to 0} \frac{h(2x + h + 3)}{h}$$

$$= \lim_{h \to 0} (2x + h + 3) = 2x + 3$$

So $f'(x) = 2x + 3$. Knowing that the derivative gives the slope of the tangent line at any point, we can quickly compute that at $x = -2$,

$$m_{\tan} = f'(-2)$$
$$= 2(-2) + 3$$
$$= -1$$

Figure 2.4.6 has a graph of f and the line tangent to f at $x = -2$.

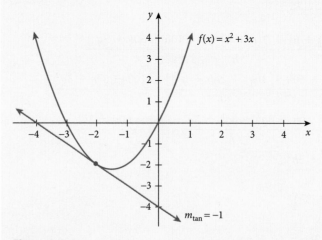

Figure 2.4.6

▶ **Try It Yourself**

Some related Exercises are 35, 37, and 47.

Example 5 communicates that the derivative of a function, in its most general form, is very useful when determining tangent line slopes or instantaneous rates of change at **any** point on the graph of the given function.

We conclude this section by analyzing when a derivative may not exist.

Nondifferentiable Functions

There are some functions that are **nondifferentiable**. That is, some functions cannot be differentiated at certain values. We conclude this section by determining where these functions are not differentiable. We begin by analyzing the **absolute value function**, $f(x) = |x|$, whose graph is shown in **Figure 2.4.7**.

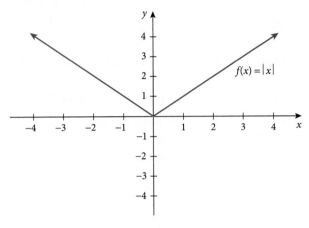

Figure 2.4.7

Recall that the definition for $f(x) = |x|$ is

$$f(x) = |x| = \begin{cases} x, & x \geq 0 \\ -x, & x < 0 \end{cases}$$

Example 6: Analyzing Nondifferentiability at a Point

OBJECTIVE 5

Analyze nondifferentiable functions.

Show that $f(x) = |x|$ is not differentiable at $x = 0$.

Perform the Mathematics

We appeal to the definition of the derivative and try to compute $f'(0)$ from the definition. This gives us

$$f'(0) = \lim_{h \to 0} \frac{f(0 + h) - f(0)}{h} = \lim_{h \to 0} \frac{f(h) - f(0)}{h}$$

$$= \lim_{h \to 0} \frac{|h| - 0}{h} = \lim_{h \to 0} \frac{|h|}{h}$$

In Section 2.1, we learned that if we try to substitute here, it produces the indeterminate form $\frac{0}{0}$. In **Table 2.4.1**, we numerically analyze this limit.

Table 2.4.1

		$h \to 0^-$				$h \to 0^+$			
h	-1	-0.1	-0.01	-0.001	0	0.001	0.01	0.1	1
$\dfrac{|h|}{h}$	-1	-1	-1	-1		1	1	1	1

Table 2.4.1 suggests that the limit does not exist. We confirm this belief algebraically by appealing to the definition of absolute value. Specifically,

$$|h| = -h, \text{ for } h < 0 \qquad \lim_{h \to 0^-} \frac{|h|}{h} = \lim_{h \to 0^-} \frac{-h}{h}$$

Limit of a constant
$$\lim_{h \to 0^-} (-1) = -1$$

Similarly,

$$|h| = h, \text{ for } h \geq 0 \qquad \lim_{h \to 0^+} \frac{|h|}{h} = \lim_{h \to 0^+} \frac{h}{h}$$

Limit of a constant
$$\lim_{h \to 0^+} (1) = 1$$

Because the **left-hand limit** does not equal the **right-hand limit**, $\lim_{h \to 0^-} \frac{|h|}{h} \neq \lim_{h \to 0^+} \frac{|h|}{h}$, we conclude that $\lim_{h \to 0} \frac{|h|}{h}$ **does not exist**. Because $\lim_{h \to 0} \frac{|h|}{h}$ does not exist, and since $f'(0) = \lim_{h \to 0} \frac{|h|}{h}$, we conclude that the derivative does not exist. This is exactly what we wanted to show—that the absolute value function is not differentiable at $x = 0$. Notice that $f(x) = |x|$ is continuous at $x = 0$. ∎

Geometrically, the reason $f(x) = |x|$ is not differentiable at $x = 0$ is because its graph consists of two lines with slopes of -1 and $+1$ that meet at the origin. See **Figure 2.4.8**.

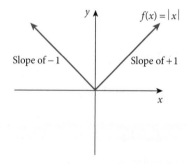

Figure 2.4.8

The two conflicting slopes make it impossible to define a single slope at the origin.

 Technology Option

To gain a different perspective on why $f(x) = |x|$ is not differentiable at $x = 0$, consider the sequence of graphs in **Figures 2.4.9(a)–(e)**. Notice that after four applications of the Zoom In command, the graph of $f(x) = |x|$ still has a corner at the origin. The graph has not "straightened out." If a function is differentiable at a point, several applications of the Zoom In command will show that the graph "straightens" out and practically becomes its own tangent line.

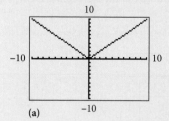

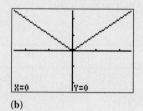

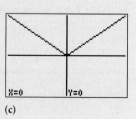

Figure 2.4.9 (a) Original graph. (b) After one Zoom In. (c) After second Zoom In.

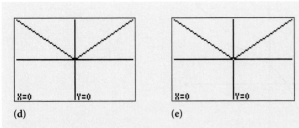

Figure 2.4.9 **(d)** After third `Zoom In`. **(e)** After fourth `Zoom In`.

Example 6 and Figure 2.4.8 show that if a graph has a **sharp turn** or a **corner** at a point, then the function is not differentiable at that point. In Example 7, we analyze another scenario in which a function is nondifferentiable.

Example 7: Analyzing Nondifferentiability at a Point

Show that $f(x) = \sqrt[3]{x}$ is not differentiable at $x = 0$.

Perform the Mathematics

By looking at the graph of $f(x) = \sqrt[3]{x}$ in **Figure 2.4.10**, it appears that the tangent line at $x = 0$ would be a vertical line.

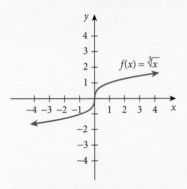

Figure 2.4.10

We know the slope of a vertical line is undefined, which leads us to believe that here, $f'(0)$ is undefined. It can be shown that the derivative of $f(x) = \sqrt[3]{x}$ is

$$f'(x) = \frac{1}{3}x^{-2/3} = \frac{1}{3x^{2/3}} = \frac{1}{3\sqrt[3]{x^2}}$$

Remembering that f' is a function, we see that the domain is $(-\infty, 0) \cup (0, \infty)$. In other words, the domain of the derivative **excludes** zero. Because $f'(0)$ is not defined, the derivative is not defined at zero, and so our function is not differentiable at zero. Notice that $f(x) = \sqrt[3]{x}$ is continuous at $x = 0$. ∎

Example 7 reminds us that the derivative is also a function. Using that fact can aid us in determining where functions are nondifferentiable. So far we have seen that a function is not differentiable at $x = c$ if the graph of the function has a sharp turn or a vertical tangent at $x = c$. The final situation for which a function is not differentiable is wherever the function is **discontinuous**. The function graphed in **Figure 2.4.11** is not differentiable at $x = c$ because it is discontinuous at $x = c$.

**Differentiability and
Nondifferentiability**

A function is not
differentiable at $x = c$ if

1. The graph of the
   function has a **sharp
   turn** or **corner** at $x = c$.
2. The graph of the
   function has a **vertical
   tangent** at $x = c$.
3. The graph of the
   function is **not
   continuous** at $x = c$.

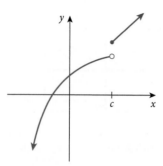

Figure 2.4.11

Both functions analyzed in Examples 6 and 7 were continuous at $x = 0$, but in both cases $f'(0)$ was undefined. At this time, we present the **theorem** shown in the margin.

A proof of this theorem is given in Appendix C. We could condense the contents of this theorem to the following:

Differentiability implies continuity, but continuity does **not** imply differentiability.

Note that Examples 6 and 7 illustrate continuous functions that are not differentiable. Examples 6 and 7, along with Figure 2.4.11, provide the meaning for **nondifferentiability**.

Example 8: Determining Points of Nondifferentiability

For what values of x is the function graphed in **Figure 2.4.12** not differentiable?

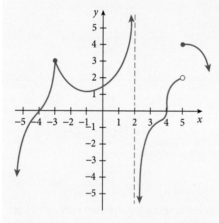

Figure 2.4.12

Perform the Mathematics

From the graph, we determine that $f(x)$ is not differentiable at

$$x = -3, \qquad \text{Sharp turn or corner}$$
$$x = 2, \qquad \text{Not continuous}$$
$$x = 4, \qquad \text{Vertical tangent}$$
$$x = 5 \qquad \text{Not continuous}$$

▶ *Try It Yourself*

Some related Exercises are 57 and 59.

Summary

We have presented some very powerful mathematical concepts in this section and the last section. The main ideas have been rates of change, specifically introducing the **instantaneous rate of change**. We calculated **derivatives** for specific values of x and for any arbitrary x. The derivative is a function, which means that it has a domain and can be evaluated at any value c in the domain. Remember when computing a derivative, whether at a specific value of c or for an arbitrary x, that we are finding a limit. Therefore, the limit techniques learned in Sections 2.1 and 2.2 can be employed.

Average Rate of Change	*Instantaneous Rate of Change*
Change over an interval	Change at a point
Slope of a secant line	Slope of a tangent line; Slope of a curve
$\dfrac{f(x + h) - f(x)}{h}$	$\lim\limits_{h \to 0} \dfrac{f(x + h) - f(x)}{h}$
Measured in units of f per unit of x	Measured in units of f per unit of x

We concluded this section by looking at **nondifferentiable functions** or, more specifically, by locating values where a function is not differentiable.

- A function is not differentiable at $x = c$ if

  1. The graph of the function has a sharp turn at $x = c$.
  2. The graph of the function has a vertical tangent at $x = c$.
  3. The function is not continuous at $x = c$.

Section 2.4 Exercises

Vocabulary Exercises

1. The notation _____ means slope of the tangent line.

2. The _____ is a function that gives an instantaneous rate of change of f at x.

3. The notation _____ is used to denote the derivative of the function f.

4. The process of determining the derivative of a function is called _____.

Skill Exercises

In Exercises 5–16, compute $f'(c)$ by using the definition of derivative $f'(c) = \lim\limits_{h \to 0} \dfrac{f(c + h) - f(c)}{h}$.

5. $f(x) = 2x + 1; c = 2$ 6. $f(x) = 3x - 4; c = 3$

7. $f(x) = x^2 - 2x; c = -2$ 8. $f(x) = x^2 - 3x; c = 2$

9. $f(x) = -x^2 + 2x; c = 4$ 10. $f(x) = -2x^2 + x; c = 4$

11. $f(x) = 2\sqrt{x}; c = 4$ 12. $f(x) = 5\sqrt{x}; c = 9$

13. $f(x) = 2.3x^2 - 0.2x + 3.2; c = 8$ 14. $f(x) = -1.2x^2 + 3.2x - 2.1; c = 7$

15. $f(x) = \dfrac{1}{x}; c = 3$ 16. $f(x) = \dfrac{3}{x}; c = 1$

In Exercises 5–16, we found the slope of the tangent line of f at $x = c$. For Exercises 17–28, use that information to determine the equation of the tangent line of f at $x = c$.

17. $f(x) = 2x + 1; c = 2$ (see Exercise 5)   18. $f(x) = 3x - 4; c = 3$ (see Exercise 6)

19. $f(x) = x^2 - 2x; c = -2$ (see Exercise 7)   20. $f(x) = x^2 - 3x; c = 2$ (see Exercise 8)

21. $f(x) = -x^2 + 2x; c = 4$ (see Exercise 9)   **22.** $f(x) = -2x^2 + x; c = 4$ (see Exercise 10)

23. $f(x) = 2\sqrt{x}; c = 4$ (see Exercise 11)   **24.** $f(x) = 5\sqrt{x}; c = 9$ (see Exercise 12)

25. $f(x) = 2.3x^2 - 0.2x + 3.2; c = 8$ (see Exercise 13)

26. $f(x) = -1.2x^2 + 3.2x - 2.1; c = 7$ (see Exercise 14)

27. $f(x) = \dfrac{1}{x}; c = 3$ (see Exercise 15)   **28.** $f(x) = \dfrac{3}{x}; c = 1$ (see Exercise 16)

In Exercises 29–44, compute the derivative function for the given function f using the definition of the derivative $f'(x) = \lim\limits_{h \to 0} \dfrac{f(x + h) - f(x)}{h}$.

29. $f(x) = 2x - 5$ **30.** $f(x) = 3x + 1$

31. $f(x) = -2x + 3$ **32.** $f(x) = -3x - 1$

33. $f(x) = x^2$ **34.** $f(x) = x^2 - 3x$

35. $f(x) = x^2 - 2x + 3$ **36.** $f(x) = x^2 - 3x + 1$

37. $f(x) = 2.1x^2 + 3.2x$ **38.** $f(x) = 1.3x^2 - 2.1x$

39. $f(x) = -2x^2 + 3x$ **40.** $f(x) = -3x^2 + 2x$

41. $f(x) = -2.3x^2 + 3.1x$ **42.** $f(x) = x^3$

43. $f(x) = x^3 + x^2$ **44.** $f(x) = \sqrt{x}$

Exercises 45–54 use the results of Exercises 29–44. Use the derivative f' to determine the slope of the tangent line to the graph of the function at the given value.

45. $f(x) = 2x - 5$ at $x = -1$ and $x = 2$ (see Exercise 29)

46. $f(x) = -3x - 1$ at $x = -1$ and $x = 2$ (see Exercise 32)

47. $f(x) = x^2 - 2x + 3$ at $x = -3$ and $x = 2$ (see Exercise 35)

48. $f(x) = x^2 - 3x + 1$ at $x = -2$ and $x = 3$ (see Exercise 36)

49. $f(x) = 2.1x^2 + 3.2x$ at $x = 0$ and $x = 2$ (see Exercise 37)

50. $f(x) = 1.3x^2 - 2.1x$ at $x = 1$ and $x = 4$ (see Exercise 38)

51. $f(x) = -2x^2 + 3x$ at $x = 3$ and $x = 6$ (see Exercise 39)

52. $f(x) = -3x^2 + 2x$ at $x = 2$ and $x = 7$ (see Exercise 40)

53. $f(x) = x^3 + x^2$ at $x = -1$ and $x = 1$ (see Exercise 43)

54. $f(x) = \sqrt{x}$ at $x = 4$ and $x = 9$ (see Exercise 44)

For each function graphed in Exercises 55–62, state the x-values for which the derivative does not exist and explain why.

55.

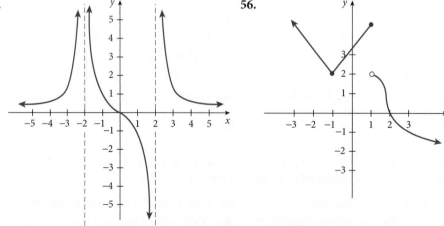

56.

57.

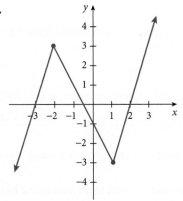

58.

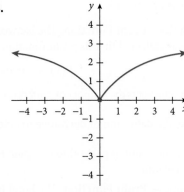

59.

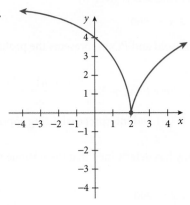

60.

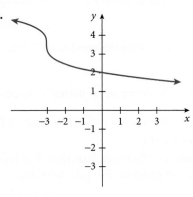

61.

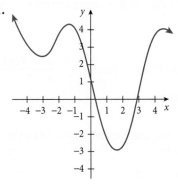

62.

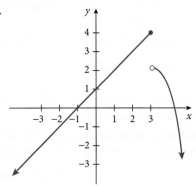

For Exercises 63–70, complete the following.

(a) Determine the domain of f.

(b) Determine the derivative.

(c) Are there any x-values in the domain of f that make the derivative undefined?

(d) For the x-values where the derivative is undefined, does the graph of the function have a sharp turn or a vertical tangent, or is it discontinuous?

63. $f(x) = \sqrt{x}$

64. $f(x) = \sqrt[4]{x}$

65. $f(x) = x^{2/3}$

66. $f(x) = x^{3/5}$

67. $f(x) = \dfrac{x^2 - 1}{x + 1}$

68. $f(x) = \dfrac{x^2 - 1}{x - 1}$

69. $f(x) = \dfrac{x^2 - 4}{x - 2}$

70. $f(x) = \dfrac{x^2 - 4}{x + 2}$

Application Exercises

71. **Economics—Profit Function:** The Network Standards Company has determined from data that it has collected that its profit function for its new E1020 routers is given by

$$P(x) = 1000x - 2x^2 \quad 0 \le x \le 300$$

where x represents the number of routers produced and sold and $P(x)$ represents the profit in dollars.

(a) Use the definition of derivative to determine $P'(x)$, where $P'(x) = \lim_{h \to 0} \dfrac{P(x+h) - P(x)}{h}$.

(b) Use the result of part (a) to compute $P'(200)$ and $P'(300)$ and interpret each using appropriate units.

72. **Economics—Profit Function:** The ForeLink Company has determined from data that it has collected that its profit function for its Legend Series golf club sets is given by

$$P(x) = 1000x - x^2 \quad 0 \le x \le 800$$

where x represents the number of club sets produced and sold and $P(x)$ represents the profit in dollars.

(a) Use the definition of derivative to determine $P'(x)$, where $P'(x) = \lim_{h \to 0} \dfrac{P(x+h) - P(x)}{h}$.

(b) Use the result of part (a) to compute $P'(400)$ and $P'(600)$ and interpret each using appropriate units.

73. **Economics—Revenue Function:** The ForeLink Company has determined that the revenue for the sale of its Junior Pro golf clubs is given by

$$R(x) = 200x - x^2 \quad 0 \le x \le 200$$

where x represents the number of golf clubs produced and sold and $R(x)$ represents the revenue in dollars.

(a) Evaluate $R(50)$ and $R(150)$, and interpret each.

(b) Compute $R'(x)$ using the definition of derivative, where $R'(x) = \lim_{h \to 0} \dfrac{R(x+h) - R(x)}{h}$.

(c) Use the result of part (b) to determine $R'(50)$ and $R'(150)$ and interpret each using appropriate units.

74. **Economics—Revenue Function:** The Network Standard Company has determined that the revenue function for their Z45h wi-fi units can be modeled by

$$R(x) = 300x - x^2 \quad 0 \le x \le 300$$

where x represents the number of units produced and sold, and $R(x)$ represents the revenue in dollars.

(a) Evaluate $R(100)$ and $R(200)$ and interpret each.

(b) Compute $R'(x)$ using the definition of derivative, where $R'(x) = \lim_{h \to 0} \dfrac{R(x+h) - R(x)}{h}$.

(c) Use the result of part (b) to determine $R'(100)$ and $R'(200)$ and interpret each using appropriate units.

75. **Economics—Cost Function:** The FrezMore Company has determined that the weekly cost of producing their Upscale line of refrigerators is given by

$$C(x) = x^2 + 15x + 1500 \quad 0 \le x \le 200$$

where x represents the number of refrigerators produced, and $C(x)$ represents the cost in dollars.

(a) Compute $C'(x)$ using the definition of derivative, where $C'(x) = \lim_{h \to 0} \dfrac{C(x+h) - C(x)}{h}$.

(b) Use the result of part (a) to determine $C'(40)$ and $C'(100)$ and interpret each using appropriate units.

76. **Economics—Cost Function:** The FrezMore Company has determined that the weekly cost of producing their Roommate line of dormitory-sized refrigerators is given by

$$C(x) = \frac{1}{2}x^2 + 3x + 15 \qquad 0 \le x \le 300$$

where x represents the number of dormitory-sized refrigerators produced and $C(x)$ represents the cost in dollars.

(a) Compute $C'(x)$ using the definition of derivative, where $C'(x) = \lim_{h \to 0} \dfrac{C(x + h) - C(x)}{h}$.

(b) Use the result of part (a) to determine $C'(40)$ and $C'(100)$ and interpret each using appropriate units.

77. **Marketing—Advertising Budget:** The number of dollars spent on advertising for a product influences the number of items of the product that will be purchased by customers. The number of Krypto-Cases sold for a new smartphone is a function of the amount of advertising spent and can be modeled by

$$N(x) = 2000 - \frac{520}{x} \qquad x > 0$$

where x represents the amount spent on advertising, measured in thousands of dollars, and $N(x)$ represents the number of cases purchased.

(a) Given $N'(x) = \dfrac{520}{x^2}$, evaluate $N(20)$ and $N'(20)$ and interpret each.

(b) Evaluate $N(36)$ and $N'(36)$, and interpret each.

(c) As the amount of dollars, x, spent on advertising increases, what happens to $N'(x)$, the rate of change in the items sold per thousands in advertising dollars spent? Is it increasing or decreasing?

(d) Graph N and N', and use the graphs to graphically verify your answer for part (c).

78. **Marketing—Advertising Budget:** The number of dollars spent on advertising for a product influences the number of items of the product that will be purchased by customers. The number of Find-it-Now GPS devices is a function of the amount of advertising spent and can be modeled by

$$N(x) = 3200 - \frac{750}{x} \qquad x > 0$$

where x represents the amount spent on advertising, measured in thousands of dollars, and $N(x)$ represents the number of devices purchased.

(a) Given that $N'(x) = \dfrac{750}{x^2}$, evaluate $N(15)$ and $N'(15)$ and interpret each.

(b) Evaluate $N(25)$ and $N'(25)$, and interpret each.

(c) As the amount in dollars, x, spent on advertising increases, what happens to $N'(x)$, the rate of change in the items sold per thousands in advertising dollars spent? Is it increasing or decreasing?

(d) Graph N and N' and use the graphs to graphically verify your answer for part (c).

79. **International Trade—Rice Exports:** The Southeast Asian country of Thailand exports more rice than any other country in the world. The rice exports from Thailand from 2000 to 2010 can be modeled by the linear function

$$f(x) = 0.16x + 7.26 \qquad 0 \le x \le 10$$

where x represents the number of years since 2000, and $f(x)$ represents the annual Thai rice export, measure in millions of metric tons. (*Source:* USDA Economic Research Service.)

(a) Use the definition of derivative to determine $f'(x)$.

(b) Evaluate and interpret $f'(5)$.

 80. International Trade—Rice Imports: According to a February 2010 report on the PBS documentary series *Frontline*, Iran used to be an exporter of rice, but now it is a net importer. The annual amount of rice imported to Iran from 2000 to 2010 can be modeled by

$$f(x) = 81.19x + 786.83 \qquad 0 \le x \le 10$$

where x represents the number of years since 2000, and $f(x)$ represents the annual Iranian rice imports, measures in thousands of metric tons. (*Source:* USDA Economic Research Service.)

(a) Use the definition of derivative to determine $f'(x)$.

(b) Evaluate and interpret $f'(8)$.

 81. Engineering—Nuclear Power Generation: Since the first commercial nuclear power plant was constructed in the United Kingdom in 1953, nuclear power has been a promising and sometimes controversial method to supply the public's electric needs. The annual amount of power generation from nuclear sources from 1990 to 2010 in the United States can be modeled by the function

$$f(x) = -0.55x^2 + 22.57x + 575.87 \qquad 0 \le x \le 20$$

where x represents the number of years since 1990, and $f(x)$ represents the amount of power generation from nuclear power plants, measured in billions of kilowatt hours (BkWh). (*Source:* U.S. Energy Information Administration.)

(a) Use the definition of derivative to determine $f'(x)$.

(b) Evaluate and interpret $f'(15)$.

 82. Engineering—Natural Gas Power Generation: Russia and the United States are the two leading countries in natural gas production, and natural gas accounts for 23.4% of the total U.S. electricity production. The annual amount of power generation from natural gas sources from 1990 to 2010 in the United States can be modeled by the function

$$f(x) = 0.67x^2 + 16.63x + 373.13 \qquad 0 \le x \le 20$$

where x represents the number of years since 1990, and $f(x)$ represents the amount of power generation from natural gas power plants, measured in billions of kilowatt hours (BkWh). (*Source:* U.S. Energy Information Administration.)

(a) Use the definition of derivative to determine $f'(x)$.

(b) Evaluate and interpret $f'(8)$.

Concept and Writing Exercises

83. Explain in a few sentences the meaning of the derivative of a function.

84. Consider a function f with a value a in the domain of f. Explain in a sentence the difference between the value of $f(a)$ and the value of $f'(a)$.

85. Consider the expression $\dfrac{(4 + h)^2 - (4)^2}{h}$. As $h \to 0$, the expression approaches the derivative of a function f at a value $x = c$. Determine the function f and the value of c.

86. Consider the expression $\dfrac{\sqrt{9 + h} - \sqrt{9}}{h}$. As $h \to 0$, the expression approaches the derivative of a function f at a value $x = c$. Determine the function f and the value of c.

In Exercises 87–89, a function f in polynomial form is given. Use the limit definition of the derivative to determine a general expression for the derivative f'.

87. $f(x) = ax + b$

88. $f(x) = ax^2 + bx + c$

89. $f(x) = ax^3 + bx^2 + cx + d$ (Hint: Note that $(x + h)^3 = x^3 + 3x^2h + 3xh^2 + h^3$)

In Exercises 90 and 91, a piecewise-defined function and an x-value are given. Complete the following.

(a) Evaluate the function at the given x-value.

(b) Use the limit definition of derivative to determine the derivative, if possible, at the indicated value.

90. $f(x) = \begin{cases} x^2, & x < 0 \\ x, & x \geq 0 \end{cases}; x = 0$ **91.** $f(x) = \begin{cases} 4 - x, & x \leq 1 \\ 3, & x > 1 \end{cases}; x = 1$

92. Consider the function $f(x) = \dfrac{1}{\sqrt{x - 3}}$. Answer the following questions about f.

     (a) State the domain of f.

     (b) Explain in a sentence why $f'(3)$ is not defined.

Section Project

Use **Figure 2.4.13** to answer the following questions.

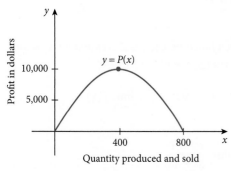

Figure 2.4.13

     (a) For what value of x is $P(x)$ a maximum?

     (b) For what value of x does $P'(x) = 0$?

     (c) What is the maximum profit?

     (d) Using interval notation, determine for what values of x is $P(x)$ increasing.

     (e) Using interval notation, determine for what values of x is $P'(x) > 0$.

     (f) Using interval notation, determine for what values of x is $P(x)$ decreasing.

     (g) Using interval notation, determine for what values of x is $P'(x) < 0$.

     (h) In your own words, describe the connection between the results in parts (a) and (b).

     (i) In your own words, describe the connection between the results in parts (d) and (e).

     (j) In your own words, describe the connection between the results in parts (f) and (g).

Section 2.1 Review Exercises

1. For $f(x) = \dfrac{1 - \sqrt{1 - 2x - x^2}}{x}$, complete the table to numerically estimate the following:

x	-0.1	-0.01	-0.001	0	0.001	0.01	0.1
$f(x)$				?			

   (a) $\lim\limits_{x \to 0^-} f(x)$ (b) $\lim\limits_{x \to 0^+} f(x)$ (c) $\lim\limits_{x \to 0} f(x)$

2. For $f(x) = (\sqrt{4 - x})^5$, complete the table to numerically estimate the following.

x	3	3.9	3.99	4	4.01	4.1	5
$f(x)$				?			

   (a) $\lim\limits_{x \to 4^-} f(x)$ (b) $\lim\limits_{x \to 4^+} f(x)$ (c) $\lim\limits_{x \to 4} f(x)$

For Exercises 3 and 4, use your calculator to graph the given function. Use the Zoom and Trace commands to graphically estimate the indicated limits. Verify your estimate numerically.

  3. $f(x) = x^3 - 2x$; $\lim\limits_{x \to 3.1} f(x)$

4. $f(x) = \sqrt{x} - \dfrac{1}{|x|}$; $\lim\limits_{x \to 2.5} f(x)$

For Exercises 5–10, determine the limits algebraically.

5. $\lim\limits_{x \to 2}(7x^3 - 10x)$

6. $\lim\limits_{x \to -1} \dfrac{x^2 - 1}{x + 1}$

7. $\lim\limits_{x \to -3} |x - 5|$

8. $\lim\limits_{x \to 0} \sqrt{36 - 8x}$

9. $\lim\limits_{x \to 10} \dfrac{x + 10}{x^2 - 100}$

10. $\lim\limits_{x \to 2.2} (x + 9)^3$

11. Use the graph of the function f to determine the following limits.

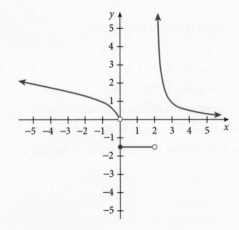

   (a) $\lim\limits_{x \to -4} f(x)$ (b) $\lim\limits_{x \to 0^-} f(x)$

   (c) $f(0)$ (d) $\lim\limits_{x \to 2} f(x)$

   (e) $\lim\limits_{x \to 3} f(x)$ (f) $f(3)$

For Exercises 12–14, determine the indicated limit.

12. $\lim\limits_{h \to 0} (x + 2h)^2$

13. $\lim\limits_{h \to 0} \dfrac{2x^2 h - 9h}{h}$

14. $\lim\limits_{h \to 0} \dfrac{6x^3 h^2 + h}{h}$

15. For $f(x) = 3x^2$, evaluate the following.

(a) $f(2 + h)$

(b) $\lim\limits_{h \to 0} \dfrac{f(2 + h) - f(2)}{h}$

16. For $f(x) = \dfrac{9}{x}$, evaluate the following.

(a) $f(4 + h)$

(b) $\lim\limits_{h \to 0} \dfrac{f(4 + h) - f(4)}{h}$

17. Business—Production Cost: Comfy Stuffed Animals, Inc. is introducing a new line of teddy bears. The total cost of producing Scare Bear (with glowing eyes) is projected to be

$$C(x) = 36{,}000 + \sqrt{10{,}000x}$$

where x represents the number of units made, and $C(x)$ represents the total cost, measured in dollars.

(a) Find and interpret $C(400)$.

(b) Find and interpret $\lim\limits_{x \to 100} C(x)$.

(c) Find and interpret $AC(x) = \dfrac{C(x)}{x}$ at a production level of 25 Scare Bears.

18. Physics—Distance: The Wave Crasher radio controlled speedboat moves away from a dock, and its distance from the dock is given by the piecewise-defined function

$$f(x) = \begin{cases} 2x^2 & 0 \le x \le 3 \\ 12x - 18 & x > 3 \end{cases}$$

where x represents the number of seconds since the boat was launched, and $f(x)$ represents the boat's distance from the dock in feet.

(a) Graph f on the closed interval $[0, 6]$.

(b) Evaluate $\lim\limits_{x \to 3^-} f(x)$.

(c) Evaluate $\lim\limits_{x \to 3^+} f(x)$.

(d) Determine $\lim\limits_{x \to 3} f(x)$.

(e) Use a complete sentence to describe the behavior of the boat during the first six seconds after its launch.

19. Baking—Sugar Content: A pastry chef in a commercial test kitchen is fine-tuning a pie-filling recipe. Colleagues acting as tasters have rated various recipes on a scale of 1 to 10. The average rating $f(x)$ can be modeled as a function of the sugar content x by

$$f(x) = 8 - \dfrac{x^2 - 48x + 512}{50}$$

where x is the number of tablespoons of sugar in the recipe, and $f(x)$ represents the taster rating.

(a) Graph f using the viewing window $[0, 40]$ by $[0, 10]$.

(b) Evaluate $f(20)$, $f(25)$, and $f(30)$, and interpret each.

(c) Evaluate and interpret $\lim\limits_{h \to 0} \dfrac{f(24 + h) - f(24)}{h}$.

(d) From the results (a), (b), and (c), how much sugar seems to be optimal?

20. Heating—Temperature Control: A furnace switches on in such a way that the temperature of the heating element is a function of time, as shown by the graph.

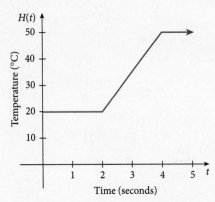

(a) Based on the graph, write a piecewise-defined function $H(t)$.

(b) Evaluate $\lim\limits_{h \to 0} \dfrac{H(5 + h) - H(5)}{h}$, if it exists.

(c) Evaluate $\lim\limits_{h \to 0} \dfrac{H(2 + h) - H(2)}{h}$, if it exists.

(d) Considering the context of the function, explain why the middle segment is not vertical.

Section 2.2 Review Exercises

21. For the function $f(x) = \dfrac{2}{(x + 3)^2}$, complete the table to numerically estimate the following limits. Use the symbols $-\infty$ and ∞ as appropriate.

x	-3.1	-3.01	-3.001	-3	-2.999	-2.99	-2.9
$f(x)$							

(a) $\lim\limits_{x \to -3^-} f(x)$ (b) $\lim\limits_{x \to -3^+} f(x)$ (c) $\lim\limits_{x \to -3} f(x)$

22. For $f(x) = \dfrac{x + 1}{x - 2}$, complete the table to numerically estimate the following limits. Use the symbols $-\infty$ and ∞ as appropriate.

x	1.9	1.99	1.999	2	2.001	2.01	2.1
$f(x)$							

(a) $\lim\limits_{x \to 2^-} f(x)$ (b) $\lim\limits_{x \to 2^+} f(x)$ (c) $\lim\limits_{x \to 2} f(x)$

For the function in Exercises 23 and 24, complete the following.

(a) Write the domain of f using interval notation.

(b) Evaluate the indicated limit.

23. $f(x) = \dfrac{1250 + 3.2x}{x}$; $\lim\limits_{x \to 0} f(x)$ 24. $f(x) = \dfrac{x + 5}{x^2 - 25}$; $\lim\limits_{x \to -5} f(x)$

For Exercises 25–28, determine the indicated limits algebraically. Then verify your answer numerically.

25. $\lim\limits_{x \to -\infty} \dfrac{x + 2}{2x - 3}$ 26. $\lim\limits_{x \to \infty} \dfrac{-2x^2 + 5x - 1}{x^2 - 13}$

27. $\lim\limits_{x \to -\infty} \dfrac{-4x^2 - 3x + 11}{8x^3 - 5}$ 28. $\lim\limits_{x \to -\infty} \dfrac{3x^4 - 27}{x + 3}$

For Exercises 29 and 30, write the equation for the horizontal asymptote of the given function in the form y = b, where b is a real number.

29. $f(x) = \dfrac{2x^2 - 9x + 9}{6x^2 + x + 11}$

30. $f(x) = \dfrac{7x + 9}{3x^2 + 2}$

For Exercises 31–33, use the graph of the function to determine if f is continuous at the given value of x. If f is not continuous, explain why it is not.

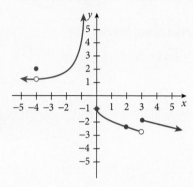

31. $x = -4$

32. $x = 0$

33. $x = 2$

34. Use the function $g(x) = \dfrac{x^2 - 9}{x + 3}$ to answer the following.

(a) Is g continuous at $x = 9$? Explain your answer.

(b) Is g continuous at $x = -3$? Explain your answer.

35. Use the function $h(x) = \begin{cases} x - 8, & x \le 2 \\ \dfrac{6}{x - 3}, & x > 2 \end{cases}$ to answer the following.

(a) Is h continuous at $x = 2$? Explain your answer.

(b) Is h continuous at $x = 3$? Explain your answer.

36. Operations Research—Reliability Rates: The annual cost of getting the reliability rate of the Overdrive automated production quality control system up to $x\%$ is given by

$$C(x) = \begin{cases} 0, & 0 \le x \le 80 \\ \dfrac{400 - 5x}{x - 100}, & 80 < x < 100 \end{cases}$$

where x represents the reliability rate as a percentage, and $C(x)$ represents the cost, measured in thousands of dollars. Evaluate the following limits and interpret the result.

(a) $\displaystyle\lim_{x \to 80} C(x)$

(b) $\displaystyle\lim_{x \to 100^-} C(x)$

37. Metaphysics—Healing Auras: The Wingmakers Company creates lithographs depicting the healing auras surrounding crystals. Producing q copies of a lithograph costs $C(x) = 238 + 5x$ dollars. Knowing that the average cost of the lithographs is given by the average cost function $AC(x) = \dfrac{C(x)}{x}$, answer the following.

(a) Evaluate and interpret $\displaystyle\lim_{x \to 40} AC(x)$.

(b) Evaluate and interpret $\displaystyle\lim_{x \to \infty} AC(x)$.

38. Farming—Crop Profits: The Duncan Peach Farm has started a new quality control program to improve the quality of their product over several years. Their records show that the annual profit follows the function

$$P(t) = \dfrac{25(t + 4)}{t + 5}$$

where t represents the time since the program took effect, and $P(t)$ represents the profit, measured in thousands of dollars.

(a) Determine the annual profits three years after the quality control program started.

(b) Evaluate and interpret $\lim\limits_{t \to \infty} P(t)$.

Section 2.3 Review Exercises

In Exercises 39–41, compute the average rate of change of f on the given closed interval.

39. $f(x) = x^3 - 6x + 1;\ [3, 7]$ **40.** $f(x) = 2\sqrt{4x};\ [4, 9]$

41. $f(x) = \dfrac{3}{x - 2};\ [6.2, 7.9]$

42. Consider the function given by the function f.

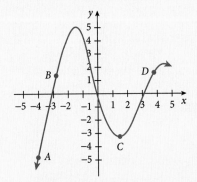

(a) Between which pairs of points is the average rate of change of f positive?

(b) Between which pairs of points is the average rate of change of f negative?

(c) Between which pairs of points is the average rate of change of f zero?

(d) At which points of f is the tangent line slope positive?

(e) At which points of f is the tangent line slope negative?

(f) At which points of f is the tangent line slope zero?

For Exercises 43 and 44, estimate the instantaneous rate of change of the function at the indicated value by using $\lim\limits_{h \to 0} \dfrac{f(x + h) - f(x)}{h}$ and completing the following tables.

h	-1	-0.1	-0.01	-0.001	-0.0001
$\dfrac{f(x+h) - f(x)}{h}$					

(table header: $h \to 0^-$)

h	1	0.1	0.01	0.001	0.0001
$\dfrac{f(x+h) - f(x)}{h}$					

(table header: $h \to 0^+$)

43. $f(x) = 2.5x;\ x = 4$ **44.** $f(x) = 3x^3 + 3x - 25;\ x = -2$

45. Find the slope of the tangent line to the graph of $f(x) = 3\sqrt{x} + 2$ at $x = 4$ by estimating $\lim\limits_{h \to 0} f(x)$ using successively smaller values of h.

Section 2.4 Review Exercises

In Exercises 46–48, compute the derivative of f at x = c by using the definition of derivative
$$f'(c) = \lim_{h \to \infty 0} \frac{f(c + h) - f(c)}{h}.$$

46. $f(x) = x^2 - 2x; c = 7$

47. $f(x) = x^3 + 1; c = 4$

48. $f(x) = \dfrac{1}{x^2}; c = 3$

49. For the function $f(x) = -3x^2 - 2x$, complete the following.

   (a) Find an equation for the line tangent to the graph of the function at $x = -1$.

   (b) Graph the function and tangent line in the same coordinate system.

In Exercises 50–53, determine f′ by using the definition of the derivative $f'(x) = \lim\limits_{h \to 0} \dfrac{f(x + h) - f(x)}{h}$.

50. $f(x) = 3x - 7$

51. $f(x) = x^4$

52. $f(x) = 3.4x^2 + 1.9x$

53. $f(x) = \dfrac{1}{x^2}$

54. Find the slope of the line tangent to the graph of $f(x) = 6x$ at $x = 4$ by first finding a general formula for the derivative.

55. Find the slope of the line tangent to the graph of $f(x) = 3x^2$ at $x = 9$ by first finding a general formula for the derivative.

56. Find the slope of the line tangent to the graph of $f(x) = \dfrac{1}{x}$ at $x = 7$ by first finding a general formula for the derivative.

57. Use the graph of the profit function P to answer the following.

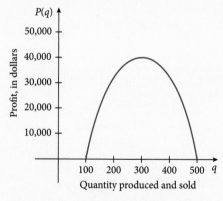

   (a) For what value of q is $P(q)$ a maximum?

   (b) For what value of q does $P'(q) = 0$?

   (c) What is the maximum profit?

In Exercises 58–61, determine if f is continuous at the given point. Then determine if the function f is differentiable at that point.

58. $f(x) = \begin{cases} \dfrac{x^2 - 9}{x - 3}, & x \neq 3 \\ 5, & x = 3 \end{cases}; \quad x = 3$

59. $f(x) = \begin{cases} \dfrac{x}{2x^2 - x}, & x \neq 0 \\ 0, & x = 0 \end{cases}; \quad x = 0$

60. $f(x) = |x - 5|; \quad x = 5$

61. $f(x) = |x - 1|; \quad x = 0$

Differentiation Techniques, the Differential, and Marginal Analysis

© Comstock/Thinkstock **(a)**

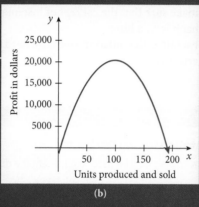

(b)

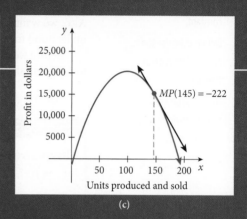

(c)

One of the factors that must be taken into consideration by business managers when determining whether or not to increase production of an item is that of marginal profit. While the goal for any successful business manager is for profit to increase, there is a point at which profit is maximized and the production of a single additional unit will result in a loss of gain in profits. Consider the case of a refrigerator manufacturer. The graph of the profit function in Figure (c) illustrates the profit lost by producing and selling the 146th refrigerator.

What We Know

In Chapter 2, we found that the instantaneous rate of change gave us the slope of the line tangent to the graph of a function. The way we found this new rate of change was through the derivative.

Where Do We Go

In this chapter, we will learn rules to allow us to compute derivatives more efficiently. We will also use the derivative to study a new quantity called the differential. We will use the differential to approximate solutions and to study some functions of business.

3.1 Derivatives of Constants, Powers, and Sums

In Section 2.4 we defined the derivative of a function and used it to find slopes of tangent lines as well as instantaneous rates of change. However, computing the derivative of a function from the definition can be quite involved. Calculus would not be very useful if all derivatives had to be calculated from the definition.

In the first three sections of this chapter, we present several **rules of differentiation**, or shortcuts if you like, that will greatly simplify differentiation. So why did we painstakingly compute derivatives via the definition if these shortcuts exist? Quite simply, these rules are *derived* from the definition! Also, we want to make sure that the *concept* of instantaneous rate of change/tangent line slope was understood for what it is … a limit.

Before we delve into differentiation rules, we need to present alternative ways that can be used to represent the derivative.

Alternative Notation for the Derivative

For $y = f(x)$, the following may be used to represent the derivative:

$$f'(x), y', \frac{dy}{dx}, \frac{d}{dx}[f(x)]$$

Each notation has its own advantage, and we will use each of these where appropriate.

Derivative of a Constant Function

Rule 1: Constant Function Rule

For any constant k, if $f(x) = k$, then $f'(x) = 0$.

NOTE: Rule 1 can be summarized by saying that the derivative of a constant is zero.

The first rule that we present shows how to differentiate a constant function. Geometrically, this rule is fairly obvious. The graph of a constant function, $f(x) = k$, is a horizontal line like the one shown in **Figure 3.1.1**. We know that the slope of a horizontal line is 0. This leads us to believe that the instantaneous rate of change at any point on the graph of a horizontal line is 0 or, in other words, for $f(x) = k$ in Figure 3.1.1, we believe that $f'(x) = 0$.

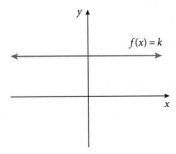

Figure 3.1.1

To show that the **Constant Function Rule** is true, we simply apply the definition of the derivative as follows. If $f(x) = k$ then we know that $f(x + h) = k$ as well. From the definition of the derivative we have

$$f'(x) = \lim_{h \to 0} \frac{f(x + h) - f(x)}{h} = \lim_{h \to 0} \frac{k - k}{h} = \lim_{h \to 0} \frac{0}{h} = \lim_{h \to 0} 0 = 0$$

Example 1: Differentiating Constant Functions

Differentiate the following.

a. $f(x) = 7$ b. $g(x) = \sqrt[3]{2}$

Perform the Mathematics

a. Since 7 is a constant, we have $f'(x) = 0$.

b. Since $\sqrt[3]{2}$ is a constant, we have $g'(x) = 0$. ∎

Power Rule

Rule 2: Power Rule

If $f(x) = x^n$, where n is any real number, then $f'(x) = nx^{n-1}$.

We now present one of the most useful differentiation rules in calculus. It is used to differentiate functions of the form $f(x) = x^n$, where n is any real number. Functions of this form are collectively called **power functions**. The definition of the derivative can be used to determine derivatives for power functions, although the algebra may become cumbersome quickly. For example, the definition of the derivative can be used to derive the following:

$$\text{For } f(x) = x^3, \text{ we have } f'(x) = 3x^2.$$
$$\text{For } f(x) = x^6, \text{ we have } f'(x) = 6x^5.$$
$$\text{For } f(x) = x^7, \text{ we have } f'(x) = 7x^6.$$

NOTES:

1. Rule 2 is equivalent to writing $\frac{d}{dx}[x^n] = nx^{n-1}$.

2. Rule 2 can be summarized by saying that to differentiate x^n, simply bring the exponent out front as a coefficient and then decrease the exponent by 1.

There appears to be a pattern with the derivatives of power functions. It appears that the power of $f(x)$ becomes the coefficient of $f'(x)$, and the power of $f'(x)$ is 1 less than the power of $f(x)$. This observation leads us to the **Power Rule**. The proof of the Power Rule is in Appendix C.

Example 2: Differentiating Power Functions

Differentiate the following functions.

a. $f(x) = x^4$ b. $g(x) = x^{1.32}$ c. $y = \sqrt{x}$ d. $g(x) = \dfrac{1}{x^3}$

Perform the Mathematics

a. Thanks to the Power Rule, we quickly compute the derivative to be

$$f'(x) = 4x^{4-1} = 4x^3$$

b. Again, using the Power Rule, we have

$$g'(x) = 1.32x^{1.32-1} = 1.32x^{0.32}$$

c. Here we need to use a little algebra before differentiating. We need to recall that $\sqrt[n]{x} = x^{1/n}$. This allows us to rewrite $y = \sqrt{x}$ as $y = x^{1/2}$. Now we can apply the Power Rule to get

$$y' = \frac{1}{2}x^{1/2-1} = \frac{1}{2}x^{-1/2}$$

Utilizing the algebraic fact that $x^{-n} = \dfrac{1}{x^n}$ allows us to write the simplified version of the derivative as

$$y' = \frac{1}{2x^{1/2}} = \frac{1}{2\sqrt{x}}$$

d. We begin by using a little algebra to rewrite the function. That is, by utilizing the algebraic fact that $x^{-n} = \dfrac{1}{x^n}$, the function can be rewritten as

$$g(x) = \frac{1}{x^3} = x^{-3}$$

Now we can apply the Power Rule and get

$$g'(x) = -3x^{-3-1} = -3x^{-4} = -\frac{3}{x^4}$$ ∎

Rule 3: Constant Multiple Rule

If $f(x) = k \cdot g(x)$, where k is any real number, then $f'(x) = k \cdot g'(x)$, assuming that g is differentiable.

NOTES:

1. Rule 3 is equivalent to writing
$$\frac{d}{dx}[k \cdot g(x)] = k \cdot \frac{d}{dx}[g(x)].$$

2. In words, the Constant Multiple Rule states that the derivative of a constant times a function is simply the constant times the derivative of the function.

OBJECTIVE 3

Differentiate a function that is multiplied by a constant.

▶ *Try It Yourself*

Some related Exercises are 13 and 14.

Constant Multiple Rule

The Constant Multiple Rule extends the Power Rule to differentiating functions that are of the form $k \cdot g(x)$, a constant times a function. The definition of the derivative can be used to derive the following:

For $f(x) = 2x^3$, we have $f'(x) = 6x^2$.

For $f(x) = 3x^5$, we have $f'(x) = 15x^4$.

For $f(x) = -2x^4$, we have $f'(x) = -8x^3$.

Again, there appears to be a pattern in these derivatives. It appears that the power of $f(x)$ gets multiplied by the coefficient of $f(x)$, and this product becomes the coefficient of $f'(x)$, while the power of $f'(x)$ is 1 less than the power of $f(x)$. This observation leads to the **Constant Multiple Rule**.

The proof of this rule is in Appendix C.

Example 3: Differentiating a Constant Times a Function

Differentiate the following.

a. $g(x) = 1.2x^5$ **b.** $y = \dfrac{1}{7x^3}$ **c.** $f(x) = \dfrac{2}{3}\sqrt[5]{x}$

Perform the Mathematics

a. Applying the Constant Multiple Rule yields

$$g'(x) = 1.2 \cdot (5x^{5-1}) = 6x^4$$

b. Rewriting the function via algebra gives

$$y = \frac{1}{7}x^{-3}$$

Now we apply the Constant Multiple Rule to get

$$y' = \frac{1}{7}(-3x^{-3-1}) = \frac{1}{7}(-3x^{-4})$$

$$= -\frac{3}{7}x^{-4} = -\frac{3}{7x^4}$$

c. We begin by rewriting the function as

$$f(x) = \frac{2}{3}x^{1/5}$$

Now we apply the Constant Multiple Rule, which gives

$$f'(x) = \frac{2}{3}\left(\frac{1}{5}x^{1/5-1}\right) = \frac{2}{15}x^{-4/5}$$

Simplifying the derivative yields

$$f'(x) = \frac{2}{15x^{4/5}} = \frac{2}{15\sqrt[5]{x^4}} \qquad \blacksquare$$

▶ *Try It Yourself*

Some related Exercises are 19 and 23.

Sum and Difference Rule

Consider two functions $f(x) = 2x^2$ and $g(x) = 3x + 1$. We can now define a new function, S, as follows.

$$S(x) = f(x) + g(x) = 2x^2 + 3x + 1$$

From the differentiation rules developed so far in this section, we know that

$$f'(x) = 4x \text{ and } g'(x) = 3$$

However, we have no rule to help us determine $S'(x)$. We can appeal to the definition of the derivative to help us compute $S'(x)$:

$$S'(x) = \lim_{h \to 0} \frac{S(x+h) - S(x)}{h}$$

$$= \lim_{h \to 0} \frac{[2(x+h)^2 + 3(x+h) + 1] - [2x^2 + 3x + 1]}{h}$$

$$= \lim_{h \to 0} \frac{[2(x^2 + 2xh + h^2) + 3x + 3h + 1] - 2x^2 - 3x - 1}{h}$$

$$= \lim_{h \to 0} \frac{2x^2 + 4xh + 2h^2 + 3x + 3h + 1 - 2x^2 - 3x - 1}{h}$$

$$= \lim_{h \to 0} \frac{4xh + 2h^2 + 3h}{h}$$

$$= \lim_{h \to 0} \frac{h(4x + 2h + 3)}{h}$$

$$= \lim_{h \to 0} (4x + 2h + 3)$$

$$= 4x + 3$$

So we have determined that $S'(x) = 4x + 3$. Recall that we defined $S(x)$ as

$$S(x) = f(x) + g(x) = 2x^2 + 3x + 1$$

Also recall that we knew that $f'(x) = 4x$ and $g'(x) = 3$. It appears that we have discovered that

$$S'(x) = 4x + 3 = f'(x) + g'(x)$$

This is not a coincidence. This illustrates a rule that handles functions that are sums or differences of other functions and is called the **Sum and Difference Rule.**

Rule 4: Sum and Difference Rule

If $h(x) = f(x) \pm g(x)$, where f and g are both differentiable functions, then $h'(x) = f'(x) \pm g'(x)$.

NOTES:

1. Rule 4 is equivalent to writing $\frac{d}{dx}[f(x) \pm g(x)] = \frac{d}{dx}[f(x)] \pm \frac{d}{dx}[g(x)]$.

2. In words, the Sum and Difference rule states that to differentiate a sum/difference of two (or more) functions, just differentiate the functions separately and add/subtract the results.

OBJECTIVE 4

Differentiate the sum and/
or difference of functions.

Example 4: Differentiating Sums and Differences

Differentiate the following functions.

a. $f(x) = 3x^2 + 2x - 1$ **b.** $g(x) = \frac{1}{2}x^3 - \frac{3}{2}x^{-1}$ **c.** $y = 3x^4 + 2\sqrt{x} - \frac{2}{x^2}$

Perform the Mathematics

a. According to the Sum and Difference Rule,

$$f'(x) = \frac{d}{dx}[3x^2 + 2x - 1] = \frac{d}{dx}[3x^2] + \frac{d}{dx}[2x] - \frac{d}{dx}[1]$$

$$= 6x + 2 - 0 = 6x + 2$$

b. Applying the Sum and Difference Rule yields

$$g'(x) = \frac{d}{dx}\left[\frac{1}{2}x^3 - \frac{3}{2}x^{-1}\right] = \frac{d}{dx}\left[\frac{1}{2}x^3\right] - \frac{d}{dx}\left[\frac{3}{2}x^{-1}\right]$$

$$= \frac{3}{2}x^2 - \left(-\frac{3}{2}x^{-2}\right) = \frac{3}{2}x^2 + \frac{3}{2}x^{-2}$$

c. First we rewrite the function as

$$y = 3x^4 + 2x^{1/2} - 2x^{-2}$$

Now we differentiate to get

$$\frac{dy}{dx} = \frac{d}{dx}[3x^4 + 2x^{1/2} - 2x^{-2}]$$

$$= \frac{d}{dx}[3x^4] + \frac{d}{dx}[2x^{1/2}] - \frac{d}{dx}[2x^{-2}]$$

$$= 12x^3 + x^{-1/2} + 4x^{-3} \qquad \blacksquare$$

▶ *Try It Yourself*
Some related Exercises are 31 and 43.

Notice in Example 4 that we not only used the Sum and Difference Rule, but we also used the Constant Function Rule, the Power Rule, and the Constant Multiple Rule. In short, we used all the rules presented in this section.

Applications

Example 5: Analyzing Greenhouse Gas Emissions

The majority of scientists support the notion that the release of greenhouse gasses since the beginning of the industrial age has had an effect on the climate and will continue to have an effect on the climate. Carbon dioxide is one such gas that is attributed to warming due to the greenhouse effect. The amount of carbon dioxide emissions by the U.S. can be modeled by

$$f(x) = 0.61x^3 - 22.57x^2 + 276.36x + 4767.7 \qquad 1 \le x \le 19$$

where x represents the number of years since 1989 and $f(x)$ represents the total carbon dioxide emissions in millions of metric tons. (*Source:* U.S. Census Bureau.) Evaluate $f'(18)$ and interpret.

Understand the Situation

We can use the differentiation techniques learned in this section to quickly determine the derivative, $f'(x)$. To evaluate $f'(18)$ we will simply substitute $x = 18$ into the derivative, $f'(x)$, for each occurrence of x. Note that $x = 18$ corresponds to the year 2007.

© LiquidLibrary

Perform the Mathematics

Using the techniques learned in this section we get

$$f'(x) = 0.61(3x^2) - 22.57(2x) + 276.36(1) + 0$$
$$f'(x) = 1.83x^2 - 45.14x + 276.36$$

To determine $f'(18)$, we substitute $x = 18$ into the derivative. This gives us

$$f'(18) = 1.83(18)^2 - 45.14(18) + 276.36$$
$$= 56.76$$

> **OBJECTIVE 5**
>
> Apply the differentiation rules to environmental science, physics, and business.

Interpret the Result

This means that in 2007, U.S. emissions of carbon dioxide into the atmosphere was increasing at a rate of 56.76 million metric tons per year. See **Figure 3.1.2**.

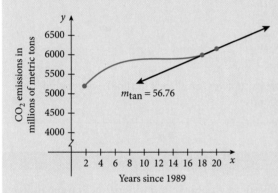

Figure 3.1.2

▷ *Try It Yourself*

Some related Exercises are 81 and 83. ■

The next example deals with what is known as a **position function** and a **velocity function**. These functions are commonly used in physics.

Example 6: Analyzing Velocity

If a coconut falls from a tree that is 75 feet tall, its height above the ground after t seconds is given by

$$s(t) = 75 - 16t^2$$

where $s(t)$ is measured in feet. This function is called a **position function** since it gives the position of the coconut above the ground as a function of time.

a. The derivative of a position function, $s'(t)$, is called the **velocity function** and is denoted by $v(t)$. Determine $v(t)$.

b. Compute $s(2)$ and $v(2)$, and interpret each.

c. When does the coconut hit the ground?

Perform the Mathematics

a. Since we have been told that $s'(t) = v(t)$, we determine $v(t)$ as follows:

$$v(t) = s'(t) = -32t$$

b. Evaluating $s(2)$ and $v(2)$, we get

$$s(2) = 75 - 16(2)^2 = 11 \text{ feet}$$
$$v(2) = -32(2) = -64 \text{ feet/sec}$$

This means that after 2 seconds, the coconut is 11 feet above the ground and is falling at a rate of 64 feet per second. Notice that the negative value of $v(2)$ indicates that the coconut is falling toward the ground.

c. The coconut hits the ground when it is 0 feet above the ground. Hence, we need to solve $s(t) = 0$. Setting $s(t) = 0$ and solving yields

$$75 - 16t^2 = 0$$
$$75 = 16t^2$$
$$\frac{75}{16} = t^2$$
$$\pm\sqrt{\frac{75}{16}} = t$$
$$t \approx \pm 2.17$$

Since -2.17 seconds does not make sense (we cannot yet travel back in time), we conclude that the coconut hits the ground in about 2.17 seconds. ∎

Technology Option

We can graphically support the result of Example 6c by graphing the position function and utilizing the `Zero` command. See **Figure 3.1.3**.

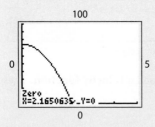

Figure 3.1.3 Result of `Zero` command.

Our final example of this section looks at how we can apply the derivative to a profit function.

Example 7: Analyzing a Profit Function

The Cool Air Company has determined that the total cost function for the production of its refrigerators can be described by

$$C(x) = 2x^2 + 15x + 1500$$

where x is the weekly production of refrigerators and $C(x)$ is the total cost in dollars. The revenue function for these refrigerators is given by

$$R(x) = -0.3x^2 + 460x$$

where x is the number of refrigerators sold and $R(x)$ is in dollars. Determine $P(20)$ and $P'(20)$ and interpret each.

Understand the Situation

We begin by determining the profit function, P, and evaluating the profit function when $x = 20$. Once we have the profit function, P, we can determine the derivative, P', and then evaluate it at $x = 20$.

Perform the Mathematics

In Chapter 1 we learned that profit function is the revenue function minus the cost function. This gives us

$$\begin{aligned} P(x) &= R(x) - C(x) \\ &= (-0.3x^2 + 460x) - (2x^2 + 15x + 1500) \\ &= -2.3x^2 + 445x - 1500 \end{aligned}$$

Evaluating the profit function when $x = 20$ gives

$$\begin{aligned} P(20) &= -2.3(20)^2 + 445(20) - 1500 \\ &= 6480 \end{aligned}$$

Since $P(x) = -2.3x^2 + 445x - 1500$, the derivative of P is

$$P'(x) = \frac{d}{dx}[-2.3x^2] + \frac{d}{dx}[445x] - \frac{d}{dx}[1500] = -4.6x + 445$$

Evaluating $P'(20)$ yields

$$\begin{aligned} P'(20) &= -4.6(20) + 445 \\ &= 353 \end{aligned}$$

Interpret the Results

This means that when the Cool Air Company makes and sells 20 refrigerators, a profit of \$6480 is realized and the profit is increasing at a rate of \$353 per refrigerator.

> ▷ **Try It Yourself**

Some related Exercises are 73 and 74.   ■

Summary

In this section we introduced four differentiation rules to allow us to compute derivatives more quickly:

Name	Rule	Example
Constant Function Rule	If $f(x) = k$, then $f'(x) = 0$.	If $f(x) = 7$, then $f'(x) = 0$.
Power Rule	If $f(x) = x^n$, then $f'(x) = nx^{n-1}$.	If $f(x) = x^8$, then $f'(x) = 8x^7$.
Constant Multiple Rule	If $f(x) = k \cdot g(x)$, then $f'(x) = k \cdot g'(x)$.	If $f(x) = 3x^4$, then $f'(x) = 12x^3$.
Sum and Difference Rule	If $h(x) = f(x) \pm g(x)$, then $h'(x) = f'(x) \pm g'(x)$.	If $h(x) = x^3 + x^2$, then $h'(x) = 3x^2 + 2x$.

Even though we now have some tools that allow us to calculate a derivative more quickly than by using the definition of the derivative, do not lose sight of what the derivative tells us: It is a limit that gives an instantaneous rate of change as well as the slope of a tangent line.

Section 3.1 Exercises

Vocabulary Exercises

1. The process of finding the derivative of a function is known as _____.

2. For a derivative f', the symbol above the function notation is read _____.

3. The derivative of a constant is _____.

4. To differentiate a power term x^n, we bring the exponent n out front as a _____ and then subtract _____ from the exponent.

5. The derivative of a constant times a function is the constant times the _____ of the function.

6. The rule that essentially states that we can differentiate functions term by term is the _____ rule.

7. A function used in physics that has the form $s(t) = -16t^2 + v_0 t + h_0$ is known as a _____ function.

8. A function used in physics that has the form $v(t) = -32t + v_0$ is known as a _____ function.

Skill Exercises

In Exercises 9–24, determine the derivative for the given single-term function. When appropriate, simplify the derivative so that there are no negative or fractional exponents. A few helpful rules from algebra are:

(i) $x^{-n} = \dfrac{1}{x^n}$ (ii) $x^{m/n} = \sqrt[n]{x^m}$

9. $f(x) = 3$ 10. $f(x) = 5$

11. $f(x) = -2$ 12. $f(x) = -7$

13. $f(x) = x^6$ 14. $f(x) = x^{10}$

15. $f(x) = -3x^4$ 16. $f(x) = -5x^7$

17. $f(x) = 2x^{2/3}$ 18. $f(x) = 3x^{4/5}$

19. $f(x) = -3x^{-1/3}$ 20. $f(x) = -2x^{-2/5}$

21. $f(x) = \dfrac{2}{3}x^4$ 22. $f(x) = \dfrac{3}{4}x^8$

23. $f(x) = -\dfrac{2}{5}x^{5/3}$ 24. $f(x) = -\dfrac{5}{6}x^{6/5}$

In Exercises 25–56, determine the derivative for the given function. When appropriate, simplify the derivative so that there are no negative or fractional exponents.

25. $f(x) = 2x^3 + 4x^2 - 7x + 1$ 26. $f(x) = 4x^3 - 3x^2 + 5x - 3$

27. $f(x) = 3x^2 - 2x + 6$ 28. $f(x) = 3x^2 + 5x - 2$

29. $f(x) = -5x^2 - 6x + 2$ 30. $f(x) = -3x^2 + 9x - 1$

31. $f(x) = -5x^3 + 7x - 5$ 32. $f(x) = -8x^3 - 8x + 3$

33. $f(x) = \dfrac{1}{2}x^3 + \dfrac{3}{5}x^2 - \dfrac{2}{3}x + \dfrac{2}{5}$ 34. $f(x) = \dfrac{2}{5}x^3 - \dfrac{2}{3}x^2 + \dfrac{1}{2}x - 5$

35. $f(x) = 1.31x^2 + 2.05x - 3.9$

36. $f(x) = 3.15x^2 - 1.13x + 5.2$

37. $f(x) = -0.2x^2 + 3.5x^3 - 0.4x^4$

38. $f(x) = 0.3x^2 - 0.67x^3 + 0.8x^4$

39. $f(x) = 1.15x^3 - 2.3x^2 + 2.53x - 7.1$

40. $f(x) = 2.35x^3 + 3.56x^2 - 63.25x + 365.3$

41. $f(x) = 3\sqrt{x} + \dfrac{1}{2}x - 5x^2$

42. $f(x) = 5\sqrt{x} - \dfrac{1}{2}x + 7x^2$

43. $f(x) = \sqrt[3]{x} + x^2 - 3x^3$

44. $f(x) = \sqrt[3]{x} - 2x^3 + 5x^4$

45. $f(x) = \sqrt[3]{x^2} - \dfrac{4}{\sqrt{x}}$

46. $f(x) = \sqrt[3]{x^2} + \dfrac{3}{\sqrt{x}}$

47. $f(x) = 2.35x^{1.35}$

48. $f(x) = 3.2x^{1.14}$

49. $f(x) = 2000 + \dfrac{5}{x^2}$

50. $f(x) = 3300 + \dfrac{13}{x^2}$

51. $f(x) = 2x^2 + \dfrac{1}{x}$

52. $f(x) = -3x^2 - \dfrac{1}{x}$

53. $f(x) = 3x^{-3/2} - 4x^{-1/2} + 5$

54. $f(x) = 4x^{-3/2} - 2x^{-1/2} + x^{-1/2}$

55. $f(x) = 2.35x^{-1/2} - 2.3x^{-2/3}$

56. $f(x) = 3.52x^{-2/5} + 3.2x^{-1/2}$

For Exercises 57–64, an algebraic function in the form $f(x) = \dfrac{g(x)}{h(x)}$ is given. Complete the following:

(a) Write the domain of the f using interval notation.

(b) Use the addition rule for fractions, $\dfrac{a+b}{c} = \dfrac{a}{c} + \dfrac{b}{c}$, to rewrite f. For example,

$$f(x) = \frac{x^3 + 3x^2 - x + 2}{x} = \frac{x^3}{x} + \frac{3x^2}{x} - \frac{x}{x} + \frac{2}{x} = x^2 + 3x - 1 + 2x^{-1}.$$

(c) Compute f' using the rules from this section.

(d) Write the domain of f' using interval notation.

57. $f(x) = \dfrac{3x^3 - 9x^2 + 4}{12}$

58. $f(x) = \dfrac{x^2 + 4x + 3}{2}$

59. $f(x) = \dfrac{2x^3 + 3x^2 - x + 3}{x}$

60. $f(x) = \dfrac{-3x^3 - 4x^2 + 2x - 7}{x}$

61. $f(x) = \dfrac{7x^4 - 50x^2 + x}{x^2}$

62. $f(x) = \dfrac{-5x^4 + 25x^2 - x}{x^2}$

63. $f(x) = \dfrac{4x^3 - 14x^2 + 3}{2\sqrt{x}}$

64. $f(x) = \dfrac{x^3 + 2x^2}{\sqrt[3]{x^2}}$

For Exercises 65–70, complete the following.

(a) Use the differentiation rules from this section to determine f'.

(b) Find the slope of the tangent line to the graph of f at the indicated x-value.

(c) Use the result of part (b) to write an equation of the line tangent to the graph of f at the indicated x-value.

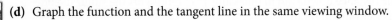

(d) Graph the function and the tangent line in the same viewing window.

65. $f(x) = x^3$ at $x = -1$

66. $f(x) = \sqrt{x}$ at $x = 4$

67. $f(x) = \dfrac{1}{x}$ at $x = 3$

68. $f(x) = \dfrac{1}{x^2}$ at $x = 3$

69. $f(x) = x^{2/3}$ at $x = 8$

70. $f(x) = x^{-1/3}$ at $x = 1$

Application Exercises

71. **Marketing—Advertising Budget:** The number of dollars spent on advertising for a product influences the number of items of the product that will be purchased by customers. The number of Krypto-Cases sold for a new smartphone is a function of the amount spent on advertising and can be modeled by

$$N(x) = 2000 - \frac{520}{x} \qquad x \geq 0.208$$

where x represents the amount spent on advertising, measured in thousands of dollars, and $N(x)$ represents the number of cases purchased.

   (a) Determine $N'(x)$.

   (b) Compute $N(10)$ and $N'(10)$ and interpret each.

72. **Marketing—Advertising Budget:** The number of dollars spent on advertising for a product influences the number of items of the product that will be purchased by customers. The number of Find-it-Now GPS devices is a function of the amount spent on advertising and can be modeled by

$$N(x) = 3600 - \frac{700}{x} \qquad x \geq 0.195$$

where x represents the amount spent on advertising, measured in thousands of dollars, and $N(x)$ represents the number of devices purchased.

   (a) Determine $N'(x)$.

   (b) Compute $N(5)$ and $N'(5)$ and interpret each.

73. **Microeconomics—Manufacturing Cost:** From past data analysis, the LaRoche Raincoat Company finds that the manufacturing cost for their Jeffery coat can be modeled by

$$C(x) = 3000 + 11x - 7\sqrt{x} + 0.03x^{3/2}$$

where x represents the number of coats produced and $C(x)$ represents the cost in dollars.

   (a) Determine $C'(x)$.

   (b) Evaluate $C(300)$ and $C'(300)$ and interpret each.

74. **Microeconomics—Manufacturing Cost:** (*continuation of Exercise 73*) Recall that for a cost function, the average cost $AC(x)$ is given by $AC(x) = \dfrac{C(x)}{x}$ and gives the per-unit cost of making a product at production level x.

   (a) For the cost function in Exercise 73, write and simplify the average cost function $AC(x)$.

   (b) Determine $AC'(x)$.

   (c) Evaluate $AC(300)$ and $AC'(300)$ and interpret each.

75. **Ecology—Sulfur Dioxide Levels:** The Sperry Company's Cantor Ridge coal-burning power plant collects air samples of its emissions at various points downwind to determine the concentration of sulfur dioxide at different distances from the plant. The amount of sulfur dioxide measured can be modeled by the function

$$f(x) = \frac{93.21}{x^2} \qquad x > 0$$

where x represents the distance downwind in miles and $f(x)$ represents the sulfur dioxide concentration in parts per million (ppm).

   (a) Determine $f'(x)$.

   (b) Evaluate $f(1)$ and $f'(1)$ and interpret each.

76. **Ecology—Sulfur Dioxide Levels:** The Sperry Company's Rangel Valley coal-burning power plant collects air samples of its emissions at various points downwind to determine the

concentration of sulfur dioxide at different distances from the plant. The amount of sulfur dioxide measured can be modeled by the function

$$f(x) = \frac{78.35}{x^2} \qquad x > 0$$

where x represents the distance downwind in miles and $f(x)$ represents the sulfur dioxide concentration in parts per million (ppm).

(a) Determine $f'(x)$.

(b) Evaluate $f(2)$ and $f'(2)$ and interpret each.

*In Exercises 77–80, a supply function s is given. A **supply function** gives the price $s(x)$ at which exactly x units of a product is supplied. Its derivative $s'(x)$ gives the instantaneous rate of change in the price of the product at supply level x.*

77. Macroeconomics—Supply Function: Balata Incorporated produces golf balls and finds that its supply function for the new Xtrah golf ball is given by

$$s(x) = 0.24x + 3.70$$

where x represents the number of dozens of golf balls supplied each month and $s(x)$ is the price per dozen, measured in dollars. Evaluate $s'(65)$ and interpret.

78. Macroeconomics—Supply Function: The Farver Company has determined that the supply function for their new Healthy Day breakfast bar is given by

$$s(x) = \frac{1}{2}x + 40$$

where x is the number of cases of breakfast bars supplied each month and $s(x)$ represents the price per case. Evaluate $s'(70)$ and interpret.

79. Macroeconomics—Supply Function: Baker's Bake Shoppe makes specialty baking pans and finds that the supply function for the candle-shaped birthday cake pan is given by

$$s(x) = 3.75x^{0.22}$$

where x is the number of baking pans supplied and $s(x)$ represents the price per pan. Evaluate $s'(31)$ and interpret.

80. Macroeconomics—Supply Function: The U-Build-It home improvement corporation determines that the supply function for their Haul-Out disposal bags is given by

$$s(x) = 6.81x^{0.33}$$

where x is the number of boxes of bags supplied and $s(x)$ represents the price per box of bags. Evaluate $s'(5)$ and interpret.

 81. Political Science—Hispanic Voters: One group of voters that has received attention recently is Hispanic Americans. The number of Hispanic voters from 1995 to 2010 can be modeled by

$$f(x) = 1.08x + 11.7 \qquad 0 \le x \le 15$$

where x represents the number of years since 1995 and $f(x)$ represents the number of Hispanic voters in millions. (*Source:* U.S. Census Bureau.) Determine $f(9)$ and $f'(9)$ and interpret each.

 82. Political Science—African American Voters: A group that political analysis has traditionally been interested in is African American voters. The number of African American voters from 1995 to 2010 can be modeled by

$$f(x) = 3.10x + 20.67 \qquad 0 \le x \le 15$$

REGISTER
HERE
TO VOTE

© Jupiterimages/Thinkstock

where x represents the number of years since 1995 and $f(x)$ represents the number of African American voters in millions. (*Source:* U.S. Census Bureau.) Determine $f(9)$ and $f'(9)$ and interpret each.

 83. Macroeconomics—European Unemployment: Suppose an economist wishes to examine the unemployment rate in the middle of the first decade of the 21st century. The unemployment rate in Bulgaria from 2000 to 2010 can be modeled by

$$f(x) = 0.08x^3 - 1.24x^2 + 3.34x + 15.37 \qquad 0 \le x \le 10$$

where x represents the number of years since 2000 and $f(x)$ represents the unemployment rate as a percentage. (*Source:* Google Public Data.) Use the model to determine the rate of change in the Bulgarian unemployment rate in 2005.

 84. Macroeconomics—European Unemployment: Suppose an economist wishes to examine the unemployment rate in the middle of the first decade of the 21st century. The unemployment rate in Latvia from 2000 to 2010 can be modeled by

$$f(x) = 0.09x^3 - 0.98x^2 + 1.63x + 13.21 \qquad 0 \le x \le 10$$

where x represents the number of years since 2000 and $f(x)$ represents the unemployment rate as a percentage. (*Source:* Google Public Data.) Use the model to determine the rate of change in the Latvian unemployment rate in 2008.

Concept and Writing Exercises

85. The constant function rule states that for any constant k, if $f(x) = k$, then $f'(x) = 0$. Sketch a graph of the constant function f and explain using the properties of slope that the derivative is zero.

*For a function f at a value $x = a$, the **normal line** of f at a is the equation of the line that is perpendicular to the tangent line at $x = a$. In Exercises 86, 87, and 88, a function f and a value a are given. Complete the following.*

(a) *Determine the equation of the tangent line at $x = a$. Call this line y_1.*

(b) *Determine the equation of the normal line at a. Recall from algebra that perpendicular lines have negative reciprocal slopes. Call this line y_2.*

 (c) *Graph f along with the lines y_1 and y_2 in the same viewing window.*

86. $f(x) = 2x + 3 \quad x = 4$ **87.** $f(x) = \sqrt{x} \quad x = 9$

88. $f(x) = x^2 - 5x + 6 \quad x = 1$

89. The Sum Rule states that $\frac{d}{dx}[f(x) + g(x)] = \frac{d}{dx}[f(x)] + \frac{d}{dx}[g(x)]$, and the Constant Multiple Rule states that, for a real number constant k, that $\frac{d}{dx}[k \cdot f(x)] = k \cdot \frac{d}{dx}f(x)$. Use these rules to prove the Difference Rule, which states that for differentiable functions f and g, $\frac{d}{dx}[f(x) - g(x)] = \frac{d}{dx}[f(x)] - \frac{d}{dx}[g(x)]$.

90. Determine the derivative for the function $f(x) = (3x^2 + 4)^2$ by squaring the binomial $3x^2 + 4$ and differentiating the resulting trinomial.

*A function f is said to be **differentiable from the left** if $\lim\limits_{h \to 0^-} \dfrac{f(x + h) - f(x)}{h}$ exists, and a function f is said to be **differentiable from the right** if $\lim\limits_{h \to 0^+} \dfrac{f(x + h) - f(x)}{h}$ exists. In Exercises 91 and 92, a piecewise-defined function is given. Complete the following.*

(a) Determine if f is differentiable from the left at $x = a$. If so, state the derivative's value.

(b) Determine if f is differentiable from the right at $x = a$. If so, state the derivative's value.

(c) Determine if $f'(a)$ exists.

91. $f(x) = \begin{cases} 2x^2 - 3, & x < 1 \\ 4, & x \geq 1 \end{cases}; \quad x = 1$

92. $f(x) = \begin{cases} x^2, & x \leq 1 \\ \sqrt{x}, & x > 1 \end{cases}; \quad x = 1$

93. Recall from algebra that a function is *even* if $f(-x) = f(x)$ for all values of x in the domain of f, and a function is *odd* if $f(-x) = -f(x)$ for all values of x in the domain of f. Show that the derivative of an even function is an odd function.

Section Project

Turbochargers were used to increase the horsepower output of cars in the Indianapolis 500 from 1952 until 1996, when they were banned from competition through 2011. The first driver to complete a lap at over 200 mph was Tom Sneva in 1978. (*Source:* Indycar.com.) The pole position speed for the Indianapolis 500 from 1978 to 2011 can be modeled by the piecewise-defined function

$$f(x) = \begin{cases} 404.23 - \dfrac{16{,}329.3}{x}, & 78 \leq x \leq 96 \\ -0.092x^2 + 19.54x - 805, & 97 \leq x \leq 111 \end{cases}$$

where x represents the number of years since 1900 and $f(x)$ represents the pole position speed in miles per hour. (*Source:* Indy500.com.) Use the model to answer the following.

© Bettmann/CORBIS

(a) Evaluate $f(96)$ and $f(97)$ and interpret each value.

(b) Evaluate $f(97) - f(96)$ and interpret in terms of the turbocharger technology.

(c) Compute the rate of change in the pole position speed at the Indianapolis 500 in 2011.

(d) With the return of turbocharged engines to the Indianapolis 500 in 2012, would we expect the value of $f'(112)$ to be positive or negative? Explain your answer in a brief sentence.

3.2 Derivatives of Products and Quotients

SECTION OBJECTIVES

1. Differentiate products using the Product Rule.

2. Apply the Product Rule to business functions.

3. Differentiate quotients using the Quotient Rule.

4. Apply Quotient Rule to medication concentration.

In Section 3.1 we learned four useful rules for computing derivatives. In this section, we illustrate how to differentiate the **product** and **quotient** of two functions. These two rules are not as simple as the ones presented in Section 3.1.

Product Rule

To differentiate the product of two functions, we use the **Product Rule**.

Rule 5: Product Rule

If $h(x) = f(x) \cdot g(x)$, where f and g are differentiable functions, then the derivative is
$h'(x) = f'(x) \cdot g(x) + f(x) \cdot g'(x)$.

NOTE: In words, the Product Rule states that we can take the derivative of the first function times the second function plus the first function times the derivative of the second function.

A proof of the Product Rule is in Appendix C.

OBJECTIVE 1

Differentiate products
using the Product Rule.

Example 1: Differentiating Products

Differentiate $h(x) = 3x^3(x^4 + 2)$ by using the Product Rule.

Perform the Mathematics

Let $f(x) = 3x^3$ and $g(x) = x^4 + 2$. By the Product Rule, we compute the derivative to be

$$h'(x) = f'(x) \cdot g(x) + f(x) \cdot g'(x)$$
$$= 9x^2(x^4 + 2) + 3x^3(4x^3)$$
$$= 9x^6 + 18x^2 + 12x^6$$
$$= 21x^6 + 18x^2$$
$$= 3x^2(7x^4 + 6) \qquad \blacksquare$$

Example 2: Differentiating Products

Differentiate the following by using the Product Rule.

a. $f(x) = (2x^2 + 4x + 5)(5x - 4)$

b. $y = \sqrt{x}(3x^3 - 4x^2 + 8x)$

Perform the Mathematics

a. Applying the Product Rule yields

$$f'(x) = (4x + 4)(5x - 4) + (2x^2 + 4x + 5)(5)$$
$$= (20x^2 - 16x + 20x - 16) + (10x^2 + 20x + 25)$$
$$= 30x^2 + 24x + 9$$

b. Before differentiating, we rewrite the function as

$$y = x^{1/2}(3x^3 - 4x^2 + 8x)$$

Now using the Product Rule we get

$$y' = \frac{1}{2}x^{-1/2}(3x^3 - 4x^2 + 8x) + x^{1/2}(9x^2 - 8x + 8)$$

Writing the derivative without negative or fractional exponents yields

$$y' = \frac{3x^3 - 4x^2 + 8x}{2\sqrt{x}} + \sqrt{x} \cdot (9x^2 - 8x + 8) \qquad \blacksquare$$

▶ *Try It Yourself*

Some related Exercises are 15 and 19.

Example 3: Differentiating Products

Differentiate $f(x) = (6x^{4/3} + 2x)(3x^{5/3} + 4x - 1)$.

Perform the Mathematics

By the Product Rule, we immediately compute the derivative to be

$$f'(x) = (8x^{1/3} + 2)(3x^{5/3} + 4x - 1) + (6x^{4/3} + 2x)(5x^{2/3} + 4) \qquad \blacksquare$$

Example 4: Determining an Equation for a Tangent Line

Determine an equation for the line tangent to $y = (x^4 - x^2)(x^3 - x + 2)$ at $x = 1$.

Perform the Mathematics

Since the derivative gives the slope of a tangent line at any point, our first task is to determine the derivative of the function by using the Product Rule. This gives

$$\frac{dy}{dx} = (4x^3 - 2x)(x^3 - x + 2) + (x^4 - x^2)(3x^2 - 1)$$

We now evaluate $\frac{dy}{dx}$ at $x = 1$ to get the slope of the tangent line. Substituting $x = 1$ gives

$$\frac{dy}{dx} = (4 \cdot 1^3 - 2 \cdot 1)(1^3 - 1 + 2) + (1^4 - 1^2)(3 \cdot 1^2 - 1) = 4$$

So, at $x = 1$, $m_{tan} = 4$. Also, when $x = 1$, we determine the point of tangency by substituting 1 for x in the original function. This gives

$$y = (1^4 - 1^2)(1^3 - 1 + 2) = 0$$

So our point of tangency is $(1, 0)$. Since we know a point on the tangent line and we know that the slope of the tangent line is 4, we use the point-slope form of a line to determine an equation for the tangent line. This results in

$$y - y_1 = m(x - x_1)$$
$$y - 0 = 4(x - 1)$$
$$y = 4x - 4$$

Figure 3.2.1 shows a graph of the function along with the tangent line.

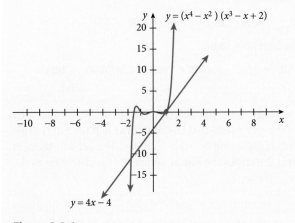

Figure 3.2.1

■

▶ *Try It Yourself*

Some related Exercises are 41 and 43.

Technology Option

In **Figure 3.2.2**, we have used a graphing calculator to graph the function in Example 4. We have also used the `Draw Tangent` command. Notice that the `Draw Tangent` command does not give the same equation for the tangent line as we found in Example 4. This is a limitation of the calculator. Even though it is very close, we needed calculus for the **exact** answer.

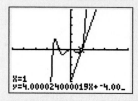

Figure 3.2.2

OBJECTIVE 2

Apply the Product Rule to business functions.

Example 5: Applying the Product Rule

Extensive market research has determined that for the next five years the price of a certain mountain bike is predicted to vary according to $p(t) = 300 - 30t + 7.5t^2$, where t is time in years and $p(t)$ is the price in dollars. The number of mountain bikes sold annually by Skinner's Bikes is expected to follow $q(t) = 3000 + 90t - 15t^2$, where $q(t)$ is the number sold and t is time in years.

a. Determine $R(t)$ and $R'(t)$.

b. Compute $R'(1)$ and interpret.

c. Compute $R'(4)$ and interpret.

Perform the Mathematics

a. Because revenue equals price times quantity, we have

$$R(t) = p(t) \cdot q(t) = (300 - 30t + 7.5t^2) \cdot (3000 + 90t - 15t^2)$$

By the Product Rule, we compute the derivative to be

$$R'(t) = \underbrace{(-30 + 15t)}_{p'(t)} \cdot \underbrace{(3000 + 90t - 15t^2)}_{q(t)} + \underbrace{(300 - 30t + 7.5t^2)}_{p(t)}\underbrace{(90 - 30t)}_{q'(t)}$$

We will not simplify $R'(t)$ so that we may see separately the effects on $R(t)$ caused by changing prices and sales. Notice in the derivative that the first term is $p'(t) \cdot q(t)$. This term describes the rate at which $R(t)$ is changing as price changes. The second term in the derivative is $p(t) \cdot q'(t)$. This term describes the rate at which $R(t)$ is changing as the number of bikes sold changes.

b. We compute $R'(1)$ to be

$$R'(1) = (-30 + 15)(3000 + 90 - 15) + (300 - 30 + 7.5)(90 - 30)$$
$$= (-15)(3075) + (277.5)(60)$$
$$= -46,125 + 16,650 = -29,475$$

The negative sign indicates that after one year, the revenue is decreasing at a rate of $29,475 per year. Notice that the effect of falling prices represented by the first term, $p'(1) \cdot q(1) = -46,125$, is greater than the effect of rising sales given by the second term, $p(1) \cdot q'(1) = 16,650$.

c. We have

$$R'(4) = (-30 + 60)(3000 + 360 - 240) + (300 - 120 + 120)(90 - 120)$$
$$= (30)(3120) + (300)(-30)$$
$$= 93,600 - 9000 = 84,600$$

Since this result is positive, we claim that after four years the revenue is increasing at a rate of $84,600 per year. Notice that the effect of rising prices represented by the first term, $p'(4) \cdot q(4) = 93,600$, is greater than the effect of falling sales given by the second term, $p(4) \cdot q'(4) = -9000$. ∎

Quotient Rule

The final differentiation rule of this section, the **Quotient Rule**, shows how to differentiate the quotient of two functions.

> **Rule 6: Quotient Rule**
>
> If $h(x) = \dfrac{f(x)}{g(x)}$, where f and g are differentiable functions, then $h'(x) = \dfrac{f'(x) \cdot g(x) - f(x) \cdot g'(x)}{[g(x)]^2}$, where $g(x) \neq 0$.

NOTE: In words, the Quotient Rule states that we can take the derivative of the numerator times the denominator minus the numerator times the derivative of the denominator, all of this over the denominator squared.

Example 6: Differentiating Quotients

OBJECTIVE 3

Differentiate Quotients using the Quotient Rule.

Differentiate the following using the Quotient Rule.

a. $h(x) = \dfrac{x + 3}{x - 2}$　　　　**b.** $y = \dfrac{x^4 - 3x}{x^2 + 1}$

Perform the Mathematics

a. Using the Quotient Rule, we have

$$h'(x) = \frac{f'(x)g(x) - f(x)g'(x)}{[g(x)]^2}$$
$$= \frac{1 \cdot (x - 2) - (x + 3) \cdot 1}{(x - 2)^2}$$
$$= \frac{x - 2 - x - 3}{(x - 2)^2} = \frac{-5}{(x - 2)^2}$$

b. Again, by the Quotient Rule, the derivative is

$$y' = \frac{f'(x)g(x) - f(x)g'(x)}{[g(x)]^2}$$
$$= \frac{(4x^3 - 3)(x^2 + 1) - (x^4 - 3x)(2x)}{(x^2 + 1)^2}$$
$$= \frac{(4x^5 + 4x^3 - 3x^2 - 3) - 2x^5 + 6x^2}{(x^2 + 1)^2}$$
$$= \frac{2x^5 + 4x^3 + 3x^2 - 3}{(x^2 + 1)^2}$$
∎

▶ **Try It Yourself**

Some related Exercises are 29 and 31.

The next example requires us to use both the Product Rule and the Quotient Rule.

Example 7: Differentiating Using the Product Rule and Quotient Rule

Differentiate $y = \dfrac{(2x + 1)(3x - 2)}{x + 1}$.

Perform the Mathematics

Notice that when applying the Quotient Rule, we will use the Product Rule when differentiating the numerator. To aid us in this computation, let $f(x) = 2x + 1$ and $g(x) = 3x - 2$. We then have

$$f'(x) = 2 \ \text{ and } \ g'(x) = 3$$

Now when applying the Quotient Rule, the derivative of the numerator is

$$\frac{d}{dx}[(2x + 1)(3x - 2)] = 2(3x - 2) + (2x + 1)3$$

$$= 6x - 4 + 6x + 3 = 12x - 1$$

Putting this all together using the Quotient Rule gives

$$\frac{dy}{dx} = \frac{(12x - 1)(x + 1) - (2x + 1)(3x - 2) \cdot 1}{(x + 1)^2}$$

$$= \frac{12x^2 + 11x - 1 - (6x^2 - x - 2)}{(x + 1)^2}$$

$$= \frac{12x^2 + 11x - 1 - 6x^2 + x + 2}{(x + 1)^2}$$

$$= \frac{6x^2 + 12x + 1}{(x + 1)^2} \qquad ■$$

▶ **Try It Yourself**

Some related Exercises are 35 and 37.

Example 8: Applying the Quotient Rule

OBJECTIVE 4

Apply the Quotient Rule to medication concentration.

Researchers have determined through experimentation that the percent concentration of a certain medication can be approximated by

$$p(t) = \frac{200t}{2t^2 + 5} - 4 \qquad [0.25, 20]$$

where t is the time in hours after administering the medication and $p(t)$ is the percent concentration. Evaluate $p'(1)$ and $p'(6)$ and interpret each.

Understand the Situation

In order to evaluate $p'(1)$ and $p'(6)$, we first need to determine $p'(t)$. Once we have $p'(t)$, we will substitute $t = 1$ and $t = 6$ to determine the values of $p'(1)$ and $p'(6)$, respectively.

© Hemera/Thinkstock

Perform the Mathematics

Utilizing the Quotient Rule, along with the Sum and Difference Rules, we compute the derivative to be

$$p'(t) = \frac{200(2t^2 + 5) - (200t)(4t)}{(2t^2 + 5)^2} = \frac{-200(2t^2 - 5)}{(2t^2 + 5)}$$

Evaluating $p'(t)$ for $t = 1$ gives

$$p'(1) = \frac{-200(2(1)^2 - 5)}{(2(1)^2 + 5)^2} \approx 12.24$$

Evaluating $p'(t)$ for $t = 6$ gives

$$p'(6) = \frac{-200(2(6)^2 - 5)}{(2(6)^2 + 5)^2} \approx -2.26$$

Interpret the Results

For $t = 1$, $p'(1) \approx 12.24$. This means that at the end of one hour, the concentration of medication is increasing at a rate of about 12.24% per hour. For $t = 6$, $p'(6) \approx -2.26$. This means that at the end of six hours, the concentration of medication was decreasing at a rate of about 2.26% per hour. See **Figure 3.2.3**.

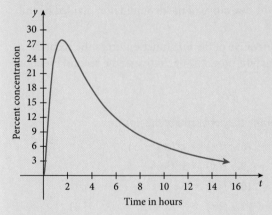

Figure 3.2.3

▷ *Try It Yourself*

Some related Exercises are 65 and 71. ∎

Summary

In this section we presented the **Product Rule** and **Quotient Rule** for computing derivatives. The differentiation rules learned so far are listed in **Table 3.2.1**. The letters k and n represent constants, whereas f and g represent differentiable functions of x.

Table 3.2.1

Name	Rule
Constant Function	$\dfrac{d}{dx}[k] = 0$
Power	$\dfrac{d}{dx}[x^n] = nx^{n-1}$
Constant Multiple	$\dfrac{d}{dx}[k \cdot f(x)] = k \cdot f'(x)$
Sum and Difference	$\dfrac{d}{dx}[f(x) \pm g(x)] = f'(x) \pm g'(x)$
Product	$\dfrac{d}{dx}[f(x) \cdot g(x)] = f'(x) \cdot g(x) + f(x) \cdot g'(x)$
Quotient	$\dfrac{d}{dx}\left[\dfrac{f(x)}{g(x)}\right] = \dfrac{f'(x) \cdot g(x) - f(x) \cdot g'(x)}{[g(x)]^2}, g(x) \neq 0$

Section 3.2 Exercises

Vocabulary Exercises

1. When we use the term *product*, we mean the result when two functions are being

   _____.

2. The quotient of two functions implies that the functions are being _____.

3. When using the Quotient Rule, we take the derivative of the numerator times the denominator, minus the numerator times the derivative of the denominator, divided by the _____ of the denominator.

4. The Product Rule states that we take the derivative of the first function times the second function _____ the first function times the derivative of the second function.

Skill Exercises

In Exercises 5–26, use the Product Rule to determine the derivative of the function.

5. $f(x) = x^2(2x + 1)$

6. $f(x) = x^2(3x - 5)$

7. $f(x) = x^3(3x^2 + 2x - 5)$

8. $f(x) = x^3(5x^2 - 6x + 3)$

9. $f(x) = 3x^4(2x^2 - 9x + 1)$

10. $f(x) = 5x^3(3x^2 - 6x + 2)$

11. $f(x) = -5x^2(3x^3 + 5x - 7)$

12. $f(x) = -7x^3(2x^3 - 3x^2 + 8)$

13. $f(x) = (3x + 4)(2x - 1)$

14. $f(x) = (4x - 1)(x + 6)$

15. $f(x) = (5x + 3)(3x^3 + 2x^2 + 1)$

16. $f(x) = (2x - 1)(x^2 - 2x + 3)$

17. $f(x) = (3x^2 - 2x + 1)(2x^2 + 5x - 7)$

18. $f(x) = (2x^2 + 5x - 1)(3x^2 - 7x + 3)$

19. $f(x) = (2\sqrt{x} + 4x - 3)(3x - 4)$

20. $f(x) = (3\sqrt{x} - 2x + 1)(5x + 2)$

21. $f(x) = (3x^{6/5} - 5x)(4x^{5/3} + 2x - 5)$

22. $f(x) = (2x^{4/3} + 3x)(-2x^{7/3} + 2x - 5)$

23. $f(x) = (3\sqrt{x} - 5)\left(2\sqrt{x} - \dfrac{1}{x^3}\right)$

24. $f(x) = (4\sqrt{x} + 2x - 6)(3\sqrt{x} + 6x)$

25. $f(x) = (x^{2/3} + x + 1)(x^{-1} + x^{-2})$

26. $f(x) = (6x^{4/3} - 2x + 3)(3x^{-1} - 4x^{-2})$

In Exercises 27–40, determine the derivative for the given function.

27. $f(x) = \dfrac{x + 2}{x + 1}$

28. $f(x) = \dfrac{3x - 4}{x - 1}$

29. $f(x) = \dfrac{4x - 3}{2x + 1}$

30. $f(x) = \dfrac{5x - 11}{3x - 4}$

31. $f(x) = \dfrac{3x^2 - 5x + 1}{5x^2 + 3x + 2}$

32. $f(x) = \dfrac{-2x^2 + 6x - 5}{3x^2 + 5x + 2}$

33. $f(x) = \dfrac{3\sqrt{x} - 5}{6x - 1}$

34. $f(x) = \dfrac{4\sqrt{x} + 3}{2x + 7}$

35. $f(x) = \dfrac{(x^2 + 2)(x - 3)}{x - 1}$

36. $f(x) = \dfrac{(x + 2)(x^3 - 3x^2 + 1)}{x - 2}$

37. $f(x) = \dfrac{(5x^4 + 2)(x^2 + 3)}{x - 4}$

38. $f(x) = \dfrac{(6x^3 - 2x^2 + 1)(3x - 5)}{2x + 1}$

39. $f(x) = \dfrac{4x^3 + 2x^2 - 3x - 5}{2}$

40. $f(x) = \dfrac{2x^3 + 3x^2 - 7x + 1}{5}$

For Exercises 41–48, complete the following:

(a) *Determine the derivative.*

(b) *Write the equation of the line tangent to the graph of the function at the indicated x-value.*

41. $f(x) = x^2(x^2 - 5)$ at $x = 1$

42. $f(x) = -3x^2(2x + 3)$ at $x = 3$

43. $f(x) = (x^2 + 1)(x^3 + 1)$ at $x = 1$

44. $f(x) = (2x - 3)(x^3 + 3)$ at $x = 5$

45. $f(x) = \dfrac{x + 2}{x - 1}$ at $x = 2$

46. $f(x) = \dfrac{x^2 + 1}{x}$ at $x = -1$

47. $f(x) = \dfrac{3x^2 - 2x}{-2x + 3}$ at $x = -1$

48. $f(x) = \dfrac{-2x^2 + 3x}{3x - 5}$ at $x = 3$

For Exercises 49–54, complete the following:

(a) *Determine the equation of the line tangent to the graph of the function at the indicated x-value.*

(b) *Graph the function and the tangent line in the same viewing window.*

(c) *Use the* dy/dx *command or the* Draw Tangent *command on your calculator to verify the result in part (a).*

 49. $f(x) = x^2(x^2 - 3)$ at $x = 2$

50. $f(x) = -2x^2(3x + 1)$ at $x = 1$

 51. $f(x) = (x^3 + 2)(x^2 + 2)$ at $x = -1$

52. $f(x) = (3x - 1)(x^2 + 1)$ at $x = -2$

53. $f(x) = \dfrac{4x^3 - 3x}{2x + 1}$ at $x = 2$

54. $f(x) = \dfrac{-2^2 + 5x}{4x - 1}$ at $x = -1$

In this section, the Product Rule is shown using the product of two functions. The rule can be extended to differentiate the product of any finite number of differentiable functions. For example, if $k(x) = f(x) \cdot g(x) \cdot h(x)$, then the derivative of k is given by $k'(x) = f'(x) \cdot g(x) \cdot h(x) + f(x) \cdot g'(x) \cdot h(x) + f(x) \cdot g(x) \cdot h'(x)$. Use this form of the Product Rule to differentiate the functions in Exercises 55–60.

55. $f(x) = (x + 1)(x - 2)(x + 5)$

56. $f(x) = x^2(x^3 - 3x^2 + 1)(x - 4)$

57. $f(x) = (x + 1)(2x^2 - 3)(3x + 4)$

58. $f(x) = (x - 4)(3x^2 - 5)(2x - 9)$

59. $f(x) = \sqrt{x}(2x - 1)(3x^2 + 2)$

60. $f(x) = \sqrt[3]{x}(3x + 1)(2x^3 - 3)$

Application Exercises

In Exercises 61–64, recall that for varying quantities produced and sold over time period t, the revenue function is $R(t) = p(t) \cdot q(t)$.

61. Finance—Market Analysis: Market analysts at the Maxwell Company have estimated that the monthly sales during the first seven months of its new MePad tablet computer can be modeled by

$$q(t) = 30t - 0.5t^2 \quad 0 \le t \le 7$$

where t represents the number of months since the MePad became available and $q(t)$ represents the number of computers sold in hundreds. The analysts also determine that the price of the MePad will vary and sets the price using the model

$$p(t) = 2200 - 34t^2 \quad 0 \le t \le 7$$

where t represents the number of months since the MePad became available and $p(t)$ represents the price of the MePad in dollars.

(a) Compute $q(3)$ and $q'(3)$ and interpret each.

(b) Compute $p(3)$ and $p'(3)$ and interpret each.

(c) Write the revenue function $R(t)$. Do not algebraically simplify the function.

(d) Compute $R(3)$ and $R'(3)$ and interpret each.

62. **Finance—Market Analysis** (*continuation of Exercise 61*): Use the revenue function from Exercise 61 to complete the following.

(a) Verify that after multiplying and simplifying $R'(t)$ in Exercise 61, we get the function $R'(t) = 68t^3 - 3060t^2 - 2200t + 66{,}000$.

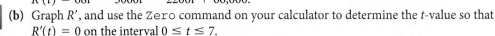 (b) Graph R', and use the `Zero` command on your calculator to determine the t-value so that $R'(t) = 0$ on the interval $0 \le t \le 7$.

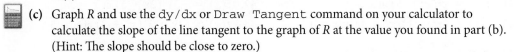 (c) Graph R and use the `dy/dx` or `Draw Tangent` command on your calculator to calculate the slope of the line tangent to the graph of R at the value you found in part (b). (Hint: The slope should be close to zero.)

(d) Is the revenue maximized or minimized at the t-value you found in part (b)? Explain your answer.

63. **Finance—Market Analysis:** The PlayPro Company has projected that the monthly sales of their Bluetooth headphones can be modeled by

$$q(t) = 30t - \frac{t^2}{2} \quad 0 \le t \le 5$$

where t represents the number of months since the headphones were initially sold and $q(t)$ represents the number of units sold in hundreds. They also project that the retail price of the headphones can be modeled by

$$p(t) = 220 - t^2 \quad 0 \le t \le 5$$

where t represents the number of months since the headphones were initially sold and $p(t)$ represents the retail price in dollars.

(a) Evaluate $q(3)$ and $q'(3)$ and interpret each.

(b) Evaluate $p(3)$ and $p'(3)$ and interpret each.

64. **Finance—Market Analysis** (*continuation of Exercise 63*): Use the sales and price functions from Exercise 63 to complete the following.

(a) Write the revenue function $R(t)$. Do not algebraically simplify the function.

(b) Compute $R(3)$ and $R'(3)$ and interpret each.

65. **Ecology—Removing Pollutants:** In planning its future operating budget, the city of Rockton hires a consulting firm to determine the cost of removing the pollutants discharged into Lake Watson. They find the cost can be projected by the model

$$C(x) = \frac{113x}{100 - x} \quad 0 \le x < 100$$

where x represents the percentage of the pollutants removed and $C(x)$ represents the cost in thousands of dollars.

(a) Use the Quotient Rule to differentiate $C(x)$.

(b) Evaluate $C(50)$ and $C'(50)$ and interpret each.

66. **Ecology—Removing Pollutants:** In planning its future operating budget, the city of Utica hires a consulting firm to determine the cost of removing the pollutants discharged into Lake Chevelle. They find the cost can be projected by the model

$$C(x) = \frac{50x}{100 - x} \qquad 0 \le x < 100$$

where x represents the percentage of the pollutants removed and $C(x)$ represents the cost in thousands of dollars.

(a) Use the Quotient Rule to differentiate $C(x)$.

(b) Evaluate $C(50)$ and $C'(50)$ and interpret each.

67. **Conservation Science—Fish Propagation:** At the request of local anglers, the Graysville game commission decides to stock Lake Mumford with bass. From past data collection, conservation scientists determine that the population of the stocked lake can be modeled by the function

$$P(t) = \frac{10(10 + 7t)}{1 + 0.02t} \qquad t \ge 0$$

where t represents the number of months since the lake was initially stocked and $P(t)$ represents the population size.

(a) How many bass were initially stocked in the lake?

(b) Determine $P'(t)$.

(c) Evaluate and interpret $P(5)$ and $P'(5)$.

68. **Conservation Science—Fish Propagation:** In order to generate income from public resources, the Crogan city council decides to stock Lake Simpson with commercial catfish. From past data collection, conservation scientists determine that the population of the stocked lake can be modeled by the function

$$P(t) = \frac{20(10 + 7t)}{1 + 0.02t} \qquad t \ge 0$$

where t represents the number of months since the lake was initially stocked and $P(t)$ represents the population size.

(a) How many catfish were initially stocked in Lake Simpson?

(b) Determine $P'(t)$.

(c) Evaluate and interpret $P(12)$ and $P'(12)$.

69. **Recreation Studies—Pleasure Craft Purchasing:** Suppose an investment firm is trying to determine which type of recreational activity shows a high rate of growth so that it can make a profit on its investment. The amount Americans spent on boats and pleasure craft annually from 1990 to 2010 can be modeled by

$$f(x) = 0.16(\sqrt{x} + 13.19)(x + 12.6) \qquad 0 \le x \le 20$$

where x represents the number of years since 1990, and $f(x)$ represents the annual amount spent on boats and pleasure craft in billions of dollars. (*Source:* U.S. Bureau of Economic Analysis.)

© PhotoDisc/Getty Images

(a) Determine $f'(x)$.

(b) Use part (a) to determine the rate at which the spending on boats and pleasure craft was increasing in 2008.

70. **Public Policy—Patents Issued:** A patent is a set of exclusive rights granted by the U.S. government to an inventor for a limited period of time in exchange for a public disclosure of an invention and is often used as a yardstick to measure both the innovativeness and entrepreneurial climate of the country. The number of patents issued in the United States annually from 1990 to 2010 can be modeled by

$$f(x) = 0.2(\sqrt{x} + 7.78)(x + 60.86) \qquad 0 \le x \le 20$$

where x represents the number of years since 1990 and $f(x)$ represents the number of patents issued, measured in thousands. (*Source:* U.S. Patent and Trademark Office.)

(a) Determine $f'(x)$.

(b) Use part (a) to determine the rate that the issuance of patents was increasing in 2007.

 71. International Economics—Unemployment in Ireland: Suppose a social worker wants to explore the number of unemployed in Ireland to see how rapidly the unemployment rate has been increasing. The number unemployed in Ireland annually from 2000 to 2010 can be modeled by

$$f(x) = \frac{2.03x - 26.34}{0.03x - 0.35} \qquad 0 \le x \le 10$$

where x represents the number of years since 2000, and $f(x)$ represents the number of unemployed in Ireland, measured in thousands. (*Source:* Google Public Data.)

(a) Determine $f'(x)$.

(b) Use part (a) to determine the rate at which the number of unemployed in Ireland was increasing in 2009.

 72. Economics—Consumer Spending: Social networking and Internet commerce have contributed to the amount of spending on video and audio products during the past two decades. The amount spent on audio and video products in the United States from 1990 to 2010 can be modeled by

$$f(x) = \frac{3.39x + 44.41}{-0.01x + 0.82} \qquad 0 \le x \le 20$$

where x represents the number of years since 1990 and $f(x)$ represents the amount spent on audio and video products, measured in billions of dollars. (*Source:* U.S. Bureau of Economic Analysis).

(a) Determine $f'(x)$.

(b) Use the results of part (a) to determine the rate at which spending on video and audio products was increasing in 2008.

 73. Macroeconomics—Corporate Tax Rates: In a July 22, 2010, Associated Press interview, Treasury Secretary Tim Geithner said, "We are likely to have to take a broader look at corporate tax reform next year," adding that it was likely to be one of the areas the fiscal commission appointed by President Obama would examine to make recommendations on deficit reduction. The amount collected in U.S. corporate taxes annually from 1990 to 2010 can be modeled by

$$f(x) = 1.65x + 13.13 \qquad 0 \le x \le 20$$

where x represents the number of years since 1990 and $f(x)$ represents the amount of corporate taxes collected in billions of dollars. The amount in corporate profits of U.S. corporations from 1990 to 2010 can be modeled by

$$g(x) = 2.45x^2 + 26.52x + 419.01 \qquad 0 \le x \le 20$$

where x represents the number of years since 1990 and $g(x)$ represents the corporate profits in billions of dollars. (*Sources:* Associated Press, U.S. Bureau of Economic Analysis.)

 (a) Form the function $h(x) = \dfrac{f(x)}{g(x)} \cdot 100, \quad 0 \le x \le 20,$ to determine the ratio of taxes per profit as a percentage. Then graph h in the viewing window $[0, 20]$ by $[0, 4]$.

(b) Determine $h'(x)$.

(c) Evaluate $h'(3)$ and $h'(18)$ and interpret each.

 74. Demographics—Immigration and Greenhouse Gases: In his book *Bleeding Hearts and Empty Promises: A Liberal Rethinks Immigration*, Philip Cafaro, associate professor of philosophy at Colorado State University, claims that immigration to the United States has contributed to higher levels of greenhouse gas emissions. (*Source:* Center for Immigration Studies.) The amount of greenhouse gas emissions annually in the United States from 1990 to 2010 can be modeled by

$$f(x) = -3.39x^2 + 122.44x + 5977 \qquad 0 \le x \le 20$$

where x represents the number of years since 1990 and $f(x)$ represents the amount of greenhouse gas emissions in millions of metric tons. (*Source:* Google Public Data.) The number of immigrants to the United States from 1990 to 2010 can be modeled by

$$g(x) = -0.03x + 1.42 \qquad 0 \le x \le 20$$

where x represents the number of years since 1990, and $g(x)$ represent the number of immigrants coming to the United States annually in millions. (*Source:* U.S. Census Bureau.)

(a) Form the function $h(x) = \dfrac{f(x)}{g(x)}$, $0 \le x \le 20$, to measure the number of tons of greenhouse gasses in the United States per immigrant.

(b) Determine $h'(x)$.

(c) Evaluate $h'(2)$ and $h'(17)$ and interpret each.

(d) Do the results in part (c) support the professor's premise? Explain.

Concept and Writing Exercises

In Exercises 75 and 76, assume that f, g, and h are differentiable functions.

75. If $k(x) = f(x) \cdot g(x) \cdot h(x)$, determine the derivative $k'(x)$ by differentiating $f(x)[g(x) \cdot h(x)]$ and apply the Product Rule twice.

76. If $k(x) = \dfrac{f(x) \cdot g(x)}{h(x)}$, use the Quotient and Product Rules to determine a general formula for $k'(x)$.

77. Let $h(x) = c \cdot f(x)$, where c represents a real number constant, and let f represent a differentiable function. Use the Product Rule to verify the Constant Multiple Rule by showing that $\frac{d}{dx}[c \cdot f(x)] = c \cdot f'(x)$.

78. Prove the Quotient Rule by showing that the difference quotient for $\dfrac{f(x)}{g(x)}$ is

$$\frac{1}{h}\left[\frac{f(x + h)}{g(x + h)} - \frac{f(x)}{g(x)}\right] = \frac{g(x)f(x + h) - f(x)g(x + h)}{g(x + h)g(x)h}. \text{ (Hint: Before computing the limit,}$$

add and subtract $g(x)f(x)$ in the numerator.)

79. Apply the Quotient Rule to the function $f(x) = \dfrac{1}{x^n}$ to show that the Power Rule applies if n is a negative integer.

80. Suppose we have a differentiable function g and know that $g'(x) = \frac{1}{x}$. If $h(x) = x \cdot (g(x) - 1)$, determine an expression for $h'(x)$.

81. Let $f(x) = \frac{1}{x}$ and $g(x) = x^4$. Use the Product Rule to determine h', where $h(x) = f(x) \cdot g(x)$. Compute the product $f'(x) \cdot g'(x)$ to show that the result is not the same as h'.

82. A form of rational function that is commonly used is the ratio of two linear functions

$$f(x) = \frac{ax + b}{cx + d} \text{ where } a, b, c \text{ and } d \text{ are real number constants. Use the Quotient Rule to}$$

determine the general form for the derivative f'.

83. Let $h(x) = (f(x))^2$. Use the Product Rule to show that $h'(x) = 2f(x)f'(x)$.

84. Let $h(x) = (f(x))^3$. Show that $h'(x) = 3(f(x))^2 f'(x)$.

Section Project

Hauser's Law is a dictum from San Francisco investment economist W. Kurt Hauser, who claims that since the Second World War, federal tax revenues have been equal to about 19.5% of the U.S. gross domestic product (GDP). (*Source:* The Hoover Institute.) The annual amount of tax receipts from 1990 to 2010 can be modeled by

$$f(x) = 1.29x^3 - 39.1x^2 + 342.99x + 1036.16 \quad 0 \le x \le 20$$

where x represents the number of years since 1990 and $f(x)$ represents the annual federal tax receipts in billions of dollars. The annual U.S. GDP from 1990 to 2010 can be modeled by

$$g(x) = 11.73x^2 + 263.06x + 5815.13 \quad 0 \le x \le 20$$

where x represents the number of years since 1990 and $g(x)$ represents the annual GDP in billions of dollars. (*Source:* U.S. Bureau of Economic Analysis).

(a) Form the function $h(x) = \dfrac{f(x)}{g(x)} \cdot 100, \quad 0 \le x \le 20$, to determine the ratio of federal tax receipts to GDP as a percentage.

(b) Graph $h(x)$ in the viewing window [0, 20] by [0, 30].

(c) During what years is the percentage less than 19.5%? During what years is the percentage greater than 19.5%?

(d) Use the Quotient Rule to determine $h'(x)$.

(e) Evaluate $h'(7)$ and $h'(19)$ and interpret each rate of change.

SECTION OBJECTIVES

1. Differentiate power functions using the Chain Rule.

2. Differentiate radical functions using the Chain Rule.

3. Differentiate rational functions using the Chain Rule.

4. Apply the Chain Rule to health science and business functions.

3.3   The Chain Rule

So far, the types of functions that we have differentiated are polynomial functions, rational functions, and power functions. But one family of functions that we have not differentiated is the **composite function family**. If the composite function $f(x) = 1000\sqrt{180 - 2x}$ models the number of college graduates surviving to x years of age, we must find a way to compute the derivative of f in order to determine the rate of change of this function. In this section, we introduce the **Chain Rule**, a powerful technique used to differentiate composite functions.

Chain Rule

We will discuss the differentiation of composite functions that have the form $h(x) = f(g(x))$. A reasonable question to ask at this point is "Can we even determine the derivative of a composite function in the first place?" To answer that question, let's take a Flashback to an example first presented in Chapter 1.

Flashback: Oil Spill Revisited

In Section 1.4 we saw that a refinery's underwater supply line ruptures, resulting in a fairly circular oil spill. The radius was modeled by

$$r(t) = 0.7t$$

where t represents the number of seconds since the spill occurred and $r(t)$ represents the radius of the oil slick in feet. The area of the spill was given by

$$A(r) = \pi r^2$$

© Rainer von Brandis/iStockphoto

where r is the radius of the slick in feet and $A(r)$ is the area of the slick in square feet. Simplify the composition $A(r(t))$, which yields $A(t)$, and find the derivative $\frac{d}{dt}[A(t)]$. Interpret the resulting function.

Perform the Mathematics

Simplifying, we get

$$A(r(t)) = A(0.7t)$$

Substituting $0.7t$ for r in the function $A(r) = \pi r^2$, we get the area of the oil slick as a function of time:

$$A(r(t)) = \pi(0.7t)^2 = 0.49\pi t^2 = A(t)$$

Differentiating this result yields

$$\frac{d}{dt}[A(t)] = \frac{d}{dt}[0.49\pi t^2] = 0.98\pi t$$

The derivative, $\frac{d}{dt}[A(t)] = 0.98\pi t$, represents the instantaneous rate of change of the area of the oil slick with respect to time. ∎

The Flashback shows that we can differentiate a composite function, yet it does not explicitly show how this is done. Let's examine the functions in the Flashback again and see whether there is another way to find the derivative of $A(t) = A(r(t))$. If we compute the derivatives of the original functions $r(t) = 0.7t$ and $A(r) = \pi r^2$, with respect to t and r respectively, we have

$$r'(t) = 0.7 \quad \text{and} \quad A'(r) = 2\pi r$$

Since we know that $A'(r(t)) = 2\pi(0.7t) = 1.4\pi t$ and $r'(t) = 0.7$, it appears that the derivative $A'(t) = \frac{d}{dt}[A(t)]$ is equivalent to

$$A'(t) = A'(r(t)) \cdot r'(t) = (1.4\pi t) \cdot (0.7) = 0.98\pi t$$

This illustrates exactly how to differentiate a composite function using the **Chain Rule**. We now state this rule for functions made up of a composition of functions f and g.

Chain Rule

If $y = f(u)$ and $u = g(x)$ are used to define $h(x)$, where $h(x) = f(g(x))$, then

$$h'(x) = f'(g(x)) \cdot g'(x)$$

provided $f'(g(x))$ and $g'(x)$ exist.

NOTE: Using Leibniz notation, the Chain Rule is equivalent to $h'(x) = \dfrac{dy}{du} \cdot \dfrac{du}{dx}$ provided that $\dfrac{dy}{du}$ and $\dfrac{du}{dx}$ exist.

To differentiate $h(x) = f(g(x))$, it appears that we can differentiate the "outside" function f, and then chain it to the derivative of the "inside" function g. Many functions that we will study have the form $h(x) = (\textit{function})^{\textit{power}}$, where the power is some real number. For these types of functions, we can rely on an extension, or corollary, of this rule, called the **Generalized Power Rule**. Observe that this rule looks much like our Power Rule presented in Section 3.1.

Generalized Power Rule

If u is a differentiable function of x and n is any real number with $f(x) = [u(x)]^n$, then

$$f'(x) = n[u(x)]^{n-1} \cdot u'(x)$$

OBJECTIVE 1

Differentiate power functions using the Chain Rule.

Example 1: Using the Generalized Power Rule

Use the Generalized Power Rule to determine derivatives for the following.

a. $f(x) = (5x^3 + 3x)^4$　　　**b.** $g(x) = (x^2 + 1)^{15}$

Perform the Mathematics

a. For the function $f(x) = (5x^3 + 3x)^4$, consider $u(x) = 5x^3 + 3x$ and $n = 4$. Applying the Generalized Power Rule gives

$$f'(x) = \frac{d}{dx}[(5x^3 + 3x)^4]$$

$$= 4(5x^3 + 3x)^{4-1} \cdot \frac{d}{dx}(5x^3 + 3x)$$

$$= 4(5x^3 + 3x)^3(15x^2 + 3)$$

b. Here, we can consider $u(x) = x^2 + 1$ and $n = 15$ and apply the generalized Power Rule to get

$$g'(x) = \frac{d}{dx}[(x^2 + 1)^{15}]$$

$$= 15(x^2 + 1)^{14} \cdot \frac{d}{dx}(x^2 + 1)$$

$$= 15(x^2 + 1)^{14}(2x) = 30x(x^2 + 1)^{14} \qquad \blacksquare$$

▶ *Try It Yourself*

Some related Exercises are 15 and 17.

Notice the power of the Generalized Power Rule in part (b) of Example 1. It would have been possible, yet not at all practical, to algebraically expand the binomial $(x^2 + 1)^{15}$ in order to apply the differentiation techniques from Section 3.1.

We can use this new technique of differentiation to readily determine derivatives of new families of functions, including the radical and rational exponent functions. To use the Generalized Power Rule with these functions, let's review how these functions can be rewritten as shown in the Toolbox.

From Your Toolbox: Radical and Rational Exponent Functions

A function of the form $f(x) = \sqrt[b]{[g(x)]^a}$ is called a *radical function*. Rewriting $f(x) = \sqrt[b]{[g(x)]^a}$ as $f(x) = [g(x)]^{a/b}$ produces what we call the *rational exponent function*.

The key in differentiating radical functions is to rewrite them in the rational exponent form so that we can apply the Generalized Power Rule. We illustrate this in Example 2.

OBJECTIVE 2

Differentiate radical functions using the Chain Rule.

Example 2: Determining Derivatives of Radical Functions

a. Use the Generalized Power Rule to determine $f'(x)$ for $f(x) = \sqrt[3]{2x - 4}$.

b. Find an equation of the line tangent to the graph of f at the point $(6, 2)$.

Perform the Mathematics

a.   Before we can differentiate, we rewrite $f(x) = \sqrt[3]{2x - 4}$ using rational exponents.

$$f(x) = \sqrt[3]{2x - 4} = (2x - 4)^{1/3}$$

Using the Generalized Power Rule with $u(x) = (2x - 4)$ and $n = \frac{1}{3}$ gives us

$$f'(x) = \frac{d}{dx}[(2x - 4)^{1/3}]$$

$$= \frac{1}{3}(2x - 4)^{1/3 - 1} \cdot \frac{d}{dx}(2x - 4) = \frac{1}{3}(2x - 4)^{-2/3}(2) = \frac{2}{3}(2x - 4)^{-2/3}$$

Writing the derivative without negative or rational exponents, this simplifies to

$$f'(x) = \frac{2}{3}(2x - 4)^{-2/3} = \frac{2}{3(2x - 4)^{2/3}} = \frac{2}{3\sqrt[3]{(2x - 4)^2}}$$

b.   Since $f'(6)$ gives the slope of the tangent line at $x = 6$, we have

$$f'(6) = \frac{2}{3\sqrt[3]{(2(6) - 4)^2}} = \frac{2}{3\sqrt[3]{64}} = \frac{2}{3 \cdot 4} = \frac{1}{6}$$

With a slope of $\frac{1}{6}$ and point $(6, 2)$, we use the point-slope form of a line to get the tangent line equation:

$$y - y_1 = m(x - x_1)$$

$$y - 2 = \frac{1}{6}(x - 6)$$

$$y = \frac{1}{6}(x - 6) + 2$$

The graphs of $f(x) = \sqrt[3]{2x - 4}$ and $y = \frac{1}{6}(x - 6) + 2$ are shown in **Figure 3.3.1**.

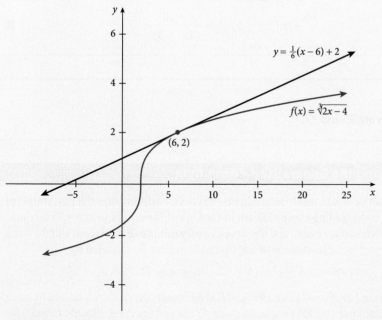

Figure 3.3.1

> ▶ **Try It Yourself**
>
> Some related Exercises are 41 and 47.

Another family of functions that can be differentiated using the Generalized Power Rule are rational functions. Even though the Generalized Power Rule can be used on any rational function, it is particularly useful on rational functions that have constants in their numerators.

OBJECTIVE 3

Differentiate rational functions using the Chain Rule.

Example 3: Determining Derivatives of Rational Functions Two Ways

For the function $h(x) = \dfrac{5}{(2x-3)^2}$, determine $h'(x)$ using the Quotient Rule and then determine $h'(x)$ using the Generalized Power Rule.

Perform the Mathematics

Using the Quotient Rule, we get

$$h'(x) = \frac{\dfrac{d}{dx}[5] \cdot (2x-3)^2 - (5) \cdot \dfrac{d}{dx}[(2x-3)^2]}{[(2x-3)^2]^2}$$

$$= \frac{0 \cdot (2x-3)^2 - 5 \cdot [2(2x-3)(2)]}{(2x-3)^4}$$

$$= \frac{-20(2x-3)}{(2x-3)^4} = \frac{-20}{(2x-3)^3}$$

When using the Generalized Power Rule, we rewrite the rational function as $h(x) = 5(2x-3)^{-2}$. With help from the Constant Multiple Rule, we determine

$$h'(x) = 5 \cdot \frac{d}{dx}[(2x-3)^{-2}]$$

$$= 5 \cdot (-2)(2x-3)^{-3} \cdot (2)$$

$$= -20(2x-3)^{-3} = \frac{-20}{(2x-3)^3} \qquad ■$$

> ▶ **Try It Yourself**
>
> Some related Exercises are 35 and 39.

Example 4: Applying the Generalized Power Rule

Even though the number of deaths due to heart disease has been declining over the past couple of decades, heart disease is the leading cause of death in the United States. About every 25 seconds, an American will have a coronary event, and about once every minute an American will die from one. The death rate caused by heart disease in the United States can be modeled by

$$f(x) = 369.2(x+1)^{-0.17} \qquad 0 \le x \le 17$$

where x represents the number of years since 1990 and $f(x)$ represents the death rate caused by heart disease measured in deaths per 100,000 people. Evaluate $f'(12)$ and interpret. (*Source:* Centers for Disease Control and Prevention.)

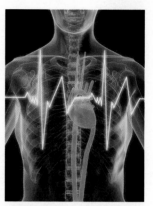

© Sebastian Kaulitzki/ShutterStock, Inc.

Understand the Situation

In order to evaluate $f'(12)$, we need to determine the derivative, $f'(x)$, using the Generalized Power Rule. Once we have the derivative, we will substitute 12 for each occurrence of x. Note that $x = 12$ corresponds to 2002.

Perform the Mathematics

Using the Generalized Power Rule, we get

$$f'(x) = \frac{d}{dx}[369.2(x + 1)^{-0.17}]$$

$$= (369.2) \cdot (-0.17)(x + 1)^{-0.17-1} \cdot (1)$$

$$= -62.764(x + 1)^{-1.17}$$

We now evaluate $f'(12)$ as follows:

$$f'(12) = -62.764(12 + 1)^{-1.17} \approx -3.12$$

Interpret the Result

This means that in 2002, the death rate caused by heart disease was decreasing at a rate of about 3.12 deaths per 100,000 people per year. See **Figure 3.3.2**.

OBJECTIVE 4

Apply the Chain Rule to health science and business functions.

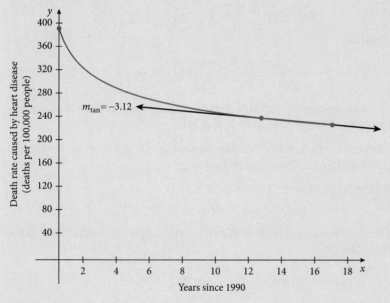

Figure 3.3.2

▶ *Try It Yourself*

Some related Exercises are 81 and 83. ∎

Our final example in this section demonstrates how the Generalized Power Rule can be used with business functions.

Example 5: Applying the Generalized Power Rule to Business Functions

The RadRadio Company produces and sells personal stereo devices. Market research has found that the price-demand function is

$$p(x) = 100 - \sqrt{x^2 + 20} \qquad 0 \le x \le 35$$

where x is the number of devices demanded in thousands and $p(x)$ represents the unit price in dollars.

a. Evaluate $p'(30)$ and interpret.   **b.** Evaluate $R'(30)$ and interpret.

Perform the Mathematics

a. To aid us in determining $p'(x)$, we write this radical function as a rational exponent function. The rational exponent form of the price-demand function is

$$p(x) = 100 - (x^2 + 20)^{1/2} \qquad 0 \le x \le 35$$

Using the Generalized Power Rule, we get the derivative

$$p'(x) = -\frac{1}{2}(x^2 + 20)^{-1/2} \cdot \frac{d}{dx}(x^2 + 20)$$

$$= -\frac{1}{2}(x^2 + 20)^{-1/2} \cdot (2x)$$

$$= -x(x^2 + 20)^{-1/2}$$

$$= \frac{-x}{(x^2 + 20)^{1/2}} = \frac{-x}{\sqrt{x^2 + 20}}$$

Evaluating $p'(30)$ yields

$$p'(30) = \frac{-30}{\sqrt{(30)^2 + 20}} = \frac{-30}{\sqrt{920}} \approx -0.99$$

This means that at a demand level of 30,000 personal stereo devices, the price is decreasing at a rate of about \$0.99 (or 99 cents) per 1000 devices sold.

b. We need to first determine the revenue function. Knowing that the revenue function is $R(x) = x \cdot p(x)$, we use the price-demand function to get

$$R(x) = x \cdot p(x) = x \cdot (100 - \sqrt{x^2 + 20})$$
$$= 100x - x\sqrt{x^2 + 20} = 100x - x(x^2 + 20)^{1/2}$$

Notice that when we determine the derivative $R'(x)$, we must apply the Product Rule to the second term of $R(x)$. This gives us

$$R'(x) = \frac{d}{dx}[100x - x(x^2 + 20)^{1/2}]$$

Sum and Difference Rule $\qquad = \dfrac{d}{dx}[100x] - \dfrac{d}{dx}[x(x^2 + 20)^{1/2}]$

Product Rule $\qquad = \dfrac{d}{dx}[100x] - \left[\dfrac{d}{dx}(x) \cdot (x^2 + 20)^{1/2} + x \cdot \dfrac{d}{dx}(x^2 + 20)^{1/2}\right]$

Chain Rule $\qquad = 100 - (1)(x^2 + 20)^{1/2} - x\left(\dfrac{1}{2}\right)((x^2 + 20)^{-1/2}(2x))$

Simplify $\qquad = 100 - (x^2 + 20)^{1/2} - x^2(x^2 + 20)^{-1/2}$

Rewrite $\qquad = 100 - \sqrt{x^2 + 20} - \dfrac{x^2}{\sqrt{x^2 + 20}}$

Evaluating $R'(x)$ at $x = 30$ produces

$$R'(30) = 100 - \sqrt{(30)^2 + 20} - \frac{(30)^2}{\sqrt{(30)^2 + 20}}$$

$$= 100 - \sqrt{920} - \frac{900}{\sqrt{920}} \approx 40.0$$

This means that when the production level is 30,000 personal stereo devices, revenue is increasing at a rate of about \$40 per thousand devices sold. ∎

Summary

In this section, we reviewed composite functions and then introduced the Chain Rule, which is a technique used to differentiate composite functions. We then focused on the Generalized Power Rule, which is an extension of the Chain Rule.

- **Chain Rule:** If $y = f(u)$ and $u = g(x)$ are used to define $h(x)$, where $h(x) = f(g(x))$, then $h'(x) = f'(g(x)) \cdot g'(x)$ provided $f'(g(x))$ and $g'(x)$ exist.

- **Generalized Power Rule:** If u is a differentiable function of x and n is any real number with $f(x) = [u(x)]^n$, then $f'(x) = n[u(x)]^{n-1} \cdot u'(x)$.

Section 3.3 Exercises

Vocabulary Exercises

1. The Chain Rule is used to differentiate _____ functions.
2. A function of the form $f(x) = \sqrt[b]{[g(x)]^a}$ is called a _____ function.
3. If n represents a real number and $f(x) = [g(x)]^n$, then we can differentiate f using the _____ _____ Rule.
4. A function of the form $f(x) = [g(x)]^{a/b}$ produces what we call a _____ exponent function.

Skill Exercises

For Exercises 5–10, differentiate using the Generalized Power Rule.

5. $f(x) = (x + 1)^2$
6. $f(x) = (x + 3)^2$
7. $f(x) = (x - 5)^3$
8. $f(x) = (x - 2)^3$
9. $f(x) = (2 - x)^2$
10. $f(x) = (5 - x)^2$

For Exercises 11–34, differentiate using the Generalized Power Rule.

11. $f(x) = (2x + 4)^3$
12. $f(x) = (3x + 3)^3$
13. $f(x) = (5 - 2x)^5$
14. $f(x) = (10 - 5x)^4$
15. $f(x) = (3x^2 + 7)^5$
16. $f(x) = (4x^2 - 3)^3$
17. $f(x) = (x^3 - 2x^2 + x)^2$
18. $f(x) = (2x^3 + 4x + 3)^3$
19. $f(x) = 3(x^3 - 4)^3$
20. $f(x) = 5(4x^2 - 10)^6$
21. $f(x) = 5(5x^2 - 3x - 1)^{10}$
22. $f(x) = 10(10x^3 + x - 9)^8$
23. $f(x) = (4x^2 - x - 4)^{55}$
24. $f(x) = (8x^2 - 2x + 5)^{94}$

25. $f(x) = (2x - 4)^{1/2}$

26. $f(x) = (7x + 6)^{1/2}$

27. $f(x) = (x^2 + 2x)^{1/3}$

28. $f(x) = (x^3 + 5x)^{1/3}$

29. $f(x) = (5x - 2)^{-2}$

30. $f(x) = (4x + 3)^{-3}$

31. $f(x) = (x^2 + 2x + 4)^{-1/2}$

32. $f(x) = (3x^2 + 5x + 6)^{-1/2}$

33. $f(x) = (3x^3 - x)^{-1/4}$

34. $f(x) = (4x^5 + 5x^3)^{-1/3}$

For the rational functions in Exercises 35–40, complete the following:

(a) *Determine the derivative using the Quotient Rule.*

(b) *Determine the derivative using the Generalized Power Rule.*

35. $f(x) = \dfrac{1}{3x + 4}$

36. $f(x) = \dfrac{1}{7x - 5}$

37. $f(x) = \dfrac{5}{(x - 2)^2}$

38. $f(x) = \dfrac{10}{(2x - 1)^3}$

39. $f(x) = \dfrac{2}{x^2 + 2x + 3}$

40. $f(x) = \dfrac{9}{x^3 + 2x + 10}$

In Exercises 41–48, determine an equation of the line tangent to the graph of f at the indicated point.

41. $f(x) = (2x - 1)^3; (1, 1)$

42. $f(x) = (3x - 4)^3; (1, -1)$

43. $f(x) = (2 - x)^4; (1, 0)$

44. $f(x) = (x^2 - 1)^4; (1, 0)$

45. $f(x) = (x^3 - 4x + 2)^4; (2, 16)$

46. $f(x) = (4x - 3)^{1/2}; (3, 3)$

47. $f(x) = (2x - 4)^{1/2}; (2, 0)$

48. $f(x) = (2x + 8)^{1/2}; (4, 4)$

Use the Generalized Power Rule to differentiate the functions in Exercises 49–56.

49. $f(x) = \sqrt{x^2 + 5}$

50. $f(x) = \sqrt{3x + 6}$

51. $f(x) = \sqrt[3]{2x - 1}$

52. $f(x) = \sqrt[3]{4x - 3}$

53. $f(x) = \dfrac{5}{\sqrt{2x - 8}}$

54. $f(x) = \dfrac{10}{\sqrt{5x + 8}}$

55. $f(x) = \dfrac{64}{\sqrt[3]{5x^2 - 6x + 3}}$

56. $f(x) = \dfrac{27}{\sqrt[3]{3x^3 + x}}$

In Exercises 57–66, use the Generalized Power Rule, along with the Product and Quotient Rules, to determine the derivative of the given functions.

57. $f(x) = x(x - 4)^3$

58. $f(x) = x(10 - x)^3$

59. $f(x) = x\sqrt{x^2 + 3x}$

60. $f(x) = x^2\sqrt{2x^2 - 11}$

61. $f(x) = \dfrac{x^3}{(3x - 8)^2}$

62. $f(x) = \dfrac{x^2}{(4x^2 - x + 5)^3}$

63. $f(x) = (x + 3)^3(2x - 1)^2$

64. $f(x) = (3x - 3)^4(2x - 2)^3$

65. $f(x) = \sqrt{\dfrac{x + 3}{x - 3}}$

66. $f(x) = \sqrt{\dfrac{2x + 1}{2x - 1}}$

In Exercises 67–72, find the derivatives using the Generalized Power Rule.

67. $f(x) = (4x^2 + 5x + 6)^{0.23}$

68. $f(x) = 3(0.7x^3 - 0.02x^2)^{0.09}$

69. $f(x) = \left(\dfrac{1}{x + 3}\right)^{-1.03}$

70. $f(x) = \left(\dfrac{1}{0.2x + 1.7}\right)^{-1.1}$

71. $f(x) = 1.44(x + 1)^{1.22}$

72. $f(x) = 67.41(x + 1)^{0.97}$

Application Exercises

73. **Actuarial Science—Survival Rates:** The Hogan Actuary Firm has determine that for residents who were born and raised in Buchannan County, the number of people surviving since World War II can be modeled by

$$f(x) = 400\sqrt{100 - x} \qquad 0 \le x \le 100$$

where x represents the age of the Buchannan County resident and $f(x)$ represents the number of residents surviving. Evaluate and interpret $f'(70)$.

74. **Market Research—TV Ratings:** During its first season, the number of viewers who watched the new television series "It Ain't Me!" can be modeled by

$$f(x) = \sqrt[3]{(50 + 2x)^2} \qquad 1 \le x \le 26$$

where x represents the number of weeks the series has been airing and $f(x)$ represents the number of viewers in millions. Evaluate and interpret $f'(13)$.

75. **Education—Enrollment Trends:** A study by the Dean of Clarksman University determines that the annual student enrollment in the Arts and Science College from 2001 to 2011 can be modeled by

$$f(x) = -\frac{10,000}{\sqrt{1 + 0.18x}} + 10,000 \qquad 1 \le x \le 11$$

where x represents the number of years since 2000 and $f(x)$ represents the student enrollment. Use the model to determine the rate of change in enrollment in the Arts and Science College in 2010.

76. **Medicine—Arteriosclerosis:** Medical researchers studying arteriosclerosis at the Giuliani Institute have found that if the radius of an examined patient's artery is currently one centimeter, the amount of fatty tissue called plaque that will build up in the artery can be modeled by the function

$$f(x) = 0.5x^2(x^2 + 10)^{-1} \qquad 0 \le x \le 10$$

where x represents the number of years since the initial examination and $f(x)$ represents the thickness of the artery wall in centimeters. Use the model to determine the rate of change in plaque thickness seven years after a patient was initially examined.

In Exercises 77 and 78, consider the following. In the early 1930s, psychologist L. L. Thurstone determined that the time needed to memorize a list of random words is given by the model

$$f(x) = ax\sqrt{x - b} \qquad x \ge b$$

where x represents the number of words on the list and $f(x)$ represents the time needed to memorize the list, measured in minutes. The constants a and b are different for each subject and are determined by pretesting.

77. **Psychology—Memorization Rates:** Erica is a freshman subject in a psychology class and learns the items on the memorization list according to the model

$$f(x) = \frac{5}{2}x\sqrt{x - 6} \qquad x \ge 6$$

   (a) Evaluate $f(20)$ and interpret.
   (b) Determine $f'(x)$.
   (c) Evaluate $f'(20)$ and interpret.

78. **Psychology—Memorization Rates:** José has completed his pretesting, and it has been determined that he learns the items on the memorization list according to the model

$$f(x) = 2x\sqrt{x - 3} \qquad x \ge 3$$

(a) Evaluate $f(12)$ and interpret.

(b) Determine $f'(x)$.

(c) Evaluate $f'(12)$ and interpret.

79. **Ecology—Chemical Run-Off:** The amount of toxic material entering Lake Formica is related to the number of years that the Bristine Chemical Company has been operating and can be modeled by

$$f(x) = \left(\frac{4}{5}x^{1/5} + 2\right)^4 \qquad 0 \le x \le 30$$

where x represents the number of years that the company has been operating and $f(x)$ represents the amount of toxic material entering the lake, measured in gallons.

(a) Evaluate $f(15)$ and interpret.

(b) Determine $f'(x)$.

(c) Evaluate $f'(15)$ and interpret.

80. **Accounting—Tax Revenue:** Suppose that the rural town of Rufusville decides to relax its zoning laws so that more land can be made eligible for commercial use. They estimate that the town's annual tax revenue after the zoning changes can be modeled by

$$f(x) = 5x\sqrt{2x + 2} \qquad x \ge 0$$

where x represents the number of years since the zoning laws were changed, and $f(x)$ represents the annual tax revenue in thousands of dollars.

(a) Evaluate $f(15)$ and interpret.

(b) Determine $f'(x)$.

(c) Evaluate $f'(15)$ and interpret.

 81. **Macroeconomics—Minimum Wage:** Suppose an economist wants to examine the rate of growth in minimum wage in certain European countries. The monthly minimum wage in Romania from 2000 to 2010 can be modeled by

$$f(x) = 2.27(2.66x + 8.11)^{1.19} \qquad 0 \le x \le 10$$

where x represents the number of years since 2000 and $f(x)$ represents the Romanian national monthly minimum wage, measured in euros per month. (*Source:* EuroStat.) Evaluate $f'(8)$ and interpret.

 82. **Macroeconomics—Minimum Wage:** Suppose an economist wants to examine the rate of growth in minimum wage in certain European countries. The monthly minimum wage in Greece from 2000 to 2010 can be modeled by

$$f(x) = 2.46(0.83x + 23.98)^{1.68} \qquad 0 \le x \le 10$$

where x represents the number of years since 2000 and $f(x)$ represents the Greek national monthly minimum wage, measured in euros per month. (*Source:* EuroStat.) Evaluate $f'(4)$ and interpret.

 83. **Women's Studies—Unmarried Labor Force:** A sociological topic that has been discussed since the end of World War II is the status of women in the labor force. The number of unmarried women in the U.S. labor force from 1970 to 2010 can be modeled by

$$f(x) = 1.11(1.57x + 17.27)^{0.66} \qquad 0 \le x \le 40$$

where x represents the number of years since 1970 and $f(x)$ represents the number of unmarried women in the U.S. labor force, measured in millions. (*Source:* U.S. Bureau of Labor Statistics.)

(a) Determine $f'(x)$.

(b) Evaluate $f(35)$ and $f'(35)$ and interpret each.

84. Women's Studies—Married Labor Force: Over the past half century, increasing numbers of married women have gone to work. The number of married women in the U.S. labor force from 1970 to 2010 can be modeled by

$$f(x) = 2.57(18.33x + 139.79)^{0.4} \quad 0 \le x \le 40$$

where x represents the number of years since 1970 and $f(x)$ represents the number of married women in the U.S. labor force, measured in millions. (*Source:* U.S. Bureau of Labor Statistics.)

(a) Determine $f'(x)$.

(b) Evaluate $f(35)$ and $f'(35)$ and interpret each.

Concept and Writing Exercises

For Exercises 85–88, suppose that g is a differentiable function and the derivative of g is $g'(x) = \frac{1}{x}$. Use the Chain Rule to differentiate the following functions. Algebraically simplify your answers.

85. $f(x) = g(x^3)$

86. $f(x) = g\left(\dfrac{1}{x}\right)$

87. $f(x) = g\left(\dfrac{3}{4\sqrt{x}}\right)$

88. $f(x) = g\left(\dfrac{1 + 2x}{1 - x}\right)$

89. Consider the function $f(x) = (3x + 5)^4(2x - 8)^3$. Write a few complete sentences that describe the strategy you would use to differentiate the function f. Do not determine the derivative f'; just explain how you would do it.

90. Let $h(x) = \dfrac{f(x)}{g(x)}$. Use the Chain Rule to derive the Quotient Rule by first rewriting h as $h(x) = f(x) \cdot g(x)^{-1}$.

91. Determine the derivative of the function $f(x) = \sqrt{x + \sqrt{x + \sqrt{x}}}$ by using the Chain Rule two times.

*In Exercises 92–94, consider the following. Suppose that the value of x and y are determined by another value, or parameter t. That is, $x = x(t)$ and $y = y(t)$. Then x and y are called **parametric equations**. We can show that the derivative of y with respect to x for parametric equations is given by $\dfrac{dy}{dt} = \dfrac{y'(t)}{x'(t)} = \dfrac{\frac{dy}{dt}}{\frac{dx}{dt}}$. Determine the derivative $\frac{dy}{dx}$ for each of the following and write your answer in terms of the independent variable t.*

92. $y = y(t) = t^2 + 1, x = x(t) = 3t^3$

93. $y = y(t) = \sqrt{t + 1}; x = x(t) = \sqrt{2t}$

94. $y = y(t) = \dfrac{1}{1 - t}; x = x(t) = \dfrac{1}{t}$

 ## Section Project

In November 2006, Mitra Toossi, an economist at the Office of Occupational Statistics and Employment Projections, stated that "peaking at 2.6 percent during the 1970s, the growth rate of the labor force has been decreasing with the passage of each decade and is expected to continue to do so in the future." (*Source:* Monthly Labor Review Online). To test the economist's remarks, we can derive a model for the size of the labor force, and then inspect its relative growth. The civilian labor force since 1980 and projected into 2016 can be modeled by the function

$$f(x) = \sqrt{ax + b} \quad 0 \le x \le 36$$

where x represents the number of years since 1980 and $f(x)$ represents the civilian labor force, measured in millions. (*Source:* U.S. Bureau of Labor Statistics.) The values a and b represent whole number constants.

(a) In 1980 the civilian labor force was 107.35 million. For the model this means that $f(0) = 107.35$. Use this information to solve for the constant b.

(b) We also know that in 2000, the civilian labor force was 140.9 million. Use this information to determine the constant a in the model $f(x) = \sqrt{ax + b}$.

(c)   Use the techniques of this section to determine $f'(x)$.

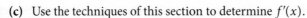

(d)   Use your calculator to make a table of values for f and f' where $Y_1 = f(x)$ and $Y_2 = f'(x)$. What observation can you make about the values of the derivative?

(e)   Assign $Y_3 = Y_2/Y_1*100$ to form the percentage rate of change function
$$g(x) = \frac{f'(x)}{f(x)} \cdot 100.$$ Evaluate this percentage rate of change function at $x = 0$, $x = 10$, $x = 20$, $x = 30$, and $x = 36$. Do these percentages support the economist's claim?

SECTION OBJECTIVES

1. Compute a differential.
2. Apply the differential to a business function.
3. Compute a linear approximation.

3.4   The Differential and Linear Approximations

How often have you heard on TV or read in a newspaper or online a statement such as, "If crime keeps growing at this rate. . . ." The problem with this statement is that rates do indeed change. This is the reason that we study calculus—to see the impact of these fluctuating rates of change. In this section, we will examine the mathematics of the statement made above, and how to use the **differential** to make short-term conclusions about applications. Then we will study the idea of **linear approximations**, which can be used to approximate the change in the dependent variable as changes in the independent variable are made.

The Differential

The differential is a tool used to study the relationship between changes in the independent and dependent variable values. So let's begin by revisiting some tools that we introduced with the difference quotient.

From Your Toolbox: The Difference Quotient

Recall that for a function f on a closed interval $[x_1, x_2]$:

- $\Delta x = x_2 - x_1$ is called the increment in x. In Chapter 2 we called this quantity h.
- $\Delta y = f(x_2) - f(x_1)$ is called the increment in y.
- The slope of a secant line, m_{sec}, is given by $m_{sec} = \dfrac{\Delta y}{\Delta x} = \dfrac{f(x + \Delta x) - f(x)}{\Delta x}$. Recall that $\dfrac{f(x + \Delta x) - f(x)}{\Delta x}$ is called the difference quotient.

Now we can establish a context for our discussion through a Flashback.

Flashback: Prison Inmate Population Revisited

In Section 3.1, we modeled the total U.S. Federal Bureau of Prisons inmate population with the function
$$f(x) = -\frac{3}{50}x^2 + \frac{361}{40}x + 56.21 \qquad 1 \le x \le 19$$

where x represents the number of years since 1989, and $f(x)$ represents the total inmate population in thousands. (*Source:* Federal Bureau of Prisons.)

a.   Evaluate $f(17) - f(16)$ and interpret this result.

b.   Determine the difference quotient with $x = 16$ and $\Delta x = 1$ and interpret. Compare to part (a).

© AbleStock

Perform the Mathematics

a.   First note that the values $x = 16$ and $x = 17$ correspond to the years 2005 and 2006, respectively. Also note that when computing $f(17) - f(16)$, we are simply finding a change in y, that is, Δy.

$$f(17) - f(16) = \left[-\frac{3}{50}(17)^2 + \frac{361}{40}(17) + 56.21 \right] - \left[-\frac{3}{50}(16)^2 + \frac{361}{40}(16) + 56.21 \right]$$

$$= 192.295 - 185.25 = 7.045$$

So from 2005 to 2006, the U.S. federal prison inmate population increased by 7.045 thousand, or 7045 inmates.

b. Computing the difference quotient with $x = 16$ and $\Delta x = 1$ gives us

$$m_{sec} = \frac{f(x + \Delta x) - f(x)}{\Delta x} = \frac{f(16 + 1) - f(16)}{1}$$

$$= \frac{f(17) - f(16)}{1} = \frac{192.295 - 185.25}{1} = 7.045$$

This means that from 2005 to 2006, the U.S. federal prison inmate population increased at an average rate of 7045 inmates per year. Notice that this result is the same as in part (a). In other words, **when $\Delta x = 1$, then** $\Delta y = 3.61 = m_{sec}$, where m_{sec} is the slope of the secant line. ∎

Now let's mathematically try the scenario that we suggested at the beginning of this section. What if the prison population continued to grow at a rate constant with that of 2005? Will this assumption give an accurate prediction of the 2006 U.S. federal prison population? Using the derivative and the tangent line, we can find out. First, we find the growth rate in 2005 using the derivative f', since we know that $f'(16)$ is the instantaneous rate of change at $x = 16$:

$$f'(x) = \frac{d}{dx}\left(-\frac{3}{50}x^2 + \frac{361}{40}x + 56.21 \right) = -\frac{3}{25}x + \frac{361}{40}$$

$$f'(16) = -\frac{3}{25}(16) + \frac{361}{40} = \frac{1421}{200} = 7.105$$

Thus, the U.S. federal prison inmate population was growing at a rate of 7105 inmates per year in 2005. Assuming that the rate of change is constant during 2005, we use the equation of the line tangent to the curve at $x = 16$ to predict the 2006 federal inmate population. Example 1 illustrates the process.

Example 1: Using the Tangent Line Equation

Determine an equation of the line tangent to the graph of the model $f(x) = -\frac{3}{50}x^2 + \frac{361}{40}x + 56.21$ when $x = 16$. On the tangent line, determine y when $x = 17$ and interpret what this means.

Perform the Mathematics

We will use the derivative to determine the slope of the tangent line. Specifically, $f'(16)$ gives us the slope of the line tangent to the graph of f at $x = 16$. The point of tangency is $(16, f(16)$. Once we have the slope of the tangent line and the point of tangency, we will use the point-slope form to determine an equation of the tangent line.

Knowing that $f(16) = -\frac{3}{50}(16)^2 + \frac{361}{40}(16) + 56.21 = 185.25$ and $f'(16) = 7.105$, we use the point-slope form of a line to determine an equation of the tangent line at $x = 16$:

$$y - y_1 = m(x - x_1)$$

$$y - 185.25 = 7.105(x - 16)$$

$$y = 7.105(x - 16) + 185.25$$

Evaluating the tangent line equation at $x = 17$ gives us

$$y = 7.105(17 - 16) + 185.25 = 7.105(1) + 185.25 = 192.355$$

This means that if the U.S. federal prison inmate population continued growing at the 2005 rate, the 2006 inmate population would be about 192.355 thousand. Using the model, the number of inmates in 2006 was $f(17) = 192.295$, meaning that there was a difference of only about 0.06 thousand, or 60 inmates. See **Figure 3.4.1**.

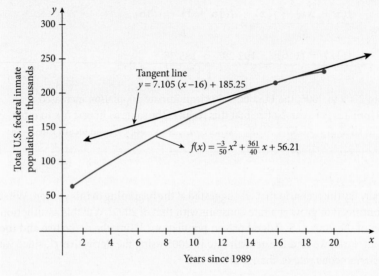

Figure 3.4.1   ■

We can make two important observations from the results of the Flashback and from Example 1:

1. From part (a) of the Flashback, we found that the actual change as x changed from 16 to 17 ($\Delta x = 1$) was $\Delta y = 7.045$ thousand prisoners. We also found that $f'(16)$, the instantaneous rate of change at $x = 16$ was 7.105 thousand inmates per year. Thus, for $\Delta x = 1$, Δy is approximately equal to $f'(16)$.

2. Example 1 showed that $f(17) = 192.295$ and the y-value on the tangent line at $x = 17$ ($y = 192.355$) are very close.

The concept of the **differential** is embedded in observation 1, while the concept of a **linear approximation** is embedded in observation 2. We will discuss the linear approximation later in this section, but before that, let's examine the **differential** and its applications.

Many disciplines think of dy as the change in y on the tangent line. We show the difference between Δy and dy in **Figure 3.4.2**.

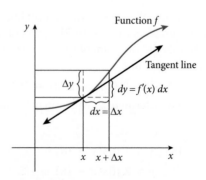

Figure 3.4.2

Example 2: Computing the Differential

For the function $f(x) = x^2 + 3x - 8$, evaluate Δy and dy for $x = 2$ and $\Delta x = dx = 0.1$.

Perform the Mathematics

We can compute the change in y, Δy, by evaluating $f(x + \Delta x) - f(x)$:

$$
\begin{aligned}
y &= f(2 + 0.1) - f(2) \\
&= f(2.1) - f(2) \\
&= ((2.1)^2 + 3(2.1) - 8) - ((2)^2 + 3(2) - 8) \\
&= 2.71 - 2 = 0.71
\end{aligned}
$$

Since $f'(x) = 2x + 3$, the differential in y for the function is

$$
\begin{aligned}
dy &= f'(x)\,dx \\
dy &= (2x + 3)\,dx
\end{aligned}
$$

So at $x = 2$ and $\Delta x = dx = 0.1$, we get

$$
dy = (2(2) + 3)(0.1) = 7(0.1) = 0.7
$$

Notice that $dy \approx \Delta y$ in this example. Again, this is true when dx is small. ∎

▶ *Try It Yourself*

Some related Exercises are 23 and 25.

Our next example illustrates how differentials can be used in business applications.

Example 3: Using Differentials in Applications

The Garland Toddler Company determined that the price-demand function for their pacifier/thermometer is given by

$$
p(x) = 15 - 0.2\sqrt{x}
$$

where x represents the quantity demanded and $p(x)$ represents the unit price in dollars.

a. Compute Δp, the actual change in price, for $x = 100$ and $\Delta x = dx = 1$.

b. Determine the differential dp for the price-demand function. Use the differential dp to approximate the change in price that would cause the quantity demanded to increase from 100 to 101 units.

Perform the Mathematics

a. Using $x = 100$ and $\Delta x = dx = 1$, we get the increment in p as

$$
\begin{aligned}
\Delta p &= p(x + \Delta x) - p(x) \\
&= p(100 + 1) - p(100) \\
&= p(101) - p(100) \\
&\approx 12.99002 - 13 \\
&= -0.00998
\end{aligned}
$$

b. For the function $p(x) = 15 - 0.2\sqrt{x}$, the differential in the dependent variable p is

$$dp = p'(x)dx = \frac{d}{dx}[15 - 0.2\sqrt{x}]dx$$

$$= -0.2\left(\frac{1}{2}x^{-1/2}\right)dx$$

$$= (-0.1x^{-1/2})dx$$

Evaluating dp when $x = 100$ and $\Delta x = dx = 101 - 100 = 1$, we get

$$dp = (-0.1(100)^{-1/2})(1)$$

$$= (-0.1)(0.1)(1) = -0.001$$

This means that as the quantity demanded changes from 100 to 101 units, the unit price of the pacifier/thermometer would decrease by about 1 cent. Notice how close the numerical solutions are in parts (a) and (b). This is further evidence that **if dx is small, then** $dy \approx \Delta y$.

Linear Approximations

To get a better picture of why linear approximations are used, let's return to the model for the U.S. federal prison inmate population:

$$f(x) = -\frac{3}{50}x^2 + \frac{361}{40}x + 56.21 \qquad 1 \le x \le 19$$

Let's investigate the graph of f around $x = 18$. We graph f in smaller and smaller intervals as shown in **Figure 3.4.3**. Notice that as we zoom in on f for values close to $x = 18$, the graph appears to straighten out and look like a line. Let's see if we can take advantage of this characteristic to approximate function values.

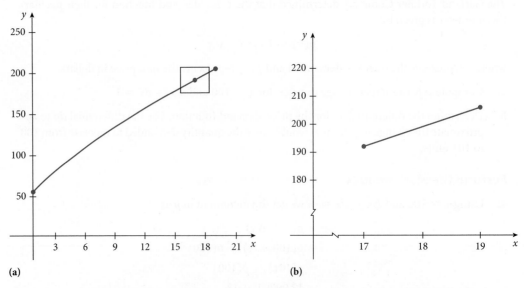

(a)

(b)

Figure 3.4.3 **(a)** We zoom in on that portion of the graph shown in the rectangle and redraw the graph shown in Figure 3.4.3(b). **(b)** On this smaller interval, the graph of f appears to be nearly linear.

Example 4: Computing a Linear Approximation

For the model of the total U.S. federal prison inmate population:

a.   Determine dy and evaluate when $x = 18$ and $\Delta x = dx = 1$.

b.   Evaluate $f(19)$ and interpret.

c.   Add $f(18)$ to the result from part (a) and compare to part (b).

Perform the Mathematics

a.   For the model $f(x) = -\frac{3}{50}x^2 + \frac{361}{40}x + 56.21$, the derivative is $f'(x) = -\frac{3}{25}x + \frac{361}{40}$. The differential in y for the model is

$$dy = f'(x)dx = \left(-\frac{3}{25}x + \frac{361}{40}\right)dx$$

Evaluating for $x = 18$ and $\Delta x = dx = 1$, the differential in y is

$$dy = f'(18) \cdot 1$$
$$= 6.865$$

b.   We evaluate $f(19)$ by substituting 19 for every occurrence of x in the function $f(x) = -\frac{3}{50}x^2 + \frac{361}{40}x + 56.21$. Note that $x = 19$ corresponds to the year 2008. Evaluating we get

$$f(x) = -\frac{3}{50}x^2 + \frac{361}{40}x + 56.21$$

$$= -\frac{3}{50}(19)^2 + \frac{361}{40}(19) + 56.21 = 206.025$$

This means that in 2008, there were about 206,025 inmates in U.S. federal prisons.

c.   Evaluating the model at $x = 18$ gives

$$f(18) = -\frac{3}{50}(18)^2 + \frac{361}{40}(18) + 56.21 = 199.22$$

Adding this result to part (a) gives us

$$f(18) + dy = f(18) + f'(18) \cdot 1 = 199.22 + 6.865 = 206.085$$

Notice how close this result is to part (b). In other words, $f(19) \approx f(18) + f'(18) \cdot 1$. ∎

Notice in Example 4 that we can rewrite $f(19)$ as $f(18 + 1)$. So it appears, for a small value of dx, that $f(18 + 1) \approx f(18) + f'(18)(1)$. This is exactly what a **linear approximation** is. We generalize this new tool in the following definition.

DEFINITION

Linear Approximation

For a differentiable function f, where $y = f(x)$, the **linear approximation** of f is given by

$$f(x + dx) \approx f(x) + dy = f(x) + f'(x)dx$$

when the value of dx is small.

Both the linear approximation and the differential will be the cornerstone when we study Marginal Analysis in Section 3.5. Let's now try a classic application of linear approximations.

Example 5: Using Linear Approximations to Make Estimates

Use a linear approximation to estimate $\sqrt[3]{63}$ and compare with the calculator value of 3.9791 (rounded to four decimal places).

Perform the Mathematics

A first, it appears that we have nothing to work with to get the differential. But, on closer inspection, we see that a number very close to $\sqrt[3]{63}$ has an integer cube root, that is, $\sqrt[3]{64} = 4$. Thus, we can use the linear approximation to make an estimate of $\sqrt[3]{63}$ by letting $f(x) = \sqrt[3]{x}$, $x = 64$, and $dx = 63 - 64 = -1$. This means that we want to determine

$$f(x + dx) \approx f(x) + f'(x)dx$$
$$f(64 + (-1)) \approx f(64) + f'(64)(-1)$$

Knowing that $f(x) = \sqrt[3]{x} = x^{1/3}$, we get $f'(x) = \dfrac{1}{3}x^{-2/3} = \dfrac{1}{3\sqrt[3]{x^2}}$. So the linear approximation is

$$\sqrt[3]{63} = f(64 + (-1)) \approx f(64) + f'(64)(-1)$$

$$= \sqrt[3]{64} + \frac{1}{3\sqrt[3]{(64)^2}}(-1)$$

$$= 4 + \frac{1}{3(16)}(-1)$$

$$= 4 - \frac{1}{48} = 3\frac{47}{48}$$

As a decimal, $3\frac{47}{48} \approx 3.9792$, whereas $\sqrt[3]{63} \approx 3.9791$. The difference using a linear approximation is only about one ten-thousandth. ∎

Summary

This section began with a discussion of the **differential in x,** denoted by dx, and the **differential in y**, where $dy = f'(x)dx$. We said that if dx is small in value, then $dy \approx \Delta y$. Then we introduced the **linear approximation** and stated that, when dx is small, $f(x + dx) \approx f(x) + dy = f(x) + f'(x)dx$.

Section 3.4 Exercises

Vocabulary Exercises

1. Recall that the expression $\dfrac{f(x + \Delta x) - f(x)}{\Delta x}$ is called the _____ quotient.

2. The _____ in x is given by $dx = \Delta x$.

3. The value Δy represents the _____ change in the dependent variable values.

4. The differential in y, denoted by dy, gives an _____ change in the dependent variable values.

5. We know that dy is approximately equal to Δy if the value of dx is relatively _____.

6. If dx is relatively small, then $f(x) + dy$ gives a _____ approximation of $f(x + dx)$.

Skill Exercises

In Exercises 7–22, find dy for the given function and write your answer in the form $dy = f'(x)dx$.

7. $f(x) = 6x$

8. $f(x) = -3x$

9. $f(x) = -3x^2 + 2x$

10. $f(x) = 7x^3 + 3x^2 - 13$

11. $f(x) = \dfrac{5}{x-1}$

12. $f(x) = \dfrac{-2}{x+3}$

13. $f(x) = \dfrac{x}{x+1}$

14. $f(x) = \dfrac{2x}{x-3}$

15. $f(x) = \sqrt{x} + \dfrac{2}{x}$

16. $f(x) = \sqrt[3]{x} - \dfrac{3}{x^2}$

17. $f(x) = \dfrac{1}{\sqrt{x}} + \sqrt[3]{x^2}$

18. $f(x) = \dfrac{1}{2\sqrt{x}} - \sqrt{x}$

19. $f(x) = \dfrac{x^2 + 1}{x^2 - 1}$

20. $f(x) = \dfrac{x^2 + 3}{x^2 - 3}$

21. $f(x) = 3x^{1.7} + 7x^{0.8} + 3$

22. $f(x) = 4x^{2.2} - 6x^{0.7} + 7$

In Exercises 23–34, evaluate Δy and dy for each function using the given x and dx values.

23. $f(x) = x^2 - 2x - 1$, $x = 2$, $\Delta x = dx = 0.1$

24. $f(x) = x^2 + 5x$, $x = 1$, $\Delta x = dx = 0.2$

25. $f(x) = 750 + 5x - 2x^3$, $x = 50$, $\Delta x = dx = 2$

26. $f(x) = 1000 + 2x - 3x^2$, $x = 100$, $\Delta x = dx = 1$

27. $f(x) = 100 - \dfrac{270}{x}$, $x = 9$, $\Delta x = dx = 0.5$

28. $f(x) = 75 - \dfrac{150}{x}$, $x = 5$, $\Delta x = dx = 0.5$

29. $f(x) = \sqrt{x}$, $x = 2$, $\Delta x = dx = 0.1$

30. $f(x) = 3\sqrt{x}$, $x = 1.5$, $\Delta x = dx = 0.1$

31. $f(x) = \dfrac{x^2 + 1}{x^2 - 1}$, $x = 2$, $\Delta x = dx = 0.1$

32. $f(x) = \dfrac{x^2 - 5}{2x^2 + 1}$, $x = 3$, $\Delta x = dx = 0.1$

33. $f(x) = 2x^2(3x^2 - 2x)$, $x = 1$, $\Delta x = dx = 0.1$

34. $f(x) = x^3(x^2 - 1)$, $x = 2$, $\Delta x = dx = 0.1$

For Exercises 35–42, use the linear approximation to estimate the value of the given number. Compare to the calculator value when rounded to four decimal places.

35. $\sqrt{26}$

36. $\sqrt{8.9}$

37. $\sqrt[3]{26}$

38. $\sqrt[3]{124}$

39. $\sqrt[4]{15.8}$

40. $\sqrt[4]{15}$

41. $\dfrac{2}{\sqrt{50}}$

42. $\dfrac{3}{\sqrt[3]{7}}$

Application Exercises

43. Geometry—Area Increase: The radius of a circle increases from an initial value of $r = 5$ inches by an amount $\Delta r = dr = 0.2$ inch. Estimate the corresponding increase in the circle's area by evaluating dA. The area of a circle is given by $A(r) = \pi r^2$.

44. **Geometry—Area Increase:** The radius of a circle increases from an initial value of $r = 8$ inches by an amount $\Delta r = dr = 0.1$ inch. Estimate the corresponding increase in the circle's area by evaluating dA. The area of a circle is given by $A(r) = \pi r^2$.

45. **Business Management—Insurance Costs:** The Dakorn Company determines that the annual cost of covering its employees' vision and dental insurance can be modeled by the function

$$f(x) = 1000 + 110\sqrt{x} \qquad x \geq 0$$

where x represents the number of employees covered and $f(x)$ represents the annual insurance cost in dollars.

(a) Determine the differential dy for the model.

(b) Suppose that the company's CEO has decided to hire four new employees in the Human Resources Department, so that the number of employees increases from 250 to 254. Use the differential to approximate the increase in cost in dental and vision insurance.

46. **Demographics—Poverty:** A study conducted by the Northern Aid Organization shows that the number of people identified as having incomes below poverty level in the northern provinces of a certain country can be modeled by

$$f(x) = 10 + 707\sqrt{x} \qquad x \geq 0$$

where x represents the population in thousands and $f(x)$ represents the number of people identified as having incomes below poverty level, measured in thousands.

(a) Determine the differential dy for the model.

(b) Suppose the population of the Prudential Province has increased from 20 thousand to 22 thousand. Use the differential to approximate the increase in number of people below poverty level.

© K. Thorsen/ShutterStock, Inc.

47. **Business Management—Production Costs:** The Paulson Motor Company estimates that the weekly cost of producing its new car, the Evadour, can be modeled by the cost function

$$C(x) = 0.22x^3 - 2.35x^2 + 14.32x + 10.22 \qquad 0 \leq x \leq 50$$

where x represents the number of Evadours produced each week and $C(x)$ represents the weekly manufacturing cost, measured in thousands of dollars.

(a) Determine the differential $dy = C'(x)dx$ for the model.

(b) Increased demand of the Evadour has resulted in a weekly increase in production from 30 to 33 cars. Use the differential to approximate the change in production cost.

48. **Business Management—Production Costs:** The A&D Publishing Company has determined that the printing cost of its new self-help book *Me and You Too* can be modeled by the cost function

$$C(x) = 0.02x^3 - 0.6x^2 + 9.15x + 98.43 \qquad 0 \leq x \leq 40$$

where x represents the number of books produced each day and $C(x)$ represents the daily production cost in thousands of dollars.

(a) Determine the differential $dy = C'(x)dx$ for the model.

(b) Favorable reviews of the book have resulted in an increase in daily production from 19 to 20 books. Use the differential to approximate the change in production cost.

49. **Marketing—Advertising Budget:** Using past records, managers at the PowerSet Company have estimated that the monthly amount spent on advertising and the sales of its Spiker volleyballs can be modeled by

$$f(x) = 120x - 2.4x^2 \qquad 0 \leq x \leq 25$$

where x represents the amount spent on advertising in thousands of dollars and $f(x)$ represents the number of volleyballs sold, in hundreds.

(a) Determine the differential dy for the model.

(b) The company's director of advertising has increased the advertising budget from $10,000 to $11,000. Use the differential to approximate the increase in sales by increasing the amount spent on advertising from $x = 10$ to $x = 11$.

50. Marketing—Advertising Budget: The PowerSet Company has estimated that the monthly amount spent on advertising and the sales of its Side-Out volleyballs can be modeled by

$$f(x) = 189.24x - 3.5x^2 \qquad 0 \le x \le 25$$

where x represents the amount spent on advertising in thousands of dollars and $f(x)$ represents the number of volleyballs sold, in hundreds.

(a) Determine the differential dy for the model.

(b) The company's director of advertising has increased the advertising budget from $10,000 to $11,000. Use the differential to approximate the increase in sales by increasing the amount spent on advertising from $x = 10$ to $x = 11$.

(c) Compute the actual change in sales $\Delta y = f(11) - f(10)$ and compare Δy to the approximation dy found in part (b).

 51. Nursing—Inpatient Healthcare: Countries have different procedures and philosophies in the way they approach acute inpatient healthcare. The country with the longest average inpatient stay is Japan. The annual average number of days an acute inpatient stays in the hospital in Japan from 2000 to 2010 can be modeled by

$$f(x) = 0.07x^2 - 1.33x + 24.8 \qquad 0 \le x \le 10$$

where x represents the number of years since 2000 and $f(x)$ represents the average number of days an acute inpatient stays in the hospital. (*Source:* Organization for Economic Cooperation and Development.)

(a) Evaluate dy for $x = 9$ and $\Delta x = dx = 1$ and interpret.

(b) Write the equation of the tangent line of f at $x = 9$ and evaluate the equation for the year 2014. Interpret this result.

52. Nursing—Inpatient Healthcare: The country with the shortest average inpatient stay is Denmark. The annual average number of days an acute inpatient stays in the hospital in Denmark from 2000 to 2010 can be modeled by

$$f(x) = 0.01x^2 - 0.11x + 3.8 \qquad 0 \le x \le 10$$

where x represents the number of years since 2000, and $f(x)$ represents the average number of days an acute inpatient stays in the hospital. (*Source*: Organization for Economic Cooperation and Development.)

(a) Evaluate dy for $x = 10$ and $\Delta x = dx = 1$ and interpret.

(b) Write the equation of the tangent line of f at $x = 10$ and evaluate the equation for the year 2014. Interpret this result.

53. Personal Finance—Bank Credit Cards: A topic that has gained attention since the mid-2000s is the growing private credit card debt in the United States. The number of bank-issued credit cards in circulation annually in the United States can be modeled by

$$f(x) = -10.8x^2 + 82.8x + 455 \qquad 0 \le x \le 10$$

where x represents the number of years since 2000 and $f(x)$ represents the number of bank-issued credit cards in circulation annually, measured in millions. (*Source:* The Nilson Report.)

(a) Evaluate dy for $x = 8$ and $\Delta x = dx = 1$ and interpret.

(b) Write the equation of the tangent line of f at $x = 8$ and evaluate the equation for the year 2012. Interpret this result.

© Deklofenak/ShutterStock, Inc.

54. Personal Finance—Oil Company Credit Cards: Many consumers use credit cards issued by gas stations in association with oil companies. The number of oil company–issued credit cards in circulation annually in the United States can be modeled by

$$f(x) = 0.67x^2 + 8.33x + 98 \qquad 0 \le x \le 10$$

where x represents the number of years since 2000 and $f(x)$ represents the number of oil company–issued credit cards in circulation annually, measured in millions. (*Source:* The Nilson Report.)

(a) Evaluate dy for $x = 9$ and $\Delta x = dx = 1$ and interpret.

(b) Write the equation of the tangent line of f at $x = 10$ and evaluate the equation for the year 2016. Interpret this result.

Concept and Writing Exercises

55. In a few sentences, explain the difference between the values of Δy and dy.

56. By using differentials, explain why 1.06 is a reasonable approximation for $(1.01)^6$.

In Exercises 57 and 58, suppose we do not know the formula for the function f, but know $f(2) = -4$ and $f'(x) = \sqrt{x^2 + 5}$. Complete the following.

57. Use linear approximation to estimate $f(1.96)$ and $f(2.04)$.

58. In a few sentences, explain whether the estimates in Exercise 57 are too large or too small.

In Exercises 59-64, let u and v represent differentiable functions and let k and n represent real number constants. Establish the following rules for working with differentials.

59. Constant Rule for Differentials: $dk = 0$.

60. Power Rule for Differentials: $du^n = nu^{n-1} \cdot du$

61. Sum Rule for Differentials: $d(u + v) = du + dv$

62. Difference Rule for Differentials: $d(u - v) = du - dv$

63. Product Rule for Differentials: $d(uv) = du \cdot v + u \cdot dv$

64. Quotient Rule for Differentials: $d\left(\dfrac{u}{v}\right) = \dfrac{du \cdot v - u \cdot dv}{v^2}$

Section Project

One way we can use differentials is to approximate the volume of Earth and its atmosphere. We know that the radius of the earth is about 3963 miles and that the formula for the volume of a sphere is given by $V(r) = \frac{4}{3}\pi r^3$. In this situation, we let r represent the radius in miles and $V(r)$ represent the volume in cubic miles.

(a) Determine an expression for dV the differential in the volume with respect to the radius r.

(b) There is no definitive border between the end of the atmosphere and the beginning of space, but many meteorologists consider the thickness of Earth's atmosphere to be about 62.1 miles. In the expression for dV, what does this value represent?

(c) Use the expression found in part (a) and the value in part (b) to determine a value for dV. Interpret the result.

SECTION OBJECTIVES

1. Determine a marginal cost function.

2. Determine a marginal profit function.

3. Determine a linear approximation.

4. Determine an average cost function.

5. Compute a marginal average cost function.

3.5   Marginal Analysis

Many decisions made by managers in business involve analyzing the effect on the dependent variable when a small change is made to a specific independent variable value. For example, a company may wish to consider changing the price of an item and examining how this change affects the revenue or profit of the product. **Marginal analysis** can be defined as the study of the amount of change in the dependent variable that results from a single unit change in an independent variable. A **unit change** means a change of a single unit. This change in the dependent variable is a direct application of our now-familiar tool—the derivative.

Marginal Analysis

Let's start this discussion of business functions by reviewing their definitions.

1. The *price-demand function* p gives us the price $p(x)$ at which people buy exactly x units of product.

2. The cost $C(x)$ of producing x units of a product is given by the *cost function*

    $$C(x) = (\text{variable costs}) \cdot (\text{units produced}) + (\text{fixed cost})$$

    Note that since variable costs are often expressed as a function, $C(x)$ may be a higher-order polynomial function.

3. The total revenue R generated by producing and selling x units of product at price $p(x)$ is given by the *revenue function*

    $$R(x) = (\text{quantity sold}) \cdot (\text{unit price}) = x \cdot p(x)$$

4. The profit P generated after producing and selling x units of a product is given by the *profit function*

    $$P(x) = \text{revenue} - \text{cost} = R(x) - C(x)$$

We referred to marginal analysis as the study of the dependent variable if the independent variable had a single unit change. Let's say that we want to study the marginal cost at a production level x, given a cost function C. We can start by determining the *actual change* in cost, denoted by ΔC, when the number of units produced is increased by 1.

$$(\text{actual change in cost}) = (\text{cost to produce } x + 1 \text{ units}) - (\text{cost to produce } x \text{ units})$$

In terms of the cost function C, this is the same as

$$\Delta C = C(x + 1) - C(x)$$

This relationship is illustrated in **Figure 3.5.1**.

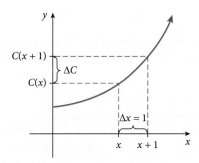

Figure 3.5.1 ΔC is the actual change in cost.

Let's try an example of computing this actual change. Suppose that the cost of producing x units of a recreation vehicle can be modeled by

$$C(x) = 100 + 60x + 3x^2$$

where x represents the number of vehicles produced and $C(x)$ is the cost in hundreds of dollars. To find ΔC for $x = 5$ and $\Delta x = 1$, we seek the difference in cost where

$$\begin{aligned}
\Delta C &= C(x + \Delta x) - C(x) \\
&= C(5 + 1) - C(5) \\
&= C(6) - C(5) \\
&= 568 - 475 = 93
\end{aligned}$$

Since $C(6)$ is the cost of producing six vehicles and $C(5)$ is the cost of producing the first five vehicles, then $\Delta C = C(6) - C(5)$ must represent the cost of producing the sixth vehicle. Thus, the cost of producing the sixth vehicle is \$9300.

An exact value can be computed by evaluating $\Delta C = C(x + \Delta x) - C(x)$, but in Section 3.4 we found that, when dx is small, $\Delta C \approx dy$, where dy is the differential in y given by $dy = C'(x)dx$. So for dx being small, we have

$$\text{(actual change in cost, } \Delta C) \approx \text{(differential in } C, dC)$$

But for marginal analysis, $dx = 1$. This gives us

$$\text{(marginal cost at production level } x) \approx \text{(differential in } C, \text{ where } dx = 1) = C'(x) \cdot 1 = C'(x)$$

Consequently, the **marginal cost function**, denoted by MC, is simply the derivative of the cost function. Also note that MC is the differential in the cost function where $dx = 1$. This is sometimes called a *unit differential*. The relationship of $C(x)$, ΔC, and $MC(x)$ is shown in **Figure 3.5.2**.

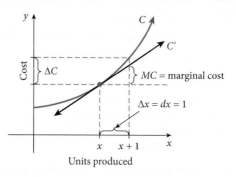

Figure 3.5.2

Example 1: Computing a Marginal Cost Function

The cost of producing x units of a certain recreational vehicle can be modeled by

$$C(x) = 100 + 60x + 3x^2$$

where x represents the number of vehicles produced and $C(x)$ is the cost in hundreds of dollars. Compute the marginal cost $MC(x) = C'(x)$. Evaluate $MC(5)$ and interpret.

Perform the Mathematics

Computing the derivative of $C(x)$ with respect to x, we get

$$MC(x) = C'(x) = \frac{d}{dx}(100 + 60x + 3x^2) = 60 + 6x$$

Evaluating the marginal cost function at $x = 5$ yields

$$MC(5) = 60 + 6(5) = 90$$

This means that the approximate cost of producing the next, or sixth, vehicle is 90 hundred, or \$9000. Earlier in this section, we showed that the *exact cost* of producing the sixth vehicle is \$9300, so the error of the marginal cost approximation is \$300. ∎

▶ *Try It Yourself*

Some related Exercises are 7 and 9.

Other Marginal Business Functions

At the beginning of this section, we stated that marginal analysis focused on the change in cost, revenue, and profit with a single unit change of the independent variable. This small change can have a significant impact on profit. This is because a change in price can cause changes in the quantity produced and sold, the cost and revenue, and the profit. See **Figure 3.5.3**.

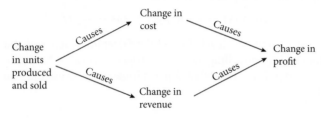

Figure 3.5.3

So it seems logical that we need to examine the marginal functions that are associated with the remainder of the business functions.

These definitions show that differentiating the business functions gives us the **marginal business functions**. Let's see how these functions can be used to affect profit.

Example 2: Using the Marginal Profit Function

OBJECTIVE 2

Determine a marginal profit function.

The FrezMore Company has determined that its cost of producing x refrigerators can be modeled by

$$C(x) = 2x^2 + 15x + 1500 \qquad 0 \le x \le 200$$

where x is the number of refrigerators produced each week and $C(x)$ represents the weekly cost in dollars. The company also determines that the price-demand function for the refrigerators is

$$p(x) = -0.3x + 460$$

a. Determine the profit function, P, for the refrigerators.

b. Determine the marginal profit function.

c. Compute $MP(60)$ and $MP(145)$ and interpret the results.

Perform the Mathematics

a. In order to determine the profit function, we must first find the revenue function. Since the price-demand function for producing and selling x refrigerators is $p(x) = -0.3x + 460$, the revenue function from selling x refrigerators is

$$R(x) = x \cdot p(x) = x(-0.3x + 460) = -0.3x^2 + 460x$$

The profit function can now be determined as

$$
\begin{aligned}
P(x) &= R(x) - C(x) \\
&= (-0.3x^2 + 460x) - (2x^2 + 15x + 1500) \\
&= -0.3x^2 + 460x - 2x^2 - 15x - 1500 \\
&= -2.3x^2 + 445x - 1500
\end{aligned}
$$

b. Differentiating the profit function to determine the marginal profit function yields

$$MP(x) = \frac{d}{dx}(-2.3x^2 + 445x - 1500) = -4.6x + 445$$

c. Evaluating the marginal profit function at $x = 60$ gives

$$MP(60) = -4.6(60) + 445 = 169$$

This means that at a production level of $x = 60$ refrigerators each week, there is about $169 profit for making and selling the 61st refrigerator.

Evaluating the marginal profit function at $x = 145$ gives

$$MP(145) = -4.6(145) + 445 = -222$$

This means that at a production level of $x = 145$ refrigerators each week, there is about $222 loss in profit for making and selling the 146th refrigerator. **Figure 3.5.4** shows that the profit at $x = 145$ is decreasing at a rate of $222 per refrigerator because the slope of the tangent line at $x = 145$ is -222.

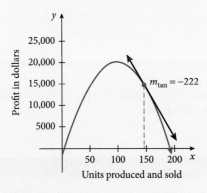

Figure 3.5.4

An extension of the marginal concept is the use of linear approximation for business functions. Recall from Section 3.4 that for a differentiable function f where $y = f(x)$, the **linear approximation** of f is given by

$$f(x + dx) \approx f(x) + dy = f(x) + f'(x)dx$$

when the value of dx is small.

In the case of marginal analysis, $dx = 1$. This means that we can write a linear approximation for a cost function with $f(x) = C(x)$ as

$$C(x + 1) \approx C(x) + C'(x)$$

or, in terms of the marginal cost function, as

$$C(x + 1) \approx C(x) + MC(x)$$

Consequently, we can apply the linear approximation concept to marginal analysis to easily make approximations for cost, revenue, and profit functions with information that is already known. This is particularly useful if the data given are tabular or some partial results are already available. Let's apply this new piece of information to determine a linear approximation for a revenue function.

OBJECTIVE 3

Determine a linear approximation.

Example 3: Computing Linear Approximations for a Revenue Function

The revenue function for the production of x refrigerators per week in Example 2 was given as $R(x) = -0.3x^2 + 460x$.

a. Compute $R(110)$ and interpret.

b. Compute $MR(110)$ and interpret.

c. Use the solution from parts (a) and (b) to get a linear approximation for $R(111)$. Compare the result to the exact value of $R(111)$.

Perform the Mathematics

a. For the revenue function $R(x) = -0.3x^2 + 460x$, we substitute $x = 110$ to get

$$R(110) = -0.3(110)^2 + 460(110) = -0.3(12{,}100) + 50{,}600$$
$$= -3630 + 50{,}600 = 46{,}970$$

So when producing and selling 110 refrigerators per week, the revenue realized is $46,970.

b. We differentiate the revenue function to determine the marginal revenue function as follows:

$$MR(x) = R'(x) = \frac{d}{dx}(-0.3x^2 + 460x)$$
$$= -0.6x + 460$$

The marginal revenue at a production level of $x = 110$ is

$$MR(110) = -0.6(110) + 460$$
$$= -66 + 460 = 394$$

This means that the revenue gained from producing and selling the 111th refrigerator is about $394.

c. To determine an approximation for $R(111)$, we use the linear approximation for the revenue function

$$R(x + 1) \approx R(x) + MR(x)$$

Using this, along with the results from parts (a) and (b), we get an approximation for $R(111)$ as follows:

$$R(111) = R(110 + 1) \approx R(110) + MR(110)$$
$$= 46{,}970 + 394 = 47{,}364$$

So our approximation for $R(111)$ is $R(111) \approx 47{,}364$. The exact revenue using the given revenue function is $R(111) = 47{,}363.70$, so the error from using the linear approximation is only $0.30! The key to the closeness of this linear approximation is that the change in the independent variable is relatively small. ■

Average Business Functions

Many times in business situations, financial reports are simplified so that the numerical results can be easily understood. Often managers are interested in the *per unit* cost of a product, which is usually easier to work with than the total cost for the production of x units of a product. For example, it is easier for a manager to think of the production costs at a video production company to be $8.25 per DVD or for a cycling company to think of it costing $215 to produce each bike.

To find this per unit, or **average**, for the cost function, we take the total cost and divide it by the number of items produced. In words,

Per unit cost = (total cost to produce x items)/(number of items produced, x)

Let's see how this works through a Flashback.

© Hemera/Thinkstock

Flashback: Fashion Mystique Revisited

In Section 1.2 we found that the cost function of copying promotional flyers for The Fashion Mystique was given by

$$C(x) = 0.10x + 5$$

where x represents the number of copies produced and $C(x)$ represents the cost of producing x flyers.

a. Compute $C(100)$ and interpret.

b. Evaluate $\dfrac{C(100)}{100}$ and interpret.

Perform the Mathematics

a. Evaluating the cost function at $x = 100$, we get

$$C(100) = 0.10(100) + 5 = 10 + 5 = 15$$

This means that the cost of producing 100 promotional flyers is $15.

b. We use the result from part (a) to evaluate the expression $\dfrac{C(100)}{100}$.

$$\frac{C(100)}{100} = \frac{15}{100} = 0.15$$

This tells us that at a production level of $x = 100$ flyers, the average cost is 15 cents per flyer. ∎

DEFINITION

Average Business Functions

The **average cost function**, AC, which gives the per unit cost of producing x items, is given by

$$AC(x) = \frac{C(x)}{x}$$

The **average profit function**, AP, which gives the per unit profit of producing and selling x items, is given by

$$AP(x) = \frac{P(x)}{x}$$

NOTE: Statistics students may recall that the average of a set of data is given by $\bar{x}$. This is why some textbooks denote the average cost function as $\overline{C}$.

The Flashback leads us to the definition of the **average business functions**.

Notice that the average revenue function is omitted from the average business functions definition. This is because the average revenue function is $AR(x) = \dfrac{R(x)}{x} = \dfrac{x \cdot p(x)}{x} = p(x)$. So we see that the average revenue function is just another name for the price function.

OBJECTIVE 4

Determine an average cost function.

Example 4: Analyzing an Average Cost Function

The Ventoux Athletic Shoe Company knows that for its Stampeder model basketball shoes, the daily cost function can be modeled by

$$C(x) = 700\sqrt{x} + 5000 \qquad 0 \le x \le 500$$

where x is the number of pairs of shoes produced daily and $C(x)$ is the daily cost in dollars.

a. Determine AC, the average cost function.

b. Evaluate and interpret $C(400)$ and $AC(400)$.

Perform the Mathematics

a. Using the definition of average cost function, we get

$$AC(x) = \frac{C(x)}{x} = \frac{700\sqrt{x} + 5000}{x}$$

b. Evaluating the cost function when $x = 400$ yields

$$C(400) = 700\sqrt{400} + 5000$$
$$= 700(20) + 5000$$
$$= 19{,}000$$

This means that the total cost of producing 400 pairs of shoes a day is $19,000.

Evaluating $AC(x)$ when $x = 400$, we get

$$AC(400) = \frac{700\sqrt{400} + 5000}{400} = 47.5$$

This means that the average cost of producing 400 pairs of shoes a day is $47.50 for each pair. ∎

▶ Try It Yourself

Some related Exercises are 25a and 26a.

A manager may also wish to apply marginal analysis techniques to the average business functions. The result is the **marginal average business functions**. These are found by computing the derivative of each of the average business functions, respectively. The result is a function that approximates the per unit cost (or profit) of producing one more item. Now let's define these marginal average business functions.

DEFINITION

Marginal Average Business Functions

The **marginal average cost function** approximates the per unit cost for producing an additional item of a product and is given by

$$MAC(x) = \frac{d}{dx}[AC(x)] = \frac{d}{dx}\left[\frac{C(x)}{x}\right]$$

The **marginal average profit function** approximates the per unit profit for producing and selling an additional item of a product and is given by

$$MAP(x) = \frac{d}{dx}[AP(x)] = \frac{d}{dx}\left[\frac{P(x)}{x}\right]$$

NOTE: To determine these marginal average business functions, we first determine the average business function and then compute the derivative, *in that order*.

Example 5: Computing a Marginal Average Cost Function

OBJECTIVE 5

Compute a marginal average cost function.

In Example 4, we found that the cost function for producing x pairs of Stampeder model basketball shoes was

$$C(x) = 700\sqrt{x} + 5000 \qquad 0 \le x \le 500$$

a. Compute the marginal average cost function MAC.

b. Evaluate $MAC(400)$. Round the result to the nearest hundredth and interpret.

Perform the Mathematics

a. In Example 4 we computed the average cost function as $AC(x) = \dfrac{700\sqrt{x} + 5000}{x}$. Before differentiating, we simplify to get

$$AC(x) = \frac{700\sqrt{x}}{x} + \frac{5000}{x} = 700x^{-1/2} + 5000x^{-1}$$

Now, differentiating this average cost function with respect to x gives us

$$MAC(x) = \frac{d}{dx}(700x^{-1/2} + 5000x^{-1})$$

$$= -350x^{-3/2} - 5000x^{-2}$$

Simplifying the rational and negative exponents, we get the marginal average cost function

$$MAC(x) = -\frac{350}{\sqrt{x^3}} - \frac{5000}{x^2}$$

b. Evaluating the result of part (a) at $x = 400$, we get

$$MAC(400) = -\frac{350}{\sqrt{(400)^3}} - \frac{5000}{(400)^2} = -0.075 \approx -0.08$$

This means that when 400 pairs of basketball shoes have been produced, the average cost per pair decreases by about $0.08 for an additional pair produced. Note that the negative sign tells us that the per unit cost is decreasing.   ■

Summary

In this section, we revisited the business functions and from them derived the marginal business functions. These functions are found by differentiation, and they approximate the cost, revenue, or profit for producing one more item of a product. Then we discussed the average business functions, which were found by taking the business function and dividing by the independent variable. Finally, we discussed the marginal average business functions, which were the derivatives of the average business functions.

Important Functions

- **Marginal cost function:** $MC(x) = C'(x)$

- **Marginal revenue function:** $MR(x) = R'(x)$

- **Marginal profit function:** $MP(x) = P'(x)$

- **Average cost function:** $AC(x) = \dfrac{C(x)}{x}$

- **Average profit function:** $AP(x) = \dfrac{P(x)}{x}$

- **Marginal average cost function:** $MAC(x) = \dfrac{d}{dx}[AC(x)]$

- **Marginal average profit function:** $MAP(x) = \dfrac{d}{dx}[AP(x)]$

Section 3.5 Exercises

Vocabulary Exercises

1. The study of the amount of change in the dependent variable that results from a single unit change in the independent variable is called _____ analysis.

2. A change of a single unit is called a _____ change.

3. The approximate cost of producing one additional unit of a product at production level x is given by the _____ _____ function.

4. The total cost divided by the number of units produced gives the _____ cost.

5. The average revenue function is another name for the _____ function.

6. To get the marginal profit function, we begin by _____ the profit function.

Skill Exercises

In Exercises 7–12, assume the cost function $C(x)$ is measured in dollars. Complete the following:

(a) *Determine the marginal cost function MC.*

(b) *Evaluate and interpret $MC(x)$ for the given production level x.*

(c) *Evaluate the actual change in cost by evaluating $C(x + 1) - C(x)$ and compare with the answer in part (b).*

7. $C(x) = 23x + 5200; x = 10$

8. $C(x) = 14x + 870; x = 12$

9. $C(x) = \dfrac{1}{2}x^2 + 12.7x + 2100; x = 11$

10. $C(x) = \dfrac{1}{2}x^2 + 27x + 1200; x = 20$

11. $C(x) = 0.2x^3 - 3x^2 + 50x + 20; x = 30$

12. $C(x) = 0.08x^3 - 2x^2 + 10x + 70; x = 90$

In Exercises 13–18, the cost function C and the price-demand function p are given. Assume that the value of $C(x)$ and $p(x)$ are in dollars. Complete the following.

(a) *Determine the revenue function R and the profit function P.*

(b) *Determine the marginal cost function MC and the marginal profit function MP.*

13. $C(x) = 5x + 500; p(x) = 6$

14. $C(x) = 12x + 4500; p(x) = 15$

15. $C(x) = \dfrac{x^2}{100} + 7x + 1000; p(x) = -\dfrac{x}{20} + 15$

16. $C(x) = \dfrac{1}{100}x^2 + \dfrac{1}{2}x + 8, p(x) = -\dfrac{x}{200} + 1$

17. $C(x) = -.0001x^3 + 4x + 100, 0 \le x \le 70; p(x) = -0.005x + 7$

18. $C(x) = -0.002x^3 + 0.01x^2 + 2x + 50, 0 \le x \le 40; p(x) = -x^{-1/2} + 5$

Application Exercises

19. **Economics—Marginal Cost:** The Country Day Company determines that the daily cost of producing its Garden King lawn tractor tires can be modeled by

$$C(x) = 100 + 40x - 0.001x^2 \qquad 0 \le x \le 300$$

where x represents the number of tires produced each day, and $C(x)$ represents the cost in dollars. Determine MC, the marginal cost function. Evaluate $MC(200)$ and interpret.

20. **Economics—Marginal Cost:** The Kelomata Company determined that the monthly cost of producing its Sun Stopper patio swings can be modeled by

$$C(x) = 15,000 + 100x - 0.001x^2 \quad 0 \le x \le 200$$

where x represents the number of patio swings manufactured monthly and $C(x)$ represents the production cost in dollars. Determine MC, the marginal cost function. Evaluate $MC(100)$ and interpret.

21. **Manufacturing—Raincoat Production:** Using historical data, the Seas Beginning Corporation determines that the daily cost of producing its Rain Forrest Ultra raincoat can be modeled by

$$C(x) = 1000 + 35x - 0.01x^2 \quad 0 \le x \le 300$$

where x represents the number of raincoats produced daily, and $C(x)$ represents the daily cost in dollars. Determine MC, the marginal cost function. Evaluate $MC(200)$ and interpret.

22. **Manufacturing—Transmission Production:** The accounting division of the Kranky Krank Company determines that its weekly cost of producing its continuous variable transmissions (CVTs) can be modeled by the cost function

$$C(x) = 55,000 + 600x - 1.25x^2 \quad 0 \le x \le 230$$

where x represents the number of CVTs produced annually and $C(x)$ represents the manufacturing costs in dollars. Determine MC, the marginal cost function. Evaluate $MC(210)$ and interpret.

23. **Manufacturing—Hard Drive Production:** The Memory Master Company makes portable hard drives for laptop computers, and its managers have determined that the cost of producing its 3-terabyte hard drive can be modeled by

$$C(x) = 10,000 + 200x - 0.2x^2 \quad 0 \le x \le 650$$

where x represents the number of hard drives produced each week and $C(x)$ represents the manufacturing cost in dollars.

  (a) Determine the marginal cost function $MC(x)$ and evaluate $MC(500)$.

  (b) If the Memory Master Company sells 500 hard drives weekly to a computer chain store for $120 each, should production be increased? Explain your answer.

24. **Economics—Marginal Analysis:** For the cost function in Exercise 19, complete the following.

  (a) Determine the average cost function AC. Evaluate and interpret $AC(200)$.

  (b) Determine MAC, the marginal average cost function. Evaluate and interpret $MAC(200)$.

25. For the cost function in Exercise 20, complete the following.

  (a) Determine the average cost function AC. Evaluate and interpret $AC(100)$.

  (b) Determine MAC, the marginal average cost function. Evaluate and interpret $MAC(100)$.

26. **Profit Analysis—Bobbleheads:** The financial planning team at the Tesch Company determines that the profit function for producing and selling its Captain U.S. bobbleheads can be modeled by

$$P(x) = -0.001x^2 + 8x - 4000 \quad 0 \le x \le 7000$$

where x represent the number of bobbleheads produced and sold and $P(x)$ represents the monthly profit in dollars.

  (a) Determine MP, the marginal profit function. Evaluate $MP(3000)$ and interpret.

  (b) If the Tesch Company is producing and selling 3000 bobbleheads per month, is profit increasing or decreasing?

27. **Profit Analysis—Radio Production:** The Hanash Corporation determines that the weekly profit from producing and selling its Jog-R-Radios can be modeled by

$$P(x) = -0.01x^2 + 12x - 2000 \quad 0 \le x \le 1000$$

where x represents the number of radios produced and sold weekly and $P(x)$ represents the weekly profit in dollars.

(a) Determine MP, the marginal profit function. Evaluate $MP(700)$ and interpret.

(b) If the Hanash Corporation is producing and selling 700 radios per week, is profit increasing or decreasing?

28. **Publishing—Magazine Publications:** By inspecting their tracking records, the telemarketing company Calls-R-Us has concluded that the monthly profit from selling magazine subscriptions can be modeled by

$$P(x) = 5x + \sqrt{x} \qquad 0 \le x \le 100$$

where x represents the number of subscriptions sold per month, and $P(x)$ represents profit in dollars. Determine MP, the marginal profit function. Evaluate $MP(55)$ and interpret.

29. **Publishing—Newspaper Profit:** Ashton is a newspaper motor courier and determines that the monthly profit from his current newspaper route can be modeled by

$$P(x) = 2x - \sqrt{x} \qquad 0 \le x \le 200$$

where x represents the number of subscribers, and $P(x)$ represents the monthly profit in dollars. Determine MP, the marginal profit function. Evaluate $MP(110)$ and interpret.

30. **Economics—Marginal Analysis:** For the profit function in Exercise 28, complete the following.

(a) Determine AP, the average profit function. Evaluate and interpret $AP(55)$.

(b) Determine MAP, the marginal profit function. Evaluate $MAP(55)$ and interpret.

31. **Economics—Marginal Analysis:** For the profit function in Exercise 29, complete the following.

(a) Determine AP, the average profit function. Evaluate and interpret $AP(110)$.

(b) Determine MAP, the marginal profit function. Evaluate $MAP(110)$ and interpret.

32. **Economics—Marginal Analysis:** The GlobalText Company makes social media–oriented mobile phones apps and determines that the fixed and variable costs to produce x apps are $1200 and $12 per app, respectively.

(a) Write the cost function C in the form $C(x) = mx + b$.

(b) Determine the marginal cost function MC and then evaluate $MC(100)$ and $MC(150)$ and interpret these results.

(c) Why are the marginal costs in part (b) equal?

33. **Economics—Marginal Analysis:** The NewJoy toy company has just produced a new Street Kings action figure set that it sells to wholesalers for $20 each. From collecting data, they determine that the cost to produce x action figures can be modeled by the function $C(x) = 0.001x^2 + 4x + 5000$.

(a) Derive and algebraically simplify the profit function $P(x)$.

(b) Evaluate $P(1000)$ and interpret.

(c) Evaluate $MP(1000)$ and interpret.

34. **Economics—Marginal Analysis:** The NewJoy toy company hires an accounting firm to audit their books and revises their price-demand and cost functions to $p(x) = 23$ and $C(x) = \frac{x^2}{95} + \frac{7}{2}x + 5500$, respectively.

(a) Derive and algebraically simplify the profit function $P(x)$.

(b) Evaluate $P(500)$ and interpret.

(c) Evaluate $MP(500)$ and interpret.

35. **Economics—Marginal Analysis:** The EZ-Craft Company determines that the price-demand function, in dollars, for their new U-Make-It picture frame is $p(x) = -\frac{x}{30} + 200$, with a cost

function of $C(x) = 60x + 72,000$, where x represents the number of frames produced and $C(x)$ represents the cost in dollars.

(a) Determine the revenue function R.

(b) Determine the profit function P. Find the smallest and largest production levels x so that the company realizes a profit. Do this by determining the smallest and largest x-values so that $R(x) > C(x)$.

(c) Evaluate $P'(3000)$ and interpret.

36. **Economics—Marginal Analysis**: Vroncom Incorporated determines that the price-demand function for their TruTouch tablet device can be modeled by $p(x) = -\frac{x}{30} + 300$. They also have determined that their fixed costs are $150,000 and variable costs are 30 dollars per device.

(a) Determine the revenue function R and then write the cost function in the form $C(x) = mx + b$.

(b) Determine the profit function P. Find the smallest and largest production levels x so that the company realizes a profit. Do this by determining the smallest and largest x-values so that $R(x) > C(x)$.

(c) Evaluate $P'(1000)$ and interpret.

37. **Economics—Marginal Analysis**: The stockholder's report for the Step-Up Company lists the following information and graph for its recently released W-Racer walking shoes.

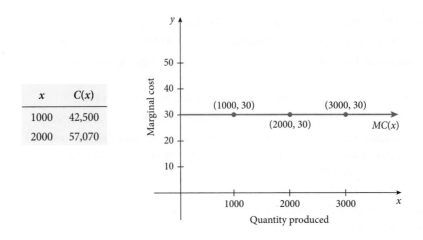

x	$C(x)$
1000	42,500
2000	57,070

(a) Use the table along with the graph of $MC(x)$ to get a linear approximation for $C(1001)$. Use the approximation $C(x + 1) \approx C(x) + MC(x)$.

(b) Repeat the procedure in part (a) to get a linear approximation for $C(2001)$.

Concept and Writing Exercises

38. Show that for a linear cost function, the marginal cost is constant.

39. Write a few sentences to explain the relationship between the marginal revenue and the marginal cost at the point where the marginal profit is zero.

In Exercises 40–43, let P(x) represent the profit for producing x units of a product and let k and c represent constants. Answer the following.

40. How do we determine the profit from producing k units of a product?

41. How do we determine the level of production so that the marginal profit will be c dollars?

42. How do we determine the marginal profit from producing k units of a product?

43. How do we find the production level so that the profit will be c dollars?

For Exercises 44–47, consider the general cost function $C(x) = -ax^2 + bx + c$ and let $R(x) = x \cdot k$ where k represents a real number constant.

44. Determine the marginal cost function MC.

45. Determine the marginal revenue function MR.

46. Write the profit function P.

47. Determine the marginal profit function MP.

Section Project

Consider the following data table for the cost and revenues at various production levels for a new brand of handheld television.

Number of Handheld TVs produced, x	Cost, C(x)	Revenue, R(x)
100	$5500	$3,050
200	$9600	$24,100
300	$13,500	$81,200
400	$17,600	$192,200

© iStockphoto/Thinkstock

Use the tabular information to complete the following.

(a) Use the regression capabilities of your graphing calculator to get a linear regression model for the cost of producing the TVs in the form

$$C(x) = ax + b \qquad 100 \le x \le 400$$

where x represents the number of TVs produced and $C(x)$ represents the cost in dollars.

(b) Use the regression capabilities of your graphing calculator to get a cubic regression model for the revenue of producing and selling the TVs in the form

$$R(x) = ax^3 + bx^2 + cx + d \qquad 100 \le x \le 400$$

where x represents the number of TVs produced and $R(x)$ represents the revenue in dollars. Round the values of a, b, c, and d to four decimal places if necessary.

(c) Determine the average cost function AC and simplify the result.

(d) Determine the marginal average cost function MAC. At a production level of $x = 150$, is the average cost increasing or decreasing?

(e) Using the results of parts (a) and (b), write and simplify the profit function $P(x)$.

(f) Determine the marginal average profit function MAP. At a production level of $x = 250$, is the average profit increasing or decreasing?

Section 3.1 Review Exercises

In Exercises 1–12, determine the derivative of the given function. When appropriate, simplify the derivative so that there are no negative or fractional exponents.

1. $f(x) = x^8$

2. $f(x) = 3x^{1/5}$

3. $f(x) = -4x^{-3/7}$

4. $f(x) = -\dfrac{4}{5}x^{7/6}$

5. $f(x) = 6x^2 + 5x - 11$

6. $f(x) = -9x^3 + x + 12$

7. $f(x) = \dfrac{1}{2}x^3 - 3x^2 + \dfrac{2}{3}x + 5$

8. $f(x) = 6.23x^2 + 1.98x - 3.34$

9. $f(x) = -7.10x^3 - 5.02x^2 + 11.19x - 16.07$

10. $f(x) = 2\sqrt{x} + 7x^2 - \dfrac{1}{2}x^3$

11. $f(x) = \sqrt[5]{x} - \sqrt{x} + \dfrac{1}{x}$

12. $f(x) = 2.08x^{3.79}$

For the functions in Exercises 13 and 14, complete the following.

(a) Write the domain of f in interval notation.

(b) Use the addition rule for fractions to rewrite and simplify f.

(c) Find $f'(x)$ using the differentiation rules.

(d) Write the domain of f' using interval notation.

13. $f(x) = \dfrac{3x^2 - 2x^2 + 6x + 2}{x}$

14. $f(x) = \dfrac{6x^4 + 25x^3 - 9x + 11}{x^2}$

For Exercises 15 and 16, complete the following.

(a) Determine the derivative for the given function using the differentiation rules.

(b) Determine the slope of the line tangent to the graph of the function at the indicated x-value.

(c) Determine an equation for the line tangent to the graph of the function at the indicated x-value.

(d) Graph the function and the tangent line in the same viewing window.

15. $f(x) = \dfrac{1}{x^2}; \qquad x = 4$

16. $f(x) = x^{3/4}; \qquad x = 16$

17. Energy Consumption—Dishwashers: The total amount of energy that the Kitchen Master dishwasher uses can be modeled by the function

$$f(t) = 6t + 9\sqrt{t} + 0.02t^{3/2} \qquad t \geq 0$$

where t represents the number of hours the dishwasher has been running and $f(t)$ represents the amount of energy spent in kilowatt-hours.

(a) Compute and simplify $f'(t)$.

(b) Suppose that an electrician wants to know how fast the dishwasher is using energy as it is turned on. Evaluate $f'\left(\dfrac{1}{3600}\right)$ and interpret this value.

18. Rocket Science—Ballistics: A rocket taking off has an altitude from the ground of $h(t) = 200t^2$, where $h(t)$ is in feet and t is in seconds. Find and interpret $h'(4)$.

19. Real Estate—Home Prices: Over the course of one year, the average price of homes for sale in a certain area can be modeled by the function

$$P(x) = -30x^2 + 2100x + 163{,}250$$

where x represents the number of weeks after January 1, and $P(x)$ represents the average home price, measured in dollars.

(a) Find the average rate of change of the price from $x = 10$ weeks to $x = 30$ weeks.

(b) Graph the function in (a) in the viewing window [0, 52] by [0, 250,000] and determine the maximum price and the week at which that price is at its maximum.

20. **Energy Consumption—Strategic Oil Reserve:** Throughout recent U.S. history, the U.S. Department of Energy has periodically released oil in its strategic oil reserves to help ease gasoline prices. In August 2011, the release had negligible effect on gas prices. "Although it helped initially to pull down prices it was probably too little," said American Automobile Administration spokesman John Townsend. (*Source: The Washington Times.*) The amount of oil in the U.S. strategic oil reserve from 1980 to 2011 can be modeled by the function

$$f(x) = 0.1x^3 - 5.24x^2 + 88.1x + 128 \qquad 0 \le x \le 31$$

where x represents the number of years since 1980, and $f(x)$ represents the amount of oil in reserve in millions of barrels. (*Source:* U.S. Energy Information Administration.)

(a) Determine $f'(x)$.

(b) How fast was the strategic oil reserve increasing in 1984?

(c) Evaluate $f'(20)$ and interpret the result.

Section 3.2 Review Exercises

In Exercises 21–28, determine the derivative of the given function.

21. $f(x) = x^3(6x^2 - 3x + 8)$

22. $f(x) = -3x^2(6x^3 - 2x^2 + 9x + 10)$

23. $f(x) = (2x^2 - 5x + 1)(3x^2 + 4x - 1)$

24. $f(x) = (3x^{1/2} + 5x)(-2x^{2/5} + x - 9)$

25. $f(x) = (5\sqrt{x} - 3x - 1)(4\sqrt{x} - 7x)$

26. $f(x) = \dfrac{3x + 2}{x - 1}$

27. $f(x) = \dfrac{2x^2 + 2x - 5}{3x^2 - x + 9}$

28. $f(x) = \dfrac{3\sqrt{x} - 2}{3x + 1}$

For Exercises 29 and 30, complete the following.

(a) Determine the derivative.

(b) Determine an equation of the line tangent to the graph of the function at the indicated x-value.

(c) Graph the function and the tangent line in the same viewing window.

(d) Use the dy/dx or Draw Tangent command on your calculator to verify the solution to part (b).

29. $f(x) = -3x^2(3x + 5); x = 2$

30. $f(x) = \dfrac{2x^2 - 3}{x}; x = 2$

Section 3.3 Review Exercises

In Exercises 31–40, differentiate using the Generalized Power Rule.

31. $f(x) = (x + 2)^3$

32. $f(x) = (x - 5)^2$

33. $f(x) = (8 - x)^3$

34. $f(x) = (4x - 3)^3$

35. $f(x) = (2x + 5)^4$

36. $f(x) = (4x^2 + 7)^5$

37. $f(x) = 3(x^2 - 5x + 3)^2$

38. $f(x) = (3x^2 + 7x - 2)^{63}$

39. $f(x) = (2x^2 - 5x + 7)^{1/3}$

40. $f(x) = (3x^2 - 9x - 4)^{-6}$

For the rational functions in Exercises 41–44, complete the following.

(a) Determine the derivative using the Quotient Rule.

(b) Determine the derivative using the Generalized Power Rule.

41. $f(x) = \dfrac{3}{2x + 9}$

42. $f(x) = \dfrac{7}{(x - 7)^2}$

43. $f(x) = \dfrac{2}{(3x + 5)^3}$

44. $f(x) = \dfrac{9}{x^2 - 6x + 18}$

In Exercises 45–48, determine an equation of the line tangent to the graph of f at the indicated ordered pair.

45. $f(x) = (5x + 3)^5$; $(-1, -32)$

46. $f(x) = (7x - 6)^3$; $(1, 1)$

47. $f(x) = (x^2 - 5x + 8)^4$; $(3, 16)$

48. $f(x) = (6x - 11)^{1/2}$; $(6, 5)$

In Exercises 49–52, use the Generalized Power Rule to differentiate the functions.

49. $f(x) = \sqrt{7x - 12}$

50. $f(x) = \sqrt[3]{8x + 1}$

51. $f(x) = \dfrac{3}{\sqrt{4x + 5}}$

52. $f(x) = \dfrac{8}{\sqrt[3]{2x^3 - 5x + 4}}$

For Exercises 53–56, use the Generalized Power Rule, along with the Product and Quotient Rules, to find the derivatives of the given functions.

53. $f(x) = x(x^2 + 5)^3$

54. $f(x) = 4x\sqrt{x^2 - 2x}$

55. $f(x) = \dfrac{3x - 7}{\sqrt{5x - 6}}$

56. $f(x) = (6x - 5)^7(3x + 2)^4$

In Exercises 57–60, determine the derivatives using the Generalized Power Rule.

57. $f(x) = (3x^2 - x + 1)^{0.67}$

58. $f(x) = \left(\dfrac{1}{8.3x - 5.7}\right)^{-2.4}$

59. $f(x) = (x^3 + x^2 + 5x + 1)^{-0.7}$

60. $f(x) = 4.96(x + 1)^{2.78}$

 61. **Corporate Finance—Oil Profits:** An economic sector that has done well in the past few decades is petroleum- and coal-based energy companies. "The earnings reflect continued leadership in operational performance during a period of strong commodity prices," said Exxon's chairman, Rex W. Tillerson. (*Source: The New York Times*). The net profits for petroleum and coal corporations from 1990 to 2011 can be modeled by

$$f(x) = (1.8x + 6.3)^{1.26} \quad 0 \le x \le 21$$

where x represents the number of years since 1990 and $f(x)$ represents the net profits for petroleum and coal corporations, measured in billions of dollars. (*Source: U.S. Census Bureau.*)

(a) Use the Chain Rule to determine $f'(x)$.

(b) Evaluate and interpret $f'(8)$.

(c) Graph f' in the viewing window $[0, 21]$ by $[0, 7]$.

(d) Use the value command to verify your answer to part (b).

 62. **Corporate Finance—Renewable Energy:** Renewable energy companies have recently attempted to benefit from the same tax advantages oil and coal companies receive. "It would be a real boon to the renewable energy industry," said John McKenna, of Hamilton Clark Securities. (*Source: The San Francisco Chronicle.*) The net profits for renewable energy companies from 1990 to 2011 can be modeled by

$$f(x) = (1.16x + 4.2)^{1.37} \quad 0 \le x \le 21$$

where x represents the number of years since 1990 and $f(x)$ represents the net profits for renewable energy companies, measured in billions of dollars. (*Source:* U.S. Census Bureau.)

(a) Use the Chain Rule to determine $f'(x)$.

(b) Evaluate and interpret $f'(8)$.

(c) Graph f' in the viewing window $[0, 21]$ by $[0, 6]$.

(d) Compare the solution to part (b) of Exercise 61. Which model was growing at a faster rate?

Section 3.4 Review Exercises

In Exercises 63–74, determine dy for the given function.

63. $f(x) = 4x + 2$

64. $f(x) = 5x^2 - 3x + 2$

65. $f(x) = \dfrac{x + 3}{x - 5}$

66. $f(x) = \dfrac{7}{x + 5}$

67. $f(x) = \sqrt{x} - \dfrac{3}{x^4}$

68. $f(x) = \dfrac{8}{x^2} + \sqrt[3]{x}$

69. $f(x) = x^4 - 2x + \sqrt[3]{x^2}$

70. $f(x) = x^3 - 5x^2 + 2x + 3$

71. $f(x) = 4x^5 - 21x + 4$

72. $f(x) = \dfrac{x^2 - 5}{x^2 + 5}$

73. $f(x) = 2x^{1.7} - 5x^{0.8} + 4$

74. $f(x) = 3x^{4.1} + 7x^{0.6} - 12$

For Exercises 75–80, evaluate Δy and dy for the given function and indicated values.

75. $f(x) = 4x^2 - x + 6; x = 3, \Delta x = dx = 0.1$

76. $f(x) = \dfrac{18}{x} + 5; x = 2, \Delta x = dx = 0.5$

77. $f(x) = \sqrt[3]{x}; x = 8, \Delta x = dx = 1.261$

78. $f(x) = 2x^3 - 7x^2 + 2x; x = 4, \Delta x = dx = 0.2$

79. $f(x) = \dfrac{x^2 + 2}{x^2 - 2}; x = 2, \Delta x = dx = 0.1$

80. $f(x) = 4x(2x + 5); x = 1, \Delta x = dx = 0.1$

For Exercises 81–84, use the linear approximation to estimate the values of the given numbers. Compare to the calculator value when rounded to four decimal places. Recall that $f(x + dx) \approx f(x) + dy$.

81. $\sqrt{65}$

82. $\sqrt{24.6}$

83. $\sqrt[4]{16.3}$

84. $\sqrt[3]{62}$

85. Advertising—T-Shirt Sales: The Wild & Wacky T-shirt company has estimated that the association between its monthly T-shirt sales and its advertising can be modeled by

$$f(x) = 90x - 2.7x^2 \qquad 0 \le x \le 8$$

where x represents the amount spent on advertising in hundreds of dollars and $f(x)$ is the number of T-shirts sold in hundreds.

(a) Determine dy.

(b) Approximate the increase in sales if the advertising is increased from \$400 to \$500.

86. Repeat Exercise 85 using the model $f(x) = 82.76x - 1.87x^2$.

87. For Exercises 85 and 86, compute the actual change Δy in sales and compare to the approximation.

88. **International Ecology—India CO$_2$ Emissions:** One country that has been aware of its increasing emission of fossil fuel emissions is India. "India's carbon dioxide emissions will increase by nearly three-fold to 3200 million metric tons by 2030," according to an economic survey that has been tabled in Parliament. (*Source: The Hindu News.*) The amount of carbon dioxide emissions from fossil fuels in India from 1990 to 2011 can be modeled by

$$f(x) = 0.91x^2 + 28x + 625.6 \qquad 0 \le x \le 21$$

where x represents the number of years since 1990 and $f(x)$ represents the amount of carbon dioxide emissions emitted annually, measured in millions of metric tons. (*Source: The International Energy Statistics Database.*)

(a) According to the model, is the 2030 estimate for CO$_2$ emissions accurate?

(b) Determine the derivative $f'(x)$.

(c) Write the equation of the tangent line of f when $x = 20$.

(d) Find the y-value on the tangent line when $x = 40$. Explain what this value means in terms of the CO$_2$ emissions.

Section 3.5 Review Exercises

For Exercises 89–94, complete the following.

(a) Determine the marginal cost function $MC(x)$.

(b) For the given production level x, evaluate $MC(x)$ and interpret.

(c) Determine the actual change in cost by evaluating $C(x + 1) - C(x)$ and compare with the answer to part (b).

89. $C(x) = 18x + 642; x = 2$

90. $C(x) = 9x + 1460; x = 27$

91. $C(x) = 26.7x + 87.4; x = 8$

92. $C(x) = \frac{1}{2}x^2 + 3x + 16; x = 15$

93. $C(x) = \frac{1}{4}x^2 + 12x + 47; x = 31$

94. $C(x) = \frac{1}{3}x^2 + 318x + 1783; x = 23$

In Exercises 95–100, the cost function C and the price–demand function p are given.

(a) Determine the revenue function R.

(b) Determine the profit function P.

(c) Differentiate P in part (b) to get the marginal profit function MP.

(d) Determine the marginal cost function MC and the marginal revenue function MR.

(e) Subtract the solutions found in part (d) to get $MR - MC$ and simplify. Compare with the result of part (c).

95. $C(x) = 7x + 250; p(x) = 11$

96. $C(x) = 14x + 1380; p(x) = 21$

97. $C(x) = \frac{1}{10}x^2 + 3x + 850; p(x) = -\frac{x}{15} + 50$

98. $C(x) = \frac{1}{50}x^2 + \frac{1}{4}x + 70; p(x) = -\frac{x}{50} + 5$

99. $C(x) = -0.01x^3 + 8x^2 + 100; p(x) = -0.005x + 10$

100. $C(x) = -0.01x^3 + 0.1x^2 + 4x + 18; p(x) = -0.6x + 15$

Manufacturing—Bicycles: For Exercises 101–105, the Wheelex Company manufactures bicycles and finds the price function for the bicycles to be

$$p(x) = -0.02x + 150$$

where x represents the number of bicycles produced and sold and p(x) is the price of the bicycle. Furthermore, the fixed and variable costs to produce x bicycles are $5600 and $85 per bicycle, respectively.

101. The cost function follows the linear form $C(x) = mx + b$. Answer the following.

    (a) Write the cost C in the linear form $C(x) = mx + b$.

    (b) Use calculus to compute the marginal cost function $MC(x) = \frac{dC}{dx}$.

102. Using the information found in Exercise 101, complete parts (a) through (d).

    (a) Evaluate $MC(250)$ and $MC(500)$ and interpret these answers.

    (b) Why are the answers in part (a) equal?

    (c) Algebraically find $AC(x)$ and simplify.

    (d) Evaluate $AC(250)$ and interpret.

103. Use the solution from part (c) of Exercise 102 to answer parts (a) and (b).

    (a) Use calculus to compute the marginal average cost function $MAC(x)$.

    (b) Evaluate $MAC(250)$ and interpret.

104. Use $p(x)$ and the cost function information given in Exercise 101 to complete parts (a) through (c).

    (a) Derive the revenue function $R(x)$.

    (b) Use calculus to compute $MR(x) = \frac{dR}{dx}$.

    (c) Evaluate $MR(500)$ and interpret.

105. Use the solution from part (a) of Exercise 101 with the solution to part (a) of Exercise 104 to complete parts (a) and (b).

    (a) Derive the profit function $P(x)$.

    (b) If the bicycles must be manufactured in lots of 250, how many bicycles should be manufactured so that the profit is as large as possible? Verify your answer by completing the table.

Produced, x	Total profit, P(x)
0	
250	
500	
750	
1000	
1250	
1500	
1750	
2000	

106. Sales Analysis—Watercolors: The Colorama Company, a paint manufacturer, has just produced a watercolor set that it sells to wholesalers for $6 each. The cost $C(x)$ to produce x watercolor sets is given by the function

$$C(x) = 0.0002x^2 + 2x + 1250$$

    (a) Algebraically derive the profit function $P(x)$ and simplify it.

    (b) Evaluate $P(3000)$ and interpret.

    (c) Differentiate to compute the marginal profit function.

    (d) Evaluate $MP(3000)$ and interpret.

(e) Use the solutions from parts (b) and (d) to get a linear approximation for the value of $P(3001)$.

(f) Compute the error of the approximation for $P(3001)$ in part (e).

107. Sales Analysis—Watercolors: The Colorama Company hires a consulting firm to assess its work and consequently revises its price and cost functions to

$$p(x) = 6.50 \text{ and } C(x) = \frac{x^2}{5500} + \frac{7}{3}x + 1500$$

Redo parts (a) to (f) in Exercise 106 using these revisions.

108. Knowing that $AC(x) = \dfrac{C(x)}{x}$, use the quotient rule to show that the marginal average cost function can be written as $MAC(x) = \dfrac{MC(x) - AC(x)}{x}$.

109. Publishing—Price/Demand: The Between the Lines Publishing Company determines that the price-demand function for a new book is $p(x) = \dfrac{-x}{500} + 20$, with fixed costs of \$12,000 and variable costs of 4.5 dollars per book.

(a) Determine the cost function C and the revenue function R.

(b) Find the profit function P.

(c) Find the smallest and largest production levels x so that the company realizes a profit. (That is, find the smallest and largest independent values so that the revenue is greater than the cost.)

(d) Compute the marginal profit function $P'(x)$.

(e) Evaluate $P'(2500)$ and interpret the result.

110. Manufacturing—Price/Demand: The Deluxe Furniture Company determines that the price–demand function for its new bookshelf is

$$p(x) = -\frac{x}{75} + 250$$

The fixed costs are \$10,000 and variable costs are 150 dollars per unit. Redo parts (a) through (e) in Exercise 109.

Finance—Stockholder Analysis: For Exercises 111–116, consider a stockholder's report that lists the information shown in the table.

q	$C(q)$	$R(q)$
100	22,830	38,000
200	30,830	72,000
300	38,830	102,000

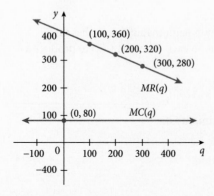

Use the table and the graphs of MC(q) and MR(q) to get a linear approximation for the given cost and revenue functions.

111. $C(101)$

112. $C(201)$

113. $C(301)$

114. $R(101)$

115. $R(201)$

116. $R(301)$

117. Manufacturing—Average Costs: Consider the following data table for the cost and revenues at various production levels for a new brand of computer printer.

Number of printers produced, x	Cost, C(x)	Revenue, R(x)
1000	243,600	296,950
2000	363,000	575,800
3000	482,800	818,550
4000	603,300	1,007,200

(a) Use your calculator to determine a linear regression model for the cost of producing the printer in the form

$$C(x) = ax + b \qquad 1 \le x \le 4$$

where x represents the number of printers produced in thousands and $C(x)$ represents the cost of production.

(b) Use your calculator to determine a cubic regression model for the revenue of producing and selling the printers in the form

$$R(x) = ax^3 + bx^2 + cx + d \qquad 1 \le x \le 4$$

where x represents the number of printers produced and sold in thousands and $R(x)$ represents the resulting revenue.

(c) Compute $MAC(x)$ and simplify the result.

(d) Evaluate $MAC(1.5)$ and interpret.

118. Use the models found in Exercise 117 parts (a) and (b), to answer parts (a) and (b).

(a) Compute $MAP(x)$ and simplify the result.

(b) Evaluate $MAP(1.5)$ and interpret the answer.

Exponential and Logarithmic Functions

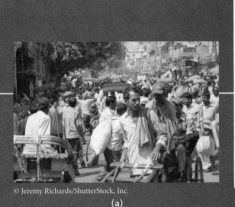

© Jeremy Richards/ShutterStock, Inc.

(a)

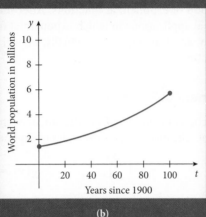

(b)

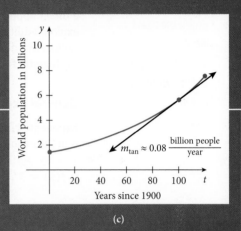

$m_{\tan} \approx 0.08 \dfrac{\text{billion people}}{\text{year}}$

(c)

While the rate of population growth has slowed in most of the industrialized world, it continues to grow at an alarming rate in many places around the world. In Figure (b) is a graph of a world population model, $p(t) = 1.419e^{0.014t}$, that shows the population in billions during the 20th century. As shown in Figure (c), finding the slope of the line tangent to the curve at $t = 99$ (that is, evaluating $p'(99)$) indicates that in 1999 the world's population was growing at a rate of about 80 million people per year.

What We Know

In the previous chapter, we learned the basic rules of differentiation and their applications. We also learned that the derivative can be used to study marginal analysis of business functions. We found that the role of the differential was central to marginal analysis.

Where Do We Go

In this chapter, we will learn how to differentiate other families of functions for which the rules we have learned so far may not apply. These include exponential and logarithmic functions.

Chapter Sections

SECTION OBJECTIVES

1. Graph a general exponential function.

2. Apply an exponential decay model.

3. Calculate compound interest.

4. Consider the number e and the exponential function.

5. Evaluate a logistic function.

4.1   Exponential Functions

Often business people make a proclamation such as, "Our sales are growing exponentially!" This may not be exactly true, but it is an effective way to communicate that the sales are growing very rapidly. In this section, we will study **general exponential functions** that typically have function values that either increase or decrease rapidly. Then we will discuss compound interest, which is a direct application of exponential functions. After introducing the irrational number e, we will explore additional applications in which exponential functions arise, such as population growth, radioactive decay, and learning curves. Finally, we will introduce a special type of function called the **logistic function**.

General Exponential Functions

Exponential functions are different from the functions studied so far in that the independent variable is in the exponent. The following are examples of exponential functions:

$$f(x) = 3^x, \qquad y = \left(\frac{1}{5}\right)^x, \qquad g(x) = 2 \cdot 4^x, \qquad f(x) = 1.6 \cdot (1.09)^x$$

DEFINITION

General Exponential Function

A **general exponential function** has the form

$$f(x) = a \cdot b^x$$

where a is a real number **constant** with $a \neq 0$, and b is a real number with $b > 0$, $b \neq 1$. Here b is called the **base**.

NOTE: Do not confuse exponential functions with power functions. For an exponential function, the base is a constant and the exponent is the independent variable. The reverse is true for power functions. For example, $f(x) = x^3$ is a power function, whereas $f(x) = 3^x$ is an exponential function.

OBJECTIVE 1

Graph a general exponential function.

Example 1: Graphing Exponential Functions

a.   Make a table of values for the general exponential functions $f(x) = 2^x$ and $g(x) = (\frac{1}{2})^x$ for $x = -5, -4, -3, \ldots, 3, 4, 5$.

b.   Graph the functions f and g and write their domain using interval notation.

Perform the Mathematics

a.   The values for the functions are shown in **Table 4.1.1**.

Table 4.1.1

x	-5	-4	-3	-2	-1	0	1	2	3	4	5
$f(x) = 2^x$	$\frac{1}{32}$	$\frac{1}{16}$	$\frac{1}{8}$	$\frac{1}{4}$	$\frac{1}{2}$	1	2	4	8	16	32
$g(x) = \left(\dfrac{1}{2}\right)^x$	32	16	8	4	2	1	$\frac{1}{2}$	$\frac{1}{4}$	$\frac{1}{8}$	$\frac{1}{16}$	$\frac{1}{32}$

b.   From the table and the graphs in **Figure 4.1.1 (a)** and (b), we see that any real number can be substituted for x, so the domain of the functions $f(x) = 2^x$ and $g(x) = (\frac{1}{2})^x$ is $(-\infty, \infty)$.

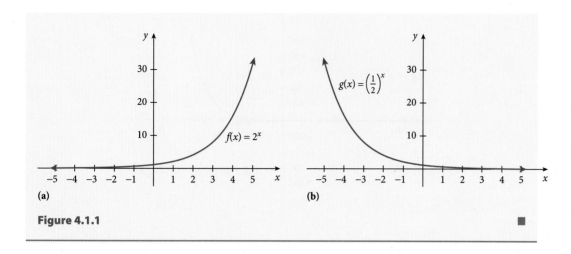

(a)

(b)

Figure 4.1.1

Notice that $f(x) = 2^x$ is increasing on its domain and $\lim\limits_{x \to \infty} 2^x = \infty$ and $\lim\limits_{x \to -\infty} 2^x = 0$. We call this kind of function an **exponential growth function**. Whereas, notice that $g(x) = \left(\frac{1}{2}\right)^x$ is decreasing on its domain, $\lim\limits_{x \to \infty} \left(\frac{1}{2}\right)^x = 0$, and $\lim\limits_{x \to -\infty} \left(\frac{1}{2}\right)^x = \infty$. We call this kind of function an **exponential decay function**. We summarize this in the **Properties of General Exponential Functions** box.

Properties of General Exponential Functions

- The domain is the set of all real numbers $(-\infty, \infty)$.
- The y-intercept is at $(0, 1)$.
- The behavior of the function is shown in the table.

Function type	Definition	Graph
Exponential growth function	$f(x) = b^x, b > 1$ $\lim\limits_{x \to \infty} b^x = \infty$ $\lim\limits_{x \to -\infty} b^x = 0$	
Exponential decay function	$f(x) = b^x, 0 < b < 1$ $\lim\limits_{x \to \infty} b^x = 0$ $\lim\limits_{x \to -\infty} b^x = \infty$	

For functions of the form $f(x) = a \cdot b^x$, the value of a can affect the end behavior. First, we see that with any value of a, $f(0) = a \cdot b^0 = a \cdot 1 = a$. This means that the y-intercept of the graph of $f(x) = a \cdot b^x$ is at $(0, a)$. To see the effect of a, compare the graphs of $f(x) = 2 \cdot 3^x$ and $g(x) = -2 \cdot 3^x$ in **Figure 4.1.2**. In Figure 4.1.2 we see that switching the sign of a from positive to negative results in a reflection about the x-axis. Now let's see how these exponential functions are used in applications.

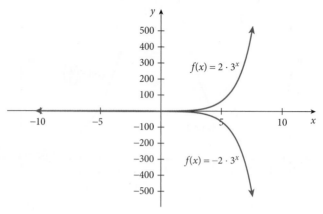

Figure 4.1.2

© iofoto/ShutterStock

OBJECTIVE 2

Apply an exponential decay model.

Example 2: Using General Exponential Decay Models

The number of disabled World War II veterans receiving compensation for service-related disabilities can be modeled by

$$f(x) = 1445.83(0.93)^x \qquad 1 \le x \le 20$$

where x represents the number of years since 1989, and $f(x)$ represents the number of World War II veterans receiving compensation, measured in thousands. (*Source:* U.S Census Bureau.)

a. Evaluate $f(11)$ and interpret.

b. Classify the function as an exponential growth or exponential decay function and describe the behavior of f as the x-values increase.

Perform the Mathematics

a. We begin by noting that $x = 11$ corresponds to the year 2000. Evaluating the function, we get

$$f(11) = 1445.83(0.93)^{11} \approx 650.773$$

This means that in 2000 there were approximately 650,773 disabled World War II veterans receiving compensation.

b. The graph of f is shown in **Figure 4.1.3**. For the model $f(x) = 1445.83(0.93)^x$, the value of $b = 0.93$ is between zero and one. This means f is an exponential decay function, which is supported by Figure 4.1.3.

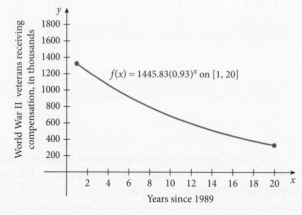

Figure 4.1.3

We observe in Figure 4.1.3 that as the x-values (representing years) increase, the function values decrease. Since x represents the number of years since 1989, this means that the farther away from 1989, the number of World War II veterans receiving compensation is decreasing. ∎

Compound Interest

A common application of the general exponential function in the banking industry is **compound interest**. To understand where the compound interest formula comes from, we begin by reviewing **simple interest**.

> **DEFINITION**
>
> **Simple Interest Formula**
>
> If P dollars is deposited in an account that earns interest at a rate of r (written as a decimal), then the interest I accumulated after t years is
>
> $$I = P \cdot r \cdot t$$
>
> Interest = Principal · rate · time

> **DEFINITION**
>
> **Compound Interest Formula**
>
> If a principal of P dollars is invested in an account earning an annual interest rate of r (in decimal form) compounded k times per year, then the amount A in the account at the end of t years is given by the **compound interest formula**
>
> $$A = P\left(1 + \frac{r}{k}\right)^{kt}$$

Compound interest means that interest is paid not only on the money deposited, called the **principal**, but also on interest that has already accumulated. Usually, a bank computes the interest earned and deposits the interest in your account. Then, the next time that the interest is computed, the bank computes interest on the original principal **plus** previously earned interest. The amount in an account that earns compound interest is given by the **compound interest formula**.

Example 3: Calculating Compound Interest

OBJECTIVE 3

Calculate compound interest.

Suppose that $20,000 is deposited into an account that yields an annual interest rate of 6.5%, compounded quarterly.

a.   How much will be in the account after three years? Round to the nearest cent.

b.   How much interest was earned at the end of the three-year period?

Perform the Mathematics

a.   Using the compound interest formula with $P = 20{,}000$, $r = 0.065$, $k = 4$, and $t = 3$ we get

$$A = P\left(1 + \frac{r}{k}\right)^{kt} = 20{,}000\left(1 + \frac{0.065}{4}\right)^{4(3)}$$

$$= 20{,}000(1.01625)^{12} \approx 24{,}268.15$$

So there will be about $24,268.15 in the account in three years.

b.   To determine the interest earned, we take the amount A and subtract the principal P from it. This gives us $24,268.15 − $20,000 = $4268.15. So about $4268.15 in interest is earned in three years. ∎

▶ **Try It Yourself**

Some related Exercises are 31 and 33.

Table 4.1.2

k	$\left(1 + \dfrac{1}{k}\right)^{k}$
1	2
10	$2.59374\ldots$
100	$2.70481\ldots$
1000	$2.71692\ldots$
10,000	$2.71815\ldots$
100,000	$2.71827\ldots$
1,000,000	$2.71828\ldots$

If we hold the values of P, r, and k constant, we can think of the compound interest formula as a function of time t. Writing this gives us the **compound interest function**: $A(t) = P(1 + \frac{r}{k})^{kt}$. Notice here that the independent variable is time, denoted with a t. For the account in Example 3, we can write the compound interest function as

$$A(t) = 20{,}000\left(1 + \frac{0.065}{4}\right)^{4t}$$

To find the value of the account after five years, we evaluate $A(5)$.

$$A(5) = 20{,}000\left(1 + \frac{0.065}{4}\right)^{4(5)} = 20{,}000(1.01625)^{20} \approx 27{,}608.40$$

This means that after five years, there will be about \$27,608.40 in the account.

The Number e

OBJECTIVE 4

Consider the number e and the exponential function.

At the beginning of this section, we saw that we could write an exponential function $f(x) = a \cdot b^x$ with any base b as long as $b > 0$, $b \neq 1$. However, the most frequently used base is the famous number e. This number is so widely used that it is found on nearly every scientific and graphing calculator. To see what this number is, let's numerically examine the expression $(1 + \frac{1}{k})^k$, where k is an independent variable. See **Table 4.1.2**.

From Table 4.1.2, it appears that as the k-values become very large, that is, $k \to \infty$, the values of the expression $(1 + \frac{1}{k})^k$ seem to be getting close to a single value called e.

Many applications in the physical, life, and social sciences use e as the base of the exponential function. As a matter of fact, this is often referred to as *the* **exponential function**.

DEFINITION

The Number e

The real number e is defined

$$e = \lim_{k \to \infty}\left(1 + \frac{1}{k}\right)^{k}$$

DEFINITION

The Exponential Function

The **exponential function** has the form

$$f(x) = ae^{bx}$$

where a and b are real number constants.

NOTE: The simplest form of the exponential function is $f(x) = e^x$. See **Figure 4.1.4**.

Properties of Exponential Functions

Given the exponential function $f(x) = ae^{bx}$, with $a > 0$ and $b \neq 0$, we have

- The domain is the set of all real numbers $(-\infty, \infty)$.
- The y-intercept is at $(0, a)$.
- If $b > 0$, then f is an exponential growth function.
- If $b < 0$, then f is an exponential decay function.

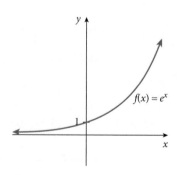

Figure 4.1.4

As with general exponential functions, exponential functions of the form $f(x) = ae^{bx}$ can also model exponential growth and exponential decay. The growth or decay **property of an exponential function** $f(x) = ae^{bx}$ is determined by the sign of the constant b.

Example 4: Applying the Exponential Decay Function

The annual per capita consumption of whole milk in the United States can be modeled by

$$f(x) = 15.85e^{-0.035x} \qquad 1 \le x \le 29$$

where x represents the number of years since 1979 and $f(x)$ represents the per capita consumption of whole milk in gallons. (*Source*: U.S. Department of Agriculture.)

a. Evaluate $f(21)$ and interpret.

b. Classify the function as an exponential growth or exponential decay function, and describe the behavior of f as the x-values increase.

Perform the Mathematics

a. First, we notice that $x = 21$ corresponds to the year 2000. Evaluating the function at $x = 21$ yields

$$f(21) = 15.85e^{-0.035(21)} = 15.85e^{-0.735} \approx 7.6$$

This means that in 2000, the annual per capita consumption of whole milk was about 7.6 gallons.

b. The graph of f is given in **Figure 4.1.5**. For the model $f(x) = 15.85e^{-0.035x}$, the value of $b = -0.035$ is less than zero, that is, $b < 0$. This means f is an exponential decay function, which is supported by Figure 4.1.5.

© Dave McAleavy/ShutterStock, Inc.

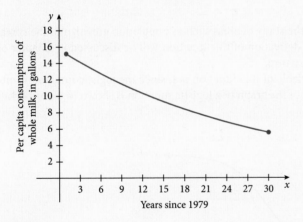

Figure 4.1.5

The exponential function $f(x) = ae^{bx}$ is often used in the sciences, as well as the business sciences. Let's suppose that a banker with a strong entrepreneurial spirit wants to offer customers an account that accumulates interest **continuously** over time. To find a formula for this kind of interest, recall that the compound interest formula is

$$A = P\left(1 + \frac{r}{k}\right)^{kt}$$

It can be shown that $\lim_{k \to \infty} \left(1 + \frac{r}{k}\right)^k = e^r$. So rewriting the compound interest formula, we get

$$A = P\left(1 + \frac{r}{k}\right)^{kt}$$

$$= P\left[\left(1 + \frac{r}{k}\right)^k\right]^t$$

DEFINITION

Continuous Compound Interest

If a principal P dollars is invested into an account earning an annual interest rate r (in decimal form) compounded continuously, then the amount A in the account at the end of t years is given by the **continuous compound interest formula**

$$A = Pe^{rt}$$

For $k \to \infty$, we can substitute e^r for the expression in the brackets and get

$$A = Pe^{rt}$$

This is the **continuous compound interest formula**.

As a function of time, we can define the **continuous compound interest function** as $A(t) = Pe^{rt}$.

Example 5: Computing Compound Interest

Suppose that we have $5000 deposited in an account that earns 5.3% interest compounded continuously. Find the continuous compound interest function for this account, $A(t)$, and evaluate $A(3)$ and interpret.

Perform the Mathematics

Knowing that $r = 5.3\% = 0.053$ and $P = 5000$, we get the function

$$A(t) = 5000e^{0.053t}$$

Evaluating $A(t)$ for $t = 3$, we get

$$A(3) = 5000e^{0.053(3)} = 5000e^{0.159} \approx \$5861.69$$

So after 3 years, there would be $5861.69 in the account. ∎

DEFINITION

Logistic Function

A **logistic function** has the form

$$f(t) = \frac{L}{1 + ae^{-kLt}}$$

where L is the limit of growth (that is, a horizontal asymptote) and a and k are real number constants.

Logistic Curves

One type of function that is used in the study of areas such as population growth and the spread of disease is the **logistic function**. The derivation of this function will be discussed in Chapter 6; for now, we simply define the **logistic function**.

Notice that the independent variable of this function is t, since the logistic function is often a function of time. The typical S-shape of the graph of a logistic function is shown in **Figure 4.1.6**.

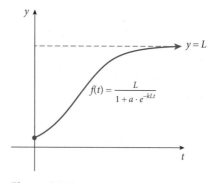

Figure 4.1.6

OBJECTIVE 5

Evaluate a logistic function.

Example 6: Using Logistic Functions

The number of people affected by a flu epidemic in the isolated town of Fall Peak (population 1800) can be modeled by

$$f(t) = \frac{1800}{1 + 359e^{-0.9t}}$$

where t represents the number of weeks since the flu was first detected and $f(t)$ represents the number of people who are affected. See **Figure 4.1.7**.

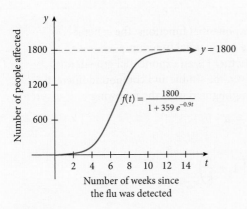

Figure 4.1.7

a. Evaluate $f(1)$ and interpret.

b. Determine $\displaystyle \lim_{t \to \infty} \frac{1800}{1 + 359e^{-0.9t}}$ and interpret.

Perform the Mathematics

a. Evaluating the function at $t = 1$, we get

$$f(1) = \frac{1800}{1 + 359e^{-0.9(1)}} = \frac{1800}{1 + 359e^{-0.9}} \approx 12.25$$

This means that, according to the model, after one week about 12 people were affected.

b. From Figure 4.1.7, it appears that this limit is 1800. That is, graphically we believe

$$\lim_{t \to \infty} \frac{1800}{1 + 359e^{-0.9t}} = 1800$$

Analytically, we determine the limit as follows.

$$\lim_{t \to \infty} \frac{1800}{1 + 359e^{-0.9t}}$$

Limit Theorems $\qquad \displaystyle = \frac{\lim\limits_{t \to \infty} 1800}{\lim\limits_{t \to \infty} (1 + 359e^{-0.9t})}$

Limit Theorems $\qquad \displaystyle = \frac{\lim\limits_{t \to \infty} 1800}{\lim\limits_{t \to \infty} 1 + \lim\limits_{t \to \infty} 359e^{-0.9t}}$

Limit Theorems $\qquad \displaystyle = \frac{\lim\limits_{t \to \infty} 1800}{\lim\limits_{t \to \infty} 1 + 359 \lim\limits_{t \to \infty} e^{-0.9t}}$

Limit Theorems $\qquad \displaystyle = \frac{1800}{1 + 359 \cdot 0}$

$$= 1800$$

This means that as time increases without bound, the number of people affected by the flu is 1800. ∎

Summary

In this section, we examined two types of exponential functions: the general exponential function $f(x) = a \cdot b^x$ and the exponential function $f(x) = ae^{bx}$. We found that the value of b determines whether f is an exponential growth or exponential decay function. Then we reviewed the simple and compound interest formulas. The definitions that this section highlighted were the following:

- **General Exponential Function:** $f(x) = a \cdot b^x$
- **Simple Interest Formula:** $I = P \cdot r \cdot t$
- **Compound Interest Formula:** $A = P\left(1 + \dfrac{r}{k}\right)^{kt}$
- **The number e:** $\lim\limits_{k \to \infty}\left(1 + \dfrac{1}{k}\right)^k = e$
- **Exponential Function:** $f(x) = ae^{bx}$
- **Continuous Compound Interest Formula:** $A = Pe^{rt}$
- **Logistic Function:** $f(t) = \dfrac{L}{1 + ae^{-kLt}}$

Section 4.1 Exercises

Vocabulary Exercises

1. The number _____ is an irrational number and can be defined as $\lim\limits_{k \to \infty}(1 + \tfrac{1}{k})^k$.

2. The _____ exponential function is given by $f(x) = a \cdot e^{kx}$.

3. The _____ exponential function is given by $f(x) = a \cdot b^x$.

4. The formula $I = Prt$ gives the _____ interest of an account.

5. If $b < 0$ in the function $f(x) = a \cdot e^{bx}$, then we call the function an exponential _____ function.

6. In the function $f(x) = a \cdot b^x$, we call b the _____.

7. A function of the form $f(t) = \dfrac{L}{1 + a \cdot e^{-kLt}}$ is called a _____ function.

8. The formula $A = Pe^{rt}$ gives _____ compound interest.

Skill Exercises

For Exercises 9–22, complete the following.

(a) Make a table of function values for $x = -5, -4, \ldots, 4, 5$.

(b) Graph and classify f as an exponential decay function or an exponential growth function.

9. $f(x) = \left(\dfrac{1}{3}\right)^x$

10. $f(x) = \left(\dfrac{1}{4}\right)^x$

11. $f(x) = 4^x$

12. $f(x) = 3^x$

13. $f(x) = (2.3)^x$

14. $f(x) = (4.5)^x$

15. $f(x) = (0.7)^x$

16. $f(x) = (0.22)^x$

17. $f(x) = e^{2x}$

18. $f(x) = e^{7x}$

19. $f(x) = e^{0.3x}$

20. $f(x) = e^{0.92x}$

21. $f(x) = e^{-1.6x}$

22. $f(x) = e^{-3.1x}$

Application Exercises

23. **Medicine—Healing Time:** Medical researchers have recently begun to use mathematical modeling to project the length of time that it takes to heal many common wounds. Suppose that

the healing of a 30-square-inch chest wound can be modeled by

$$f(x) = 30\left(\frac{4}{5}\right)^x \quad x \ge 0$$

where x represents the number of days since the wound occurred and $f(x)$ represents the size of the wound in square inches.

(a) Evaluate $f(7)$ and interpret.

(b) Compute the slope of the secant line for $x = 1$ and $\Delta x = 7$ and interpret.

(c) Determine $\lim\limits_{x \to \infty} f(x)$ and interpret.

24. **Consumer Mathematics—Sales Projections:** The use of recreational utility vehicles (RUVs) is an off-road adventure activity that has gained in popularity in recent years. The Desert Fox Dune Buggy Company launches a sales blitz for its new Family Buggy RUV and offers a 30% discount during the first week of sales. The projected percentage of customers who respond to this sale can be modeled by

$$f(x) = 70 - 100\left(\frac{3}{5}\right)^x \quad 1 \le x \le 7$$

where x represents the number of days since the sales blitz and $f(x)$ represents the number of customers who respond to the sale.

(a) Evaluate $f(1)$ and interpret.

(b) Compute the slope of the secant line for $x = 2$ and $\Delta x = 2$ and interpret.

(c) Determine $\lim\limits_{x \to \infty} f(x)$ and interpret.

 25. **Civil Engineering—Traffic Fatalities:** Vision Zero is a traffic safety program started in Sweden in the late 1990s that emphasized ethics, responsibility, and safety mechanisms in order to reduce the number of traffic deaths in European countries. (*Source:* Association of Air Medical Services.) The annual number of traffic fatalities in France from 2000 to 2010 can be modeled by

$$f(x) = 8505(0.91)^x \quad 0 \le x \le 10$$

where x represents the number of years since 2000, and $f(x)$ represents the number of traffic fatalities.

(a) Evaluate and interpret $f(5)$.

(b) Compute the slope of the secant line of f for $x = 5$ and $\Delta x = 4$ and interpret.

(c) From analyzing the exponential model, was the Vision Zero program effective in France? Explain.

 (d) Graph the model in the viewing window $[0,10]$ by $[2000,4000]$. Graph $Y1 = 8505*.91^x$ and $Y2 = 3312$, then use the `Intersect` command to determine the year there were 3312 traffic fatalities in France.

 26. **Civil Engineering—Traffic Fatalities:** The Vision Zero program mentioned in Exercise 25 was also implemented in Germany. The annual number of traffic fatalities in Germany from 2000 to 2010 can be modeled by

$$f(x) = 7580(0.94)^x \quad 0 \le x \le 10$$

where x represents the number of years since 2000 and $f(x)$ represents the number of traffic fatalities.

(a) Evaluate and interpret $f(4)$.

(b) Compute the slope of the secant line of f for $x = 4$ and $\Delta x = 5$ and interpret.

(c) From analyzing the exponential model, was the Vision Zero program effective in Germany? Explain.

(d) Graph the model in the viewing window $[0,10]$ by $[4000,8000]$. Graph $Y1 = 7580*.94^x$ and $Y2 = 5563$, then use the `Intersect` command to determine the year there were 5563 traffic fatalities in Germany.

27. **Marketing—Direct Mail Advertising:** Information from a report titled "The Power of Direct Marketing" claims that direct mail marketing is growing faster than total advertising spending and the U.S. economy as a whole. (*Source:* Direct Marketing Association.) The annual amount spent on direct marketing in the United States from 2000 to 2010 can be modeled by

$$f(x) = 43.7(1.04)^x \quad 0 \le x \le 10$$

where x represents the number of years since 2000 and $f(x)$ represents the annual amount spent on direct marketing, measured in billions of dollars. (*Source:* U.S. Statistical Abstract.)

(a) Evaluate and interpret $f(1)$.

(b) Compute the slope of the secant line of f for $x = 1$ and $\Delta x = 6$ and interpret.

(c) From analyzing the exponential model, is the amount spent on direct marketing increasing or decreasing? Explain.

28. **Marketing—Internet Advertising:** A report by the Interactive Advertising Bureau claims that the growth of advertising on the Internet has begun to outpace the growth of TV advertising. (*Source:* TechCrunch.) The annual amount spent on Internet advertising in the United States from 2000 to 2010 can be modeled by

$$f(x) = 4.66(1.12)^x \quad 0 \le x \le 10$$

where x represents the number of years since 2000 and $f(x)$ represents the annual amount spent on Internet advertising, measured in billions of dollars. (*Source:* U.S. Statistical Abstract.)

(a) Evaluate and interpret $f(2)$.

(b) Compute the slope of the secant line of f for $x = 2$ and $\Delta x = 5$ and interpret.

(c) From analyzing the exponential model, is the amount spent on Internet advertising increasing or decreasing? Explain.

Banking—Computing Simple Interest: For Exercises 29–38, compute the desired interest amount.

29. Use the simple interest formula to find the total interest earned for an account in which $500 is deposited at a 4% interest rate for six years.

30. Use the simple interest formula to find the total interest earned for an account in which $2000 is deposited at a 6.5% interest rate for four years.

31. If Kevin deposits $3000 into an account that yields 6% interest compounded annually, how much will be in the account after three years?

32. If Shauna deposits $10,000 into an account that yields 6.5% interest compounded semiannually, how much will be in the account after four years?

33. If Lucy deposits $1800 into an account that yields 7.1% interest compounded monthly, how much will be in the account after five years?

34. If Lana deposits $800 into an account that yields 7% interest compounded annually, how much will be in the account after eight years?

35. If Joe deposits $2000 into an account that yields 8% annual interest, how much will be in the account after six years if the interest is compounded

    (a) annually?   (b) semiannually?   (c) quarterly?

36. If Mrs. Johanson deposits $5000 into an account that yields 6.5% annual interest, how much will be in the account after ten years if the interest is compounded

    (a) annually?   (b) monthly?   (c) weekly (52 times a year)?

37. If the Klein family deposits $15,000 into an account that yields 5.9% annual interest, how much will be in the account after 16 years if the interest is compounded

    (a) annually?   (b) monthly?   (c) weekly (52 times a year)?

38. If the LaDukes deposit $8000 into an account that yields 6.1% annual interest, how much will be in the account after ten years if the interest is compounded

    (a) annually?   (b) semiannually?   (c) quarterly?

Banking—Computing Compound Interest: For Exercises 39–42, use the compound interest function $A(t) = P\left(1 + \frac{r}{k}\right)^{kt}$ to complete the following.

(a) *Evaluate $A(5)$ and interpret the result.*

(b) *Determine how much in interest was made after the end of the five-year period.*

39. $5500 is invested at 5.5% interest compounded quarterly.

40. $6400 is invested at 6.5% interest compounded quarterly.

41. $10,000 is invested at 7.25% interest compounded monthly.

42. $7500 is invested at 6.25% interest compounded monthly.

43. Personal Banking—Doubling Time: Suppose Elizabeth deposits $4000 at 5.75% annual interest compounded monthly.

   (a) Write the compound interest function A for the given information.

   (b) Graph A in the viewing window $[0, 16]$ by $[4000, 10,000]$.

   (c) Use your calculator to graphically find the amount of time that it takes the account to accumulate a total balance of $8000. This is called the *doubling time* of the account.

44. Personal Banking—Doubling Time: Now assume that Elizabeth deposits $6000 at a 5.75% annual interest rate compounded monthly.

   (a) Write and graph the compound interest function A for the given information.

   (b) Find the doubling time of this new account. Compare your answer to part (c) of Exercise 43.

45. Biological Science—Bacterial Growth: Under certain conditions, the spread of the *E. coli* strain of bacteria can be modeled by

$$f(t) = N_0 e^{0.23t} \qquad t \geq 0$$

where t represents the time in minutes and $f(t)$ represents the size of the culture after time t. If the size of the culture begins at $N_0 = 1,200,000$, answer the following.

   (a) Evaluate $f(5)$ and interpret.

   (b) Determine the slope of the secant line for $t = 0$ and $\Delta t = 10$ and interpret.

46. Medicine—Drug Absorption: The absorption of a drug is determined by several factors, including its physicochemical properties and method of administration. The amount of a new experimental drug that is in a patient's bloodstream can be modeled by

$$f(t) = 5e^{-0.3t} \qquad t \geq 0$$

where t represents the time since the drug was administered in hours and $f(t)$ represents the amount of the drug in the bloodstream, measured in milligrams.

   (a) Evaluate $f(2)$ and interpret.

   (b) Determine the slope of the secant line for $t = 2$ and $\Delta t = 3$ and interpret.

47. Ecology—Atlantic Fish Catches: During the 1950s and 1960s, the global expansion of fisheries was coupled with ever-growing catches at a rate so rapid that they tended to exceed fish population growth. (*Source:* Institute of Development Research.) The annual amount of fish caught by Norwegians from 2000 to 2010 in the northwest Atlantic Ocean can be modeled by

$$f(x) = 30{,}111e^{-0.22x} \qquad 0 \leq x \leq 10$$

where x represents the number of years since 2000 and $f(x)$ represents the amount of fish caught in tons. (*Source:* EuroStat.)

   (a) Classify the model as an exponential growth or exponential decay model.

   (b) Evaluate $f(2)$ and interpret.

   (c) Compute the slope of the secant line for $x = 2$ and $\Delta x = 6$. Does this result support the Institute's claim?

48. Ecology—Mediterranean Fish Catches: The annual amount of fish caught by Slovenians from 2000 to 2010 in the Mediterranean Sea can be modeled by

$$f(x) = 19.46e^{-0.13x} \qquad 0 \le x \le 10$$

where x represents the number of years since 2000 and $f(x)$ represents the amount of fish caught in tons. (*Source:* EuroStat.)

(a) Classify the model as an exponential growth or exponential decay model.

(b) Evaluate $f(2)$ and interpret.

(c) Compute the slope of the secant line for $x = 2$ and $\Delta x = 6$. Comparing this result to Exercise 47(c), which country has the greatest decrease in fish catches, Norway or Slovenia?

 49. Political Economics—Food Stamp Program: Fueled by rising unemployment and food prices, the number of Americans on food stamps has been steadily increasing. The number of participants in the federal food stamp program from 2000 to 2010 can be modeled by

$$f(x) = 17.73e^{0.06x} \qquad 0 \le x \le 10$$

where x represents the number of years since 2000 and $f(x)$ represents the number of participants in the food stamp program, measured in millions. (*Source:* U.S. Department of Agriculture.)

(a) Classify f as an exponential growth or exponential decay model.

(b) Evaluate $f(3)$ and interpret.

(c) Compute the slope of the secant line for $x = 2$ and $\Delta x = 7$ and interpret.

 50. Political Economics—School Breakfast Program: The national School Breakfast Program (SBP) was established by Congress as a pilot program in 1966 in areas where children had long bus rides to school and in areas where many mothers were in the workforce. It became a permanent entitlement program in 1975 to assist schools in providing nutritious morning meals to the nation's children. The annual number of participants in the SBP from 1990 to 2010 can be modeled by

$$f(x) = 4.42e^{0.05x} \qquad 0 \le x \le 20$$

where x represents the number of years since 1990 and $f(x)$ represents the number of SBP participants in millions of students. (*Source:* U.S. Department of Agriculture.)

(a) Classify f as an exponential growth or exponential decay model.

(b) Evaluate $f(3)$ and interpret.

(c) Compute the slope of the secant line for $x = 12$ and $\Delta x = 7$. From 2002 to 2009, which program has had a more rapid rise in participants, the food stamp program (consult Exercise 49(c)) or the SBP?

Banking—Compound Interest: In Exercises 51–56, use the continuous compound interest function to evaluate $A(7)$ and interpret.

51. $1000 is invested at a $5\frac{1}{2}$% interest rate.   **52.** $800 is invested at a 4.75% interest rate.

53. $20,000 is invested at a 5.9% interest rate.   **54.** $10,000 is invested at a 7.1% interest rate.

55. $8000 is invested at a 8.1% interest rate.   **56.** $75,000 is invested at a $5\frac{3}{4}$% interest rate.

57. Business—Sales: Suppose that the proportion of game systems that have the latest 3-D WOW technology can be modeled by

$$f(t) = \frac{0.9}{1 + 4e^{-0.3t}} \qquad t \ge 0$$

where t represents the number of months since the 3-D technology has been available on the market and $f(t)$ represents the proportion of game systems sold with the 3-D technology.

(a) Evaluate $f(1)$ and interpret.

(b) Compute the secant line slope for $t = 1$ and $\Delta t = 5$ and interpret.

(c) Determine $\lim_{t \to \infty} f(t)$ and interpret.

58. Public Health—Measles Spread: Suppose that the number of students getting the measles in the Grover local school system can be modeled by

$$f(t) = \frac{200}{1 + 200.34e^{-t}} \qquad t \geq 0$$

where t represents the number of days since the outbreak began, and $f(t)$ represents the number of students getting the measles.

(a) Evaluate $f(4)$ and interpret.

(b) Compute the secant line slope for $t = 4$ and $\Delta t = 5$ and interpret.

(c) Determine $\lim_{t \to \infty} f(t)$ and interpret.

59. Conservation Science—Population Analysis: Suppose that conservationists wish to repopulate a rare breed of lizard by transporting a seed colony to a wildlife preserve. They estimate that the protected lizard population will grow according to the logistics model

$$f(t) = \frac{1000}{1 + 121.51e^{-0.72t}} \qquad t \geq 0$$

where t represents the number of years since the seeding and $f(t)$ represents the lizard population.

(a) What was the initial size of the seed colony?

(b) According to the model, what will the lizard population be in 8 years?

(c) As time goes on, what is the projected population of the lizards in the colony?

60. Biology—Bacterial Growth: In a biological experiment, suppose that a seed colony of paramecia was placed in a Petri dish along with a nutritional medium. The number of paramecia in the dish can then be modeled by

$$f(t) = \frac{380}{1 + 181.22e^{-2.1t}} \qquad t \geq 0$$

where t represents the number of days since the start of the experiment and $f(t)$ represents the paramecium population.

(a) How many paramecia were initially placed in the Petri dish?

(b) According to the model, how many paramecia will be in the Petri dish after two days?

(c) As time goes on, what is the limiting size of the paramecium population in the dish?

61. Epidemiology—AIDS Cases: One of the primary applications of logistic curves is to model the spread of disease. During the early 1980s acquired immune deficiency syndrome (AIDS) was introduced to the U.S. population, and the number of cases grew rapidly during that decade. The annual number of new AIDS cases in the United States from 1980 to 1995 can be modeled by

$$f(t) = \frac{46820}{1 + 246e^{-0.76t}} \qquad 0 \leq t \leq 15$$

where t represents the number of years since 1980, and $f(t)$ represents the number of new AIDS cases. (*Source:* Centers for Disease Control and Prevention.)

(a) Use the model to determine the number of new AIDS cases in the United States in 1986.

(b) Compute the slope of the secant line for $x = 6$ and $\Delta x = 5$ and interpret.

(c) Determine $\lim_{t \to \infty} f(t)$ and interpret.

62. Demographics—U.S. Population: The logistic function is often used to express the demographic transition, which is the portion of population growth reflected by a transition from high birth and death rates to low birth and death rates. The population of the United States from 1900 to 2010 can be modeled by

$$f(t) = \frac{659}{1 + 7.22e^{-0.017t}} \qquad 0 \leq t \leq 110$$

where t represents the number of years since 1900 and $f(t)$ represents the U.S. population, measured in millions. (*Source: U.S. Census Bureau.*)

(a) Graph the function f in Y_1 using the viewing window $[0,110]$ by $[80,320]$ along with the line $Y_2 = 192$. Use the `Intersect` command to determine the year that the U.S. population was 192 million.

(b) Find the slope of the secant line for $x = 64$ and $\Delta x = 15$ and interpret.

(c) Determine $\lim\limits_{t\to\infty} f(t)$ and interpret.

Concept and Writing Exercises

63. Write a sentence explaining the difference between a general exponential function and a natural exponential function.

64. Suppose that a student claims that the number e is equal to $\frac{27,180}{9999}$ and you check on your calculator and the fraction seems to be the same as e^1. Does that prove that $e = \frac{27,180}{9999}$? Explain your answer.

65. Use the properties of exponents to show that $3^{x-2} = \frac{1}{9} \cdot 3^x$.

66. Use the properties of exponents to show that $32(2^x) = 2^{x+5}$.

67. Graph $f(x) = 2^x$ and $g(x) = 3^x$ in the viewing window $[-3, \ 5]$ by $[0, \ 10]$ and use the `Trace` command to solve the inequality $f(x) > g(x)$. Write your answer using interval notation.

For Exercises 68–72, consider the graph of the function $f(x) = b^x$. Use the graph to answer the following.

68. State the domain and range of f.

69. From inspecting the graph, is the value of b greater than one or between zero and one? Explain your answer.

70. What is $\lim\limits_{x\to-\infty} f(x)$?

71. Determine the function values $f(-1)$, $f(0)$, and $f(1)$.

72. What is $\lim\limits_{x\to\infty} f(x)$?

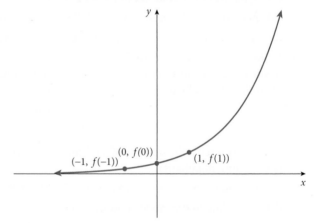

Section Project

Investors compare accounts that compound interest differently by using an **effective rate**. For example, an interest rate of 6.25% compounded monthly produces the same amount of interest as an account having a 6.43% interest rate compounded annually. Here, 6.43% is called the effective rate and 6.25% is called the nominal rate. The effective rate is given by

$$\text{Effective rate} = \left(1 + \frac{r}{k}\right)^k - 1$$

where r represents the nominal rate in decimal form, and k represents the number of times the interest is compounded each year. Compute the effective rates for the following accounts and interpret the results.

(a) A 4% nominal rate compounded weekly (52 times a year).

(b) A 7.5% nominal rate compounded monthly.

(c) A 6.1% nominal rate compounded daily (365 times a year).

(d) A 5.75% nominal rate compounded semiannually.

For interest compounded continuously, the effective rate is given by

$$\text{Effective rate} = e^r - 1$$

(e) A 4% nominal rate compounded continuously.

(f) A 7.5% nominal rate compounded continuously.

(g) Knowing that $e^x = \lim\limits_{x \to \infty} (1 + \frac{1}{x})^x$, derive the formula for the effective rate that is compounded continuously.

4.2 Logarithmic Functions

SECTION OBJECTIVES

1. Determine if two functions are inverses of each other.

2. Use properties of logarithms.

3. Apply a logarithmic model.

In the previous section, we studied exponential functions and their properties. Now we want to develop some mathematical tools that undo the process of exponentiation. This undoing is accomplished with **logarithms**. To understand logarithms, let's quickly review the process of evaluating composition of functions first shown in Section 1.4.

Example 1: Evaluating Composite Functions

Consider the functions $f(x) = \dfrac{x + 1}{2}$ and $g(x) = 2x - 1$. Evaluate $f(g(7))$ and $g(f(7))$.

Perform the Mathematics

Starting from the inside of the expression $f(g(7))$ and working our way out, we first evaluate

$$g(7) = 2(7) - 1 = 13$$

To evaluate $f(g(7))$, we substitute 13 for $g(7)$ to get

$$f(g(7)) = f(13) = \frac{13 + 1}{2} = \frac{14}{2} = 7$$

In the case of $g(f(7))$, we first start by evaluating $f(7)$ to get

$$f(7) = \frac{7 + 1}{2} = \frac{8}{2} = 4$$

So, to evaluate $g(f(7))$, we get

$$g(f(7)) = g(4) = 2(4) - 1 = 7 \qquad \blacksquare$$

Inverse Functions

Notice in Example 1 that $f(g(7))$ and $g(f(7))$ were the same. Was this a coincidence, or would this be true for any value of x? Well, to answer this question, let's compute $f(g(x))$ and $g(f(x))$.

For $f(x) = \dfrac{x + 1}{2}$ and $g(x) = 2x - 1$, we determine $f(g(x))$ as

$$f(g(x)) = f(2x - 1)$$
$$= \frac{(2x - 1) + 1}{x}$$
$$= \frac{2x}{x}$$
$$= x$$

Likewise, for $g(f(x))$ we have

$$g(f(x)) = g\left(\frac{x + 1}{2}\right)$$
$$= 2\left(\frac{x + 1}{2}\right) - 1$$
$$= x + 1 - 1$$
$$= x$$

DEFINITION

One-to-One Function

A function f is a **one-to-one function** if, for elements a and b in the domain of f,

$$a \neq b \text{ implies that}$$
$$f(a) \neq (b)$$

Horizontal Line Test

If every horizontal line intersects the graph of a function f in no more than one point, then f is a one-to-one function

So it appears that for any value of x, we have $f(g(x)) = x = g(f(x))$. This illustrates the definition of **inverse functions**. Before we present the formal definition of inverse functions, we must note that all inverse functions have a special property: they are all **one-to-one functions**.

Finding these values for some functions can be rather difficult, so there is another way to determine if a function is 1-1. It is called the **horizontal line test**.

Now that we know what kind of functions have inverses, we present the definition of an **inverse function**.

DEFINITION

Inverse Function

Two one-to-one functions f and g are **inverses** of each other if

$$(f(g(x)) = x \text{ for every } x\text{-value in the domain of } g$$

and

$$(g(f(x)) = x \text{ for every } x\text{-value in the domain of } f$$

NOTES:

1. Only one-to-one functions have inverse functions.
2. An inverse function will "undo" what the function has done.

A special notation is often used for the inverse function. If g is the inverse of f, then g may be written as f^{-1} (read "f inverse"). Do *not* confuse this with a negative exponent. $f^{-1}(x)$ does **not** represent $\frac{1}{f(x)}$; it represents the inverse function of f. **Figure 4.2.1** shows the general relationship between a function and its inverse function.

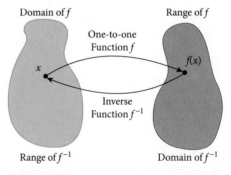

Figure 4.2.1

OBJECTIVE 1

Determine if two functions are inverses of each other.

Example 2: Showing That Two Functions Are Inverses

Use the definition of inverse functions to show that $f(x) = 3x - 2$ and $g(x) = \dfrac{x+2}{3}$ are inverses of each other.

Perform the Mathematics

We need to show that $f(g(x)) = x$ and $g(f(x)) = x$. Completing the first composition, we get

$$f(g(x)) = f\left(\frac{x+2}{3}\right)$$
$$= 3\left(\frac{x+2}{3}\right) - 2$$
$$= x + 2 - 2 = x$$

Completing the second composition yields

$$g(f(x)) = g(3x - 2)$$
$$= \frac{(3x - 2) + 2}{3}$$
$$= \frac{3x}{3} = x$$

Since we have shown that $f(g(x)) = x$ and $g(f(x)) = x$, we conclude f and g are inverses of each other. See **Figure 4.2.2**.

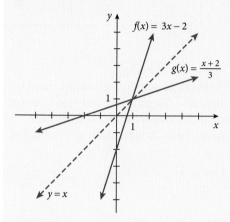

Figure 4.2.2

Figure 4.2.2 shows that there is a **graphical relationship between two inverse functions**.

Logarithmic Functions

Now we want to look at a special family of inverse functions. Their importance can be seen in the Flashback that follows.

> **Graphical Relationship Between Inverse Functions f and g**
>
> If f is a one-to-one function, then the graph of its inverse function g is a reflection about the line $y = x$.

Flashback: Disabled World War II Veterans Revisited

In Section 4.1, we saw that the number of disabled World War II veterans receiving compensation for service-related disabilities can be modeled by

$$f(x) = 1445.83(0.93)^x \qquad 1 \leq x \leq 20$$

where x represents the number of years since 1989 and $f(x)$ represents the number of World War II veterans receiving compensation, measured in thousands. (*Source*: U.S. Census Bureau.) Use tables to numerically estimate the time x when the number of disabled World War II veterans in the United States dipped below 900,000. Write the answer accurate to the half-year.

Perform the Mathematics

We start by making a table of values for $x = 1, 2, 3, \ldots$. The result is shown in **Table 4.2.1**. Numerically, we see that at some time between the values of $x = 6$ and $x = 7$, the $f(x)$-value dipped below 900. To get a better estimate, let's make another table using increments on 0.5 for x. This output is shown in **Table 4.2.2**. This table tells us that the $f(x)$-value dipped below 900 in the interval $6.5 < x < 7$. Since $f(6.5) \approx 902.1$, our estimate accurate to the 0.5 is $x = 6.5$.

Table 4.2.1		**Table 4.2.2**	
x	$f(x)$	x	$f(x)$
1	1344.6	6	935.44
2	1250.5	6.5	902.1
3	1163	7	869.96
4	1081.6		
5	1005.8		
6	935.44		
7	869.96		

DEFINITION

Logarithm

We say that

$$y = \log_b x \text{ if and only if}$$
$$b^y = x$$

where $b > 0$, $b \neq 1$, and $x > 0$. ($\log_b x$ is read "the log of x, base b" or "the log base b of x.")

We could continue to get a more accurate solution to the Flashback by continuing with this numerical method or even by using the calculator's `Intersect` command. To get an accurate solution algebraically, we need a way to undo the exponential function. This is why we study the **logarithm**. The inverse association between logarithm and exponent is illustrated in the **definition of a logarithm**.

This definition tells us, for example, that

$$6^2 = 36 \quad \text{means} \quad \log_6 36 = 2$$

or

$$3^{-2} = \frac{1}{9} \quad \text{means} \quad \log_3\left(\frac{1}{9}\right) = -2$$

To illustrate that the logarithmic function is an inverse function, let's recall the table of values of $f(x) = 2^x$ that was given in Section 4.1. These values are repeated on the left-hand side of **Table 4.2.3**. Now, if we use the definition of the logarithm, we obtain the values on the right-hand side of Table 4.2.3. For example, for $x = \frac{1}{32}$, we see that since $\frac{1}{32} = 2^{-5}$, we can say that $\log_2\left(\frac{1}{32}\right) = -5$.

Table 4.2.3

x	$f(x) = 2^x$	x	$g(x) = \log_2 x$
-5	$\frac{1}{32}$	$\frac{1}{32}$	-5
-4	$\frac{1}{16}$	$\frac{1}{16}$	-4
-3	$\frac{1}{8}$	$\frac{1}{8}$	-3
-2	$\frac{1}{4}$	$\frac{1}{4}$	-2
-1	$\frac{1}{2}$	$\frac{1}{2}$	-1
0	1	1	0
1	2	2	1
2	4	4	2
3	8	8	3
4	16	16	4
5	32	32	5

The table shows that the x- and y-coordinates of $f(x) = 2^x$ and $g(x) = \log_2 x$ are interchanged. Thus, the graph of $g(x) = \log_2 x$ is a reflection of the graph of $f(x) = 2^x$ about the line $y = x$. See **Figure 4.2.3**.

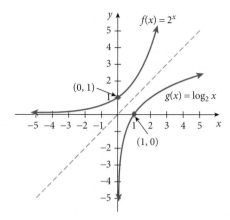

Figure 4.2.3

DEFINITION

Logarithmic Function

The **logarithmic function** is given by

$$f(x) = \log_b x$$

where $b > 0$, $b \neq 1$, and the domain is $(0, \infty)$.

So it seems that the logarithm undoes exponentiation. That is, $f(x) = b^x$ and $g(x) = \log_b x$ **are inverse functions**. Table 4.2.3 and Figure 4.2.3 also suggest that, since the range of the exponential function is the set of positive reals, the domain of the logarithmic function is also the set of positive reals. In other words, we cannot take the logarithm of zero or a negative number. This is all summarized in the **definition of a logarithmic function.**

The definition states that the base b of a logarithm can be any positive number not equal to 1, but two bases are the most widely used. These two bases give the **common and natural logarithm.**

The **properties of logarithms** are much like those for exponents. Some of these properties are proved in Appendix C and some are left for you to prove in the Exercises.

DEFINITION

Common and Natural Logarithm

1. The **common logarithm** is $\log_{10} x$ and is denoted by $\log x$.

2. The **natural logarithm** is $\log_e x$ and is denoted by $\ln x$.

Properties of Logarithms

Let m and n be positive numbers. Then the following are true:

1. $\log_b(mn) = \log_b m + \log_b n$

2. $\log_b\left(\dfrac{m}{n}\right) = \log_b m - \log_b n$

3. $\log_b m^n = n \cdot \log_b m$

4. $\log_b 1 = 0$

5. $\log_b b = 1$

6. $b^{\log_b x} = x$

7. $\log_b b^x = x$

8. $\log_b m = \log_b n$ if and only if $m = n$

Example 3: Using the Properties of Logarithms

Use the properties of logarithms to rewrite the following.

a. $\log(6 \cdot 17)$ **b.** $\log_5\left(\dfrac{10}{3}\right)$ **c.** $\log_6\sqrt{11}$

Perform the Mathematics

a. Using Property 1, we get $\log(6 \cdot 17) = \log 6 + \log 17$.

b. Property 2 tells us that we can rewrite $\log_5(\frac{10}{3})$ as $\log_5 10 - \log_5 3$.

c. Before we apply Property 3, we rewrite $\sqrt{11}$ as $11^{1/2}$,

$$\log_6\sqrt{11} = \log_6 11^{1/2} = \frac{1}{2}\log_6 11 \qquad \blacksquare$$

OBJECTIVE 2

Use properties of logarithms.

▶ **Try It Yourself**

Some related Exercises are 27 and 29.

In our study of calculus, we primarily use the **natural logarithm**. Since the natural logarithm is key to solving exponential equations, we note that all properties of logarithms are true for $\ln x$. These properties allow us to *take the log* of both sides of an equation so that the independent variable is no longer in the exponent. We outline this procedure to solve a general exponential equation $y = a \cdot b^x$ in Example 4.

Example 4: Solving an Exponential Equation

In the Flashback, we numerically approximated the solution to the exponential equation $900 = 1445.83(0.93)^x$. Use the properties of logarithms to solve the equation, and round the solution to the nearest hundredth.

Perform the Mathematics

We start solving the equation $900 = 1445.83(0.93)^x$ by dividing both sides by 1445.83.

$$900 = 1445.83(0.93)^x$$

Divide by 1445.83.
$$\frac{900}{1445.83} = (0.93)^x$$

Take the ln of both sides.
$$\ln\left(\frac{900}{1445.83}\right) = \ln(0.93)^x$$

Property 3.
$$\ln\left(\frac{900}{1445.83}\right) = x \cdot \ln(0.93)$$

Divide by $\ln(0.93)$
$$\frac{1}{\ln(0.93)} \cdot \ln\left(\frac{900}{1445.83}\right) = x$$

$$6.53 \approx x$$

Note that no rounding takes place until the final step. Using a calculator and rounding earlier in the solving process can throw off the final result. ∎

Applications with Logarithmic Functions

Many phenomena that we study start with rapid growth, and then the growth begins to level off. This type of behavior is well suited to the **logarithmic model** of the form $f(x) = a + b\ln x$, where a and b are constants. Since we cannot take the logarithm of a negative number, the domain of this function is the set of positive numbers $(0, \infty)$ or a subset of this set. In practice, we usually start with an x-value of 1.

Example 5: Applying a Logarithmic Model

OBJECTIVE 3

Apply a logarithmic model.

The total enrollment in public elementary and secondary schools in the United States can be modeled by

$$f(x) = 47.04 + 1.12\ln x \qquad 1 \le x \le 9$$

where x represents the number of years since 1999, and $f(x)$ represents the total enrollment in public elementary and secondary schools in the United States in millions. (*Source*: U.S. National Center for Education Statistics.) Evaluate $f(8)$ and interpret.

Understand the Situation

To evaluate $f(8)$, we substitute 8 for every occurrence of x. We also note that $x = 8$ corresponds to 2007.

Perform the Mathematics

Evaluating, we get

$$f(8) = 47.04 + 1.12 \ln 8 \approx 49.37$$

Interpret the Result

This means that in 2007, the total enrollment in public elementary and secondary schools in the United States was about 49.37 million students.

▷ *Try It Yourself*

Some related Exercises are 51 and 52.

■

Summary

This section featured several new topics, including the definition and properties of inverse functions. Then we used this information to discover that the exponential function and logarithmic functions were inverses.

- Two **one-to-one** functions f and g are **inverses** of each other if $f(g(x)) = x$ for every x-value in the domain of g and $g(f(x)) = x$ for every x-value in the domain of f.

- We say that $y = \log_b x$ if and only if $b^y = x$, where $b > 0$, $b \neq 1$, $x > 0$.

- The logarithmic model is $f(x) = a + b \ln x$.

Properties of Logarithms

1. $\log_b(mn) = \log_b m + \log_b n$ **2.** $\log_b\left(\dfrac{m}{n}\right) = \log_b m - \log_b n$

3. $\log_b m^n = n \cdot \log_b m$

4. $\log_b 1 = 0$

5. $\log_b b = 1$

6. $b^{\log_b x} = x$

7. $\log_b b^x = x$

8. $\log_b m = \log_b n$ if and only if $m = n$

Section 4.2 Exercises

Vocabulary Exercises

1. A function is a _____ if it passes the horizontal line test.
2. If $f(g(x)) = x = g(f(x))$ for every value in the domains of f and g, then the functions are

   _____.

3. If $f^{-1}(x) = g(x)$, then the graphs of the functions are a _____ about the line $y = x$.
4. In the logarithmic function $f(x) = \log_b x$, the value b is called the _____.
5. The base for the common logarithm is _____.
6. The base for the natural logarithm is _____.

Skill Exercises

In Exercises 7–10, functions f and g are given. Evaluate the following.

7. $f(x) = 3x$, $g(x) = 6x$. Evaluate $f(g(3))$ and $f(g(-2))$.
8. $f(x) = 3x - 1$, $g(x) = x^2 + 1$. Evaluate $g(f(2))$ and $g(f(3))$.

9. $f(x) = 3x^2 - x, g(x) = 6x + 1$. Evaluate $f(g(0))$ and $g(f(0))$.

10. $f(x) = \sqrt{2x}, g(x) = 3x^2$. Evaluate $f(g(2))$ and $g(f(2))$.

For Exercises 11–18, determine the composition of functions $g(f(x))$ and $f(g(x))$.

11. $f(x) = 3x - 1$; $g(x) = 8x + 2$

12. $f(x) = 6x - 9$; $g(x) = 7x + 5$

13. $f(x) = 5x - 3$; $g(x) = x^2 + 3x + 4$

14. $f(x) = 3x^2 + x + 5$; $g(x) = x - 1$

15. $f(x) = x^3$; $g(x) = \dfrac{1}{x}$

16. $f(x) = \dfrac{2}{x}$; $g(x) = x + 3$

17. $f(x) = \sqrt{x + 1}$; $g(x) = x^2 + 2$

18. $f(x) = \sqrt{x + 2}$; $g(x) = 4x - 3$

For Exercises 19–26, one-to-one functions f and g are given. Show that f and g are inverses of each other.

19. $f(x) = 8x$; $g(x) = \dfrac{1}{8}x$

20. $f(x) = \dfrac{3}{4}x$; $g(x) = \dfrac{4}{3}x$

21. $f(x) = 7x - 10$; $g(x) = \dfrac{x + 10}{7}$

22. $f(x) = \dfrac{x - 3}{4}$; $g(x) = 4x + 3$

23. $f(x) = x^3 + 6$; $g(x) = \sqrt[3]{x - 6}$

24. $f(x) = x^5 - 9$; $g(x) = \sqrt[5]{x + 9}$

25. $f(x) = \sqrt{x - 1}$; $g(x) = x^2 + 1$ for $x \geq 0$

26. $f(x) = \sqrt[4]{x - 1}$; $g(x) = x^4 + 1$ for $x \geq 0$

In Exercises 27–34, use the properties of logarithms to rewrite the following as a sum and/or difference of logarithms.

27. $\log_2\left(\dfrac{3}{5}\right)$

28. $\log_6\left(\dfrac{4}{7}\right)$

29. $\log(8 \cdot 20)$

30. $\log(17 \cdot 11)$

31. $\ln\sqrt{26}$

32. $\ln\sqrt[3]{11}$

33. $\log_3\dfrac{4\sqrt{3}}{9}$

34. $\log_{16}\dfrac{8\sqrt[3]{6}}{11}$

For Exercises 35–40, rewrite the equation in the logarithmic form $y = \log_b x$.

35. $2^5 = 32$

36. $3^4 = 81$

37. $2^{-3} = \dfrac{1}{8}$

38. $5^{-2} = \dfrac{1}{25}$

39. $e^1 = e$

40. $e^0 = 1$

For Exercises 41–46, solve the exponential equation algebraically and round the solution to hundredths place.

41. $8 \cdot 2^x = 51$

42. $6 \cdot 9^x = 126$

43. $2e^x = 62$

44. $4e^x = 81$

45. $1.21(0.3)^x = 42$

46. $0.33(1.27)^x = 58$

Application Exercises

47. **Physics—Balloon Inflation:** Suppose that a spherical balloon is being inflated with the radius increasing at an average rate of 1.3 inches per second.

(a) Write the radius $r(t)$ as a function of time t. (Assume that when $t = 0$, then $r = 0$.)

(b) The volume of a sphere is a function of its radius and is given by $V(r) = \frac{4}{3}\pi r^3$. Simplify $V(t) = V(r(t))$ and interpret the meaning of this composition.

(c) Evaluate $V(t)$ when $t = 6$ and interpret.

48. Pharmacy—Pill Dissolution: The Top Drug Company has developed a new type of pill in the shape of a ball. The spherical pill is dropped into a glass of water and dissolves. Tests show that when the 2-centimeter pill is dropped into the water, its radius decreases at an average rate of 0.003 centimeters per second.

(a) Write the radius $r(t)$ as a function of time t. In this case, $r(t) = 2$ when $t = 0$.

(b) Knowing that the surface area of a sphere as a function of its radius is $S(r) = 4\pi r^2$, simplify $S(t) = S(r(t))$ and interpret the meaning of this composition.

(c) Evaluate $S(t)$ when $t = 20$ and interpret.

49. Agriculture Economics—Federal Farm Subsidies: Traditionally, farmers have benefitted from government farm subsidies to account for fluctuating commodity process. But some believe that these subsidies have contributed to obesity in the United States. "What we've had is a cheap feed grain policy, or a cheap calorie policy, and that's been pretty consistent from farm bill to farm bill over the last 30-odd years," said David Wallinga, advisor for the Agriculture and Trade Policy Institute. (*Source:* National Public Radio.) The amount of direct federal subsidies to farms from 2000 to 2011 can be modeled by

$$h(x) = (1351x + 1355)^{0.27} \qquad 0 \le x \le 11$$

where x represents the number of years since 2000 and $h(x)$ represents the annual amount of direct federal subsidies to farms in billions of dollars. (*Source:* U.S. Department of Agriculture.)

(a) Write two functions f and g so that $f(g(x)) = h(x)$.

(b) Evaluate $h(11) - h(1)$ and interpret.

(c) Determine m_{sec} for $x = 1$ and $\Delta x = 10$ and interpret.

50. Consumer Psychology—Shrub Care Spending: Changes in personal economics have resulted in homeowners spending more time taking care of their own landscaping. "We found that homeowners should be prepared to dedicate an average of 6.5 hours per week working outdoors during growing season," said Peter Sawchuk, project leader at *Consumer Reports* (*Source: Consumer Reports.*) The amount spent on professional shrub care from 2000 to 2011 can be modeled by

$$h(x) = (-2.78x + 59.4)^{1.8} \qquad 0 \le x \le 11$$

where x represents the number of years since 2000 and $h(x)$ represents the amount spent annually on shrub care, measured in millions of dollars. (*Source:* The National Gardening Association.)

(a) Write two functions f and g so that $f(g(x)) = h(x)$.

(b) Evaluate $h(6) - h(1)$ and interpret.

(c) Determine m_{sec} for $x = 1$ and $\Delta x = 5$ and interpret.

51. Travel and Tourism—Foreign Travel: Political unrest, higher fuel prices, and shorter vacation times resulted in a leveling off of foreign travel over the past decade. "The flattening of total per capita travel over so many countries has never been experienced," wrote Stanford University professors Adam Millard-Ball and Lee Schippe. (*Source: National Geographic Daily News.*) The amount Americans spent on travel to other countries from 2001 to 2011 can be modeled by

$$f(x) = 81.3 + 9.8\ln x \qquad 1 \le x \le 11$$

where x represents the number of years since 2000 and $f(x)$ represents the annual amount Americans spent on foreign travel, measured in billions of dollars. (*Source:* U.S. Department of Commerce.)

(a) Evaluate and interpret $f(10)$.

(b) How much did the amount Americans spend on foreign travel change from 2001 to 2006?

(c) Determine the average rate of change in American spending on foreign travel from 2001 to 2006.

52. **Consumerism—Reading and Entertainment Spending:** The convenience of e-readers has resulted in a gradual increase in spending in reading and personal entertainment. "Driven largely by the surging digital revolution, the consumer migration to digital has continued at an ever faster pace," said Stefanie Kane, of PricewaterhouseCoopers. (*Source: The New York Times.*) The average amount each American adult spent on reading and entertainment from 2001 to 2011 can be modeled by

$$f(x) = 1998 + 314.8\ln x \qquad 1 \le x \le 11$$

where x represents the number of years since 2000 and $f(x)$ represents the average amount spent on reading and entertainment, measured in dollars each year. (*Source:* Consumer Expenditure Survey.)

(a) Evaluate $f(11)$ and interpret.

(b) How much did the amount that American adults spend on reading and entertainment change from 2006 to 2011?

(c) Determine the average rate of change in American spending on reading and entertainment from 2006 to 2011.

53. **Consumerism—Broadway Plays:** Throughout the past decade, theater goers have increasingly patronized Broadway. "The economic impact on New York City reminds us of how important Broadway is to tourism," said Charlotte St. Martin, Executive Director of The Broadway League. (*Source:* The Broadway League.) The annual gross ticket sales for Broadway plays from 2001 to 2011 can be modeled by

$$f(x) = 556 + 169\ln x \qquad 1 \le x \le 11$$

where x represents the number of years since 2000 and $f(x)$ represents the gross ticket sales, measured in millions of dollars. (*Source:* The Broadway League.)

(a) Solve the equation $f(x) = 828$ for x. Round x to the nearest whole number and interpret.

(b) On average, how fast did spending on Broadway plays grow each year from 2001 to 2005?

54. **Health—Gym Membership Spending:** Boredom and sameness have contributed to a stabilization of participation in commercial gymnasiums. "They say your body gets bored, but I was getting bored too, and it got a lot harder to make myself go to the gym," says Erin McEvoy, an administrative assistant. (*Source: The Wall Street Journal.*) The amount spent annually on commercial gyms and fitness centers from 2001 to 2011 can be modeled by the function

$$f(x) = 25.8 + 3.45\ln x \qquad 1 \le x \le 11$$

where x represents the number of years since 2000 and $f(x)$ represents the amount spent on gyms and fitness centers, measured in billions of dollars. (*Source:* U.S. Bureau of Economic Analysis.)

(a) Make a table of values for $x = 1, 2, \ldots, 10$. During what year did the amount spent on gyms and fitness center memberships exceed 30 billion dollars?

(b) On average, how much did the amount spent on gyms and fitness center memberships increase each year from 2005 to 2011?

Concept and Writing Exercises

55. Explain the difference between a logarithmic function and an exponential function.

56. A common logarithm has a base of 10 so that $\log_{10} x$ is written as $\log x$. Explain the difference between a common logarithm and a natural logarithm.

For Exercises 57–60, let a, b and c represent positive real numbers. Rewrite each of the following as the sum and/or difference of logarithms.

57. $\log\left(\dfrac{ab}{c}\right)$

58. $\log(a \cdot b)^c$

59. $\ln\left(\dfrac{a\sqrt{c}}{b}\right)$

60. $\ln\sqrt[c]{ab}$

Section Project

Even though the United States is often called "breadbasket to the world," it has increased its agricultural imports for such commodities as soybeans. "Soybean oil producers will purchase more imported beans because they have been able to raise prices and increase their profit margins," said Zhang Xingchao from Jingyi Futures Co. (*Source: China Daily.*) The annual amount of agricultural imports from China to the United States from 1991 to 2011 can be modeled by

$$f(x) = -8.23 + 880.6 \ln x \quad 1 \le x \le 21$$

where x represents the number of years since 1990 and $f(x)$ represents the annual agricultural imports from China, measured in millions of dollars. (*Source:* U.S. Department of Agriculture). Use the logarithmic function to answer the following.

(a) Evaluate $f(10)$ and interpret.

(b) For the function f, compute the difference quotient for $x = 10$ and $\Delta x = 10$ and interpret.

Another country that is using agriculture to its benefit is Brazil. "Brazil is poised to become the fourth-largest economy in the world by 2040, surpassing Japan, France and even the United Kingdom," said David Sterman of Jesup & Lamont Securities. (*Source:* Street Authority.) The annual amount of agricultural imports from Brazil to the United States from 1991 to 2011 can be modeled by the function

$$g(x) = 1107e^{0.04x} \quad 1 \le x \le 21$$

where x represents the number of years since 1990 and $g(x)$ represents the annual agricultural imports from Brazil, measured in millions of dollars. (*Source:* U.S. Department of Agriculture.) Use the exponential function to answer the following.

(c) Evaluate $f(10) - g(10)$ and interpret.

(d) For the function g, compute the difference quotient for $x = 10$ and $\Delta x = 10$ and interpret.

(e) Compare the results from parts (b) and (d). What conclusion can you draw from these results?

4.3 Derivatives of Exponential Functions

In this section, we discuss how to differentiate the **exponential function**, $f(x) = e^x$, and the **general exponential function**, $f(x) = b^x$. The exponential function has applications in business as well as in the social and life sciences. One application that was introduced in Section 4.1 is exponential growth and decay. Here we analyze the rate of change of the growth or decay. To discuss these applications, in particular when determining rates of change, we need to learn the derivative. As in Section 2.4, we determine the derivative using the definition of the derivative.

Derivatives of Exponential Functions with Base *e*

An important function that has not yet been differentiated is the exponential function with base e, $f(x) = e^x$. To determine this derivative, we return to the definition of the derivative. From the definition, if $f(x) = e^x$, then

$$f'(x) = \lim_{h \to 0} \frac{f(x + h) - f(x)}{h}$$

$$= \lim_{h \to 0} \frac{e^{x+h} - e^x}{h}$$

Using properties of exponents, we get

$$f'(x) = \lim_{h \to 0} \frac{e^x \cdot e^h - e^x}{h}$$

$$= \lim_{h \to 0} \frac{e^x(e^h - 1)}{h}$$

SECTION OBJECTIVES

1. Differentiate an exponential function involving e^x.

2. Differentiate an exponential function involving $e^{f(x)}$.

3. Apply the derivative of an exponential function to an exponential growth model.

4. Differentiate an exponential function involving b^x.

5. Differentiate an exponential function involving $b^{f(x)}$.

6. Apply the derivative of an exponential function to an exponential decay model.

Since we are computing the limit with respect to h, e^x can be treated as a constant. This gives

$$f'(x) = \lim_{h \to 0}\left(e^x \cdot \frac{e^h - 1}{h}\right)$$

$$= e^x \cdot \lim_{h \to 0}\frac{e^{h-1}}{h}$$

But what is $\lim\limits_{h \to 0}\frac{e^h - 1}{h}$? If we simply substitute zero for h, we get the indeterminate form $\frac{0}{0}$. To find this limit, we rely on our numerical and graphical methods.

From the graph of $\frac{e^h - 1}{h}$ in **Figure 4.3.1**, we believe that $\lim\limits_{h \to 0}\frac{e^h - 1}{h} = 1$. Checking numerically, **Table 4.3.1** shows the values of $\frac{e^h - 1}{h}$ for h-values close to zero. From the graph and the table, we believe that

$$\lim_{h \to 0}\frac{e^h - 1}{h} = 1$$

We can now take the limit of the difference quotient and finally determine the derivative for $f(x) = e^x$.

$$f'(x) = e^x \cdot \lim_{h \to 0}\frac{e^h - 1}{h} = e^x \cdot 1 = e^x$$

This surprising result means that the **derivative of the exponential function** $f(x) = e^x$ is itself. (Up to now, the only other function that had a property like this is the trivial function $f(x) = 0$.)

Derivative of the Exponential Function

The **derivative** of the **exponential function** $f(x) = e^x$ is

$$f'(x) = e^x$$

NOTE: This rule is equivalent to writing $\frac{d}{dx}[e^x] = e^x$.

Table 4.3.1

h	$\dfrac{e^h - 1}{h}$
-0.1	0.9516
-0.001	0.9995
-0.0001	0.9999
0	
0.0001	1.0001
0.001	1.0005
0.1	1.0517

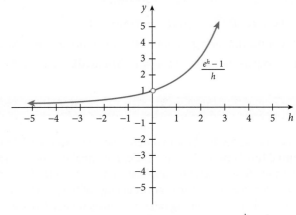

Figure 4.3.1 Graphical determination that $\lim\limits_{h \to 0}\dfrac{e^h - 1}{h} = 1$.

OBJECTIVE 1

Differentiate an exponential function involving e^x.

Example 1: Differentiating Exponential Functions

Determine derivatives for the following functions.

a. $y = \dfrac{x^2}{e^x}$ **b.** $g(x) = x^4 \cdot e^x$

Perform the Mathematics

a. Applying the Quotient Rule, we get

$$y' = \frac{d}{dx}\left[\frac{x^2}{e^x}\right] = \frac{\frac{d}{dx}[x^2] \cdot e^x - x^2 \cdot \frac{d}{dx}[e^x]}{(e^x)^2}$$

$$= \frac{2xe^x - x^2e^x}{e^{2x}} = \frac{e^x(2x - x^2)}{e^{2x}} = \frac{2x - x^2}{e^x}$$

b. Applying the Product Rule for this function gives

$$g'(x) = \frac{d}{dx}[x^4 \cdot e^x] = \frac{d}{dx}[x^4] \cdot e^x + x^4 \cdot \frac{d}{dx}[e^x]$$

$$= 4x^3 e^x + x^4 e^x \qquad \blacksquare$$

The derivative of e^x by itself has limited value. But the **Chain Rule** applied to the **exponential function** is very useful, as it lends itself to many applications. Let's add this extension to our list of differentiation rules.

> **Chain Rule for the Exponential Function**
>
> If f is a differentiable function of x, the derivative of the composition of functions $h(x) = e^{f(x)}$ is given by
>
> $$h'(x) = e^{f(x)} \cdot f'(x)$$

Example 2: Differentiating Composite Exponential Functions

> **OBJECTIVE 2**
>
> Differentiate an exponential function involving $e^{f(x)}$.

Determine derivatives for the following functions.

a. $h(x) = e^{3x-3}$ **b.** $g(x) = e^{6x-x^6}$

Perform the Mathematics

a. In the composite form $h(x) = e^{f(x)}$, $f(x) = 3x - 3$.

$$h'(x) = \frac{d}{dx}[e^{3x-3}] = e^{3x-3} \cdot \frac{d}{dx}[3x - 3]$$

$$= e^{3x-3} \cdot 3 = 3e^{3x-3}$$

b. Applying the Chain Rule for exponential functions gives

$$g'(x) = \frac{d}{dx}[e^{6x-x^6}] = e^{6x-x^6} \cdot \frac{d}{dx}[6x - x^6]$$

$$= e^{6x-x^6} \cdot (6 - 6x^5) \qquad \blacksquare$$

▶ **Try It Yourself**

Some related Exercises are 25 and 29.

The properties that we studied in earlier chapters can be applied to the exponential functions as well. This is illustrated in Example 3.

Example 3: Applying an Exponential Growth Model

The world's population during the 20th century closely followed the model

$$p(t) = 1.419 e^{0.014t} \qquad 0 \le t \le 100$$

> **OBJECTIVE 3**
>
> Apply the derivative of an exponential function to an exponential growth model.

where t represents the number of years since 1900 and $p(t)$ represents the world's population in billions of people. Evaluate and interpret $p'(10)$ and compare to $p'(99.)$ (*Source:* U.S. Census Bureau.)

Perform the Mathematics

Differentiating the population model, we get

$$p'(t) = \frac{d}{dt}[1.419 e^{0.014t}] = 1.419 \frac{d}{dt}[e^{0.014t}]$$

$$= 1.419 e^{0.014t} \cdot \frac{d}{dt}[0.014t]$$

$$= 1.419 e^{0.014t} \cdot 0.014 \approx 0.02 e^{0.014t}$$

The year 1910 corresponds to $t = 10$. Evaluating $p'(10)$ gives

$$p'(10) = 0.02e^{(0.014)(10)} \approx 0.023 \frac{\text{billion people}}{\text{year}}$$

This means that in 1910, the world's population was increasing at a rate of about 0.023 billion (or 23 million) people per year.

The year 1999 corresponds to $t = 99$. Evaluating $p'(99)$ yields

$$p'(99) = 0.02e^{(0.014)(99)} \approx 0.08 \frac{\text{billion people}}{\text{year}}$$

This means that in 1999 the world's population was increasing at a rate of about 0.08 billion (that is, 80 million) people per year. The growth rate in 1999 was nearly 3.5 times the growth rate in 1910. See **Figure 4.3.2**.

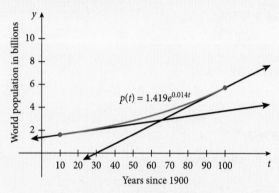

Figure 4.3.2 Growth rate in 1999 was about 3.5 times the growth rate in 1910 as shown by tangent lines.

Derivative of General Exponential Functions

One type of mathematical model we have not yet differentiated is the **general exponential function**, $f(x) = b^x$. Remember that for these functions, b is any positive number other than 1. To find this derivative, let's return to a property that we covered in Section 4.2, specifically the property that since the exponential and logarithmic functions are inverses, for $x > 0$, we have

$$e^{\ln x} = x$$

This property means that we can substitute the value of b for x and write $b = e^{\ln b}$. Thus,

$$b^x = (e^{\ln b})^x = e^{\ln b \cdot x}$$

So we can rewrite the general exponential function $f(x) = b^x$ as $f(x) = e^{\ln b \cdot x}$. Now we can differentiate using the Chain Rule for the exponential function to get

$$f'(x) = \frac{d}{dx}[b^x]$$

$$= \frac{d}{dx}[e^{\ln b \cdot x}]$$

$$= e^{\ln b \cdot x} \cdot \frac{d}{dx}[\ln b \cdot x]$$

Since b is a constant, so is $\ln b$, which gives us

$$f'(x) = e^{\ln b \cdot x} \cdot \ln b$$

Rewriting b for $e^{\ln b}$ gives

$$f'(x) = b^x \cdot \ln b$$

Derivative of the General Exponential Function

If $f(x) = b^x$, with $b > 0$ and $b \neq 1$, then the **derivative** of the **general exponential function** is

$$f'(x) = b^x \cdot \ln b$$

Example 4: Determining Derivatives of General Exponential Functions

OBJECTIVE 4

Differentiate an exponential function involving b^x.

Determine the derivatives for the given functions.

a. $f(x) = 10^x$

b. $g(x) = \dfrac{3^x}{4^x}$

Perform the Mathematics

a. Here we use the general exponential function differentiation rule with $b = 10$.

$$f'(x) = \frac{d}{dx}[10^x]$$

$$= 10^x \cdot \ln 10$$

b. Instead of using the Quotient Rule, we can use the properties of exponents to rewrite $g(x) = \frac{3^x}{4^x}$ as $g(x) = \left(\frac{3}{4}\right)^x$. Now we have a general exponential function with $b = \frac{3}{4}$. So the derivative is

$$g'(x) = \frac{d}{dx}\left[\left(\frac{3}{4}\right)^x\right]$$

$$= \left(\frac{3}{4}\right)^x \cdot \ln\left(\frac{3}{4}\right) \qquad \blacksquare$$

It seems natural at this point to extend our differentiation capabilities of the general exponential function by adding in the **Chain Rule for general exponential functions**.

> **Chain Rule for the General Exponential Function**
>
> If f is a differentiable function of x, and $b > 0, b \neq 1$, the derivative of the composition of functions $h(x) = b^{f(x)}$ is given by
>
> $$h'(x) = b^{f(x)} \cdot \ln b \cdot f'(x)$$

Example 5: Differentiating Composite General Exponential Function

OBJECTIVE 5

Differentiate an exponential function involving $b^{f(x)}$.

Determine the derivative for $g(x) = 5^{9x-5}$.

Perform the Mathematics

In the form $h(x) = b^{f(x)}$, we have $b = 5$ and $f(x) = 9x - 5$. So the derivative is

$$g'(x) = \frac{d}{dx}[5^{9x-5}]$$

$$= 5^{9x-5} \cdot \ln 5 \cdot \frac{d}{dx}[9x - 5]$$

$$= 5^{9x-5} \cdot \ln 5 \cdot 9 \qquad \blacksquare$$

▶ *Try It Yourself*

Some related Exercises are 35 and 39.

> **DEFINITION**
>
> **Rewriting the General Exponential Model**
>
> The general exponential model $f(x) = a \cdot b^x$ can be written in the exponential form
>
> $$g(x) = a \cdot e^{kx}$$
>
> where $k = \ln b$.

Applications

Recall that models such as $p(t) = 1.419e^{0.014t}$ from Example 3 are called **exponential growth models**. Before we look at one final application, we would like to point out that, since $f(x) = b^x$ can be written as $f(x) = e^{\ln b \cdot x}$, any **general exponential model** may be written as shown to the right.

For example, the exponential model $f(x) = 3.1(1.92)^x$ can be written as $f(x) = 3.1e^{(\ln 1.92) \cdot x} \approx 3.1e^{0.65x}$. This rewriting technique is useful for modeling, since many calculators will model data in the general exponential form $f(x) = a \cdot b^x$.

Flashback: World War II Veterans Revisited

The number of disabled World War II veterans receiving compensation for service-related disabilities can be modeled by

$$f(x) = 1445.83(0.93)^x \qquad 1 \le x \le 20$$

OBJECTIVE 6

Apply the derivative of an exponential function to an exponential decay model.

where x represents the number of years since 1989, and $f(x)$ represents the number of World War II veterans receiving compensation, measured in thousands. (*Source:* U.S. Census Bureau.) Determine $f'(11)$ and interpret.

Perform the Mathematics

The derivative is computed to be

$$f'(x) = \frac{d}{dx}[1445.83(0.93)^x]$$

$$= 1445.83 \cdot (0.93)^x \cdot \ln(0.93)$$

Evaluating the derivative at $x = 11$ gives us

$$f'(11) = 1445.83 \cdot (0.93)^{11} \cdot \ln(0.93) \approx -47.23$$

Since $x = 11$ corresponds to the year 2000, we conclude that in 2000 the number of disabled World War II veterans receiving compensation for service-related disabilities was decreasing at a rate of about 47,230 veterans per year.   ■

▶ *Try It Yourself*

Some related Exercises are 45 and 55.

Summary

In this section, we examined the derivatives of exponential functions. The differentiation rules for the exponential function, the general exponential function, and the Chain Rule for both were determined as follows:

- $\dfrac{d}{dx}[e^x] = e^x$

- $\dfrac{d}{dx}[e^{f(x)}] = e^{f(x)} \cdot f'(x)$

- $\dfrac{d}{dx}[b^x] = b^x \cdot \ln b$

- $\dfrac{d}{dx}[b^{f(x)}] = b^x \cdot \ln b \cdot f'(x)$

Please note that in the Chain Rule, f is a differentiable function of x.

Section 4.3 Exercises

Vocabulary Exercises

1. The exponential function $f(x) = e^x$ and its derivative are _____.

2. If $h(x) = e^{f(x)}$, then we can determine the derivative of h using the _____ Rule.

3. For $f(x) = b^x$, the derivative f' represents the derivative of the _____ exponential function.

4. For the _____ _____ _____ function $h(x) = b^{f(x)}$, the derivative is $h'(x) = b^{f(x)} \cdot \ln b \cdot f'(x)$.

5. If $f(x) = a \cdot b^x$ is expressed in the form $g(x) = a \cdot e^{(\ln b)x}$, then we are _____ the general exponential model.

6. Rewriting the general exponential model is useful when a _____ will only model data of the form $f(x) = a \cdot b^x$.

Skill Exercises

In Exercises 7–22, use the differentiation rule of the natural exponential function to determine the derivative of the given function.

7. $f(x) = 7e^x$

8. $f(x) = 10e^x$

9. $f(x) = 2x(4 + e^x)$

10. $f(x) = 10x(e^x + 40)$

11. $f(x) = \dfrac{10}{5 - e^x}$

12. $f(x) = \dfrac{15}{2 + e^x}$

13. $f(x) = 4x^2 e^x$

14. $f(x) = 10x^2 e^x$

15. $f(x) = \sqrt{12 - e^x}$

16. $f(x) = \sqrt[3]{e^x + 5}$

17. $f(x) = \dfrac{e^x - 10}{x^3 - 1}$

18. $f(x) = \dfrac{x^2 + 5}{2 - e^x}$

19. $f(x) = \dfrac{e^x + 1}{e^x - 1}$

20. $f(x) = \dfrac{4 - e^x}{4 + e^x}$

21. $f(x) = 2xe^x - x$

22. $f(x) = 4x^2 e^x - e^x$

In Exercises 23–30, determine the derivative of the given function.

23. $f(x) = e^{2x-1}$

24. $f(x) = e^{6x+9}$

25. $f(x) = e^{\sqrt{x}}$

26. $f(x) = e^{\sqrt[3]{x}}$

27. $f(x) = 5x \cdot e^{2x}$

28. $f(x) = 2x^3 \cdot e^{6x^2}$

29. $f(x) = \dfrac{e^{x-1}}{e^{x+1}}$

30. $f(x) = \dfrac{e^{2+x}}{e^{2-x}}$

In Exercises 31–36, determine the derivative of the given function.

31. $f(x) = 10^x$

32. $f(x) = 4^x$

33. $f(x) = \dfrac{5^x}{15^x}$

34. $f(x) = \dfrac{8^x}{2^x}$

35. $f(x) = x^3 \cdot 0.3^x$

36. $f(x) = x^2 \cdot 0.2^x$

In Exercises 37–42, determine the derivative of the given function.

37. $f(x) = 10^{x+3}$

38. $f(x) = 2^{9-x}$

39. $f(x) = 9^{1/x}$

40. $f(x) = 4^{\sqrt{x}}$

41. $f(x) = x \cdot e^x - 5^{2x}$

42. $f(x) = 2x \cdot e^{3x} + 4^{2x-1}$

Application Exercises

43. Biology—Tumor Growth: Suppose a veterinarian wants to determine how fast a tumor grows in a lab specimen when exposed to a new pesticide. The growth of the tumor in specimen can be modeled by

$$f(x) = 2.1e^{0.2x} \qquad x > 0$$

where x represents the number of days since exposure to the pesticide, and $f(x)$ represents the diameter of the tumor in millimeters.

(a) Determine $f'(x)$.

(b) Evaluate and interpret $f'(3)$.

44. Agriculture—Crop Yield: Farmers at AgriCorp want to determine the rate of change in crop yield after adding the popular HB-73 herbicide. From past data, they find the total yield of the crop after adding the herbicide can be modeled by

$$f(x) = 1500e^{1.5x} \qquad x \geq 0$$

where x represents the number of growing seasons since the herbicide was first used, and $f(x)$ represents the crop yield in bushels.

(a) Determine $f'(x)$.

(b) Evaluate and interpret $f'(5)$.

45. Zoology—Fish Population: A zoologist is writing a dissertation on the impact of the fish population in Lake Maben since the Bellmore Company built a factory nearby. Initial sampling shows that the fish population can be modeled by

$$p(t) = 12 \cdot (0.8)^t \qquad 0 \leq t \leq 10$$

where t represents the number of years since the factory opened, and $p(t)$ represents the fish population in hundreds.

(a) Determine $p'(t)$.

(b) Evaluate and interpret $p'(1)$ and compare to $p'(8)$.

46. Immunology—Spread of the Flu: The director of health services at Deter University is pondering the termination of its flu immunization program for budgetary reasons, but first wishes to know how rapidly the flu can spread among the student population. From studying the university's infirmary records, his staff determined that the number of students getting the flu can be modeled by

$$f(x) = 7000(0.92)^x \qquad 0 \leq x \leq 12$$

where x represents the number of years since the immunization program began, and $f(x)$ represents the number of student who contract the flu each year.

(a) Determine $f'(x)$.

(b) Evaluate and interpret $f'(2)$ and compare to $f'(11)$.

47. Finance—Compound Interest: Suppose that $2000 is invested in an account that earns 6.5% interest compounded continuously. The amount accumulated in the account after t years can be modeled by

$$A(t) = 2000e^{0.065t} \qquad t \geq 0$$

(a) Determine $A'(t)$.

(b) Evaluate $A(5)$ and $A'(5)$ and interpret each.

48. Finance—Account Analysis (*continuation of Exercise 47*): Use the function A from Exercise 47. Remember that t represents the independent variable.

(a) Graph A in the viewing window [0,10] by [2000,4000].

(b) Use the `Intersect` command to determine how many years it takes for the amount in the account to exceed $3000.

(c) Use the `dy/dx` command to approximate $A'(2)$.

49. Finance—Compound Interest: Suppose that $5000 is invested in an account that earns 5% interest compounded continuously. The amount accumulated in the account after t years can be modeled by

$$A(t) = 5000e^{0.05t} \qquad t \geq 0$$

(a) Determine $A'(t)$.

(b) Evaluate $A(9)$ and $A'(9)$ and interpret each.

50. Finance—Account Analysis (*continuation of Exercise 49*): Use the function A from Exercise 49. Remember that t represents the independent variable.

(a) Graph A in the viewing window $[0,10]$ by $[5000,8500]$.

(b) Use the `Intersect` command to determine how many years it takes for the amount in the account to exceed $7000.

(c) Use the `dy/dx` command to approximate $A'(2)$.

51. Population Geography—Asylum Applicants in Germany: Since 2000, German regulations have required airlines and other transportation companies to ensure that their passengers have the necessary travel documents to enter Germany. (*Source:* U.S. Committee for Refugees and Immigrants.) These new regulations have resulted in a decrease in the number seeking asylum. The number of refugees seeking asylum annually in Germany from 2000 to 2010 can be modeled by

$$f(x) = 91{,}957e^{-0.2x} \qquad 0 \leq x \leq 10$$

where x represents the number of years since 2000 and $f(x)$ represents the number of asylum applicants. (*Source:* EuroStat.)

(a) Determine $f'(x)$.

(b) Use the derivative to determine the rate of change in asylum applicants in 2002.

52. Population Geography—Asylum Applicants in Ireland: Applications for asylum have been decreasing in Ireland. Minister for Justice, Equality and Law Reform Dermot Ahern explained the decrease, stating, "Many asylum applicants are economic migrants. Therefore it is possible that economic conditions in Ireland may have been a factor in the reduction in asylum application numbers." (*Source: The Irish Examiner.*) The number of refugees seeking asylum annually in Ireland from 2000 to 2010 can be modeled by

$$f(x) = 12{,}150e^{-0.17x} \qquad 0 \leq x \leq 10$$

where x represents the number of years since 2000 and $f(x)$ represents the number of asylum applicants. (*Source:* EuroStat.)

(a) Determine $f'(x)$.

(b) Use the derivative to determine the rate of change in asylum applicants in 2008.

53. Sociology—Social Networking: One of the entities that is truly experiencing exponential growth is the social networking site called Facebook. Peter Thiel, one of Facebook's early investors, said "People underestimate the power of the exponential growth we are seeing." (*Source:* ZDnet.) The number of Facebook users from 2005 to 2010 can be modeled by

$$f(x) = 1.4e^{0.8x} \qquad 5 \leq x \leq 10$$

where x represents the number of years since 2000 and $f(x)$ represents the number of Facebook users in millions. (*Source:* Crunchbase.)

(a) Determine $f'(x)$.

(b) Use the derivative to determine the rate of change in Facebook users in 2009.

54. Business—Social Networking Revenue: The social networking site Facebook's revenue also exhibits exponential growth. Early investor Peter Thiel said, "It's not difficult to see the immense revenue opportunities in front of Facebook if it can keep the wheels from falling off." (*Source:* ZDnet.) The annual revenue generated by Facebook from 2005 to 2010 can be modeled by

$$f(x) = 2.78e^{0.6x} \qquad 5 \le x \le 10$$

where x represents the number of years since 2000 and $f(x)$ represents the Facebook revenue in millions of dollars. (*Source:* Crunchbase.)

(a) Determine $f'(x)$.

(b) Use the derivative to determine the rate of change in Facebook revenue in 2009.

55. Popular Culture—Motion Picture Revenue: In spite of the proliferation of media inside the home, the motion picture industry has continued steady growth through the 21st century. The annual revenues from motion pictures from 2004 to 2010 can be modeled by

$$f(x) = 47(1.08)^x \qquad 4 \le x \le 10$$

where x represents the number of years since 2000 and $f(x)$ represents the revenues from motion pictures, measured in millions. (*Source:* U.S. Census Bureau.)

(a) Determine $f'(x)$.

(b) In which year was the rate of change in motion picture revenue greater, 2004 or 2008? Explain.

56. Journalism—Online News Consumption: A significant change in news consumption from the 20th to the 21st century is how the delivery of news has transitioned from newspaper to online format. The percentage of Hispanic American adults who say that they regularly get their news online can be modeled by

$$f(x) = 24(1.04)^x \qquad 0 \le x \le 10$$

where x represents the number of years since 2000 and $f(x)$ represents the percentage of Hispanic American adults who say that they regularly get their news online. (*Source:* Pew Internet and American Life Project.)

(a) Determine $f'(x)$.

(b) In which year was the rate of change in the percentage of Hispanics getting news online greater, 2003 or 2007? Explain.

Concept and Writing Exercises

57. If $x \ge 0$, show that $e^x \ge x + 1$. To do this, let $f(x) = e^x - (x + 1)$ and show that the graph of f is increasing for x-values.

58. Use the result of Exercise 57 to show that $e^x \ge x + 1 + \frac{x^2}{2}$ for $x \ge 0$.

59. Graph the functions $f(x) = x^{10}$ and $g(x) = e^x$. For what values of x is $g(x) > f(x)$?

60. Graph the functions $f(x) = x^5$ and $g(x) = 5^x$. Which function grows more rapidly for large values of x? Use this result to determine $\lim\limits_{x \to \infty} \dfrac{x^5}{5^x}$.

61. The number e can be defined as a limit by writing $e = \lim\limits_{n \to \infty} \left(1 + \frac{1}{n}\right)^n$. Let r represent a fixed real number constant. Use the limit definition of e to show that $\lim\limits_{n \to \infty} \left(1 + \frac{r}{n}\right)^{n/r} = e$.

62. Use the result from Exercise 61 to show that $\lim\limits_{n \to \infty} \left(1 + \frac{r}{n}\right)^{nt} = e^{rt}$.

63. Use the result of Exercise 62 to derive the continuous compound interest formula. That is, show that $\lim\limits_{n \to \infty} P\left(1 + \frac{r}{n}\right)^{nt} = Pe^{rt}$.

64. Let f represent a function whose derivative is $f'(x) = \frac{1}{x}$. Let g represents a differentiable function with the property that $g(x) = f(e^x)$. Determine the derivative $g'(x)$.

In Exercises 65 and 66, use the fact that $e^0 = 1$ to determine the limits.

65. $\displaystyle\lim_{x\to 0} \frac{e^x - 1}{x}$

66. $\displaystyle\lim_{x\to 0} \frac{e^{2x} - 1}{x}$

 Section Project

An ongoing concern for college students is the escalating cost of tuition and fees. "All of the colleges . . . are looking carefully at their budgets. As to whether it would push some over the edge, I don't know yet. They have cut back on the number of class sections and many may well have to eliminate summer session," said California Chancellor Jack Scott. (*Source: The Los Angeles Times.*) The average cost of tuition and fees for four-year colleges in the United States from 1980 to 2010 can be modeled by

$$f(x) = (0.79 - 0.001x)^{-100/3} \qquad 0 \le x \le 30$$

where x represents the number of years since 1980 and $f(x)$ represents the annual cost of tuition and fees for four-year colleges in dollars. (*Source: U.S. National Center for Education Statistics.*) The average cost of tuition and fees for two-year colleges in the United States from 1980 to 2010 can be modeled by the exponential function

$$g(x) = 2668.23(1.03)^x \qquad 0 \le x \le 30$$

where x represents the number of years since 1980 and $f(x)$ represents the annual cost of tuition and fees for two-year colleges in dollars. (*Source: U.S. National Center for Education Statistics.*)

(a) Graph the functions f and g using the viewing window suggested in the accompanying graph.

(b) Determine $f'(x)$ then evaluate and interpret $f'(18)$.

(c) Determine $g'(x)$ then evaluate and interpret $g'(18)$. Compare the results to those in part (b).

(d) Now form the function $h(x) = f(x) - g(x)$ using the models f and g above. What does this function h represent?

 (e) Graph the function h and use the dy/dx command to estimate $h'(18)$. How does this solution compare to that in part (c)? What differentiation rule does this result illustrate?

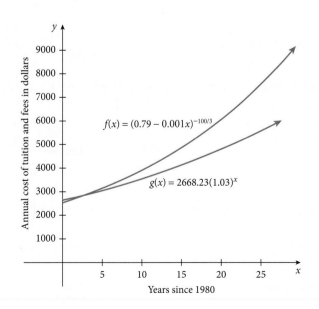

4.4 Derivatives of Logarithmic Functions

Now we wish to focus our attention on a function introduced in Section 4.2, the **logarithmic function**. This function has applications in business, social, and life sciences. To discuss its application in calculus, in particular for determining rates of change, we first must learn its derivative. Since the logarithmic function is not a polynomial function, we must develop a new rule for its derivative.

Derivative of the Natural Logarithm Function

To set the stage for studying the derivative of the natural logarithm function, let's review some of its properties that we presented in Section 4.2 in the following Toolbox.

From Your Toolbox: Properties of Logarithms

- The natural logarithm is $\log_e x$ and is denoted $\ln x$ for $x > 0$.
- Let m and n be positive numbers. Then the following are true:

  1. $\ln(mn) = \ln m + \ln n$ **2.** $\ln\left(\dfrac{m}{n}\right) = \ln m - \ln n$

  3. $\ln m^n = n \cdot \ln m$ **4.** $\ln 1 = 0$

  5. $\ln e = 1$ **6.** $e^{\ln x} = x$

  7. $\ln e^x = x$

To determine the derivative of $f(x) = \ln x$, we cannot rely on differentiation rules learned so far. To determine the derivative, we return to the definition of the derivative. This gives us

$$f'(x) = \lim_{h \to 0} \frac{f(x + h) - f(x)}{h}$$

$$= \lim_{h \to 0} \frac{\ln(x + h) - \ln x}{h}$$

Property 2 of logarithms $$f'(x) = \lim_{h \to 0} \frac{\ln\left(\dfrac{x + h}{x}\right)}{h}$$

Simplifying $\dfrac{x + h}{x}$ to $1 + \dfrac{h}{x}$ $$= \lim_{h \to 0} \frac{\ln\left(1 + \dfrac{h}{x}\right)}{h}$$

We now let $k = \frac{x}{h}$. Notice that $h \to 0$ implies that $k \to \infty$ (for $h > 0$). We also note that if $k = \frac{x}{h}$, then $\frac{1}{k} = \frac{h}{x}$ and $h = \frac{x}{k}$. So rewriting the limit in terms of k gives us

$k = \dfrac{x}{h}$ implies $h = \dfrac{x}{k}$ $$= \lim_{k \to \infty} \frac{\ln\left(1 + \dfrac{1}{k}\right)}{\dfrac{x}{k}}$$

Dividing by $\dfrac{x}{k}$ is same as multiplying by $\dfrac{k}{x}$ $$= \lim_{k \to \infty} \frac{k}{x}\left[\ln\left(1 + \dfrac{1}{k}\right)\right]$$

Rewriting $\dfrac{k}{x}$ as $\dfrac{1}{x} \cdot k$ $$= \lim_{k \to \infty} \frac{1}{x} \cdot \left[k\ln\left(1 + \dfrac{1}{k}\right)\right]$$

Property 3 of logarithms $$= \lim_{k \to \infty} \frac{1}{x} \cdot \left[\ln\left(1 + \dfrac{1}{k}\right)^k\right]$$

Derivative of the Natural Logarithm Function

For $f(x) = \ln x$, with $x > 0$, the **derivative of the natural logarithm function**, is given by

$$f'(x) = \frac{1}{x}$$

NOTE: This is equivalent to writing $\dfrac{d}{dx}[\ln x] = \dfrac{1}{x}$.

In Section 4.1 we numerically saw that $\lim\limits_{k \to \infty} (1 + \frac{1}{k})^k = e$. Using this information we continue and get

$$= \frac{1}{x} \cdot \ln e$$

Property 5 of logarithms

$$= \frac{1}{x}$$

We have shown that the **derivative of the natural logarithm function** $f(x) = \ln x$ is $f'(x) = \frac{1}{x}$.

Example 1: Determining Derivatives of Natural Logarithmic Functions

OBJECTIVE 1

Determine the derivative of a natural logarithmic function.

Compute derivatives for the following functions.

a. $f(x) = 2\ln x$ **b.** $g(x) = \ln x^3$ **c.** $y = 7 - 4\ln x$

Perform the Mathematics

a. We can apply the Constant Multiple Rule here to get

$$f'(x) = \frac{d}{dx}[2\ln x] = 2 \cdot \frac{d}{dx}[\ln x]$$

$$= 2 \cdot \frac{1}{x} = \frac{2}{x}$$

b. Before we differentiate, we first take advantage of logarithm property 3, $\ln m^n = n \cdot \ln m$, and rewrite $g(x) = \ln x^3$ as $g(x) = 3\ln x$. Differentiation gives us

$$g'(x) = \frac{d}{dx}[\ln x^3] = \frac{d}{dx}[3\ln x]$$

$$= 3 \cdot \frac{d}{dx}[\ln x] = 3 \cdot \frac{1}{x} = \frac{3}{x}$$

c. For this derivative, we use the Sum and Difference Rule. This gives

$$y' = \frac{d}{dx}[7 - 4\ln x] = \frac{d}{dx}[7] - \frac{d}{dx}[4\ln x]$$

$$= 0 - 4 \cdot \frac{1}{x} = \frac{-4}{x}$$ ∎

▶ *Try It Yourself*

Some related Exercises are 9 and 15.

Example 1c illustrates a special form that we will frequently use for modeling data. The function $y = 7 - 4\ln x$ is an example of the **natural logarithm model** because it has the form $f(x) = a + b\ln x$, where a and b represent constants. Example 2 illustrates how to calculate the rate of change of this type of function.

Example 2: Differentiating and Interpreting a Natural Logarithm Model

OBJECTIVE 2

Apply the derivative of a natural logarithmic function.

The percentage of women who are college graduates in the United States who completed 4 years of college or more can be modeled by

$$f(x) = 23.34 + 2.22\ln x \qquad 1 \le x \le 10$$

© Andresr/ShutterStock, Inc.

where x represents the number of years since 1999, and $f(x)$ represents the percentage of women who are college graduates in the United States who completed 4 years of college or more. (*Source*: U.S. Census Bureau.) Evaluate $f'(6)$ and interpret.

Understand the Situation

We need to first determine the derivative and then evaluate it at $x = 6$. Note that $x = 6$ corresponds to the year 2005.

Perform the Mathematics

The derivative of the model is

$$f'(x) = \frac{d}{dx}[23.34 + 2.22\ln x]$$

$$= \frac{d}{dx}[23.34] + \frac{d}{dx}[2.22\ln x]$$

$$= 0 + 2.22 \cdot \frac{1}{x} = \frac{2.22}{x}$$

Evaluating $f'(6)$, we get

$$f'(6) = \frac{2.22}{6} = 0.37$$

Interpret the Result

This means that in 2005, the percentage of women who were college graduates who completed 4 years of college or more was increasing at a rate of 0.37% per year.

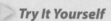 *Try It Yourself*

Some related Exercises are 49 and 53.

Sometimes we need to differentiate functions that involve more than just $\ln x$. Let's say that the argument of the logarithm function was another function, which we will call $u = g(x)$. If we let $f(x) = \ln x$, then we can form the composite function $h(x) = f(g(x)) = f(u) = \ln u$. We can use the Chain Rule to determine the derivative of $h(x) = f(g(x))$. Since $f'(x) = \frac{1}{x}$ and $g'(x) = u' = \frac{du}{dx}$, we have

$$h'(x) = f'(g(x)) \cdot g'(x)$$

$$= \frac{1}{g(x)} \cdot g'(x) = \frac{g'(x)}{g(x)}$$

Chain Rule for the Natural Logarithmic Function

If g is a differentiable function of x and the range of g is $(0, \infty)$, the derivative of the composition of functions $h(x) = \ln[g(x)]$ is given by

$$h'(x) = \frac{1}{g(x)} \cdot g'(x) = \frac{g'(x)}{g(x)}$$

NOTE: This rule is equivalent to writing $\frac{d}{dx}[\ln(g(x))] = \frac{1}{g(x)} \cdot \frac{d}{dx}[g(x)] = \frac{1}{g(x)} \cdot g'(x) = \frac{g'(x)}{g(x)}$.

Example 3: Differentiating the Natural Logarithm Function with the Chain Rule

Determine the derivative of each function.

a. $y = \ln(x^4 - 2x)$ **b.** $h(x) = \ln\sqrt{6x - 1}$ **c.** $f(x) = (\ln x)^3$

Determine the derivative of a natural logarithmic function using the Chain Rule.

Perform the Mathematics

a. Here the inside, or g, function is $g(x) = (x^4 - 2x)$. So applying the Chain Rule gives us

$$y' = \frac{1}{x^4 - 2x} \cdot \frac{d}{dx}[x^4 - 2x]$$

$$= \frac{1}{x^4 - 2x} \cdot (4x^3 - 2)$$

$$= \frac{4x^3 - 2}{x^4 - 2x}$$

b. Here, we begin by using some algebra to rewrite the function:

$$h(x) = \ln\sqrt{6x - 1} = \ln(6x - 1)^{1/2} = \frac{1}{2}\ln(6x - 1)$$

Now we differentiate using the Chain Rule:

$$h'(x) = \frac{d}{dx}\left[\frac{1}{2}\ln(6x - 1)\right]$$

$$= \frac{1}{2} \cdot \frac{d}{dx}[\ln(6x - 1)]$$

$$= \frac{1}{2} \cdot \frac{1}{6x - 1} \cdot \frac{d}{dx}[6x - 1]$$

$$= \frac{1}{2} \cdot \frac{1}{6x - 1} \cdot 6 = \frac{3}{6x - 1}$$

c. This is a little different from the previous two parts in that we must apply the Generalized Power Rule to get the derivative:

$$f'(x) = \frac{d}{dx}[(\ln x)^3]$$

$$= 3(\ln x)^2 \cdot \frac{d}{dx}[\ln x]$$

$$= 3(\ln x)^2 \cdot \frac{1}{x}$$

$$= \frac{3(\ln x)^2}{x} \qquad \blacksquare$$

Derivative of the General Logarithmic Function

1. For the general logarithmic function $f(x) = \log_b x$ with $b > 0, b \neq 1$,

$$f'(x) = \frac{1}{\ln b} \cdot \frac{1}{x}$$

2. If g is a differentiable function, where the range of g is $(0, \infty)$, then the **Chain Rule** for the general logarithmic function $f(x) = \log_b[g(x)]$ is

$$f'(x) = \frac{1}{\ln b} \cdot \frac{1}{g(x)} \cdot g'(x)$$

The difference between the functions in Example 1b and Example 3c is an important one. In Example 1b, $g(x) = \ln x^3$ means that x is taken to the third power and *then* the logarithm is taken. In Example 3c, $f(x) = (\ln x)^3$ means that the logarithm of x is taken first and *then* the result is cubed. The parentheses are used to make the difference clear.

Derivatives of General Logarithmic Functions

The most frequently used logarithm function for applications is the natural logarithm function $f(x) = \ln x$. But what about the derivatives of the logarithm function with other bases? In other words, what is $\frac{d}{dx}[\log_b x]$? The box to the right shows how to **differentiate a general logarithmic function**. The derivation of this result is left to you in the Exercises.

OBJECTIVE 4

Determine the derivative of a general logarithmic function.

Example 4: Determining Derivatives of General Logarithmic Functions

Differentiate the following.

a. $y = \log_3 x$

b. $f(x) = \log(x^3 + 9)$

Perform the Mathematics

a. Here we have a general logarithmic function with base 3. So the derivative is

$$y' = \frac{d}{dx}[\log_3 x] = \frac{1}{\ln 3} \cdot \frac{1}{x} = \frac{1}{x \ln 3}$$

b. Since the base is not written, we know that this is a common logarithm with base 10. Applying the Chain Rule, with $g(x) = x^3 + 9$, gives us

$$f'(x) = \frac{d}{dx}[\log(x^3 + 9)]$$

$$= \frac{1}{\ln 10} \cdot \frac{1}{x^3 + 9} \cdot \frac{d}{dx}[x^3 + 9]$$

$$= \frac{1}{\ln 10} \cdot \frac{1}{x^3 + 9} \cdot 3x^2 = \frac{1}{\ln 10} \cdot \frac{3x^2}{x^3 + 9}$$ ∎

▶ *Try It Yourself*

Some related Exercises are 33 and 37.

Summary

In this section we learned how to differentiate the natural logarithmic function, as well as the logarithmic function with any base.

- **Derivative of the natural logarithm function:** For $f(x) = \ln x$, $f'(x) = \frac{1}{x}$.

- **Chain Rule for the natural logarithm function:** If $h(x) = \ln[g(x)]$, then
  $h'(x) = \frac{1}{g(x)} \cdot g'(x) = \frac{g'(x)}{g(x)}$.

- **Derivative of the general logarithmic function:** For the general logarithmic function $f(x) = \log_b x$ with $b > 0$, $b \neq 1$, $f'(x) = \frac{1}{\ln b} \cdot \frac{1}{x}$. The **Chain Rule** for the general logarithmic function $f(x) = \log_b[g(x)]$ is $f'(x) = \frac{1}{\ln b} \cdot \frac{1}{g(x)} \cdot g'(x)$.

Section 4.4 Exercises

Vocabulary Exercises

1. A function of the type $f(x) = \ln x$ is called the _____ logarithm function.

2. The base of $f(x) = \ln x$ is the number _____.

3. We call a function of the type $f(x) = a + b\ln x$ a natural logarithm _____.

4. When we determine the derivative $\frac{d}{dx}[\ln(f(x))]$, where f is a differentiable function, we must use the _____ Rule.

5. A function of the type $f(x) = \log_b x$ is a _____ logarithmic function.

6. The domain of $f(x) = \log_b x$ is the set of all _____ real numbers.

Skill Exercises

In Exercises 7–16, determine the derivative for the following functions.

7. $f(x) = 5\ln x$ **8.** $f(x) = -8\ln x$

9. $f(x) = \ln x^6$ **10.** $f(x) = \ln x^4$

11. $f(x) = 4x^3 \cdot \ln x$ **12.** $f(x) = 12x^3 \cdot \ln x$

13. $f(x) = \dfrac{3x^5}{\ln x}$ **14.** $f(x) = \dfrac{12}{\ln x}$

15. $f(x) = 10 - 12\ln x$ **16.** $f(x) = -2 + 8\ln x$

In Exercises 17–30, determine the derivative of the following functions.

17. $f(x) = \ln(x + 7)$ **18.** $f(x) = \ln(2 - x)$

19. $f(x) = \ln(2x - 5)$ **20.** $f(x) = \ln(3x + 4)$

21. $f(x) = \ln(x^2 + 3)$ **22.** $f(x) = \ln(3x^3 - 11)$

23. $f(x) = \ln(\sqrt{2x + 5})$ **24.** $f(x) = \ln(\sqrt[3]{4x + 2})$

25. $f(x) = (\ln x)^6$ **26.** $f(x) = (\ln x)^4$

27. $f(x) = \sqrt{x} \cdot \ln(\sqrt{x})$ **28.** $f(x) = 4x^5 \cdot \ln(3x^3)$

29. $f(x) = \dfrac{x^2 + 2x + 3}{\ln(x + 5)}$ **30.** $f(x) = \dfrac{4x^3 - x + 2}{\ln(x + 7)}$

For Exercises 31–40, determine the derivative of the following functions.

31. $f(x) = \log_{10} x$ **32.** $f(x) = \log_5 x$

33. $f(x) = 6\log_3 x$ **34.** $f(x) = 11\log_4 x$

35. $f(x) = x^2\log_9 x$ **36.** $f(x) = 2x^5\log_8 x$

37. $f(x) = \log_2(5x + 3)$ **38.** $f(x) = \log_5(3x + 9)$

39. $f(x) = \log_{10}\left(\dfrac{x + 3}{x^2 + 1}\right)$ **40.** $f(x) = \log_2\left(\dfrac{x^3}{x^2 - 1}\right)$

For Exercises 41–48, determine an equation of the line tangent to the graph of the function at the indicated point.

41. $f(x) = \ln x$; $(2, \ln 2)$ **42.** $f(x) = \ln x$; $(1, 0)$

43. $f(x) = \ln\sqrt{2x - 1}$; $(1, 0)$ **44.** $f(x) = \ln(3x)$; $(2, \ln 6)$

45. $f(x) = 4x^3 \cdot \ln x$; $(1, 0)$ **46.** $f(x) = 12x^3 \cdot \ln x$; $(2, 96\ln 2)$

47. $f(x) = (\ln x)^6$; $(e, 1)$ **48.** $f(x) = \ln x^6$; $(e, 6)$

Application Exercises

49. Biology—Bacterial Growth: Suppose a research assistant in biology finds in an experiment that at low temperatures, the growth of a certain bacteria culture can be modeled by

$$f(t) = 750 + 12\ln t \qquad t \geq 1$$

where t represents the number of hours since the start of the experiment and $f(t)$ represents the number of bacteria present.

(a) Determine $f'(t)$.

(b) Determine the number of bacteria present and the rate of growth after 12 hours.

50. **Urban Planning—Zoning:** Suppose the city of Plantersville has enacted new zoning laws in order to curb the city's growing population. In an effort to evaluate their laws, the zoning commission has determined from census records that the city's population can be modeled by

$$P(x) = 10{,}000 + 100\ln x \qquad x \geq 1$$

where x represents the number of years since the zoning laws were enacted and $P(x)$ represents the city's population.

(a) Determine $P'(x)$.

(b) Determine the city's population and rate of growth 20 years after the zoning laws were enacted.

51. **Medicine—Drug Popularity:** Suppose that prescription drug companies have found that the popularity of the new drug Vectrum has plateaued in recent years. User data has shown that annual number of prescriptions written for the drug can be modeled by

$$f(x) = 150 + 5\log_2 x \qquad x \geq 1$$

where x represents the number of years that the drug has been on the market and $f(x)$ represents the number of prescriptions written annually in thousands.

(a) Determine $f'(x)$.

(b) Evaluate $f'(2)$ and $f'(10)$ and interpret each.

52. **Education—Student Hygiene:** The school district of Molisburg started an educational campaign in an attempt to curb the increase in head lice found in the elementary school population. In part of a program to evaluate the effectiveness of the campaign, the school nurse's records show that the number of children who contracted lice annually can be modeled by

$$f(t) = 200 + 8\log_3 t \qquad t \geq 1$$

where t represents the number of years since the educational campaign began and $f(t)$ represents the number of children who contracted lice.

(a) Determine $f'(t)$.

(b) Evaluate $f'(2)$ and $f'(7)$ and interpret each.

 53. **Political Economics—U.S. Postal Service:** In order to regain revenue lost to third-party couriers, the U.S. Postal Service adopted a flat parcel shipping rate during the mid-2000s. The annual USPS revenue from 1991 to 2010 can be modeled by

$$f(x) = 38.16 + 12.22\ln x \qquad 1 \leq x \leq 20$$

where x represents the number of years since 1990 and $f(x)$ represents the USPS revenue in billions of dollars. (*Source:* U.S. Postal Service.)

(a) Determine $f'(x)$.

(b) According to the model, how fast was the USPS revenue growing in 2009 as compared to 1999?

© Steve Krull/iStockphoto

 54. **Government Services—USPS Mail Delivery:** The steadily growing popularity of fax machines, texting, and e-mail has reduced the amount of mail handled by the U.S. Postal Service. "The replacement of letter mail and business-transaction mail by electronic alternatives continues to cause downward pressure on mail volume." (*Source:* CNNMoney.) The number of pieces of mail handled annually by the USPS from 1991 to 2010 can be modeled by

$$f(x) = 165.53 + 16.5\ln x \qquad 1 \leq x \leq 20$$

where x represents the number of years since 1990 and $f(x)$ represents the number of pieces of mail handled by the USPS measured in billions. (*Source:* U.S. Postal Service.)

(a) Determine $f'(x)$.

(b) According to the model, how fast was the USPS mail volume growing in 2008 as compared to 1998?

55. **Law Enforcement—New Zealand Prison Population:** A country in which the prison population has grown gradually is New Zealand. In June 2005, the Parliament passed a new Corrections Act, whose provisions were scheduled to come into force in mid-2005. The objectives of the act were to eliminate private management of prisons, establish individual management plans for prisoners, and make prisoners' minimum entitlements more consistent with U.N. standards. (*Source:* U.S. State Department.) The population of New Zealand prisons from 2001 to 2010 can be modeled by

$$f(x) = 4480 + 3986 \log x \qquad 1 \le x \le 10$$

where x represents the number of years since 2000 and $f(x)$ represents the number incarcerated. (*Source:* Statistics New Zealand.)

(a) Determine $f'(x)$.

(b) Compute $f'(2)$ and compare the rate to $f'(9)$. From inspecting the rates, was the Corrections Act effective in reducing the prison population?

56. **Law Enforcement—Croatia Prison Population:** Overcrowding, lack of natural light, and prisoner abuse are among the concerns documented concerning Croatian prisons by the United Nations High Commission for Refugees. (*Source:* UNHCR.org.) The population of prisons in Croatia from 2001 to 2010 can be modeled by

$$f(x) = 2286 + 1833 \log x \qquad 1 \le x \le 10$$

where x represents the number of years since 2000 and $f(x)$ represents the number incarcerated. (*Source:* EuroStat.)

(a) Determine $f'(x)$.

(b) Evaluate $f(7)$ and $f'(7)$ and interpret each.

Concept and Writing Exercises

57. Determine the derivative of the function $f(x) = \ln(\ln(x))$.

For Exercises 58–60, consider the function $f(x) = \ln|x|$.

58. Graph f and write the domain using interval notation.

59. For what x-values is f not differentiable?

60. Use the definition of absolute value to rewrite f and write a formula for f'.

61. Use the definition of derivative to prove that $\lim\limits_{x \to 0} \dfrac{\ln(1+x)}{x} = 1$.

For Exercises 62–64, consider the functions $f(x) = \ln x$ and $g(x) = x^{1/10}$.

62. Graph f and g then estimate the values where $f(x) > g(x)$.

63. Let $h(x) = \dfrac{\ln x}{x^{1/10}}$ then determine $\lim\limits_{x \to \infty} h(x)$.

64. Determine the value of k so that $h(x) < \frac{1}{10}$ whenever $x < k$.

65. Use the Chain Rule to prove that if $f(x) = \ln(g(x))$, then $f'(x) = \dfrac{1}{g(x)} \cdot g'(x)$.

66. Let $f(x) = x \cdot \ln x$. Determine the instantaneous rate of change of f with respect to x when $x = e$.

Section Project

A topic that has been in the news lately is energy conservation and the future of energy generation. Oil tycoon T. Boone Pickens has been touting the use of natural gas in the United States as a "bridge" source of energy until new sources such as hydrogen fuel cells are developed. The annual world production of natural gas from 1991 to 2010 can be modeled by the logarithmic function

$$f(x) = 68.05 + 11.97 \ln x \qquad 1 \le x \le 20$$

where x represents the number of years since 1990, and $f(x)$ represents the world production of natural gas in trillions of cubic feet. (*Source:* U.S. Energy Information Administration.)

 (a) Graph f in the viewing window [0, 20] by [0, 110].

(b) Determine the derivative $f'(x)$.

(c) Evaluate and interpret $f'(19)$.

(d) Write the equation of the tangent line at $x = 19$.

The annual U.S. production of natural gas from 1991 to 2010 can be modeled by the cubic function

$$g(x) = 0.003x^3 - 0.1x^2 + 0.87x + 17 \qquad 1 \le x \le 20$$

where x represents the number of years since 1990 and $f(x)$ represents the U.S. production of natural gas in trillions of cubic feet. (*Source:* U.S. Energy Information Administration.)

 (e) Graph g in the viewing window [0, 20] by [0, 110].

(f) Determine the derivative $g'(x)$.

(g) Evaluate and interpret $g'(19)$.

(h) Write the equation of the tangent line at $x = 19$.

Now form a function $w(x) = f(x) - g(x)$ using models f and g above.

(i) What does the function w represent?

(j) If the world and U.S. natural gas output stay at the 2009 rates, will the United States ever surpass the rest of the world in natural gas production? If so, when? What are the drawbacks of this kind of estimate?

Now form a rational function $h(x) = \dfrac{g(x)}{f(x)} \cdot 100$ using the models f and g above.

(k) What does this model represent?

(l) Evaluate and interpret $h(19)$.

Section 4.1 Review Exercises

For Exercises 1–4, complete the following.

(a) Complete the following table.

x	−5	−4	−3	−2	−1	0	1	2	3	4	5
$f(x)$											

(b) Graph f.

(c) Classify the function as exponential growth or decay.

1. $f(x) = \left(\dfrac{2}{5}\right)^x$

2. $f(x) = 2.3^x$

3. $f(x) = e^{0.8x}$

4. $f(x) = e^{-2.7x}$

5. Sales Analysis—Grocery Stores: The SuperValue Grocery chain announced a two-week sale in conjunction with the grand opening of its new location. The percentage of customers responding to the sale can be modeled by

$$f(x) = 65 - 80(0.71)^x$$

where x is the day of the sale, and $f(x)$ represents the percentage of customers who respond to the sale.

(a) Evaluate $f(1)$ and interpret.

(b) Find the slope of the secant line for $x = 4$ and $\Delta x = 3$ and interpret.

(c) When does the percentage of customers responding to the sale first exceed 50%?

6. Personal Banking—Compound Interest: Suppose that Paul deposits $1700 into an account that pays interest at a rate of 5.3% compounded monthly. How much will be in the account after 7 years?

7. Personal Banking—Compound Interest: Suppose that Yushu deposits $20,000 into an account that pays interest at a rate of 7.5%. How much will be in the account after 12 years if the interest is compounded

(a) Annually?

(b) Monthly (12 times a year)?

(c) Weekly (52 times a year)?

8. Personal Banking—Compound Interest: Suppose that $8000 is invested at a 6.75% interest rate compounded quarterly (four times a year).

(a) Write the compound interest function

$$A(t) = P\left(1 + \frac{r}{n}\right)^{nt}$$

for the given information and interpret.

(b) Evaluate $A(6)$ and interpret.

(c) Calculate how much interest was made at the end of the 6-year period.

9. Personal Banking—Compound Interest: Suppose that Frederick deposits $5000 at a 6.2% interest rate compounded monthly.

(a) Write the compound interest function A for the given information.

(b) Graph A in the viewing window [0, 15] by [5000, 15,000].

(c) Use your calculator to graphically find the doubling time of the account (that is, the amount of time that it takes the account to accumulate a total balance of $10,000).

10. **Demographics—City Population:** The population of Granaco City can be modeled by

$$f(x) = 64{,}000e^{0.04x} \qquad 0 \le x \le 20$$

where x represents the number of years since 1980 and $f(x)$ represents the number of people living in Granaco City.

(a) Evaluate $f(4)$ and interpret.

(b) Find the slope of the secant line for $x = 5$ and $\Delta x = 3$ and interpret.

(c) In what year did the population first exceed 110,000?

11. Suppose that $2500 is invested at a 5.7% interest rate compounded continuously.

(a) Write the continuous compound interest function $A(t) = Pe^{rt}$ for the given information.

(b) Evaluate $A(4)$ and interpret.

(c) Use your calculator to graphically find the doubling time of the account.

12. **Personal Banking—Compound Interest:** For each account, compute the effective rate, and interpret the result.

(a) An 8.2% nominal rate compounded monthly.

(b) A 6.7% nominal rate compounded continuously.

13. **Wildlife Conservation—Leopard Population:** The population of leopards in a certain region is given by the logistic model

$$f(t) = \frac{160}{1 + 42.29e^{-0.1x}} \qquad 0 \le x \le 60$$

where x represents the number of years since 1940 and $f(x)$ represents the number of leopards.

(a) Evaluate $f(35)$ and interpret.

(b) Compute the difference quotient for $x = 47$ and $\Delta x = 3$ and interpret.

Section 4.2 Review Exercises

For Exercises 14 and 15, let $f(x) = x^2 - 3x$ and $g(x) = 2x - 5$. Evaluate each expression.

14. $f(g(4))$

15. $g(f(-2))$

For Exercises 16 and 17, complete the following.

(a) State the domain of each of the functions f and g.

(b) Algebraically simplify $f(g(x))$.

(c) Algebraically simplify $g(f(x))$.

16. $f(x) = 5x + 3; \ g(x) = x^2$ 17. $f(x) = \sqrt{3x - 2}; \ g(x) = 3x^2 + 6$

For Exercises 18–21, determine the functions f and g so that $f(g(x)) = h(x)$ for each composite function h; then check your answer. There is more than one correct answer.

18. $h(x) = (x^2 + 2)^5$ 19. $h(x) = x^2 + 5$

20. $h(x) = \sqrt{6 - x}$ 21. $h(x) = \dfrac{3x - 4}{3x + 10}$

22. Graph the function $g(x) = \frac{1}{x}$, and use the horizontal line test to determine if g is a one-to-one function.

For the functions in Exercises 23 and 24, complete the following.

(a) Show that $f(g(x)) = x$.

(b) Graph f, g, and the line $y = x$ in the same Cartesian plane.

23. $f(x) = 4 - 3x$; $g(x) = \dfrac{-x + 4}{3}$ **24.** $f(x) = \sqrt[3]{x + 5}$; $g(x) = x^3 - 5$

For Exercises 25 and 26, rewrite the equation in logarithmic form, $y = \log_b x$.

25. $3^5 = 243$ **26.** $10^{-4} = 0.0001$

For Exercises 27 and 28, use the properties of logarithms to rewrite the expression.

27. $\ln 18^3$ **28.** $\log_2 \dfrac{5^2}{7}$

For Exercises 29 and 30, solve the exponential equations algebraically and round the solution to the hundredths place.

29. $5 \cdot 4^x = 65$ **30.** $6.3(1.7)^x = 24$

Section 4.3 Review Exercises

For Exercises 31–34, determine the derivatives of the following functions.

31. $f(x) = 3e^x + 5$ **32.** $f(x) = \sqrt{e^x - 5}$

33. $f(x) = 7x^3 e^x - 3e^x$ **34.** $f(x) = e^{2x - 13}$

For Exercises 35 and 36, complete the following.

(a) Determine the tangent line equation for the function at the given point.

(b) Check by graphing the function and the tangent line in the same viewing window.

35. $f(x) = 3e^x$; $(2, 3e^2)$ **36.** $f(x) = e^x - e^{-x}$; $(0, 0)$

In Exercises 37–42, determine the derivatives of the general exponential functions.

37. $f(x) = 4^x$ **38.** $f(x) = \dfrac{3^x}{15^x}$

39. $f(x) = x^4 \cdot 0.7^x$ **40.** $f(x) = \sqrt[3]{10^x}$

41. $f(x) = 5^{x^2 - 1}$ **42.** $f(x) = 0.4^{\sqrt{x}}$

43. Agriculture—Peruvian Imports: The American appetite for fresh produce has resulted in a rise in agricultural imports from Latin American countries. One country that has benefited from this is Peru, which supplements California's supply of avocados. "California doesn't affect us, as demand and consumption in the U.S. are very high," said Arturo Medina, general manager of the Association of Prohass. (*Source:* Fresh Plaza.) The amount of agricultural exports from Peru to the United States from 1990 to 2011 can be modeled by the exponential function

$$f(x) = 56.53e^{0.14x} \qquad 0 \le x \le 21$$

where x represents the number of years since 1990, and $f(x)$ represents the annual amount of agricultural exports from Peru to the United States, measured in millions of dollars. (*Source:* U.S. Department of Agriculture.)

(a) Determine $f'(x)$.

(b) Evaluate $f'(21)$ and interpret.

44. Agriculture—Peruvian Imports: For the model in Exercise 43, write the equation of the line tangent to the graph of f at the point $(21, f(21))$, then evaluate the tangent line equation when $x = 30$. Interpret the meaning of this ordered pair.

Section 4.4 Review Exercises

For Exercises 45–50, determine the derivatives of the functions.

45. $f(x) = -6\ln x$

46. $f(x) = \ln x^5$

47. $f(x) = 4x^5 \cdot \ln x$

48. $f(x) = \ln(3x^2 - 5)$

49. $f(x) = (\ln x)^{-2}$

50. $f(x) = \dfrac{x^2 + 5x - 2}{\ln(x + 4)}$

For Exercises 51–54, determine the derivatives of the functions.

51. $f(x) = 3\log_5 x$

52. $f(x) = 4x^3 \log_2 x$

53. $f(x) = \log_{10}(6x - 5)$

54. $f(x) = \log_4\left(\dfrac{x^2 + 5}{2x - 3}\right)$

For Exercises 55 and 56, complete the following.

(a) Determine the tangent line equation for the function at the given point.

(b) Check by graphing the function and the tangent line in the same viewing window.

55. $f(x) = \ln x^4$; $(e^4, 16)$

56. $f(x) = \ln\sqrt{2x + 3}$; $(3, \ln 3)$

For Exercises 57 and 58, determine the derivatives of the given functions.

57. $f(x) = 3xe^{5x} - \ln(x + 4)$

58. $f(x) = \log_7 3x - \log_3 7x$

59. Educational Funding—College Donations: Many colleges and universities benefit from significant contributions from local philanthropists. "There are a group of people who give to have their names on buildings, and that's the easiest place to have your name on a building," says Edward H. Merrin, an art dealer who tops the donor list at Tufts University. (*Source: The Chronicle of Philanthropy.*) The annual amount of donations to college and universities from 1991 to 2011 can be modeled by

$$f(x) = 7.7 + 10.54\ln x \qquad 1 \leq x \leq 21$$

where x represents the number of years since 1990 and $f(x)$ represents the amount of donations to colleges and universities, measured in billions of dollars. (*Source:* Giving USA Foundation.)

(a) Determine $f'(x)$.

(b) Evaluate $f'(20)$ and interpret.

60. Educational Funding—College Donations: For the model in Exercise 59, write the equation of the tangent line of f at the point $(20, f(20))$ then evaluate the tangent line equation when $x = 30$. Interpret the meaning of this ordered pair.

Applications of the Derivative

© NoDerog/iStockphoto **(a)**

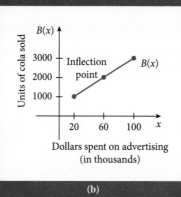
(b)

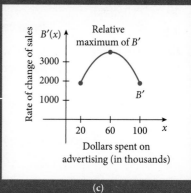
(c)

To gain domination in the lucrative soft drink market, cola makers can spend a fortune on advertising. But how are decisions made on how much money to spend on advertising? In Figure (b), $B(x)$ represents the number of units of cola sold after spending x thousand dollars on advertising. Figure (c) shows that the rate of change of sales $B'(x)$ is maximized at $x = 60$. At $x = 60$ on the graph of $B(x)$, we find what is called an inflection point. In economics, this point is also called the *point of diminishing returns* and indicates where the rate of change of sales started to decrease.

What We Know

In Chapter 4, we continued our study of differentiation and how the derivative gives an instantaneous rate of change. We also examined exponential and logarithmic functions and applications utilizing the power of the derivative.

Where Do We Go

In this chapter, we will see how the rate of change concept and derivative tell us the behavior of a function: specifically, where we can find extrema and intervals of concavity. We will also see many applications of extrema and concavity. This chapter will also analyze further applications of the derivative including elasticity of demand as well as related rates.

Chapter Sections

5.1   First Derivatives and Graphs

In this section we examine more deeply the relationship that exists between a function and its derivative. Our study will reveal how the sign of a derivative, whether it is positive or negative, on an interval indicates whether the function is increasing or decreasing. Also, we will see how the derivative can help locate the **relative extrema** for a function.

Intervals of Increase and Decrease for Functions

So far, we have looked at the graphs of many different functions. The graphs of functions generally have portions that are **increasing** and portions that are **decreasing**, such as the graph in **Figure 5.1.1**.

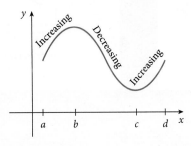

Figure 5.1.1

In the following Flashback, we review how to determine intervals where a quadratic function is increasing and where it is decreasing.

Flashback: Paper and Paperboard Waste Generation Revisited

Earlier in the text, we saw that the total paper and paperboard waste that is generated in the United States can be modeled by

$$f(x) = -0.15x^2 + 3.38x + 69.38 \qquad 1 \le x \le 19$$

where x represents the number of years since 1989 and $f(x)$ represents the total paper and paperboard waste generated in millions of tons. See **Figure 5.1.2**. (*Source:* U.S. Statistical Abstract.)

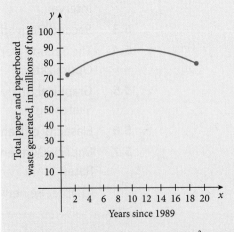

Figure 5.1.2 Graph of $f(x) = -0.15x^2 + 3.38x + 69.38$

Determine intervals where f is increasing and decreasing. Round to the nearest hundredth and interpret.

Perform the Mathematics

Since the graph of the model is a parabola, to determine intervals where the model is increasing or decreasing, we must find the coordinates of the vertex, the maximum point on the graph. In Chapter 1 it was shown that the x-coordinate of the vertex can be determined algebraically by using $x = -\frac{b}{2a}$. Using this formula, the x-coordinate is

$$x = -\frac{b}{2a} = -\frac{3.38}{2(-0.15)} = -\frac{3.38}{-0.3} \approx 11.27$$

The y-coordinate is found by substituting $x = 11.27$ back into $f(x)$. This gives

$$f(11.27) = -0.15(11.27)^2 + 3.38(11.27) + 69.38 \approx 88.42$$

So we conclude that f is increasing on the interval $(1, 11.27)$ and decreasing on the interval $(11.27, 19)$. Rounding 11.27 to the nearest year gives a value of 11. Thus, the total paper and paperboard waste that is generated in the United States increased from 1990 to 2000 and then it decreased from 2000 to 2008. There was a peak of about 88.42 million tons of paper and paperboard waste generated, and it occurred in 2000. ∎

Technology Option

You can verify the results in the Flashback by graphing the model on your calculator and using the `Maximum` command.

In the Flashback, we found intervals where f is increasing and where it is decreasing. Another way to find these intervals is to analyze the sign of f'. Since $f'(x)$ gives the rate of change of f at x, f' can be called a **rate function**. In Example 1, we observe the relationship between the graph of f and the sign of f'.

Example 1: Analyzing the Behavior of a Function and Its Derivative

OBJECTIVE 1

Analyze the relationship between the sign of f' and the behavior of f.

The graphs of $f(x) = x^3 - 12x + 4$ and its derivative $f'(x) = 3x^2 - 12$ are shown in **Figures 5.1.3** and **5.1.4**. Use the graphs to determine where:

a. f is increasing

b. $f' > 0$

c. f is decreasing

d. $f' < 0$

e. f is constant

f. $f' = 0$

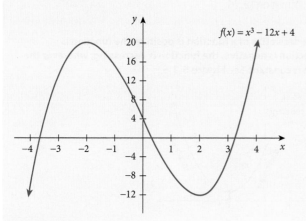

Figure 5.1.3

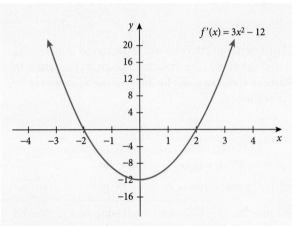

Figure 5.1.4

Perform the Mathematics

a. f is increasing on $(-\infty, -2) \cup (2, \infty)$.

b. Recall that a function is greater than zero whenever its graph is *above* the x-axis. Thus, $f' > 0$ on $(-\infty, -2) \cup (2, \infty)$, since this is where the graph of f' is above the x-axis.

c. f is decreasing on $(-2, 2)$.

d. Recall that a function is less than zero whenever its graph is *below* the x-axis. Thus, $f' < 0$ on $(-2, 2)$, since this is where the graph of f' is below the x-axis.

e. f is constant at $x = -2$ and $x = 2$.

f. Recall that a function is equal to zero whenever its graph *crosses* the x-axis. Thus, $f'(x) = 0$ at $x = -2$ and $x = 2$, since this is where the graph of f' crosses the x-axis. ∎

Notice in Example 1 that there appears to be a connection between our answers to parts (a) and (b) as well as parts (c) and (d) and parts (e) and (f). Pairing the results in this manner leads us to make the following observation about **increasing and decreasing functions**, which is true for any function and its derivative.

DEFINITION

Increasing and Decreasing Functions

On an open interval (a, b) on which f is differentiable and continuous:

1. If $f'(x) > 0$ for all x in (a, b), then f is increasing on (a, b).

2. If $f'(x) < 0$ for all x in (a, b), then f is decreasing on (a, b).

3. If $f'(x) = 0$ for all x in (a, b), then f is constant on (a, b).

NOTE: In English this means that **wherever the derivative of a function is positive, the function is increasing; wherever the derivative of a function is negative, the function is decreasing; wherever the derivative of a function is zero, the function is constant**. See **Figure 5.1.5**.

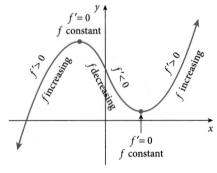

Figure 5.1.5

Knowing that the derivative tells us the slope of a tangent line at a point, the preceding should come as no surprise. Basically, wherever tangent lines have a positive slope, $f'(x) > 0$, we see that a function is increasing (**Figure 5.1.6**), and wherever tangent lines have a negative slope, $f'(x) < 0$, we see that a function is decreasing (**Figure 5.1.7**). Also, where we have **horizontal tangent lines**, $f'(x) = 0$, we say that the function is constant (**Figure 5.1.8**).

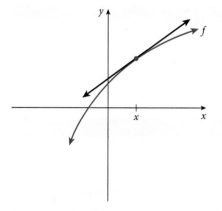

Figure 5.1.6 Tangent line has positive slope; that is, $f'(x) > 0$.

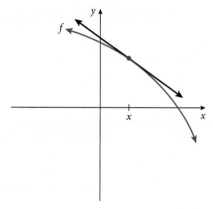

Figure 5.1.7 Tangent line has negative slope; that is, $f'(x) < 0$.

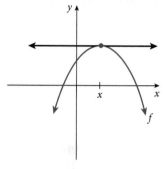

Figure 5.1.8 Tangent line has zero slope; that is, $f'(x) = 0$.

Relative Extrema and Critical Values

In Figure 5.1.5, notice that the graph has a *valley* and a *hill*. In mathematics, the hill is called a **relative maximum**, while the valley is called a **relative minimum**. By using the word **relative** we mean **relative to points nearby**. (Later in this chapter, we will discuss *absolute maximum* and *absolute minimum*.) These terms are expressed more formally in the **relative extrema** definition.

The plural of relative maximum is relative maxima and the plural of relative minimum is relative minima. Collectively, the maxima and minima are called **extrema**.

> **DEFINITION**
>
> **Relative Extrema**
>
> If f is a continuous function on an open interval containing c, then the point $(c, f(c))$:
>
> - Is a **relative maximum** if $f(c) \geq f(x)$ for all x in the interval.
> - Is a **relative minimum** if $f(c) \leq f(x)$ for all x in the interval.

Example 2: Locating Relative Extrema

Determine where the graphs in **Figure 5.1.9** and **5.1.10** have a relative maximum and a relative minimum.

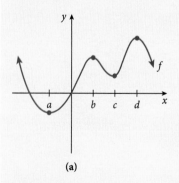

(a)

Figure 5.1.9

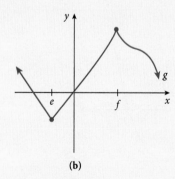

(b)

Figure 5.1.10

Perform the Mathematics

a. f has a relative maximum at $x = b$ and $x = d$, and f has a relative minimum at $x = a$ and $x = c$.

b. g has a relative maximum at $x = f$ and a relative minimum at $x = e$. ∎

In Example 2(a), the relative extrema occurred where the graph has horizontal tangent lines, that is, where $f'(x) = 0$. In Example 2(b), the relative extrema occurred where the graph had a corner or made a sharp turn, that is, where $f'(x)$ is undefined. We call points on a curve where $f'(x) = 0$ or where $f'(x)$ is undefined the **critical points** of the function. The x-coordinate of the critical points are called **critical values** or **critical numbers**.

OBJECTIVE 2

Determine critical values.

DEFINITION

Critical Values

A **critical value** for f is an x-value in the domain of f for which $f'(x) = 0$ **or** $f'(x)$ is undefined.

Example 3: Determining Critical Values

Determine the critical values for $f(x) = 5x^3 + 4x^2 - 12x - 25$.

Understand the Situation

The key to locating critical values is to first determine the derivative. Once we have the derivative, we simply determine value(s) for x for which $f'(x) = 0$ or $f'(x)$ is undefined.

Perform the Mathematics

The derivative is

$$f'(x) = 15x^2 + 8x - 12$$

Since $f'(x)$ is defined for all x, the only critical values occur where $f'(x) = 0$. So we need to solve

$$15x^2 + 8x - 12 = 0$$
$$(5x + 6)(3x - 2) = 0$$
$$x = -\frac{6}{5} \text{ or } x = \frac{2}{3}$$

Interpret the Results

So the critical values occur at $x = -\frac{6}{5}$ and $x = \frac{2}{3}$. See **Figure 5.1.11**.

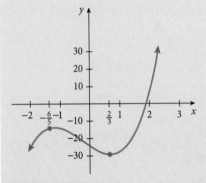

Figure 5.1.11

▷ *Try It Yourself*

Some related Exercises are 21 and 23.

Example 4: Determining Critical Values

Determine the critical values for the following functions.

a. $f(x) = \sqrt[3]{x}$ **b.** $y = x^3$

Perform the Mathematics

a. Again, our first step is to compute the derivative. This yields

$$f(x) = \sqrt[3]{x} = x^{1/3}$$

$$f'(x) = \frac{1}{3}x^{-2/3} = \frac{1}{3x^{2/3}} = \frac{1}{3\sqrt[3]{x^2}}$$

Here we notice that $f'(x)$ is undefined at $x = 0$ and that $f'(x)$ cannot equal zero; that is, $f'(x) = 0$ has no solution. So we conclude that the only critical value is $x = 0$. See **Figure 5.1.12**.

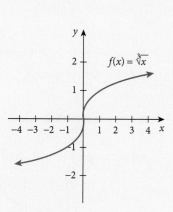

Figure 5.1.12   **Figure 5.1.13**

b. The derivative is $\frac{dy}{dx} = 3x^2$ and it is defined for all x. We see that $\frac{dy}{dx} = 0$ at $x = 0$, which means that the only critical value is $x = 0$. See **Figure 5.1.13**. ■

Notice that in Example 3, our critical values produced a relative maximum and a relative minimum, whereas in Example 4, our critical values produced neither a relative maximum nor a relative minimum. What this means is that **every relative extreme occurs at a critical value, but not every critical value produces a relative extreme**. So how do we know if a critical value is going to produce a relative maximum or a relative minimum? The answer to this question is found in the First Derivative Test.

First Derivative Test

The **First Derivative Test** pulls together all of the ideas in this section.

First Derivative Test

Let $x = c$ be a critical value of a function f that is continuous on an open interval containing $x = c$. The point $(c, f(c))$ can be called:

1. A **relative minimum** if $f' < 0$ to the left of c and $f' > 0$ to the right of c.
2. A **relative maximum** if $f' > 0$ to the left of c and $f' < 0$ to the right of c.

Recalling that the sign of the derivative tells us the increasing-decreasing behavior of the function, we can construct a **sign diagram** that communicates the **First Derivative Test** in a more visual manner. In the following sign diagrams, the bottom row gives the sign of f' and the top row gives the behavior of f.

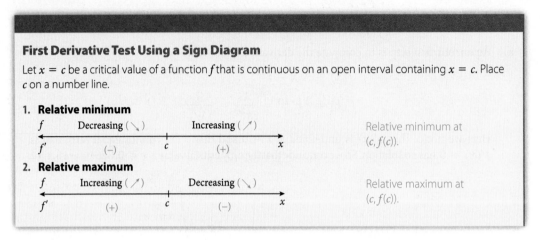

First Derivative Test Using a Sign Diagram

Let $x = c$ be a critical value of a function f that is continuous on an open interval containing $x = c$. Place c on a number line.

1. **Relative minimum**

   Relative minimum at $(c, f(c))$.

2. **Relative maximum**

   Relative maximum at $(c, f(c))$.

NOTE: If f' has the same sign on both sides of $x = c$, then f has neither a relative maximum nor a relative minimum at $x = c$.

Figure 5.1.14 (a–c) illustrates the First Derivative Test when $x = c$ is a critical value, where $f'(c) = 0$, that is, a horizontal tangent.

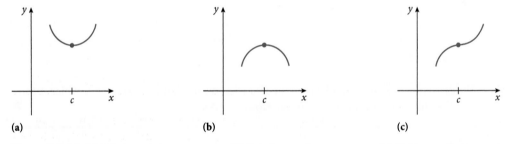

(a)   **(b)**   **(c)**

Figure 5.1.14 **(a)** Relative minimum at $x = c$. **(b)** Relative maximum at $x = c$. **(c)** Neither a relative maximum nor a relative minimum at $x = c$.

Figure 5.1.15 (a–c) illustrates the First Derivative Test when $x = c$ is a critical value, where $f'(c)$ is undefined.

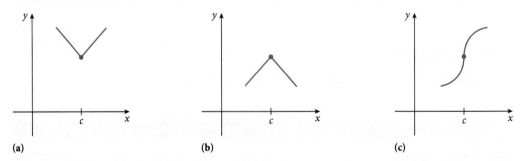

(a)   **(b)**   **(c)**

Figure 5.1.15 **(a)** Relative minimum at $x = c$. **(b)** Relative maximum at $x = c$. **(c)** Neither relative maximum nor relative minimum at $x = c$.

To understand why the sign diagram and the method that we will use in constructing sign diagrams works, we need to quickly review **continuous functions**. Suppose that a function f is continuous on the open interval $(2, 7)$, and $f(x) \neq 0$ for any x in $(2, 7)$. If $f(x)$ is positive for some x, say $f(3) = 4$, then $f(x)$ is positive for every x in $(2, 7)$. Why? Well, suppose we assume that $f(6)$ is negative, say $f(6) = -3$. Since f is **continuous**, the graph of f could not connect the points $(3, 4)$ and $(6, -3)$ without crossing the x-axis. See **Figure 5.1.16**.

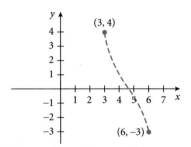

Figure 5.1.16

But recall that $f(x) = 0$ at the point where the graph of f crosses the x-axis. Since we agreed that $f(x) \neq 0$ for any x in $(2, 7)$, this assumption has been contradicted, and thus f cannot be negative in $(2, 7)$. This property of continuous functions is stated in **Theorem 5.1**, which we will use frequently in this chapter.

> **Theorem 5.1**
>
> If f is continuous and $f(x) \neq 0$ for any x in the interval (a, b), then f cannot change sign on (a, b).

Example 5: Applying the First Derivative Test

OBJECTIVE 3

Use the First Derivative Test.

For $f(x) = 5x^3 + 4x^2 - 12x - 25$, determine intervals where f is increasing and where f is decreasing by making a sign diagram. Locate all relative extrema.

Perform the Mathematics

To find intervals where f is increasing and where it is decreasing, we need to determine the critical values for this function. In order to find the critical values, we must first determine the derivative. In Example 3, we determined that the derivative is $f'(x) = 15x^2 + 8x - 12$ and the critical values are

$$x = -\frac{6}{5} \text{ and } x = \frac{2}{3}$$

To determine intervals where the function is increasing and decreasing, we place the critical values on a number line representing the x-axis in **Figure 5.1.17**.

Figure 5.1.17

Since f' is a polynomial function, it is continuous on the open intervals determined by the critical values. Thanks to Theorem 5.1, f' can change sign only at the critical values. We can determine the sign of the derivative within each interval by evaluating $f'(x)$ for some x within the interval. We call such x-values **test numbers**. In the interval $(-\infty, -\frac{6}{5})$, we choose $x = -2$ as our test number.

$$f'(-2) = 15(-2)^2 + 8(-2) - 12 = 32$$

Since $f'(-2) = 32$, which is positive, we conclude that on the interval $(-\infty, -\frac{6}{5})$, $f(x) = 5x^3 + 4x^2 - 12x - 25$ is *increasing*. Choosing $x = 0$ as our test number in the interval $(-\frac{6}{5}, \frac{2}{3})$ gives us

$$f'(0) = 15(0)^2 + 8(0) - 12 = -12$$

So $f'(0) = -12$, a negative result. In a similar fashion, we select $x = 1$ as our test number in the interval $(\frac{2}{3}, \infty)$ and determine that

$$f'(1) = 15(1)^2 + 8(1) - 12 = 11$$

So $f'(1) = 11$, a positive number. We now list the signs of the derivative on the sign diagram in **Figure 5.1.18**.

Figure 5.1.18

Now, using our knowledge of how the sign of the derivative tells us the behavior of the function, we complete the sign diagram in **Figure 5.1.19**. This sign diagram indicates that f is increasing on $(-\infty, -\frac{6}{5}) \cup (\frac{2}{3}, \infty)$ and decreasing on $(-\frac{6}{5}, \frac{2}{3})$. To find the relative extrema, we use the First Derivative Test with a sign diagram. Since f is increasing on $(-\infty, -\frac{6}{5})$ and then decreasing on $(-\frac{6}{5}, \frac{2}{3})$, we conclude that there is a relative maximum at $x = -\frac{6}{5}$. Since f is decreasing on $(-\frac{6}{5}, \frac{2}{3})$ and then increasing on $(\frac{2}{3}, \infty)$, we conclude that there is a relative minimum at $x = \frac{2}{3}$. Specifically, we have a relative maximum at the point $(-\frac{6}{5}, -\frac{337}{25})$ and a relative minimum at the point $(\frac{2}{3}, -\frac{803}{27})$.

Figure 5.1.19   ■

Locating Relative Extrema

To locate relative extrema of a function, find the critical values of the function and apply the First Derivative Test.

Example 5 demonstrates the process we can use to **locate relative extrema**.

OBJECTIVE 4

Apply the First Derivative Test.

Example 6: Applying the First Derivative Test

For $f(x) = x + \frac{1}{x}$, determine intervals where f is increasing and decreasing. Locate all relative extrema.

Perform the Mathematics

To determine critical values, we need to first determine the derivative.

$$f(x) = x + \frac{1}{x} = x + x^{-1}$$

$$f'(x) = 1 - x^{-2} = 1 - \frac{1}{x^2}$$

Next, we set the derivative equal to zero and solve.

$$1 - \frac{1}{x^2} = 0$$

$$1 = \frac{1}{x^2}$$

$$x^2 = 1$$

$$x = \pm 1$$

Notice that, although the derivative is undefined at $x = 0$, we do not list $x = 0$ as a critical value, because $x = 0$ is not in the domain of f. So the only critical values are $x = 1$ and $x = -1$. We proceed as in Example 5 and construct the sign diagram in **Figure 5.1.20**.

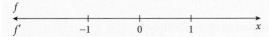

Figure 5.1.20

Notice that we placed $x = 0$, where the derivative was undefined, in our sign diagram. We did this to take advantage of Theorem 5.1. In the interval $(-\infty, -1)$, we select the test number -2; in $(-1, 0)$, we select $-\frac{1}{2}$; in $(0, 1)$, we pick $\frac{1}{2}$; and in $(1, \infty)$, we choose 2. (Notice that in each of these open intervals our function is continuous, thus allowing us to use Theorem 5.1.) Substituting these test numbers into the derivative yields

$$x = -2 \qquad\qquad x = -\frac{1}{2} \qquad\qquad x = \frac{1}{2} \qquad\qquad x = 2$$

$$f'(-2) = 1 - \frac{1}{(-2)^2} \quad f'\left(-\frac{1}{2}\right) = 1 - \frac{1}{\left(-\frac{1}{2}\right)^2} \quad f'\left(\frac{1}{2}\right) = 1 - \frac{1}{\left(\frac{1}{2}\right)^2} \quad f'(2) = 1 - \frac{1}{(2)^2}$$

$$= 1 - \frac{1}{4} \qquad\qquad = 1 - \frac{1}{\frac{1}{4}} \qquad\qquad = 1 - \frac{1}{\frac{1}{4}} \qquad\qquad = 1 - \frac{1}{4}$$

$$= \frac{3}{4} \qquad\qquad\qquad = -3 \qquad\qquad\qquad = -3 \qquad\qquad\qquad = \frac{3}{4}$$

Finally, using these results and our knowledge of how the sign of the derivative determines the behavior of the function, we complete the sign diagram as shown in **Figure 5.1.21**.

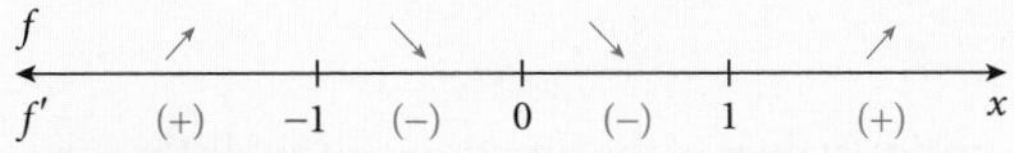

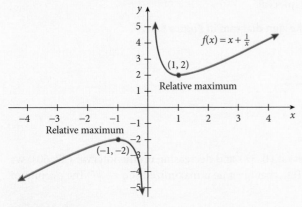

Figure 5.1.21

So $f(x) = x + \frac{1}{x}$ is increasing on the intervals $(-\infty, -1) \cup (1, \infty)$ and decreasing on the intervals $(-1, 0) \cup (0, 1)$. Using the sign diagram and the First Derivative Test, we have a relative maximum at $x = -1$ and a relative minimum at $x = 1$. Specifically, we have a relative maximum at $(-1, -2)$ and a relative minimum at $(1, 2)$. See **Figure 5.1.22**.

Figure 5.1.22

> **Try It Yourself**
>
> Some related Exercises are 37 and 41.

Now let's see how the First Derivative Test can be used in an application.

Example 7: Determining Maximum Revenue from Marginal Revenue

The owner of Everything Fit to Print has received a shipment of new computer tutorial software. The manufacturer of the software claims that their market research has determined that a retail price of $74 maximizes revenue. The manufacturer supplied the graph of the marginal revenue function shown in **Figure 5.1.23**. Determine what value of q, quantity, will maximize revenue and determine the maximum revenue.

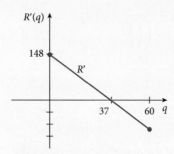

Figure 5.1.23

Perform the Mathematics

To find the relative maximum, we determine the critical values of the function and apply the First Derivative Test. Since the critical values of R occur where $R'(q) = 0$, we know from the graph of R' that $q = 37$ is a critical value. Now we determine the sign of R' to the left and to the right of this critical value.

- The graph of R' is above the q-axis on the interval $(0, 37)$. This means that R' is positive and thus R is increasing on this interval.

- The graph of R' is below the q-axis on the interval $(37, 60)$. This means that R' is negative and thus R is decreasing on this interval.

We use this information to construct the sign diagram in **Figure 5.1.24**.

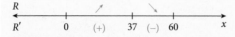

Figure 5.1.24

Since revenue is increasing on the interval $(0, 37)$ and decreasing on the interval $(37, 60)$, we conclude by using the First Derivative Test, that revenue is maximized at $q = 37$. The maximum revenue is then simply

$$p \cdot q = 74 \cdot 37 = \$2738$$

Summary

In this section we saw how the sign of the derivative tells us the behavior of a function. Specifically, on an open interval:

- If $f'(x) > 0$ for all x in (a, b), then f is increasing on (a, b).

- If $f'(x) < 0$ for all x in (a, b), then f is decreasing on (a, b).

- If $f'(x) = 0$ for all x in (a, b), then f is constant on (a, b).

We defined **critical values** as x-values in the domain of the function such that $f'(x) = 0$ or $f'(x)$ is undefined. To find relative extrema, determine critical values and apply the **First Derivative Test**. Use the following process:

1. Determine the derivative and use it to find critical values.

2. Make a sign diagram to determine intervals of increase and decrease.

3. Use the sign diagram and the First Derivative Test to locate relative extrema.

4. Use the calculator as an effective tool to verify your work.

Section 5.1 Exercises

Vocabulary Exercises

1. Whenever the derivative of a continuous function is negative, the function is _____.

2. Whenever the derivative of a continuous function is positive, the function is _____.

3. Whenever the derivative of a continuous function is zero, the function is _____.

4. An x-value in the domain of f for which $f'(x) = 0$ or $f'(x)$ is undefined is a _____ value.

5. If the derivative of a continuous function is negative to the left of $x = c$ and positive to the right of c, then $(c, f(c))$ is a relative _____.

6. If the derivative of a continuous function is positive to the left of $x = c$ and negative to the right of c, then $(c, f(c))$ is a relative _____.

Skill Exercises

In Exercises 7–12, write the intervals where the function f is increasing, where it is decreasing, and where it is constant.

7.
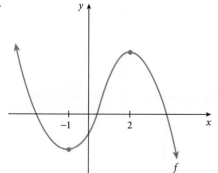

8.
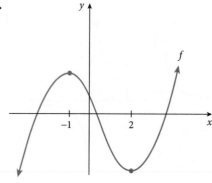

9.

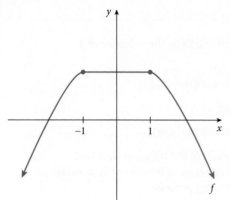

10.

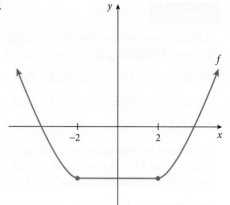

11.

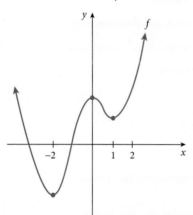

12.

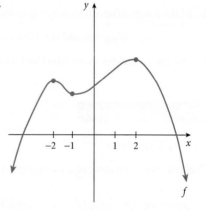

For Exercises 13–18, the graph of a function f is given. Complete the following.

(a) Determine intervals where the derivative is positive.

(b) Determine intervals where the derivative is negative.

(c) List any *x*-values where the derivative is undefined or where the derivative is equal to zero.

13.

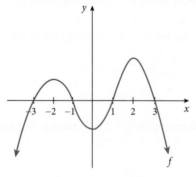

14.

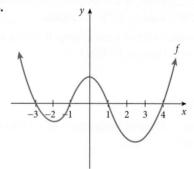

15.

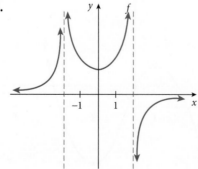

16.

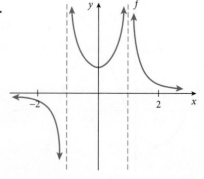

17.

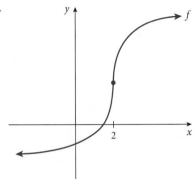

18.

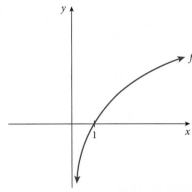

For Exercises 19–34, determine the critical values for each function.

19. $f(x) = 2x - 3$

20. $f(x) = -3x + 5$

21. $f(x) = x^2 - 5x + 6$

22. $f(x) = 2x^2 + 3x - 5$

23. $f(x) = \dfrac{1}{3}x^3 + x^2 - 15x + 3$

24. $f(x) = -x^3 - 3x^2 + 45x - 5$

25. $f(x) = \sqrt{2x + 1}$

26. $f(x) = \sqrt{3x - 7}$

27. $f(x) = e^x$

28. $f(x) = e^{3x}$

29. $f(x) = x + \ln x$

30. $f(x) = x - \ln x$

31. $f(x) = \sqrt[3]{x + 1}$

32. $f(x) = \sqrt[3]{x - 1}$

33. $f(x) = \dfrac{x - 2}{2x - 3}$

34. $f(x) = \dfrac{2x + 1}{5 - 3x}$

Exercises 35–44, use the results of Exercises 21–32. For each function complete the following.

(a) Determine the intervals where the function is increasing and decreasing.

(b) Write the *x*-value associated with any relative extrema.

35. The function in Exercise 21.

36. The function in Exercise 22.

37. The function in Exercise 23.

38. The function in Exercise 24.

39. The function in Exercise 25.

40. The function in Exercise 26.

41. The function in Exercise 29.

42. The function in Exercise 30.

43. The function in Exercise 31.

44. The function in Exercise 32.

In Exercises 45–50, a derivative f′ is given. Use the derivatives to determine the x-values where the function f has a relative maximum and/or a relative minimum.

45. $f'(x) = -2x + 5$

46. $f'(x) = -2x - 3$

47. $f'(x) = 3x(x + 1)$

48. $f'(x) = -2x(x - 4)$

49. $f'(x) = 2(x + 1)^2(x - 1)(x + 3)^3$

50. $f'(x) = 4(x - 1)(x + 1)^2(x + 2)^3$

For Exercises 51–54, sketch a graph of a continuous function that satisfies the given data.

51.

x	$f(x)$	$f'(x)$
0	3	−1
1	2	−1
2	1	Undefined
3	2	1
4	3	1
5	4	1

52.

x	$f(x)$	$f'(x)$
0	2	−2
1	1	−1
2	0.25	0
3	1.25	1
4	2.5	1.5
5	2	−0.5

53.

x	$f(x)$	$f'(x)$
0	0	Undefined
1	1	$\frac{1}{2}$
2	1.4	0.35
3	1.7	0.29
4	2	0.25
5	2.2	0.22

54.

x	$f(x)$	$f'(x)$
0	−7	12
1	0	3
2	1	0
3	2	3
4	9	12
5	28	27

Application Exercises

55. Macroeconomics—Maximum Revenue: The Network Standards Company determines that the revenue from their popular 56N cable modem can be modeled by

$$R(x) = 300x - x^2 \quad 0 \le x \le 300$$

where x represents the number of modems produced and sold, and $R(x)$ represents the revenue in dollars.

(a) Determine the intervals where R is increasing and decreasing.

(b) Determine the relative maximum and determine each coordinate.

56. Macroeconomics—Maximum Revenue: Linguini's Pizza Palace is starting an all-you-can-eat pizza buffet from 5:00 to 9:00 PM on Friday evenings. The district manager conducts a survey of local residents and determines that the price-demand function for the buffet can be modeled by

$$p(x) = -0.02x + 8.3 \quad x \ge 0$$

where x represents the quantity demanded and $p(x)$ represents the price in dollars.

(a) Use the model to determine the price if the demand is 250. Round your answer to the nearest cent.

(b) Determine the revenue function $R(x)$ for the problem situation.

(c) Determine the intervals where R is increasing and where R is decreasing.

(d) Determine the relative maximum and interpret each coordinate.

57. Macroeconomics—Maximum Revenue: Linguini's Pizza Palace is introducing a new line of specialty pizzas, such as chicken and garlic pizza. Taste tests of this type of pizza have produced a price-demand function that can be modeled by

$$p(x) = 15.22e^{-0.015x} \quad x \ge 0$$

where x represents the quantity demanded and $p(x)$ represents the price in dollars.

(a) Use the model to determine the price if the demand is 55. Round your answer to the nearest cent.

(b) Determine the revenue function $R(x)$ for the problem situation.

(c) Determine the intervals where R is increasing and where R is decreasing.

(d) Determine the relative maximum. Round the x- and y-coordinates to the nearest hundredth, and interpret each coordinate.

58. Medical Science—Drug Concentration: Suppose researchers are investigating the duration and concentration of the experimental drug Protaine and find that the concentration of the medication during the first 20 hours after administration can be modeled by

$$p(t) = \frac{230t}{t^2 + 6t + 9} \quad 0 \le t \le 20$$

where t represents the time since the drug was administered in hours and $p(t)$ represents the concentration as a percentage.

(a) Determine $p'(t)$ and determine the critical value(s).

(b) Determine the intervals where p is increasing and decreasing.

(c) Determine the relative maximum and interpret each coordinate.

59. **Medical Science—Drug Concentration:** Suppose researchers are investigating the duration and concentration of the experimental drug Duremia and find that the concentration of the medication during the first 20 hours after it had been administered can be modeled by

$$p(t) = \frac{200t}{2t^2 + 5} - 4 \qquad \frac{1}{4} \le t \le 20$$

where t represents the time since the drug had been administered in hours and $p(t)$ represents the concentration as a percentage.

(a) Determine $p'(t)$ and determine the critical value(s).

(b) Determine the intervals where p is increasing and decreasing.

(c) Determine the relative maximum and interpret each coordinate.

60. **Macroeconomics—Minimizing Average Cost:** A consulting firm hired for the Coffee Clutch Company determines that the cost to produce their Bottomless Cup coffee cups can be modeled by

$$C(x) = 50 + x + \frac{x^2}{40} \qquad 0 \le x \le 100$$

where x represents number of cups made daily and $C(x)$ represents the cost in dollars.

(a) Determine the average cost function AC and find the critical value(s) of AC.

(b) Determine the intervals where AC is increasing and where it is decreasing.

(c) Determine the relative minimum for AC and interpret each coordinate.

61. **Macroeconomics—Minimizing Average Cost:** The Virtual Buddy Company seeks to minimize the daily cost of producing their digital pet called Scamper. The finance department at the company has analyzed the data and determines the cost can be modeled by

$$C(x) = 150 + 3x + \frac{2x^2}{30} \qquad 0 \le x \le 200$$

where x represent the number of Scampers made daily and $C(x)$ represents the cost in dollars.

(a) Determine the average cost function AC and find the critical value(s) for AC.

(b) Determine the intervals where AC is increasing and where it is decreasing.

(c) Determine the relative minimum for AC and interpret each coordinate.

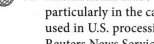 62. **International Trade—Chicken Exports:** Annual U.S. exports of chicken were affected, particularly in the case of Russia, because Russia cited concerns about a chlorine rinse used in U.S. processing plants to kill pathogens that can cause food poisoning. (*Source: Reuters News Service.*) The annual U.S. exports of chicken from 1990 to 2010 can be modeled by

$$f(x) = -1.34x^2 + 30x + 28.72 \qquad 0 \le x \le 20$$

where x represents the number of years since 1990 and $f(x)$ represents the amount of U.S. chicken exports, measured in millions of pounds. (*Source: U.S. Department of Agriculture.*)

(a) Determine the intervals where the function is increasing and decreasing.

(b) Determine the relative maximum and interpret each coordinate.

© Sandeep Subba/iStockphoto

 63. **Applied Economics—Hog Production:** Pig and hog production has been curtailed in the state of Minnesota in the past few years because of new anti–corporate farming laws that prohibit corporations from owning farmland. (*Source:* University of Minnesota Department

of Applied Economics.) The annual hog production in Minnesota from 2005 to 2010 can be modeled by

$$f(x) = -199.5x^2 + 2935.5x - 8969 \qquad 5 \leq x \leq 10$$

where x represents the number of years since 2000 and $f(x)$ represents the Minnesotan hog production in millions of pounds. (*Source: U.S. Department of Agriculture.*)

(a) Determine the intervals where the function is increasing and decreasing.

(b) Determine the relative maximum and interpret each coordinate.

64. **Linguistics—Learning English:** In a move that surprised some educators in 2008, Polish digital channel TVN Lingua used three seasons of Great Britain's TV reality show *Big Brother*, which aired from 2001 to 2003, to help viewers learn English as a "living language." (*Source: The London Daily Mail.*) The annual percentage of Polish students learning English from 2000 to 2010 can be modeled by

$$f(x) = -0.06x^3 + 0.36x^2 + 0.79x + 88.62 \qquad 0 \leq x \leq 10$$

where x represents the number of years since 2000 and $f(x)$ represents the percentage of Polish students learning English. (*Source: EuroStat.*)

(a) Determine $f'(x)$.

(b) Determine the x-value in the domain where $f'(x) = 0$.

(c) Based on the answer in part (b), determine the year that the percentage of Polish students learning English was at its maximum. What was the maximum percentage?

65. **Travel and Tourism—Greek Tourism:** Greece's post-Olympic tourism surge coming to an end, along with rising inflation, has had an impact on tourism to Greece. "Things look restrained," said George Drakopoulos, director-general of the Association of Greek Tourism Enterprises. (*Source: SperoNews.*) The annual number of tourists coming to Greece from 2000 to 2010 can be modeled by

$$f(x) = 21.39x^2 - 18.1x + 4309 \qquad 0 \leq x \leq 10$$

© AbleStock

where x represents the number of years since 2000 and $f(x)$ represents the number of tourists coming to Greece in thousands. (*Source: EuroStat.*)

(a) Determine the x-value in the domain where $f'(x) = 0$.

(b) Based on the answer in part (a), determine the year that the number of tourists coming to Greece was at its minimum. What was the minimum number of tourists?

66. **Business—Airline Revenues:** Airline passenger revenues have increased in recent years due in part to à la carte fees for passengers to add perks. "There certainly are perks that will now become available to passengers who normally would not qualify for them based on the amount of flying they do," says Jay Sorensen, a consultant and expert on airlines' so-called ancillary revenue that they draw from fees. "Network airlines have realized that they have an attractive asset called 'services' they provide to elite travelers, and they've learned they can sell these on an à la carte basis to any consumer." (*Source: USA Today.*) The annual passenger revenue from 1995 to 2010 can be modeled by

$$f(x) = 0.11x^3 - 3.7x^2 + 39.5x - 48.18 \qquad 5 \leq x \leq 20$$

where x represents the number of years since 1990 and $f(x)$ represents the airline revenues in billions of dollars. (*Source: U.S. Census Bureau.*)

(a) Determine $f'(x)$.

(b) Determine the x-values in the domain where $f'(x) = 0$.

(c) Based on the answer in part (b), determine the year that the airline revenue was at a relative maximum and the year that the airline revenue was at a relative minimum.

67. **Business—Computer Programming Revenue:** The introduction of a new generation of video game systems from companies such as Sony, Microsoft, and Nintendo has created an increased demand for software programmers. The annual revenue for computer programming services from 2000 to 2010 can be modeled by

$$f(x) = 1.36x^2 - 8.21x + 70.06 \qquad 0 \le x \le 10$$

where x represents the number of years since 2000 and $f(x)$ represents the revenue for computer programming services in billions of dollars. (*Source:* U.S. Census Bureau.)

(a) Determine $f'(x)$.

(b) Determine the x-values in the reasonable domain where $f'(x) = 0$.

(c) Based on the answer to part (b), determine the year that the revenue for computer programming services was at a relative minimum.

Concept and Writing Exercises

68. Let a and b represent two nonzero real numbers. Show, for $1 < a < b$ that $a + \frac{1}{a} < b + \frac{1}{b}$. Hint: Prove that the function $f(x) = x + \frac{1}{x}$ is increasing for $x \ge 1$.

69. Show that if $x > 0$, then $x + \frac{1}{x} \ge 2$. Use the hint from Exercise 68.

70. Consider the cubic function $f(x) = x^3 + bx^2 + cx + 2$. Find the value for b and c so that f has a relative maximum at $x = -3$ and f has a relative minimum at $x = 1$.

71. Consider the cubic function $f(x) = ax^3 + bx^2 + cx + d$. Determine coefficients of f so that the function has a relative minimum at $(1, 0)$ and a relative maximum at $(-2, 3)$.

72. Let f and g represent differentiable functions defined on the open interval (a, b). Show that if the functions f and g are positive and increasing on the interval (a, b), then the sum of the functions $(f + g)$ is also increasing on (a, b).

73. Let f and g represent differentiable functions defined on the open interval (a, b). Show that if the functions f and g are positive and increasing on the interval (a, b), then the product of the functions (fg) is also increasing on (a, b).

74. Sketch a graph of the function f that satisfies the following conditions.

   (i) $f(1) = f(4) = 2$

   (ii) $f'(1) = f'(4) = 0$

   (iii) $f'(x) > 0$ on $(-\infty, 1)$ and on $(4, \infty)$; $f'(x) < 0$ on $(1, 4)$.

75. Consider a quadratic function of the form $f(x) = ax^2 + bx + c$ where $a \ne 0$ and a, b, and c represent real numbers. Use the derivative f' to determine the x-coordinate where a horizontal tangent line occurs.

For Exercises 76–78, recall that profit equals revenue minus cost. That is, $P(x) = R(x) - C(x)$.

76. At production level x show that $P(x)$ is maximized when $R'(x) = C'(x)$: that is, when marginal revenue equals marginal cost.

77. For all x on a production interval (a, b), show that profit is increasing when $R'(x) > C'(x)$: that is, profit is increasing when marginal revenue is greater than marginal cost.

78. For all x on a production interval (a, b), show that profit is decreasing when $R'(x) < C'(x)$: that is, profit is decreasing when marginal revenue is less than marginal cost.

 **Section Project**

In the *Time* magazine article "Finding the Man Who Started the Global Recession," Douglas McIntyre writes, "Someone who took out a subprime loan in 2003 is the 'patient zero' who began the great recession. He was supposed to pay his mortgage for ten years, then sell his home. When his mortgage reset in 2006, he defaulted." (*Source:* www.time.com.) The annual number of built-for-sale housing starts in the United States from 2001 to 2010 is shown in the table (*Source:* U.S. Census Bureau.)

Years since 2000	Housing starts (in thousands)
1	919
2	999
3	1120
4	1240
5	1358
6	1121
7	760
8	589
9	553
10	549

Use the data to answer the following.

(a) From inspecting the table, during what year was the number of housing starts at its maximum?

(b) From 2006 to 2007, what was the decrease in the housing starts? In other words, compute the average rate of change in housing starts from 2006 to 2007.

 (c) Use the regression capabilities of your graphing calculator to determine a quadratic model for the data in the form $f(x) = ax^2 + bx + c$, $1 \le x \le 10$, where the value of a, b, and c are rounded to two decimal places.

(d) Differentiate f then determine the x-value for the maximum of the function. Is this the same value as you got in part (a)? How do you account for the difference?

(e) Evaluate $f'(6)$ and compare your answer to part (b). How do you account for this difference?

SECTION OBJECTIVES

1. Determine absolute extrema.

2. Apply absolute extrema to egg consumption.

3. Apply absolute extrema to optimize area.

4. Apply absolute extrema to optimize medication concentration.

5. Apply absolute extrema to maximize profit.

5.2 Optimizing Functions on a Closed Interval

In Section 5.1 we used the first derivative and the First Derivative Test to determine the location of the *relative maximum* and *relative minimum* (plural is *relative extrema*) for a function. In this section we go one step further and determine the location of the **absolute maximum** and **absolute minimum** for a function. Intuitively, these points are on the "highest hill" and in the "lowest valley." Once we know where these **absolute extrema** are located, we will start to apply our new knowledge to a wide array of applications. Thus, we begin our voyage into what is known as **optimization**. We begin by taking a look at the tools used to find the absolute maximum and absolute minimum values on a closed interval.

Absolute Extrema on a Closed Interval

In Section 5.1 we stated that the word *relative* in *relative maximum* and *relative minimum* meant *relative to points nearby*. An **absolute maximum** is the largest value that a function attains on its domain or, stated another way, the highest point on the graph. Similarly, an **absolute minimum** is the smallest value that the function attains on its domain or, stated another way, the lowest point on the graph. For any given function, the **absolute extrema** may or may not exist. The graphs in **Figure 5.2.1** show the possibilities for some functions that are defined for all values of x.

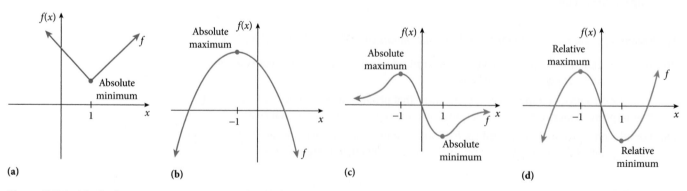

Figure 5.2.1 **(a)** Absolute minimum at $x = 1$. No absolute maximum. **(b)** Absolute maximum at $x = -1$. No absolute minimum. **(c)** Absolute maximum at $x = -1$. Absolute minimum at $x = 1$. **(d)** Relative maximum at $x = -1$. Relative minimum at $x = 1$. No absolute extrema.

These graphs suggest that absolute extrema may occur at critical values, that is, where $f'(x) = 0$ or where $f'(x)$ is undefined, which is quite similar to what we did in Section 5.1. Notice that **Figure 5.2.1**(d) indicates that a critical value does not necessarily give an absolute extrema.

Now, if we consider any continuous function on a *closed interval* $[a, b]$, the function will have an absolute maximum and an absolute minimum on the interval. In fact, these absolute extrema will be located either (1) at critical values that are within the interval (a, b), or (2) at the endpoints of the interval $[a, b]$. **Figures 5.2.2** through **5.2.4** illustrate some possibilities.

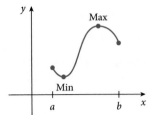

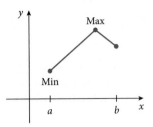

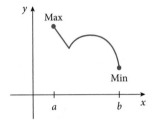

Figure 5.2.2 Absolute extrema occur at critical points.

Figure 5.2.3 Absolute maximum at critical point. Absolute minimum at endpoint.

Figure 5.2.4 Absolute extrema occur at endpoints.

The prior discussion and figures lead us to the **Extreme Value Theorem**.

The Extreme Value Theorem tells us when an absolute maximum and absolute minimum are guaranteed to exist. Since we know that absolute extrema occur at critical values within the open interval or at the endpoints of the interval, we supply the following procedure to **locate these absolute extrema**. When we determine the absolute extrema for a function, we are said to be **optimizing** the function.

> **Extreme Value Theorem**
>
> If a function is continuous on a closed interval $[a, b]$, then the function must have both an absolute maximum and an absolute minimum on $[a, b]$.

> **Four-step Process for Locating the Absolute Extrema of *f* on [*a*, *b*]**
>
> 1. Verify that *f* is continuous on $[a, b]$.
> 2. Determine the critical values of *f* in (a, b).
> 3. Evaluate $f(x)$ at the critical values and at the endpoints of the interval.
> 4. The largest value that you obtain from step 3 is the absolute maximum of *f* on $[a, b]$, and the smallest value that you obtain from step 3 is the absolute minimum of *f* on $[a, b]$.

Example 1: Determining Absolute Extrema

Determine the absolute extrema for $f(x) = 2x^3 + 3x^2 - 12x + 1$ on $[-3, 2]$.

Perform the Mathematics

Step 1: Since *f* is a polynomial function, we know that it is continuous on the closed interval $[-3, 2]$. By the Extreme Value Theorem, it must have an absolute maximum and an absolute minimum.

Step 2: To determine the critical values, we first need the derivative

$$f'(x) = 6x^2 + 6x - 12$$

Since $f'(x)$ is defined for all *x*, the only critical values are where $f'(x) = 0$. So we need to solve

$$6x^2 + 6x - 12 = 0$$
$$6(x^2 + x - 2) = 0$$
$$6(x + 2)(x - 1) = 0$$
$$x = -2 \text{ or } x = 1$$

So the critical values are $x = -2$ and $x = 1$.

OBJECTIVE 1

Determine absolute extrema.

Step 3: Here, we evaluate $f(x)$ at the critical values and the endpoints.

Critical values:	$x = -2$	$f(-2) = 2(-2)^3 + 3(-2)^2 - 12(-2) + 1 = 21$
	$x = 1$	$f(1) = 2(1)^3 + 3(1)^2 - 12(1) + 1 = -6$
Endpoints:	$x = -3$	$f(-3) = 2(-3)^3 + 3(-3)^2 - 12(-3) + 1 = 10$
	$x = 2$	$f(2) = 2(2)^3 + 3(2)^2 - 12(2) + 1 = 5$

Step 4: Since 21 is the largest value from step 3, we conclude that the absolute maximum is 21, and it occurs at $x = -2$. The smallest value from step 3 is -6, so -6 is the absolute minimum and it occurs at $x = 1$. See **Figure 5.2.5**.

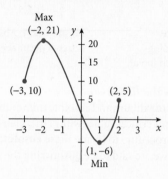

Figure 5.2.5

Example 2: Determining Absolute Extrema

Determine the absolute extrema for $f(x) = 3x^4 + 4x^3 - 36x^2 + 1$ on $[-1, 1]$.

Perform the Mathematics

Step 1: Since f is a polynomial function, we know that it is continuous on the closed interval $[-1, 1]$. By the Extreme Value Theorem, it must have an absolute maximum and an absolute minimum.

Step 2: To determine the critical values, we first need the derivative

$$f'(x) = 12x^3 + 12x^2 - 72x$$

Since $f'(x)$ is defined for all x, the only critical values are where $f'(x) = 0$. So we need to solve

$$12x^3 + 12x^2 - 72x = 0$$
$$12x(x^2 + x - 6) = 0$$
$$12x(x + 3)(x - 2) = 0$$
$$x = 0, x = -3, \text{ or } x = 2$$

So the critical values are $x = 0$, $x = -3$, and $x = 2$.

Step 3: Since $x = 0$ is the only critical value in the interval $[-1, 1]$, we evaluate $f(x)$ at this critical value and at the endpoints.

Critical value:	$x = 0$	$f(0) = 3(0)^4 + 4(0)^3 - 36(0)^2 + 1 = 1$
Endpoints:	$x = -1$	$f(-1) = 3(-1)^4 + 4(-1)^3 - 36(-1)^2 + 1 = -36$
	$x = 1$	$f(1) = 3(1)^4 + 4(1)^3 - 36(1)^2 + 1 = -28$

Step 4: The absolute maximum is 1, and it occurs at $x = 0$. The absolute minimum is -36, and it occurs at $x = -1$. See **Figure 5.2.6**.

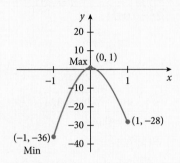

Figure 5.2.6 ■

▶ *Try It Yourself*

Some related Exercises are 9 and 11.

Applications

We now turn our attention to problems where we can maximize revenue, profit, area, or volume, as well as minimize costs, average costs, and pollution. This is a small sample of the types of problems that we may encounter. The examples that follow supply a general strategy for solving any optimization problem on a closed interval.

 Example 3: Optimizing Egg Consumption

A staple of breakfast for many generations in the United States has been the egg. The annual per capita consumption of eggs in the United States can be modeled by

$$f(x) = -0.014x^3 + 0.769x^2 - 11.34x + 282.03 \qquad 1 \le x \le 29$$

where x is the number of years since 1979 and $f(x)$ is the annual per capita consumption in the number of eggs. (*Source:* U.S. Department of Agriculture.) Determine the absolute extrema on $[1, 29]$ and interpret. Give the answer to the nearest year.

Understand the Situation

Even though our function models a real-world situation, we still employ the four-step process.

Perform the Mathematics

Step 1: Since we have a polynomial function, it is continuous on the interval $[1, 29]$. Thus, the Extreme Value Theorem applies, and we can expect to find an absolute maximum and an absolute minimum on the interval.

Step 2: To determine the critical values, we need the derivative, which is

$$f'(x) = -0.042x^2 + 1.538x - 11.34$$

The derivative is defined for all x, so the only critical values are where $f'(x) = 0$. So we need to solve

$$-0.042x^2 + 1.538x - 11.34 = 0$$

Using the quadratic formula gives us

$$x = \frac{-b \pm \sqrt{b^2 - 4ac}}{2a} = \frac{-1.538 \pm \sqrt{(1.538)^2 - 4(-0.042)(-11.34)}}{2(-0.042)}$$

$$= \frac{-1.538 \pm \sqrt{0.460324}}{-0.084}$$

So the two solutions are

$$x = \frac{-1.538 + \sqrt{0.460324}}{-0.084} \quad \text{or} \quad x = \frac{-1.538 - \sqrt{0.460324}}{-0.084}$$

$$x \approx 10.232 \qquad\qquad\qquad x \approx 26.387$$

Step 3: Both critical values are in the interval $[1, 29]$, so we now evaluate the function at the critical values and at the endpoints of the interval. (We will round to the nearest thousandth.)

Critical values: $x = 10.232$

$$f(10.232) = -0.014(10.232)^3 + 0.769(10.232)^2 - 11.34(10.232) + 282.03$$

$$\approx 231.511$$

$$x = 26.387$$

$$f(26.387) = -0.014(26.387)^3 + 0.769(26.387)^2 - 11.34(26.387) + 282.03$$

$$\approx 261.02$$

Endpoints: $x = 1$

$$f(1) = -0.014(1)^3 + 0.769(1)^2 - 11.34(1) + 282.03 \approx 271.445$$

$$x = 29$$

$$f(29) = -0.014(29)^3 + 0.769(29)^2 - 11.34(29) + 282.03 \approx 258.453$$

Step 4: We have 231.511 as the absolute minimum and it occurs at $x = 10.232$. Rounding to the nearest year gives $x = 10$. The absolute maximum is 271.445 and it occurs at $x = 1$.

Interpret the Result

From 1980 to 2008, the absolute maximum of annual egg consumption per capita in the United States was about 271 eggs per person, and it occurred in 1980. The absolute minimum of annual egg consumption per capita in the United States was about 232 eggs per person and it occurred in 1989.

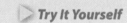 **Try It Yourself**

Some related Exercises are 47 and 51.   ■

In many optimization problems, we must first determine the function that we are trying to optimize. Example 4 illustrates how we handle these types of problems.

Example 4: Optimizing Area

OBJECTIVE 3

Apply absolute extrema to optimize area.

One of the authors needs to build a rectangular enclosure for his beloved beagle. He has 200 feet of fence. Determine the dimensions of the rectangle that will make the area of the enclosure as large as possible.

Understand the Situation

Two rectangular regions are shown in **Figure 5.2.7** that each use 200 feet of fencing. Our goal is to find the one rectangle that has the maximum area.

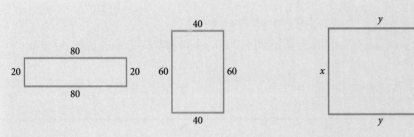

Figure 5.2.7   **Figure 5.2.8**

Instead of continuing in a "guess-and-check" fashion, let's consider a generic rectangle with a width of x and a length of y, as shown in **Figure 5.2.8**. The area, A, of this generic rectangle is given by

$$A = x \cdot y$$

We want to maximize the area, A, but it is written as a function of two variables. To use the tools we currently have, we need to write $A = x \cdot y$ as a function of *one* variable. Since we have 200 feet of fence, we conclude that the perimeter, P, of the rectangular enclosure is 200 feet. That is,

$$P = x + y + x + y = 200$$
$$2x + 2y = 200$$
$$2y = 200 - 2x$$
$$y = 100 - x$$

We can substitute this into our area equation and get $A = x \cdot (100 - x)$ or, more precisely

$$A(x) = 100x - x^2$$

Since this enclosure is for an animal that must be able to move around in it and turn around in it, a reasonable domain is $10 \leq x \leq 90$. So we now have the situation of determining the absolute maximum of

$$A(x) = 100x - x^2 \text{ on } [10, 90]$$

We now employ the four-step process just as we did in Examples 1, 2, and 3.

Perform the Mathematics

Step 1: Since A is a polynomial function, it is continuous on the stated interval, which means that the Extreme Value Theorem applies.

Step 2: Next, we determine the critical values. The derivative of $A(x)$ is

$$A'(x) = 100 - 2x$$

Since $A'(x)$ is defined for all x, the only critical value is where $A'(x) = 0$. Solving this equation yields

$$100 - 2x = 0$$
$$x = 50$$

Step 3: Evaluating $A(x)$ at the critical value and endpoints results in

Critical value: $x = 50$ $A(50) = 100(50) - (50)^2 = 2500$

Endpoints: $x = 10$ $A(10) = 100(10) - (10)^2 = 900$

             $x = 90$ $A(90) = 100(90) - (90)^2 = 900$

Step 4: The largest number from step 3 is 2500 and it occurs at $x = 50$. So the dimensions of the rectangular enclosure that maximize the area for the beagle are

$$x = 50$$
$$y = 100 - x = 100 - 50 = 50$$

Interpret the Result

This means that a rectangle with a length of 50 feet and width of 50 feet maximizes the area.

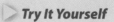 **Try It Yourself**

Some related Exercises are 31 and 33.   ∎

Notice that in Example 4 we needed to do some preliminary work *before* we could use our four-step process. Example 4 suggests the following **strategy for solving applied optimization problems.**

Strategy for Solving Applied Optimization Problems on a Closed Interval

1. **Understand the Situation.** This includes:
   - Read the question carefully and, if possible, sketch a picture that represents the problem.
   - Select variables to represent the quantity to be maximized or minimized and all other unknowns.
   - Write an equation for the quantity to be maximized or minimized. If necessary, eliminate extra variables so that the quantity to be optimized is a function of one variable.
2. **Perform the Mathematics.** Here, this means apply the four-step process to determine the absolute extrema.
3. **Interpret the Result.** This includes writing a sentence that answers the question posed in the problem.

In Example 5 we return to a problem encountered in the Section 5.1 Exercises.

OBJECTIVE 4

Apply absolute extrema to optimize medication concentration.

Example 5: Determining a Maximum Medication Concentration

Researchers have determined through experimentation that the percent concentration of a certain medication t hours after it has administered can be approximated by

$$p(t) = \frac{230t}{t^2 + 6t + 9} \qquad 1 \le t \le 20$$

where $p(t)$ is the percent concentration. How many hours after the administration of this medication is the concentration at a maximum? What is the maximum concentration? See **Figure 5.2.9.**

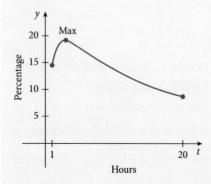

Figure 5.2.9

Perform the Mathematics

Since we are given the function to be maximized, we can immediately use the four-step process.

Step 1: The function p is continuous for all values of t except $t = -3$. But, since $t = -3$ is not in the stated interval, p is continuous on the stated interval. Thus, the Extreme Value Theorem guarantees an absolute maximum.

Step 2: By the Quotient Rule, we compute the derivative to be

$$p'(t) = \frac{230(t^2 + 6t + 9) - 230t(2t + 6)}{(t^2 + 6t + 9)^2} = \frac{-230t^2 + 2070}{(t^2 + 6t + 9)^2} = \frac{-230(t^2 - 9)}{(t^2 + 6t + 9)^2}$$

Notice that $p'(t)$ is undefined at $t = -3$, but, since this value is not in the stated interval, we do not consider it. We now solve $p'(t) = 0$, which gives

$$\frac{-230(t^2 - 9)}{(t^2 + 6t + 9)^2} = 0$$

Since the only time that a fraction can equal zero is when the numerator equals zero, we have

$$-230(t^2 - 9) = 0$$
$$t^2 - 9 = 0$$
$$t^2 = 9$$
$$t = \pm 3$$

Disregarding $t = -3$, we have a critical value at $t = 3$.

Step 3: Evaluating $p(t)$ at the critical value $t = 3$ and at the endpoints and rounding to the nearest thousandth gives

$$\text{Critical value: } \quad t = 3 \quad p(3) = \frac{230(3)}{(3)^2 + 6(3) + 9} \approx 19.17$$

$$\text{Endpoints: } \quad t = 1 \quad p(1) = \frac{230(1)}{(1)^2 + 6(1) + 9} \approx 14.38$$

$$t = 20 \quad p(20) = \frac{230(20)}{(20)^2 + 6(20) + 9} \approx 8.70$$

Step 4: The largest result from step 3 is 19.17 and it occurs at $t = 3$. This means that 3 hours after administration of this medication, the percent concentration is maximized at about 19.17%. ∎

▶ *Try It Yourself*

Some related Exercises are 37 and 41.

In our final example, we return to the functions of business, as well as determining an equation for a line.

Example 6: Maximizing Profit

Only Beef, a local sandwich store, estimates that it can sell 500 roast beef sandwiches per week if it sets the price at $2.50, but its weekly sales will rise by 50 sandwiches for each $0.05 decrease in price. The company has fixed costs each week of $525.00, and variable costs for making a roast beef sandwich are $0.55. Let x represent the number of roast beef sandwiches made and sold each week. Determine the value of x that maximizes weekly profit, assuming that $500 \leq x \leq 1500$.

Understand the Situation

We want to maximize profit, $P(x)$. Recall that $P(x) = R(x) - C(x)$, $R(x) = $ price $\cdot$ quantity, and $C(x) = $ fixed costs $+$ variable costs. Since the fixed costs are $525 and x is the number of sandwiches made and sold each week, the variable costs are $0.55x$. Hence,

$$C(x) = \text{fixed costs } + \text{ variable costs}$$
$$C(x) = 525 + 0.55x$$

Table 5.2.1

x, number of sandwiches	$p(x)$, price of sandwich
500	2.50
550	2.45
600	2.40
650	2.35

To determine $R(x)$, we need price, $p(x)$. We know that when $x = 500$ price is $2.50 and for each $0.05 *decrease* in price, weekly sales *increase* by 50. To aid us in analyzing this situation, we make **Table 5.2.1** with this information.

Notice that, for each change in x of 50, $p(x)$ changes by 0.05. As x increases at a constant rate, $p(x)$ seems to be decreasing at a constant rate, which means that we should be able to model $p(x)$ with a **linear function**. We can write

$$\frac{\text{Change in } p(x)}{\text{Change in } x} = \frac{-0.05}{50} = -0.001$$

This tells us that the slope of the line is $m = -0.001$. Using the point-slope form of a line with the point $(500, 2.50)$ gives

$$y - y_1 = m(x - x_1)$$
$$y - 2.50 = -0.001(x - 500)$$
$$y = -0.001(x - 500) + 2.50$$
$$y = -0.001x + 3$$

So the price function is given by

$$p(x) = -0.001x + 3$$

The revenue function is then given by

$$R(x) = \text{price} \cdot \text{quantity}$$
$$R(x) = p(x) \cdot x$$
$$R(x) = (-0.001x + 3) \cdot x = -0.001x^2 + 3x$$

Now that we have $R(x)$ and $C(x)$, profit, $P(x)$, is given by

$$P(x) = R(x) - C(x)$$
$$P(x) = -0.001x^2 + 3x - (525 + 0.55x) = -0.001x^2 + 2.45x - 525$$

We have determined the function that is to be optimized. In other words, our task is to determine the absolute maximum for

$$P(x) = -0.001x^2 + 2.45x - 525 \text{ on } [500, 1500]$$

Perform the Mathematics

Step 1: Since P is a polynomial function, it is continuous on the stated interval and the Extreme Value Theorem guarantees an absolute maximum.

Step 2: The derivative is

$$P'(x) = -0.002x + 2.45$$

Since $P'(x)$ is defined for all x, the only critical value is where $P'(x) = 0$. Solving this equation gives

$$-0.002x + 2.45 = 0$$
$$x = 1225$$

Step 3: Evaluating $P(x)$ at the critical value $x = 1225$ and the endpoints results in

Critical value: $\quad x = 1225 \quad P(1225) = -0.001(1225)^2 + 2.45(1225) - 525 = 975.625$

Endpoints: $\quad x = 500 \quad\;\; P(500) = -0.001(500)^2 + 2.45(500) - 525 = 450$

$\qquad\qquad\quad x = 1500 \quad P(1500) = -0.001(1500)^2 + 2.45(1500) - 525 = 900$

Step 4: The largest value from step 3 is 975.625 and it occurs at $x = 1225$.

Interpret the Result

To maximize weekly profit, Only Beef should sell 1225 roast beef sandwiches each week.

 **Try It Yourself**

Some related Exercises are 43 and 45.

In Example 6, the price function, p, was determined from the information given in the problem. We recommend that you check its validity. One way of doing this is to substitute values into the price function that correspond with the table. For example, we could substitute $x = 500$ or $x = 600$ into the price function to verify that we get a price of $2.50 and $2.40, respectively. This type of checking is an invaluable tool in helping to secure the correct result.

Summary

The main concept in this section was **optimizing**, that is, determining absolute extrema for a function on a closed interval. The **Extreme Value Theorem** stated that if f is a continuous function on a closed interval, then f is guaranteed to have an absolute maximum and an absolute minimum. These absolute extrema occur at either the endpoints of the interval or at a critical value that is in the interior of the interval. We then presented a **four-step process** to locate these absolute extrema and the **Strategy for Solving Applied Optimization Problems on a Closed Interval**.

Section 5.2 Exercises

Vocabulary Exercises

1. The largest value that a function attains on its domain is the function's absolute _____.

2. The smallest value that a function attains on its domain is the function's absolute _____.

3. We usually find absolute extrema on functions defined on a _____ interval.

4. When we determine the absolute extrema for a function, the function is being

_____.

5. The _____ Value Theorem guarantees the existence of absolute extrema on a closed interval.

6. When we use the Four-Step Process for locating absolute extrema, we must first verify that the function is _____ on a closed interval.

Skill Exercises

In Exercises 7–22, determine the absolute extrema of each function on the indicated interval.

7. $f(x) = x^2 - 2x - 7;\ [-2, 1]$

8. $f(x) = 2x^2 + 3x - 1;\ [-1, 2]$

9. $f(x) = x^3 - 2x^2 - 5x + 6;\ [-2, 2]$

10. $f(x) = x^3 + 4x^2 + x - 6;\ [-2, 0]$

11. $f(x) = x^3 - 3x^2;\ [-1, 3]$

12. $f(x) = x^3 - 12x;\ [0, 4]$

13. $f(x) = x^4 - x^3 + 5;\ [-2, 2]$

14. $f(x) = 2x^3 - 6x^2 + 4;\ [-1, 4]$

15. $f(x) = (2x^2 - 1)^4;\ [0, 2]$

16. $f(x) = (3x + 1)^3;\ [-2, 1]$

17. $f(x) = \sqrt[3]{x};\ [-1, 1]$

18. $f(x) = \sqrt[3]{x^2};\ [-1, 8]$

19. $f(x) = \dfrac{1}{x - 2};\ [0, 1]$

20. $f(x) = \dfrac{x}{x - 2};\ [3, 5]$

21. $f(x) = \dfrac{1}{x^2 + 1};\ [1, 4]$

22. $f(x) = \dfrac{1}{x^2 + 1};\ [-1, 1]$

In Exercises 23–26, use your graphing calculator to graph the derivative of the function, and then use the Zero *command to determine the rational critical values of f. Use this information to determine the absolute extrema of each function on the indicated interval.*

23. $f(x) = x^4 - 15x^2 - 10x + 24;\ [-3, 3]$

24. $f(x) = x^4 + 2x^3 - 13x^2 - 14x + 24;\ [0, 4]$

25. $f(x) = \dfrac{1}{2}x^4 - \dfrac{7}{3}x^3 + x^2 + 3x - 1;\ [-2, 4]$

26. $f(x) = \dfrac{1}{2}x^4 - \dfrac{7}{4}x^3 + \dfrac{3}{2}x + 2;\ [-1, 3]$

Application Exercises

27. **Business Management—Maximizing Profit:** By using past records, the owner of the Sleep Cheap Motel has determined a profit function based on room rates. The CEO of the franchise has dictated that the room rate must be no less than $40 and not more than $60 per night. The profit can be modeled by

$$P(x) = -x^2 + 92x - 180 \qquad 40 \le x \le 60$$

where x represents the room rate and $P(x)$ represents the daily profit.

(a) What is the room rate that maximizes the daily profit?

(b) What is the maximum daily profit?

28. **Business Management—Maximizing Revenue:** The Wax and Wick Company sells jumbo-sized holiday candles with 10 in each box. The price demand function for the candles is given by

$$p(x) = 102 - 3x$$

where x represents the number of boxes, and $p(x)$ represents the price in dollars per box.

(a) Determine the revenue function R.

(b) How many boxes must be produced and sold to maximize revenue?

(c) What is the maximum revenue?

29. **Bacteriology—Pool Maintenance:** The Cement Pond is the local swimming facility for residents of the Point King gated community. Periodically the pond is treated for *rotavirus,* which is a common cause of gastroenteritis. Suppose the pool maintenance crew finds that the concentration of rotavirus during a two-week period can be modeled by

$$f(x) = 15x^2 - 210x + 750 \qquad 0 \le x \le 14$$

where x represents the number of days since the pool was treated and $f(x)$ represents the concentration of rotavirus per cubic centimeter.

   (a) How many days after the treatment will the rotavirus concentration be minimized?

   (b) What is the minimum concentration?

30. **Bacteriology—Pool Maintenance:** The Camp-n-Swim campgrounds has a small pool for recreational swimming. Periodically the pool is treated for *rotavirus,* which is a common cause of gastroenteritis. Suppose the pool maintenance crew finds that the concentration of rotavirus during a two-week period can be modeled by

$$f(x) = 3x^2 - 42x + 150 \qquad 0 \le x \le 14$$

where x represents the number of days since the pool was treated and $f(x)$ represents the concentration of rotavirus per cubic centimeter.

   (a) How many days after the treatment will the rotavirus concentration be minimized?

   (b) What is the minimum concentration?

31. **Construction—Dog Enclosure:** In Example 4, we learned that one of the authors needs to build a rectangular enclosure for his beloved beagle. His wife suggested that if the enclosure was built so that one side was along the outside wall of the house, the 200 feet of fencing would make an even bigger enclosure. Assume that the author has 200 feet of fencing, and the side along the wall needs no fencing.

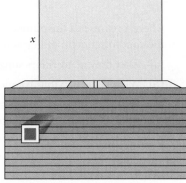

   (a) Write a function $A(x)$ for the area of the enclosure as a function of x, the sides of the enclosure perpendicular to the house. See the diagram.

   (b) For what value of x is the function A maximized?

   (b) What is the maximum area?

32. **Construction—Playground:** Suppose the Recreation Department in Crown Point has been authorized to construct a rectangular playground whose area is 10,000 square feet. Because of city ordinance, the playground must be enclosed on all four sides with fencing and be at least 50 feet wide. Let x represent the width of the playground and let y represent the length.

   (a) Write a function P for the perimeter of the playground, where x represents the playground's width for $50 \le x \le 200$.

   (b) For what value of x is the function P minimized?

   (c) What is the minimum perimeter?

33. **Horticulture—Gardening:** Suppose that one of the authors recently purchased a new home, and he has an extra acre of land to grow a garden. He wants the garden to be rectangular and to have an area of 1440 square feet. The design of the garden is to incorporate an eight-foot walkway on the north and south sides and a five-foot walkway on the east and west sides.

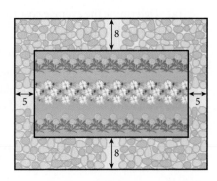

   (a) Write a function A for the total area of the garden and the walkways, where x represents the length of the garden and y represents the width. See the diagram. Assume that $10 \le x \le 100$.

(b) For what value of x is the function A maximized? What is the corresponding y value?

(c) What is the maximum area?

34. **Marketing Science—Publishing Costs:** The marketing research department of *Shank*, the quarterly magazine for beginning golfers, has determine that the price-demand function for the magazine can be modeled by

$$p(x) = 2.75 - 0.01x \qquad 0 \le x \le 275$$

where x represents the number of magazines printed and sold in a quarter and $p(x)$ represents the price of the magazine in hundreds of dollars. Further analysis of the data shows that the cost of printing, distributing, and advertising the magazine can be modeled by

$$C(x) = 0.003x^2 + 0.5x + 5 \qquad x \ge 0$$

where x represents the number of magazine produced and $C(x)$ represents the cost in hundreds of dollars.

(a) Determine the level of sales that maximizes the profit.

(b) Determine the price p that the magazine should sell at in order to maximize profit.

(c) What is the maximum profit?

35. **Biochemistry—Drug Concentration:** Suppose that scientists have determined that the concentration of the experimental drug Malpeprin in a patient's bloodstream can be modeled by

$$f(x) = \frac{2.5x}{2x^2 + 7x + 4} \qquad 0 \le x \le 8$$

where x represents the number of hours since the drug was administered and $f(x)$ represents the concentration in milligrams per cubic centimeter.

(a) After how many hours since administering the drug is its concentration at a maximum?

(b) What is the maximum concentration?

36. **Biochemistry—Drug Concentration:** Suppose that scientists have determined that the concentration of the experimental drug Conuprien in a patient's bloodstream can be modeled by

$$f(x) = \frac{5.3x}{x^2 + 4x + 5} \qquad 0 \le x \le 8$$

where x represents the number of hours since the drug was administered and $f(x)$ represents the concentration in milligrams per cubic centimeter.

(a) After how many hours since administration of the drug is its concentration at a maximum?

(b) What is the maximum concentration?

37. **Fiscal Research—Average Cost:** Financial planners at the Double D Corporation have determined that the cost to produce its Roxie fleece boot and backpack combo can be modeled by

$$C(x) = 0.03x^3 - 2.4x^2 + 73.16x + 102.27 \qquad 0 \le x \le 55$$

where x represent the number of combos produced each day and $C(x)$ represents the daily cost in dollars.

(a) Determine the average cost function AC.

(b) Determine the daily production level x that minimizes the average cost.

(c) Determine the average cost and total cost at the production level found in part (b).

38. **Fiscal Research—Maximizing Revenue:** For the Roxie fleece boot and backpack combo in Exercise 37, suppose that the Double D Corporation gathers data for the price of a combo and the number of combos demanded each day and finds that the price-demand function for the product is given by

$$p(x) = 0.14x^2 - 15.52x + 561.21 \qquad 5 \le x \le 55$$

where x represents the number of combos demanded each day, and $p(x)$ represents the price that consumers are willing to pay for the combos in dollars.

(a) Determine the revenue function R for the fleece boot and backpack combos.

(b) Determine the maximum revenue.

(c) Determine the demand that maximizes revenue.

(d) Determine the price of the combos that maximizes revenue.

39. **Fiscal Research—Maximizing Profit** (*continuation of Exercises 37 and 38*): Using the cost function in Exercise 37 and the revenue function from Exercise 38, complete the following.

(a) Write the profit function P.

(b) Determine the maximum profit for the combos.

(c) Find the demand that maximizes profit.

(d) Find the price of the combos that maximizes profit.

40. **Hematology—Hematocrit Levels:** Hematocrit is a medical term that represents the percentage of red blood cells in a blood sample. In general, men with a hematocrit less than 41% and women with a hematocrit less than 36% are considered anemic. (*Source:* National Anemia Action Council.) Blood viscosity is a measure of the resistance of blood to flow. As hematocrit increases, so does the blood viscosity. If we consider water to have a relative viscosity of 1, then blood has a viscosity of about 3 or 4. In one model, the relationship between hematocrit and viscosity is given by

$$f(x) = 0.0015x^2 - 0.019x + 1.563 \qquad 10 \le x \le 80$$

where x represents the hematocrit and $f(x)$ represents the relative viscosity. Determine the absolute minimum and absolute maximum of the function and interpret each.

41. **Business Management—Maximizing Profit:** The Boogie-All-Night Company manufactures black lights for night clubs. Suppose the company's business software reveals that the monthly fixed costs are $100 and the variable costs are $12 per light. The software also projects that 20 lights can be sold each month if the price for a light is $20, and two more lights can be sold for each decrease of $1 in the price. Let x represents the number of lights produced and sold each month.

(a) Determine the monthly cost function C.

(b) Determine the monthly revenue function R.

(c) Determine the production level x that maximizes the monthly profit, assuming that $10 \le x \le 40$.

42. **Business Management—Maximizing Profit:** The Boogie-All-Night Company also manufactures floor-to-ceiling lava lamps. The monthly fixed cost for the lamps is $1000, and the variable costs are $60 for each lamp. The company estimates that 100 lamps can be sold each month at a price of $100, and that seven more lamps can be sold for each decrease of $5 in the price. Let x represent the number of lamps produced each month.

(a) Determine the monthly cost function C.

(b) Determine the monthly revenue function R.

(c) Determine the production level x that maximizes the monthly profit, assuming that $50 \le x \le 200$.

© Dimitris Christou/iStockphoto

43. **Business Management—Maximizing Profit:** The Tool Shack Company manufactures drill presses. The account supervisor at the company determines that the monthly fixed costs are $15,000, and the variable costs are $180 for each drill press produced. The supervisor also estimates that 300 presses can be sold each month for $280 each, and that 20 more units can be sold for each decrease of $14 in price. Let x represent the number of drill presses produced and sold each month.

(a) Determine the monthly cost function C.

(b) Determine the monthly revenue function R.

(c) Determine the production level x that maximizes the monthly profit, assuming that $200 \le x \le 580$.

44. Business Management—Maximizing Revenue: The VirtualPet Company makes handheld virtual pets. The marketing department is confident that it can sell 3000 of its popular Piggy Pal virtual pets per week at a price of $10 each. The company's fiscal officer believes that reducing the price by $0.17 each will increase the weekly sales by 180. Let x represent the number of Piggy Pals manufactured and sold each week.

(a) Determine the weekly revenue R.

(b) Determine the value of x that maximizes weekly revenue, assuming that
$3000 \leq x \leq 10{,}000$.

 45. Law Enforcement—Alarm System Sales: In an August 10, 2010, article, Nicholas J. Halmond, managing director of instantspy.net, said that he believes there is a significant connection between economic downturn and alarm system sales, adding, "I believe that in times of hardship, people are much more likely to commit crimes, poverty increases and so do the chances of suffering a home invasion." (*Source:* www.einpresswire.com.) The annual revenue from alarm systems from 2000 to 2010 can be modeled by

$$f(x) = 32.78x^2 - 35.7x + 2776 \qquad 0 \leq x \leq 10$$

where x represents the number of years since 2000, and $f(x)$ represents the revenue from alarm system sales in millions of dollars. (*Source:* U.S. Census Bureau.) Determine the absolute minimum and the absolute maximum of f on the closed interval $[0, 10]$ and interpret.

 46. Nutrition—Spending on Meat Products: *New York Times* food critic Mark Bittman suggested that Americans should decrease their spending on meat, saying, "To suggest that Americans eat 50% less meat, it's not enough of a cut, but it's a start. It would seem absurd, but that's exactly what should happen." (*Source:* www.ted.com.) The average per capita annual amount spent on meat in the United States from 1990 to 2010 can be modeled by

$$f(x) = -1.12x^2 + 26.46x + 660 \qquad 0 \leq x \leq 20$$

where x represents the number of years since 1990 and $f(x)$ represents the per capita annual amount spent on meat, measured in dollars. (*Source:* U.S. Census Bureau.) Determine the absolute minimum and the absolute maximum of f on the closed interval $[0, 20]$ and interpret.

 47. Marketing—Local Radio Advertising: Revenues from local radio ads have been decreasing during the past decade, and in response, the National Association of Broadcasters asked for a government mandate that would force cell phone manufacturers to offer FM radio on their devices. Gary Shapiro, chief executive of the Consumer Electronics Association, says the cell phone industry is "completely, inalterably opposed to this." (*Source:* TechNewsWorld.) The annual revenue from local radio advertising from 1990 to 2010 can be modeled by

$$f(x) = -0.98x^3 - 18.73x^2 + 1071x + 6621 \qquad 0 \leq x \leq 20$$

where x represents the number of years since 1990 and $f(x)$ represents the local radio advertising, measured in millions of dollars. (*Source:* Universal McCann.)

(a) Differentiate f and use the quadratic formula to determine the critical values.

(b) Use the result of part (a) to determine the absolute maximum and absolute minimum of f on the interval $[0, 20]$. Interpret the coordinates.

 48. Marketing—Yellow Pages Advertising: Local Yellow Pages books once dominated local advertising dollars, but electronic means have limited the book's effectiveness. AIM advertising analyst Joe Michaud of the AIM consulting group, said in a recent client report that the shift to the Internet for information "represents an end to the era of big printed books filled with everyone from auto mechanics to zydeco sellers. People have opted for the desktop and the mobile device." (*Source:* WebProNews.) The annual revenue from local Yellow Pages advertising from 1990 to 2010 can be modeled by

$$f(x) = -1.48x^3 + 27.83x^2 + 199x + 7795 \qquad 0 \leq x \leq 20$$

where x represents the number of years since 1990 and $f(x)$ represents the local Yellow Page advertising, measured in millions of dollars. (*Source:* Universal McCann.)

(a) Differentiate f and use the quadratic formula to determine the critical values.

(b) Use the result of part (a) to determine the absolute maximum and absolute minimum of f on the interval $[0, 20]$. Interpret the coordinates.

Concept and Writing Exercises

49. Consider a quadratic function $f(x) = ax^2 + bx + c$ defined on a closed interval $[c, d]$. What is the least number of absolute extrema that f can have on $[c, d]$? What is the greatest number? Explain.

50. Consider a linear function of the form $f(x) = mx + b$ defined on a closed interval $[c, d]$. What is the least number of absolute extrema that f can have on $[c, d]$? What is the greatest number? Explain.

51. Make a graph to determine the absolute extrema of $f(x) = |2x - 6|$ on $[0, 5]$.

52. Make a graph to determine the absolute extrema of $f(x) = |x^2 - 25|$ on $[3, 6]$.

For Exercises 53 and 54, consider the function $f(x) = \sqrt{9 - x^2} + 3$.

53. Determine the domain of f.

54. Find and classify the absolute extrema of f on its domain.

Section Project

Suppose that Lisa's Bakery has collected data on the price of a dozen brownies, and the number sold each day. The data is shown in **Table 5.2.2**.

Table 5.2.2

Brownies demanded (in dozens)	Price (in Dollars)
14	4.00
19	3.75
21	3.50
23	3.25
26	3.00
31	2.75
35	2.50

A linear price demand function that models the data in the table is given by

$$p(x) = -0.07x + 5.04 \qquad 14 \le x \le 40$$

where x represents the number of brownies demanded, in dozens, and $p(x)$ represents the price at which customers buy x brownies per day, measured in dollars.

(a) Use the price-demand function to determine a revenue function R for the brownies.

(b) Determine the maximum revenue on the interval $[14, 40]$, and find the price and the demand that will yield the maximum revenue.

(c) Each dozen brownies costs the bakery $1.50 to make. Use this information to determine the profit function P.

(d) Determine the maximum profit on the interval $[14, 40]$ and find the price and the demand that will yield the maximum profit.

(e) The price demand function for the data can also be modeled by the exponential function

$$p(x) = 5.64e^{-0.023x} \qquad 14 \le x \le 40$$

where x represents the number of brownies demanded, in dozens, and $p(x)$ represents the price at which customers buy x brownies per day, measured in dollars. Determine a revenue function R for the brownies.

(f) Use the result of part (e) to determine the maximum revenue on the interval $[14, 40]$, and find the price and the demand that will yield the maximum revenue.

(g) Knowing that each dozen brownies cost the bakery $1.50 to make, use the revenue function from part (f) to determine the profit function P.

(h) Determine the maximum profit on the interval $[14, 40]$, and find the price and the demand that will yield the maximum profit.

(i) If you owned Lisa's Bakery, which model would you use to determine the price in order to maximize profit: the linear model from part (a), or the exponential model from part (e)? Explain your answer.

SECTION OBJECTIVES

1. Compute higher-order derivatives.

2. Determine intervals of concavity.

3. Locate inflection points.

4. Apply inflection point and interpret its meaning.

5.3 Second Derivatives and Graphs

In Section 5.1, we saw how the derivative of a function, f', allowed us to determine where the function f is increasing and where it is decreasing. Recall that $f'(x)$ tells us an *instantaneous rate of change* at x and that f' is itself a function. Since the derivative is a function, we can compute the derivative of $f'(x)$ and use it to determine where f' is increasing and decreasing. The derivative of a derivative is called the **second derivative**.

We just stated that the second derivative tells us where the derivative is increasing and decreasing. This analysis will also give us more information about the behavior of the original function. Specifically, it will give us the intervals of **concavity** and **inflection points**, and other characteristics of the graphs of functions. We will interpret these concepts in the context of rates of change. Before we go much further, let's consider **higher-order derivatives**.

Higher-Order Derivatives

If f is some function, $f'(x)$ is the derivative at x that gives the instantaneous rate of change for f at x or, equivalently, the slope of a tangent line at x. We know that f' is used to tell where f is increasing and where it is decreasing. Since f' is itself a function, computing the derivative of f' tells us where the derivative is increasing and where it is decreasing. The derivative of f', denoted f'' and read "f double prime," is called the **second derivative**. Table 5.3.1 gives some different notations for the second derivative.

Table 5.3.1

Function	First derivative	Second derivative
$f(x)$	$f'(x)$	$f''(x)$
y	y'	y''
y	$\dfrac{dy}{dx}$	$\dfrac{d^2y}{dx^2}$

For most applications, we have no need for a derivative beyond the second derivative. However, any derivative beyond the first derivative is called a **higher-order derivative**.

OBJECTIVE 1

Compute higher-order derivatives.

Example 1: Computing Higher-Order Derivatives

Determine the first three derivatives for the following functions.

a. $f(x) = 2x^5 + 6x^3 - 7x + 1$

b. $y = e^{2x}$

Perform the Mathematics

a. Applying the rules for differentiation yields

$$f'(x) = 10x^4 + 18x^2 - 7$$

$$f''(x) = \frac{d}{dx}[10x^4 + 18x^2 - 7] = 40x^3 + 36x$$

$$f'''(x) = \frac{d}{dx}[40x^3 + 36x] = 120x^2 + 36$$

b. Again, applying the rules for differentiation gives us

$$y' = 2e^{2x}$$

$$y'' = \frac{d}{dx}[2e^{2x}] = 4e^{2x}$$

$$y''' = \frac{d}{dx}[4e^{2x}] = 8e^{2x}$$

∎

▶ Try It Yourself

Some related Exercises are 9 and 15.

As we can see in Example 1, computing higher-order derivatives is no different from just computing a derivative.

Flashback: First Derivative Test Revisited

Determine intervals where $f(x) = 5x^3 + 4x^2 - 12x - 25$ is increasing and where it is decreasing. Also, locate any relative extrema.

Perform the Mathematics

To find intervals where f is increasing and where it is decreasing, we need to determine the critical values for this function. As we saw in Section 5.1, Example 5, the derivative is given by $f'(x) = 15x^2 + 8x - 12$, and the critical values are $x = -\frac{6}{5}$ and $x = \frac{2}{3}$. Placing the critical values on a number line and choosing test points produces the sign diagram in **Figure 5.3.1**.

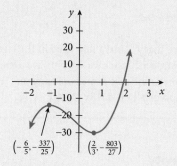

Figure 5.3.1

Figure 5.3.2

The sign diagram indicates that $f(x) = 5x^3 + 4x^2 - 12x - 25$ is increasing on $(-\infty, -\frac{6}{5}) \cup (\frac{2}{3}, \infty)$ and decreasing on $(-\frac{6}{5}, \frac{2}{3})$. To locate the relative extrema, we use the first derivative test using a sign diagram. This gives a relative maximum at $(-\frac{6}{5}, -\frac{337}{25})$ and a relative minimum at $(\frac{2}{3}, -\frac{803}{27})$, as shown in **Figure 5.3.2**.

∎

Recall that, since f' is also a function, we should be able to compute the **second derivative**, f'' and use it to determine the behavior of the derivative, f'. Later we will see how this information is used to determine the shape of the graph of f. First, let's see how we can use f'' to find information about f'.

Example 2: Using f'' to Graph f'

For the function in the Flashback, compute $f''(x)$ and use it to determine where f' is increasing and decreasing.

Perform the Mathematics

In the Flashback, we computed the derivative of $f(x) = 5x^3 + 4x^2 - 12x - 25$ to be $f'(x) = 15x^2 + 8x - 12$. So the second derivative is

$$f''(x) = \frac{d}{dx}[15x^2 + 8x - 12] = 30x + 8$$

Since $f''(x)$ is defined for all x, the only critical value for f' is when $f''(x) = 0$. So we solve

$$30x + 8 = 0$$

$$x = -\frac{8}{30} = -\frac{4}{15}$$

So the only critical value is $x = -\frac{4}{15}$. We employ the same process from Section 5.1 and reviewed in the Flashback: We place $x = -\frac{4}{15}$ on a number line, select some test numbers, and make a sign diagram. See **Figure 5.3.3**.

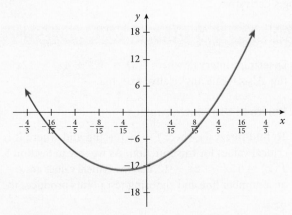

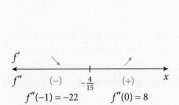

Figure 5.3.3

Figure 5.3.4 Graph of $f'(x) = 15x^2 + 8x - 12$.

Notice that f'' and f' are used in the sign diagram. The process is exactly the same as what we did in Section 5.1, since we simply have a function and its derivative listed here. From the sign diagram, we determine that f' is decreasing on $(-\infty, -\frac{4}{15})$ and increasing on $(-\frac{4}{15}, \infty)$. **Figure 5.3.4** shows a graph of $f'(x) = 15x^2 + 8x - 12$. ∎

Concavity

In Chapter 2 we saw how the derivative gives an instantaneous rate of change of a function at any point. Example 2 showed how the second derivative tells us where the first derivative is increasing and where it is decreasing. Since the first derivative gives an instantaneous rate of change, the second derivative tells us the increasing or decreasing behavior of the instantaneous rate of change. Knowing intervals where the instantaneous rate of change, f', is increasing and intervals where it is decreasing then tells us the type of **concavity** that the graph of a function, f, has on the interval. **Concavity** is important in our continuing analysis of functions and their graphs.

Notice in **Figure 5.3.5** that, on an interval where the graph of a function is concave up, tangent lines to the curve for any point in the interval are below the curve. Also, on an interval where the graph of a function is concave down, tangent lines to the curve for any point in the interval are above the curve.

DEFINITION

Concavity

On an open interval (a, b) where f is differentiable:

1. If f' is increasing, then the graph of f is **concave up.**
2. If f' is decreasing, then the graph of f is **concave down.**

Concave up: Tangent lines lie below the curve.

Concave down: Tangent lines lie above the curve.

Figure 5.3.5

Now let's return to our work in the Flashback and Example 2. Recall that

$$f(x) = 5x^3 + 4x^2 - 12x - 25 \qquad \text{(see **Figure 5.3.6**)}$$
$$f'(x) = 15x^2 + 8x - 12 \qquad \text{(see **Figure 5.3.7**)}$$
$$f''(x) = 30x + 8 \qquad \text{(see **Figure 5.3.8**)}$$

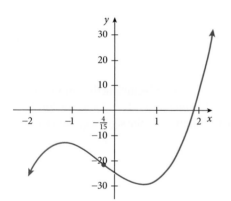

Figure 5.3.6

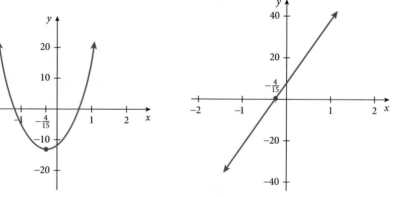

Figure 5.3.7

Figure 5.3.8

The Flashback, Example 2, and Figures 5.3.6 through 5.3.8 illustrate the following:

- On the interval $\left(-\infty, -\frac{4}{15}\right)$, $f'' < 0$ (Figure 5.3.8) and f' is decreasing (Figure 5.3.7 and Example 2).
- On the same interval $\left(-\infty, -\frac{4}{15}\right)$, the graph of f is concave down. See Figure 5.3.6.
- On the interval $\left(-\frac{4}{15}, \infty\right)$, $f'' > 0$ (Figure 5.3.8) and f' is increasing (Figure 5.3.7 and Example 2).
- On the same interval $\left(-\frac{4}{15}, \infty\right)$, the graph of f is concave up. See Figure 5.3.6.

These observations can be condensed into the following **Tests for Concavity**.

Tests for Concavity

For a function f whose second derivative exists on open interval (a, b):

1. If $f''(x) > 0$ for all x on (a, b), then the graph of f is concave up on (a, b).
2. If $f''(x) < 0$ for all x on (a, b), then the graph of f is concave down on (a, b).

NOTE: In Section 5.1 we saw how the sign of f' determines where f is increasing or decreasing. Here we need to determine the sign of f'' in order to determine the concavity of the graph of f. Hence, the process outlined in Section 5.1 will be used here, as shown in Example 3.

OBJECTIVE 2

Determine intervals of
concavity.

Example 3: Determining Intervals of Concavity

Determine intervals where the graph of $f(x) = x^3 + 3x^2 - 4$ is concave up and where the graph is concave down.

Perform the Mathematics

To determine the concavity, we need to determine the sign of the second derivative. We compute the second derivative to be

$$f'(x) = 3x^2 + 6x$$
$$f''(x) = 6x + 6$$

We note that f'' is defined for all values of x, so, using the same process as in Section 5.1, we need to determine where $f''(x) = 0$. Solving this equation yields

$$6x + 6 = 0$$
$$x = -1$$

We place $x = -1$ on a number line and make a sign diagram in **Figure 5.3.9** using the same steps as in Section 5.1.

Figure 5.3.9

Selecting $x = -2$ as a test number from the interval $(-\infty, -1)$ and substituting this into $f''(x)$ gives $f''(-2) = 6(-2) + 6 = -6$. Selecting $x = 0$ from the interval $(-1, \infty)$ and substituting into $f''(x)$ gives $f''(0) = 6(0) + 6 = 6$. Putting this information on the sign diagram yields **Figure 5.3.10**.

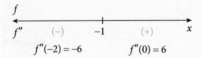

Figure 5.3.10

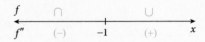

Figure 5.3.11

Knowing that the sign of the second derivative tells us the concavity of the graph, we complete the sign diagram as shown in **Figure 5.3.11**. We use $\cap$ to mean concave down and $\cup$ to mean concave up. The graph of $f(x) = x^3 + 3x^2 - 4$ is concave down on $(-\infty, -1)$ and concave up on $(-1, \infty)$.

▶ *Try It Yourself*

Some related Exercises are 23 and 25.

The graph of $f(x) = x^3 + 3x^2 - 4$ is shown in **Figure 5.3.12**. At the point $(-1, -2)$, the graph switches concavity from concave down to concave up.

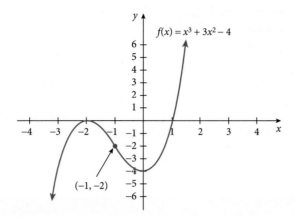

Figure 5.3.12

Determining Intervals of Concavity

To determine intervals of concavity of the graph of a function f, we determine the second derivative f'' and apply the Tests for Concavity.

A point on the graph where concavity changes from concave up to concave down (or from concave down to concave up) is called an **inflection point**. It is also worth noting that the x-coordinate of an inflection point is a critical value for f'.

Before continuing to Example 4, we offer a procedure for **determining intervals of concavity** of the graph of a function.

Example 4: Determining Intervals of Concavity

OBJECTIVE 3

Locate inflection points.

Determine intervals where the graph of $f(x) = x^4$ is concave up and where it is concave down. Also, locate any inflection points.

Perform the Mathematics

We begin by computing the second derivative to be

$$f'(x) = 4x^3$$
$$f''(x) = 12x^2$$

Since $f''(x)$ is defined for all x, just as in Example 3, we need to solve $f''(x) = 0$. This results in

$$12x^2 = 0$$
$$x = 0$$

We place $x = 0$ on a number line and make a sign diagram as in **Figure 5.3.13**.

Figure 5.3.13   **Figure 5.3.14**

Selecting $x = -1$ as a test number from the interval $(-\infty, 0)$ and substituting this into $f''(x)$ gives $f''(-1) = 12(-1)^2 = 12$. Selecting $x = 1$ from the interval $(0, \infty)$ and substituting into $f''(x)$ gives $f''(1) = 12(1)^2 = 12$. Knowing that the sign of the second derivative tells the concavity of the graph, we complete the sign diagram as shown in **Figure 5.3.14**.

We conclude that the graph of $f(x) = x^4$ is concave up on $(-\infty, 0)$ and concave up on $(0, \infty)$. Since there is no change in concavity, the graph of the function $f(x) = x^4$ has no inflection points. ∎

Example 3 showed that $f''(x)$ may equal zero at an inflection point. There is only one other possibility for an inflection point. If a continuous function has an inflection point at $x = d$, then either $f''(d) = 0$ or $f''(d)$ is undefined. We use the following process to **locate inflection points**.

Locating Inflection Points

1. Determine the values $x = d$ where $f''(d) = 0$ or where $f''(d)$ is undefined.
2. Place these values on a number line and make a sign diagram.
3. The point $(d, f(d))$ is an inflection point if f'' changes sign at $x = d$ **and** if $x = d$ is in the domain of f.

NOTE: Example 4 indicates that not every value of x that satisfies $f''(x) = 0$ produces an inflection point. It is worth noting one more time that the process outlined in Examples 3 and 4 is similar to the process that we used in Section 5.1 to locate relative extrema.

Example 5: Locating Inflection Points

Determine any inflection points for the graph of $f(x) = 5x^3 + 4x^2 - 12x - 25$.

Perform the Mathematics

This is the same function from the Flashback and Example 2. In Example 2 we determined that $f''(x) = 30x + 8$ and $f''(x) = 0$ at $x = -\frac{4}{15}$. Placing $x = -\frac{4}{15}$ on a number line and constructing a sign diagram gives **Figure 5.3.15**.

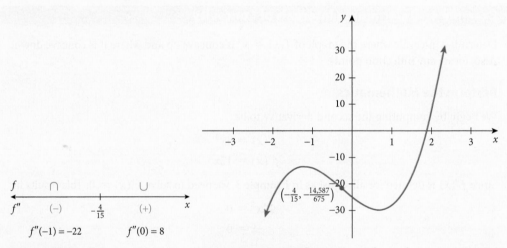

Figure 5.3.15

Figure 5.3.16 Graph of $f(x) = 5x^3 + 4x^2 - 12x - 25$ has an inflection point at $(\frac{-4}{15}, \frac{-14{,}587}{675})$.

The sign diagram indicates that the graph of f changes concavity at $x = -\frac{4}{15}$. So the point $(-\frac{4}{15}, f(-\frac{4}{15}))$, which is the same as $(-\frac{4}{15}, -\frac{14{,}587}{675})$, is an inflection point. See **Figure 5.3.16**. ∎

▶ *Try It Yourself*

Some related Exercises are 43(c) and 45(c).

Inflection Points, Rates of Change, and Applications

In this section, we have learned that the second derivative tells us the intervals where the rate of change is increasing and where it is decreasing. In addition, we know that:

1. f' gives the intervals where f is increasing and where it is decreasing.
2. f'' gives the intervals where the graph of f is concave up and where it is concave down.

We summarize the results of all this information in **Table 5.3.2**.

Table 5.3.2

First and Second Derivative	Shape of Graph of f	Behavior of f
$f' > 0$ implies f increasing $f'' > 0$ implies f concave up		f increasing at a faster rate
$f' > 0$ implies f increasing $f'' < 0$ implies f concave down		f increasing at a slower rate
$f' < 0$ implies f decreasing $f'' < 0$ implies f concave down		f decreasing at a faster rate
$f' < 0$ implies f decreasing $f'' > 0$ implies f concave up		f decreasing at a slower rate

Example 6: Maximizing a Rate of Change

The marketing research department for the Spritz Cola company analyzed data on the number of units of a certain cola that sold after spending x dollars on advertising. They estimate that the company will sell $B(x)$ units of a diet cola after spending x thousand dollars on advertising, according to

$$B(x) = -\frac{1}{3}x^3 + 60x^2 - 110x + 5200 \qquad 20 \le x \le 100$$

a. Determine where the rate of change of sales is increasing and where it is decreasing.

b. What level of spending maximizes the rate of change of sales?

Perform the Mathematics

a. In order to determine where the rate of change of sales is increasing and where it is decreasing, we must first determine the rate of change of sales, $B'(x)$. Differentiating $B(x)$ gives us

$$B'(x) = -x^2 + 120x - 110$$

To determine where B' is increasing and decreasing, we compute $B''(x)$, the derivative of $B'(x)$ and get

$$B''(x) = -2x + 120$$

Since $B''(x)$ is defined for all x, we solve $B''(x) = 0$ and get

$$-2x + 120 = 0$$
$$x = 60$$

Making a sign diagram and analyzing B' and B'' gives **Figure 5.3.17**. From this diagram we conclude that the rate of change of sales, B', is increasing on the interval $(20, 60)$ and decreasing on the interval $(60, 100)$.

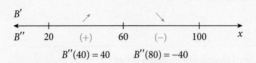

Figure 5.3.17

b. The sign diagram indicates that a spending level of 60, which is $60,000, for advertising maximizes the rate of change of sales. **Figure 5.3.18** shows graphs of B and B' that demonstrate this result.

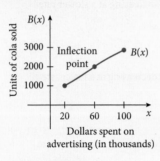

Figure 5.3.18   ■

Example 6 and Figure 5.3.18 nicely illustrate the key concepts of this section. That is, where B' is increasing, the graph of B is concave up, and where B' is decreasing, the graph of B is concave down. Also, where the rate of change of sales is maximized, at $x = 60$, the graph of B has an inflection point. In Example 6, the inflection point is also known as the **point of diminishing returns**. At the inflection point in Example 6, concavity changes from concave up to concave down. As we analyzed in Example 6, the point of diminishing returns is exactly where the rate of change of sales started to decrease.

Example 7: Interpreting Concavity and Inflection Point

OBJECTIVE 4

Apply inflection point and interpret its meaning.

Since the late 1980s, some women have been delaying starting a family until they reach the age of 30. The number of births to U.S. women who were 30 to 34 years old can be modeled by

$$f(x) = -0.05x^3 + 1.46x^2 - 5.05x + 890.39 \qquad 1 \le x \le 18$$

where x is the number of years since 1989 and $f(x)$ is the number of births to women who were 30 to 34 years old, measured in thousands. (*Source:* U.S. National Center for Health Statistics.) Locate and interpret the inflection point.

Understand the Situation

To locate the inflection point, we employ the process outlined for Locating Inflection Points. This means that we must determine the second derivative.

Perform the Mathematics

Differentiation gives us

$$f'(x) = -0.15x^2 + 2.92x - 5.05$$
$$f''(x) = -0.30x + 2.92$$

Since f'' is defined for all x, we must solve $f''(x) = 0$. This gives

$$-0.30x + 2.92 = 0$$
$$x = 9.73 \qquad \text{rounded to the nearest hundredth}$$

We place $x = 9.73$ on a number line and construct a sign diagram as in **Figure 5.3.19**.

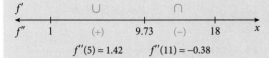

Figure 5.3.19

So the graph of f is concave up on $(1, 9.73)$ and concave down on $(9.73, 18)$. Since the graph of f changes from concave up to concave down at $x = 9.73$, we conclude that there is an inflection point at $(9.73, f(9.73))$ or, specifically, at $(9.73, 933.42)$. Note that both coordinates have been rounded to the nearest hundredth.

Interpret the Result

The graph of f is concave up on the interval $(1, 9.73)$ which means that the rate of change of births is *increasing* on this interval. The graph of f is concave down on the interval $(9.73, 18)$ which means that the rate of change of births is *decreasing* on this interval. Hence, we interpret the inflection point as indicating that the rate of change of births to U.S. women who were 30 to 34 years old was maximized at $(9.73, 933.42)$, or some time in 1998, at about 933,420 births per year.

▷ **Try It Yourself**

Some related Exercises are 75 and 77.   ■

Summary

In this section, we saw how the second derivative can be used to determine the increasing and decreasing behavior of the rate of change, or derivative, of a function. We examined the intervals where the graph of a function is concave up and where it is concave down, and we saw that a point where a graph switches concavity is called an **inflection point**. To determine the **concavity** of the graph of a function on an interval, apply the Test for Concavity.

Section 5.3 Exercises

Vocabulary Exercises

1. If we compute the derivative of the function $f'(x)$, we get the _____ derivative.

2. Generally speaking, if we continue to take derivatives of a function successively, we get the function's _____-_____ derivatives.

3. If the derivative of a function is decreasing on an open interval (a, b), then we know that the graph of the function is _____ _____ on that interval.

4. If the values of $f''(x)$ are positive for all x-values in an open interval (a, b), then we know that the graph of the function is _____ _____ on (a, b).

5. Values where $f''(x) = 0$ are candidates for _____ points.

6. If the second derivative of a sales function changes from positive values to negative values at a value x, we say that $(x, f(x))$ is a point of _____ _____.

Skill Exercises

In Exercises 7–16, compute the first three derivatives for the given function.

7. $f(x) = -4x^5 - 6x^3 + 7x$

8. $f(x) = 5x^4 + 3x^2 - 7x + 1$

9. $f(x) = 7x^3 - 3x^2 + 4x + 5$

10. $f(x) = -8x^3 - 7x^2 + 5x + 6$

11. $f(x) = e^x$

12. $f(x) = e^{x^2}$

13. $f(x) = \sqrt{x}$

14. $f(x) = \sqrt[3]{x}$

15. $f(x) = \ln x$

16. $f(x) = \ln 2x$

In Exercises 17–22, the graph of a function f is given.

(a) Determine the intervals where the graph of f is concave up and where it is concave down.

(b) Determine intervals where the derivative of the function is increasing and where the derivative is decreasing.

17.

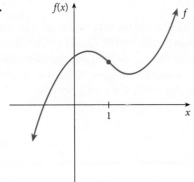

18.

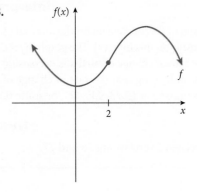

19.

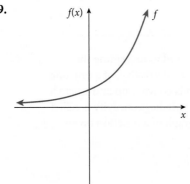

20.

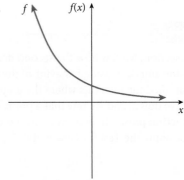

21.

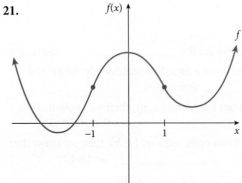

22.

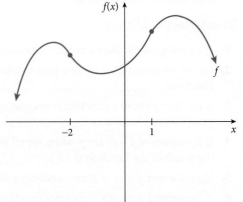

In Exercises 23–40, determine the intervals where the graph of the given function is concave up and where it is concave down. Then identify any inflection points.

23. $f(x) = x^3 + 6x^2 + 18x - 5$

24. $f(x) = x^3 - 6x^2 + 6x - 3$

25. $f(x) = -12x^3 + 6x^2 - 24x - 11$

26. $f(x) = -3x^3 + 5x^2 - 2x + 4$

27. $f(x) = 2x^2 + 3x + 1$

28. $f(x) = 3x^2 - 2x - 5$

29. $f(x) = -3x^2 + 3x - 2$

30. $f(x) = -2x^2 - 5x + 3$

31. $f(x) = x + \dfrac{2}{x}$

32. $f(x) = x + \dfrac{5}{x}$

33. $f(x) = x^{2/3}$

34. $f(x) = (x - 1)^{2/3}$

35. $f(x) = e^{2x}$

36. $f(x) = e^{1.2x}$

37. $f(x) = e^{-2x}$

38. $f(x) = e^{-1.5x}$

39. $f(x) = \ln(x + 1)$

40. $f(x) = \ln(x - 3)$

For each function in Exercises 41–54, complete the following.

(a) Determine the intervals where the function is increasing and where it is decreasing.

(b) Identify any relative extrema.

(c) Determine the intervals where the graph of the function is concave up and where it is concave down.

(d) Identify any inflection points.

41. $f(x) = 2x^2 + 3x + 1$

42. $f(x) = 3x^2 - 2x - 5$

43. $f(x) = x^3 + 6x^2 - 15x - 5$

44. $f(x) = x^3 - 6x^2 - 15x + 3$

45. $f(x) = -3x^3 + 5x^2 + 2x - 9$

46. $f(x) = -2x^3 + 3x^2 + 6x - 5$

47. $f(x) = x + \dfrac{5}{x}$

48. $f(x) = x + \dfrac{2}{x}$

49. $f(x) = (x - 1)^{2/3}$

50. $f(x) = (x + 2)^{2/3}$

51. $f(x) = e^{3x}$

52. $f(x) = e^{-3x}$

53. $f(x) = \ln(x + 3)$

54. $f(x) = \ln(x - 1)$

For Exercises 55–58, sketch a graph of a function that satisfies the given conditions.

55. Domain: $(-\infty, \infty)$
Range: $(-\infty, 5]$
Continuous on $(-\infty, \infty)$
$f' > 0$ on $(-\infty, 2)$
$f' < 0$ on $(2, \infty)$
$f'' < 0$ on $(-\infty, \infty)$

56. Domain: $(-\infty, \infty)$
Range: $[5, \infty)$
Continuous on $(-\infty, \infty)$
$f' < 0$ on $(-\infty, -3)$
$f' > 0$ on $(-3, \infty)$
$f'' > 0$ on $(-\infty, \infty)$

57. Domain: $(-\infty, \infty)$
Range: $(-\infty, \infty)$
Continuous on $(-\infty, \infty)$
$f' > 0$ on $(-\infty, -1) \cup (3, \infty)$
$f' < 0$ on $(-1, 3)$
$f'' < 0$ on $(-\infty, 1)$
$f'' > 0$ on $(1, \infty)$

58. Domain: $(-\infty, \infty)$
Range: $(-\infty, \infty)$
Continuous on $(-\infty, \infty)$
$f' > 0$ $(-\infty, 0) \cup (0, \infty)$
$f'(0)$ is undefined
$f'' > 0$ on $(-\infty, 0)$
$f'' < 0$ on $(0, \infty)$

Application Exercises

59. Microeconomics—Production Cost: The Beagle Works Company has determined that the cost function for their Beagle Buddy chew toy is given by

$$C(x) = x^3 - 6x^2 + 13x + 10$$

where x represents the number of items produced and $C(x)$ represents the cost in hundreds of dollars.

(a) Evaluate $C(5)$ and $C'(5)$ and interpret each.

(b) Determine the intervals where the marginal cost is increasing and where it is decreasing. Determine the relative minimum of the marginal cost function.

(c) Determine the inflection point for the graph of C.

60. Microeconomics—Production Cost: The Chug-a-Mug Company has determined that the cost function for their Bottomless Pit Coffee Cup is given by

$$C(x) = x^3 - 6x^2 + 15x$$

where x represents the number of cups produced and $C(x)$ represents the cost in hundreds of dollars.

(a) Evaluate $C(5)$ and $C'(5)$ and interpret each.

(b) Determine the intervals where the marginal cost is increasing and where it is decreasing. Determine the relative minimum of the marginal cost function.

(c) Determine the inflection point for the graph of C.

61. Microeconomics—Marginal Profit: The financial planning department of Cool Air Refrigerator Company has determined that the monthly profit for their Polar Pod can be modeled by

$$P(x) = -2.3x^3 + 445x^2 - 1500x - 200$$

where x represents the number of Polar Pods produced and sold and $C(x)$ represents the monthly profit in dollars.

(a) Evaluate $P(35)$ and $P'(35)$ and interpret each.

(b) Determine the intervals where the marginal profit is increasing and where it is decreasing. Determine the relative extremum of the marginal profit function.

62. Microeconomics—Marginal Profit (*Continuation of Exercise 61*):

(a) Determine the inflection point for the graph of P.

(b) Explain why the relative extremum for the marginal profit and the inflection point for the graph of P have the same x-value.

63. Microeconomics—Marginal Profit (*Continuation of Exercise 61*): For the profit function P, explain why the intervals where the marginal profit is increasing and where it is decreasing and the intervals where the graph of P is concave up and concave down are the same.

64. Marketing—Product Sales: The Sucrado Cola Company determines that the sales of *Screamin'*, their new energy drink, can be modeled by

$$S(x) = -\frac{5}{2}x^3 + 112.5x^2 + 150x + 10{,}000 \qquad 10 \leq x \leq 30$$

where x represents the amount spent on advertising measured in millions of dollars and $S(x)$ represents the total sales of the energy drink. Locate the point of diminishing returns for $S(x)$ and interpret its meaning. Recall that this point is given by the inflection point for the graph of S.

65. Marketing—Product Sales: The Big Cola Company has determined that the total sales of its new diet cola *MinaMax* can be modeled by the function

$$S(x) = -2x^3 + 90x^2 - 1200x + 10{,}000 \qquad 10 \leq x \leq 25$$

where x represents the amount spent on advertising measured in millions of dollars and $S(x)$ represents the total sales of the diet cola. Locate the point of diminishing returns for $S(x)$ and interpret its meaning. Recall that this point is given by the inflection point for the graph of S.

66. **Physiology—Arterial Pressure:** The pressure of blood circulating in the arteries is called arterial pressure. Suppose that a medical researcher determines that the blood flow through an artery can be modeled by

$$f(x) = 0.267e^{0.0256x} \qquad 20 \le x \le 120$$

where x represents the arterial pressure, measured in millimeters of mercury (mmHg) and $f(x)$ represents the blood flow, measured in milliliters per minute (mL/min).

(a) Evaluate $f(50)$ and $f'(50)$ then interpret each.

(b) Show that the graph of f is concave up on the interval $[20, 120]$. Explain what this means.

67. **Physiology—Arterial Pressure:** The pressure of blood circulating in the arteries is called arterial pressure. Suppose that a medical researcher determines that the blood flow through an artery can be modeled by

$$f(x) = 0.278(1.026)^x \qquad 20 \le x \le 120$$

where x represents the arterial pressure, measured in millimeters of mercury (mmHg) and $f(x)$ represents the blood flow measured in milliliters per minute (mL/min).

(a) Evaluate $f(50)$ and $f'(50)$ then interpret each.

(b) Show that the graph of f is concave up on the interval $[20, 120]$. Explain what this means.

68. **Psychology—Male Blood Volume:** Suppose that after collecting data from a sample of subjects, medical researchers have determined that the volume of blood in males can be modeled by

$$f(x) = -10{,}822 + 3800\ln x \qquad 40 \le x \le 90$$

where x represents the weight of the male in kilograms (kg) and $f(x)$ represents the blood volume in milliliters (mL).

(a) Evaluate $f(60)$ and $f'(60)$ then interpret each.

(b) Show that the graph of f is concave down on the interval $[40, 90]$. Explain what this means.

69. **Psychology—Cardiac Index:** The cardiac index is a cardiovascular value that relates the cardiac output to body surface area and is often determined by the age of the person. The cardiac index can be modeled by

$$f(x) = \frac{7.644}{\sqrt[4]{x}} \qquad 10 \le x \le 80$$

where x represents the age of the patient in years and $f(x)$ represents the cardiac output per square meter of body surface area, measured in $\dfrac{\text{liters per minute}}{\text{square meters}}$.

(a) Evaluate $f(20)$ and $f'(20)$ then interpret each.

(b) Determine the intervals where f is increasing and where it is decreasing.

(c) Show that the graph of f is concave up on the interval $[10, 80]$, and explain what this means.

70. **Microeconomics—Sales Curve:** The accounting office of the Trim Pill Company has tracked the sales of its new FastFit quick loss pill kit and has determined that the sales for the first year can be modeled by

$$f(x) = \frac{5000}{1 + 2e^{-0.2x}} \qquad 0 \le x \le 52$$

where x represents the number of weeks since the product went on the market and $f(x)$ represents the total number of pill kits sold.

(a) Evaluate $f(26)$ and $f'(26)$ then interpret each.

(b) Determine $f''(x)$.

(c) Graph $f''(x)$ on the interval $[0, 52]$ and then use the `Zero` command to help determine the intervals where f is concave up and where it is concave down.

(d) Assuming that advertising increases sales, use the result of part (c) to decide after how many weeks it would be an ideal time to start an advertising campaign.

 71. Ecological Studies—U.S. Carbon Monoxide Emissions: In recent years, the levels of carbon monoxide (CO) emissions have decreased in the United States. "The large decline in emissions was driven by the economic downturn, combined with an ongoing trend toward a less energy-intensive economy and a decrease in the carbon-intensity of the energy supply," said Richard Newell, head of EIA, a data-analysis section of the Department of Energy. (*Source: USA Today.*) The annual amount of CO emissions from 1970 to 2010 can be modeled by

$$f(x) = -0.03x^2 - 2.22x + 260 \qquad 0 \le x \le 40$$

where x represents the number of years since 1970 and $f(x)$ represents the amount of CO emissions, measured in millions of tons. (*Source: U.S. Environmental Protection Agency.*)

(a) How fast were the CO emissions decreasing in 1980? How fast in 2005?

(b) According to the model, would you say that the CO emissions are decreasing at a decreasing rate, decreasing at an increasing rate, or decreasing at a constant rate? Explain your answer.

 72. Ecological Studies—International Carbon Dioxide Emissions: Over the past decade, emissions of carbon dioxide (CO_2) emissions have increased. Jerry Hatfield, a plant physiologist with the U.S. Department of Agriculture, said that the consequences can be positive and negative agriculturally. "Corn and soybean plants are likely to grow and mature faster but will be more subject to crop failures from spikes in summer temperatures." (*Source: The New York Times*). The annual international carbon dioxide emissions from 2000 to 2010 can be modeled by

$$f(x) = 0.11x^3 - 1.93x^2 + 12.77x + 349 \qquad 0 \le x \le 10$$

where x represents the number of years since 2000 and $f(x)$ represents the annual carbon dioxide emissions in parts per million (ppm). (*Source: Science Daily.*)

(a) How fast were the CO_2 emissions increasing in 2001? In 2009?

(b) According to the model, would you say that the CO_2 emissions are increasing at a decreasing rate, increasing at an increasing rate, or increasing at a constant rate? Explain your answer.

© AbleStock

 73. Personal Finance—Revolving Debt: Revolving debt is constantly renewed as you pay it off. It is basically an open credit account. Credit cards are the most common type of credit card debt. For college students, revolving debt (from credit cards) has recently exceeded total nonrevolving debt (from student loans). Mark Kantrowitz, publisher of FinAid.org, compared the growth in education debt to cooking a lobster: "The increase in total student debt occurs slowly but steadily, so by the time you notice that the water is boiling, you're already cooked." (*Source: The University of Houston Daily Cougar*). The total annual amount of revolving debt in the United States from 1990 to 2010 can be modeled by

$$f(x) = 0.16x^3 - 4.92x^2 + 78.39x + 239 \qquad 0 \le x \le 20$$

where x represents the number of years since 1990 and $f(x)$ represents the total consumer revolving debt in billions of dollars. (*Source: Board of Governors, Federal Reserve System.*)

(a) Determine the year(s) that the revolving debt was increasing at a rate of 28.71 billions of dollars per year.

(b) Does the model have an inflection point in the reasonable domain $0 \leq x \leq 20$? Show work to verify your answer.

 74. **Personal Finance—Nonrevolving Debt:** Nonrevolving debt is the type of debt that is designed so that once a loan is paid off, if you want another loan, you have to reapply. This type of debt could be from home mortgages, car loans, and so on. Mark Fleming, chief economist with First American CoreLogic, states that this type of debt "is a significant drag on both the housing market and on economic growth. It is driving foreclosures and decreasing mobility for millions of homeowners." (*Source: The Philadelphia Bulletin.*) The total annual amount of nonrevolving debt in the United States from 1990 to 2010 can be modeled by

$$f(x) = 0.39x^3 + 12.37x^2 - 36.72x + 570 \qquad 0 \leq x \leq 20$$

where x represents the number of years since 1990 and $f(x)$ represents the total consumer nonrevolving debt in billions of dollars. (*Source: Board of Governors, Federal Reserve System.*)

(a) Determine the year that the nonrevolving debt was increasing at a rate of 17.44 billions of dollars per year.

(b) Does the model have an inflection point in the reasonable domain $0 \leq x \leq 20$? Show work to verify your answer.

 75. **Economics—Research and Development Spending:** The global recession has resulted in reduced research and development spending in Japan. Sadao Nagaoka, faculty fellow at Japan's Research Institute of Economy Trade and Industry, warns of the repercussions of this reduction, concluding: "A significant decline in R&D investment due to financial constraints resulting from the recession would not only stifle effective demand in the short term, but would also decelerate the speed of creating and applying new knowledge. The overall result would be a lower long-term growth rate." (*Source:* www.rieti.go.jp.) The annual amount spent on research and development in the business sector in Japan from 2000 to 2010 can be modeled by

$$f(x) = -0.14x^3 + 2.05x^2 - 10.79x + 110.8 \qquad 0 \leq x \leq 10$$

where x represents the number of years since 2000 and $f(x)$ represents the amount spent on research and development by the Japanese business sector, measured in billions of euros. (*Source: EuroStat.*)

(a) Graph f' and use the graph to show that f is decreasing on its reasonable domain.

(b) During what year was the decrease in R&D spending at its least? To find this year, determine the relative maximum of f' on the interval $0 \leq x \leq 10$.

 76. **Education—Research and Development in Higher Education:** Another area in which research and development spending has been decreasing in Japan is in the area of higher education. The annual amount spent per capita on research and development in higher education in Japan from 2000 to 2010 can be modeled by

$$f(x) = -0.31x^3 + 4.49x^2 - 25.9x + 179 \qquad 0 \leq x \leq 10$$

where x represents the number of years since 2000 and $f(x)$ represents the amount spent per capita on research and development in higher education in euros per inhabitant. (*Source: EuroStat.*)

(a) Graph f' and use the graph to show that f is decreasing on its reasonable domain.

(b) During what year was the decrease in higher education R&D spending at its least? To find this year, determine the relative maximum of f' on the interval $0 \leq x \leq 10$.

Concept and Writing Exercises

77. In Chapter 1 we saw that the graph of the quadratic function $f(x) = ax^2 + bx + c$, where a, b, and c are real numbers and $c \neq 0$, is concave up if $a > 0$ and concave down if $a < 0$. Compute the second derivative of f to show that this is true.

78. Consider the cubic function $f(x) = ax^3 + bx^2 + cx + d$. What conditions for the values of a, b, and c must be met so that f is always increasing on its domain?

79. Consider a cubic function of the form $f(x) = ax^3 + bx^2 + cx + d$ where a, b, c, and d are real numbers and $a \neq 0$. Show that the x-coordinate of the inflection point of f is given by
$$x = -\frac{b}{3a}.$$

80. Consider the function $f(x) = a\sqrt{x} + \dfrac{b}{\sqrt{x}}$. Determine the values of a and b so that f has an inflection point at $(4, 13)$.

81. Consider the cubic function f that has real number zeros at c_1, c_2 and c_3. That is, the function can be written in the form $f(x) = a(x - c_1)(x - c_2)(x - c_3)$ for some real number a, $a \neq 0$. Show that f has an inflection point at $x = \frac{1}{3}(c_1 + c_2 + c_3)$.

In Exercises 82 and 83, suppose that $f'(x) > 0$ and $g'(x) > 0$ for all x in the domain of these functions. Determine what additional conditions are necessary (if any) to ensure that the following are true.

82. The composition of the functions $f(g(x))$ is increasing for all x.

83. The product of functions $(f \cdot g)(x)$ is increasing for all x.

In Exercises 84 and 85, suppose that $f''(x) > 0$ and $g''(x) > 0$ for all x in the domain of these functions. Determine what additional conditions are necessary (if any) to ensure that the following are true.

84. The composition of the functions $f(g(x))$ is concave up for all x.

85. The product of functions $(f \cdot g)(x)$ is concave up for all x.

86. Consider the quartic function $f(x) = x^4$. Show that $f''(x) = 0$, at the origin, but the origin is not an inflection point.

87. Consider the function $f(x) = x \cdot |x|$. Show that the origin is an inflection point of f. (Hint: Rewrite f as a piecewise-defined function.)

 Section Project

After decreasing during the first half of the decade in the 2000s, the number of births by teenage women began to increase. "We could be reaching a place where further decreases are harder to achieve," said Stephanie Ventura, chief of reproductive statistics branch at the National Center for Health Statistics. (*Source:* The Cable News Network.) The number of births to women 20 and under in the United States from 1980 to 2010 can be modeled by the quartic function

$$f(x) = 0.0062x^4 - 0.327x^3 + 5.58x^2 - 31.5x + 552.2 \qquad 0 \leq x \leq 30$$

where x represents the number of years since 1980 and $f(x)$ represents the number of births to women 20 and under, measured in thousands. (*Source:* U.S. National Center for Health Statistics.)

(a) Graph f in the window $[0, 30]$ by $[400, 600]$.

(b) Determine $f'(x)$. Evaluate $f'(9)$ and interpret.

 (c) Use $f'(x)$ to determine the intervals where f is increasing and where it is decreasing.

(d) Determine the intervals where the rate of change of f is increasing and where it is decreasing.

(e) Determine the intervals where the graph of f is concave up and where it is concave down.

(f) Use the result of part (c) to determine the relative extrema of f and interpret each coordinate.

(g) Use the result of part (e) to determine the inflection points of f and interpret the coordinates.

5.4 The Second Derivative Test and Optimization

SECTION OBJECTIVES

1. Use Second Derivative Test to determine relative extrema.

2. Use Second Derivative Test to determine absolute minimum.

3. Apply Second Derivative Test to maximize volume.

4. Apply Second Derivative Test to maximize harvest.

5. Apply Second Derivative Test to minimize inventory costs.

In Section 5.2 we learned that if a function is continuous on a closed interval, then it has an absolute maximum and an absolute minimum. We also saw how to locate the absolute extrema by a four-step process. The key in Section 5.2 was that we considered functions on *closed intervals*. In this section we consider functions on **open** and **half-open** intervals. This means that we cannot use the four-step process and must rely on something else. What we tend to rely on when optimizing a function on an open (or half-open) interval is a graph of the function, a table of values, and the **Second Derivative Test**. What we use depends on the number of critical values on the open interval.

The Second Derivative Test

In Section 5.1 we saw how the First Derivative Test was used to locate relative extrema. At this time, we explore another method for locating relative extrema called the **Second Derivative Test**.

Example 1: Determining Relative Extrema

Determine the critical values of the function shown in **Figure 5.4.1**. Classify each as giving a relative maximum or a relative minimum, and determine the concavity of the graph of f at each relative extrema.

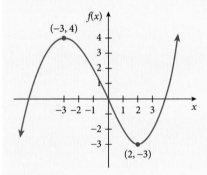

Figure 5.4.1

Perform the Mathematics

Since the graph of f has no breaks or sharp turns, $f'(x)$ is defined for all x. It appears that $f'(x) = 0$ at $x = -3$ and $x = 2$. So the critical values are $x = -3$ and $x = 2$. We have a relative maximum at $x = -3$, specifically at $(-3, 4)$, and a relative minimum at $x = 2$, specifically at $(2, -3)$. From the graph, it appears that at the relative maximum $(-3, 4)$ the graph is **concave down**, whereas at the relative minimum $(2, -3)$ the graph is **concave up**. ■

The Second Derivative Test

For a function whose second derivative exists on an open interval containing c and has a critical value $x = c$ where $f'(c) = 0$, the point $(c, f(c))$ is a

1. **Relative minimum** if $f''(c) > 0$

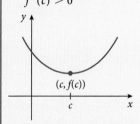

2. **Relative maximum** if $f''(c) < 0$

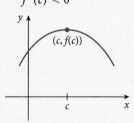

The Second Derivative Test fails if $f''(c) = 0$ or if $f''(c)$ is undefined. In this case, the First Derivative Test may be used.

In Example 1, since the graph is concave down at the relative maximum $(-3, 4)$, we know that $f''(-3) < 0$. Since the graph is concave up at the relative minimum $(2, -3)$, we know that $f''(2) > 0$. This observation is formally given as the **Second Derivative Test**.

Example 2: Using the Second Derivative Test to Locate Extrema

Use the Second Derivative Test to locate the relative extrema for

$$f(x) = x^3 + 3x^2 - 9x + 5$$

Perform the Mathematics

To apply the Second Derivative Test, we must first determine the critical values for f, which means that we need to determine $f'(x)$.

$$f'(x) = 3x^2 + 6x - 9$$

Since $f'(x)$ is defined for all x, the only critical values are where $f'(x) = 0$. Solving $f'(x) = 0$ yields

$$3x^2 + 6x - 9 = 0$$
$$3(x^2 + 2x - 3) = 0$$
$$3(x + 3)(x - 1) = 0$$
$$x = -3 \quad \text{or} \quad x = 1$$

So the critical values are $x = -3$ and $x = 1$. We now compute $f''(x)$ to be

$$f''(x) = 6x + 6$$

We evaluate $f''(x) = 6x + 6$ at each critical value and obtain

$$f''(-3) = 6(-3) + 6 = -12$$
$$f''(1) = 6(1) + 6 = 12$$

Since $f''(-3) < 0$, there is a relative maximum at $x = -3$. Since $f''(1) > 0$, there is a relative minimum at $x = 1$. **Figure 5.4.2** shows a graph of f.

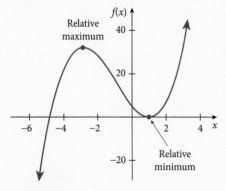

Figure 5.4.2   ■

▶ *Try It Yourself*

Some related Exercises are 9 and 11.

Absolute Extrema on Open Intervals

Recall that in Section 5.1 we used the Extreme Value Theorem to find the absolute maximum and absolute minimum on a closed interval. The key to this theorem was that it told us on a *closed interval*, a function is *guaranteed* to have an absolute maximum and an absolute minimum. If the interval is not closed, there is no guarantee that absolute extrema exist.

Figures 5.4.3 through **5.4.5** demonstrate that when the interval is open, we may or may not have absolute extrema. Probably one of the best methods to determine if a function has absolute extrema on an open interval is to look at the graph of the function. However, when the function in question is continuous and has only one critical value in the indicated interval, we can employ the **Second Derivative Test** to determine if an absolute maximum or an absolute minimum exists.

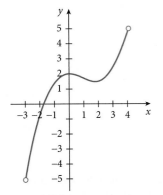

Figure 5.4.3 No absolute extrema on $(-3, 4)$.

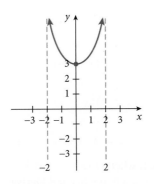

Figure 5.4.4 Absolute minimum, but no absolute maximum on $(-2, 2)$.

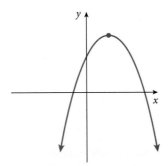

Figure 5.4.5 Absolute maximum, but no absolute minimum on $(-\infty, \infty)$.

The Second Derivative Test may be used on any interval (closed, open, or half-open) that contains only one critical value. So when we have a function, f, that is continuous on an interval and $x = c$ is the only critical value in the interval, we can modify the Second Derivative Test from earlier in this section to locate **absolute extrema** as follows:

Second Derivative Test for Absolute Extrema

For a continuous function, f, on *any* interval (open, closed, or half-open), and if $x = c$ is the **only** critical value in the interval where $f'(c) = 0$, if $f''(c)$ exists, then the point $(c, f(c))$ is

1. an **absolute minimum** if $f''(c) > 0$ **2.** an **absolute maximum** if $f''(c) < 0$

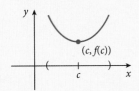   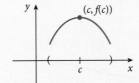

If $f''(c) = 0$, then this test fails.

Example 3: Determining an Absolute Extrema on an Interval

Determine the absolute minimum for $f(x) = 3x + \frac{27}{x}$ on $(0, \infty)$.

Perform the Mathematics

Since $(0, \infty)$ is an open interval, the techniques of Section 5.2 do not apply here. Applying the new techniques of this section requires us to first determine the derivative. Rewriting the function and then differentiating gives us

$$f(x) = 3x + \frac{27}{x} = 3x + 27x^{-1}$$
$$f'(x) = 3 - 27x^{-2} = 3 - \frac{27}{x^2}$$

OBJECTIVE 2

Use Second Derivative Test to determine absolute minimum.

Now, according to the Second Derivative Test for Absolute Extrema, we need to find the critical value $x = c$, where $f'(c) = 0$. Setting the derivative equal to zero and solving gives

$$3 - \frac{27}{x^2} = 0$$

$$\frac{3x^2 - 27}{x^2} = 0$$

$$\frac{3(x^2 - 9)}{x^2} = 0$$

$$\frac{3(x + 3)(x - 3)}{x^2} = 0$$

$$x = 3 \quad \text{or} \quad x = -3$$

So the only critical value in $(0, \infty)$, where $f'(x) = 0$, is $x = 3$. Since there is only one critical value that makes $f'(x) = 0$, we can apply the Second Derivative test for Absolute Extrema. To apply this requires the second derivative:

$$f'(x) = 3 - 27x^{-2}$$

$$f''(x) = 54x^{-3} = \frac{54}{x^3}$$

We now evaluate the second derivative at the critical value and get

$$f''(3) = \frac{54}{(3)^3} = \frac{54}{27} = 2$$

Since $f''(3) = 2$, $f''(3) > 0$, we conclude by the Second Derivative Test for Absolute Extrema that the absolute minimum for f on $(0, \infty)$ occurs at $x = 3$, and the absolute minimum is $f(3) = 3(3) + \frac{27}{3} = 18$. See **Figure 5.4.6**.

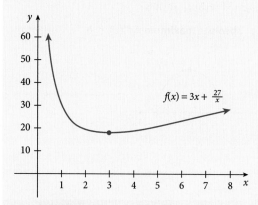

Figure 5.4.6

▶ *Try It Yourself*

Some related Exercises are 23 and 25.

Applications

In contrast to the problems in Section 5.2, the problems in this section have intervals that are not closed. Hence, we need the techniques presented in this section.

Example 4: Maximizing Volume

Lydia would like to make an open-top box out of a square piece of corrugated cardboard that measures 12 inches on each side. To do this, she needs to cut small squares from each corner and then turn up the edges. Determine the dimensions of the resulting box if its volume is to be a maximum.

Understand the Situation

We begin the solution by sketching the square piece of corrugated cardboard and labeling the length and width of the cutout pieces with an x. See **Figure 5.4.7**.

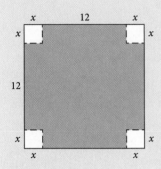

Figure 5.4.7

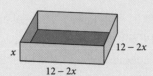

Figure 5.4.8

The formula for volume of a box is

$$V = l \cdot w \cdot h$$

Volume is what we want to maximize, but as it is currently written, we do not have the tools in our calculus arsenal to solve it. We need to represent the volume as a function of a single variable. To do this, we now draw a figure of what the box would look like. See **Figure 5.4.8**. The height, h, of the box is the same as x. The length, l, of our box is given by $(12 - 2x)$. This is due to the fact that the 12-inch square had x inches cut out from each corner. Hence, $12 - x - x$ results in the length being $(12 - 2x)$. By a similar argument, the width, w, of the box also measures $(12 - 2x)$. So we now have for the volume of this box

$$V = l \cdot w \cdot h$$
$$V(x) = (12 - 2x) \cdot (12 - 2x) \cdot x = (12 - 2x)^2 \cdot x$$

Perform the Mathematics

Now that we have the volume as a function of a single variable, we can apply techniques that are in our calculus arsenal to solve the problem. But first we need to consider the interval, or reasonable domain, for V. Since we want to make a box, $x \neq 0$, because $x = 0$ corresponds to making no cuts at all. Also, $x < 6$, since a square cutout of 6 inches on each side would produce four square pieces of cardboard and we could not make a box. (There would be nothing left to turn up!) So we claim that the length and width of the cutout, x, is in the interval $0 < x < 6$. Thus, we need to maximize

$$V(x) = (12 - 2x)^2 \cdot x \text{ on } (0, 6)$$

The interval is an open interval, so we need the techniques of this section to solve it. We first determine the critical values by computing the derivative using the Product and Chain Rules as follows:

$$V'(x) = 2(12 - 2x)(-2) \cdot x + (12 - 2x)^2 \cdot 1$$
$$= -4x(12 - 2x) + (12 - 2x)^2$$
$$= (12 - 2x)[-4x + (12 - 2x)]$$
$$= (12 - 2x)(-6x + 12)$$

Setting the derivative equal to zero and solving gives

$$(12 - 2x)(-6x + 12) = 0$$

$$x = 6 \text{ or } x = 2$$

The only critical value that is **in** the interval $(0, 6)$ is $x = 2$. We now use the Product Rule to determine $V''(x)$:

$$V''(x) = \frac{d}{dx}[(12 - 2x)(-6x + 12)]$$

$$= -2(-6x + 12) + (12 - 2x)(-6)$$

$$= 12x - 24 - 72 + 12x = 24x - 96$$

Evaluating $V''(x)$ at the critical value $x = 2$ yields

$$V''(2) = 24(2) - 96 = -48$$

Since $V''(2) = -48$, $V''(2) < 0$, we have an absolute maximum at $x = 2$. Thus, the dimensions of the box that has maximum volume is

Length: $12 - 2x = 12 - 2(2) = 12 - 4 = 8$
Width: $12 - 2x = 12 - 2(2) = 12 - 4 = 8$
Height: $x = 2$

Interpret the Result

So a length of 8 inches, width of 8 inches, and a height of 2 inches maximize the volume at 128 cubic inches.

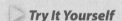 **Try It Yourself**

Some related Exercises are 44 and 45. ■

Example 4 suggests the following procedure when solving applied optimization problems on **an open interval**.

Strategy for Solving Applied Optimization Problems on an Open Interval

1. **Understand the Situation.** This includes:
   - Read the question carefully and, if possible, sketch a picture that represents the problem.
   - Select variables to represent the quantity to be maximized or minimized and all other unknowns.
   - Write an equation for the quantity to be maximized or minimized and determine the interval (reasonable domain) over which the function is to be optimized. If necessary, eliminate extra variables so that the quantity to be optimized is a function of one variable.

2. **Perform the Mathematics.** If the interval is open (or half-open), you must use the techniques of this section. If the interval is closed, you can use the techniques of Section 5.2.

3. **Interpret the Result.** This includes writing a sentence that answers the question posed in the problem.

In many settings, it is very easy to become too greedy. For example, if a farmer plants too much corn on an acre of ground, the yield can actually be reduced. If a restaurant has too many tables and they are too close together, patrons may actually stay away. If an accounting firm places too many accountants in an office, their productivity may decrease. Calculus and optimization can allow us to handle the problem of overcrowding, as illustrated in Example 5.

Example 5: Maximizing a Harvest

Laurie's Lemons has determined that the annual yield per lemon tree is fairly constant at 320 pounds when the number of trees per acre is 50 or fewer. The owner of Laurie's Lemons would like to maximize the annual yield per acre. To do this, she wants to plant more lemon trees. Research has shown that for each additional tree over 50, the annual yield per tree decreases by 4 pounds due to overcrowding. How many trees should be planted on each acre to maximize the annual yield from an acre?

Perform the Mathematics

The total yield for an acre is represented by

$$\text{Total yield} = (\text{number of trees per acre}) \cdot (\text{yield per tree})$$

Let x represent the number of trees per acre. To aid us in determining the yield per tree, we make the chart shown in **Table 5.4.1**.

Table 5.4.1

Number of trees	Yield Per Tree	Total Yield
10	320	3200
20	320	6400
40	320	12,800
50	320	16,000
51	$320 - 4(51 - 50) = 316$	16,116
52	$320 - 4(52 - 50) = 312$	16,224
53	$320 - 4(53 - 50) = 308$	16,324
x	$320 - 4(x - 50)$	$x \cdot [320 - 4(x - 50)]$

The "yield per tree" column was unchanged until we hit 51 trees per acre. *Each* additional tree beyond 50 results in a decrease in yield of 4 pound per tree. After listing a few values beyond 50, we generalized by placing an x in for the number of trees and followed the pattern that was found for 51, 52, and 53 trees. If we now let $TY(x)$ be the total yield in pounds, we can represent this scenario by

$$TY(x) = x \cdot [320 - 4(x - 50)]$$

The reasonable domain for TY is [50, 130). We do not include $x = 130$ since a quick check shows that at $x = 130$ the yield would be zero! So our problem is to maximize

$$TY(x) = x \cdot [320 - 4(x - 50)] \quad \text{on} \quad [50, 130]$$

The derivative is computed, after algebraically simplifying $TY(x)$, to be

$$\begin{aligned}
TY(x) &= x \cdot [320 - 4(x - 50)] \\
&= x \cdot [320 - 4x + 200] \\
&= x \cdot [520 - 4x] \\
&= 520x - 4x^2 \\
TY'(x) &= 520 - 8x
\end{aligned}$$

Determining the critical values x such that $TY'(x) = 0$ gives

$$520 - 8x = 0$$

$$x = 65$$

We compute the second derivative to be

$$TY''(x) = -8$$

Evaluating the second derivative at $x = 65$ gives

$$TY''(65) = -8$$

So $TY''(65) < 0$, which means, according to the techniques of this section, that at $x = 65$ we have an absolute maximum. So Laurie's Lemons should plant 65 lemon trees per acre to maximize the total yield. ■

Try It Yourself

Some related Exercises are 53 and 55.

The method illustrated in Example 5 is used in any situation where increasing one item decreases total output by some amount. For example, adding additional work stations in a machine shop may decrease the output per station. This scenario would follow the Example 5 process.

Sustainable Harvest

Our next example shows how to **maximize a sustainable harvest**. If one is in the business of harvesting fish (possibly by aquaculture), animals, or even trees, the ability to resist overharvesting is fundamental to the long-term success of the business. Overharvesting can inhibit the ability of the population to reproduce and sustain itself. This is illustrated in Example 6.

Example 6: Maximizing a Sustainable Harvest

Sussex County allows hunters to shoot deer during a limited open hunting season. The length of the season is carefully determined by the State Department of Natural Resources to ensure a harvest that is sustainable year after year. A state biologist has determined that, in Sussex County, the deer population after one year is given by

$$f(x) = 2.1x - 0.001x^2$$

where x is the original population of the deer measured in hundreds. Determine the optimal population size and the yearly kill that it will sustain.

© PhotoDisc/Getty Images

Understand the Situation

We start by determining $h(x)$, the yearly harvest. The harvest is simply the difference between the deer population, $f(x)$, after one year and the original population, x. That is,

$$
\begin{aligned}
h(x) &= f(x) - x \\
&= 2.1x - 0.001x^2 - x \\
&= 1.1x - 0.001x^2
\end{aligned}
$$

Since we have no idea of the population size, the interval on which we consider $h(x)$ is $(0, \infty)$. (We do know that the population cannot be negative!) We wish to maximize $h(x)$ using the techniques of this section.

Perform the Mathematics

Computing the derivative gives us

$$h'(x) = 1.1 - 0.002x$$

The critical value x, such that $h'(x) = 0$, is found to be

$$1.1 - 0.002x = 0$$
$$x = 550$$

The second derivative is $h''(x) = -0.002$. Evaluating the second derivative at the critical value gives

$$h''(550) = -0.002$$

Thus, we have a maximum at $x = 550$.

Interpret the Result

So the population of deer in Sussex County should grow to 55,000. It will then sustain a yearly harvest of

$$h(550) = 1.1(550) - 0.001(550)^2 = 302.5$$

that is, a yearly sustainable harvest of 30,250 deer per year.

▷ Try It Yourself

Some related Exercises are 57 and 59. ∎

Inventory Costs

A successful retail store must pay attention to the size of its inventory. Overstocking can lead to extra interest costs, excessive warehouse rental charges, and possibly the danger of the product becoming obsolete or damaged. On the other hand, too small an inventory involves additional paperwork in reordering and extra delivery charges. There is also the danger of running out of stock.

Let's take a look at a real inventory problem. Records from previous years show that Linger's Luxury Office Furniture sells 300 executive desks a year. To plan their inventory accurately, their analysts assume that the desks sell steadily throughout the year. They could order these desks in lots of size 300, 150, or 100 or in general lots of size x. Regardless of the lot size, on average this store will have $\frac{x}{2}$ executive desks in stock on which it must pay **inventory costs**. See **Figure 5.4.9**.

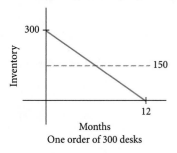
One order of 300 desks

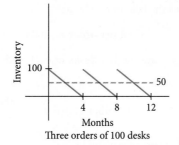
Three orders of 100 desks

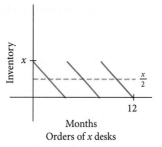
Orders of x desks

Figure 5.4.9

Ordering all 300 at once could result in high **storage costs**, whereas reordering several times throughout the year could result in high **reorder costs**. In Example 7 we determine the best lot size that minimizes the total of storage costs and reorder costs.

Example 7: Minimizing Inventory Costs

Linger's Luxury Office Furniture expects to sell 300 executive desks a year. Each desk costs the store $400, and there is a fixed charge of $800 per order. If it costs $200 to store an executive desk for a year, how large should each order be, and how often should orders be placed to minimize the inventory costs?

Perform the Mathematics

We begin by letting x represent the lot size, the number in each order. Our total inventory costs are represented by

$$\text{Total costs} = \text{storage costs} + \text{reorder costs}$$

We begin by determining the storage costs. We assume that the desks sell steadily throughout the year and that Linger's reorders x more when the stock is depleted. The inventory throughout the year would look like the graph of inventory as a function of the months in a year, as shown in **Figure 5.4.10**.

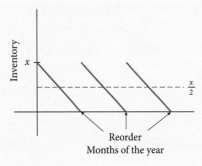

Figure 5.4.10

Since the number in stock varies from x to 0, we consider an *average inventory* of $\frac{x}{2}$. Since it costs $200 to store a desk for a year, we determine the storage costs as follows:

$$\text{Storage costs} = (\text{storage per item}) \cdot (\text{average number of items})$$
$$= 200 \cdot \left(\frac{x}{2}\right) = 100x$$

The next piece to determine is the reorder costs. In general, reorder costs are determined by

$$\text{Reorder costs} = (\text{cost per order}) \cdot (\text{number of orders})$$

Since each desk costs $400, we know that ordering x desks costs $400x$. There is a fixed charge of $800 per order, which gives the cost per order as

$$\text{Cost per order} = 400x + 800$$

A yearly supply is 300 desks, and with x desks in each order, we find the number of orders in one year to be

$$\text{Number of orders} = \frac{300}{x}$$

This means that our reorder costs are given by

$$\text{Reorder costs} = (\text{cost per order}) \cdot (\text{number of orders})$$
$$= (400x + 800) \cdot \frac{300}{x}$$

Thus our total inventory costs, $IC(x)$, are given by

$$\text{Total costs} = \text{storage costs} + \text{reorder costs}$$

$$IC(x) = 100x + (400x + 800) \cdot \frac{300}{x} \qquad 0 < x \leq 300$$

$$IC(x) = 100x + 120{,}000 + \frac{240{,}000}{x}$$

To minimize $IC(x)$, we employ the methods of this section. The derivative is

$$IC(x) = 100x + 120{,}000 + \frac{240{,}000}{x}$$

$$IC(x) = 100x + 120{,}000 + 240{,}000x^{-1}$$

$$IC'(x) = 100 - 240{,}000x^{-2} = 100 - \frac{240{,}000}{x^2}$$

We now find the critical value x, where $IC'(x) = 0$:

$$100 - \frac{240{,}000}{x^2} = 0$$

$$100 = \frac{240{,}000}{x^2}$$

$$x^2 = 2400$$

$$x \approx \pm 48.99$$

Since x represents the lot size, we round to the nearest integer and get $x = \pm 49$. We now reject $x = -49$ (since one cannot order a negative number of items) and begin to verify that $x = 49$ minimizes inventory costs. Computing the second derivative gives

$$IC''(x) = 480{,}000x^{-3} = \frac{480{,}000}{x^3}$$

Evaluating the second derivative at the critical value $x = 49$ gives

$$IC''(49) = \frac{480{,}000}{(49)^3} \approx 4.0799$$

Since $IC''(49) > 0$, we conclude that inventory costs, $IC(x)$, are minimized at $x = 49$. If there are 49 desks per order, the yearly total of 300 would require $\frac{300}{49} \approx 6.1224$ orders per year. Since this number of orders per year is impossible, it seems reasonable to round x to 50 (the number of orders would then be 6) and answer the question as follows: To minimize inventory costs, each lot size should have 50 desks with orders placed 6 times per year. ∎

> **Try It Yourself**

Some related Exercises are 61 and 63.

Summary

The main theme in this section was determining absolute extrema on intervals that are not closed. We illustrated how the second derivative handles this case, provided that there is only one critical value, $x = c$, on the interval where $f'(c) = 0$.

Specifically, we presented the **Second Derivative test for Absolute Extrema** and the **Strategy for Solving Applied Optimization Problems on an Open Interval**.

- For a continuous function, f, on *any* interval (open, closed, or half-open), and if $x = c$ is the **only** critical value in the interval where $f'(c) = 0$, if $f''(c)$ exists, then the point $(c, f(c))$ is (1) an **absolute minimum** if $f''(c) > 0$ or (2) an **absolute maximum** if $f''(c) < 0$. If $f''(c) = 0$, then this test fails.

Section 5.4 Exercises

Vocabulary Exercises

1. An interval of the form (a, b) or $a < x < b$ is called an _____ interval.

2. An interval of the form $[a, b)$ or $(a, b]$ is called a _____ _____ interval.

3. At the critical value $x = c$, if $f'(c) = 0$ and $f''(c) < 0$, then the point $(c, f(c))$ is an absolute _____ .

4. At the critical value $x = c$, if $f'(c) = 0$ and $f''(c) > 0$, then the point $(c, f(c))$ is an absolute _____ .

5. The Second Derivative Test is used to determine absolute _____ .

6. At the critical value $x = c$, if $f'(c) = 0$ and $f''(c) = 0$, then the Second Derivative Test _____ .

Skill Exercises

In Exercises 7–22, use the Second Derivative Test to locate any relative extrema, if they exist. In Exercises 15 and 16, use your calculator to graph the derivative and then use the Zero *command to approximate the solutions to $f'(x) = 0$.*

7. $f(x) = 3x^2 - 2x - 3$

8. $f(x) = -2x^2 + 3x + 2$

9. $f(x) = x^3 - 2x^2 - 13x - 10$

10. $f(x) = x^3 + 3x^2 - x - 3$

11. $f(x) = \frac{1}{3}x^3 + \frac{5}{2}x^2 + 6x - 2$

12. $f(x) = \frac{1}{3}x^3 - \frac{1}{2}x^2 - 6x + 2$

13. $f(x) = x^3 + \frac{3}{2}x^2 - 6x - 3$

14. $f(x) = -x^3 + 3x^2 - 3x + 5$

15. $f(x) = x^4 + x^3 - 7x^2 - x + 6$

16. $f(x) = x^4 - 3x^3 - 8x^2 + 12x + 16$

17. $f(x) = 3x^4 - 24x^2 + 16$

18. $f(x) = 2x^4 - 36x^2 + 16$

19. $f(x) = \frac{1}{x^2 + 1}$

20. $f(x) = \frac{1}{x^2 - 1}$

21. $f(x) = 3x^6 + 9x^4 - 5$

22. $f(x) = -\frac{1}{3}x^6 - 2x^4 + 3$

In Exercises 23–30, determine the absolute minimum (if any) of f on the indicated interval.

23. $f(x) = 2x + \frac{6}{x}$; $(0, 10)$

24. $f(x) = 4x + \frac{2}{x}$; $(0, 10)$

25. $f(x) = 4x - 3 + \frac{2}{x}$; $(0, 20)$

26. $f(x) = 2x - 5 + \frac{6}{x}$; $(0, 20)$

27. $f(x) = 3x^2 - 2 + \frac{6}{x^2}$; $(0, 5)$

28. $f(x) = 2x^2 - 3 + \frac{2}{x^2}$; $(0, 5)$

29. $f(x) = 3x^2 - 1 + \frac{2}{x^2}$; $(0, 10)$

30. $f(x) = 4x^2 - 2 + \frac{3}{x^2}$; $(0, 4)$

In Exercises 31–38, determine the absolute maximum (if any) of f on the indicated interval.

31. $f(x) = -2x - \dfrac{9}{x}$; $(0, 20)$

32. $f(x) = -3x - \dfrac{2}{x}$; $(0, 20)$

33. $f(x) = -3x - \dfrac{5}{x}$; $(0, 10)$

34. $f(x) = -2x - \dfrac{3}{x}$; $(0, 10)$

35. $f(x) = 3 - 2x^2 - \dfrac{3}{x^2}$; $(0, 10)$

36. $f(x) = 2 - 3x^2 - \dfrac{2}{x^2}$; $(0, 7)$

37. $f(x) = 2 - 3x - \dfrac{2}{x^2}$; $(0, 10)$

38. $f(x) = 3 - 2x - \dfrac{5}{x}$; $(0, 10)$

Application Exercises

39. Construction—Fence Building: Melissa has 400 feet of fencing with which to enclose two adjacent lots as shown in **Figure 5.4.11**.

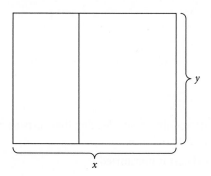

Figure 5.4.11

(a) Write a function $A(x)$ that represents the total area of the lots in terms of the length x and write an interval for the possible values of x.

(b) Determine the dimensions x and y that maximize the total area.

(c) What is the maximum area?

40. Construction—Fence Building: Melissa has decided to enclose two adjacent lots along a canal in such a way that one side requires no fencing, as shown in the figure. She has 400 feet of fencing in which to enclose the lots.

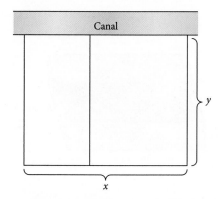

(a) Write a function $A(x)$ that represents the total area of the lots in terms of the length x and write an open interval for the possible values of x.

(b) Determine the dimensions x and y that maximize the total area.

(c) What is the maximum area?

41. Construction—Track Building: A 1320-foot athletic track encloses a rectangular region and the adjoining semicircular ends, as shown in **Figure 5.4.12**.

Figure 5.4.12

(a) Determine the dimensions x and y of the rectangle of maximum area.

(b) What is the maximum area?

42. Agriculture—Pest Prevention: Juan is an avid gardener and wishes to enclose two identical rectangular plots, each with an area of 1500 square feet, as shown in **Figure 5.4.13**.

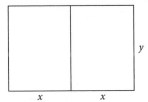

Figure 5.4.13

To keep raccoons out of his garden, the outer boundary requires heavy fencing that costs $8.50 per foot. The fence for the middle partition costs only $3.20 per foot.

(a) Determine the dimensions x and y so that the total cost is minimized.

(b) Determine the minimum cost.

43. Agriculture—Pest Prevention: Richelle is an avid gardener and wishes to enclose two identical rectangular plots, each with an area of 1200 square feet. (Consult Figure 5.4.13 here, too.) To keep skunks and rabbits out of her garden, the outer boundary requires heavy fencing that costs $6 per foot. The fence for the middle partition costs only $4 per foot.

(a) Determine the dimensions x and y so that the total cost is minimized.

(b) Determine the minimum cost.

44. Manufacturing—Constructing a Tray: The Brilliant Butler Company wishes to design a serving tray that will contain the most fluids in case of a spill. The serving tray is designed from a 6- by 16-inch piece of tin by cutting identical squares from the corners and folding up the flaps as shown in **Figure 5.4.14**.

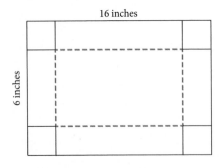
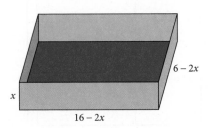

Figure 5.4.14

(a) Write a function $V(x)$ that gives the volume of the serving tray in terms of its height x. Write an open interval for the domain of the function.

(b) Determine the height, length, and width of the tray that maximizes the volume.

(c) What is the maximum volume?

45. Manufacturing—Constructing a Toy Box: The Rinky-Dinky Toy Company makes small farm animal figures and wishes to design a toy box that optimizes the number of figures that can fit inside. An open-top box to carry the toys is made from a 10- by 12-inch piece of cardboard by butting identical squares and then folding up and adjoining the flaps as in shown in **Figure 5.4.15**.

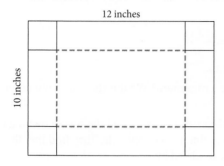
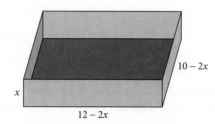

Figure 5.4.15

(a) Write a function $V(x)$ that gives the volume of the box in terms of its height x. Write an open interval for the domain of the function.

(b) Determine the height, length, and width of the box that maximizes the volume.

(c) What is the maximum volume?

For Exercises 46 and 47, consider the following scenario. The Parcel Packing Company is designing new shipping containers for international shipping and is using an open-top box design with a square base with lengths x and various heights y as shown in **Figure 5.4.16**. Recall that the surface area is equal to the area of the base plus the area of the four sides.

Figure 5.4.16

46. Commercial Design—Making Shipping Boxes: The Parcel Packing Company's SuperBox is made to carry multiple small items or large items such as personal safes and electronic equipment. The SuperBox has a volume of 6000 cubic inches. The material on the sides costs \$0.20 per square inch, and the material for the base is stronger and costs \$0.30 per square inch.

(a) Determine the dimensions of the SuperBox that minimize the cost.

(b) Compute the minimum cost.

47. Commercial Design—Making Shipping Boxes: The Parcel Packing Company's MightyMini box is made to carry small items or multiple items such as jewelry safely. The MightyMini box has a volume of 343 cubic inches. The material on the sides costs \$0.02 per square inch, and the material for the base is stronger and costs \$0.04 per square inch.

(a) Determine the dimensions of the MightyMini box that minimize the cost.

(b) Compute the minimum cost.

48. Marketing—Minimizing Average Cost: The marketing research department of the golfing magazine *Shank* has determined that the cost of printing, advertising, and distributing the magazine can be modeled by

$$C(x) = 0.003x^2 + 0.5x + 5 \ (0, \infty)$$

where x represents the number of magazines printed and sold each quarter in hundreds, and $C(x)$ represents the cost in hundreds of dollars.

(a) Write and simplify the average cost function $AC(x) = \dfrac{C(x)}{x}$.

(b) Determine the production level x that minimizes the average cost.

49. **Personal Finance—Total Car Costs:** *Car-Do-It* Magazine has examined the cost of owning the new NeoCommuter car, including the service and maintenance contracts. They find that the cost of owning the car that sells for $11,000 can be modeled by

$$C(x) = 100x^2 + 400x + 11,000 \ (0, 30)$$

where x represents the number of years since the car's purchase and $C(x)$ represents the owner's cost in dollars.

   (a) Determine the average cost function $AC(x) = \dfrac{C(x)}{x}$.

   (b) Determine the value of x that minimizes $AC(x)$.

   (c) After how many years will the average cost be a minimum? What is the minimum annual cost of owning the car?

50. **Personal Finance—Total Car Costs:** *Car-Do-It* Magazine has examined the cost of owning the new AeroCrossover car, including the service and maintenance contracts. They find that the cost of owning the car that sells for $15,000 can be modeled by

$$C(x) = 100x^2 + 600x + 15,000 \ (0, 30)$$

where x represents the number of years since the car's purchase and $C(x)$ represents the owner's cost in dollars.

   (a) Determine the average cost function $AC(x) = \dfrac{C(x)}{x}$.

   (b) Determine the value of x that minimizes $AC(x)$.

   (c) After how many years will the average cost be a minimum? What is the minimum annual cost of owning the car?

51. **Managerial Science—Determining Room Rates:** The Cost-Saver Cruise Inn motel has 200 rooms. The manager finds that the motel is filled when he sets the room rates at $50 per night. The records show that for each $1.50 increase in the room rate, two fewer rooms are rented.

   (a) Write the revenue function $R(x) = x \cdot p(x)$ for the problem situation and include the half-open interval that represents the function's reasonable domain.

   (b) Determine the room rate that maximizes the motel's nightly revenue. What is the maximum nightly revenue?

52. **Managerial Science—Incorporating Maintenance Costs:** Suppose that for the Cost-Saver Cruise Inn motel described in Exercise 51, it costs $4.50 to clean and maintain each room that is occupied.

   (a) Write the cost, C, and profit, P, functions for the motel where x represents the number of rooms that are occupied. Write the half-open interval that represents the reasonable domain of the profit function.

   (b) Determine the room rate that maximizes the motel's nightly profit. What is the maximum nightly profit?

53. **Agriculture—Maximizing Yield:** Dudley's Delicious Apple Farm has recorded their total annual yield and tree density and has determined that their apple tree yield is fairly constant at 352 pounds per tree when the number of trees per acre is 55 or fewer. They also find that because of overcrowding, for each additional tree over 55, the annual yields decreases by five pounds. How many trees need to be planted per acre to maximize the annual yield?

54. **Agriculture—Maximizing Yield:** The Pear Paradise Farm has recorded their total annual yield and tree density and has determined that their pear tree yield is fairly constant at 160 pounds per tree when the number of trees per acre is 40 or fewer. They also find that because of

overcrowding, for each additional tree over 40, the annual yields decreases by two pounds. How many trees need to be planted per acre to maximize the annual yield?

55. **Agriculture—Maximizing Yield:** The Orange Works Farm has recorded their total annual yield and tree density and has determined that their orange tree yield is fairly constant at 270 pounds per tree when the number of trees per acre is 30 or fewer. They also find that because of overcrowding, for each additional tree over 30, the annual yields decreases by three pounds. How many trees need to be planted per acre to maximize the annual yield?

56. **Agriculture—Maximizing Yield:** Walt's Walnut Grove has recorded their total annual yield and tree density and has determined that their walnut tree yield is fairly constant at 50 pounds per tree when the number of trees per acre is 30 or fewer. They also find that because of overcrowding, for each additional tree over 30, the annual yields decreases by 1.25 pounds. How many trees need to be planted per acre to maximize the annual yield?

57. **Environmental Studies—Maximum Sustainable Harvest:** In Sumber County, hunters are permitted to shoot deer during a limited hunting season from November 10th to December 5th. The duration of the hunting season is carefully determined by the state's Department of Natural Resources to ensure a harvest that is sustainable year after year. A state biologist has determined that the yearly growth curve for the deer population can be modeled by

$$f(x) = 1.5x - 0.002x^2$$

where x represents the original deer population in thousands. Determine the optimal population size and maximum yearly kill that the population will sustain.

58. **Environmental Studies—Maximum Sustainable Harvest:** In Poling County, hunters are permitted to shoot elk during a limited hunting season from August 28th to September 6th. The duration of the hunting season is carefully determined by the state's Department of Natural Resources to ensure a harvest that is sustainable year after year. A state biologist has determined that the yearly growth curve for the elk population can be modeled by

$$f(x) = 1.75x - 0.003x^2$$

where x represents the original elk population in hundreds. Determine the optimal population size and maximum yearly kill that the population will sustain.

59. **Environmental Studies—Maximum Sustainable Harvest:** In Hammon County, trappers are permitted to trap cottontail rabbits during a season from September 18th to February 28th. The duration of the trapping season is carefully determined by the state's Department of Natural Resources to ensure a harvest that is sustainable year after year. A state biologist has determined that the yearly growth curve for the cottontail rabbit population can be modeled by

$$f(x) = 1.3x - 0.0003x^2$$

where x represents the original rabbit population in hundreds. Determine the optimal population size and maximum yearly kill that the population will sustain.

60. **Environmental Studies—Maximum Sustainable Harvest:** In Kroner County, hunters are permitted to shoot pheasant during a limited hunting season from November 5th to January 9th. The duration of the hunting season is carefully determined by the state's Department of Natural Resources to ensure a harvest that is sustainable year after year. A state biologist has determined that the yearly growth curve for the pheasant population can be modeled by

$$f(x) = 2.2x - 0.01x^2$$

where x represents the original pheasant population in hundreds. Determine the optimal population size and maximum yearly kill that the population will sustain.

In Exercises 61–69, complete the following.

Find the lot size and how often orders should be placed to minimize inventory costs. Round to the nearest lot size that produces an integer value of the number of orders placed.

61. **Inventory Management—Minimizing Inventory Costs:** Linger's Luxury Office Furniture expects to sell 400 junior executive desks a year. Each desk costs the store $300 and there is a fixed charge of $600 per order. If it costs $200 to store a junior executive desk for a year, how large should each order be and how often should orders be placed to minimize the inventory costs?

62. **Inventory Management—Minimizing Inventory Costs:** Linger's Luxury Office Furniture expects to sell 200 presidential desks a year. Each desk costs the store $600 and there is a fixed charge of $600 per order. If it costs $300 to store a presidential desk for a year, how large should each order be and how often should orders be placed to minimize the inventory costs?

63. **Inventory Management—Minimizing Inventory Costs:** The VideoPhile Camera Store expects to sell 480 video camera carrying cases in a year. Each carrying case costs the store $26 and there is a fixed charge of $30 per order. If it costs $1 to store a carrying case for a year, how large should each order be and how often should orders be placed to minimize inventory costs?

64. **Inventory Management—Minimizing Inventory Costs:** The VideoPhile Camera Store described in Exercise 63 has recently found a new facility to store the carrying cases. In this new facility, it costs $0.50 to store a carrying case for a year. Given this new amount, rework Exercise 63 to determine how large each order should be and how often orders should be placed to minimize inventory costs.

65. **Inventory Management—Minimizing Inventory Costs:** Sandkuhl's Appliances expects to sell 2500 gas stoves in a year. Each gas stove costs the store $500, and there is a fixed charge of $20 per order. If it costs $10 to store a gas stove for a year, how large should each order be and how often should orders be placed to minimize inventory costs?

66. **Inventory Management—Minimizing Inventory Costs:** Sandkuhl's Appliances expects to sell 2400 microwave ovens in a year. Each microwave oven costs the store $120, and there is a fixed charge of $54 per order. If it costs $9.00 to store a microwave oven for a year, how large should each order be and how often should orders be placed to minimize inventory costs?

67. **Inventory Management—Minimizing Inventory Costs:** Beagle's Department Store expects to sell 1200 pairs of blue jeans in a year. Each pair of blue jeans costs the store $12, and there is a fixed charge of $100 per order. If it costs $4 to store a pair of blue jeans for a year, how large should each order be and how often should orders be placed to minimize the inventory costs?

68. **Inventory Management—Minimizing Inventory Costs:** Gene's New and Used Cars expects to sell 300 cars in a year. On average, the cars cost $8500 each, and there is a fixed charge of $750 per order. If it costs $1500 to store a car for a year, how large should each order be and how often should orders be placed to minimize the inventory costs?

69. **Inventory Management—Minimizing Inventory Costs:** Simpson's Shoes expects to sell 1000 pairs of a certain running shoe in a year. Each pair of shoes costs the store $15, and there is a fixed charge of $150 per order. If it costs $6 to store a pair of shoes for a year, how large should each order be and how often should orders be placed to minimize inventory costs?

Exercises 70–72 examine the aspect of manufacturing known as **production runs**. *These Exercises can be handled in a manner similar to Exercises 61–69.*

70. **Production Management—Production Runs:** A publisher estimates that the annual demand for a book will be 5000 copies. Each book costs $17 to print, and setup costs are $1800 for each printing. If storage costs are $3 per book per year, determine how many books should be printed per run and how many printings will be needed in order to minimize costs.

71. **Production Management—Production Runs:** A manufacturer estimates the demand for a new CD-ROM game to be 8000 per year. Each CD-ROM game costs $14 to produce, and it costs $800 in setup costs. If a CD-ROM game can be stored for a year at a cost of $3, how many should be produced at a time and how many production runs will be needed in order to minimize costs?

72. **Production Management—Production Runs:** A toy manufacturer estimates the demand for a toy car to be 50,000 per year. Each toy car costs $3 to make, plus setup costs of $500 for each production run. If it costs $2 to store a toy car for a year, how many toy cars should be manufactured at a time and how many production runs will be needed in order to minimize costs?

Concept and Writing Exercises

73. Explain why $(c, f(c))$ is a relative maximum if $f''(c) < 0$, provided $f'(c) = 0$.

74. In the March 2010 article "Increasing Rate of IRS Threats Spurred by Plane Crash?" tax specialist Manny Davis stated, "In 2009 threats on the IRS have increased and in the past few weeks the IRS has been receiving threats at an increasing rate." (*Source:* BackTaxesHelp.com). In terms of applied differential calculus, what is wrong with Mr. Davis's claim about the IRS threats?

For Exercises 75–78, assume that a, b, and c represent positive real numbers.

75. Determine the absolute minimum of $f(x) = ax + \frac{b}{x}$ on the interval $(0, \infty)$.
76. Determine the absolute maximum of $f(x) = -ax - \frac{b}{x}$ on the interval $(-\infty, 0)$.
77. Determine the absolute minimum of $f(x) = ax - b + \frac{c}{x}$ on the interval $(0, \infty)$.
78. Determine the absolute maximum of $f(x) = a - bx - \frac{c}{x^2}$ on the interval $(0, \infty)$.

Section Project

The source of the data for this project is the Federal Express website (www.fedex.com).

(a) Federal Express states that any package that exceeds 165 inches in length plus girth is to be considered U.S. Domestic Freight. See **Figure 5.4.17**. Assuming that the front of the package is a square, determine the largest volume package that Federal Express will accept and not declare it to be U.S. Domestic Freight.

(b) Federal Express states that the maximum length plus girth for its FedEx Overnight Freight and its FedEx 2Day Freight shipments is 300 inches. See Figure 5.4.17.

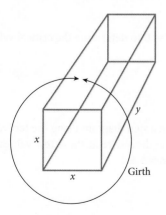

Figure 5.4.17

Assuming that the front of the package is a square and the maximum length plus girth is 300 inches, determine the dimensions that will maximize the volume and determine the maximum volume.

1. Analyze a function and its graph.
2. Apply graphical analysis to shrimp capture.

5.5   Graphical Analysis and Curve Sketching

In Section 5.1, we learned how the rate function f' can be used to determine the behavior of a continuous function f. To find relative extrema, determine the critical values of f and use the First Derivative Test. Then in Section 5.2, we learned how to optimize functions on a closed interval, which was another application of the derivative.

In Section 5.3, we learned how the second derivative f'' gives important information about the behavior of the function f. To find intervals of concavity for a function f, determine the second derivative f'' and apply the Tests for Concavity. Then in Section 5.4, we learned the Second Derivative Test and used it to optimize functions on *any* interval (open or half-open) that had only one critical value in the interval.

Synthesis

In this section, we pull together all of the concepts studied so far in Chapter 5 to continue to analyze functions. We will analyze more in depth the graph of a function, and we will continue to look at locating and interpreting relative extrema and inflection points.

OBJECTIVE 1

Analyze a function and its graph.

Example 1: Analyzing a Function and Its Graph

Consider $f(x) = 5x^4 - x^5$.

a.   Determine intervals where f is increasing and where it is decreasing and locate all relative extrema.

b.   Determine intervals where the graph of f is concave up and where it is concave down.

c.   Locate any inflection points.

d.   Using the information gathered in parts (a)–(c), sketch a graph of f and label the points found in parts (a) and (c).

Perform the Mathematics

a.   To determine intervals where f is increasing and where it is decreasing, as well as locating any relative extrema, we determine the critical values of f and apply the First Derivative Test. The derivative is

$$f'(x) = 20x^3 - 5x^4$$

Since $f'(x)$ is defined for all x, we need to solve $f'(x) = 0$ to determine the critical values.

$$20x^3 - 5x^4 = 0$$
$$5x^3(4 - x) = 0$$
$$x = 0 \text{ or } x = 4$$

We place the critical values on a number line and make a sign diagram using the test values $x = -1, x = 1$, and $x = 5$. Recall that we evaluate the derivative at these test values and place a sign of the result in our sign diagram as in **Figure 5.5.1**.

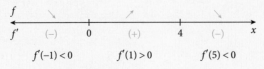

Figure 5.5.1

From the sign diagram in Figure 5.5.1, we have that $f(x) = 5x^4 - x^5$ is decreasing on $(-\infty, 0) \cup (4, \infty)$ and increasing on $(0, 4)$. There is a relative minimum at $x = 0$, specifically, $(0, 0)$, and a relative maximum at $x = 4$, specifically, $(4, 256)$.

b. To find intervals of concavity, we determine the second derivative and apply the Tests for Concavity. The second derivative is

$$f''(x) = 60x^2 - 20x^3$$

Since $f''(x)$ is defined for all x, we need to solve $f''(x) = 0$.

$$60x^2 - 20x^3 = 0$$
$$20x^2(3 - x) = 0$$
$$x = 0 \text{ or } x = 3$$

We place $x = 0$ and $x = 3$ on a number line and make a sign diagram using the test values $x = -1, x = 1,$ and $x = 5$. Recall that we evaluate the second derivative at these test values and place the sign of the result in our sign diagram as shown in **Figure 5.5.2**.

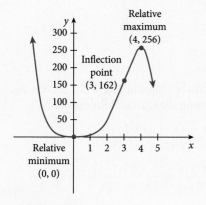

Figure 5.5.2

So the graph of $f(x) = 5x^4 - x^5$ is concave up on $(-\infty, 0) \cup (0, 3)$ and concave down on $(3, \infty)$.

c. Recall that at an inflection point the sign of f'' changes. Since the sign of f'' changes at $x = 3$, the only inflection point occurs when $x = 3$. Specifically, it is at the point $(3, 162)$.

d. **Figure 5.5.3** is a sketch of the function incorporating all the information gathered in parts (a)–(c).

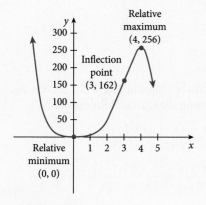

Figure 5.5.3

▶ *Try It Yourself*

Some related Exercises are 19 and 21.

▶ **Technology Option**

When doing the graphical analysis of a function, such as we did in Example 1, we can use a graphing calculator as a final step to confirm our sketch. In doing so, we notice how the calculus is helpful in determining an appropriate viewing window. For example, in Example 1 part (a), we determined that the function has a relative maximum at $(4, 256)$ and a relative minimum at $(0, 0)$. This tells us that a viewing window of $[-10, 10]$ by $[-50, 300]$ should show the important features of the graph. See **Figure 5.5.4**.

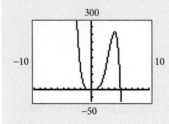

Figure 5.5.4

Graphical Analysis of Function f

1. Use $f'(x)$ to:
   - Determine critical values of f.
   - Construct a sign diagram to determine intervals where f is increasing and where it is decreasing.
   - Locate relative extrema by the First Derivative Test.
2. Use $f''(x)$ to:
   - Construct a sign diagram to determine intervals where the graph of f is concave up and where it is concave down.
   - Locate inflection points.

Before proceeding to Example 2, let's summarize the steps employed in Example 1 for the **graphical analysis of a function**.

Example 2: Analyzing a Function and Its Graph

Consider $f(x) = \dfrac{x - 1}{2x - 3}$.

a. Determine intervals where f is increasing and where it is decreasing. Also, locate any relative extrema.

b. Determine intervals where the graph of f is concave up and where it is concave down. Also, locate any inflection points.

c. Identify any vertical and horizontal asymptotes.

d. Sketch a graph of f.

Perform the Mathematics

a. To determine intervals where the function is increasing and where it is decreasing, we need the derivative. Differentiating this function using the Quotient Rule gives

$$f'(x) = \frac{1 \cdot (2x - 3) - (x - 1) \cdot 2}{(2x - 3)^2}$$

$$= \frac{2x - 3 - 2x + 2}{(2x - 3)^2} = \frac{-1}{(2x - 3)^2}$$

Notice that $f'(x)$ is undefined at $x = \frac{3}{2}$ since this value makes the denominator zero. Even though $x = \frac{3}{2}$ makes the derivative undefined, it is not a critical value, because $x = \frac{3}{2}$ is not in the domain of f. Also notice that there are no values of x such $f'(x) = 0$, that is,

$$\frac{-1}{(2x - 3)^2} = 0$$

has no real solution. Thus, f has no critical values. However, we include $x = \frac{3}{2}$ on a number line when constructing our sign diagram so that we can exploit the power of Theorem 5.1. Consult the Toolbox for a reminder on Theorem 5.1.

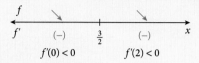

If f is continuous and $f(x) \neq 0$ for any x in the interval (a, b), then f cannot change sign on (a, b).

Figure 5.5.5 has the result of constructing a sign diagram.

$$f \qquad\qquad \searrow \qquad\qquad\qquad \searrow$$

$$f' \qquad (-) \qquad \tfrac{3}{2} \qquad (-) \qquad\qquad x$$

$$\quad f'(0) < 0 \qquad\qquad f'(2) < 0$$

Figure 5.5.5

So $f(x) = \dfrac{x - 1}{2x - 3}$ is decreasing on $(-\infty, \tfrac{3}{2}) \cup (\tfrac{3}{2}, \infty)$. Utilizing the First Derivative Test using a sign diagram indicates that f has no relative extrema.

b. To find intervals of concavity, we determine the second derivative and apply the Test for Concavity. We have from part (a) that the first derivative is $f'(x) = \dfrac{-1}{(2x - 3)^2}$. To compute $f''(x)$, we rewrite $f'(x)$ as

$$f'(x) = -1(2x - 3)^{-2}$$

and employ the Power and Chain Rules. This gives us

$$f''(x) = -1[-2(2x - 3)^{-3} \cdot 2]$$

$$= 4(2x - 3)^{-3}$$

$$= \frac{4}{(2x - 3)^3}$$

As in part (a), $f''(x)$ is undefined at $x = \tfrac{3}{2}$, and $f''(x) = 0$ has no real solution. Thus, the only value that we place on the number line is $x = \tfrac{3}{2}$. See **Figure 5.5.6**.

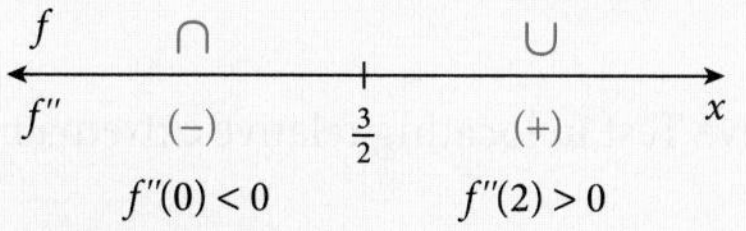

$$f \qquad\qquad \cap \qquad\qquad\qquad \cup$$

$$f'' \qquad (-) \qquad \tfrac{3}{2} \qquad (+) \qquad\qquad x$$

$$\quad f''(0) < 0 \qquad\qquad f''(2) > 0$$

Figure 5.5.6

So the graph of $f(x) = \dfrac{x - 1}{2x - 3}$ is concave down on $(-\infty, \tfrac{3}{2})$ and concave up on $(\tfrac{3}{2}, \infty)$. Even though the graph changes concavity at $x = \tfrac{3}{2}$, as stated earlier, this number is not in the domain of f; hence, there are no inflection points for the graph of f.

c. Utilizing information from Section 2.2, since $\displaystyle\lim_{x \to (3/2)^-} \frac{x - 1}{2x - 3} = -\infty$ and $\displaystyle\lim_{x \to (3/2)^+} \frac{x - 1}{2x - 3} = \infty$, we have a vertical asymptote at $x = \tfrac{3}{2}$. Also from Section 2.2, we determine the horizontal asymptote by computing

$$\lim_{x \to \infty} \frac{x - 1}{2x - 3} = \lim_{x \to \infty} \frac{\dfrac{x}{x} - \dfrac{1}{x}}{\dfrac{2x}{x} - \dfrac{3}{x}}$$

$$= \lim_{x \to \infty} \frac{1 - \dfrac{1}{x}}{2 - \dfrac{3}{x}} = \frac{1 - 0}{2 - 0} = \frac{1}{2}$$

Similarly,

$$\lim_{x \to -\infty} \frac{x-1}{2x-3} = \frac{1}{2}$$

So the horizontal asymptote is $y = \frac{1}{2}$.

d. **Figure 5.5.7** shows a graph of f using the information gathered in parts (a)–(c).

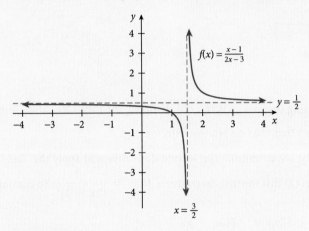

Figure 5.5.7

▶ *Try It Yourself*

Some related Exercises are 29 and 31.

Applications

Example 3 revisits the power of using the Second Derivative Test in locating relative extrema in some applications.

OBJECTIVE 2

Apply graphical analysis to shrimp capture.

Example 3: Analyzing Domestic Shrimp Catch

The oil spill in the Gulf of Mexico in 2010 highlighted how important the shrimp industry is to a state such as Louisiana. The total pounds of shrimp caught in the United States can be modeled by

$$f(x) = -1.86x^3 + 31.39x^2 - 154.03x + 456.99 \qquad 1 \le x \le 9$$

where x represents the number of years since 1999, and $f(x)$ is the total amount of shrimp caught in the United States in millions of pounds. (*Source:* U.S. National Oceanic and Atmospheric Administration.) Use the Second Derivative Test to locate any relative extrema and interpret each coordinate.

Perform the Mathematics

To apply the Second Derivative Test, we must first determine the critical values for f. This means that we need to determine $f'(x)$. We compute $f'(x)$ to be

$$f'(x) = -5.58x^2 + 62.78x - 154.03$$

Since the derivative is defined for all x, the only critical value is when $f'(x) = 0$. Hence, we must solve

$$-5.58x^2 + 62.78x - 154.03 = 0$$

We solve this quadratic equation by using the quadratic formula. This gives

$$x = \frac{-b \pm \sqrt{b^2 - 4ac}}{2a} = \frac{-62.78 \pm \sqrt{(62.78)^2 - 4(-5.58)(-154.03)}}{2(-5.58)}$$

$$= \frac{-62.78 \pm \sqrt{503.3788}}{-11.16}$$

So our two solutions are, rounded to the nearest hundredth,

$$x = \frac{-62.78 + \sqrt{503.3788}}{-11.16} \text{ or } x = \frac{-62.78 - \sqrt{503.3788}}{-11.16}$$

$$x = 3.62 \qquad\qquad x = 7.64$$

Each value is within the interval $[1, 9]$, so both are critical values that we will analyze with the Second Derivative Test. We now compute the second derivative to be

$$f''(x) = -11.16x + 62.78$$

Evaluating $f''(x)$ at $x = 3.62$ gives us

$$f''(3.62) = -11.16(3.62) + 62.78 = 22.3808$$

Since $f''(3.62) > 0$, this means that we have a relative minimum at $x = 3.62$, specifically at the point $(3.62, 222.51)$. The y-coordinate has been rounded to the nearest hundredth. This means that in 2002, the total shrimp catch in the United States reached a relative minimum of about 222.51 million pounds.

Evaluating $f''(x)$ at our other critical value $x = 7.64$ gives us

$$f''(7.64) = -11.16(7.64) + 62.78 = -22.4824$$

Since $f''(7.64) < 0$, this means that we have a relative maximum at $x = 7.64$, specifically at the point $(7.64, 282.97)$. The y-coordinate has been rounded to the nearest hundredth. This means that in 2006, the total shrimp catch in the United States reached a relative maximum of about 282.97 million pounds. ∎

Example 4: Analyzing Domestic Shrimp Catch

In Example 3 we saw that the total pounds of shrimp caught in the United States can be modeled by

$$f(x) = -1.86x^3 + 31.39x^2 - 154.03x + 456.99 \qquad 1 \le x \le 9$$

where x represents the number of years since 1999 and $f(x)$ is the total amount of shrimp caught in the U.S. in millions of pounds. (*Source:* U.S. National Oceanic and Atmospheric Administration.) Locate any inflection points and interpret each coordinate.

Perform the Mathematics

To find any inflection points, we need to determine intervals of concavity and apply the Test for Concavity. From Example 3 we determined that the second derivative is

$$f''(x) = -11.16x + 62.78$$

The second derivative is defined for all x, so we must solve $f''(x) = 0$. This gives

$$-11.16x + 62.78 = 0$$

$$x = 5.63 \text{ rounded to the nearest hundredth}$$

Placing this value on a number line, we construct the sign diagram shown in **Figure 5.5.8**. We use $x = 2$ and $x = 7$ as our test values. Evaluating the second derivative at these test values gives the completed sign diagram as shown in Figure 5.5.8. From Figure 5.5.8 we see that the graph of f changes concavity at $x = 5.63$. So we have an inflection point at $(5.63, 252.84)$. (The y-coordinate has been rounded to the nearest hundredth.) So we conclude that the rate at which the total shrimp catch in the United States was increasing the greatest occurred a little more than halfway through 2004. See **Figure 5.5.9**.

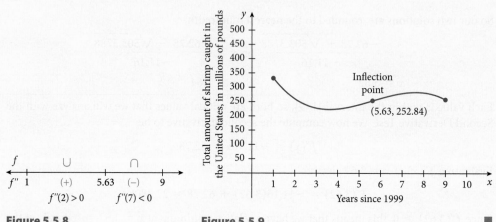

Figure 5.5.8   **Figure 5.5.9**

▶ *Try It Yourself*

Some related Exercises are 53 and 55.

Summary

Section 5.5 has brought together many concepts studied in Section 5.1 and 5.3. For a function f:

$f'(x)$	$f''(x)$
Gives the slope of a tangent line at any point	Gives the increasing or decreasing behavior of the rate of change
Gives an instantaneous rate of change	Determines concavity for the graph of f
Determines critical values	Determines inflection points of the graph of f
Determines intervals of increase or decrease for f	
Determines relative extrema	

Section 5.5 Exercises

Vocabulary Exercises

1. When analyzing a graph of a function f, we can begin by solving the equation $f(x) = 0$ to determine the _____ of the function.

2. When performing graphical analysis of a function, we begin by computing the _____ derivative.

3. We use a sign diagram of f' to determine where the graph of f is _____ and _____.

4. A sign diagram of f' is also used to determine the relative _____ of f.

5. We use a sign diagram of f'' to determine where the graph of f is _____ _____ and _____ _____.

6. If we are given a rational function, our graphical analysis would include finding the horizontal and vertical _____.

Skill Exercises

*For Exercises 7–12, refer to the graph of f shown in **Figure 5.5.10** for analysis.*

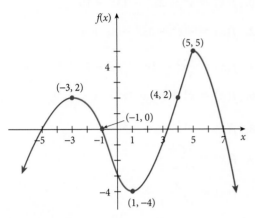

Figure 5.5.10

7. Determine intervals where f is positive and where f is negative.

8. Determine intervals where f is increasing and where f is decreasing.

9. Determine intervals where f' is positive and where f' is negative.

10. Determine intervals where f' is increasing and where f' is decreasing.

11. Determine intervals where the graph of f is concave up and where the graph of f is concave down.

12. Locate any relative extrema, and locate any inflection points.

*For Exercises 13–18, refer to the graph of f shown in **Figure 5.5.11** for analysis.*

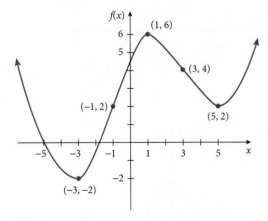

Figure 5.5.11

13. Determine intervals where f is positive and where f is negative.

14. Determine intervals where f is increasing and where f is decreasing.

15. Determine intervals where f' is positive and where f' is negative.

16. Determine intervals where f' is increasing and where f' is decreasing.

17. Determine intervals where the graph of f is concave up and where the graph of f is concave down.

18. Locate any relative extrema, and locate any inflection points.

In Exercises 19–38, sketch the graph of the given function using the techniques in this section. Label all relative extrema, inflection points, and asymptotes. In Exercises 25–28, use your calculator to graph the derivative and then use the Zero *command to approximate the solutions to* $f'(x) = 0$.

19. $f(x) = x^3 + 3x^2 - x - 3$

20. $f(x) = x^3 - 2x^2 - 13x - 10$

21. $f(x) = \frac{1}{3}x^3 - \frac{1}{2}x^2 - 6x + 2$

22. $f(x) = -x^3 + 3x^2 - 3x + 5$

23. $f(x) = x^3 - 6x^2 + 5$

24. $f(x) = \frac{1}{3}x^3 - 3x^2 + 5x - 2$

 25. $f(x) = x^4 - 3x^3 - 8x^2 + 12x + 16$

 26. $f(x) = x^4 + x^3 - 7x^2 - x + 6$

 27. $f(x) = x^4 - 2x^2 + 1$

28. $f(x) = x^4 - 9x^2 + 2$

29. $f(x) = \dfrac{x - 1}{x + 2}$

30. $f(x) = \dfrac{3x + 1}{5x - 2}$

31. $f(x) = \dfrac{1}{x^2 + 1}$

32. $f(x) = \dfrac{x}{x^2 - 1}$

33. $f(x) = 0.2x + 40 + \dfrac{20}{x}$

34. $f(x) = 0.3x + 20 + \dfrac{10}{x}$

35. $f(x) = x - \ln x$

36. $f(x) = e^x - x$

37. $f(x) = \sqrt[3]{x^2}$

38. $f(x) = \sqrt[3]{(x - 1)^2}$

In Exercises 39–42, the graph of f'' is given. Use the graph to sketch a graph of (a) f' and (b) f. There are many correct answers.

39.

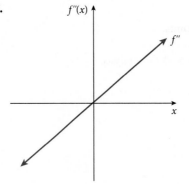

40.

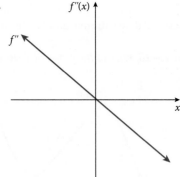

41.

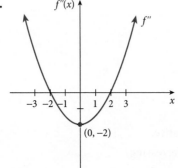

42.

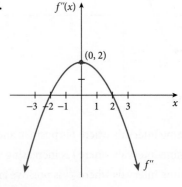

Application Exercises

43. Microeconomics—Revenue Analysis: The marketing research department for Music Time, a manufacturer of integrated amplifiers, used a large metropolitan area to test market their new product. They determined that price, $p(x)$, in dollars per unit, and quantity demanded per week, x, was approximated by

$$p(x) = 216 - 0.08x^2 \qquad 0 \le x \le 50$$

So the weekly revenue can be approximated by

$$R(x) = p(x) \cdot x = 216x - 0.08x^3 \qquad 0 \le x \le 50$$

(a) Compute $R'(x)$ and use it to determine intervals where R is increasing and where it is decreasing.

(b) Determine intervals where the graph of R is concave up and where it is concave down.

44. Microeconomics—Cost Analysis: Suppose that Music Time, the manufacturer in Exercise 43, has a weekly cost function in dollars given by

$$C(x) = 20x + 1000$$

(a) Determine P, the weekly profit function.

(b) Using the techniques of this section, graph P. Label all extrema and inflection points, if they exist.

45. Microeconomics—Profit Analysis: The Hanash Corporation determines that the weekly profit from producing and selling its Jog-R-Radio can be modeled by

$$P(x) = -0.01x^2 + 12x - 2000 \qquad 0 \le x \le 1000$$

where x represents the number of radios produced and sold each week and $P(x)$ is the weekly profit in dollars.

(a) Recall that the average profit function, denoted AP, is given by $AP(x) = \dfrac{P(x)}{x}$. Determine $AP(x)$ and simplify algebraically.

(b) Using the techniques of this section, sketch a graph of AP on the interval $(0, 1000]$. Label all extrema and inflection points.

46. Microeconomics—Profit Analysis *(Continuation of Exercise 45):*

(a) Determine the marginal profit function $MP(x)$ and simplify algebraically.

 (b) Graph both MP and AP (from Exercise 45) in the same coordinate system on the interval $(0, 1000]$.

(c) Algebraically determine the point where the two graphs intersect.

(d) Compare the x-coordinate of the result from part (c) with the x-coordinate of the relative maximum for AP. Are they the same?

47. Microeconomics—Profit Analysis *(Continuation of Exercises 45 and 46):* Fill in the blanks illustrating an important fact from economics:

The maximum average profit occurs when _____ *is equal to* _____ .

48. Economics—Average Cost: The Digital Pet Company has determined that its daily cost, in dollars, for producing x virtual pets is given by

$$C(x) = 150 + 3x + \frac{x^2}{15} \qquad 0 \le x \le 200$$

(a) Determine the average cost function $AC(x) = \dfrac{C(x)}{x}$.

(b) Using the techniques of this section, sketch a graph of AC on the interval $(0, 200]$. Label all extrema and inflection points.

49. Economics—Marginal Analysis (*Continuation of Exercise 48*):

  (**a**) Determine the marginal cost function $MC(x)$.

  (**b**) Graph both MC and AC (from Exercise 48) in the same coordinate system on the interval $(0, 200]$.

  (**c**) Algebraically determine the point where the two graphs intersect.

  (**d**) Compare the x-coordinate of the result from part (c) with the x-coordinate of the relative minimum for AC. Are they the same?

50. (*Continuation of Exercises 48 and 49*): Fill in the blanks illustrating an important fact from economics.

> *The minimum average cost occurs when* _____ *is equal to* _____.

51. Personal Finance—New Car Purchase: Miranda's New and Used Cars has a special program for individuals who purchase a new car. At the time of purchase, buyers can also purchase a maintenance and service contract with the dealership. This contract covers all recommended servicing of the vehicle. Tony just purchased a new car for \$11,000. The cost for the maintenance and service contract for his new car is \$500 the first year and increases \$200 per year thereafter, as long as Tony owns the car.

    Using the techniques of regression, he found that the total cost of the car (excluding items not covered by the service contract such as gasoline) after t years is given by

$$C(t) = 100t^2 + 400t + 11,000$$

  (**a**) Determine a function for the average cost per year where $AC(t) = \dfrac{C(t)}{t}$.

  (**b**) When is the average cost per year a minimum? Round your answer to the nearest tenth.

  (**c**) What is the minimum average cost per year rounded to the nearest dollar?

52. Personal Finance—New Car Purchase: Elisa just purchased a new car at Miranda's New and Used Cars. (See Exercise 51.) The cost was \$15,000 and the cost for the maintenance and service contract for her new car is \$700 the first year and increases \$200 per year thereafter, as long as Elisa owns the car.

    Using the techniques of regression, she found that the total cost of the car (excluding items not covered by the service contract such as gasoline) after t years is given by

$$C(t) = 100t^2 + 600t + 15,000$$

  (**a**) Determine a function for the average cost per year where $AC(t) = \dfrac{C(t)}{t}$.

  (**b**) When is the average cost per year a minimum? Round to the nearest tenth.

  (**c**) What is the minimum average cost per year rounded to the nearest dollar?

53. Hematology—Male Blood Volume: For males, blood volume in milliliters can be approximated by

$$f(x) = -10{,}822 + 3800 \ln x \qquad 40 \le x \le 90$$

where x represents the body weight in kilograms (kg) and $f(x)$ represents the blood volume in milliliters (mL).

  (**a**) Show that f is increasing on the interval $[40, 90]$. How can one interpret this function's behavior?

  (**b**) Show that the graph of f is concave down on the interval $[40, 90]$.

54. Hematology—Female Blood Volume: For females, blood volume can be approximated by

$$f(x) = -11{,}456 + 3888 \ln x \qquad 40 \le x \le 60$$

where x represents the body weight in kilograms (kg) and $f(x)$ represents the blood volume in milliliters (mL).

  (**a**) Show that f is increasing on the interval $[40, 60]$. How can one interpret this function's behavior?

  (**b**) Show that the graph of f is concave down on the interval $[40, 60]$.

55. **Biology—Heat Loss:** The movement of air is known as *convection*, and the removal of heat from the human body by convection air currents is called *heat loss by convection*. If we let x represent wind velocity, in miles per hour, then $f(x)$ represents the percent of total heat loss by convection and is approximated by

$$f(x) = 12 + 16.694\ln(x + 1) \qquad 0 \le x \le 60$$

   (a) Compute $f(20)$ and $f'(20)$ and interpret each.
   (b) Show that f is increasing on the interval $[0, 60]$.
   (c) Show that the graph of f is concave down on the interval $[0, 60]$.
   (d) Sketch a graph of f on the interval $[0, 60]$.

Concept and Writing Exercises

For Exercises 56–59, fill in the blanks. When we say "trace a graph," we mean to follow it from left to right.

56. If we trace a graph and it is increasing at a decreasing rate, we say it is concave _____.

57. If we trace a graph and it is decreasing at a decreasing rate, we say it is concave _____.

58. If we trace a graph and it is decreasing at an increasing rate, we say it is concave _____.

59. If we trace a graph and it is increasing at an increasing rate, we say it is concave _____.

60. Can an inflection point be determined by only using f'? Explain why or why not.

61. State in your own words the meaning of the second derivative in terms of instantaneous rate of change.

Section Project

A worry of many in central Europe is hazardous landfill waste. "It is expected that till the end of 2007 the Act on Environmental Burden will be adopted, that will solve problems of closed landfills where above all hazardous industrial wastes were landfilled," said Milena Okoličányiová of the Slovak Environmental Agency. (*Source:* Centre of Waste and Environmental Management.) The amount of waste landfilled per capita annually in Slovakia from 2000 to 2011 can be modeled by

$$f(x) = 0.36x^3 - 4.64x^2 + 21.06x + 195.11 \qquad 0 \le x \le 11$$

where x represents the number of years since 2000 and $f(x)$ represents the amount of waste that is landfilled per capita, measured in kilograms per person. (*Source:* Eurostat.) Use the function to analyze the following.

   (a) Graph f and $y = 251$ in the same viewing window and use the `Intersect` command to determine the year when the amount of waste was about 251 kilograms per person.
   (b) Determine the derivative f'.
   (c) Find the intervals where f is increasing. Are there any intervals where it is decreasing?
   (d) Determine the second derivative f''.
   (e) Find the coordinates of the inflection point.
   (f) Use the answer from part (e) to determine the years in which the amount of waste was increasing at a decreasing rate.
   (g) Use the answer from part (e) to determine the years in which the amount of waste was increasing at an increasing rate.

5.6 Elasticity of Demand

In Chapter 3, we discussed the concept of marginal analysis, which focused on the effect that a small change in the quantity produced and sold had on cost, revenue, and profit. Now we wish to know what effect a change in price has on the quantity demanded. Let's consider the following scenario. In the 1980s and early 1990s, Japanese microprocessor manufacturers flooded the technology market with low-price microchips in an attempt to force other overseas manufacturers out of business. This huge increase in microchips caused lower prices. The low prices, in turn, stimulated consumer demand, but reduced manufacturers' revenues dramatically. **Elasticity of demand** is a mathematical tool that can be used to measure the impact that a change in price has on the demand for a product. The term **elasticity** generally refers to how sensitive the demand is to a change in price.

Arc Elasticity

Let's begin by looking at some examples of elasticity. We call a product **elastic** if a small change in price produces a significant change in demand. The furniture market is an example of demand that is elastic. This market is sensitive to changes in price, as furniture shoppers tend to seek value and quality. On the other hand, we call a product **inelastic** if a change in price generally does not affect demand. The heating oil market is an example of a commodity that is inelastic. Since homeowners who use heating oil must have the product in the winter months, changes in price have little effect on the demand.

Generally, there are two types of elasticity. The first type that economists commonly use is **arc elasticity**. This type of elasticity is computed when the actual changes in price and quantity are known and a price-demand function may not be given. Arc elasticity measures the ratio of relative change in quantity to the relative change in price

$$(\text{Arc elasticity}) = -\frac{(\text{relative change in quantity demanded})}{(\text{relative change in price})}$$

$$= -\frac{\dfrac{\Delta q}{q}}{\dfrac{\Delta p}{p}}$$

The relative changes are simply the change in the quantity or price divided by their original value. The reason for the negative sign is purely a convention used by economists.

DEFINITION

Elastic and Inelastic Products

1. If small changes in the unit price of a product result in significant changes in demand, the product is **elastic.**

2. If small changes in the unit price of a product do not result in significant changes in demand, the product is **inelastic.**

DEFINITION

Arc Elasticity

For finite changes in quantity and price, the **arc elasticity,** denoted by E_a, is

$$E_a = -\frac{\dfrac{\Delta q}{q}}{\dfrac{\Delta p}{p}} = -\frac{\dfrac{q_2 - q_1}{q_1}}{\dfrac{p_2 - p_1}{p_1}}$$

where q_1 and q_2 are the original and new quantities, respectively, and p_1 and p_2 are the original and new prices, respectively.

OBJECTIVE 1

Determine an arc elasticity.

Example 1: Determining an Arc Elasticity

The GreenLawn Company can sell 300 mulching lawn mowers each week when the unit price of the mowers is $200. They find through a promotion that when the price is reduced by 10% to $180, the number sold rises by 20% to 360 mowers each week. Determine the arc elasticity.

Perform the Mathematics

Here we have the original quantity $q_1 = 300$ and the original price $p_1 = 200$. We also have the new quantity $q_2 = 360$ and the new price $p_2 = 180$. We will use the values to compute the arc elasticity as follows:

$$E_a = -\frac{\dfrac{q_2 - q_1}{q_1}}{\dfrac{p_2 - p_1}{p_1}} = -\frac{\dfrac{360 - 300}{300}}{\dfrac{180 - 200}{200}} = -\frac{\dfrac{60}{300}}{\dfrac{-20}{200}} = -\frac{0.2}{-0.1} = 2$$

So the arc elasticity, E_a, is 2. We observe that $E_a > 1$ and, in general, in order for this to happen, the percent change in sales exceeds the percent change in price. When this happens, we say that the demand is **elastic**. ∎

Try It Yourself

Some related Exercises are 7 and 39.

Arc elasticity is applicable when we have finite changes in quantity and price. However, we cannot determine the elasticity at a single quantity demanded. For this, we need the *instantaneous* change in price and quantity. To compute this type of elasticity, we need the derivative. This results in a new type of elasticity called **point elasticity**.

Point Elasticity

To determine this new type of elasticity, we need to rewrite the price-demand function. Recall that the **price-demand function** p gives us the price $p(x)$ at which people buy exactly x units of a product. Since the price-demand function is always decreasing on its domain, we know that the function is one-to-one. Thus, we can solve the price-demand function for x to get what economists call the **demand function**. This function plays a key role in determining point elasticity.

A graph of a typical demand function is shown in **Figure 5.6.1**. Notice that the price p is now on the horizontal axis and the demand x is on the vertical axis. To determine the demand function, we simply need to solve the price-demand function for x. Since p will become the independent variable, we will express $p(x)$ simply as p. Example 2 illustrates.

> **DEFINITION**
>
> **Demand Function**
>
> The **demand function**, denoted by $d(p)$, is given by
>
> $$x = d(p)$$
>
> The demand function gives the quantity x of a product demanded by consumers at price p.

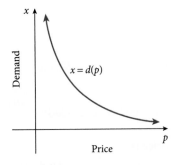

Figure 5.6.1 Graph of demand function $x = d(p)$.

OBJECTIVE 2

Determine a demand
function given
price-demand.

Example 2: Determining the Demand Function

Determine the demand function $x = d(p)$ for the price-demand function

$$p = \frac{1000}{x^2} \qquad 10 \le x \le 50$$

Perform the Mathematics

We can get the demand function by solving $p = \dfrac{1000}{x^2}$ for x.

$$p = \frac{1000}{x^2}$$

$$x^2 \cdot p = 1000$$

$$x^2 = \frac{1000}{p}$$

$$x = \sqrt{\frac{1000}{p}}$$

So the demand function is $x = d(p) = \sqrt{\dfrac{1000}{p}}$. Notice that we use only the positive square root, since both price and demand are represented by positive values. ∎

▶ **Try It Yourself**

Some related Exercises are 15 and 17.

To determine point elasticity, suppose that for a demand function the unit price of a product is increased by h dollars to $p + h$ dollars, as shown in **Figure 5.6.2**.

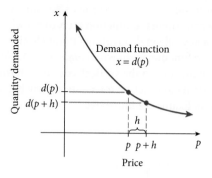

Figure 5.6.2

This causes the quantity demanded to decrease from $d(p)$ to $d(p + h)$ units. So we see that the relative change in demand is

$$(\text{relative change in quantity demanded}) = \frac{d(p + h) - d(p)}{d(p)}$$

The corresponding change in unit price is

$$(\text{relative change in price}) = \frac{(p + h) - p}{p} = \frac{h}{p}$$

Knowing from the beginning of this section that elasticity is the ratio of relative change in quantity to the relative change in price, we get

$$-\frac{(\text{relative change in quantity demanded})}{(\text{relative change in price})} = -\frac{\dfrac{d(p+h) - d(p)}{d(p)}}{\dfrac{h}{p}}$$

As the change in price becomes small, that is, $h \to 0$, we get the point elasticity formula:

$$(\text{Point elasticity}) = \lim_{h \to 0}\left[-\frac{\dfrac{d(p+h) - d(p)}{d(p)}}{\dfrac{h}{p}} \right]$$

$$= \lim_{h \to 0}\left[-\frac{\dfrac{d(p+h) - d(p)}{h}}{\dfrac{d(p)}{p}} \right] = \left[-\frac{\displaystyle\lim_{h \to 0}\dfrac{d(p+h) - d(p)}{h}}{\dfrac{d(p)}{p}} \right]$$

$$= \frac{-d'(p)}{\dfrac{d(p)}{p}} = -\frac{p \cdot d'(p)}{d(p)}$$

We will simply call this type of elasticity the **elasticity of demand**.

DEFINITION

Elasticity of Demand

Given the demand function $x = d(p)$, the **elasticity of demand**, denoted by $E(p)$, of a product at a price p is given by

$$E(p) = \frac{-p \cdot d'(p)}{d(p)}$$

Example 3: Determining the Elasticity of Demand

Determine the elasticity of demand $E(p)$ for each price-demand function.

a. $p = 12 - 0.4x$

b. $p = \dfrac{50}{2 - x}$

OBJECTIVE 3

Determine an elasticity of demand (point elasticity).

Perform the Mathematics

a. First, we must solve this price-demand function for x to get the demand function, $x = d(p)$.

$$p = 12 - 0.4x$$
$$p - 12 = -0.4x$$
$$\frac{p - 12}{-0.4} = x$$
$$-2.5p + 30 = x = d(p)$$

Now, we find $d'(p)$.

$$d'(p) = \frac{d}{dp}[-2.5p + 30] = -2.5$$

We can now determine the elasticity of demand as

$$E(p) = \frac{-p \cdot d'(p)}{d(p)} = \frac{-p(-2.5)}{-2.5p + 30} = \frac{2.5p}{-2.5p + 30}$$

b. Solving $p = \dfrac{50}{2 - x}$ for x, we get

$$p = \frac{50}{2 - x}$$

$$p(2 - x) = 50$$

$$2p - xp = 50$$

$$-xp = 50 - 2p$$

$$x = \frac{50 - 2p}{-p} = -\frac{50}{p} + 2 = 2 - \frac{50}{p} = 2 - 50p^{-1} = d(p)$$

Next, we determine $d'(p)$.

$$d'(p) = \frac{d}{dp}[2 - 50p^{-1}] = 50p^{-2} = \frac{50}{p^2}$$

This gives the elasticity of demand $E(p)$ as

$$E(p) = \frac{-p \cdot d'(p)}{d(p)} = \frac{-p\left(\dfrac{50}{p^2}\right)}{2 - \dfrac{50}{p}}$$

$$= \frac{\dfrac{-50}{p}}{\dfrac{2p - 50}{p}} = \frac{-50}{2p - 50} \qquad\blacksquare$$

Notice that the value of $E(p)$ has no units, since the units of demand are canceled out when computing the ratio of the relative changes. Also notice that the point elasticity function requires only one price to compute the elasticity of demand. The interpretation of the values of $E(p)$ is summarized in **Table 5.6.1**.

Table 5.6.1 Interpreting the $E(p)$ value

If the E(p) value is:	Then the demand is:	And this means a change in price will cause:
Between 0 and 1	Inelastic	Relatively small changes in demand
Greater than 1	Elastic	Relatively large changes in demand
Equal to 1	Unitary	A relatively equal change in demand

OBJECTIVE 4

Interpret an elasticity of demand.

Example 4: Interpreting the Elasticity of Demand

A school's junior business club is having its annual raffle. Data collected from past raffles indicate that the demand function for the tickets follows the model

$$x = d(p) = 36 - p^2$$

where p represents the price of a ticket and $x = d(p)$ represents the number of tickets each member sells each day.

a. Find the elasticity of demand $E(p)$.

b. Evaluate $E(3)$ and $E(4)$ and interpret the results.

Perform the Mathematics

a. We first determine $d'(p)$ and get

$$d(p) = 36 - p^2$$
$$d'(p) = -2p$$

Using the definition of elasticity of demand yields

$$E(p) = \frac{-p \cdot d'(p)}{d(p)} = \frac{-p \cdot (-2p)}{36 - p^2} = \frac{2p^2}{36 - p^2}$$

b. Evaluating $E(p)$ at $p = 3$, we get

$$E(3) = \frac{2(3)^2}{36 - (3)^2} = \frac{18}{27} = \frac{2}{3}$$

Since $E(3) = \frac{2}{3} < 1$, this means that the demand for the raffle ticket is **inelastic** at a price of $3 a ticket. A small increase in price will cause a negligible change in demand.

Evaluating $E(p)$ at $p = 4$, we get

$$E(4) = \frac{2(4)^2}{36 - (4)^2} = \frac{32}{20} = \frac{8}{5}$$

Since $E(4) = \frac{8}{5} > 1$, this means that the demand for the raffle ticket is **elastic** at a price of $4 a ticket. So a small increase in price will cause a significant change in demand. ∎

▶ *Try It Yourself*

Some related Exercises are 23 and 25.

In Example 4, we found that at a price of $3, the demand for a ticket was inelastic, so demand is only negligibly sensitive to price changes. Increasing prices slightly will increase the revenue $R(x)$, because the higher price will turn away relatively few buyers. Conversely, when the price was $4, the demand was elastic, so the demand is sensitive to price changes. Thus, to increase revenue in this case, the club should slightly lower its price. The lower price will increase demand for tickets and will generate more revenue, which will offset the lower price. Furthermore, if the point elasticity was $E(p) = 1$, then prices should not be changed, because an increase in price will result in exactly the same percentage decrease in demand. Thus, when $E(p) = 1$, revenue is at a maximum.

Since $R'(p)$ is the rate of change in the revenue as the price p changes, the equation $R'(p) = d(p)(1 - E(p))$ expresses this relationship between $R'(p)$ and $d(p)$. This is summarized in **Table 5.6.2**.

Table 5.6.2 Elasticity-Revenue Relationship

If the demand is:	Then E(p) is:	And R'(p) is:	So the revenue is:
Elastic	Greater than 1	Negative	Decreasing
Inelastic	Less than 1	Positive	Increasing
Unitary	1	Zero	At a maximum

From Table 5.6.2 and **Figure 5.6.3**, we observe the following:

1. If demand is elastic, $E(p) > 1$, then prices should be lowered to increase revenue.
2. If demand is inelastic, $E(p) < 1$, then prices should be raised to increase revenue.
3. If demand is unitary, $E(p) = 1$, then the revenue is at its maximum and the price should not be changed.

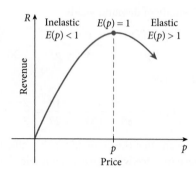

Figure 5.6.3

OBJECTIVE 5

Determine price adjustments.

Example 5: Determining Price Adjustment to Increase Revenue

The PowerTi Company produces neckties and sells them at a price of $15. The demand function for the ties is given by

$$x = d(p) = 170{,}000\sqrt{20 - p}$$

where p is the price of a necktie and $d(p)$ is the demand for the ties.

a. Determine $E(15)$ and state whether the manufacturer should lower or raise the price to increase revenue.

b. Set $E(p) = 1$ and solve for p to determine at what price the revenue is greatest.

Perform the Mathematics

a. First, we need the elasticity of demand $E(p)$. Computing the derivative of the demand function $d(p) = 170{,}000\sqrt{20 - p} = 170{,}000(20 - p)^{1/2}$, we get

$$d(p) = 170{,}000\sqrt{20 - p} = 170{,}000(20 - p)^{1/2}$$

$$d'(p) = 170{,}000\left(\frac{1}{2}\right)(20 - p)^{-1/2}(-1) = \frac{-85{,}000}{\sqrt{20 - p}}$$

Now, using this result in the point elasticity of demand formula gives

$$E(p) = \frac{-p \cdot d'(p)}{d(p)} = \frac{-p\left(\dfrac{-85{,}000}{\sqrt{20 - p}}\right)}{170{,}000\sqrt{20 - p}}$$

$$= \frac{85{,}000p}{\sqrt{20 - p}} \cdot \frac{1}{170{,}000\sqrt{20 - p}} = \frac{0.5p}{20 - p}$$

So at a price of $p = 15$, we get

$$E(15) = \frac{0.5(15)}{20 - 15} = \frac{7.5}{5} = 1.5$$

$E(15) = 1.5 > 1$, the demand is relatively elastic and the prices should be lowered to increase revenue.

b. To find the price that maximizes revenue, we need to find the value of p that makes $E(p) = 1$. So, solving this equation, we get

$$E(p) = 1$$

$$\frac{0.5p}{20 - p} = 1$$

$$0.5p = 20 - p$$

$$1.5p = 20$$

$$p = \frac{20}{1.5} \approx 13.33$$

Thus, to maximize revenue, the price of a necktie should be $13.33. ∎

▶ *Try It Yourself*

Some related Exercises are 49 and 51.

Summary

In this section, we discussed a topic that is specific to economics, **elasticity of demand**. We first computed elasticity for finite changes in price and demand using the **arc elasticity** formula. We saw that products could be classified as **elastic**, **inelastic**, or **unitary**. Then we examined elasticity for instantaneous changes in price and quantity, called **point elasticity**. Finally, we examined the relationship between the **elasticity of demand** and maximizing revenue.

- **Arc elasticity:** $E_a = -\dfrac{\dfrac{q_2 - q_1}{q_1}}{\dfrac{p_2 - p_1}{p_1}}$

- **Point elasticity:** $E(p) = \dfrac{-p \cdot d'(p)}{d(p)}$

Section 5.6 Exercises

Vocabulary Exercises

1. The sensitivity that a change in demand is to a change in price is called _____.
2. If a change in price does not affect demand, we call a product _____.
3. For finite changes in quantity and price, we can compute _____ elasticity.
4. The _____ function gives the quantity x of a product demanded by consumers at price p.
5. For continuous changes in quantity and price, we can compute _____ elasticity.
6. If the revenue of a product is at its maximum, we call the demand _____.

Skill Exercises

In Exercises 7–12, compute the arc elasticity E_a.

7. $q_1 = 50, q_2 = 55, p_1 = 110, p_2 = 100$ 8. $q_1 = 10, q_2 = 13, p_1 = 8, p_2 = 5$

9. $q_1 = 300, q_2 = 260, p_1 = 40, p_2 = 45$ 10. $q_1 = 5000, q_2 = 5800, p_1 = 60, p_2 = 54$

11. $q_1 = 5700, q_2 = 5600, p_1 = 390, p_2 = 400$ 12. $q_1 = 420, q_2 = 419, p_1 = 389, p_2 = 395$

In Exercises 13–22, the price-demand function p is given. Write the demand function $x = d(p)$.

13. $p = 600 - 100x$ 14. $p = 100 - x$

15. $p = \dfrac{300}{x^2} + 10$ 16. $p = \dfrac{700}{2x + 1}$

17. $p = 12 - 0.04x$ 18. $p = \sqrt{250 - x}$

19. $p = \sqrt{300 - x^2}$ 20. $p = 1200 - x^2$

21. $p = 100e^{-0.1x}$ 22. $p = 250e^{-0.3x}$

In Exercises 23–38, the demand function $x = d(p)$ and the price p are given. Determine the point elasticity of demand $E(p)$, then classify the demand as elastic, inelastic, or unitary.

23. $d(p) = 220 - 5p;\ p = 10$ 24. $d(p) = 50 - 4p;\ p = 5$

25. $d(p) = 200 - p^2;\ p = 8$ 26. $d(p) = 250 - p^2;\ p = 10$

27. $d(p) = \dfrac{200}{p};\ p = 15$ 28. $d(p) = \dfrac{1000}{7p};\ p = 20$

29. $d(p) = \sqrt{150 - 3p};\ p = 30$ 30. $d(p) = \sqrt{100 - 2p};\ p = 25$

31. $d(p) = \dfrac{100}{p^2};\ p = 30$ 32. $d(p) = \dfrac{300}{p^3};\ p = 20$

33. $d(p) = 4500e^{-0.02p};\ p = 200$ 34. $d(p) = 6500e^{-0.04p};\ p = 100$

35. $d(p) = 100 \cdot \ln(1000 - 10p);\ p = 19$ 36. $d(p) = 150 \cdot \ln(100 - 2p);\ p = 5$

37. $d(p) = 100e^{-0.05p};\ p = 40$ 38. $d(p) = 3000e^{-0.03p};\ p = 100$

Application Exercises

39. **Microeconomics—Product Demand:** The 4Com Company determines that the demand function for their mouse pad is

$$x = d(p) = 20 - 2p$$

where p is the unit price of the mouse pad and $d(p)$ is the quantity demanded in hundreds.

(a) Determine $d(5)$ and $d(6)$ and interpret.

(b) Using $q_1 = d(5)$, $q_2 = d(6)$, $p_1 = 5$, and $p_2 = 6$, determine the arc elasticity E_a and interpret.

40. **Microeconomics—Product Demand:** The WetLands Company determines that the demand function for their rain poncho is

$$x = d(p) = 50 - 5p$$

where p is the unit price of the rain poncho and $d(p)$ is the quantity demanded in thousands.

(a) Determine $d(5)$ and $d(7)$ and interpret.

(b) Using $q_1 = d(5)$, $q_2 = d(7)$, $p_1 = 5$, and $p_2 = 7$, determine the arc elasticity E_a and interpret.

41. **Microeconomics—Ticket Elasticity:** The 725th Street Movie Theater finds that it can sell 320 tickets per day when the price of each ticket is $8.50. During a promotion week, the theater finds that by reducing the ticket price to $7.25, the number of tickets sold increases to 410 tickets per day. Determine the arc elasticity and interpret.

42. **Microeconomics—Photo Elasticity:** The HappyFace Portrait Studio sells 325 family packs of photos monthly at a unit price of $49. When they offer a sale price of $39, the number of family packs increases to 330 sold per month. Determine the arc elasticity and interpret.

43. **Microeconomics—Jewelry Elasticity:** The JemJam Jewelry Store finds that they sell 52 pairs of emerald earrings per week at a price of $149 each. Under new management, the price is increased to $159 and the number sold decreases to 45 pairs per week. Determine the arc elasticity and interpret.

44. **Microeconomics—Trinket Elasticity:** The Fleetwood Company sells 600 stained glass window ornaments monthly at a unit price of $19. When they increased the price to $24 in order to absorb higher fixed costs, they found that the monthly sales decreased to 450 ornaments. Determine the arc elasticity and interpret.

45. **Transportation—Ticket Demand:** Currently, about 1800 people ride the MetroTram local public transportation system per day and pay $4 for each ticket. The number of people x willing to take the public transportation at price p is given by the demand function

$$x = 600(5 - p^{1/2})$$

   (a) Is the demand elastic or inelastic at the current $4 ticket price?

   (b) Should the ticket price be raised or lowered to increase revenue?

46. **Transportation—Ticket Demand:** The EconoCommute bus system charges 65 cents for each ride and serves approximately 57,800 riders each day. The demand function for x riders at p cents for each ride is given by

$$x = 2000\sqrt{900 - p}$$

   (a) Is the demand elastic or inelastic at the current 65-cent fare?

   (b) Should the rider's price be raised or lowered to increase revenue?

47. **Microeconomics—Clothing Demand:** The SilkTop clothing manufacturer sells blouses at $15 each and calculates the demand function for the blouses to be

$$d(p) = 60 - 3p$$

where p is the price of a blouse and $d(p)$ represents the demand in hundreds.

   (a) Is the demand elastic or inelastic at the current $15 price?

   (b) Should the blouse price be raised or lowered to increase revenue?

48. **Microeconomics—Cigarette Elasticity:** Suppose that the demand function for a certain brand of cigarettes is

$$d(p) = 4.5p^{-0.73}$$

where p is the price of a pack of cigarettes and $d(p)$ is the demand in thousands.

   (a) Determine the point elasticity of demand $E(p)$.

   (b) Show that the demand is inelastic for any price $p > 0$. Interpret what this means in terms of raising cigarette prices through manufacturer and tax price increases.

49. **Microeconomics—Ice Cream Elasticity:** The IceDream Company determines that the demand function for their frozen yogurt is

$$d(p) = 50 - 2p$$

where p is the price, in dollars, of a quart of frozen yogurt and $d(p)$ represents the demand in hundreds.

   (a) Determine $E(4)$ and state whether the manufacturer should lower or raise the price to increase revenue.

   (b) Set $E(p) = 1$ and solve for p to determine at what price the revenue is greatest.

50. **Microeconomics—Backpack Elasticity:** The PackIt Company determines that the demand function for their lightweight daypack is given by

$$d(p) = -p^2 + 400$$

where p is the unit price, in dollars, of a daypack and $d(p)$ is the demand.

(a) Determine $E(10)$ and state whether the manufacturer should lower or raise the price to increase revenue.

(b) Set $E(p) = 1$ and solve for p to determine at what price the revenue is greatest.

51. **Microeconomics—Night Light Elasticity:** The EverGlo Company determines that the demand for their new night light is given by

$$d(p) = 2 - \frac{p^2}{5}$$

where p is the unit price of a night light, in dollars, and $d(p)$ is the demand in thousands.

(a) Determine $E(1.50)$ and state whether the manufacturer should lower or raise the price to increase revenue.

(b) Set $E(p) = 1$ and solve for p to determine at what price the revenue is greatest.

52. **Microeconomics—Potato Chip Elasticity:** The SuperChip Company determines that the demand function for their gourmet potato chips is given by

$$d(p) = 3 - 0.1p - 0.1p^2$$

where p is the unit price, in dollars, of a bag of potato chips and $d(p)$ is the demand in thousands.

(a) Determine $E(4)$ and state whether the manufacturer should lower or raise the price to increase revenue.

(b) Set $E(p) = 1$ and solve for p to determine at what price the revenue is greatest.

53. **Microeconomics—TV Elasticity:** The MicroTV Company determines that the demand for their 3-inch color TV is

$$d(p) = 100 \cdot \ln(150 - p)$$

where p is the unit price, in dollars, of a 3-inch color TV and $d(p)$ is the demand in hundreds.

(a) Determine $E(100)$ and state whether the manufacturer should lower or raise the price to increase revenue.

(b) Set $E(p) = 1$ and solve for p to determine at what price the revenue is greatest.

Concept and Writing Exercises

54. Explain in your own words what the term *elasticity* means in economics.

55. Explain in terms of calculus how unitary elasticity corresponds to maximized revenue.

For Exercises 56–59, a demand function is given. Determine a function $E(p)$ for the point elasticity of demand. Assume that a and b are fixed constants.

56. $d(p) = a - bp$

57. $d(p) = a - p^2$

58. $d(p) = \dfrac{a}{p^2}$

59. $d(p) = ae^{-bp}$

Section Project

Some who study the theory of economics want to generalize the trends that they see relative to demand functions. To begin to understand how this is done, answer the following:

(a) If the demand for a product is $d(p) = 300p^{-0.3}$, show that the elasticity of demand $E(p)$ is a constant.

(b) Consider the demand function $d(p) = c \cdot p^{-n}$, where c and n are fixed constants. Show that the elasticity $E(p)$ is also a fixed constant.

(c) If the demand for a product is $d(p) = \dfrac{100}{e^{0.2p}}$, show that the elasticity of demand $E(p)$ is a linear function.

(d) Consider the demand function $d(p) = \dfrac{a}{e^{cp}}$, where a and c are fixed constants. Show that the elasticity $E(p)$ is a linear function of the form $E(p) = cp$.

5.7   Implicit Differentiation and Related Rates

SECTION OBJECTIVES

1. Use implicit differentiation to determine $\frac{dy}{dx}$.

2. Apply implicit differentiation to price-demand.

3. Solve a related rates situation.

4. Apply related rates to a geometric situation.

5. Apply related rates to a revenue situation.

Until now, all the graphs of functions that we have differentiated are ones that have passed the vertical line test. But what about other graphs like the one in **Figure 5.7.1**? We see that these graphs have tangent lines, too.

In this section we will learn a new method of differentiation that can be used to find derivatives and, subsequently, rates of change of graphs that may not pass the vertical line test. This method is called **implicit differentiation**. Then we examine applications called **related rates** that use implicit differentiation.

Implicit Differentiation

Thus far, all function that we have studied have the form $y = f(x)$. Algebraically, this means that we can solve for y *explicitly* in terms of x, such as $y = x^3 + 3x^2 - 1$. However, not all equations are expressed in this form. For example, for an equation such as

$$xy^4 + y^2 - xy + 3x^2 - 5 = 0$$

solving for y for this implicit function would be both difficult and impractical. The natural question is, "How do we find the derivative $\frac{dy}{dx}$ in this case?" The key is to use **implicit differentiation**. Since this differentiation method is really an extension of the Chain Rule, let's recall exactly what this rule states.

From Your Toolbox: The Chain Rule

If $y = f(u)$ and $u = g(x)$ are used to define $h(x)$, where $h(x) = f(g(x))$, then

$$h'(x) = f'(g(x)) \cdot g'(x)$$

provided that $f'(g(x))$ and $g'(x)$ exist.

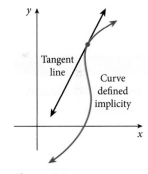

y

Tangent line

Curve defined implicity

x

Figure 5.7.1

For a function like $y = (x^3 + 3)^4$, we can use the Chain Rule to determine the derivative because our inside function, $g(x) = x^3 + 3$, is given. But what if we were asked to determine

$$\frac{d}{dx}[f(x)^4]?$$

Notice here that our inside function is not explicitly given; it is just represented as $f(x)$. We could still perform the differentiation using the Chain Rule, as follows:

$$\frac{d}{dx}[f(x)^4] = 4 \cdot [f(x)^3] \cdot f'(x)$$

Now if we replace $f(x)$ with y and $f'(x)$ with $\frac{dy}{dx}$, we have

$$\frac{d}{dx}[f(x)^4] = 4 \cdot [f(x)^3] \cdot f'(x)$$

$$\frac{d}{dx}[y^4] = 4 \cdot [y^3] \cdot \frac{dy}{dx} = 4y^3\frac{dy}{dx}$$

This suggests the following **Implicit Differentiation Rule**:

Implicit Differentiation Rule for y^n

If n is any real number and y is differentiable, then

$$\frac{d}{dx}[y^n] = n \cdot y^{n-1} \cdot \frac{dy}{dx}$$

This rule allows us to determine derivatives of expressions that involve powers of x, as well as powers of y. The process we use follows.

Determine $\frac{dy}{dx}$ When Using Implicit Differentiation

1. Use the usual differentiation rules for terms in x.
2. Use implicit differentiation for terms in y.
3. Solve for $\frac{dy}{dx}$.

OBJECTIVE 1

Use implicit differentiation to determine $\frac{dy}{dx}$.

Example 1: Determining Derivatives of Expressions Containing y^n

Use implicit differentiation to determine $\frac{dy}{dx}$ for the following.

a. $y^3 - 4x^2 = 7$

b. $x^{1/3} + y^{1/3} - 1 = 0$

Perform the Mathematics

a. We start by differentiating each term with respect to x.

$$\frac{d}{dx}[y^3] - \frac{d}{dx}[4x^2] = \frac{d}{dx}[7]$$

To differentiate $\frac{d}{dx}[y^3]$, we use the Implicit Differentiation Rule for y^n to get

$$3y^2 \cdot \frac{dy}{dx} - 8x = 0$$

Solving for $\frac{dy}{dx}$ yields

$$3y^2 \cdot \frac{dy}{dx} = 8x$$

$$\frac{dy}{dx} = \frac{8x}{3y^2}$$

b.   Again, we begin by differentiating each term with respect to x.

$$\frac{d}{dx}[x^{1/3}] + \frac{d}{dx}[y^{1/3}] - \frac{d}{dx}[1] = \frac{d}{dx}[0]$$

$$\frac{1}{3}x^{-2/3} + \frac{1}{3}y^{-2/3} \cdot \frac{dy}{dx} - 0 = 0$$

When writing the equation with positive exponents and in radical form, we get

$$\frac{1}{3\sqrt[3]{x^2}} + \frac{1}{3\sqrt[3]{y^2}} \cdot \frac{dy}{dx} = 0$$

Solving for $\frac{dy}{dx}$ gives

$$\frac{1}{3\sqrt[3]{x^2}} + \frac{1}{3\sqrt[3]{y^2}} \cdot \frac{dy}{dx} = 0$$

$$\frac{1}{3\sqrt[3]{y^2}} \cdot \frac{dy}{dx} = -\frac{1}{3\sqrt[3]{x^2}}$$

$$\frac{dy}{dx} = -\frac{3\sqrt[3]{y^2}}{3\sqrt[3]{x^2}}$$

$$\frac{dy}{dx} = -\sqrt[3]{\frac{y^2}{x^2}}$$

■

▶ **Try It Yourself**

Some related Exercises are 13 and 15.

Notice in Example 1 that the derivatives of these implicit functions are written in terms of both x and y.

Example 2: Determining $\frac{dy}{dx}$ Implicitly

Determine $\frac{dy}{dx}$ for the following.

a. $4e^y = x^2$ **b.** $2x + 1 = \sqrt{2 - y^2}$

Perform the Mathematics

a.   We begin by differentiating each term with respect to x.

$$\frac{d}{dx}[4e^y] = \frac{d}{dx}[x^2]$$

$$4e^y \frac{dy}{dx} = 2x$$

Solving for $\frac{dy}{dx}$ yields

$$\frac{dy}{dx} = \frac{2x}{4e^y} = \frac{x}{2e^y}$$

b.   Again, we begin by differentiating each term with respect to x.

$$\frac{d}{dx}[2x] + \frac{d}{dx}[1] = \frac{d}{dx}[\sqrt{2 - y^2}]$$

$$\frac{d}{dx}[2x] + \frac{d}{dx}[1] = \frac{d}{dx}[(2 - y^2)^{1/2}]$$

$$2 + 0 = \frac{1}{2}(2 - y^2)^{-1/2} \cdot (-2y)\frac{dy}{dx}$$

$$2 = -y(2 - y^2)^{-1/2}\frac{dy}{dx}$$

$$2 = \frac{-y}{\sqrt{2 - y^2}}\frac{dy}{dx}$$

Solving for $\frac{dy}{dx}$ gives

$$\frac{dy}{dx} = \frac{-2\sqrt{2 - y^2}}{y}$$

■

> ### Try It Yourself
>
> Some related Exercises are 27 and 31.

Sometimes we encounter situations that contain mixed terms, such as x^2y^3. To apply our implicit differentiation technique to these mixed terms, we must apply the Product Rule and then the Implicit Differentiation Rule for y^n when we differentiate the factor containing y^n. For example, suppose we want to determine $\frac{d}{dx}[x^2y^3]$. Using the Product Rule, we get

$$\frac{d}{dx}[x^2y^3] = \frac{d}{dx}[x^2] \cdot y^3 + x^2 \cdot \frac{d}{dx}[y^3]$$

$$= 2xy^3 + x^2 \cdot 3y^2\frac{dy}{dx}$$

Example 3: Determining $\frac{dy}{dx}$ of an Expression Containing x and y Terms

Use implicit differentiation to determine $\frac{dy}{dx}$ for $x^2y^2 - 7y^3 - 5 = 0$.

Perform the Mathematics

Differentiating each term with respect to x gives

$$\frac{d}{dx}[x^2y^2] - \frac{d}{dx}[7y^3] - \frac{d}{dx}[5] = \frac{d}{dx}[0]$$

Continuing, we need to use the Product Rule for the $\frac{d}{dx}[x^2y^2]$ term, which gives us

$$\frac{d}{dx}[x^2] \cdot y^2 + x^2 \cdot \frac{d}{dx}[y^2] - \frac{d}{dx}[7y^3] - \frac{d}{dx}[5] = \frac{d}{dx}[0]$$

$$2xy^2 + x^2 \cdot 2y\frac{dy}{dx} - 21y^2\frac{dy}{dx} - 0 = 0$$

$$2xy^2 + 2x^2y\frac{dy}{dx} - 21y^2\frac{dy}{dx} = 0$$

Now we need to solve for $\frac{dy}{dx}$.

$$2xy^2 + 2x^2y\frac{dy}{dx} - 21y^2\frac{dy}{dx} = 0$$

$$2x^2y\frac{dy}{dx} - 21y^2\frac{dy}{dx} = -2xy^2$$

$$\frac{dy}{dx}(2x^2y - 21y^2) = -2xy^2$$

$$\frac{dy}{dx} = \frac{-2xy^2}{2x^2y - 21y^2}$$ ∎

▶ *Try It Yourself*

Some related Exercises are 19 and 25.

When finding an equation for the line tangent to the graph of an implicit function, we use a procedure that is much the same as what we have always done. The only difference is that for implicit functions, we need both the x- and y-coordinates for the tangent point, since many derivatives are expressed in terms of both x and y. When we are evaluating a derivative of an implicit expression by replacing both x- and y-values, we write $\frac{dy}{dx}\big|_{(x,y)}$.

Example 4: Determining an Equation of a Line Tangent to the Graph of an Implicit Function

Consider the equation $x^2 + y^2 = 25$. This equation represents the graph of a circle centered at the origin and has a radius of 5, as shown in **Figure 5.7.2**.

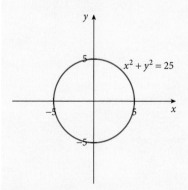

Figure 5.7.2

a.   Use implicit differentiation to determine $\frac{dy}{dx}$ for $x^2 + y^2 = 25$. Evaluate $\frac{dy}{dx}\big|_{(4,\,-3)}$ and interpret.

b.   Write an equation of the line tangent to the graph of $x^2 + y^2 = 25$ at the point $(4, -3)$.

Perform the Mathematics

a.   Using the Implicit Differentiation Rule for y^n, we get

$$\frac{d}{dx}[x^2] + \frac{d}{dx}[y^2] = \frac{d}{dx}[25]$$

$$2x + 2y\frac{dy}{dx} = 0$$

$$2y\frac{dy}{dx} = -2x$$

$$\frac{dy}{dx} = \frac{-2x}{2y} = -\frac{x}{y}$$

To evaluate $\frac{dy}{dx}\big|_{(4,\,-3)}$, we replace 4 for x and -3 for y in the derivative $\frac{dy}{dx} = -\frac{x}{y}$ to get

$$\frac{dy}{dx}\bigg|_{(4,\,-3)} = -\frac{4}{-3} = \frac{4}{3}$$

This means that the slope of the line tangent to the graph of $x^2 + y^2 = 25$ at the point $(4, -3)$ is $m_{\text{tan}} = \frac{4}{3}$.

b.   Using the slope of $\frac{4}{3}$ found in part (a) and the point-slope form of a line, we find that the tangent line equation at the point $(4, -3)$ is

$$y - y_1 = m(x - x_1)$$

$$y - (-3) = \frac{4}{3}(x - 4)$$

$$y + 3 = \frac{4}{3}(x - 4)$$

$$y + 3 = \frac{4}{3}x - \frac{16}{3}$$

$$y = \frac{4}{3}x - \frac{25}{3}$$

The graph of $x^2 + y^2 = 25$ with the tangent line $y = \frac{4}{3}x - \frac{25}{3}$ is shown in **Figure 5.7.3**.

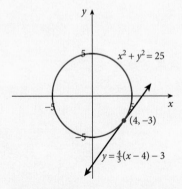

Figure 5.7.3

We can also use implicit differentiation in applications. In the next example, we are given a price-demand equation in terms of both demand x and price $p(x)$. To avoid confusion, we will denote the price in such applications by just writing p.

Example 5: Applying Implicit Differentiation

Suppose that the VeeCam Company determines that the price-demand equation for their economy tripod is given by

$$p + 2xp + x^2 = 125 \qquad 0 < x \leq 100$$

where x represents the demand for the tripod in thousands and p represents the price in dollars. Use implicit differentiation to determine $\frac{dp}{dx}$. Evaluate and interpret $\frac{dp}{dx}\big|_{(2.5, 19.5)}$.

Perform the Mathematics

To differentiate $p + 2xp + x^2 = 125$ with respect to x implicitly, we need to use the Product Rule for the term $2xp$:

$$\frac{d}{dx}[p] + \frac{d}{dx}[2xp] + \frac{d}{dx}[x^2] = \frac{d}{dx}[125]$$

$$\frac{d}{dx}[p] + \frac{d}{dx}[2x] \cdot p + 2x \cdot \frac{d}{dx}[p] + \frac{d}{dx}[x^2] = \frac{d}{dx}[125]$$

$$\frac{dp}{dx} + 2p + 2x\frac{dp}{dx} + 2x = 0$$

Solving for $\frac{dp}{dx}$ gives

$$\frac{dp}{dx} + 2x\frac{dp}{dx} = -2x - 2p$$

$$\frac{dp}{dx}(1 + 2x) = -2x - 2p$$

$$\frac{dp}{dx} = \frac{-2x - 2p}{1 + 2x} = \frac{-2(x + p)}{1 + 2x}$$

Evaluating the derivative when $x = 2.5$ and $p = 19.5$ yields

$$\frac{dp}{dx}\bigg|_{(2.5, 19.5)} = \frac{-2(2.5 + 19.5)}{1 + 2(2.5)} = \frac{-2(22)}{6} \approx -7.33$$

This means that at a demand level of 2.5 thousand tripods and a price of $19.50, the price is decreasing at a rate of $7.33 per thousand tripods. ∎

Related Rates

To begin to understand the concept behind related rates, let's suppose that the TastyMix Company determines that the daily revenue function, R, for their cupcake mix is given by

$$R(x) = 40x - x^2$$

where x represents the number of boxes of mix produced and sold in hundreds. But instead of having a price-demand equation giving price as a function of x, they find that the amount produced and sold is a function of *time t* or, mathematically, $x = x(t)$. Can we write an expression for $\frac{dR}{dt}$ even if we do not know how the quantity, x, or the revenue, R, are related to t? The answer again lies with the Chain Rule. Since the quantity x is a function of time t and the revenue $R(x)$ is a function of the quantity x, we can write

$$\begin{pmatrix} \text{Change in revenue} \\ \text{with respect to time} \end{pmatrix} = \begin{pmatrix} \text{Change in revenue} \\ \text{with respect to amount} \\ \text{produced and sold} \end{pmatrix} \cdot \begin{pmatrix} \text{Change in amount} \\ \text{produced and sold} \\ \text{with respect to time} \end{pmatrix}$$

$$\frac{dR}{dt} = \frac{dR}{dx} \cdot \frac{dx}{dt}$$

Since $\frac{dR}{dx} = \frac{d}{dx}(40x - x^2) = 40 - 2x$, we get

$$\frac{dR}{dt} = (40 - 2x) \cdot \frac{dx}{dt}$$

$$\frac{dR}{dt} = 40\frac{dx}{dt} - 2x\frac{dx}{dt}$$

This equation is called a **rate of change equation**. It expresses the rate of change of revenue with respect to time, $\frac{dR}{dt}$, in terms of $\frac{dx}{dt}$, the rate of change in the amount produced and sold with respect to time. Hence, we say that $\frac{dR}{dt}$ and $\frac{dx}{dt}$ are **related rates**. Thus, if we know the rate $\frac{dx}{dt}$, then we can find the rate $\frac{dR}{dt}$ using the rate of change equation.

Example 6: Determining a Rate of Change Equation

Determine the rate of change equation by implicitly differentiating $p^2 = x^3 - 300x$ with respect to time, t.

Perform the Mathematics

Using implicit differentiation, we $\frac{d}{dt}$ each term to get

$$\frac{d}{dt}[p^2] = \frac{d}{dt}[x^3] - \frac{d}{dt}[300x]$$

$$2p \cdot \frac{dp}{dt} = 3x^2\frac{dx}{dt} - 300\frac{dx}{dt} \qquad\qquad \blacksquare$$

▶ **Try It Yourself**

Some related Exercises are 49 and 51.

In the related rates applications that follow, it is necessary to determine a rate of change equation, using implicit differentiation, as we did in Example 6. We then solve for one of the rates, given some information, as illustrated in Example 7.

OBJECTIVE 3

Solve a related rates situation.

Example 7: Solving General Related Rates Applications

Assuming that x and p are both functions of t, determine $\frac{dp}{dt}$ for $p^2 = x^3 - 300x$, given $\frac{dx}{dt} = 5$, $x = 2$, and $p = 3$.

Perform the Mathematics

In Example 6 we determined that the rates $\frac{dp}{dt}$ and $\frac{dx}{dt}$ were related by the equation

$$2p \cdot \frac{dp}{dt} = 3x^2 \frac{dx}{dt} - 300\frac{dx}{dt}$$

Substituting the given values and solving for $\frac{dp}{dt}$ gives

$$2(3)\frac{dp}{dt} = 3(2)^2(5) - 300(5)$$

$$6\frac{dp}{dt} = 60 - 1500$$

$$6\frac{dp}{dt} = -1440$$

$$\frac{dp}{dt} = -\frac{1440}{6} = -240 \qquad \blacksquare$$

Applications

The applications of related rates are varied. To aid us in **solving related rates applications**, we offer the following as guidelines.

> ### Solving Related Rates Applications
>
> 1. Sketch a diagram, if it can be helpful.
> 2. Write down all the variables, along with the rates that are given. Be sure to include proper units.
> 3. Write an equation that relates the variables given.
> 4. Assuming that the variables are functions of time, t, differentiate the equation implicitly with respect to time, t.
> 5. Solve for the unknown rate.

Example 8: Applying Related Rates in Geometry Applications

OBJECTIVE 4

Apply related rates to a geometric situation.

The FreshDay Company delivers its bread in electric-powered trucks. The delivery area is restricted by the range of the trucks to a radius r miles from the company bakery. Through improvements in battery efficiency, the company finds that the radius for the delivery area is increasing at a rate of $3\frac{\text{miles}}{\text{year}}$. At the time that the radius is 50 miles, how fast is the delivery area increasing?

Understand the Situation

For clarity, we will follow the process in the box preceding this example.

Perform the Mathematics

1. We sketch a diagram for this application as shown in **Figure 5.7.4**.

2. Next, we can label t as the time in years, r as the radius in miles, and A as the delivery area in square miles. When the radius is $r = 50$, we know that the radius is increasing at a rate of $\frac{dr}{dt} = 3\frac{\text{miles}}{\text{year}}$.

Figure 5.7.4

3. The radius and the area are related by the equation for a circle, $A = \pi r^2$, where A is the area in square miles and r is the radius in miles.

4. Since r is changing with respect to time t, which means that A is a function of time t, we differentiate the equation $A = \pi r^2$ implicitly with respect to t. This gives,

$$\frac{d}{dt}[A] = \frac{d}{dt}[\pi r^2]$$

$$\frac{dA}{dt} = 2\pi r \frac{dr}{dt}$$

5. Since we want to know how fast the delivery area is increasing, we seek $\frac{dA}{dt}$ when $r = 50$ and $\frac{dr}{dt} = 3$. Solving for the unknown rate $\frac{dA}{dt}$ yields

$$\frac{dA}{dt} = 2\pi r \frac{dr}{dt}$$

$$= 2\pi(50)(3) = 300\pi \approx 942.48$$

Interpret the Result

This means that when the radius is 50 miles and the radius of the delivery area is increasing at a rate of 3 miles per year, the delivery area is increasing at a rate of about 942.48 square miles per year.

▷ **Try It Yourself**

Some related Exercises are 65 and 69.   ◼

Many times, the equation for a related rates application is not given directly. In the previous example, we saw that the geometry formula for the area of a circle was necessary to describe the delivery area. Now let's see how related rates are used in business applications.

OBJECTIVE 5

Apply related rates to a revenue situation.

Example 9: Applying Related Rates in Business Applications

Past records of the TechTop Company determine that the revenue for the number of software suites produced and sold is given by

$$R = 90x - x^2$$

where x is the number of units produced and sold daily and R is the generated revenue in dollars. The company also finds that the software is selling at a rate of five suites per day. How fast is the revenue changing when 40 suites are being produced and sold?

Perform the Mathematics

Again, we will use the following steps for solving related rates applications.

1. In this application, a diagram is not helpful.

2. We have the number of units sold, x, the revenue generated, R, and the time, t. We know that the suites are selling at a rate of $5\frac{\text{suites}}{\text{day}}$, so $\frac{dx}{dt} = 5$ when $x = 40$.

3. In this case, we are given the revenue equation $R = 90x - x^2$.

4. Since we are being asked how revenue, R, is changing with respect to time, t, we differentiate the equation for R with respect to time t.

$$\frac{d}{dt}[R] = \frac{d}{dt}[90x - x^2]$$

$$\frac{dR}{dt} = 90\frac{dx}{dt} - 2x\frac{dx}{dt}$$

5.   Since we want to know how fast the revenue is changing, we seek $\frac{dR}{dt}$ when $\frac{dx}{dt} = 5$ and $x = 40$. This gives us

$$\frac{dR}{dt} = 90\frac{dx}{dt} - 2x\frac{dx}{dt}$$

$$= 90(5) - 2(40)(5) = 50$$

So when 40 suites are produced and sold daily, the revenue is increasing at a rate of $50 per day.   ∎

Summary

In this section, we focused on two main topics. We saw that **implicit differentiation** was a method that could be used to determine derivatives of expressions that were written in terms of x, as well as in terms of both x and y. We introduced a new rule for implicit differentiation. Next, we turned our attention to **related rates**. By using implicit differentiation, we found that we could differentiate two quantities with respect to time, t, a third variable. We saw applications of related rates that used geometric formulas and also applications in the business sciences that used our five-step process for solving related rates applications.

- **Implicit Differentiation Rule for y^n:** $\dfrac{d}{dx}[y^n] = n \cdot y^{n-1} \cdot \dfrac{dy}{dx}$

- **Solving Related Rates Applications:**

  1. Sketch a diagram, if it can be helpful.
  2. Write down all the variables, along with the rates that are given. Be sure to include proper units.
  3. Write an equation that relates the variables given.
  4. Assuming that the variables are functions of time, t, differentiate the equation implicitly with respect to time, t.
  5. Solve for the unknown rate.

Section 5.7 Exercises

Vocabulary Exercises

1. Implicit differentiation is a technique used to determine derivatives for equations that may not pass the _____ line test.

2. To differentiate implicitly, we make use of the _____ Rule.

3. We make use of the implicit differentiation technique if we are given an equation in which we may not be able to solve for _____.

4. When evaluating a derivative of an implicit expression, we need given values for _____ and for _____.

5. An equation of the form $x^2 + y^2 = r^2$ is the equation of a _____.

6. For an equation of the form $x^2 + y^2 = r^2$, the value of r represents the _____.

Skill Exercises

For Exercises 7–16, use implicit differentiation to determine $\frac{dy}{dx}$.

7. $2x + y = 5$

8. $x + 3y = 0$

9. $x + 3y^2 = 4$

10. $7x^2 - 5y^2 - 100 = 0$

11. $2x^2 + 2y^2 = 32$

12. $3x^2 + y^2 = 81$

13. $5x^3 + y^3 - 4x = 0$

14. $4x^2 + 2y^3 = x^3$

15. $x^{1/4} - y^{1/4} = 1$

16. $x^{2/3} - y^{2/3} = 4$

For Exercises 17–26, use implicit differentiation to determine $\frac{dy}{dx}$.

17. $x^2 + xy = 6$

18. $xy + y^2 = 4$

19. $x^3 - y^3 + 12xy = 0$

20. $2x^3 + 3xy + y^3 = 0$

21. $(x + y)^2 - 1 = 7x^2$

22. $(y + 3)^2 = x^2 - 5$

23. $y \cdot \ln x = 10 - y$

24. $x + \ln y + 11 = 0$

25. $xe^y + x^2 = y^2$

26. $e^x y + y = 4x$

For Exercises 27–32, use implicit differentiation to determine $\frac{dy}{dx}$.

27. $5^y = x^3$

28. $6^y = 2x^4$

29. $10^{y-2} = x - 3$

30. $7^{3y-1} = x + 1$

31. $\sqrt[3]{y} = x^2 + 6$

32. $\sqrt{y} = 3x^4 - 9$

In Exercises 33–38, complete the following.

(a) Use implicit differentiation to determine $\frac{dy}{dx}$.

(b) Solve the equation explicitly for y and differentiate to determine $\frac{dy}{dx}$.

33. $x + 2y = 6$

34. $3x - 4y = 12$

35. $xy + 4 = x^4$

36. $xy = 2$

37. $\dfrac{x}{y} - x^2 = 1$

38. $\dfrac{1}{x} + \dfrac{1}{y} = 3$

For Exercises 39–46, determine an equation of the line tangent to the graph of the given equation at the indicated point.

39. $x^2 + y^2 = 13;\ (3, 2)$

40. $x^3 + y^3 = 9;\ (1, 2)$

41. $4x^2 + 9y^2 = 36;\ (0, 2)$

42. $x^2 + y^2 = 25;\ (3, 4)$

43. $x \ln y = 2x^3 - 2y;\ (1, 1)$

44. $x + \ln y - 2y^2 = 0;\ (2, 1)$

45. $x^2 + y^2 = e^y;\ (1, 0)$

46. $4e^y + y = x^2;\ (2, 0)$

In Exercises 47–54, determine the rate of change equation by implicitly differentiating each of the following with respect to time t.

47. $2x + 3y = 20$

48. $x^2 + 2y = 11$

49. $x^2 - 3y = 1$

50. $y = x^2 - 3x + 5$

51. $x^2 + y^2 = 5x$

52. $x^3 - y^3 = 10y$

53. $5xy + y^4 = x$

54. $xy = 4x + 5y + 9$

For Exercises 55–60, determine the indicated rate. Assume that x and y are both functions of t.

55. Find $\dfrac{dy}{dt}$ for $xy = 7$, given $x = 7$, $y = 1$, and $\dfrac{dx}{dt} = 2$.

56. Find $\dfrac{dx}{dt}$ for $x^2 + 9y^2 = 18$, given $x = 3$, $y = 1$, and $\dfrac{dy}{dt} = 10$.

57. Find $\dfrac{dx}{dt}$ for $y^2 + x = 3$, given $x = 2$, $y = 1$, and $\dfrac{dy}{dt} = 2$.

58. Find $\dfrac{dy}{dt}$ for $y = x^2 - 2x + 3$, given $x = 5$, $y = 18$, and $\dfrac{dx}{dt} = 3$.

59. Find $\dfrac{dy}{dt}$ for $x^2 + y^2 = 25$, given $x = 3$, $y = -4$, and $\dfrac{dx}{dt} = 2$.

60. Find $\dfrac{dx}{dt}$ for $y^2 + xy - 3x = -1$, given $x = 1$, $y = -2$, and $\dfrac{dy}{dt} = -2$.

Application Exercises

61. The Crabtree Company determines that the price-demand equation for their mini picture frames is given by

$$p + x^2 = 150$$

where x represents the demand for the frames in hundreds and p represents the price in dollars.

(a) Use implicit differentiation to determine $\frac{dp}{dx}$.

(b) Evaluate and interpret $\frac{dp}{dx}\big|_{(11,\,29)}$.

62. Suppose that Pottery Plus determines that the price-demand equation for their glazed presidential inauguration mugs is given by

$$\frac{1}{2}p + \sqrt{x} = 30$$

where x represents the demand for the mugs and p represents the price in dollars.

(a) Use implicit differentiation to determine $\frac{dp}{dx}$.

(b) Evaluate and interpret $\frac{dp}{dx}\big|_{(400,\,20)}$.

63. The WoodedLife Company has determined that the price-demand equation for their new single-occupant tent is given by

$$px + 2x = 1000 \qquad 10 \le x \le 30$$

where x represents the demand for the tents in hundreds and p represents the price in dollars.

(a) Use implicit differentiation to determine $\frac{dp}{dx}$.

(b) Evaluate and interpret $\frac{dp}{dx}\big|_{(20,\,48)}$.

64. The CorpStyle Company determines that the price-demand equation for their all-silk power ties is given by

$$x^2 = 1000 - p^2 \qquad 10 \le x \le 30$$

where x represents the demand for the ties in thousands, and p represents the price in dollars.

(a) Use implicit differentiation to determine $\frac{dp}{dx}$.

(b) Evaluate and interpret $\frac{dp}{dx}\big|_{(18,\,26)}$.

65. The area of a circle with radius r is increasing at a rate of 12 square inches per minute. Find the rate at which the radius is increasing when the radius is 2 inches.

66. The area of a circle with radius r is increasing at a rate of 5 square inches per minute. Find the rate at which the radius is increasing when the radius is 0.8 inch.

67. The area of a square with sides x inches is increasing at a rate of 10 square inches per minute. Find the rate at which a side is increasing when the sides are 3 inches.

68. The area of a square with sides x inches is increasing at a rate of 25 square inches per minute. Find the rate at which a side is increasing when the sides are 4 inches.

69. A circular pan is being heated and the radius of the heated area increases at a rate of 0.02 inch per minute. Find the rate at which the heated area is increasing when the radius is 8 inches.

70. A 5-foot-tall man is walking away from a 20-foot street lamp at a speed of 6 feet per second. How fast is the tip of his shadow moving along the ground?

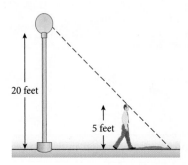

20 feet

5 feet

71. A 6-foot-tall woman is walking away from a 24-foot street lamp at a speed of 8 feet per second. How fast is the tip of her shadow moving along the ground?

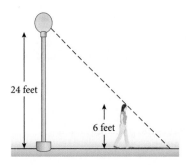

24 feet

6 feet

72. The SoftSkirt Company determines that the monthly revenue for a new style of skirt is given by

$$R = 60 - \frac{1}{2}x^2$$

where x is the number of skirts produced and sold in hundreds and R is the revenue, in thousands of dollars, generated from the skirts.

(a) Differentiate the revenue function with respect to time t to get $\frac{dR}{dt}$.

(b) Determine the rate of change in the revenue with respect to time at a production level of $x = 3$ and $\frac{dx}{dt} = 20$ hundred skirts per month.

73. The SharpSuit Company determines that monthly revenue for a new type of casual suit is given by

$$R = 250x - \frac{2}{5}x^2$$

where x is the number of suits produced and sold and R is the revenue in dollars.

(a) Differentiate the revenue function with respect to time t to get $\frac{dR}{dt}$.

(b) Determine the rate of change in the revenue with respect to time at a production level of $x = 100$ and $\frac{dx}{dt} = 200$ suits per month.

74. A 24-foot ladder is leaning against the wall as shown in the accompanying figure. Assume that the values of x and y are related by the Pythagorean Theorem equation

$$x^2 + y^2 = 24^2$$

(a) Differentiate each side of the equation with respect to time t.

(b) If the base of the ladder is sliding away from the wall at a rate of 2 feet per second, find the rate at which the top of the ladder is sliding down the wall when the top of the ladder is 12 feet above the ground.

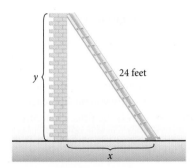

75. A 20-foot ladder is leaning against the wall as shown in the accompanying figure. Assume that the values of x and y are related by the Pythagorean Theorem equation

$$x^2 + y^2 = 20^2$$

(a) Differentiate each side of the equation with respect to time t.

(b) If the base of the ladder is sliding away from the wall at a rate of 3 feet per second, find the rate at which the top of the ladder is sliding down the wall when the top of the ladder is 8 feet above the ground.

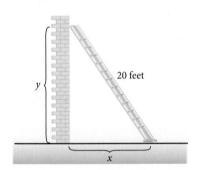

76. A surgeon has a patient with a tumor in the shape of a sphere. Assume that the volume of the tumor is given by the equation

$$V = \frac{4}{3}\pi r^2$$

where r is the radius of the tumor measured in millimeters and V is its volume, measured in cubic millimeters.

(a) Differentiate each side of the equation with respect to time t.

(b) If tests show that the radius of the tumor is growing at a rate of 0.3 millimeter per week, determine the rate at which the volume is changing when the radius is 15 millimeters.

77. Assume that the number of bass in a pond is related to the level of polychlorinated biphenyls (or PCBs) in the pond. (PCBs are a group of industrial chemicals used in plasticizers, fire retardants, and other materials.) The bass population is modeled by

$$y = \frac{2500}{1 + x}$$

where x represents the PCB level in parts per million (ppm) and y represents the number of bass in the pond.

(a) Differentiate each side of the equation with respect to time t.

(b) If the level of PCBs is increasing at rate of 40 ppm per year, find the rate at which y is changing when there are 100 bass in the pond.

78. The price-demand equation for oranges in the town of Southburg is given by

$$p = \frac{40{,}000}{x^{1.5}}$$

where x is the demand for oranges in bushels per month and p is the wholesale price per bushel. The local farmer's market finds that the demand for oranges is currently 900 bushels each month and is increasing at a rate of 100 bushels per month. How fast is the price changing?

Concept and Writing Exercises

79. Contrast explicit and implicit differentiation in a couple of sentences.

80. Explain why we need both the x- and y-values given when we evaluate an implicit derivative at a given point.

For Exercises 81 and 82, let a and b represents real numbers where $a \neq 0$ and $b \neq 0$. Determine $\frac{dy}{dx}$ for the following.

81. $x^a + y^b = abxy$

82. $\dfrac{x^2}{a^2} + \dfrac{y^2}{b^2} = 1$

For Exercises 83 and 84, a, b, and k represent real numbers. Determine the rate of change equation by implicitly differentiating each of the following with respect to time t.

83. $ax + by = k$

84. $axy + y = k$

 Section Project

The strategic planning department of the Higbrize Company is given the data in **Table 5.7.1** on the cost of producing a new lapel pager.

Table 5.7.1

Units Produced, x	Cost, C
1500	14,725
2500	18,125
3500	21,725
4000	23,600
5500	29,525
7000	35,900

The department is asked to answer the following scenario: The current production level is 7500 pagers and is increasing at a rate of 50 units per day. How is the average cost changing?

(a) Use the regression features of your calculator to determine a linear cost model for the lapel pagers.

(b) Determine the average cost function AC.

(c) Differentiate AC with respect to time t.

(d) What do the numbers 7500 and 50 represent in terms of production x and time t?

(e) Use related rates to find the rate of change for the average cost with respect to time and interpret.

Chapter 5   Review Exercises

Section 5.1 Review Exercises

For each function graphed in Exercises 1–4, complete the following.

(a) Determine where the function is increasing.

(b) Determine where the function is decreasing.

(c) Determine where the function is constant.

1.

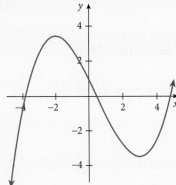

2.

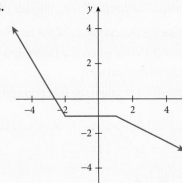

3.

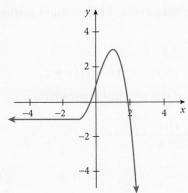

4.

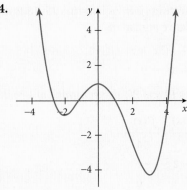

For the derivatives graphed in Exercises 5–8, complete the following.

(a) List intervals where the function is increasing.

(b) List intervals where the function is decreasing.

(c) List where the function is constant.

5.

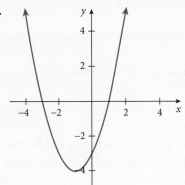

6.

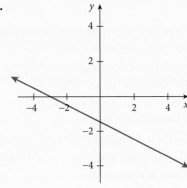

7.

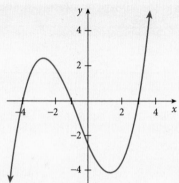

8.

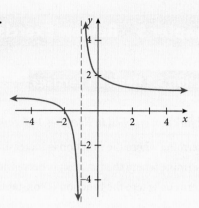

For each function in Exercises 9–14, complete the following.

(a) Determine the critical values of the function.

(b) Determine intervals where the function is increasing and decreasing.

(c) Locate the relative extrema.

 (d) Use a calculator to graph the function and verify your results to parts (a) and (b).

9. $f(x) = \dfrac{1}{2}x - 4$ **10.** $f(x) = -x^2 + 3x + 8$

11. $f(x) = x^3 + 3x^2 - 24x + 20$ **12.** $f(x) = \sqrt{9 - x}$

13. $f(x) = (3x - 2)^4$ **14.** $f(x) = \dfrac{x + 4}{3 - x}$

Use the derivatives given in Exercises 15–18 to determine the x-values where f has a relative maximum and/or a relative minimum.

15. $f'(x) = -3x + 8$ **16.** $f'(x) = 5x(x - 1)$

17. $f'(x) = x^2 - 9$ **18.** $f'(x) = (x - 4)^2(x - 1)(x + 2)^3(x + 5)$

For Exercises 19 and 20, sketch a graph of a continuous function that satisfies the given data.

19.

x	f(x)	f'(x)
0	1	−4
1	−2	−2
2	−3	0
3	−2	2
4	1	4
5	6	6

20.

x	f(x)	f'(x)
0	3	17
1	12	2
2	9	−7
3	0	−10
4	−9	−7
5	−12	2

21. Microeconomics—Shampoo Prices: The LuvYerPet Company is introducing a new brand of dog shampoo. The company has determined that the demand for the product can be modeled by

$$p(x) = 4.5e^{-0.03x}$$

where x represents the quantity demanded in thousands of containers of dog shampoo and $p(x)$ represents the price in dollars.

(a) Use the model to determine the price if there is demand for 30,000 containers of dog shampoo (that is, $x = 30$).

(b) Determine R, the revenue in thousands of dollars, as a function of the quantity demanded in thousands of containers.

(c) Use the techniques of Section 5.1 to determine intervals where R is increasing and decreasing.

(d) Determine the relative maximum. Round the x- and y-coordinates to the nearest hundredth and interpret.

 (e) Graph R in the viewing window $[0, 100]$ by $[0, 100]$ and verify your work.

22. **Pharmacology—Drug Concentration:** Researchers have determined through experimentation that the percent concentration of a certain medicine during the first 20 hours after it has been administered can be approximated by

$$p(t) = \frac{150t}{t^2 + 4t + 4} \qquad 0 \le t \le 20$$

where t is the time in hours after administration of the medication and $p(t)$ is the percent concentration.

(a) Compute $p'(t)$.

(b) Determine the critical values.

(c) Determine intervals where p is increasing and decreasing.

(d) Determine the relative maximum and interpret.

23. **Communications—Phone Prices:** A manufacturer of telephones has the following costs when producing x telephones in one day.

Fixed costs	$800
Unit production costs	$15
Equipment maintenance	$\dfrac{x^2}{18}$
Reasonable domain	$0 < x < 500$

Therefore, the total cost of manufacturing x telephones in one day is given by

$$C(x) = 800 + 15x + \frac{x^2}{18} \qquad 0 < x < 500$$

(a) Determine the average total cost, AC.

(b) Determine the critical value(s) for AC.

(c) Determine intervals where AC is increasing and decreasing.

(d) Determine the relative minimum and interpret.

24. **Public Economics—Social Security Contributions:** Many have been concerned with the future of the Social Security system, in spite of its current solvency. "Right now, there is actually a surplus. The problem is not with Social Security, but with the rest of the budget," stated Jacob Lew, director of the Office of Management and Budget. (*Source: The Bedford Journal.*) The annual contributions to the Social Security system from 1990 to 2011 can be modeled by

$$f(x) = -0.06x^3 + 2x^2 + 0.93x + 270.76 \qquad 0 \le x \le 21$$

where x represents the number of years since 1990 and $f(x)$ represents the net contributions to Social Security, measured in billions of dollars. (*Source: Social Security Administration.*)

(a) Determine the critical values. Be sure to consider the reasonable domain.

(b) Determine the intervals where f is increasing and where it is decreasing.

(c) Are there any relative extrema? Explain your answer.

Section 5.2 Review Exercises

In Exercises 25–30, determine the absolute extrema of each function on the indicated interval.

25. $f(x) = x^2 - 5x + 3; \ [0, 10]$

26. $f(x) = x^3 + 6x^2 + 9x; \ [-5, 5]$

27. $f(x) = x^3 - 6x; \ [0, 5]$

28. $f(x) = \frac{1}{5}x^5 - \frac{5}{3}x^3 + 4x; \ [-5, 5]$

29. $f(x) = \sqrt{x^3}; \ [0, 5]$

30. $f(x) = x + \frac{4}{x}; \ [-10, -5]$

31. **Construction—Fence Building:** Mike has 60 feet of fencing that he intends to use to build a rectangular enclosure for his Scottish terrier. He plans to build the enclosure against one side of his house, so the fencing is needed on just three sides of the enclosure. Determine the dimensions that will make the enclosure have the largest possible area.

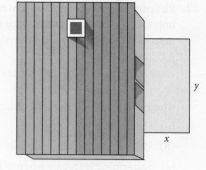

32. **Construction—Pool Building:** Your friend's family is planning to build a swimming pool. The water surface is to have an area of 1200 square feet. The pool will also have an 8-foot-wide walkway on the east and west sides and a 6-foot walkway on the north and south sides. Let x be the length of the pool (east to west), and let y be the width (north to south). Given that $10 \le x \le 100$, find the dimensions that will minimize the *total* area of the swimming pool and walkways.

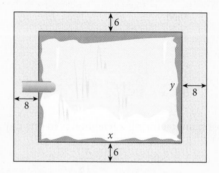

33. **Fitness—Weight Training:** The GruntWorks Company analyzed the production costs for its new weight-training equipment and determined that its weekly cost function can be given by

$$C(x) = 0.001x^3 - 0.6x^2 + 217x + 7200 \qquad 50 \le x \le 550$$

where x represents the number of units produced each week and $C(x)$ is the weekly cost in dollars.

(a) Determine the average cost function, AC.

(b) Determine the production level that minimizes the average cost.

(c) Determine the average cost and the total cost at the production level found in part (b).

34. **Fitness—Weight Training:** For the weight-training equipment in Exercise 33, the GruntWorks Company gathered data for the price of a unit and the number of units demanded per week. The price-demand function that models these data is given by

$$p(x) = -0.002x^2 + 1.5x - 1.7 \qquad 50 \le x \le 550$$

where x represents the number of units demanded each week and $p(x)$ is the price, in dollars, that consumers pay for each unit to buy exactly x units per week.

(a) Determine R, the weekly revenue as a function of x.

(b) Determine the maximum weekly revenue.

(c) Determine the demand that maximizes weekly revenue.

(d) Determine the price that maximizes weekly revenue.

35. **Fitness—Weight Training** *(Continuation of Exercise 34):* Using the functions C and R from Exercises 33 and 34, respectively, complete the following.

(a) Determine P, the profit as a function of x.

(b) Determine the maximum weekly profit.

(c) Determine the demand that maximizes weekly profit.

(d) Determine the price that maximizes weekly profit.

36. International Trade—Aerospace Exports: Some think that American economic challenges can be countered by increasing exports to meet increasing overseas demands on high-tech products. "Manufacturers in the American South are a major beneficiary of globalization, with aerospace and commodity companies exporting around the world," said Jeffrey Lacker, president of the Federal Reserve Bank of Richmond. (*Source:* Bloomberg Businessweek.) The total exports of aerospace products from 2000 to 2011 can be modeled by

$$f(x) = 0.39x^2 - 1.1x + 21.51 \qquad 0 \le x \le 11$$

where x represents the number of years since 2000 and $f(x)$ represents the annual exports of aerospace products, measured in billions of dollars. (*Source:* Aerospace Industries Association.) Determine the absolute minimum and maximum of f on its reasonable domain and interpret the coordinates.

Section 5.3 Review Exercises

In Exercises 37–40, compute the first three derivatives of the given function.

37. $f(x) = x^3 - 7x^2 + 5$

38. $f(x) = 1.3x^5 + 2.7x^4 - 6.1x^2 + 5.8x$

39. $f(x) = \sqrt[3]{x - 5}$

40. $f(x) = x^2 + 5x - \ln 3x$

For each function in Exercises 41–47, complete the following.

(a) Determine intervals where the function is concave up and where it is concave down.

(b) Determine any inflection points.

41. $f(x) = x^3 - x^2 + 5x - 7$

42. $f(x) = x^2 - 18x + 15$

43. $f(x) = 12 - \sqrt{x}$

44. $f(x) = \sqrt{25 - x^2}$

45. $f(x) = 3x - \dfrac{5}{x}$

46. $f(x) = e^{-0.7x}$

47. $f(x) = e^{5x}$

For the functions in Exercises 48–51, complete the following.

(a) Determine intervals where the function is increasing and where it is decreasing.

(b) Determine the relative extrema.

(c) Determine intervals where the graph of the function is concave up and where it is concave down.

(d) Determine any inflection points.

48. $f(x) = 3x^2 + 2x - 8$

49. $f(x) = 2x^3 + 3x^2 - 36x$

50. $f(x) = 2x + \dfrac{18}{x}$

51. $f(x) = \ln(2 - x)$

52. Sketch the graph of a function that satisfies the following conditions.

Domain:	$(-\infty, \infty)$
Range:	$[2, \infty)$
Continuous:	$(-\infty, \infty)$
$f'(x) > 0$	$(-3, 0) \cup (3, \infty)$
$f'(x) < 0$	$(-\infty, -3) \cup (0, 3)$
Undefined	$f'(0)$ is undefined
$f''(x) > 0$	$(-\infty, 0) \cup (0, \infty)$

53. Economics—Marginal Profit: The Sweet Truth Cookie Company has determined that its daily profit is given by

$$P(x) = -0.5x^3 + 3x^2 + 68x - 133$$

where x is the number of hundreds of cookies baked and sold and $P(x)$ is the daily profit in dollars.

(a) Determine $P(4)$ and $P'(4)$ and interpret each.

(b) Determine intervals where P is increasing and where it is decreasing. Determine the relative maximum and interpret each coordinate.

(c) Determine intervals where the *marginal profit* is increasing and where it is decreasing. What can you say about the graph of P on these intervals?

(d) Determine the inflection point for the graph of P.

54. **Hematology—Bird Blood Flow:** Suppose that blood flow through the artery of a certain bird species can be modeled by

$$f(x) = 0.317e^{0.0103x} \qquad 20 \leq x \leq 120$$

where x represents arterial pressure, measured in millimeters of mercury, and $f(x)$ is blood flow, measured in milliliters per minute.

(a) Evaluate $f(40)$ and interpret.

(b) Evaluate $f'(40)$ and interpret.

(c) Show that the graph of f is concave up on the interval $[20, 120]$. Explain what this means.

 55. **Conservation—Goat Population:** Suppose that the number of mountain goats in a certain region can be modeled by

$$f(x) = \frac{1400}{1 + 43e^{-0.09x}}$$

where $f(x)$ represents the number of goats t years after the goats were first introduced to the region.

(a) Graph f in the viewing window $[0, 100]$ by $[0, 1500]$.

(b) Compute $f(84)$ and $f'(84)$ and interpret each.

(c) Determine $f''(x)$.

(d) On the interval $[0, 100]$, determine intervals where the graph of f is concave up and where it is concave down. To do this, graph f'' and use the `Zero` command.

(e) When is the population of goats growing fastest?

Section 5.4 Review Exercises

56. **Construction—Window Building:** A window has the shape of a rectangle with an adjoining semicircle, as shown. The perimeter of the window is 12 feet.

(a) Determine the dimensions x and y if the area of the window is as large as possible.

(b) What is the maximum area?

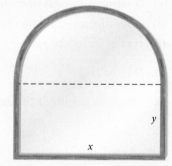

57. **Publishing—Average Cost:** A newspaper publisher has determined that the daily cost of printing, distributing, and advertising is

$$C(x) = 0.0015x^2 + 0.6x + 4.5 \qquad x > 0$$

where x represents the daily circulation, in thousands, and $C(x)$ is the cost in hundreds of dollars.

(a) Determine AC the average cost function.

(b) Determine the value of x that minimizes $AC(x)$.

58. **Construction—Box Making:** An open-top box is made by cutting identical squares from the corners of a 15-inch by 24-inch piece of cardboard and then folding up the flaps. Determine the dimensions of the box of maximum volume.

59. **Management—Inventory Costs:** The IzataFax store expects to sell 800 fax machines in a year. Each fax machine costs the store $80, and there is a fixed cost of $24 per order. If it costs $62

to store a fax machine for a year, how large should each order be, and how often should orders be placed to minimize inventory costs? Round to the nearest lot size that produces an integer value for the number of orders placed each year.

60. Management—Inventory Costs: Freddy's Fruit Orchard has determined that the annual yield per fruit tree is 150 pounds per tree when the number of trees per acre is 35 or fewer. For each additional tree over 35, the annual yield per tree decreases by 4 pounds because of overcrowding. How many trees should be planted on each acre to maximize the annual yield from an acre?

61. Business—Shipping Costs: The United Parcel Service states that a lightweight package (under 30 pounds) is *oversize* if the sum of its length and girth is 84 inches. (*Source:* United Parcel Service.) Assuming that the front of the package is a square and the maximum length plus girth is 84 inches, determine the dimensions that will maximize the volume without requiring the customer to pay the oversize charge. What is the maximum volume?

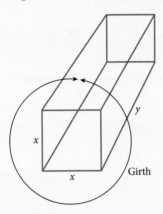

Section 5.5 Review Exercises

Use the graph of f shown to answer Exercises 62–68.

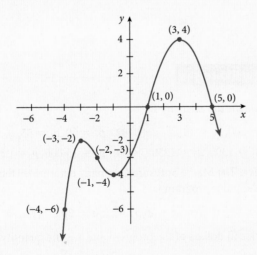

62. Determine intervals where f is positive and where f is negative.

63. Determine intervals where f is increasing and where f is decreasing.

64. Determine intervals where f' is positive and where f' is negative.

65. Determine intervals where f' is increasing and where f' is decreasing.

66. Determine intervals where the graph of f is concave up and where the graph of f is concave down.

67. Locate any relative extrema.

68. Locate any inflection points.

In Exercises 69–72, use the Second Derivative Test to locate any relative extrema, if they exist.

69. $f(x) = x^2 + 12x + 5$

70. $f(x) = x^3 - 5x^2 - 8x$

71. $f(x) = 3x^4 - 3x^3 - 3x^2$

72. $f(x) = \dfrac{1}{x^2 + 9}$

In Exercises 73–78, sketch the graph of the given function. Label all relative extrema, inflection points, and any asymptotes.

73. $f(x) = x^3 - 9x^2 + 4$

74. $f(x) = 2x^3 - \dfrac{3}{2}x^2 - 9x + 5$

75. $f(x) = x^4 - 14x^2 - 24x$

76. $f(x) = \dfrac{1}{x^2 - 4}$

77. $f(x) = 0.4x + 16 + \dfrac{8}{x}$

78. $f(x) = e^x - 4x$

In Exercises 79 and 80, the graph of f'' is given. Use the graphs to sketch graphs of f' and f. There are many correct answers.

79.

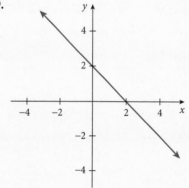

80.

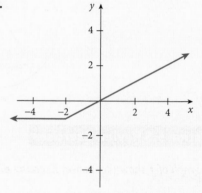

Section 5.6 Review Exercises

In Exercises 81–84, compute the arc elasticity E_a.

81. $p_1 = 18, p_2 = 20, q_1 = 200, q_2 = 195$

82. $p_1 = 50, p_2 = 55, q_1 = 80, q_2 = 70$

83. $p_1 = 630, p_2 = 650, q_1 = 1300, q_2 = 1250$

84. $p_1 = 15{,}000, p_2 = 15{,}250, q_1 = 75, q_2 = 73$

85. Economics—Elasticity: The Macro Software Company determines that the price-demand function for a new FileFinder program is

$$x = d(p) = 350 - 2p$$

where p is the unit price in dollars of the program and x is the quantity demanded in thousands.

(a) Determine $d(8)$ and $d(9)$ and interpret.

(b) Using $q_1 = d(8)$, $q_2 = d(9)$, $p_1 = 8$, and $p_2 = 9$, determine the arc elasticity E_a and interpret.

86. Economics—Elasticity: The Thai Way Buffet Restaurant typically has 45 customers for lunch each day when the price for lunch is \$6.95. During a special promotion week, the price is reduced to \$5.50 and the number of lunch customers grows to 58 per day. Determine the arc elasticity and interpret.

In Exercises 87–90, the price-demand function p(x) is given. Determine the demand function x = d(p).

87. $p = 18 - 0.3x$

88. $p = \dfrac{12}{x} + 5$

89. $p = \sqrt{500 - x^2}$

90. $p = 800e^{-0.02x}$

In Exercises 91–94, the demand function x = d(p) and the price p are given.

(a) Determine the point elasticity function $E(p)$.

(b) Classify the demand as elastic, inelastic, or unitary.

91. $d(p) = 180 - 5p;\ p = 20$

92. $d(p) = 400 - p^2;\ p = 12$

93. $d(p) = \sqrt{120 - p};\ p = 100$

94. $d(p) = 18\ln(120 - 3p);\ p = 5$

95. Economics—Elasticity: An amusement park charges \$15 per admission and serves approximately 6400 patrons per day. The demand function for the x patrons at p dollars per admission is given by

$$x = 130(35 - p)^{1.3}$$

(a) Determine the point elasticity function $E(p)$.

(b) Is the demand elastic or inelastic at the current \$15 admission price?

(c) Should the ticket price be raised or lowered in order to increase revenue?

96. Economics—Elasticity: The FlyingThyme Watch Company determines that the demand for a new watch is given by

$$d(p) = 35\ln(150 - 5p)$$

where p is the unit price, in dollars, of a watch and $d(p)$ is the demand in thousands.

(a) Determine $E(25)$ and state whether the manufacturer should raise or lower the price in order to increase revenue.

(b) Set $E(p) = 1$ and solve for p to determine the watch price that maximizes revenue.

Section 5.7 Review Exercises

For Exercises 97–104, use implicit differentiation to determine $\dfrac{dy}{dx}$.

97. $3x - 5y = 7$

98. $9x^2 + 4x^2 = 36$

99. $5x^6 + y^3 = x^2$

100. $x^{1/3} - y^{1/3} = 8$

101. $y^3 - x^2y = 5$

102. $x^2 + 6xy + 9y^2 = 4$

103. $x\ln y = y + 4$

104. $\ln(x + y) = e^x + 3y$

For Exercises 105–108, complete the following.

(a) Use implicit differentiation to determine $\dfrac{dy}{dx}$.

(b) Solve the equation explicitly for y and determine $\dfrac{dy}{dx}$.

105. $2x + 3y = 6$

106. $7x - xy = 5x^2$

107. $x^3y - y = 2x + 2$

108. $\dfrac{3}{x} - \dfrac{4}{y} = 5$

For Exercises 109–112, determine the equation of the tangent line for the given equation at the indicated point.

109. $x^2 - y^2 = 9;\ (5, 4)$

110. $x^2y + y^2x = 12;\ (3, -4)$

111. $2x\ln y = 4x^2 - 4y;\ (1, 1)$

112. $3x + xe^y = 4 + y;\ (1, 0)$

113. **Accounting—Price-Demand:** The BriteWite Light Company has determined that the price-demand equation for a new lamp is

$$px + 10x = 600 \qquad 10 \le x \le 25$$

where x represents the demand for lamps in thousands and p represents the price of one lamp in dollars.

   (a) Use implicit differentiation to determine $\frac{dp}{dx}$.

   (b) Evaluate and interpret $\frac{dp}{dx}\big|_{(15, 30)}$.

In Exercises 114–117, determine the rate of change equation by implicitly differentiating each side of the equation with respect to time t.

114. $x^2 - 3y = 4$

115. $x^3 + y^3 = 6x$

116. $2xy + 5x = 18y$

117. $xy - y^3 = 2x$

For Exercises 118–121, assume that x and y are both functions of t and determine the indicated rate.

118. Determine $\frac{dy}{dt}$ for $2x + y^2 = 7$, given $y = 3$ and $\frac{dx}{dt} = 12$.

119. Determine $\frac{dx}{dt}$ for $xy + 4y - 3x = 7$, given $x = 1$ and $\frac{dy}{dt} = 4$.

120. Determine $\frac{dx}{dt}$ for $y = 2x + 3^{x-4}$, given $x = 4$ and $\frac{dy}{dt} = -6$.

121. Determine $\frac{dy}{dt}$ for $x^2 + 2xy + y^2 = 25$, given $x = 2$, $y = 3$, and $\frac{dx}{dt} = 11$.

122. **Economics—Revenue Analysis:** The Zoot Soot Chimney Service has determined that its monthly revenue is given by

$$R = 150x - \frac{x^2}{30}$$

where x is the number of chimneys cleaned and R is the revenue in dollars.

   (a) Differentiate the revenue function with respect to time t.

   (b) Determine the rate of change in the revenue if $x = 45$ and $\frac{dx}{dt} = 3$ chimneys per month.

123. **Physics—Balloon Volume:** The surface area of a spherical balloon is given by the equation

$$S = 4\pi r^2$$

where r is the radius of the balloon and S is the volume.

   (a) Differentiate the equation with respect to time t.

   (b) If the balloon is being blown up so that the radius is increasing at a rate of 0.7 centimeters per second, determine the rate at which the surface area is changing when the radius is 8 centimeters.

Integration

© Steveheap/Dreamstime.com **(a)**

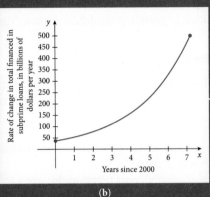

(b)

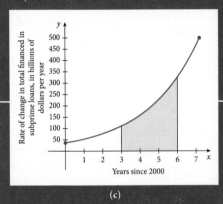

(c)

Subprime loans are offered to individuals who do not qualify for prime-rate loans based on their credit score. The proliferation of these types of loans in the early 2000s was generally regarded as the trigger for the banking crisis in 2008. Based on data gathered at the Center for Responsible Learning, the graph of Figure (b) shows the rate of change in the total financed in subprime loans, in billions of dollars per year, from 2000 to 2007. In Figure (c), the shaded area represents the total increase in subprime loans, in billions of dollars, from 2003 to 2006.

What We Know

Our focus thus far has been on the study of the derivative and instantaneous rates of change. Among other things, we have seen how we can use the derivative to determine maximum and minimum values on open and closed intervals.

Where Do We Go

We now wish to discover techniques for undoing the process of differentiation. This technique, called antidifferentiation, can be used to recover functions if the rate of change function is given. We will then study the Fundamental Theorem of Calculus, which is the glue that holds differential calculus together.

Chapter Sections

6.1 The Indefinite Integral

6.2 Area, the Definite Integral, and the Fundamental Theorem of Calculus

6.3 Integral Calculus and Total Accumulation

6.4 Integration by *u*-Substitution

6.5 Integrals That Yield Logarithmic and Exponential Functions

Chapter Review Exercises

CHAPTER 6 Integration

SECTION OBJECTIVES

1. Verify an antiderivative.

2. Evaluate an indefinite integral using the Power Rule.

3. Evaluate an indefinite integral using the Sum/ Difference Rule.

4. Evaluate an indefinite integral of a polynomial function.

5. Recover a function from its rate function.

6. Recover a profit function given its marginal profit function.

6.1 The Indefinite Integral

Until now, we have focused almost exclusively on how to determine the derivative and interpret the resulting rate of change function. For example, we found the derivative of a profit function and learned that this derivative is a rate of change function that measures the marginal profit. But what if we are given a **rate of change function** (that is, a rate function or derivative) and wish to find the original function? In this section, we will learn the basics of **integration**, a process that finds the original function, given its derivative, by reversing the differentiation. Then we will consider applications of integration.

The Indefinite Integral

To see how differentiation can be reversed, let's start with a common rate function from business. The WorkSharp Company determines that the marginal cost function for their designer suspenders is

$$MC(x) = C'(x) = 2x \qquad 0 \le x \le 10$$

where x is the number of designer suspenders produced in thousands, and $MC(x)$ is the marginal cost in thousands of dollars.

To find the company's cost function, we need to determine a cost function with the property that $\frac{d}{dx}[C(x)] = MC(x)$. In words, we seek a cost function so that when we take its derivative, we get $MC(x) = 2x$. A function such as this is referred to as an **antiderivative** of the given function MC. To gain insight into how this antidifferentiation process is done, let's review our basic differentiation rules shown in the Toolbox.

From Your Toolbox: Basic Differentiation Rules

For differentiable functions f and g with real numbers n and k, we have:

1. **Power Rule:** If $f(x) = x^n$, then $f'(x) = nx^{n-1}$.

2. **Constant Function Rule:** If $f(x) = k$, then $f'(x) = 0$

3. **Sum and Difference Rule:** If $h(x) = f(x) \pm g(x)$, then $h'(x) = f'(x) \pm g'(x)$

Since the WorkSharp Company's marginal cost function is $MC(x) = 2x$, we need an expression whose derivative is $2x$. Since the term $2x$ is linear (first degree), the Power Rule suggests that we need a quadratic function (second degree) for the antiderivative. Notice that if we choose the quadratic x^2, then $\frac{d}{dx}[x^2] = 2x$. This is just what we wanted! So is $C(x) = x^2$ the cost function we want? Not quite, because there are many functions whose derivatives are $2x$, including $x^2 + 3$, $x^2 + \frac{11}{2}$, and $x^2 - 10$. In fact, if C is any real number constant, then $x^2 + C$ is the **antiderivative** that we seek. We call C an **arbitrary constant**. To determine the WorkSharp Company's actual cost function, it would appear that we need a little more information. Before continuing, let's summarize what we have done so far.

DEFINITION

Antiderivative

We call F an **antiderivative** of f on an interval if $F'(x) = f(x)$ for all x in the interval. In other words,

$$\frac{d}{dx}[F(x)] = f(x)$$

We call $F(x) + C$, where C is any real number constant, the **general antiderivative** of f on an interval if

$$\frac{d}{dx}[F(x) + C] = f(x)$$

for all x in the interval.

NOTE: The process of determining an antiderivative is called **antidifferentiation**. Antidifferentiation is the **inverse operation** of differentiation.

Our definition provides an easy way to determine if a function is an antiderivative of another function. Example 1 illustrates how we can do this.

Example 1: Verifying General Antiderivatives

Determine if the function F is the general antiderivative of the function f.

a. $F(x) = \frac{2}{3}x^{3/2} + 4x + C$; $f(x) = \sqrt{x} + 4$

b. $F(x) = 2x^4 - x + C$; $f(x) = \frac{2}{3}x^3 - 1$

Perform the Mathematics

For each, we need to determine if $\frac{d}{dx}[F(x)] = f(x)$.

a. Differentiating, we get

$$\frac{d}{dx}[F(x)] = \frac{d}{dx}\left[\frac{2}{3}x^{3/2} + 4x + C\right] = \frac{2}{3} \cdot \frac{3}{2}x^{1/2} + 4 + 0$$

$$= x^{1/2} + 4 = \sqrt{x} + 4 = f(x)$$

Since $\frac{d}{dx}[F(x)] = f(x)$, we conclude that $F(x) = \frac{2}{3}x^{3/2} + 4x + C$ is the general antiderivative of $f(x) = \sqrt{x} + 4$.

b. Again, we compute the derivative of $F(x)$ and get

$$\frac{d}{dx}[F(x)] = \frac{d}{dx}[2x^4 - x + C] = 8x^3 - 1 \neq f(x)$$

Since $\frac{d}{dx}[F(x)] \neq f(x)$, we conclude that $F(x) = 2x^4 - x + C$ is **not** the general antiderivative of $f(x) = \frac{2}{3}x^3 - 1$. ∎

OBJECTIVE 1

Verify an antiderivative.

Another way to represent the general antiderivative of a function f is by

$$\int f(x)\,dx$$

which is called the **indefinite integral** of f.

With this new notation, we can write the result of Example 1, part (a), as

$$\int (\sqrt{x} + 4)\,dx = \frac{2}{3}x^{3/2} + 4x + C$$

Now let's look for a method to determine the indefinite integral for an expression involving x^n. First, we will consider the expression x^3. We know from the Power Rule for differentiation that the antiderivative of x^3 must be fourth degree. Since we know that $\frac{d}{dx}[x^4] = 4x^3$, then it follows that $\frac{1}{4} \cdot \frac{d}{dx}[x^4] = \frac{1}{4} \cdot 4x^3 = x^3$. Thus,

$$\int x^3\,dx = \frac{1}{4}x^4 + C$$

Notice in the antiderivative $\frac{1}{4}x^4 + C$ that the power of x is one more than the integrand x^3. Also, notice that the antiderivative includes a constant multiple that is the reciprocal of the exponent 4.

Earlier, we said that antidifferentiation is the inverse operation of differentiation. Here we are using the inverse operations of the Power Rule for differentiation to find an antiderivative. Instead of *subtracting* 1 from the exponent, we *add* 1, and instead of *multiplying* by the constant, we *divide*. We call this rule the **Power Rule for Integration**.

DEFINITION

Indefinite Integral

If $F'(x) = f(x)$ for all x, then

$$\int f(x)\,dx = F(x) + C$$

is called the **indefinite integral** of $f(x)$. In this notation, $\int$ is the **integral sign**, $f(x)$ is the **integrand**, and C is any real number constant. dx tells us the variable of integration.

NOTE: If we can determine the indefinite integral of a function, we say that the function is **integrable**. The process of determining a general antiderivative is called **integration**.

Power Rule for Integration

For any real number n, where $n \neq -1$, the indefinite integral of x^n is

$$\int x^n dx = \frac{1}{n+1} x^{n+1} + C$$

NOTE: The restriction that n cannot be -1 is because a value of -1 for n would make the denominator of the coefficient zero. In Section 6.5, we will learn the rule for the indefinite integral

$$\int x^{-1} dx = \int \frac{1}{x} dx.$$

OBJECTIVE 2

Evaluate an indefinite integral using the Power Rule.

Example 2: Using the Power Rule for Integration

Determine the following indefinite integrals.

a. $\displaystyle\int x^8 dx$ **b.** $\displaystyle\int \sqrt[4]{x}\, dx$ **c.** $\displaystyle\int \frac{1}{x^5} dx$

Perform the Mathematics

a. Applying the Power Rule for Integration, we get

$$\int x^8 dx = \frac{1}{8+1} x^{8+1} + C = \frac{1}{9} x^9 + C$$

Notice that the coefficient $\frac{1}{9}$ is the reciprocal of the exponent 9.

b. Here we use the rules of algebra to write the integrand in the form x^n. Since $\sqrt[4]{x} = x^{1/4}$, we can write

$$\int \sqrt[4]{x}\, dx = \int x^{1/4} dx = \left[\frac{1}{\frac{1}{4}+1} \right] x^{1/4+1} + C$$

$$= \frac{4}{5} x^{5/4} + C$$

Notice that the coefficient $\frac{4}{5}$ is the reciprocal of the exponent $\frac{5}{4}$.

c. Again, the rules of algebra come into play when rewriting the integrand. Here we can write $\frac{1}{x^5}$ as x^{-5} and then compute

$$\int \frac{1}{x^5} dx = \int x^{-5} dx = \frac{1}{-5+1} x^{-5+1} + C$$

$$= \frac{1}{-4} x^{-4} + C$$

$$= -\frac{1}{4} \cdot \frac{1}{x^4} + C$$

$$= -\frac{1}{4x^4} + C$$

Notice that the coefficient of $-\frac{1}{4}$ is the reciprocal of the exponent -4. ∎

▶ **Try It Yourself**

Some related Exercises are 17, 21, and 23.

The next rule involves integration of a constant. To find $\int 5dx$, we need a function whose derivative is 5. Since we know that $\frac{d}{dx}(5x) = 5$, we can write $\int 5dx = 5x + C$. This suggests the **Constant Rule for Integration**.

Some of the properties of integration are much the same as for limits and differentiation. Consider the integral $\int (2x - 5)dx$. We know that $\int 2xdx = x^2 + C$, and we can show from the Constant Rule for Integration that $\int 5dx = 5x + C$. So it seems natural that the integral of the linear function $2x - 5$ is $\int (2x - 5)dx = x^2 - 5x + C$, which you should check using differentiation. We could have written this as

$$\int (2x - 5)dx = \int 2xdx - \int 5dx = x^2 - 5x + C$$

which illustrates the **Sum and Difference Rule for Integration**.

Some may be concerned that since there are two integrals there must be two constants of integration C. But here we are simply adding two constants to get a new constant, which we call C.

> **Constant Rule for Integration**
>
> If k is any real number, then the indefinite integral of k is
> $$\int kdx = kx + C$$

Sum and Difference Rule for Integration

For integrable functions f and g, we have

$$\int [f(x) \pm g(x)]dx = \int f(x)dx \pm \int g(x)dx$$

NOTE: This rule means that we are allowed to determine indefinite integrals term by term.

Example 3: Using the Sum and Difference Rule for Integration

Apply the Sum and Difference Rule for Integration to determine the indefinite integrals.

a. $\displaystyle\int (x^2 + 3)dx$ **b.** $\displaystyle\int (\sqrt[3]{x} + 5)dx$

> **OBJECTIVE 3**
>
> Evaluate an indefinite integral using the Sum/Difference Rule.

Perform the Mathematics

a.   The Sum and Difference Rule for Integration yields

$$\int (x^2 + 3)dx = \int x^2dx + \int 3dx = \frac{1}{3}x^3 + 3x + C$$

b.   We begin by writing the first term of the integrand in the form x^n to get

$$\int (\sqrt[3]{x} + 5)dx = \int (x^{1/3} + 5)dx$$

Applying the Sum and Difference Rule for Integration gives us

$$\int (x^{1/3} + 5)dx = \int x^{1/3}dx + \int 5dx$$

$$= \frac{3}{4}x^{4/3} + 5x + C \qquad ∎$$

Our next rule addresses functions with coefficients other than 1. In Chapter 3 we learned that

$$\frac{d}{dx}[k \cdot f(x)] = k \cdot \frac{d}{dx}[f(x)]$$

For example,

$$\frac{d}{dx}[3x^2] = 3 \cdot \frac{d}{dx}[x^2] = 3 \cdot 2x = 6x$$

Since integration is the inverse operation of differentiation, we know that

$$\int 6x\,dx = 3x^2 + C$$

But how would we get the general antiderivative if we did not already know it? We could try "pulling out" the constant much as we did with the constant multiple differentiation rule.

$$\int 6x\,dx = 6 \cdot \int x\,dx = 6 \cdot \frac{1}{2}x^2 + C = 3x^2 + C$$

This illustrates that when we are integrating a constant times a function, we can put the constant in front of the integral sign and integrate the function normally. More formally, this is known as the **Constant Multiple Rule for Integration**.

Armed with these integration rules, we can find the indefinite integral of any polynomial function. For example,

$$\int (x^2 - 3x + 4)\,dx = \int x^2\,dx - 3\int x\,dx + \int 4\,dx$$

$$= \frac{1}{3}x^3 - 3 \cdot \frac{1}{2}x^2 + 4x + C$$

$$= \frac{1}{3}x^3 - \frac{3}{2}x^2 + 4x + C$$

Example 4 shows that we do not necessarily have to integrate a function with respect to x.

Constant Multiple Rule for Integration

Given any real number constant k and integrable function f,

$$\int k \cdot f(x)\,dx = k \cdot \int f(x)\,dx$$

OBJECTIVE 4

Evaluate an indefinite integral of a polynomial function.

Example 4: Integrating a Polynomial Function

Determine $\int (-2t^3 + 3t + 5)\,dt$.

Perform the Mathematics

Here the dt indicates that we are integrating the integrand with respect to t. The same rules that we have developed in this section for the independent variable x apply here as well. Performing the integration gives us

$$\int (-2t^3 + 3t + 5)\,dt = \int -2t^3\,dt + \int 3t\,dt + \int 5\,dt$$

$$= -2\int t^3\,dt + 3\int t\,dt + \int 5\,dt$$

$$= -2 \cdot \frac{1}{4}t^4 + 3 \cdot \frac{1}{2}t^2 + 5t + C$$

$$= -\frac{1}{2}t^4 + \frac{3}{2}t^2 + 5t + C$$ ∎

Recovering Functions from Rate Functions

Through much of our remaining study of calculus, we will study the properties of **rate functions**.

Recovering functions from rate functions is very useful in many applications in which the rate of change is given. For example, let's say that we have a derivative $f'(x) = x^2$ and we wish to know the function from which it came, that is, f. By integrating, we know that

$$f(x) = \int x^2 dx = \frac{1}{3}x^3 + C$$

Each value of C produces a different function and a different graph, as shown in **Figure 6.1.1**.

DEFINITION

Rate Function

A function that gives an instantaneous rate of change is a **rate function.** A rate function may be considered a derivative, since the derivative gives us an instantaneous rate of change.

NOTE: We use *rate function* and *rate of change function* interchangeably. Also, in applications, pay close attention to the units given for a rate function.

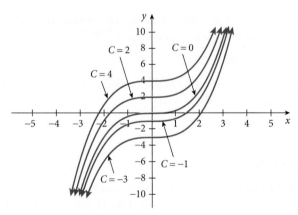

Figure 6.1.1 Graphs of $f(x) = \frac{1}{3}x^3 + C$ for $C = -3, -1, 0, 2, 4$.

Now suppose we know that an ordered pair, say $(3, 11)$, is on the graph of f. By replacing $f(x)$ with 11 and x with 3, we can write an equation to find the value of the integration constant C.

$$f(x) = \frac{1}{3}x^3 + C$$

$$11 = \frac{1}{3}(3)^3 + C$$

$$11 = \frac{1}{3} \cdot 27 + C$$

$$11 = 9 + C$$

$$2 = C$$

So, for the given derivative $f'(x) = x^2$ and point $(3, 11)$ on the graph of f, the corresponding function f is $f(x) = \frac{1}{3}x^3 + 2$. This is illustrated in **Figure 6.1.2**.

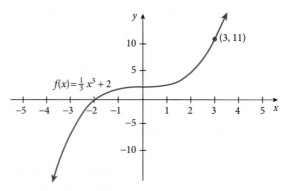

Figure 6.1.2

OBJECTIVE 5

Recover a function from its rate function.

Example 5: Recovering a Function from a Rate Function

The HonCo marketing firm finds the rate of change of total sales of a product featured on an infomercial can be modeled by

$$R(t) = 15\sqrt[3]{t} \qquad 0 \le t \le 20$$

where t represents the number of months that the infomercial has aired and $R(t)$ represents the rate of change of total sales of the product measured in $\frac{\text{thousand units}}{\text{month}}$.

a. Evaluate $R(10)$ and interpret.

b. Knowing that there were 500 units sold in trial marketing just as the infomercial began airing, find a total sales function S for the product.

c. Evaluate $S(10)$ and interpret.

Perform the Mathematics

a. First, note that R is a rate function because it measures the rate of change in total sales with respect to time. In other words, R is the derivative of some function. Evaluating the sales rate function at $t = 10$ yields

$$R(10) = 15\sqrt[3]{10} \approx 32.32$$

This means that in the 10th month of airing the infomercial, the total sales of the product were increasing at a rate of about 32,320 units per month.

b. Integrating the sales rate function R with respect to time t recovers the total sales function S. Proceeding with the integration gives us

$$S(t) = \int R(t)dt = \int 15\sqrt[3]{t}\,dt = \int 15t^{1/3}dt$$

$$= 15\int t^{1/3}dt$$

$$= 15 \cdot \frac{3}{4}t^{4/3} + C$$

$$= \frac{45}{4}\sqrt[3]{t^4} + C$$

So we now know that the sales function has the general form $S(t) = \frac{45}{4}\sqrt[3]{t^4} + C$. Since we know that 500 units were sold when the infomercial began airing, this means that, when $t = 0$, $S(t) = 0.5$. Substituting these values in our general form, we get

$$S(t) = \frac{45}{4}\sqrt[3]{t^4} + C$$

$$0.5 = \frac{45}{4}\sqrt[3]{(0)^4} + C$$

$$0.5 = 0 + C$$

$$0.5 = C$$

Thus, the total sales function is

$$S(t) = \frac{45}{4}\sqrt[3]{t^4} + 0.5$$

c. Evaluating the sales function at $t = 10$, we get

$$S(10) = \frac{45}{4}\sqrt[3]{(10)^4} + 0.5 \approx 242.87$$

This means that, during the 10th month of airing the infomercial, total sales were 242,870 units of the product. ■

In Example 5, the rate function R is measured in $\frac{\text{thousand units}}{\text{month}}$, and the recovered sales function is measured in thousands of units. **Generally, for a rate function, the units are given in $\frac{\text{dependent units}}{\text{independent unit}}$, and the function recovered is measured with dependent units.**

Recovering Business Functions

We have seen that given a rate function, we can recover the function from which it came. The rate functions, or derivatives, associated with cost C, revenue R, and profit P function are the **marginal functions**, denoted by MC, MR, and MP, respectively. Let's review these functions in the following Toolbox.

From Your Toolbox: Marginal Analysis

Let x be the number of units produced and sold of a certain product in a specific time interval.

1. The **Marginal Cost Function** $MC(x) = C'(x)$ is the approximate cost of producing one additional unit at a production level x.
2. The **Marginal Revenue Function** $MR(x) = R'(x)$ is the approximate loss or gain in revenue from producing and selling one additional unit at production level x.
3. The **Marginal Profit Function** $MP(x) = P'(x)$ is the approximate loss or gain in profit from producing and selling one additional unit at a production level x.

Recovering business functions involves integrating their corresponding marginal functions and evaluating at some given value. We illustrate in Example 6.

Example 6: Recovering a Profit Function

OBJECTIVE 6

Recover a profit function given its marginal profit function.

The Best Dressed Clothing Company finds that its marginal profit MP is linear and has the form $MP(q) = mq + b$, where m and b are constants. The company gets about $171 additional profit from producing and selling the 101st sport coat and $169 in additional profit from producing and selling the 151st sport coat in each production run.

a. Determine the marginal profit function MP.

b. Knowing that the company gets $11,300 profit from 150 sport coats, find the profit function P.

Perform the Mathematics

a. Since the marginal profit function is linear, we start by finding the slope m of the linear function. From the definition of marginal profit, we know that $MP(100) = 171$ and $MP(150) = 169$. So the slope of the marginal profit function is

$$m = \frac{\text{change in } MP(q)}{\text{change in } q} = \frac{169 - 171}{150 - 100} = -\frac{2}{50} = -\frac{1}{25}$$

Now we can use the point-slope form of a line with the point $(q, MP(q)) = (100, 171)$:

$$y - y_1 = m(x - x_1)$$

$$y - 171 = -\frac{1}{25}(x - 100)$$

$$y - 171 = -\frac{1}{25}x + 4$$

$$y = -\frac{1}{25}x + 175$$

Replacing x with q and y with $MP(q)$ gives us

$$MP(q) = -\frac{1}{25}q + 175$$

So the marginal profit function is $MP(q) = -\frac{1}{25}q + 175$.

b. To recover the profit function, P, we must integrate MP with respect to q. This gives us

$$P(q) = \int MP(q)dq = \int \left(-\frac{1}{25}q + 175\right)dq$$

$$= -\frac{1}{25} \cdot \frac{1}{2}q^2 + 175q + C$$

$$= -\frac{1}{50}q^2 + 175q + C$$

To find the value of C, we use the fact that when $q = 150$, $P(q) = 11{,}300$. So substituting $q = 150$ and $P(q) = 11{,}300$ gives us

$$11{,}300 = -\frac{1}{50}(150)^2 + 175(150) + C$$

$$11{,}300 = -\frac{1}{50}(22{,}500) + 26{,}250 + C$$

$$11{,}300 = -450 + 26{,}250 + C$$

$$11{,}300 = 25{,}800 + C$$

$$-14{,}500 = C$$

Thus, the recovered profit function is $P(q) = -\frac{1}{50}q^2 + 175q - 14{,}500$. ∎

Summary

This section began with a discussion of the **antiderivative** and quickly led us to the **indefinite integral**. We examined how to integrate powers of function, sums and differences of functions, and simple radical functions. Then we discussed how to recover a function when its rate function is given. We used this method to recover the business functions from the marginal business functions.

Important Integration Rules:

- **Power Rule:** $\int x^n dx = \dfrac{1}{n+1}x^{n+1} + C, n \neq -1$

- **Constant Rule:** $\int k\,dx = kx + C$

- **Sum and Difference Rule:** $\int [f(x) \pm g(x)]\,dx = \int f(x)\,dx \pm \int g(x)\,dx$

- **Constant Multiple Rule:** $\int k \cdot f(x)\,dx = k \cdot \int f(x)\,dx$

Section 6.1 Exercises

Vocabulary Exercises

1. If $F'(x) = f(x)$ for all x in an interval, we call F an _____ of f.
2. If $\frac{d}{dx}[F(x) + C] = f(x)$ for all x in an interval, we call C an _____ _____.
3. If $F'(x) = f(x)$ for all x in an interval, we call $\int f(x)\,dx = F(x) + C$ the _____ _____ of f.
4. The process of determining a general antiderivative is called _____.
5. A function that gives an instantaneous rate of change is called a _____ function.
6. $\int x^n dx = \dfrac{1}{n+1}x^{n+1} + C, n \neq -1$, is called the _____ Rule for Integration.

Skill Exercises

In Exercises 7–16, determine if the function F is the general antiderivative of f.

7. $F(x) = 6x + C;\ f(x) = 6$

8. $F(x) = \dfrac{1}{2}x + C;\ f(x) = \dfrac{1}{2}$

9. $F(x) = x^8 + C;\ f(x) = x^7$

10. $F(x) = \dfrac{1}{2}x^3 + C;\ f(x) = x^2$

11. $F(t) = 8t + \dfrac{t^2}{2} + C;\ f(t) = 8 + t$

12. $F(t) = 7t + et + C;\ f(t) = 7 + e$

13. $F(x) = x + C;\ f(x) = 0$

14. $F(x) = C;\ f(x) = 0$

15. $F(x) = \dfrac{x^{11}}{11} + C;\ f(x) = x^{10}$

16. $F(x) = \dfrac{5}{6x^6} + C;\ f(x) = \dfrac{5}{x^7}$

For Exercises 17–26, use the Power Rule for Integration to determine the indefinite integrals.

17. $\int x^4 dx$

18. $\int x^9 dx$

19. $\int x^{2.31} dx$

20. $\int x^{0.27} dx$

21. $\int \dfrac{1}{t^3} dt$

22. $\int \dfrac{1}{t^{11}} dt$

23. $\int \sqrt[4]{x^5}\,dx$

24. $\int \sqrt{x^5}\,dx$

25. $\int \dfrac{1}{\sqrt[3]{x}} dx$

26. $\int \dfrac{1}{\sqrt{x}} dx$

In Exercises 27–50, determine the indefinite integral.

27. $\int 0.4x^{0.6} dx$

28. $\int 0.5x^7 dx$

29. $\int (2x + 3) dx$

30. $\int (5x - 2) dx$

31. $\int \left(\dfrac{2}{3}x + 4\right) dx$

32. $\int \left(\dfrac{1}{2}t + 5\right) dt$

33. $\int (3t^2 + 2t + 10) dt$

34. $\int (4x^4 + 5x - 6) dx$

35. $\int (1 - 2x^2 + 3x^3) dx$

36. $\int (5 - 2x + x^3) dx$

37. $\int (6.21x^2 + 0.03x - 4.01) dx$

38. $\int (0.03x^2 - 0.21x + 4.02) dx$

39. $\int \left(\dfrac{1}{x^2} - \dfrac{3}{x^3}\right) dx$

40. $\int \left(\dfrac{3}{x^4} + 6x^5\right) dx$

41. $\int (3 + 2\sqrt{x}) dx$

42. $\int (\sqrt{x} - 3x^{3/2}) dx$

43. $\int \left(\dfrac{3}{x^3} + 2x^{3/2} - 4\right) dx$

44. $\int \left(x^{5/2} - \dfrac{4}{x^5} - \sqrt{x}\right) dx$

45. $\int \left(\dfrac{3t^3 - 2t}{6t}\right) dt$

46. $\int \left(\dfrac{4x^4 - 5x^3}{x^2}\right) dx$

47. $\int (0.1z^{-3} + 2z^{-2} + z^3) dz$

48. $\int (5x^{-5} + 3x^{-4}) dx$

49. $\int (2t^{0.13} + 5) dt$

50. $\int (0.1x^{0.318} + 7x) dx$

For Exercises 51–64, solve the given rate function using the given value. Note that $f(a) = b$ corresponds to the ordered pair (a, b).

51. $f'(x) = -2;\ f(0) = 4$

52. $f'(x) = 7;\ f(0) = 1$

53. $f'(x) = 5x;\ f(0) = 0$

54. $f'(x) = 3x;\ f(0) = 0$

55. $f'(x) = 2x - 3;\ f(0) = 4$

56. $f'(x) = 5 - 6x;\ f(0) = 6$

57. $f'(t) = 500 - 0.05t;\ f(0) = 40$

58. $f'(x) = 4x^2 - 6x;\ f(1) = 0$

59. $f'(x) = 2x^{-2} + 3x^{-3} - 1;\ f(1) = 2$

60. $f'(x) = \dfrac{2x^4 - x}{x^3};\ f(1) = 10$

61. $f'(t) = \dfrac{5t + 2}{\sqrt{t}};\ f(0) = 1$

62. $f'(t) = -\dfrac{10}{t^2} + t;\ f(1) = 20$

63. $f'(t) = \dfrac{1 - t^4}{t^3};\ f(1) = 4$

64. $f'(t) = \dfrac{\sqrt{t^3} - t}{(\sqrt{t})^3};\ f(9) = 4$

Application Exercises

65. Economics—Profit Function: The Krystal King jewelry store knows that the marginal profit for producing and selling x Krystal Kluster necklaces is given by the linear function

$$MP(x) = 40 - 0.05x$$

where $MP(x)$ is measured in dollars per necklace.

(a) Knowing that $P(0) = 0$, recover the profit function P.

(b) Use the solution from part (a) to determine the total profit when 200 Krystal Kluster necklaces are produced and sold.

66. Economics—Profit Function: The Pack & Plod Company determines the marginal profit function for producing and selling x tablet computer hardcases is given by

$$MP(x) = -0.20x^2 + 40$$

where $MP(x)$ is measured in dollars per hardcase.

(a) Knowing that $P(0) = -10$, recover the profit function P.

(b) Use the solution from part (a) to determine the total profit from producing and selling 150 hardcases.

67. Economics—Revenue Function: The marginal revenue function for the FrontRide Bus Company is given by.

$$MR(x) = 0.000045x^2 - 0.03x + 3.75 \qquad 0 \le x \le 500$$

where x represents the number of passengers each afternoon.

(a) Knowing that $R(0) = 0$, recover the revenue function R.

(b) Determine the price function p, knowing that $p(x) = \frac{R(x)}{x}$.

(c) What should the price be when the demand is 100 passengers?

68. Economics—Revenue Function: The marginal revenue function for the BlackDay Sunglasses Company is given by

$$MR(x) = 30 - 0.0003x^2 \qquad 0 \le x \le 540$$

where x represents the number of sunglasses produced and sold daily.

(a) Knowing that $R(50) = 1487.50$, recover the revenue function R.

(b) Determine the price function p, knowing that $p(x) = \frac{R(x)}{x}$.

(c) What should the price of the sunglasses be when the demand is 250 pairs daily?

69. Economics—Average Cost: The Banter Banner Company knows that the marginal average cost of producing and selling x of their 5-meter promotional banners is given by

$$MAC(x) = -\frac{100}{x^2} \qquad x > 0$$

(a) Knowing that it costs $2.50 per banner to make 100 banners, recover the average cost function AC.

(b) Knowing that $AC(x) = \frac{C(x)}{x}$, determine the cost function C for the banners.

(c) Using the cost function from part (b), evaluate $C(100)$ and interpret.

70. Economics—Average Cost: The marginal average cost of producing x QuickVid digital video cameras is given by

$$MAC(x) = 0.03x^2 - 0.04x + 5 \qquad 0 < x \le 50$$

(a) Knowing that it costs $422 per camera to produce 20 cameras, recover the average cost function AC.

(b) Knowing that $AC(x) = \frac{C(x)}{x}$, determine the cost function C for the banners.

(c) Using the cost function from part (b), evaluate $C(20)$ and interpret.

 71. Political Science—Hispanic Voters: One group of voters that has received attention recently is voter trends of Hispanic Americans. From 1995 to 2010, the number of Hispanic voters has increased steadily at a rate of 1.08 million voters per year. (*Source:* U.S. Census

© Digital Vision/Thinkstock.com

Bureau.) We can express the annual increase in number of Hispanic voters using the rate function

$$f'(x) = 1.08 \qquad 0 \le x \le 15$$

where x represent the number of years since 1995.

(a) Knowing that there were 11.7 million Hispanic voters in 1995, recover $f(x)$.

(b) Interpret the meaning of the function found in part (a).

(c) Use the solution to part (a) to determine the number of Hispanic voters in 2006.

 72. Political Science—African-American Voters: A group that political analysis has tradition-ally been interested in is African American voters. From 1995 to 2010, the number of African American voters has increased steadily at a rate of 3.10 million voters per year. (*Source:* U.S. Census Bureau.) We can express the annual increase in the number of African-American voters using the rate function

$$f'(x) = 3.10 \qquad 0 \le x \le 15$$

where x represent the number of years since 1995.

(a) Knowing that there were 20.67 million African-American voters in 1995, recover $f(x)$.

(b) Interpret the meaning of the function found in part (a).

(c) Use the solution to part (a) to determine the number of African-American voters in 2004.

 73. Public Health—Botulism Cases: Botulism is a rare but serious illness caused by *Clostridium botulinum* bacteria. The bacteria may enter the body through wounds, or they may live in improperly canned or preserved food. From 1980 to 2010, the rate of change in the number of cases of botulism can be modeled by the rate function

$$f'(x) = 0.22x - 0.48 \qquad 0 \le x \le 30$$

where x represents the number of years since 1980. (*Source:* U.S. Centers for Disease Control and Prevention.)

(a) Evaluate $f'(20)$ and interpret.

(b) Knowing that there were 88 cases of botulism in 1980, recover f.

(c) Use the solution in part (b) to find the number of botulism cases in the United States in 2000.

 74. Information Technology—Programming Revenue: A profession that has seen changes in the past decade is custom programming for business and industry. The rate of change in revenue (in millions of dollars per year) for custom programming services from 2000 to 2010 can be modeled by

$$R'(x) = 2.72x - 8.21 \qquad 0 \le x \le 10$$

where x represents the number of years since 2000, and $R'(x)$ represents the rate of change in revenue in millions of dollars per year. (*Source:* American Factfinder.)

(a) Evaluate $R'(3)$ and interpret.

(b) Knowing that the total revenue from custom computer programming in 2000 was 70.06 million dollars, recover R, the revenue function.

(c) Use the solution of part (b) to find the revenue generated from custom computer program-ming in 2008.

75. Macroeconomics—European Unemployment: Suppose an economist wishes to examine the European unemployment rate in the middle of the first decade of the 21st century. The rate of change in unemployment in Bulgaria from 2000 to 2010 can be modeled by the rate function

$$f'(x) = 0.24x^2 - 2.48x + 3.34 \qquad 0 \le x \le 10$$

where x represents the number of years since 2000 and $f'(x)$ is the rate of change in unemployment rate in percent per year.

(a) Knowing that the unemployment rate in Bulgaria in 2000 was 15.27%, recover f, the model for the Bulgarian unemployment rate.

(b) Use the solution to part (a) to determine the Bulgarian unemployment rate in 2005.

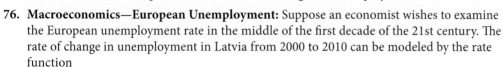 **76. Macroeconomics—European Unemployment:** Suppose an economist wishes to examine the European unemployment rate in the middle of the first decade of the 21st century. The rate of change in unemployment in Latvia from 2000 to 2010 can be modeled by the rate function

$$f'(x) = 0.27x^2 - 1.96x + 1.63 \qquad 0 \le x \le 10$$

where x represents the number of years since 2000 and $f'(x)$ is the rate of change in the unemployment rate in percent per year.

(a) Knowing that the unemployment rate in Latvia in 2000 was 13.21%, recover f, the model for the Latvian unemployment rate.

(b) Use the solution to part (a) to determine the Latvian unemployment rate in 2005.

Concept and Writing Exercises

77. Let $f'(x) = k$ be a rate function where k represents a real number constant, and let $f(a) = b$. Recover f.

78. Let $f'(x) = mx + b$ be a rate function where m and b represent real number constants. Knowing that $f(c) = d$, recover the function f.

79. Suppose a marginal average cost function has the form $MAC(x) = ax + b - \frac{d}{x^2}$ and the fixed costs are k dollars (that is, $C(0) = k$). Recover the average cost function AC and determine the cost function C.

80. Suppose a marginal average profit function has the form $MAP(x) = a\sqrt{x} + b + \frac{C}{x^2}$ and that $P(0) = -k$. Recover the average profit function AP and determine the profit function P.

81. The graph of a linear rate function f' contains the points $(-3, 1)$ and $(4, 0)$. Knowing that $f(2) = 6$ recover f.

82. A quadratic rate function has the form $f'(x) = x^2 + bx + c$ and contains the points $(2, 9)$ and $(3, 17)$. Knowing that $f(0) = 0$ recover f.

 ## Section Project

The Luv-n-Care Shop produces handcrafted art. The management of Luv-n-Care tracks its marginal costs for producing x units of art at various production levels. The data are shown in **Table 6.1.1**.

(a) Use the regression capability of your calculator to compute a power regression model of the form $MC(x) = a \cdot x^b$.

(b) Use the model from part (a) to evaluate and interpret $MC(45)$.

(c) Evaluate $MC(40)$ using the model and compare to the marginal cost given in the table. Explain why there is a difference in the marginal cost values.

(d) If the fixed costs are set at $1500 (that is, $C(0) = 1500$), recover the cost function C.

(e) Use the solution of part (d) to evaluate and interpret $C(45)$.

Table 6.1.1

Items Produced, x	Marginal Cost, $MC(x)$
10	50
20	56
30	62
40	67
50	73
60	77

6.2   Area, the Definite Integral, and the Fundamental Theorem of Calculus

In Chapter 2, we learned how to complete a mathematical task that, up until then, was seemingly impossible—to find the slope of a curve at a single point. The calculus tool we used to solve this problem was the *derivative*. Now we must try to solve yet another problem that challenged mathematicians for years: specifically, to find the area A under a curve given by a function $y = f(x)$ between two endpoints a and b or, as we also say, on a **closed interval** $[a, b]$. An illustration of our challenge is shown in **Figure 6.2.1**.

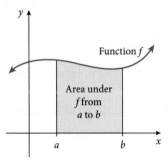

Figure 6.2.1

Finding the Area under a Curve Numerically: The Left Sum Method

To begin to tackle the problem of finding the area under a curve, we rely on area formulas learned in geometry, such as the area of a rectangle. Recall that, for a rectangle with base b and height h, the area of the rectangle is given by

$$\text{Area} = \text{base} \cdot \text{height} = b \cdot h$$

To convince ourselves that finding a definite area under a curve is possible, we will start with a continuous function $f(x) = x^2 + 1$ on a closed interval $[0, 2]$, as shown in **Figure 6.2.2**.

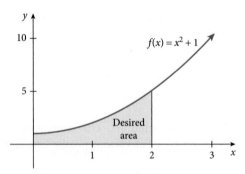

Figure 6.2.2 Area under $f(x) = x^2 + 1$ from $x = 0$ to $x = 2$.

Since we know how to find the area of a rectangle, we can approximate the area under the curve by "slicing up" the whole area into thin rectangular strips, computing the area of each strip, and then adding up the areas. Let's say that we want to start with $n = 4$ of these rectangular strips. We start the process by dividing the interval $[a, b]$ into four equal parts. These parts are called **subintervals**, and the width of each is called the **step size**, denoted by Δx. To get the step size, we simply divide the length of the closed interval $[a, b]$ by the number of subintervals n.

DEFINITION

Step Size

For a closed interval $[a, b]$ divided into n equally spaced subintervals, the **step size**, denoted by Δx, is given by

$$\Delta x = \frac{\text{length of interval } [a, b]}{\text{number of subintervals}} = \frac{b - a}{n}$$

For our function $f(x) = x^2 + 1$ on interval $[0, 2]$, if we choose four equally spaced subintervals, we get a step size of $\Delta x = \frac{2 - 0}{4} = \frac{1}{2}$. For the interval $[0, 2]$ and step size $\Delta x = \frac{1}{2}$, the endpoints of each subinterval have x-coordinates as shown in **Table 6.2.1**.

Table 6.2.1

x-Coordinate	x-Coordinate on $[0, 2]$ for $n = 4$
x_0	0
x_1	$0 + \dfrac{1}{2} = \dfrac{1}{2}$
x_2	$\dfrac{1}{2} + \dfrac{1}{2} = 1$
x_3	$1 + \dfrac{1}{2} = \dfrac{3}{2}$
x_4	$\dfrac{3}{2} + \dfrac{1}{2} = 2$

Now the coordinates in Table 6.2.1 are the x-coordinates of the endpoints of the bases of four rectangles. We can use either the left endpoint of each subinterval or the right endpoint of each subinterval to determine the height of the rectangles. Picking the left endpoints of each subinterval to determine the heights of rectangles is shown in **Figure 6.2.3(a)**, and choosing the right endpoints of each subinterval to determine the heights of rectangles is shown in Figure 6.2.3(b).

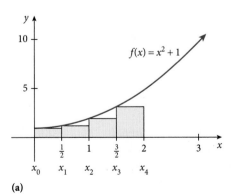

(a)

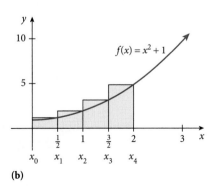

(b)

Figure 6.2.3 **(a) Left** endpoints used for the heights of rectangles. **(b) Right** endpoints used for the heights of rectangles.

OBJECTIVE 1

Approximate the area under a curve using rectangles.

For our first approximation for the area under the curve $f(x) = x^2 + 1$ on $[0, 2]$, we will use the left endpoint of each subinterval to determine the height of the rectangles. The height is found by evaluating the function $f(x) = x^2 + 1$ at the left endpoint of each of the four subintervals. We include these values in **Table 6.2.2**.

Table 6.2.2

x-Coordinate	x-Coordinate on [0, 2] for n = 4	Rectangle Height $f(x_i)$
x_0	0	$f(x_0) = f(0) = (0)^2 + 1 = 1$
x_1	$\dfrac{1}{2}$	$f(x_1) = f\left(\dfrac{1}{2}\right) = \left(\dfrac{1}{2}\right)^2 + 1 = \dfrac{5}{4}$
x_2	1	$f(x_2) = f(1) = (1)^2 + 1 = 2$
x_3	$\dfrac{3}{2}$	$f(x_3) = f\left(\dfrac{3}{2}\right) = \left(\dfrac{3}{2}\right)^2 + 1 = \dfrac{13}{4}$

So it appears that, using these four rectangles, the approximate area A under the curve $f(x) = x^2 + 1$ on $[0, 2]$ is

$$A \approx \left(\begin{array}{c}\text{area of first}\\\text{rectangle}\end{array}\right) + \left(\begin{array}{c}\text{area of second}\\\text{rectangle}\end{array}\right) + \left(\begin{array}{c}\text{area of third}\\\text{rectangle}\end{array}\right) + \left(\begin{array}{c}\text{area of fourth}\\\text{rectangle}\end{array}\right)$$

$$= \Delta x \cdot f(x_0) + \Delta x \cdot f(x_1) + \Delta x \cdot f(x_2) + \Delta x \cdot f(x_3)$$

$$= \frac{1}{2}(1) + \frac{1}{2}\left(\frac{5}{4}\right) + \frac{1}{2}(2) + \frac{1}{2}\left(\frac{13}{4}\right)$$

$$= \frac{1}{2} + \frac{5}{8} + 1 + \frac{13}{8}$$

$$= \frac{15}{4} \text{ square units or } 3.75 \text{ un}^2$$

This area is shown in **Figure 6.2.4**. We call this way of approximating the area under a curve the **Left Sum Method**.

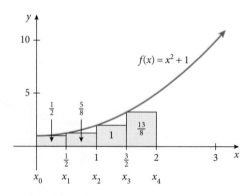

Figure 6.2.4

Left Sum Method

Given a continuous function $y = f(x)$ defined on a closed interval $[a, b]$, we compute the **Left Sum** using n equally spaced subintervals by the following steps:

1. Compute the step size Δx by calculating $\Delta x = \dfrac{b - a}{n}$.

2. Find the x-coordinates $x_0, x_1, x_2, \ldots, x_{n-1}$ of the left endpoints of each subinterval by starting with $x_0 = a$ and successively adding the step size Δx.

3. Compute the Left Sum:

$$A = \begin{pmatrix} \text{area of first} \\ \text{rectangle} \end{pmatrix} + \begin{pmatrix} \text{area of second} \\ \text{rectangle} \end{pmatrix} + \begin{pmatrix} \text{area of third} \\ \text{rectangle} \end{pmatrix} + \cdots + \begin{pmatrix} \text{area of } n\text{th} \\ \text{rectangle} \end{pmatrix}$$

$$= \Delta x \cdot f(x_0) + \Delta x \cdot f(x_1) + \Delta x \cdot f(x_2) + \cdots + \Delta x \cdot f(x_{n-1})$$

Using **summation notation**, we could write the Left Sum Method formula in step 3 as

$$A = \sum_{i=0}^{n-1} \Delta x f(x_i) \quad \text{or} \quad A = \sum_{i=0}^{n-1} f(x_i)\, \Delta x$$

This method is sometimes called **rectangular approximation** and is a straightforward way to estimate the area under a curve.

The Right Sum Method

Another way to approximate area using rectangles is to use the *right* endpoint of each subinterval to determine the height of the rectangle. Example 1 illustrates this.

Example 1: Approximating Area Using the Right Sum Method

For the function $f(x) = x^2 + 1$ on the interval $[0, 2]$, and $n = 4$:

a.   Determine the height of each rectangle using the right endpoint of each subinterval.

b.   Use the heights to approximate the area under the graph of $f(x) = x^2 + 1$ on the interval $[0, 2]$.

Perform the Mathematics

a.   Here we use the step size $\Delta x = \frac{1}{2}$. We get the heights shown in **Table 6.2.3**.

Table 6.2.3

x-Coordinate	x-Coordinate on $[0, 2]$ for $n = 4$	Rectangle Height $f(x_i)$
x_1	$0 + \dfrac{1}{2} = \dfrac{1}{2}$	$f(x_1) = f\left(\dfrac{1}{2}\right) = \left(\dfrac{1}{2}\right)^2 + 1 = \dfrac{5}{4}$
x_2	$\dfrac{1}{2} + \dfrac{1}{2} = 1$	$f(x_2) = f(1) = 2$
x_3	$1 + \dfrac{1}{2} = \dfrac{3}{2}$	$f(x_3) = f\left(\dfrac{3}{2}\right) = \dfrac{13}{4}$
x_4	$\dfrac{3}{2} + \dfrac{1}{2} = 2$	$f(x_4) = f(2) = 5$

b.   Now, approximating the area using the right endpoints of subintervals gives us

$$A \approx \begin{pmatrix} \text{area of first} \\ \text{rectangle} \end{pmatrix} + \begin{pmatrix} \text{area of second} \\ \text{rectangle} \end{pmatrix} + \begin{pmatrix} \text{area of third} \\ \text{rectangle} \end{pmatrix} + \begin{pmatrix} \text{area of fourth} \\ \text{rectangle} \end{pmatrix}$$

$$= \Delta x \cdot f(x_1) + \Delta x \cdot f(x_2) + \Delta x \cdot f(x_3) + \Delta x \cdot f(x_4)$$

$$= \frac{1}{2}\left(\frac{5}{4}\right) + \frac{1}{2}(2) + \frac{1}{2}\left(\frac{13}{4}\right) + \frac{1}{2}(5)$$

$$= \frac{5}{8} + 1 + \frac{13}{8} + \frac{5}{2}$$

$$= \frac{23}{4} \text{ square units or } 5.75 \text{ un}^2 \qquad \blacksquare$$

▶ *Try It Yourself*

Some related Exercises are 13 and 15.

Example 1 illustrated the **Right Sum Method**.

Right Sum Method

Given a continuous function $y = f(x)$ defined on a closed interval $[a, b]$, we compute the **Right Sum** using n equally spaced subintervals by the following steps:

1. Compute the step size Δx by calculating $\Delta x = \dfrac{b - a}{n}$.

2. Find the x-coordinates $x_1, x_2, x_3, \ldots, x_n$ of the right endpoints of each subinterval by starting with $x_1 = a + \Delta x$ and successively adding the step size Δx.

3. Compute the Right Sum:

$$A = \begin{pmatrix} \text{area of first} \\ \text{rectangle} \end{pmatrix} + \begin{pmatrix} \text{area of second} \\ \text{rectangle} \end{pmatrix} + \begin{pmatrix} \text{area of third} \\ \text{rectangle} \end{pmatrix} + \cdots + \begin{pmatrix} \text{area of } n\text{th} \\ \text{rectangle} \end{pmatrix}$$

$$= \Delta x \cdot f(x_1) + \Delta x \cdot f(x_2) + \Delta x \cdot f(x_3) + \cdots + \Delta x \cdot f(x_n)$$

Using **summation notation**, we could write the Right Sum Method formula in step 3 as

$$A = \sum_{i=1}^{n} \Delta x f(x_i) \quad \text{or} \quad A = \sum_{i=1}^{n} f(x_i) \Delta x$$

The approximations using the two methods are compared in **Figure 6.2.5** through **Figure 6.2.8**. Notice that in order to get better approximations for the area under the curve, we could add more and more rectangles and compute their sums.

Left Sum Method

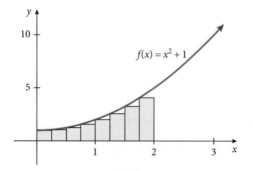

Figure 6.2.5 $n = 8$ rectangles. Area ≈ 4.1875

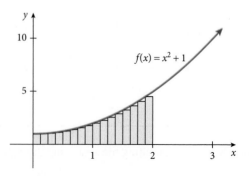

Figure 6.2.6 $n = 16$ rectangles. Area ≈ 4.421875

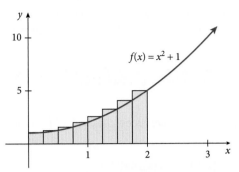

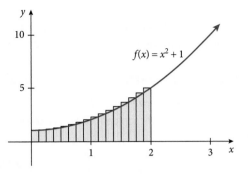

Right Sum Method

Figure 6.2.7 $n = 8$ rectangles. Area ≈ 5.1875 **Figure 6.2.8** $n = 16$ rectangles. Area ≈ 4.921875

Technology Option

The shortcoming of adding more and more rectangles is that computing these approximations by hand can quickly become taxing. Fortunately, our calculators serve not only as graphing devices, but also as programmable handheld computers that complete these computational tasks in seconds. Following are programs for the Left and Right Sum Methods. You may need to consult your calculator manual for programming your model of calculator.

Left Sum and Right Sum Method Calculator Programs

These programs approximate the area under the graph of a continuous function f on a closed interval $[a, b]$ for n rectangles. The function $y = f(x)$ must be entered in Y_1 before the program is executed. When the program executes, the user is prompted to enter a left endpoint A, a right endpoint B, and number of rectangles N.

```
PROGRAM:LEFTSUM          PROGRAM:RTSUM
Prompt A                 Prompt A
Prompt B                 Prompt B
Prompt N                 Prompt N
(B-A)/N→D                (B-A)/N→D
0→T                      0→T
For(I,0,N-1)             For(I,1,N)
A+I*D→X                  A+I*D→X
T+D*Y₁→T                 T+D*Y₁→D
End                      End
Disp "LEFT SUM APPROX",T Disp "RIGHT SUM APPROX",T
```

Using the Left and Right Sum Calculator Programs, we approximate the area under the graph of $f(x) = x^2 + 1$ on the interval $[0, 2]$ in **Table 6.2.4**.

Table 6.2.4

N *Rectangles*	LEFTSUM *Approximation*	RTSUM *Approximation*
5	3.92	5.52
10	4.28	5.08
100	4.6268	4.7068
1000	4.662668	4.670668
10,000	4.66626668	4.66706668

As the number of rectangles gets larger and larger, that is, $n \to \infty$, the approximations approach the actual area. The following **formula**, which results from letting $n \to \infty$, was derived by Georg

Riemann, who was a disciple of the famous German mathematician Karl Friedreich Gauss. Riemann also proved that the end result is the same, even if the rectangles have different widths. This is why we name the width of the rectangle Δx_i and not just Δx.

DEFINITION

Area Under a Curve Formula

If f is a continuous function defined on a closed interval $[a, b]$, then the area under the graph of f from a to b is

$$\begin{pmatrix} \text{Area under the graph} \\ \text{of } f \text{ from } a \text{ to } b \end{pmatrix} = f(x_1)\Delta x_1 + f(x_2)\Delta x_2 + f(x_3)\Delta x_3 + \cdots$$

or, alternatively,

$$= \lim_{n \to \infty} \sum_{i=1}^{n} f(x_i) \cdot \Delta x_i$$

The Definite Integral

In the Area Under a Curve Formula, we saw that an alternative way to represent the area under f from a to b is

$$\lim_{n \to \infty} \sum_{i=1}^{n} f(x_i) \cdot \Delta x_i$$

We can also use an integral that has a beginning and ending value, called a **definite integral**, to represent this limit.

DEFINITION

Definite Integral

For a continuous function f, the **definite integral** of f on the interval $[a, b]$ is

$$\int_a^b f(x)\,dx = \lim_{n \to \infty} \sum_{i=1}^{n} f(x_i) \cdot \Delta x_i$$

where a and b are the **limits of integration**.

NOTE: The definite integral is often referred to as a **limit of a sum**. Also, a definite integral can be used to represent the area under a curve.

OBJECTIVE 2

Write a definite integral to represent an area.

Example 2: Applying the Definition of Definite Integral

Write a definite integral to represent the shaded area in **Figure 6.2.9**.

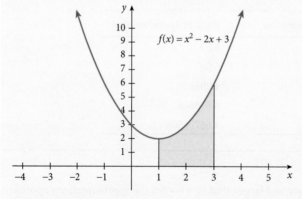

Figure 6.2.9

Perform the Mathematics

The shaded area is below the graph of the function $f(x) = x^2 - 2x + 3$ on the interval $[1, 3]$, which means that $a = 1$ and $b = 3$. We represent the shaded area as

$$\int_1^3 (x^2 - 2x + 3)\,dx$$

∎

In Section 6.1 we saw several properties of the indefinite integral. Many of the **properties** that we first saw in that section apply here, too.

Properties of the Definite Integral

1. The properties of the indefinite integral also apply to the definite integral.

    - $\displaystyle\int_a^b k \cdot f(x)\,dx = k\int_a^b f(x)\,dx$

    - $\displaystyle\int_a^b [f(x) \pm g(x)]\,dx = \int_a^b f(x)\,dx \pm \int_a^b g(x)\,dx$

2. If f is continuous on $[a, b]$ and there is a value c between a and b (see **Figure 6.2.10**), then

    $$\int_a^b f(x)\,dx = \int_a^c f(x)\,dx + \int_c^b f(x)\,dx$$

    This property, called the **Interval Addition Rule** for definite integrals, means that we can "break up" the limits of integration at a point $x = c$ that is between a and b.

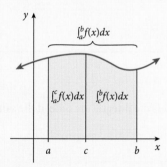

Figure 6.2.10 $\int_a^b f(x)\,dx = \int_a^c f(x)\,dx + \int_c^b f(x)\,dx$.

3. $\displaystyle\int_a^a f(x)\,dx = 0$. This property tells us that the area under a single point on a curve is zero.

The Interval Addition Rule is particularly useful when integrating piecewise-defined functions. Its use is demonstrated in Example 3.

Example 3: Applying the Interval Addition Rule for Definite Integrals

OBJECTIVE 3

Apply the Interval Addition Rule to a piecewise-defined function.

Use the Interval Addition Rule to write a definite integral to represent the shaded region for the piecewise-defined function

$$f(x) = \begin{cases} x + 1, & x \le 2 \\ -2x + 7, & x > 2 \end{cases}$$

A graph of the piecewise-defined function is given in **Figure 6.2.11**.

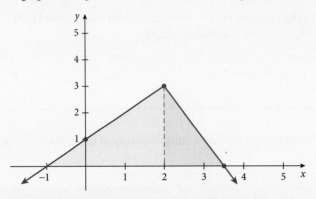

Figure 6.2.11 Graph of $f(x) = \begin{cases} x + 1, & x \le 2 \\ -2x + 7, & x > 2 \end{cases}$

Perform the Mathematics

We must break up the integral into two pieces. On the interval $[-1, 2]$ we integrate the expression $x + 1$, and on the interval $[2, \frac{7}{2}]$ we integrate $-2x + 7$. Since the "switch" in the expressions occurs at $x = 2$, and 2 is between -1 and $\frac{7}{2}$, we can use the Interval Addition Rule to write

$$\int_{-1}^{7/2} f(x)dx = \int_{-1}^{2} (2x + 1)dx + \int_{2}^{7/2} (-2x + 7)dx$$ ∎

> ### Try It Yourself
> Some related Exercises are 27(a) and 29(a).

Fundamental Theorem of Calculus

So far, we have seen that the area under a curve formula and the definite integral are indeed the same. Mathematically, this means that

$$\lim_{n \to \infty} \sum_{i=1}^{n} f(x_i) \Delta x_i = \int_{a}^{b} f(x)dx$$

As you can imagine, without the aid of calculator (or computer) programs, it can be difficult to find the area under a curve by computing the left or right sum for thousands of rectangles. Then why is the Area under a Curve Formula so important? It turns out that the founders of calculus discovered an amazing relationship between the area under a curve and the definite integral. To understand this relationship, let's reconsider an example from Section 6.1.

Flashback: Indefinite Integrals Revisited

In Example 3 part (a) in Section 6.1, we computed $\int (x^2 + 3)dx$ to be

$$F(x) = \frac{1}{3}x^3 + 3x + C$$

a.   For $F(x) = \frac{1}{3}x^3 + 3x + C$, evaluate $F(6) - F(3)$.

b. Write a definite integral to represent the area under the graph of $f(x) = x^2 + 3$ on the interval $[3, 6]$. The Left Sum approximation for this definite integral is 71.9595045 for $n = 1000$ rectangles. Compare this to the result from part (a).

Perform the Mathematics

a. Evaluating $F(6) - F(3)$, we get

$$F(6) - F(3) = \left[\frac{1}{3}(6)^3 + 3(6) + C\right] - \left[\frac{1}{3}(3)^3 + 3(3) + C\right]$$

$$= [72 + 18 + C] - [9 + 9 + C]$$

$$= 72$$

b. From the definition of the definite integral, we get

$$\int_3^6 (x^2 + 3)\,dx$$

The Left Sum approximation for this definite integral, 71.9595045, is very close to the result from part (a). ∎

The Flashback shows that the difference $F(6) - F(3)$ is very close to the approximation for the area under the curve on the interval $[3, 6]$. This result is no accident. In fact, we can show that

$$\begin{pmatrix} \text{Area under} \\ f \text{ from } a \text{ to } b \end{pmatrix} = \lim_{n \to \infty} \sum_{i=1}^{n} f(x_i)\,\Delta x_i = F(b) - F(a)$$

Since $\displaystyle\lim_{n \to \infty} \sum_{i=1}^{n} f(x_i)\,\Delta x_i = \int_a^b f(x)\,dx$, it appears that we have

$$\int_a^b f(x)\,dx = F(b) - F(a)$$

where F is any antiderivative of f. If we can find an antiderivative of f, we can employ one of the most powerful theorems in all of mathematics, the **Fundamental Theorem of Calculus**. A proof of this theorem is in Appendix C.

OBJECTIVE 4

The Fundamental Theorem of Calculus.

Fundamental Theorem of Calculus

If f is a continuous function defined on a closed interval $[a, b]$ and F is an antiderivative of f, then

$$\int_a^b f(x)\,dx = F(b) - F(a)$$

where

1. $\int_a^b f(x)\,dx$ is the **definite integral** of f on the interval $[a, b]$.

2. a and b are the **limits of integration**.

Before continuing, let's list a few **key points of the Fundamental Theorem of Calculus.**

Key Points of the Fundamental Theorem of Calculus

1. It is important to understand that $\int_a^b f(x)\,dx = F(b) - F(a)$ gives the area under the graph of f from $x = a$ to $x = b$ is a special case of the Fundamental Theorem of Calculus. This is true when $f(x) \geq 0$ everywhere on $[a, b]$. See **Figure 6.2.12**.

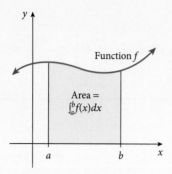

Figure 6.2.12

2. The difference $F(b) - F(a)$ is also written as $F(x)\big|_a^b$.

OBJECTIVE 5

Use the Fundamental Theorem of Calculus to evaluate definite integrals.

Example 4: Using the Fundamental Theorem of Calculus

Evaluate the definite integrals using the Fundamental Theorem of Calculus.

a. $\displaystyle\int_{-2}^{0} (x^2 - 3x)\,dx$

b. $\displaystyle\int_{1}^{5} \frac{1}{\sqrt{x}}\,dx$

Perform the Mathematics

a. $\displaystyle\int_{-2}^{0} (x^2 - 3x)\,dx = \left(\frac{1}{3}x^3 - \frac{3}{2}x^2\right)\Big|_{-2}^{0}$

$$= \left(\frac{1}{3}(0)^3 - \frac{3}{2}(0)^2\right) - \left(\frac{1}{3}(-2)^3 - \frac{3}{2}(-2)^2\right)$$

$$= (0 - 0) - \left(-\frac{8}{3} - 6\right)$$

$$= \frac{8}{3} + \frac{18}{3} = \frac{26}{3}$$

b. $\displaystyle\int_{1}^{5} \frac{1}{\sqrt{x}}\,dx = \int_{1}^{5} x^{-1/2}\,dx = (2x^{1/2})\big|_1^5 = 2\sqrt{x}\,\big|_1^5$

$$= 2\sqrt{5} - 2\sqrt{1} = 2\sqrt{5} - 2 \qquad \blacksquare$$

▶ **Try It Yourself**

Some related Exercises are 39 and 41.

Technology Option

We can evaluate definite integrals on a graphing calculator using the `fnInt` command. To check our work in Example 4a, we begin by entering $f(x) = x^2 - 3x$ into Y_1. **Figure 6.2.13** shows what must be entered into the calculator, and **Figure 6.2.14** shows the result when we execute the command.

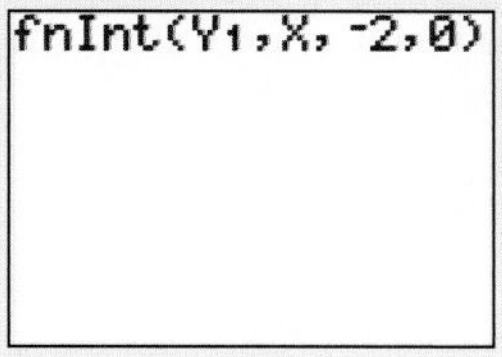

fnInt(Y1,X,-2,0)

Figure 6.2.13

fnInt(Y1,X,-2,0)
8.666666667
Ans▸Frac
26/3

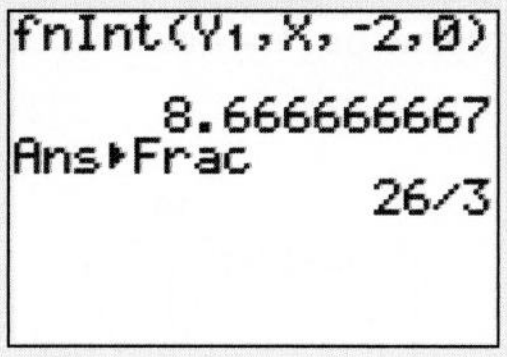

Figure 6.2.14

To see how the Fundamental Theorem of Calculus can be used to determine an area, let's take another look at Example 2.

Example 5: Determining an Exact Area

Determine the exact area shaded in **Figure 6.2.15**.

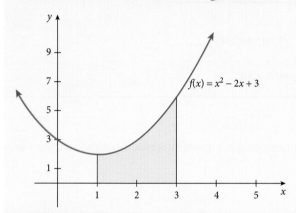

y
9
7
5
3
1
f(x) = x² – 2x + 3
1
2
3
4
5
x

Figure 6.2.15

Perform the Mathematics

As we stated in Example 2, the corresponding definite integral for the shaded area is $\int_1^3 (x^2 - 2x + 3)\,dx$. Evaluating this integral yields

$$\begin{aligned}\int_1^3 (x^2 - 2x + 3)\,dx &= \left(\frac{1}{3}x^3 - x^2 + 3x\right)\Bigg|_1^3 \\ &= \left[\frac{1}{3}(3)^3 - (3)^2 + 3(3)\right] - \left[\frac{1}{3}(1)^3 - (1)^2 + 3(1)\right] \\ &= (9 - 9 + 9) - \left(\frac{1}{3} - 1 + 3\right) \\ &= 9 - \frac{7}{3} = \frac{20}{3}\ \text{un}^2\end{aligned}$$

So the area under the graph of the function $f(x) = x^2 - 2x + 3$ on the interval $[1, 3]$ is $\frac{20}{3}$ square units. ■

◄► **Technology Option**

We can evaluate definite integrals on a graphing calculator using the $\int f(x)\,dx$ command. To check our work in Example 5, we enter $f(x) = x^2 - 2x + 3$ into Y_1 and graph the function in the standard viewing window. See **Figure 6.2.16**. From the graphing window, we execute the $\int f(x)\,dx$ command. In **Figure 6.2.17** we enter the lower limit of integration, and in **Figure 6.2.18** we enter the upper limit of integration. **Figure 6.2.19** shows the result of executing the $\int f(x)\,dx$ command.

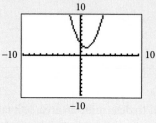

Figure 6.2.16

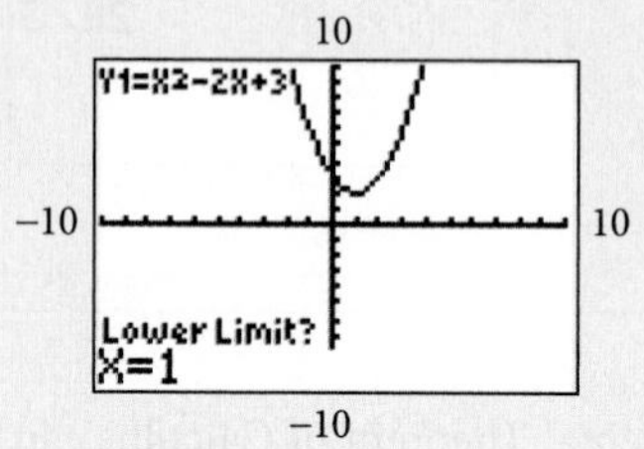

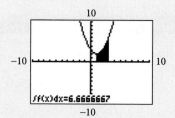

Figure 6.2.17

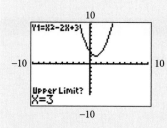

Figure 6.2.18

Figure 6.2.19

The **sign of the definite integral** depends not only on the limits of integration, but also on the orientation of the function relative to the x-axis.

Sign of $\int_a^b f(x)\,dx$

1. If $f(x) > 0$ on $[a, b]$, then $\int_a^b f(x)\,dx > 0$. Thus, if the graph of f is above the x-axis for all the values of x in $[a, b]$, then the definite integral is positive.

2. If $f(x) < 0$ on $[a, b]$, then $\int_a^b f(x)\,dx < 0$. Thus, if the graph of f is below the x-axis for all values of x in $[a, b]$, then the definite integral is negative.

Example 6: Evaluating Definite Integrals

Consider the cubic function $f(x) = x^3 - x^2 - 4x + 4$. See **Figure 6.2.20**. Evaluate $\int_{-2}^{1} f(x)\,dx$ and compare to $\int_{1}^{2} f(x)\,dx$.

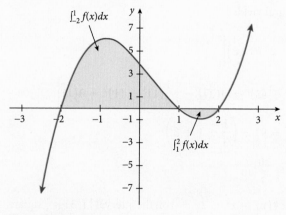

Figure 6.2.20

Perform the Mathematics

Since the graph of f is above the x-axis on $[-2, 1]$, we expect $\int_{-2}^{1} f(x)\,dx$ to be positive. Evaluating $\int_{-2}^{1} f(x)\,dx$, we get

$$\int_{-2}^{1} f(x)\,dx = \int_{-2}^{1} (x^3 - x^2 - 4x + 4)\,dx$$

$$= \left(\frac{1}{4}x^4 - \frac{1}{3}x^3 - 2x^2 + 4x \right)\Big|_{-2}^{1}$$

$$= \left[\frac{1}{4}(1)^4 - \frac{1}{3}(1)^3 - 2(1)^2 + 4(1) \right] - \left[\frac{1}{4}(-2)^4 - \frac{1}{3}(-2)^3 - 2(-2)^2 + 4(-2) \right]$$

$$= \frac{23}{12} - \left(-\frac{28}{3} \right)$$

$$= \frac{45}{4}$$

Since the graph of f is below the x-axis on $[1, 2]$, we expect $\int_{1}^{2} f(x)\,dx$ to be negative. Evaluating $\int_{1}^{2} f(x)\,dx$, we get

$$\int_{1}^{2} f(x)\,dx = \int_{1}^{2} (x^3 - x^2 - 4x + 4)\,dx$$

$$= \left(\frac{1}{4}x^4 - \frac{1}{3}x^3 - 2x^2 + 4x \right)\Big|_{1}^{2}$$

$$= \left[\frac{1}{4}(2)^4 - \frac{1}{3}(2)^3 - 2(2)^2 + 4(2) \right] - \left[\frac{1}{4}(1)^4 - \frac{1}{3}(1)^3 - 2(1)^2 + 4(1) \right]$$

$$= \frac{4}{3} - \left(\frac{23}{12} \right)$$

$$= -\frac{7}{12}$$

▶ *Try It Yourself*

Some related Exercises are 69 and 71.

Summary

In this section we saw that the area under a curve could be approximated by using the Left and Right Sum Methods. We also said that, as the number of rectangles becomes infinitely large, we get the definite integral. The definite integral was defined as

$$\int_{a}^{b} f(x)\,dx = \lim_{n \to \infty} \sum_{i=1}^{n} f(x_i)\,\Delta x_i$$

We then concluded the section by introducing the Fundamental Theorem of Calculus, and we saw how this theorem can be used to evaluate definite integrals. We then used the Fundamental Theorem of Calculus to determine the exact area of a region under a curve.

- If f is a continuous function defined on a closed interval $[a, b]$ and F is an antiderivative of f, then $\int_{a}^{b} f(x)\,dx = F(b) - F(a)$.

- If $f(x) > 0$ on $[a, b]$, then $\int_{a}^{b} f(x)\,dx > 0$.

- If $f(x) < 0$ on $[a, b]$, then $\int_{a}^{b} f(x)\,dx < 0$.

Section 6.2 Exercises

Vocabulary Exercises

1. If a closed interval is divided into n equal subintervals, the width of each subinterval is called the _____ _____.

2. For a function f, $\int_a^b f(x)\,dx$ is called a _____ integral.

3. Given $\int_a^b f(x)\,dx$, we call a and b the _____ of integration.

4. The rectangular approximation $A = \sum\limits_{i=1}^{n} \Delta x \cdot f(x_i)$ is called the _____ Sum Method.

5. The rectangular approximation $A = \sum\limits_{i=0}^{n-1} \Delta x \cdot f(x_i)$ is called the _____ Sum Method.

6. The equation $\int_a^b f(x)\,dx = F(b) - F(a)$ where F is an antiderivative of f illustrates the _____ _____ of Calculus.

Skill Exercises

In Exercises 7–12, determine the step size Δx for the given interval $[a, b]$ and the number of subintervals n.

7. $[0, 2]$; $n = 4$

8. $[0, 5]$; $n = 5$

9. $[1, 4]$; $n = 6$

10. $[3, 5]$; $n = 4$

11. $[-1, 2]$; $n = 6$

12. $[-3, 5]$; $n = 28$

In Exercises 13–18, a function f, a closed interval $[a, b]$, and a number of equally spaced subintervals n are given. Complete the following.

(a) Calculate by hand the Left Sum to approximate the area under the graph of f on $[a, b]$.

(b) Calculate by hand the Right Sum to approximate the area under the graph of f on $[a, b]$.

13. $f(x) = 3x$; $[1, 4]$; $n = 6$

14. $f(x) = 2x$; $[1, 2]$; $n = 4$

15. $f(x) = 2x^2$; $[0, 4]$; $n = 4$

16. $f(x) = x^3$; $[0, 2]$; $n = 4$

17. $f(x) = 4 - x^2$; $[-1, 2]$; $n = 6$

18. $f(x) = x^2 + 2$; $[-2, 2]$; $n = 4$

In Exercises 19–24, copy the table and use the LEFTSUM and RTSUM graphing calculator programs to approximate the area under the graph of f on $[a, b]$ for the given number of rectangles.

N Rectangles	LEFTSUM Approximation	RTSUM Approximation
10		
100		
1000		

 19. $f(x) = 3x^2$; $[-2, 0]$

 20. $f(x) = \dfrac{x^2}{2}$; $[0, 9]$

21. $f(x) = x^3 - 3x$; $[0, 4]$

22. $f(x) = x^3 + 2x$; $[0, 2]$

 23. $f(x) = 0.2x^2 + 1.3x + 2.3$; $[0, 5]$

24. $f(x) = 0.3x^3 - 2.7x + 4.1$; $[0, 3]$

For Exercises 25–30, complete the following.

(a) Write a definite integral that represents the shaded area.

(b) Use a known geometric formula to compute the shaded area.

25.

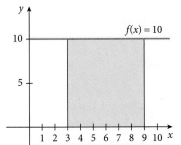

26.

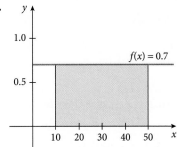

27.

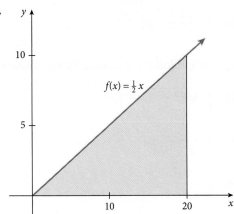

28.

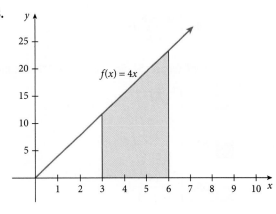

29.

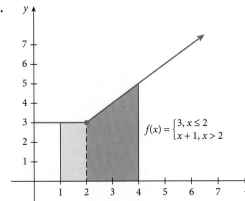

30.

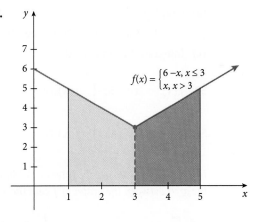

For Exercises 31–50, use the Fundamental Theorem of Calculus to evaluate the definite integrals.

31. $\displaystyle\int_1^4 3x\,dx$

32. $\displaystyle\int_1^2 2x\,dx$

33. $\displaystyle\int_0^2 5\,dx$

34. $\displaystyle\int_0^5 7\,dx$

35. $\displaystyle\int_3^5 (2x - 1)\,dx$

36. $\displaystyle\int_3^6 (4x - 2)\,dx$

37. $\displaystyle\int_0^4 2x^2\,dx$

38. $\displaystyle\int_0^2 x^3\,dx$

39. $\displaystyle\int_{-1}^2 (4 - x^2)\,dx$

40. $\displaystyle\int_{-2}^2 (x^2 + 2)\,dx$

41. $\int_1^4 8\sqrt{x}\,dx$

42. $\int_1^2 (6x + \sqrt{x})\,dx$

43. $\int_1^4 (x^3 - 3x)\,dx$

44. $\int_0^2 (x^3 + 2x)\,dx$

45. $\int_0^5 (0.2x^2 + 1.3x + 2.3)\,dx$

46. $\int_0^3 (0.3x^2 - 2.7x + 4.1)\,dx$

47. $\int_{-2}^3 (5 + x - 6x^2)\,dx$

48. $\int_1^4 (x^2 - 4x - 3)\,dx$

49. $\int_4^9 \dfrac{x - 3}{\sqrt{x}}\,dx$

50. $\int_{-2}^{-1} \dfrac{2x - 7}{x^3}\,dx$

For Exercises 51–56, use the FNINT *command on your graphing calculator to evaluate the definite integral.*

51. $\int_1^4 \sqrt{x^2 + 3}\,dx$

52. $\int_2^7 \sqrt{x^2 + 3x + 1}\,dx$

53. $\int_3^7 \dfrac{1}{x - 2}\,dx$

54. $\int_4^{10} \dfrac{3}{x - 3}\,dx$

55. $\int_2^8 \dfrac{2x}{x^2 + 1}\,dx$

56. $\int_2^6 \dfrac{3x^2}{x^3 + 1}\,dx$

For Exercises 57–60, an absolute value function f is given. Complete the following.

(a) Graph f.

(b) Rewrite f as a piecewise-defined function.

(c) Integrate f as indicated.

57. $f(x) = |x - 3|;\ \int_{-2}^5 f(x)\,dx$

58. $f(x) = |x + 6|;\ \int_{-10}^0 f(x)\,dx$

59. $f(x) = |x - 4|;\ \int_{-2}^6 f(x)\,dx$

60. $f(x) = |x - 2|;\ \int_0^9 f(x)\,dx$

In Exercises 61–66, use the Interval Addition Rule to evaluate the indicated definite integral to determine the area between the graph of f and the x-axis.

61. $\int_{-3}^4 f(x)\,dx$ where $f(x) = \begin{cases} -x, & x < 0 \\ x, & x \geq 0 \end{cases}$

62. $\int_{-3}^3 f(x)\,dx$ where $f(x) = \begin{cases} -x, & x \leq 1 \\ x - 2, & x > 1 \end{cases}$

63. $\int_{-1}^2 f(x)\,dx$ where $f(x) = \begin{cases} 1 - x^2, & x < 0 \\ x + 1, & x \geq 0 \end{cases}$

64. $\int_0^5 f(x)\,dx$ where $f(x) = \begin{cases} x^2, & x < 3 \\ x + 6, & x \geq 3 \end{cases}$

65. $\int_0^5 f(x)\,dx$ where $f(x) = \begin{cases} 10 - x^2, & x \leq 2 \\ 6, & x > 2 \end{cases}$

66. $\int_{-4}^0 f(x)\,dx$ where $f(x) = \begin{cases} 8, & x < -2 \\ x^2 + 4, & x \geq -2 \end{cases}$

For Exercises 67–72, a function f and an interval [a, b] are given. Complete the following.

(a) Determine the *x*-intercepts of the graph of the function *f* on the interval $[a, b]$.

(b) Integrate each on the interval $[a, x\text{-intercept}]$ and on the interval $[x\text{-intercept}, b]$.

(c) Interpret each integral from part (b) graphically. See Example 6.

67. $f(x) = x + 2;\ [-5, 5]$

68. $f(x) = x - 5;\ [0, 10]$

69. $f(x) = 2\sqrt{x} - 4;\ [0, 9]$

70. $f(x) = \sqrt{x} - 1;\ [0, 4]$

71. $f(x) = x^2 - 9;\ [-1, 4]$

72. $f(x) = x^2 - 4;\ [-2, 4]$

Concept and Writing Exercises

73. Write a sentence on how rectangular approximations can be used to approximate the area under a continuous function *f* on a closed interval $[a, b]$.

74. Explain the difference between using the Right Sum and Left Sum Methods to approximate the area under a curve. Does the Left Sum Method always yield an approximation less than the Right Sum Method?

For Exercises 75–78, let a and b, represent real numbers. Evaluate to write a formula for the following definite integrals.

75. $\displaystyle\int_{0}^{b} k\,dx,\ k \text{ is a real}$

76. $\displaystyle\int_{a}^{0} x^k\,dx,\ k \neq -1$

77. $\displaystyle\int_{a}^{0} \sqrt[k]{x}\,dx,\ k \text{ is an integer},\ k \geq 2$

78. $\displaystyle\int_{0}^{b} \frac{1}{x^k}\,dx,\ k \neq 1$

Section Project

For parts (a) through (d), graph each function and evaluate each definite integral.

(a) $f(x) = x^2 + 1;\ \displaystyle\int_{-3}^{0} f(x)\,dx;\ \int_{0}^{3} f(x)\,dx;\ \int_{-3}^{3} f(x)\,dx$

(b) $f(x) = x^4 + 1;\ \displaystyle\int_{-2}^{0} f(x)\,dx;\ \int_{0}^{2} f(x)\,dx;\ \int_{-2}^{2} f(x)\,dx$

(c) $f(x) = x^3;\ \displaystyle\int_{-2}^{0} f(x)\,dx;\ \int_{0}^{2} f(x)\,dx;\ \int_{-2}^{2} f(x)\,dx$

(d) $f(x) = x^5;\ \displaystyle\int_{-2}^{0} f(x)\,dx;\ \int_{0}^{2} f(x)\,dx;\ \int_{-2}^{2} f(x)\,dx$

(e) The functions in parts (a) and (b) are called **even** functions. Note that the graph of an even function is symmetric with respect to the *y*-axis. For an even function *f*, what is the relationship between $\int_{-a}^{0} f(x)\,dx$ and $\int_{-a}^{a} f(x)\,dx$?

(f) The functions in parts (c) and (d) are called **odd** functions. Note that the graph of an odd function is symmetric with respect to the origin. For an odd function *f*, what is the value of $\int_{-a}^{a} f(x)\,dx$?

6.3 Integral Calculus and Total Accumulation

In Section 6.2 we learned how to approximate area under a curve using rectangles in the Left Sum and Right Sum Method. We introduced the **definite integral** and used it to represent area under a curve. We then presented one of the most remarkable results in mathematics, **The Fundamental Theorem of Calculus**. The Fundamental Theorem of Calculus supplied us with a way to evaluate definite integrals and determine an **exact** answer. In turn, this allowed us to determine areas exactly.

In this section, we focus on why determining an area under a curve is important for us. Specifically, we focus on determining the area under a graph of a **rate of change function**, or simply a **rate function**, and interpreting what it gives us, a **total accumulation**. We will encounter many varied applications of this concept in the remainder of Chapter 6 as well as Chapter 7. Determining the area under the graph of a rate of change function is a powerful application of the calculus and specifically The Fundamental Theorem of Calculus.

The Integral as a Continuous Sum

In Section 6.2 we stated that the definite integral can be thought of as a limit of a sum. The definite integral can also be used to represent area under the graph of a function. When a definite integral is used to represent the area under the graph of a rate of change function (which we will also call a rate function), evaluating it using The Fundamental Theorem of Calculus produces a **continuous sum of a rate function over a closed interval**. To see what we mean by this statement, as well as why this observation is so important, let's consider the SafeMate Company. The SafeMate Company manufactures home emergency kits and has determined that the cost of manufacturing these kits is given by

$$C(x) = 0.02x^2 + 3x + 800 \qquad 0 \le x \le 200$$

where x represents the number of emergency kits produced daily, and $C(x)$ is the cost, in dollars. In order to calculate the additional cost of increasing production from 100 to 160 emergency kits, we can compute

$$C(160) - C(100)$$
$$= [0.02(160)^2 + 3(160) + 800] - [0.02(100)^2 + 3(100) + 800]$$
$$= 1792 - 1300 = 492$$

So it costs an additional \$492 to increase production from 100 to 160 units. Now let's examine this scenario using the **marginal cost function**. We know that the marginal cost function is $C'(x) = MC(x) = 0.04x + 3$. Let's determine the area under the graph of MC on the interval $[100, 160]$. See **Figure 6.3.1**.

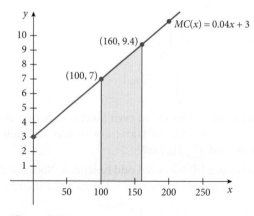

Figure 6.3.1

We can represent the area shaded in Figure 6.3.1 with the definite integral $\int_{100}^{160} (0.04x + 3)\,dx$. We determine the exact area by utilizing the Fundamental Theorem of Calculus as follows.

$$\int_{100}^{160} (0.04x + 3)\,dx = \left(\frac{0.04}{2}x^2 + 3x \right)\Big|_{100}^{160}$$

$$= \left[\frac{0.04}{2}(160)^2 + 3(160) \right] - \left[\frac{0.04}{2}(100)^2 + 3(100) \right]$$

$$= 992 - 500$$

$$= 492$$

Notice that this is exactly the same result that we had when we computed the additional cost of increasing production from 100 to 160 emergency home kits. Also notice that the marginal cost function, MC, is a type of **rate function** since it gives the instantaneous rate of change in costs.

Interpreting the area under the graph of a rate function is a powerful tool in applied calculus. To ensure that we understand the power and meaning of this tool, let's try another scenario in Example 1.

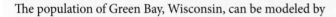

Example 1: Analyzing Population Growth

The population of Green Bay, Wisconsin, can be modeled by

$$F(x) = 2.544x + 279.948 \qquad 1 \le x \le 10$$

where x is the number of years since 1999, and $F(x)$ is the population in thousands. (*Source*: U.S. Census Bureau.)

a. Evaluate and interpret $F(7) - F(2)$.

b. Evaluate $\int_2^7 F'(x)\,dx$ and compare to part (a).

Perform the Mathematics

a. Since $F(7)$ will give us the population of Green Bay in 2006 and $F(2)$ will give us the population of Green Bay in 2001, computing $F(7) - F(2)$ gives us the total increase in Green Bay's population from 2001 to 2006. Note that these populations are in thousands. We compute $F(7) - F(2)$ as

$$F(7) - F(2) = [2.544(7) + 279.948] - [2.544(2) + 279.948] = 12.72$$

This means that from 2001 to 2006, the total increase in Green Bay's population was about 12.72 thousand people, or 12,720 people.

b. Since $F(x) = 2.544x + 279.948$ we have $F'(x) = 2.544$. So here we get

$$\int_2^7 F'(x)\,dx = \int_2^7 2.544\,dx$$

$$= 2.544x\big|_2^7$$

$$= 2.544(7) - 2.544(2)$$

$$= 12.72$$

This is exactly the same result as in part (a). Notice that the definite integral $\int_2^7 2.544\,dx$ corresponds to the shaded region in **Figure 6.3.2**. This illustrates that the area under the graph of a rate function can be interpreted as giving a total accumulation.

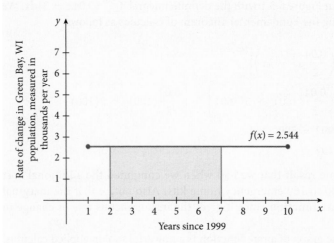

Figure 6.3.2   ■

The beginning of the section, along with Example 1, illustrates an important **interpretation of the area under the graph of a rate of change function.**

Table 6.3.1 illustrates some of the uses of this property.

Table 6.3.1

Diagram	Accumulation
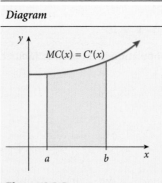 **Figure 6.3.3**	$\int_a^b MC(x)\,dx = $ Total increase in cost of producing from a to b units.
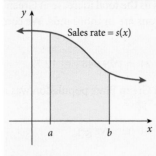 **Figure 6.3.4**	$\int_a^b s(x)\,dx = $ The total increase in sales from time a to b.
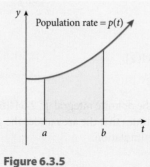 **Figure 6.3.5**	$\int_a^b p(t)\,dt = $ The total increase in population from time a to b.

Interpreting the Area Under the Graph of a Rate Function

Let f be a rate function on the interval $[a, b]$ where $[a, b]$ is in the reasonable domain of f. Then the area under the graph of f on $[a, b]$ gives the total accumulation of F from a to b, where F is an antiderivative of f.

NOTES:

1. The area under the graph of f on $[a, b]$ is given by $\int_a^b f(x)\,dx$.
2. We evaluate $\int_a^b f(x)\,dx$ by using the Fundamental Theorem of Calculus. That is, $\int_a^b f(x)\,dx = F(b) - F(a)$, where F is an antiderivative.

Example 2: Interpreting a Total Accumulation

A phrase that many Americans heard when having groceries packed for them was "Would you prefer paper or plastic?" The rate of change in total paper and paperboard waste generated in the United States can be modeled by

$$f(x) = -0.3x + 3.38 \qquad 1 \le x \le 19$$

where x represents the number of years since 1989, and $f(x)$ represents the rate of change in paper and paperboard waste generated measured in $\dfrac{\text{million of tons}}{\text{year}}$. (*Source*: U.S. Environmental Protection Agency.)

a. Graph f and shade the area that represents the increase in paper and paperboard waste generated from 1991 to 1999.

b. Write a definite integral to determine the shaded area in part (a). Use the Fundamental Theorem of Calculus to evaluate the definite integral and interpret.

OBJECTIVE 1

Interpret a total accumulation.

Perform the Mathematics

a. Since we need to find the area that represents the years 1991 ($x = 2$) to 1999 ($x = 10$), we shade below the graph of f and above the x-axis on the interval $[2, 10]$. The desired region is shown in **Figure 6.3.6**.

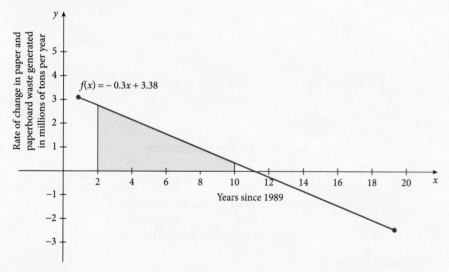

Figure 6.3.6

b. The definite integral that represents the shaded region in **Figure 6.3.6** is

$$\int_{2}^{10} (-0.3x + 3.38)\,dx$$

Evaluating this definite integral gives us

$$\int_{2}^{10} (-0.3x + 3.38)\,dx = \left(\frac{-0.3}{2}x^2 + 3.38x \right)\Bigg|_{2}^{10}$$

$$= \left[\frac{-0.3}{2}(10)^2 + 3.38(10) \right] - \left[\frac{-0.3}{2}(2)^2 + 3.38(2) \right]$$

$$= 18.8 - 6.16$$

$$= 12.64$$

This means that the total increase in paper and paperboard waste generated in the United States from 1991 to 1999 was about 12.64 million tons. ∎

▶ *Try It Yourself*

Some related Exercises are 7 and 9.

Example 3: Interpreting the Area Under the Graph of a Rate Function

OBJECTIVE 2

Interpret area under the graph of a rate function.

It has been documented that the earning power of an individual increases with more education. Since the 1970s, more and more women have been enrolling in college in the United States and increasing their own earning power. The rate of change in women enrolled in college can be modeled by

$$f(x) = 0.03x^2 - 0.22x + 0.75 \qquad 1 \le x \le 10$$

where x represents the number of years since 1998, and $f(x)$ represents the rate of change in women enrolled in college in the United States measured in millions per year. (*Source:* U.S. Census Bureau.) Determine the area of the shaded region in **Figure 6.3.7** and interpret this area as a total accumulation.

© Dmitry Kalinovsky/ShutterStock, Inc.

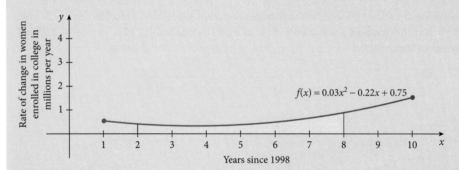

Figure 6.3.7

Understand the Situation

We first note that the interval $[2, 8]$ corresponds to the years 2000 to 2006. The area of the shaded region is represented by the definite integral $\int_2^8 (0.03x^2 - 0.22x + 0.75)\,dx$. We integrate and use the Fundamental Theorem of Calculus to evaluate this definite integral.

Perform the Mathematics

Performing the integration gives us

$$\int_2^8 (0.03x^2 - 0.22x + 0.75)\,dx$$

$$= \left(\frac{0.03}{3}x^3 - \frac{0.22}{2}x^2 + 0.75x \right) \Bigg|_2^8$$

$$= [0.01(8)^3 - 0.11(8)^2 + 0.75(8)] - [0.01(2)^3 - 0.11(2)^2 + 0.75(2)]$$

$$= 4.08 - 1.14$$

$$= 2.94$$

Interpret the Result

This means that the total increase in female enrollment in college in the United States from 2000 to 2006 was about 2.94 million.

▷ *Try It Yourself*

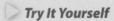

Some related Exercises are 37 and 38.   ∎

In our next example, we determine the area bounded by a rate function where the region is below the *x*-axis.

Example 4: Integrating a Rate Function

OBJECTIVE 3

Evaluate and interpret the definite integral of a rate function.

In the 1980s, the federal government launched a campaign against illegal drugs called the "War on Drugs." In recent years, some have started to debate the merits of the "War on Drugs" and its associated costs. The rate of change in pounds of marijuana per arrest in the United States can be modeled by

$$f(x) = 12.12x^2 - 145.08x + 402.78 \qquad 1 \le x \le 10$$

where *x* represents the number of years since 1999, and $f(x)$ represents the rate of change in pounds of marijuana per arrest per year. (*Source*: U.S. Drug Enforcement Administration.) See **Figure 6.3.8**.

a. Evaluate $\int_8^{10} f(x)\,dx$ and interpret.

b. Evaluate $\int_5^7 f(x)\,dx$ and interpret.

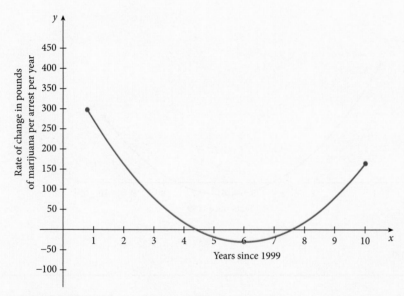

Figure 6.3.8

Perform the Mathematics

a. Since $f(x)$ is a rate function, this integration will give us a total accumulation from 2007 ($x = 8$) to 2009 ($x = 10$). Performing the integration and using the Fundamental Theorem of Calculus gives us

$$\int_8^{10} f(x)\,dx = \int_8^{10} (12.12x^2 - 145.08x + 402.78)\,dx$$

$$= (4.04x^3 - 72.54x^2 + 402.78x) \Big|_8^{10}$$

$$= [4.04(10)^3 - 72.54(10)^2 + 402.78(10)] - [4.04(8)^3 - 72.54(8)^2 + 402.78(8)]$$

$$= 165.64$$

This means that from 2007 to 2009 the total increase in pounds of marijuana per arrest in the U.S. was 165.64 pounds.

b.   Here, performing the integration and using the Fundamental Theorem of Calculus gives us

$$\int_5^7 f(x)\,dx = \int_5^7 (12.12x^2 - 145.08x + 402.78)\,dx$$

$$= (4.04x^3 - 72.54x^2 + 402.78x)\Big|_5^7$$

$$= [4.04(7)^3 - 72.54(7)^2 + 402.78(7)] - [4.04(5)^3 - 72.54(5)^2 + 402.78(5)]$$

$$= -54.68$$

This means that from 2004 to 2006 the total **decrease** in pounds of marijuana per arrest in the United States was 54.68 pounds. See **Figure 6.3.9**.

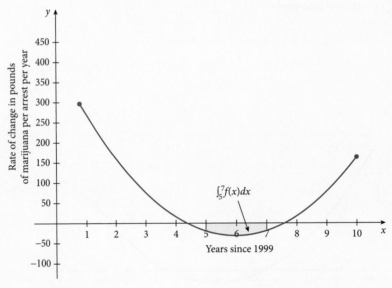

Figure 6.3.9

> **Try It Yourself**
>
> Some related Exercises are 39 and 40.

Notice in Example 4(b) that the definite integral produced a negative number. As a result, the interpretation was that it represented a total **decrease**. This reinforces that the total accumulation is not always a positive number giving us a total increase.

The integration of a rate function to get a total accumulation can also be applied to problems in the physical sciences. Before we go further, recall that for a position function s, the **velocity** of an object at time t is given by $v(t) = \frac{d}{dt}[s(t)]$. This means that velocity, v, can be considered a rate function.

So to determine the total movement (or **displacement**) of an object given its velocity, we can use a definite integral and the Fundamental Theorem of Calculus.

Example 5: Finding Total Displacement of an Object

OBJECTIVE 4

Determine total displacement.

The velocity of an object is given by

$$v(t) = 8t + 3$$

where t is the time in seconds, and $v(t)$ is the velocity measured in $\frac{\text{feet}}{\text{second}}$.

a. Evaluate $v(10)$ and interpret.

b. Determine the total movement, or displacement, of the object between 2 and 8 seconds.

Perform the Mathematics

a. Evaluating, we have

$$v(10) = 8(10) + 3 = 83$$

This means that after 10 seconds, the velocity of the object is 83 $\frac{\text{feet}}{\text{second}}$.

b. Since $v(t)$ is a rate function, we determine the total displacement by evaluating $\int_2^8 v(t)\,dt$.

Evaluating the definite integral gives us

$$\int_2^8 v(t)\,dt = \int_2^8 (8t + 3)\,dt = (4t^2 + 3t)\Big|_2^8$$

$$= [4(8)^2 + 3(8)] - [4(2)^2 + 3(2)]$$

$$= 280 - 22 = 258$$

This means that from 2 to 8 seconds, the object moved a total of 258 feet. ■

▶ **Try It Yourself**

Some related Exercises are 23 and 27.

Summary

This section focused on interpreting the area under the graph of a rate of change function. This concept is encountered regularly throughout the remainder of Chapter 6 and Chapter 7.

Interpreting the Area Under the Graph of a Rate Function

Let f be a rate function on the interval $[a, b]$, where $[a, b]$ is in the reasonable domain of f. Then the area under the graph of f on $[a, b]$ gives the total accumulation of F from a to b, where F is an antiderivative of f. The area under the graph of f on $[a, b]$ is given by $\int_a^b f(x)\,dx$. We evaluate $\int_a^b f(x)\,dx$ by using the Fundamental Theorem of Calculus. That is, $\int_a^b f(x)\,dx = F(b) - F(a)$, where F is an antiderivative.

Section 6.3 Exercises

Vocabulary Exercises

1. The Fundamental Theorem of Calculus produces a _____ sum of a rate function over a closed interval.

2. The area under the graph of f on $[a, b]$ gives the _____ _____ of F from a to b, where F is an antiderivative of f.

3. For a given marginal cost function, $\int_a^b MC(x)\,dx$ gives the total _____ in cost of production from a to b units.

4. For a given rate of change in sales function s, $\int_a^b s(x)\,dx$ gives the _____ _____ in sales from a to b units.

5. $\int_a^b p(t)\,dt$ gives the total increase in population from time a to b for a given _____ function p.

6. For a velocity function v, the definite integral $\int_a^b v(t)\,dt$ provides the total _____ of an object t from time a to time b.

Skill Exercises

For Exercises 7–10, complete the following.

(a) Write a definite integral to determine the shaded area.

(b) Evaluate the definite integral from part (a) and interpret the result.

7.

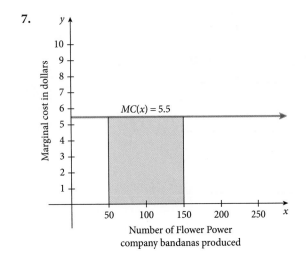

Number of Flower Power
company bandanas produced

8.

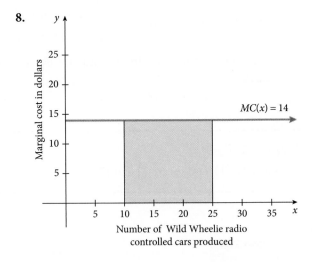

Number of Wild Wheelie radio
controlled cars produced

9.

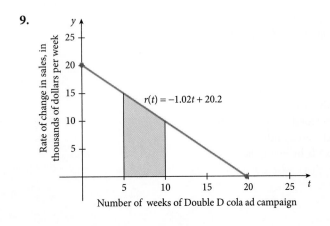

Number of weeks of Double D cola ad campaign

10.

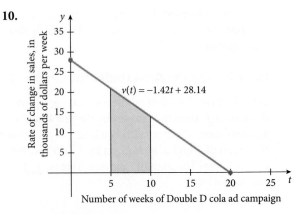

Number of weeks of Double D cola ad campaign

11. The Tuffy Toe Company determines that the marginal cost function for their new walking shoe is given by

$$MC(x) = C'(x) = 50 - 0.8x \qquad 0 \le x \le 80$$

where x represents the number of shoes produced per shift, and MC represents the marginal cost, measured in dollars per shoe.

   (a) Graph MC on its reasonable domain and shade the area that represents the total cost of producing the first 50 pairs of shoes.

   (b) Write a definite integral to determine the shaded area in part (a), then evaluate and interpret the integral.

12. The TinyTot Company determines that the marginal cost of producing their new action figure is given by

$$MC(x) = C'(x) = 4 - 0.02x \qquad 0 \le x \le 100$$

where x represents the number of action figures made daily, and MC represents the marginal cost, measured in dollars per figure.

   (a) Graph MC on its reasonable domain and shade the area that represents the total cost of producing the first 30 action figures.

   (b) Write a definite integral to determine the shaded area in part (a), then evaluate and interpret the integral.

13. The WiredWorld Audio Company determines that the marginal revenue function for producing and selling a waterproof headset radio is given by

$$MR(x) = R'(x) = 70 + x \qquad 0 \le x \le 70$$

where x represents the number of headsets produced and sold in thousands daily, and MR represents the marginal revenue, measured in dollars per headset radio.

   (a) Evaluate and interpret $\int_{10}^{40} MR(x)\,dx$.

   (b) Sketch the region you computed in part (a).

14. Using market analysis, the SatSet Satellite TV Company determines that the marginal revenue function for their Trio receiver dish is given by

$$MR(x) = R'(x) = 50 + 0.4x^3 \qquad 0 \le x \le 40$$

where x represents the number of receiver dishes produced and sold, and MR represents the marginal revenue in dollars per receiver dish.

   (a) Evaluate and interpret $\int_{20}^{30} MR(x)\,dx$.

   (b) Sketch the region you computed in part (a).

15. The marginal revenue function for the FrontRide Bus Company is given by

$$MR(x) = 0.000045x^2 - 0.03x + 3.75 \qquad 0 \le x \le 500$$

where x represents the number of bus tickets sold, and $MR(x)$ is the marginal revenue measured in dollars per ticket. Evaluate $\int_{0}^{200} MR(x)\,dx$ and interpret.

16. The daily marginal revenue function for the BlackDay Sunglasses Company is given by

$$MR(x) = 30 - 0.0003x^2 \qquad 0 \le x \le 540$$

where x represents the number of sunglasses produced and sold, and $MR(x)$ is the marginal revenue measured in dollars per sunglasses. Evaluate $\int_{0}^{300} MR(x)\,dx$ and interpret.

17. The ScandiTrac Company determines that their marginal profit function for producing and selling a new economy model of cross-country ski machine at a mall is given by

$$MP(x) = P'(x) = 0.3x^2 + 0.2x \qquad 0 \le x \le 30$$

where x is the number of machines produced and sold, and MP is the marginal profit measured in dollars per ski machine. Determine the resulting increase in profit by changing the production and sales of the ski machines from 10 to 20 units.

18. The See-The-Fine-Print Company determines that their marginal profit function for producing and selling over-the-counter reading glasses at a regional store is given by

$$MP(x) = P'(x) = 0.0015x^2 - 0.01x \qquad 0 \leq x \leq 60$$

where x is the number of reading glasses sold, and MP is the marginal profit measured in dollars. Determine the resulting increase in profit by changing the production and sales of the reading glasses from 30 to 40 units.

19. The marginal average cost function for producing x promotional banners is given by

$$MAC(x) = \frac{d}{dx}AC(x) = -\frac{100}{x^2} \qquad x > 0$$

where x represents the number of banners produced. Compute $\int_{10}^{30} MAC(x)\,dx$ and use the result to fill in the blanks for the interpretation. *This means that as the number of banners* _____ *from 10 to 30, the total* _____ *in average cost is* _____.

20. The marginal average cost function for producing x QuickVid digital cameras is given by

$$MAC(x) = \frac{d}{dx}AC(x) = 0.03x^2 - 0.04x + 5 \qquad 0 \leq x \leq 50$$

where x represents the number of digital cameras produced. Compute $\int_{0}^{25} MAC(x)\,dx$ and use the result to fill in the blanks for the interpretation. *This means that as the number of cameras* _____ *from 0 to 25, the total* _____ *in average cost is* _____.

In Exercises 21–26, the velocity function $v(t)$ for an object is given, measured in feet per second.

(a) Evaluate $v(t)$ at the given point $t = t_1$.

(b) Evaluate and interpret $\int_{t_1}^{t_2} v(t)\,dt$.

21. $v(t) = 4t$; $t_1 = 2$, $t_2 = 8$ 22. $v(t) = 0.3t$; $t_1 = 10$, $t_2 = 20$

23. $v(t) = 3t + 6$; $t_1 = 1$, $t_2 = 5$ 24. $v(t) = 2t + 10$; $t_1 = 3$, $t_2 = 5$

25. $v(t) = -9.8t + 60$; $t_1 = 1$, $t_2 = 5$ 26. $v(t) = -9.8t + 100$; $t_1 = 2$, $t_2 = 8$

27. The velocity of an object is given by

$$v(t) = -32t + 88 \qquad t \geq 0$$

where t is the time in seconds and $v(t)$ is the velocity measured in feet per second.

(a) Evaluate $v(1)$ and interpret.

(b) Determine the total displacement of the object in the first two seconds.

28. The velocity of an object is given by

$$v(t) = -32t + 120 \qquad t \geq 0$$

where t is the time in seconds and $v(t)$ is the velocity measured in feet per second.

(a) Evaluate $v(1)$ and interpret.

(b) Determine the total displacement of the object in the first three seconds.

Application Exercises

29. **Sports Science—Golf Ball Design:** The United States Golf Association mandates that one of the requirements for golf balls to be sanctioned for tournament play is that they must have an initial velocity of no more than 255 feet per second. (*Source:* RandA.org.) This means that the balls must conform to the velocity function

$$v(t) = -32t + 255 \qquad t > 0$$

where t represents the number of seconds since the ball has been stuck, and $v(t)$ represents the ball's velocity in feet per second.

© PhotoDisc/Getty Images

   (a) Use the function to determine the velocity of a golf ball three seconds after it has been it struck.

   (b) Evaluate $\int_0^3 v(t)\,dt$ and interpret.

30. **School Science—Plastic Bottle Rockets:** One common experiment for physics students is to shoot off a rocket made from a 2-liter plastic bottle. Experiments show that the initial velocity of such a rocket is 124 feet per second. (*Source:* www.mrp3.com.) Consequently, we can show that the velocity function for such rocket is given by the function

$$v(t) = -32t + 124 \qquad t > 0$$

where t represents the number of seconds since the rocket lifted off, and $v(t)$ represents the rocket's velocity in feet per second.

   (a) Use the function to determine the velocity of a rocket 2.5 seconds after it has been it launched.

   (b) Evaluate $\int_0^{2.5} v(t)\,dt$ and interpret.

31. **Political Science—Freshman Political Attitudes:** *The American Freshman* tracks interests, affiliations, and trends of incoming college freshmen. From 1980 to 2010, the rate of change in the percentage of incoming freshmen who classify their political orientation as liberal can be modeled by the function

$$f(x) = 0.02x - 0.03 \qquad 0 \le x \le 30$$

where x represents the number of years since 1980, and $f(x)$ represents the rate of change in the percentage of freshmen who classify their political orientation as liberal. (*Source:* gseis.ucla.edu.) Use the model to determine the amount of increase or decrease in the percentage of college freshmen who classify themselves as liberal between 2000 and 2010.

32. **Political Science—Freshman Political Attitudes:** *The American Freshman* tracks interests, affiliations, and trends of incoming college freshmen. From 1980 to 2010, the rate of change in the percentage of incoming freshmen who classify their political orientation as conservative can be modeled by the function

$$f(x) = 0.09 \qquad 0 \le x \le 30$$

where x represents the number of years since 1980, and $f(x)$ represents the rate of change in the percentage of freshmen who classify their political orientation as conservative. Use the model to determine the amount of increase or decrease in the percentage of college freshmen who classify themselves as conservative between 2000 and 2010.

33. **Engineering—Nuclear Power Generation:** Since the first commercial nuclear power plant was constructed in the United Kingdom in 1953, nuclear power has been a promising and

© AbleStock

sometimes controversial method to supply the public's electric needs. The rate of change in the annual amount of power generation from nuclear sources from 1990 to 2010 in the United States can be modeled by the function

$$f(x) = -1.1x + 22.57 \qquad 0 \le x \le 20$$

where x represents the number of years since 1990, and $f(x)$ represents the rate of change in the amount of power generation from nuclear power plants, measured in billions of kilowatt hours per year (BkWh/yr). (*Source:* U.S. Energy Information Administration.) Use this rate model to determine the total amount of increase or decrease in the amount of nuclear power generation between 1995 and 2005.

 34. **Engineering—Natural Gas Power Generation:** Russia and the United States are the two leading countries in natural gas production, and natural gas accounts for 23.4% of total U.S. electricity production. The rate of change in the amount of power generation from natural gas sources from 1990 to 2010 in the United States can be modeled by the function

$$f(x) = 1.34x + 16.63 \qquad 0 \le x \le 20$$

where x represents the number of years since 1990, and $f(x)$ represents the rate of change in the amount of power generation from natural gas power plants, measured in billions of kilowatt hours per year (BkWh/yr). (*Source:* U.S. Energy Information Administration).

(a) Use this rate model to determine the total amount of increase or decrease in the amount of natural gas power generation between 1995 and 2005.

(b) Consulting the result of Exercise 33, during the 1995–2005 time period, which source of power had the largest change: nuclear power or natural gas power?

 35. **Macroeconomics—European Unemployment:** Suppose an economist wishes to study the unemployment rate in the middle of the first decade of the 21st century. The rate of change in unemployment rate in Bulgaria from 2000 to 2010 can be modeled by

$$f(x) = 0.24x^2 - 2.48x + 3.34 \qquad 0 \le x \le 10$$

where x represents the number of years since 2000, and $f(x)$ represents the rate of change in unemployment rate measured in percent per year. (*Source:* Google Public Data.)

(a) Evaluate and interpret $f(6)$.

(b) Evaluate and interpret $\int_5^6 f(x)\,dx$.

 36. **Macroeconomics—European Unemployment:** Suppose an economist wishes to examine the unemployment rate in the middle of the first decade of the 21st century. The rate of change in the unemployment rate in Latvia from 2000 to 2010 can be modeled by

$$f(x) = 0.27x^2 - 1.96x + 1.63 \qquad 0 \le x \le 10$$

where x represents the number of years since 2000, and $f(x)$ represents the rate of change in the unemployment rate measured in percent per year. (*Source:* Google Public Data.)

(a) Evaluate and interpret $f(6)$.

(b) Evaluate and interpret $\int_5^6 f(x)\,dx$.

(c) Consulting Exercise 35, from 2005 to 2006, which country had the greatest change in unemployment rate: Bulgaria or Latvia?

 37. **Demographics—State Population Change:** In 2012, Michigan Governor Rick Snyder discussed the consequences in population change in the state, saying, "What happens if you have a large part of your population not paying taxes on your earnings and they're actually reasonably high users of services?" (*Source:* Bloomberg Businessweek.) A graph of the rate of change in the population of Michigan starting from 2000 and projected to 2030 is shown in the figure. (*Source:* U.S. Census Bureau.) Knowing that the rate function is $p(x) = -0.002x + 0.06, 0 \le x \le 30$, compute the shaded area and interpret.

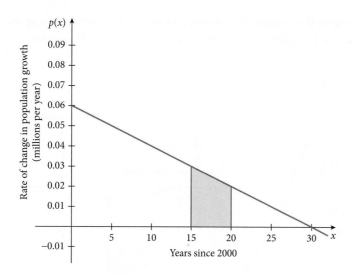

38. Demographics—State Population Change: A state experiencing rapid growth is Georgia. "Wal-Mart, the State Patrol, new high schools, and new homes are being built left and right," said Columbia County Deputy Administrator Scott Johnson. (*Source:* WRDW-TV.) A graph showing the rate of change in the population of Georgia starting from 2000 and projected to 2030 is shown in the figure. (*Source:* U.S. Census Bureau.) Knowing that the rate function is $p(x) = -0.001x + 0.14, 0 \leq x \leq 30$, compute the shaded area and interpret.

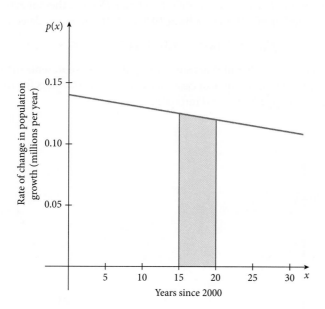

39. Public Health—Smoker Rate: One change in public health over the last few decades has been the decline in cigarette smoking. "The ban on public smoking seems to help people quit and we have begun to turn the tide in lung cancer," said Dr. Norman H. Edelman, scientific consultant for the American Lung Association. (*Source:* www.healthaim.com.) The rate of change in the percentage of smokers in the United States from 1990 to 2010 is shown in the figure and is given by the function

$$p'(t) = 0.009t^2 - 0.2t + 0.43 \qquad 0 \leq t \leq 20$$

where t represents the number of years since 1990, and $p'(t)$ represents the rate of change in percentage of smokers, measured in percent per year. Evaluate $\int_{10}^{15} p'(t)\,dt$ and interpret.

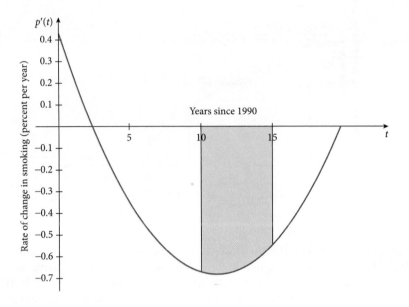

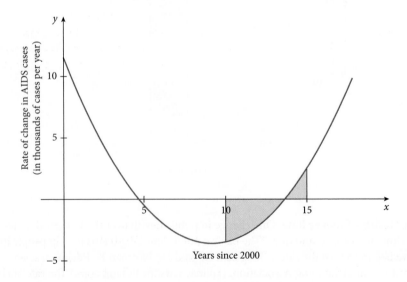

40. **Public Health—AIDS Cases:** A positive public health improvement during the last few decades has been the reduction in AIDS cases. "Our results show a strong and significant association between increased highly active antiretroviral therapy and decreased number of new AIDS diagnoses per year," said Dr. Julio Montaner. (*Source:* themoneytimes.com.) The rate of change in the number of AIDS cases from 1990 to 2010 can be modeled by

$$p'(x) = 0.18x^2 - 3.3x + 11.52 \qquad 0 \le x \le 20$$

where x represents the number of years since 1990, and $p'(x)$ represents the rate of change in AIDS cases, measured in thousands of cases per year. (*Source:* Centers for Disease Control and Prevention.) Evaluate $\int_{10}^{15} p'(x)\,dx$ and interpret.

In Exercises 41–44, you are given a rate function that, at the present time, you are unable to anti-differentiate. (Later in this chapter we will discover techniques that will allow us to determine an antiderivative.) Consequently, you will need to use the FNINT *command on your calculator to compute the result.*

41. The CustomKey Company determines that the marginal cost of producing sterling silver key fobs is given by

$$MC(x) = \frac{10x}{\sqrt{x^2 + 10,000}} \qquad 0 \le x \le 100$$

where x represents the number of fobs produced daily, and $MC(x)$ is the marginal cost in dollars. Use your calculator to approximate the integral $\int_0^{75} MC(x)\,dx$ and interpret.

42. The TightNut Company determines that the marginal cost of producing center punches is given by

$$MC(x) = 0.003x\sqrt{x + 1} \qquad 0 \le x \le 50$$

where x represents the number of center punches produced in a work shift, and $MC(x)$ represents the marginal cost in dollars. Use your calculator to approximate the integral $\int_{35}^{48} MC(x)\,dx$ and interpret.

43. Government Economics—NEH Funding: The National Endowment for the Humanities (NEH) is an independent grant-making agency of the U.S. federal government used to support research, education, and public programs in the humanities. Concerning funding cuts to the organization, NEH chairman Jim Leach said, "I think that when humanities funding is on the line, it really comes down to what our values are as a society." (*Source: The George Washington Hatchet.*) The rate of change in the amount of federal research grants to the NEH from 1990 to 2010 can be modeled by the rational function

$$f(x) = \frac{-871.41}{(x + 21.36)^2} \qquad 0 \le x \le 20$$

where x represents the number of years since 1990, and $f(x)$ represents the rate of change in the amount of federal research grants to the NEH, measured in millions of dollars per year. (*Source: U.S. National Endowment for the Humanities.*) Compare $\int_0^{10} f(x)\,dx$ to $\int_{10}^{20} f(x)\,dx$. During which decade, 1990 to 2000 or 2000 to 2010, was the amount of increase or decrease the greatest?

44. Government Economics—NEA Funding: Established by the U.S. Congress in 1965, the National Endowment for the Arts (NEA) is a public agency that provides support for excellence in the arts and leadership in art education. The NEA came under scrutiny when 2008 Republican vice-presidential candidate Sarah Palin said, "The National Endowment for the Arts, National Endowment for the Humanities, all those kind of frivolous things that government shouldn't be in the business of funding with tax dollars." (*Source: Los Angeles Times.*) The rate of change in the number of federal grants awarded by the NEA from 1990 to 2010 can be modeled the rational function

$$f(x) = \frac{-57,857,845}{35x + 408} \qquad 0 \le x \le 20$$

where x represents the number of years since 1990, and $f(x)$ represents the rate of change in the number of grants awarded per year. (*Source: U.S. Endowment for the Arts Annual Report.*) Compare $\int_0^{10} f(x)\,dx$ to $\int_{10}^{20} f(x)\,dx$. During which decade, 1990 to 2000 or 2000 to 2010, was the amount of increase or decrease the greatest?

Concept and Writing Exercises

45. Suppose that the marginal cost function for a certain product is constant at $MC(x) = k$, $0 \le x \le b$. Determine the total cost of producing b units of the product.

46. Write a few sentences in your own words about how we interpret the area under the graph of a rate function.

47. Suppose the marginal revenue function for a certain product is

$$MR(x) = 48 - 4x \qquad 0 \le x \le 10$$

where x represents the units of product produced and sold, and MR is the change in revenue measured in dollars per unit. Determine the number of units n that need to be produced so that the revenue is \$216. Find n by solving the equation $\int_0^n MR(x)\,dx = 216$.

48. Suppose the marginal cost function for a certain product is

$$MC(x) = -\frac{3}{4}x + 6 \qquad 0 \le x \le 10$$

where x represents the units of product produced, and MC is the change in cost measured in dollars per unit. Determine the number of units n that need to be produced so that the cost is zero.

49. **Energy Policy—Renewable Energy in England:** England has used a policy of Fit-in-Tariffs to promote the increase of renewable energy resources in the country. The rate of change in the percentage of energy produced by renewable sources from 1990 to 2011 can be modeled by

$$f(x) = 0.20 \qquad 0 \le x \le 21$$

where x represents the number of years since 1990, and $f(x)$ represents the rate of change in the percentage of energy produced by renewable sources measured in percent per year. (*Source:* EuroStat.) Use the rate function to determine the year that the percentage exceeded 3%. That is, determine the smallest whole number n so that $\int_0^n f(x)\,dx \ge 3$.

50. **Energy Policy—Renewable Energy in Ireland:** Another country in the British Isles that has been addressing energy needs is Ireland. Natural Resources minister Noel Dempsey said, "By the end of 2010, we will have more than doubled the capacity of new clean green technologies connected to the electricity network." (*Source:* www.renewableenergyworld.com.) The rate of change in the percentage of energy produced by renewable sources from 1990 to 2011 can be modeled by

$$f(x) = 0.21 \qquad 0 \le x \le 21$$

where x represents the number of years since 1990, and $f(x)$ represents the rate of change in the percentage of energy produced by renewable sources measure in percent per year. Use the rate function to determine the year that the percentage exceeded 3%.

 Section Project

The Federal Family Education Loan (FFEL) Program was the second largest of the U.S. higher education loan programs, with Direct Student Loans being first. The FFEL was initiated by the Higher Education Act of 1965 and was funded through a public/private partnership administered at the state and local level. In 2009, President Barack Obama called for an end to the FFEL program, calling it "a wasteful and inefficient system of taxpayers . . . paying banks a premium to act as middlemen—a premium that costs the American people billions of dollars each year . . . a premium we cannot afford." (*Source:* White House Office of the Press Secretary.) The rate of change in the funding of the FFEL program from 2000 to 2011 can be modeled by

$$f(x) = -1.24x + 8.94 \qquad 0 \le x \le 11$$

where x represents the number of years since 2000, and $f(x)$ represents the rate of change in FFEL funding, measured in billions of dollars per year. (*Source:* USDE Office of Postsecondary Education.)

(a) Knowing that the funding for the FFEL program in 2000 was 22.7 billion dollars, integrate to recover the function F that represents the annual funding of the FFEL program.

(b) Evaluate $F(9) - F(7)$ and interpret the difference.

(c) In 2007, the FFEL funding was 56.7 billion dollars, and by 2009, the amount had decreased to 51.3 billion dollars. Compute the amount of change in funding from 2007 to 2009.

(d) Use the answers from part (b) and part (c) to determine the percentage error in the estimate of the change by computing $\dfrac{F(9) - F(7) - \text{actual change}}{\text{actual change}} \cdot 100\%$.

(e) Provide an explanation as to why the solution to part (d) is relatively large.

6.4   Integration by *u*-Substitution

The rules of integration we have learned so far apply to functions that are straightforward. We have found that polynomial functions and simple root functions are easy to integrate. But what if we need to integrate a function such as $\int x\sqrt[3]{4x^2 + 1}\,dx$? We really have no way to perform the integration with the rules that we have established. In this section we will learn how to rewrite an integral that contains a composite function so that it can be expressed in a simplified form that can easily be integrated. This is called the **method of *u*-substitution**. We will see how this method is used to determine both indefinite and definite integrals and the effect that *u*-substitution has on the limits of integration. We will also examine applications that take advantage of this technique.

u-Substitution with Indefinite Integrals

We begin by reviewing the definition of the differential in the following Toolbox.

From Your Toolbox: Differential in *y*

The **differential in *y*,** denoted by dy, of the dependent variable is given by

$$dy = f'(x)dx$$

Using the definition of the differential, if we substitute *y* with *u*, we have $u = f(x)$, then $\frac{du}{dx} = f'(x)$, so the **differential in *u*** is $du = f'(x)dx$. To see how this can help us with integration, let's consider the indefinite integral $\int 4(4x + 3)^5 dx$. We could integrate by expanding $4x + 3$, but we would have to raise this binomial to the fifth power! To avoid all this unnecessary algebra, let's see if a substitution in the integrand will make it simpler. If we let $u = 4x + 3$ and compute the differential in *u*, we get

$$u = 4x + 3$$

$$\frac{du}{dx} = 4$$

$$du = 4dx$$

Notice that we now have an expression in terms of *u* for every part of the integrand. Since $u = 4x + 3$ and $du = 4dx$, we can rewrite the integral $\int \overbrace{4}^{du}(\underbrace{4x + 3}_{u})dx$ in terms of *u* as

$$\int u^5 du$$

Now we can use the Power Rule for Integration to find the integral with respect to *u* as

$$\int u^5 du = \frac{1}{6}u^6 + C$$

Since we are given the integral originally in terms of *x*, and knowing that $u = 4x + 3$, we can resubstitute and write the solution as

$$\int 4(4x + 3)^5 dx = \frac{1}{6}(4x + 3)^6 + C$$

We learned in Section 6.1 that **the derivative of the indefinite integral yields the integrand**. Hence, we can check our answer by differentiating the result via the Chain Rule.

$$\frac{d}{dx}\left[\frac{1}{6}(4x + 3)^6 + C\right] = \frac{1}{6} \cdot 6(4x + 3)^5 \cdot \frac{d}{dx}[4x + 3] + 0$$

$$= (4x + 3)^5 \cdot 4 = 4(4x + 3)^5$$

Now let's generalize by stating the **Power Rule with *u*-substitution**.

Power Rule with *u*-Substitution

If f is a differentiable function of x and $u = f(x)$, then we can write an integral of the form $\int [f(x)]^n f'(x) dx$ as $\int u^n du$ with

$$\int u^n du = \frac{1}{n+1} u^{n+1} + C \qquad n \neq -1$$

where $du = f'(x) dx$.

OBJECTIVE 1

Use the Power Rule with *u*-substitution.

Example 1: Using the Power Rule with *u*-Substitution

Use the method of *u*-substitution to evaluate the given indefinite integrals.

a. $\int \sqrt{x + 2}\, dx$ **b.** $\int 2(2x + 1)^3 dx$

Perform the Mathematics

a.   For this integration, we can let u be the radicand and write $u = x + 2$. Differentiating to get an expression for du, we find that

$$\frac{du}{dx} = 1$$

$$du = dx$$

Since $du = dx$, the substitution is

$$\int \sqrt{x + 2}\, dx = \int \sqrt{u}\, du$$

$$= \int u^{1/2} du = \frac{2}{3} u^{3/2} + C$$

Since $u = x + 2$, we resubstitute and get

$$\int \sqrt{x + 2}\, dx = \frac{2}{3}(x + 2)^{3/2} + C$$

b.   Here we will let u be the expression in the parentheses and write $u = 2x + 1$. Differentiating with respect to x, we find that

$$\frac{du}{dx} = 2$$

$$du = 2dx$$

Rewriting the integral in terms of u and integrating yields

$$\int 2(2x + 1)^3 dx = \int u^3 du = \frac{1}{4} u^4 + C$$

Resubstituting gives

$$\int 2(2x + 1)^3 dx = \frac{1}{4}(2x + 1)^4 + C \qquad \blacksquare$$

▶ *Try It Yourself*

Some related Exercises are 9 and 11.

In words, the Power Rule with *u*-substitution states that if an integrand can be expressed as an expression raised to a power, and the derivative of the expression is another factor in the integrand, we can apply the *u*-substitution. This form of the integral

$$\int \left[(\text{expression})^{\text{power}} \cdot \frac{d}{dx}(\text{expression}) \right] dx$$

occurs more frequently than one might think. As we have seen in the first example, we commonly represent *u* as an expression raised to a power or as a radicand.

Example 2: Integrating Using a *u*-Substitution

Use the Power Rule with *u*-substitution to integrate $\int \sqrt[3]{(4x^3 + x)}(12x^2 + 1)dx$.

Perform the Mathematics

To get $\int \sqrt[3]{(4x^3 + x)}(12x^2 + 1)dx$ in the form $\int u^n du$, we can let *u* be the radicand of the cube root and get

$$u = 4x^3 + x$$

So the differential in *u* is

$$du = (12x^2 + 1)dx$$

Writing the integral with respect to *u* and integrating yields

$$\int \sqrt[3]{(4x^3 + x)}(12x^2 + 1)dx = \int \sqrt[3]{u}\,du$$

$$= \int u^{1/3} du = \frac{3}{4}u^{4/3} + C$$

Knowing that $u = 4x^3 + x$, we can write the antiderivative in terms of *x* as

$$\int \sqrt[3]{(4x^3 + x)}(12x^2 + 1)dx = \frac{3}{4}(4x^3 + x)^{4/3} + C$$

$$= \frac{3}{4}\sqrt[3]{(4x^3 + x)^4} + C \qquad \blacksquare$$

Sometimes we must alter the differential in *u* to get it to match the remaining part of the integrand. We can do this by multiplying or dividing both sides of the differential by a constant. Let's say that we wish to integrate $\int x^2 \sqrt{x^3 + 9}\,dx$. If we let *u* represent the radicand, we have $u = x^3 + 9$. Computing *du*, we find that

$$du = 3x^2 dx$$

But we want *du* to just be $x^2 dx$. We can get this desired expression by just multiplying both sides of the differential by $\frac{1}{3}$. This gives us

$$du = 3x^2 dx$$

$$\frac{1}{3}du = x^2 dx$$

Now, writing the original integral in terms of *u*, we get

$$\int x^2 \sqrt{x^3 + 9}\,dx = \int \frac{1}{3}\sqrt{u}\,du = \frac{1}{3}\int u^{1/2} du$$

$$= \frac{1}{3} \cdot \frac{2}{3}u^{3/2} + C = \frac{2}{9}u^{3/2} + C$$

In terms of x, we get, by substitution,

$$\int x^2 \sqrt{x^3 + 9}\, dx = \frac{2}{9}(x^3 + 9)^{3/2} + C$$

OBJECTIVE 2

Adjust u-substitution by a constant.

Example 3: Adjusting a Constant with u-substitution

Evaluate the indefinite integral $\displaystyle\int \frac{1}{(4x - 1)^3}\, dx$.

Perform the Mathematics

Here we let $u = 4x - 1$, which gives us the differential in u as $du = 4dx$. Since we want $du = dx$, we can divide both sides of the differential by 4 to get

$$\frac{1}{4}du = dx$$

So the integral in terms of u becomes

$$\begin{aligned}
\int \frac{1}{(4x - 1)^3}\, dx &= \int \frac{1}{u^3} \cdot \frac{1}{4}\, dx \\
&= \frac{1}{4}\int u^{-3}\, du \\
&= \frac{1}{4} \cdot \frac{1}{-2}u^{-2} + C \\
&= -\frac{1}{8} \cdot \frac{1}{u^2} + C \\
&= -\frac{1}{8(4x - 1)^2} + C \qquad \blacksquare
\end{aligned}$$

Tips on u-Substitution

1. The correct choice of u is usually a radicand, a denominator, or an expression in parentheses.

2. Compute the differential in u; $du = f'(x)\,dx$.

3. Multiply or divide du by a constant, if needed, to make the remaining part of the integral match the differential in u.

4. After the u-substitution, no factors can contain x.

5. Integrate and rewrite the integral in terms of the original independent variable (usually x).

Let's summarize what we have learned so far and list some **tips on integration by u-substitution**.

Sometimes we must use a "trick" to complete the substitution. If the expression for u is linear, and so is the remainder of the integrand, we can solve the expression for u in terms of x. We demonstrate this technique in Example 4.

Example 4: Solving the u-Expression for x to Complete a Substitution

Determine the indefinite integral $\int x(x - 7)^2\, dx$.

Perform the Mathematics

If we assign u to the expression in parentheses, we have

$$u = x - 7$$
$$du = dx$$

This doesn't help because the desired du should be xdx. However, since $u = x - 7$, we can write

$$x = u + 7$$

Now, making the substitution, we write

$$\int x(x-7)^2 dx = \int (u+7)u^2 du$$

$$= \int (u^3 + 7u^2)\, du$$

$$= \frac{1}{4}u^4 + \frac{7}{3}u^3 + C$$

$$= \frac{1}{4}(x-7)^4 + \frac{7}{3}(x-7)^3 + C \qquad \blacksquare$$

> ▶ *Try It Yourself*
>
> Some related Exercises are 35 and 37.

u-Substitution with Definite Integrals

We also can apply the integration techniques of *u*-substitution to definite integrals in one of two ways. We illustrate these two methods in Example 5.

---

u-Substitution with a Definite Integral
1. Define *u*, perform the integration with respect to *du*, and rewrite the answer in terms of *x*. Evaluate the definite integral using the original limits of integration.
2. Define *u* and rewrite the integrand **and** the limits of integration in terms of the variable *u*. Integrate with respect to *u* and evaluate the integral using the new limits of integration.

Example 5: Evaluating a Definite Integral by *u*-Substitution

Evaluate $\displaystyle\int_0^4 \frac{x}{\sqrt{x^2+9}}\,dx$.

OBJECTIVE 3

Evaluate a definite integral by *u*-substitution.

Perform the Mathematics

By letting $u = x^2 + 9$, we have $du = 2x\,dx$, so $\frac{1}{2}du = x\,dx$. We can now proceed in either of the ways outlined in the *u*-substitution with a definite integral box. We will use both ways to illustrate how to use each, as well as to show that each way yields the same result.

1. $$\int \frac{x}{\sqrt{x^2+9}}\,dx = \int \frac{1}{\sqrt{u}} \cdot \frac{1}{2}\,du$$

$$= \frac{1}{2}\int \frac{1}{\sqrt{u}}\,du = \frac{1}{2}\int u^{-1/2}\,du$$

$$= \frac{1}{2} \cdot 2u^{1/2} = \sqrt{u} = (x^2+9)^{1/2}$$

Now we use the Fundamental Theorem of Calculus with the original limits to get

$$\int_0^4 \frac{x}{\sqrt{x^2+9}}\,dx = (x^2+9)^{1/2}\Big|_0^4 = (25)^{1/2} - (9)^{1/2} = 2$$

2. Since we assign $u = x^2 + 9$, we have

$$\text{If } x = 0, \text{ then } u = 0^2 + 9 = 9.$$
$$\text{If } x = 4, \text{ then } u = 4^2 + 9 = 25.$$

Thus, after we do the substitution, we can rewrite the limits as follows:

$$\int_0^4 \frac{x}{\sqrt{x^2 + 9}} dx = \frac{1}{2} \int_9^{25} \frac{1}{\sqrt{u}} du = \frac{1}{2} \int_9^{25} u^{-1/2} du$$

$$= u^{1/2} \Big|_9^{25} = (25)^{1/2} - (9)^{1/2} = 2$$

Notice that either way, we get the same value for the definite integral.   ■

▶ *Try It Yourself*

Some related Exercises are 41 and 53.

Applications

The *u*-substitution technique allows us to examine new families of applications. One of these is recovering business functions.

<table>
<tr><td>

OBJECTIVE 4

Use *u*-substitution to recover a cost function.

</td><td>

Example 6: Using *u*-Substitution to Recover Business Functions

A manager at the Black Box microprocessor manufacturing company finds through data gathered in research that the marginal cost function for a certain type of automobile computer chip made at the facility is given by

$$MC(x) = 6x\sqrt{x^2 + 11}$$

where *x* represents the number of auto computer chips produced each hour, and $MC(x)$ represents the marginal cost. The manager also knows that it costs $1932 to manufacture five chips.

</td></tr>
</table>

a.   Recover the cost function, *C*. To avoid any confusion, we will call the arbitrary constant *d*.

b.   Determine the fixed costs.

Perform the Mathematics

a.   Recall from Section 6.1 that $C(x) + d = \int MC(x) dx$, where *d* is a constant. So we must integrate $\int 6x\sqrt{x^2 + 11} dx$. To evaluate $\int 6x\sqrt{x^2 + 11} dx$, it is reasonable to let $u = x^2 + 11$, making

$$du = 2x dx$$

However, the remaining part of the integrand is $6x dx$. This means we must multiply the differential in *u* by 3 to get the parts to match.

$$du = 2x dx$$
$$3du = 6x dx$$

Our substitution is

$$\int 6x\sqrt{x^2 + 11} dx = \int \sqrt{u} \cdot 3du = 3 \int \sqrt{u} du$$

$$= 3 \int u^{1/2} du = 3 \cdot \left(\frac{2}{3}\right) u^{3/2} + d = 2u^{3/2} + d$$

So, in terms of *x*, we get the partially recovered cost function as

$$C(x) = 2(x^2 + 11)^{3/2} + d = 2\sqrt{(x^2 + 11)^3} + d$$

where d is a constant. To find d, we can use the initial value that when $x = 5$, $C(x) = 1932$ to get

$$C(x) = 2\sqrt{(x^2 + 11)^3} + d$$
$$1932 = 2\sqrt{(5^2 + 11)^3} + d$$
$$1932 = 2\sqrt{46,656} + d$$
$$1932 = 432 + d$$
$$1500 = d$$

So the recovered cost function is $C(x) = 2\sqrt{(x^2 + 11)^3} + 1500$.

b. When the cost function was represented by a polynomial function, the fixed cost was the same as the constant. But here, the cost function involves a radical, so we need to use the definition of the cost function in its literal sense and find the cost of producing zero items. This means that the fixed costs are found by evaluating $C(x)$ at $x = 0$. This gives us

$$C(x) = 2\sqrt{(x^2 + 11)^3} + 1500$$
$$C(0) = 2\sqrt{(0^2 + 11)^3} + 1500$$
$$= 2\sqrt{1331} + 1500 \approx 1572.97$$

The fixed costs are about \$1572.97. ■

As in the previous section, we can also compute the continuous sum of a rate function using the *u*-substitution technique. But now we can integrate composite rate functions, as illustrated in Example 7.

Example 7: Integrating a Rate Function Using *u*-Substitution

OBJECTIVE 5

Use *u*-substitution to integrate a rate function.

Media consultants for the new local magazine *Rave!* have projected that the number of subscriptions will grow during the first five years at a rate given by

$$S(t) = \frac{1000}{(1 + 0.3t)^{3/2}} \qquad 0 \le t \le 60$$

where t is the number of months since the magazine's first issue, and $S(t)$ is the rate of change in the number of subscriptions measured in $\dfrac{\text{subscriptions}}{\text{month}}$. Evaluate and interpret $\displaystyle\int_0^6 S(t)dt$.

Understand the Situation

Since S is a rate function, integration gives us a total accumulation. Specifically, integrating from $t = 0$ to $t = 6$ gives us the total number of subscriptions after 6 months.

Perform the Mathematics

Here we need to integrate $\displaystyle\int_0^6 \frac{1000}{(1 + 0.3t)^{3/2}}dt$. If we let $u = 1 + 0.3t$, then $du = 0.3dt$, so $\frac{1}{0.3}du = dt$. We opt to change the limits of integration and get

If $t = 0$, then $u = 1 + 0.3(0) = 1$
If $t = 6$, then $u = 1 + 0.3(6) = 2.8$

So in terms of u, the integral is

$$\int_0^6 \frac{1000}{(1+0.3t)^{3/2}} dt = 1000 \int_0^6 \frac{1}{(1+0.3t)^{3/2}} dt$$

$$= 1000 \int_1^{2.8} \frac{1}{u^{3/2}} \cdot \frac{1}{0.3} du = \frac{1000}{0.3} \int_1^{2.8} \frac{1}{u^{3/2}} du$$

$$= \frac{1000}{0.3} \int_1^{2.8} u^{-3/2} du = \frac{1000}{0.3} (-2 \cdot u^{-1/2}) \Big|_1^{2.8}$$

$$= -\frac{2000}{3} [(2.8)^{-1/2} - (1)^{-1/2}] \approx 2682.57$$

Interpret the Result

This means that at the end of the first six months, the number of subscriptions is estimated to be about 2683.

 **Try It Yourself**

Some related Exercises are 69(a) and 71(b). ■

Summary

This section concentrated on the frequently used integration technique called **u-substitution**. We saw that the technique was useful in a variety of applications.

- **Power Rule with u-substitution:** $\int u^n du = \frac{1}{n+1} u^{n+1} + C \qquad n \neq -1$

- **Tips on u-substitution:**
  1. The correct choice of u is usually a radicand, a denominator, or an expression in parentheses.
  2. Compute the differential in u; $du = f'(x)dx$.
  3. Multiply or divide du by a constant, if needed, to make the remaining part of the integral match the differential in u.
  4. After the u-substitution, no factors can contain x.
  5. Integrate and rewrite the integral in terms of the original independent variable (usually x).

Section 6.4 Exercises

Vocabulary Exercises

1. Integration by u-substitution is a method for integrating _____ functions.
2. If $u = f(x)$ so that $\frac{du}{dx} = f'(x)$, then du represents the _____ in u.
3. The method of substitution is often used when the integrand has the form of an expression raised to a real number _____.
4. When choosing the expression for u, we normally select a denominator, an expression in parentheses, or a _____.
5. After performing a u-substitution, no factors can contain the variable _____.
6. We can evaluate definite integrals using u-substitution by changing the _____ of integration.

Skill Exercises

For Exercises 7–18, use u-substitution to evaluate the indefinite integrals. Check your answer by differentiating the solution.

7. $\displaystyle\int 3(3x + 4)^2\,dx$

8. $\displaystyle\int 5(5x + 3)^2\,dx$

9. $\displaystyle\int 8x(4x^2 + 1)^3\,dx$

10. $\displaystyle\int 6x(3x^2 - 5)^3\,dx$

11. $\displaystyle\int 2x\sqrt{x^2 - 1}\,dx$

12. $\displaystyle\int 10x\sqrt[3]{5x^2 - 11}\,dx$

13. $\displaystyle\int (3x^2 - 2)(x^3 - 2x)\,dx$

14. $\displaystyle\int (4x - 1)(2x^2 - x)\,dx$

15. $\displaystyle\int \frac{3x^2}{\sqrt[3]{x^3 - 5}}\,dx$

16. $\displaystyle\int \frac{8}{\sqrt{8x + 9}}\,dx$

17. $\displaystyle\int \frac{4x}{(2x^2 + 3)^3}\,dx$

18. $\displaystyle\int \frac{3x^2 - 2}{(x^3 - 2x)^2}\,dx$

In Exercises 19–34, use u-substitution to evaluate the indefinite integrals.

19. $\displaystyle\int (4x + 7)^4\,dx$

20. $\displaystyle\int (12x - 1)^9\,dx$

21. $\displaystyle\int x(x^2 - 3)^5\,dx$

22. $\displaystyle\int x^2(x^3 - 1)^3\,dx$

23. $\displaystyle\int \frac{x^2}{\sqrt{x^3 - 5}}\,dx$

24. $\displaystyle\int \frac{x^3}{\sqrt{x^4 - 4}}\,dx$

25. $\displaystyle\int \frac{5x}{(10x^2 - 4)^2}\,dx$

26. $\displaystyle\int \frac{x^2}{(3x^3 - 8)^5}\,dx$

27. $\displaystyle\int \frac{1}{x^2}\sqrt{1 - x^{-1}}\,dx$

28. $\displaystyle\int \frac{1}{\sqrt{x}}\sqrt[3]{1 + \sqrt{x}}\,dx$

29. $\displaystyle\int (x - 7)^{10}\,dx$

30. $\displaystyle\int (6x - 10)^5\,dx$

31. $\displaystyle\int \frac{x + 1}{(x^2 + 2x + 5)^5}\,dx$

32. $\displaystyle\int \frac{2x + 3}{(2x^2 + 6x - 10)^4}\,dx$

33. $\displaystyle\int (6x + 9)(x^2 + 3x - 10)\,dx$

34. $\displaystyle\int (20x + 20)(x^2 + 2x - 9)^2\,dx$

For Exercises 35–40, evaluate the integrals by solving the u-expression to complete the substitution.

35. $\displaystyle\int x\sqrt{x + 1}\,dx$

36. $\displaystyle\int x\sqrt[3]{x - 2}\,dx$

37. $\displaystyle\int x(x - 1)^5\,dx$

38. $\displaystyle\int 2x(x + 1)^3\,dx$

39. $\displaystyle\int \frac{3x}{\sqrt{x - 1}}\,dx$

40. $\displaystyle\int \frac{x}{\sqrt[3]{x + 2}}\,dx$

In Exercises 41–52, evaluate the definite integrals. Do not change the limits of integration.

41. $\displaystyle\int_0^4 \frac{2x}{\sqrt{x^2 + 9}}\,dx$

42. $\displaystyle\int_{-2}^0 \frac{1}{(x - 4)^2}\,dx$

43. $\displaystyle\int_0^2 x(x^2 - 2)^3\,dx$

44. $\displaystyle\int_0^1 3x(x^2 - 1)^2\,dx$

45. $\displaystyle\int_0^4 (3x-5)^2\,dx$

46. $\displaystyle\int_0^2 (2x+3)^3\,dx$

47. $\displaystyle\int_{-2}^1 \sqrt{1-x}\,dx$

48. $\displaystyle\int_{-5}^0 \sqrt[3]{2-x}\,dx$

49. $\displaystyle\int_0^1 \frac{x}{(1+3x^2)^2}\,dx$

50. $\displaystyle\int_{-1}^1 \frac{2x}{(1+x^2)^3}\,dx$

51. $\displaystyle\int_2^{10} \frac{1}{\sqrt{x-1}}\,dx$

52. $\displaystyle\int_0^7 \frac{1}{\sqrt[3]{x+1}}\,dx$

For Exercises 53–64, evaluate the definite integrals. Change the limits of integration to u-limits.

53. $\displaystyle\int_0^3 \frac{x}{(x^2+1)^2}\,dx$

54. $\displaystyle\int_0^1 \frac{8x}{(2x^2+1)^3}\,dx$

55. $\displaystyle\int_1^4 \frac{2x+1}{(x^2+x-1)^2}\,dx$

56. $\displaystyle\int_1^3 \frac{2x+3}{(x^2+3x)^3}\,dx$

57. $\displaystyle\int_0^2 \frac{3x^2}{(1+x^3)^5}\,dx$

58. $\displaystyle\int_{-1}^0 \frac{x^3}{(2-x^4)^4}\,dx$

59. $\displaystyle\int_0^2 3x^2\sqrt{x^3+1}\,dx$

60. $\displaystyle\int_0^1 2x^3\sqrt{x^4+4}\,dx$

61. $\displaystyle\int_{-1}^1 x(x^2-1)^3\,dx$

62. $\displaystyle\int_{-2}^0 x^2(x^3-2)\,dx$

63. $\displaystyle\int_2^{10} \frac{3}{\sqrt{5x-1}}\,dx$

64. $\displaystyle\int_0^4 \frac{x}{\sqrt{x^2+9}}\,dx$

In Exercises 65–68, determine the exact area of the shaded region.

65.

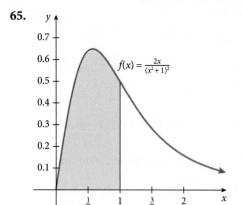

$f(x) = \dfrac{2x}{(x^2+1)^2}$

66.

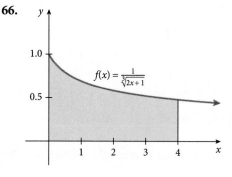

$f(x) = \dfrac{1}{\sqrt[3]{2x+1}}$

67.

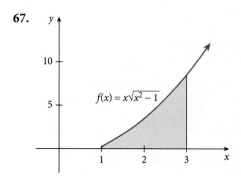

$f(x) = x\sqrt{x^2-1}$

68.

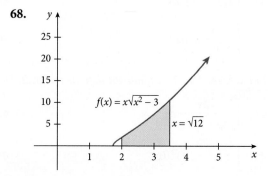

$f(x) = x\sqrt{x^2-3}$

$x = \sqrt{12}$

Application Exercises

69. Microeconomics—Marginal Cost: The CustomKey Company determines that the marginal cost of producing sterling silver key fobs is given by

$$MC(x) = \frac{10x}{\sqrt{x^2 + 10,000}} \qquad 0 \le x \le 100$$

where x represents the number of fobs produced daily, and $MC(x)$ is the marginal cost in dollars.

(a) Evaluate $MC(75)$ and interpret.

(b) Evaluate $\int_0^{75} MC(x)\,dx$ and interpret.

(c) Knowing that the cost to produce 75 fobs is $3250, recover the cost function C.

(d) Determine the fixed costs.

70. Microeconomics—Marginal Cost: The TightNut Company determines that the marginal cost of producing center punches is given by

$$MC(x) = 0.003x\sqrt{x + 1} \qquad 0 \le x \le 50$$

where x is the number of center punches produced in a work shift, and $MC(x)$ is the marginal cost in dollars.

(a) Evaluate and interpret $MC(48)$.

(b) Evaluate $\int_{35}^{48} MC(x)\,dx$ and interpret.

(c) Knowing that it costs $23.40 to produce 24 center punches, recover the cost function C.

(d) Determine the fixed costs.

71. Microeconomics—Marginal Profit: The marginal profit function for seasonal flags produced by the WaveFree Company is given by

$$MP(x) = \frac{2x}{\sqrt{2x^2 - 400}} \qquad 15 \le x \le 105$$

where x is the number of flags produced and sold monthly, and $MP(x)$ is the marginal profit in hundreds of dollars.

(a) Determine the marginal profit at a production level of 70 flags produced and sold monthly.

(b) Evaluate $\int_{20}^{100} MP(x)\,dx$ and interpret.

(c) Knowing that the WaveFree Company breaks even when 20 flags are sold, that is, $P(x) = 0$ when $x = 20$, recover the profit function P.

72. Microeconomics—Marginal Profit: The FemTouch Accessory Manufacturer determines their marginal profit function for decorative scarves is

$$MP(x) = \frac{4}{\sqrt[3]{12x - 200}} \qquad 20 \le x \le 150$$

where x is the number of scarves produced and sold daily, and $MP(x)$ is the marginal profit in hundreds of dollars.

(a) Determine the marginal profit at a production level of 100 scarves produced and sold daily.

(b) Evaluate $\int_{20}^{100} MP(x)\,dx$ and interpret.

(c) Knowing that the FemTouch accessory manufacturer breaks even when 100 units are sold, that is, $P(x) = 0$ when $x = 100$, recover the profit function P.

73. **Physics—Ballistics:** Suppose that the velocity of an object is given by the function

$$v(t) = \frac{t}{\sqrt{t^2 + 9}} \qquad t \geq 0$$

where t is the time in seconds, and $v(t)$ is the velocity in feet per second.

(a) Determine the total distance moved between 3 and 5 seconds.

(b) Knowing that when $t = 4$ seconds, $s(t) = 8$ feet, determine the position function s.

74. **Physics—Ballistics:** Suppose that the velocity of an object is given by the function

$$v(t) = \frac{2\sqrt{t + 1} + 1}{\sqrt{t + 1}} \qquad t \geq 0$$

where t is the time in seconds, and $v(t)$ is the velocity in feet per second.

(a) Determine the total distance moved between 4 and 16 seconds.

(b) Knowing that when $t = 3$ seconds, $s(t) = 16$ feet, determine the position function s.

Concept and Writing Exercises

75. Write a sentence on why we sometimes need to use the u-substitution technique.

76. Explain the differences between the two methods of evaluating definite integrals when using the u-substitution technique.

For Exercises 77–80, use the u-substitution technique to integrate the following indefinite integrals. Assume that a, b, and n represent real numbers with n > 0.

77. $\displaystyle\int a(ax + b)^n dx$

78. $\displaystyle\int 2ax(ax^2 + b)^n dx$

79. $\displaystyle\int \frac{ax}{(2ax^2 + b)^n} dx, n \neq 1$

80. $\displaystyle\int (ax + b)^n dx$

Section Project

Conservationists find that the rate of growth of carp in the first year in a stocked pond can be modeled by

$$f(x) = \frac{10x}{\sqrt{x + 1}} \qquad 0 \leq x \leq 12$$

where x is the number of months since the pond was stocked, and $f(x)$ is the rate of growth in the number of carp in the pond, measured in carp per month.

(a) Evaluate and interpret $f(6)$. Round your answer to the nearest whole number.

(b) Evaluate and interpret $\int_0^{12} f(x)dx$. Round your answer to the nearest whole number.

(c) Knowing that the pond was stocked with 100 carp, determine $F(x)$, the number of carp in the pond after x months.

(d) Evaluate and interpret $F(5)$. Round your answer to the nearest whole number.

(e) Determine an equation of the line tangent to the graph of F when $x = 6$. Evaluate the tangent line equation at the value of $x = 12$ and interpret.

In Chapter 4, we discussed the derivatives of logarithmic and exponential functions, along with their associated applications. Now we turn our attention to **integrals** that involve **exponential and logarithmic functions**. Just as we need special rules to differentiate this family of functions, we also need an additional set of integration rules. We will see how to use the *u*-substitution technique with these functions and check out applications that lend themselves to these functions.

Integrals That Yield Logarithmic Functions

Before examining the role that the logarithm plays in integration, let's review the applicable differentiation rules in the following Toolbox.

From Your Toolbox: Derivatives of Logarithms

1. $\dfrac{d}{dx}[\ln x] = \dfrac{1}{x}$

2. If f is a differentiable function of x, then, by the Chain Rule,

$$\frac{d}{dx}[\ln [f(x)]] = \frac{1}{f(x)} \cdot f'(x)$$

Remember that the Power Rule for Integration stated that

$$\int x^n dx = \frac{1}{n+1}x^{n+1} + C, \text{ provided that } n \neq -1.$$

To determine the integral of x^n when $n = -1$, we must integrate $\int x^{-1}dx = \int \frac{1}{x}dx$. Since $\frac{d}{dx}[\ln x] = \frac{1}{x}$, we can conclude that $\int \frac{1}{x}dx = \ln x + C$. But we need to be careful here, because the natural logarithm function is defined only for positive numbers. Let's say that x is negative, in which case $(-x)$ is positive. Then, by the Chain Rule for the logarithm function,

$$\frac{d}{dx}[\ln (-x)] = \frac{1}{-x} \cdot \frac{d}{dx}[-x] = \frac{1}{-x}(-1) = \frac{1}{x}$$

So we see that $\int \frac{1}{x}dx = \ln (-x) + C$ if x is negative. Instead of writing two integration rules for positive and negative integrands, we can use the absolute value to write one neat **integration formula for $\frac{1}{x}$**.

Integration Formula for $\frac{1}{x}$

For the function $f(x) = \frac{1}{x}$, the indefinite integral is given by

$$\int \frac{1}{x}dx = \ln|x| + C$$

NOTE: For $a > 1$, $\int_1^a \frac{1}{x}dx = \ln|x|\big|_1^a = \ln a - \ln 1 = \ln a - 0 = \ln a$. This means that the area under the graph of $f(x) = \frac{1}{x}$ on the interval $[1, a]$ is the same as the value of $\ln a$. Some books define the natural logarithm function this way.

The relationship between the functions $f(x) = \frac{1}{x}$ and $g(x) = \ln x$ is shown in **Figure 6.5.1**.

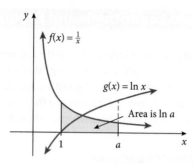

Figure 6.5.1 $\int_1^a \frac{1}{x} dx = \ln a$.

Evaluate a definite integral of the form $\int_a^b \frac{k}{x} dx$.

Example 1: Integrating Functions of the Form $f(x) = \frac{k}{x}$

Evaluate the following definite integrals.

a. $\displaystyle\int_{-6}^{-2} \frac{1}{x} dx$

b. $\displaystyle\int_1^5 \frac{6}{x} dx$

Perform the Mathematics

a.   Using the integration formula for $\frac{1}{x}$, we get

$$\int_{-6}^{-2} \frac{1}{x} dx = \ln|x| \Big|_{-6}^{-2} = \ln|-2| - \ln|-6| = \ln 2 - \ln 6 \approx -1.10$$

b.   We begin by factoring out the constant 6 and get

$$\int_1^5 \frac{6}{x} dx = 6 \int_1^5 \frac{1}{x} dx = 6 \ln|x| \Big|_1^5$$

$$= 6\ln|5| - 6\ln|1| = 6\ln 5 - 6\ln 1$$

$$= 6\ln 5 - 0 = 6\ln 5 \approx 9.66 \qquad \blacksquare$$

Integration Formula for $\frac{1}{u}$

If u is a differentiable function of x, then

$$\int \frac{1}{u} du = \ln|u| + C$$

Since we rarely must integrate $f(x) = \frac{1}{x}$, the usefulness of the integration formula for $f(x) = \frac{1}{x}$ is limited. However, the *u-substitution form of the formula* is quite useful.

Integrate using *u*-substitution that yields a logarithmic function.

Example 2: Integrating using u-Substitution

Determine the following indefinite integrals.

a. $\displaystyle\int \frac{1}{t+2} dt$

b. $\displaystyle\int \frac{4x}{x^2 + 5} dx$

Perform the Mathematics

a.   We can use a *u*-substitution here with $u = t + 2$. Computing the differential in u yields $du = dt$. So in terms of u, the integral becomes

$$\int \frac{1}{t+2} dt = \int \frac{1}{u} du = \ln|u| + C$$

In terms of t, the result is

$$\int \frac{1}{t+2} dt = \ln|t+2| + C$$

b. Here, we will let $u = x^2 + 5$, which means $du = 2xdx$. Since we want the expression containing dx to include $4x$, we can multiply both sides by 2 to get $2du = 4xdx$. In terms of u, the integral becomes

$$\int \frac{4x}{x^2 + 5}dx = \int \frac{1}{u} \cdot 2du = 2\int \frac{1}{u}du = 2\ln|u| + C$$

Resubstituting to get the result in terms of x yields

$$\int \frac{4x}{x^2 + 5}dx = 2\ln|x^2 + 5| + C = 2\ln(x^2 + 5) + C$$

Notice that, since the expression $x^2 + 5$ is positive for all values of x, we can remove the absolute value sign. ∎

▶ *Try It Yourself*

Some related Exercises are 17 and 21.

Many applications that involve integrating a rate function use our new integration rule.

Example 3: Integrating a Rate Function That Yields a Logarithmic Function

OBJECTIVE 3

Integrate a rate function that yields a logarithmic function.

Since running a series of first-come, first-served promotions, the FineHomes Furniture Store has found that its sales rate during its three-month sales drive can be modeled by

$$s(t) = \frac{10}{t} + 2 \qquad 1 \le t \le 12$$

where t represents the number of weeks that the promotion has been running and $s(t)$ is the sales rate measured in $\dfrac{\text{thousands of dollars}}{\text{week}}$. Determine the total increase in sales generated from the first to the fifth week.

Understand the Situation

Since we are given a rate function, we need to compute $\int_1^5 s(t)dt$ to obtain the total increase in sales generated from the first to the fifth week.

Perform the Mathematics

Evaluating the definite integral, we get

$$\int_1^5 s(t)dt = \int_1^5 \left(\frac{10}{t} + 2 \right)dt = \int_1^5 \frac{10}{t}dt + \int_1^5 2dt$$

$$= 10\int_1^5 \frac{1}{t}dt + 2\int_1^5 dt = \left[10\ln|t| + 2t \right] \Big|_1^5$$

$$= \left[10\ln|5| + 2(5) \right] - \left[10\ln|1| + 2(1) \right]$$

$$= 10\ln 5 + 10 - 0 - 2 = 10\ln 5 + 8$$

$$\approx 24.09$$

Interpret the Result

Interpret the Result

This means that from weeks 1 to 5 of the promotion, the total increase in sales was about $24,090.

▶ *Try It Yourself*

Some related Exercises are 69(b) and 73(b). ∎

Integration of Exponential Functions

Now we turn our attention from integrals yielding logarithmic functions to those involving exponential functions. Let's consult the following Toolbox to recall some facts on derivatives of exponential functions.

From Your Toolbox: Derivatives of Exponential Functions

1. $\dfrac{d}{dx}[e^x] = e^x$.

2. If f is a differentiable function of x, then by the Chain Rule,
$$\frac{d}{dx}[e^{f(x)}] = e^{f(x)} \cdot f'(x)$$

Integration Formula for e^x

For the exponential function $f(x) = e^x$, we have
$$\int e^x dx = e^x + C$$

From the first fact reviewed in the Toolbox, along with the fact that differentiation and integration are inverse operations, we get the following **integration rule for e^x**.

OBJECTIVE 4

Evaluate a definite integral of the form $\int_a^b e^x dx$.

Example 4: Integration of Functions Containing e^x

Determine the definite integral $\int_0^4 (2 - e^x)\,dx$.

Perform the Mathematics

Integrating, we get
$$\int_0^4 (2 - e^x)\,dx = \int_0^4 2\,dx - \int_0^4 e^x dx$$
$$= (2x - e^x)\big|_0^4 = [2(4) - e^4] - [2(0) - e^0]$$
$$= 8 - e^4 - 0 + 1 = 9 - e^4 \approx -45.60 \qquad ∎$$

Integration Formula for e^u

If u is a differentiable function of x, then
$$\int e^u du = e^u + C$$

In practice, few integrands are simply e^x. We can easily extend the methods of u-substitution to the exponential function to get an **integration formula for e^u**.

Example 5: Integration of Functions Containing e^u

OBJECTIVE 5

Integrate using
u-substitution that yields
an exponential function.

Determine the given indefinite integrals.

a. $\displaystyle\int 10xe^{x^2}dx$

b. $\displaystyle\int (2x + 1)e^{2x^2 + 2x}dx$

Perform the Mathematics

a. Since the integration formula for e^u suggests letting u be the expression in the exponent, we let $u = x^2$, so $du = 2xdx$. Multiplying by 5, we get

$$5du = 10xdx$$

So the indefinite integral is

$$\int 10xe^{x^2}dx = \int e^u 5du = 5\int e^u du = 5e^u + C$$

In terms of x, this gives

$$\int 10xe^{x^2}dx = 5e^{x^2} + C$$

b. Again, if we let u be the expression in the exponent, we have $u = 2x^2 + 2x$, and $du = (4x + 2)dx$. Since we want the du expression to contain $2x + 1$, we multiply each side by $\frac{1}{2}$ to get

$$\frac{1}{2}du = \frac{1}{2}(4x + 2)dx = (2x + 1)dx$$

So our indefinite integral is

$$\int (2x + 1)e^{2x^2 + 2x}dx = \int e^u \cdot \frac{1}{2}du$$

$$= \frac{1}{2}\int e^u du = \frac{1}{2}e^u + C$$

In terms of x, this is

$$\int (2x + 1)e^{2x^2 + 2x}dx = \frac{1}{2}e^{2x^2 + 2x} + C \qquad \blacksquare$$

▶ **Try It Yourself**

Some related Exercises are 41 and 43.

In many applications, we are asked to integrate functions of the form $f(x) = e^{kx}$, where k is a real number constant. To determine an **integration formula for e^{kx}**, we use a u-substitution with $u = kx$. With this u-substitution, we then have $du = kdx$, and this gives $\frac{1}{k}du = dx$. So, in terms of u, the integral becomes

$$\int e^{kx}dx = \frac{1}{k}\int e^u du$$

$$= \frac{1}{k}e^u + C = \frac{1}{k}e^{kx} + C$$

**Integration
Formula for e^{kx}**

If k is a nonzero real
number coefficient of x,
then for the exponential
function $f(x) = e^{kx}$,

$$\int e^{kx}dx = \frac{1}{k}e^{kx} + C$$

OBJECTIVE 6

Integrate a rate function that yields an exponential function.

Example 6: Applying the Integration Formula for e^{kx}

Over the past decade, insurance companies have stepped up their advertising of bundling together the different types of insurance, such as life, health, and car, as a way for the consumer to realize savings through the bundling with one company. The rate of change in the number of life insurance companies in the United States that also sell accident and health insurance can be modeled by

$$f(x) = -63.07e^{-0.047x} \qquad 1 \le x \le 7$$

where x represents the number of years since 2001, and $f(x)$ represents the rate of change in the number of life insurance companies that also sell accident and health insurance measured in number per year. The graph of f is shown in **Figure 6.5.2**. (*Source:* U.S. Statistical Abstract.)

a. Evaluate $f(3)$ and interpret.

b. Use the model to estimate the total increase or decrease in the number of life insurance companies that also sell accident and health insurance from 2004 to 2008.

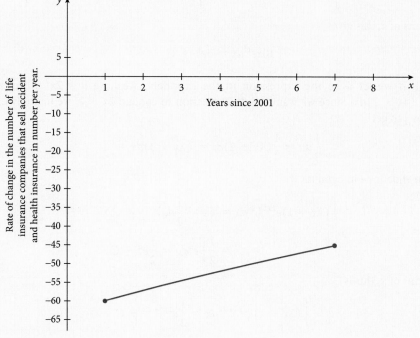

Figure 6.5.2

Perform the Mathematics

a. First we note that $x = 3$ corresponds to 2004. Evaluating $f(3)$ gives us

$$f(3) = -63.07e^{-0.047(3)} = -63.07e^{-0.141} \approx -54.78$$

This means that in 2004, the number of life insurance companies in the United States that also sell accident and health insurance was decreasing at a rate of about 55 companies per year.

b. Since function f is a rate function, and since the years 2004 and 2008 correspond to the independent variable values $x = 3$ and $x = 7$, respectively, we need to integrate $\int_3^7 f(x)\,dx$ to determine the total increase or decrease. Performing this integration gives us

$$\int_3^7 -63.07e^{-0.047x}\,dx = -63.07\int_3^7 e^{-0.047x}\,dx$$

We now use the integration formula for e^{kx} to get

$$-63.07 \int_3^7 e^{-0.047x} dx = -63.07 \cdot \frac{1}{-0.047} e^{-0.047x} \Big|_3^7$$

$$= \frac{63.07}{0.047} e^{-0.047(7)} - \frac{63.07}{0.047} e^{-0.047(3)}$$

$$= \frac{63.07}{0.047} e^{-0.329} - \frac{63.07}{0.047} e^{-0.141}$$

$$\approx -199.74$$

This means that from 2004 to 2008, the total decrease in the number of life insurance companies in the United States that also sell accident and health insurance was about 200 companies. ∎

> **Try It Yourself**
>
> Some related Exercises are 71(b) and 75(b).

Integrating General Exponential Functions

Thus far we have discussed how to integrate functions of the type $f(x) = e^{kx}$. But what about general exponential functions such as $f(x) = 2^x$ or $g(x) = 3^{0.6x}$? Surely they can be integrated, too. Some calculus courses derive a formula specifically for these functions, but we will rely on techniques already learned.

In Chapter 4 we learned that the general exponential function $f(x) = a \cdot b^x$ can be rewritten in the exponential form as $g(x) = a \cdot e^{kx}$, where $k = \ln b$. To determine the indefinite integral whose integrand has the form b^x, such as $\int 2^x dx$, we rewrite the integrand in the exponential form e^{kx}, where $k = \ln b$. Thus, we can rewrite $\int 2^x dx$ as $\int e^{(\ln 2) \cdot x} dx$ and then integrate using the integration formula for e^{kx}.

Example 7: Integrating a General Exponential Function

OBJECTIVE 7

Integrate a general exponential function.

Evaluate the definite integral $\displaystyle\int_{-2}^1 \left(\frac{3}{4}\right)^x dx$.

Perform the Mathematics

We begin by rewriting the integrand as $e^{\ln(3/4) \cdot x}$ and use the Integration Formula for e^{kx}.

$$\int_{-2}^1 \left(\frac{3}{4}\right)^x dx = \int_{-2}^1 e^{\ln(3/4) \cdot x} dx$$

$$= \frac{1}{\ln\left(\dfrac{3}{4}\right)} \cdot e^{\ln(3/4) \cdot x} \Big|_{-2}^1$$

Rewriting $e^{\ln(3/4)\cdot x}$ as $\left(\dfrac{3}{4}\right)^{x}$ yields

$$\frac{1}{\ln\left(\dfrac{3}{4}\right)}\left[\left(\dfrac{3}{4}\right)^{x}\right]\Bigg|_{-2}^{1} = \frac{1}{\ln\left(\dfrac{3}{4}\right)}\left[\left(\dfrac{3}{4}\right)^{1} - \left(\dfrac{3}{4}\right)^{-2}\right]$$

$$= \frac{1}{\ln\left(\dfrac{3}{4}\right)}\left(\dfrac{3}{4} - \dfrac{16}{9}\right) = \frac{1}{\ln\left(\dfrac{3}{4}\right)}\left(-\dfrac{37}{36}\right) = -\frac{37}{36\ln\left(\dfrac{3}{4}\right)} \approx 3.57 \quad \blacksquare$$

Many times we are given data that are always increasing or always decreasing on an interval. This kind of data is said to be **monotonic**. We can model these data with the general exponential mathematical model $y = a \cdot b^{x}$, where a is any nonzero real number and $b > 0$, $b \neq 1$. To integrate functions of the type $\int a \cdot b^{x}dx$, we factor out the constant a, and then we can determine $\int b^{x}dx$.

OBJECTIVE 8

Integrate a rate function that involves a general exponential function.

Example 8: Integrating a General Exponential Function Application

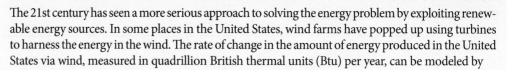

The 21st century has seen a more serious approach to solving the energy problem by exploiting renewable energy sources. In some places in the United States, wind farms have popped up using turbines to harness the energy in the wind. The rate of change in the amount of energy produced in the United States via wind, measured in quadrillion British thermal units (Btu) per year, can be modeled by

$$f(x) = 0.002(1.219)^{x} \qquad 1 \leq x \leq 20$$

where x is the number of years since 1989, and $f(x)$ is the rate of change in the amount of energy produced in the United States via wind measured in quadrillion British thermal units per year. (*Source:* U.S. Energy Information Administration.)

a. Evaluate $\displaystyle\int_{11}^{20} 0.002(1.219)^{x}dx$ and interpret.

b. Knowing that in 1995, 0.03 quadrillion British thermal units of energy was produced via wind, recover F, the amount of energy produced via wind function.

c. Evaluate $F(16)$ and interpret.

Perform the Mathematics

a. Once again, we are given a rate of change function. Integrating the rate of change function will give us a total accumulation, that is, a total increase or total decrease. Before integrating, we note that the limits of integration, $x = 11$ and $x = 20$, correspond to the years 2000 and 2009, respectively. Proceeding with the integration gives us

$$\int_{11}^{20} 0.002(1.219)^{x}dx = 0.002\int_{11}^{20}(1.219)^{x}dx = 0.002\int_{11}^{20} e^{[\ln(1.219)]\cdot x}dx$$

$$= \frac{0.002}{\ln(1.219)}\left[e^{[\ln(1.219)\cdot x]}\right]\Bigg|_{11}^{20}$$

$$= \frac{0.002}{\ln(1.219)}\left[(1.219)^{x}\right]\Bigg|_{11}^{20}$$

$$= \frac{0.002}{\ln(1.219)}\left[(1.219)^{20} - (1.219)^{11}\right]$$

$$\approx 0.441$$

This means that from 2000 to 2009, the total increase in energy produced in the United States via wind was about 0.441 quadrillion British thermal units.

b. To recover the amount of energy produced via wind function F, we begin by evaluating the indefinite integral $\int f(x)\,dx$. We then use the information that in 1995 $(x = 6)$, $F(x) = 0.03$ to determine the constant of integration. Notice that evaluating the indefinite integral, $\int 0.002(1.219)^x dx$ will yield the same antiderivative as in part (a), except here we have an arbitrary constant C. This means we should have

$$F(x) = \int 0.002(1.219)^x dx = 0.002 \int e^{[\ln(1.219)]\cdot x} dx$$

$$= \frac{0.002}{\ln(1.219)}[e^{[\ln(1.219)\cdot x]}] + C$$

$$= \frac{0.002}{\ln(1.219)}[(1.219)^x] + C$$

We now use the information that when $x = 6$, $F(x) = 0.03$ to determine C.

$$F(x) = \frac{0.002}{\ln(1.219)}[(1.219)^x] + C$$

$$0.03 = \frac{0.002}{\ln(1.219)}[(1.219)^6] + C$$

$$0.03 - \frac{0.002}{\ln(1.219)}[(1.219)^6] = C$$

$$-0.003 \approx C$$

So the amount of energy produced via wind in the United States can be modeled by

$$F(x) = \frac{0.002}{\ln(1.219)}[(1.219)^x] - 0.003 \qquad 1 \le x \le 20$$

where x is the number of years since 1989, and $F(x)$ is the amount of energy produced via wind in the United States measured in quadrillion British thermal units.

c. Using the result from part (b), we substitute $x = 16$ and find that

$$F(x) = \frac{0.002}{\ln(1.219)}[(1.219)^x] - 0.003$$

$$F(16) = \frac{0.002}{\ln(1.219)}[(1.219)^{16}] - 0.003$$

$$\approx 0.237$$

This means that in 2005, about 0.237 quadrillion British thermal units of energy were produced in the United States via the wind. ∎

Summary

In this section we analyzed and applied integration formulas for the exponential and logarithmic functions.

- **Integration formula for $\frac{1}{x}$:** $\displaystyle\int \frac{1}{x}\,dx = \ln|x| + C$

- **Integration formula for $\frac{1}{u}$:** $\displaystyle\int \frac{1}{u}\,du = \ln|u| + C$

- **Integration formula for e^x:** $\int e^x dx = e^x + C$

- **Integration formula for e^u:** $\int e^u du = e^u + C$

- **Integration formula for e^{kx}:** $\int e^{kx} dx = \frac{1}{k}e^{kx} + C$

Section 6.5 Exercises

Vocabulary Exercises

1. The function $F(x) = \ln|x|$ is called the _____ logarithm function.
2. The natural logarithm function $F(x) = \ln|x| + C$ is defined by the integral $\int f(x)dx$ where $f(x) = $ _____.
3. The value of $\ln a$ can be represented by the area under $g(x) = \frac{1}{x}$ on the closed interval _____.
4. The function $f(x) = e^x$ is referred to as the _____ exponential function.
5. The integral of the function $f(x) = e^x$ is $F(x) = $ _____.
6. A function that is always increasing or always decreasing is classified as _____.

Skill Exercises

For Exercises 7–16, evaluate the indefinite and definite integrals.

7. $\int \dfrac{2}{x} dx$

8. $\int \dfrac{5}{x} dx$

9. $\int \dfrac{-2}{3x} dx$

10. $\int \dfrac{-7}{2x} dx$

11. $\int \dfrac{1 + e}{x} dx$

12. $\int \dfrac{e^2 - 1}{x} dx$

13. $\int_1^3 \dfrac{3}{x} dx$

14. $\int_1^8 \dfrac{5}{x} dx$

15. $\int_2^5 \dfrac{1}{2}x^{-1} dx$

16. $\int_5^8 \dfrac{2}{5}x^{-1} dx$

For Exercises 17–30, use u-substitution to evaluate the integrals.

17. $\int \dfrac{1}{2x + 3} dx$

18. $\int \dfrac{1}{7 - 5x} dx$

19. $\int \dfrac{x}{x^2 + 1} dx$

20. $\int \dfrac{2x}{x^2 + 3} dx$

21. $\int \dfrac{x - 2}{x^2 - 4x + 9} dx$

22. $\int \dfrac{3x^2 - 3}{x^3 - 3x - 7} dx$

23. $\int \dfrac{x^3}{x^4 + 10} dx$

24. $\int \dfrac{3x^2}{x^3 - 4} dx$

25. $\int \dfrac{1}{x \cdot \ln 2x} dx$

26. $\int \dfrac{3}{x \cdot \ln 6x} dx$

27. $\int_1^2 \dfrac{2}{\sqrt{x}(\sqrt{x} + 4)} dx$

28. $\int_1^2 \dfrac{\ln x}{x} dx$

29. $\int_1^3 \dfrac{2}{3x - 2} dx$

30. $\int_{-1}^0 \dfrac{1}{4 - 5x} dx$

In Exercises 31–40, evaluate the indefinite and definite integrals.

31. $\displaystyle\int 2e^x\,dx$

32. $\displaystyle\int 6e^x\,dx$

33. $\displaystyle\int (e^x - 1)\,dx$

34. $\displaystyle\int (e^x - 3)\,dx$

35. $\displaystyle\int (4e^x + x - 1)\,dx$

36. $\displaystyle\int (x^3 - 3e^x + x)\,dx$

37. $\displaystyle\int_0^1 (e^x - 1)\,dx$

38. $\displaystyle\int_0^1 (6 - e^x)\,dx$

39. $\displaystyle\int_1^2 \frac{e^x + 4}{2}\,dx$

40. $\displaystyle\int_1^2 \frac{3 - e^x}{6}\,dx$

In Exercises 41–58, evaluate the indefinite and definite integrals by u-substitution.

41. $\displaystyle\int e^{4x+1}\,dx$

42. $\displaystyle\int e^{8x}\,dx$

43. $\displaystyle\int 2xe^{x^2+1}\,dx$

44. $\displaystyle\int xe^{x^2}\,dx$

45. $\displaystyle\int \frac{e^x}{e^x + 2}\,dx$

46. $\displaystyle\int \frac{e^x - e^{-x}}{e^x + e^{-x}}\,dx$

47. $\displaystyle\int \frac{3}{e^x(1 - e^{-x})}\,dx$

48. $\displaystyle\int x^2 e^{x^3}\,dx$

49. $\displaystyle\int \frac{e^{\sqrt{x}}}{\sqrt{x}}\,dx$

50. $\displaystyle\int \frac{(e^x + 1)^2}{e^x}\,dx$

51. $\displaystyle\int_1^3 e^{-4x}\,dx$

52. $\displaystyle\int_0^1 e^{2x+3}\,dx$

53. $\displaystyle\int_0^1 \frac{1 + e^x}{e^x}\,dx$

54. $\displaystyle\int_1^2 \left(e^{2x} - \frac{2}{x}\right)dx$

55. $\displaystyle\int xe^{x^2}\,dx$

56. $\displaystyle\int 3x^5 e^{x^6}\,dx$

57. $\displaystyle\int e^{4x}(e^{4x} + 4)\,dx$

58. $\displaystyle\int e^x(e^x + 10)\,dx$

For Exercises 59–68, evaluate the indefinite and definite integrals.

59. $\displaystyle\int 5^x\,dx$

60. $\displaystyle\int 10^x\,dx$

61. $\displaystyle\int 5\left(\frac{1}{2}\right)^x dx$

62. $\displaystyle\int 2\left(\frac{5}{8}\right)^x dx$

63. $\displaystyle\int 7^{-x}\,dx$

64. $\displaystyle\int 10^{-x}\,dx$

65. $\displaystyle\int_1^2 10^{3x}\,dx$

66. $\displaystyle\int_1^5 5^{-2x}\,dx$

67. $\displaystyle\int_{-1}^1 2^{3x-1}\,dx$

68. $\displaystyle\int_1^{\sqrt{2}} x \cdot 2^{x^2}\,dx$

Application Exercises

69. Management—Sales Rates: The Top-2-Bottom Dress Store has a clearance sale to remove old inventory. Past records show that the rate of sales is given by

$$S(t) = \frac{40}{t} \qquad 1 \le t \le 8$$

where t represents the number of days that the sale has been running, and $S(t)$ represents the sales rate, measured in dresses per day.

(a) Evaluate $S(2)$ and interpret.

(b) Determine the total increase in the number of dresses sold between days 2 and 4.

70. **Management—Sales Rates:** The Silver Spur Gun Shop runs a last-chance sale based on the sales rate model

$$S(t) = \frac{82}{t} \qquad 1 \le t \le 10$$

where t represents the number of days that the sale has been running, and $S(t)$ represents the sales rate, measured in guns per day.

(a) Evaluate $S(4)$ and interpret.

(b) According to the model, if the store owner must sell 150 guns in the first five days to break even, will the owner do so?

71. **Microeconomics—Marginal Cost:** The FlowStop Company determines that the marginal cost for their new line of faucet parts is given by

$$MC(x) = 1.50 + 0.04e^{0.02x}$$

(a) Evaluate $MC(20)$ and interpret.

(b) If the number of units of the part produced increases from 100 to 150, what is the total increase in cost?

72. **Population—Growth Rate:** The population of a small town is growing at a rate given by

$$R(t) = 250e^{0.04t}$$

where t represents the number of years from the present, and $R(t)$ represents the town's population growth rate, measured in people per year.

(a) Evaluate and interpret $\int_0^3 R(t)\,dt$.

(b) A civil planner estimates that 3000 more people can move into the town without negatively affecting the quality of life. How many more years will it take for the population to reach that number?

 73. **Real Estate—Mobile Homes:** The housing crisis of the late 2000s not only affected traditional houses, but mobile homes as well. "With slimmer paychecks, people are wary to take out loans for mobile and manufactured homes, even though they are generally more affordable than stick-built housing," said James Hargraves, owner of Penestar Homes. (*Source: The Shelby Star.*) The rate of change in the number of mobile homes manufactured in the southern United States from 2001 to 2011 can be modeled by

$$f(x) = \frac{-54.1}{x} \qquad 1 \le x \le 11$$

where x represents the number of years since 2000, and $f(x)$ represents the rate of change in the number of mobile homes manufactured, measured in thousands per year. (*Source:* U.S. Census Bureau.)

(a) Evaluate $f(5)$ and interpret.

(b) Integrate to determine the total decline in Southern mobile home manufacturing from 2005 to 2010.

 74. **Real Estate—Mobile Homes:** Another region stricken by the decline of mobile home sales is the Midwest. The rate of change in the number of mobile homes manufactured in the midwestern United States from 2001 to 2011 can be modeled by

$$f(x) = \frac{-19.2}{x} \qquad 1 \le x \le 11$$

where x represents the number of years since 2000, and $f(x)$ represents the rate of change in the number of mobile homes manufactured, measured in thousands per year. (*Source:* U.S. Census Bureau.)

(a) Evaluate $f(1)$ and interpret.

(b) Integrate to determine the total decline in Midwestern mobile home manufacturing from 2005 to 2010.

75. Political Economics—Food Stamp Program: Fueled by rising unemployment and food prices, the number of Americans on food stamps has been steadily increasing. The number of participants in the federal food stamp program from 2000 to 2010 can be modeled by

$$f(x) = 1.06e^{0.06x} \qquad 0 \le x \le 10$$

where x represents the number of years since 2000, and $f(x)$ represents the rate of change in the number of participants in the food stamp program, measured in millions per year. (*Source:* U.S. Department of Agriculture.)

(a) How rapidly was the number of food stamp recipients increasing in 2008?

(b) Evaluate $\int_0^8 f(x)\,dx$ and interpret.

76. Ecology—Atlantic Fish Catches: The global expansion of fisheries in the 1950s and 1960s was coupled with ever-growing catches at a rate so rapid that they tended to exceed fish population growth. (*Source:* Institute of Development Research.) The rate of change in the annual amount of fish caught by Norwegians from 2000 to 2010 in the northwest Atlantic Ocean can be modeled by

$$f(x) = -6624.42e^{-0.22x} \qquad 0 \le x \le 10$$

© iStockphoto

where x represents the number of years since 2000, and $f(x)$ represents the rate of change in the amount of fish caught in tons per year. (*Source:* EuroStat.)

(a) Evaluate $f(3)$ and interpret.

(b) Evaluate $\int_3^5 f(x)\,dx$ and interpret.

77. Finance—Subprime Loans: Subprime loans are offered to individuals who do not qualify for prime-rate loans based on their credit score. The proliferation of the loans in the early 2000s was generally regarded as the trigger for the banking crisis in 2008. "Because of the subprime mortgage debacle, we see no chance that a strong rebound in new-home sales will be a key driver of broader economic growth any time soon," said chief U.S. economist Ian Shepherdson. (*Source:* Money Morning.) The rate of change in the total amount financed in subprime loans for housing from 2000 to 2007 can be modeled by the rate function

$$f(x) = 38.37 \cdot (1.43)^x \qquad 0 \le x \le 7$$

where x represents the number of years since 2000, and $f(x)$ represents the rate of change in the total financed in subprime loans, measured in billions of dollars per year. (*Source:* Center for Responsible Lending.)

(a) Use the model to determine how rapidly the total in subprime loans was increasing in 2003.

(b) Evaluate and interpret $\int_3^6 f(x)\,dx$.

Concept and Writing Exercises

78. Explain in a sentence why we cannot use the power rule to differentiate the exponential function. In other words, why is $\frac{d}{dx}(e^x) \neq e^{x-1}$?

79. Explain in a sentence why we cannot use the Power Rule of Integration to determine $\int \frac{1}{x}\,dx$.

In Exercises 80–83, integrate knowing that that a, b, and k represent real number constants where $k > 1$.

80. $\displaystyle\int \frac{a}{x}\,dx$ **81.** $\displaystyle\int \frac{ax^k}{x^{k+1}+b}\,dx$ **82.** $\displaystyle\int x^{k-1}e^{x^k}\,dx$ **83.** $\displaystyle\int k^{ax+b}\,dx$

Section Project

Sub-Saharan Africa refers to the area of the African continent that lies south of the Sahara. This area has seen tremendous economic growth since the beginning of the 21st century. For example, "Nigeria's economic growth in relation to other countries has been on a steady path. In 2009 when most countries showed negative growth, the country had 6.9 per cent of growth," said Minister of Trade and Investment Olusegun Aganga. (*Source:* Leadership Nigeria.) The rate of change in real gross domestic product (GDP) per capita in sub-Saharan African countries from 2000 and projected through 2030 can be modeled by

$$f(x) = 28.7(1.05)^x \qquad 0 \le x \le 30$$

where x represents the number of years since 2000, and $f(x)$ represents the rate of change in per capita GDP, measured in U.S. dollars per year. (*Source:* International Monetary Fund World Economic Outlook.)

(a) Evaluate and interpret $f(3)$ and compare to $f(17)$.

(b) Divide $\dfrac{f(17)}{f(3)}$. By what factor did the per capita GDP increase between these years?

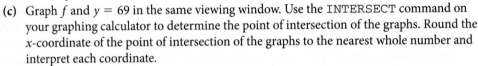 (c) Graph f and $y = 69$ in the same viewing window. Use the INTERSECT command on your graphing calculator to determine the point of intersection of the graphs. Round the x-coordinate of the point of intersection of the graphs to the nearest whole number and interpret each coordinate.

(d) Evaluate $\int_0^{10} f(x)\,dx$ and compare to $\int_{20}^{30} f(x)\,dx$. Divide the second integral value by the first. By what factor did the per capita GDP increase between these two decades?

(e) Knowing that the sub-Saharan per capita GDP was 574 U.S. dollars in 2000, integrate to recover a function of the form

$$F(x) = ae^{kx} \qquad 0 \le x \le 30$$

where x represents the number of years since 2000, and $F(x)$ represents the sub-Saharan per capita GDP, measured in USD. Round a and k to three decimal places.

(f) Divide $\dfrac{F(30)}{F(12)}$ and interpret.

Section 6.1 Review Exercises

In Exercises 1–4, determine if the function F is an antiderivative of the function f.

1. $F(x) = x^2 + C;\ f(x) = x$
2. $F(x) = x^3 - 5x + C;\ f(x) = 3x^2 - 5$
3. $F(x) = \dfrac{1}{5}x^5 + \dfrac{1}{4}x^4 + \dfrac{1}{3}x^3 + C;\ f(x) = x^4 + x^3 + x^2$
4. $F(x) = \dfrac{2}{x^2} + 4x + C;\ f(x) = \dfrac{6}{x^4} + 4$

In Exercises 5–8, use the Power Rule for Integration to find the indefinite integrals.

5. $\displaystyle\int x^7\,dx$

6. $\displaystyle\int x^{5.13}\,dx$

7. $\displaystyle\int \dfrac{1}{x^5}\,dx$

8. $\displaystyle\int \dfrac{1}{\sqrt[4]{x^3}}\,dx$

For Exercises 9–12, determine the indefinite integral.

9. $\displaystyle\int (0.5x^7 - 6x)\,dx$

10. $\displaystyle\int (6x^5 + 4x^4 - 3)\,dx$

11. $\displaystyle\int (x^2 + 0.4x - \sqrt{x^5})\,dx$

12. $\displaystyle\int \dfrac{5x^2 + 3x}{x}\,dx$

In Exercises 13 and 14, the rate function f' is shown. Use the graph to determine the intervals where its corresponding function F is increasing or decreasing, and where the graph of F is concave up or down and where the relative maximum and minimum values occur.

13.

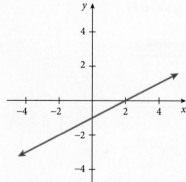

14.

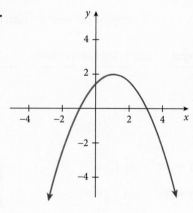

For Exercises 15–18, solve the given problem given the rate function and the given value. Note that $f(a) = b$ corresponds to the ordered pair (a, b).

15. $f'(x) = 3x - 4;\ f(0) = 11$

16. $\dfrac{dy}{dx} = x^2 - 5x;\ y = 10$ when $x = 6$

17. $f'(x) = \sqrt{x};\ (x, y) = (9, -5)$

18. $P'(x) = 0.5 + 0.15\sqrt[3]{x^4};\ P(27) = 3.1$

19. **Microeconomics—Marginal Profit:** Consider the marginal profit (in dollars per unit) for producing q units of a product given by the linear function

$$MP(q) = 48 - 0.03q$$

(a) Knowing that $P = 1600$ when $q = 100$, recover the profit function P.

(b) Use your solution from part (a) to find the total profit realized from selling 500 units of product.

20. **Microeconomics—Marginal Revenue:** The monthly marginal revenue function for the Byrne Rubber Tire Company is given by

$$MR(x) = R'(x) = -0.3x^2 - 1.4x + 35.2$$

where x is the number in thousands of tires produced and sold, and $MR(x)$ is the marginal revenue in thousands of dollars per thousand tires.

(a) Knowing that $R(x) = 0$ when $x = 0$, recover the revenue function R.

(b) Find the price-demand function p for the tires. Recall that the general formula for the revenue function is

$$R(x) = (\text{quantity produced}) \cdot (\text{price of each unit}) = x \cdot p(x)$$

(c) What should the price be when the demand is 6000 tires (that is, $x = 6$)?

Section 6.2 Review Exercises

In Exercises 21–24, determine the step size Δx for the given interval $[a, b]$ and number of subintervals n.

21. $[0, 9]; \ n = 27$

22. $[-5, 5]; \ n = 10$

23. $[7, 11]; \ n = 32$

24. $[-4, 9]; \ n = 39$

In Exercises 25–28, calculate (a) the Left Sum and (b) the Right Sum to approximate the area under the graph of f on the closed interval $[a, b]$ using n equally spaced subintervals.

25. $f(x) = 5x, \ [0, 5], \ n = 5$

26. $f(x) = 3x - 8, \ [6, 8], \ n = 4$

27. $f(x) = x^2, \ [0, 3], \ n = 6$

28. $f(x) = 3x^2 + 4, \ [-1, 1], \ n = 4$

In Exercises 29–32, complete the table for each of the functions f on $[a, b]$.

N Rectangles	LEFTSUM *Approximation*	RTSUM *Approximation*
10		
100		
1000		

 29. $f(x) = 5x^2 - 3; \ [0, 8]$

30. $f(x) = \sqrt{x - 2}; \ [3, 5]$

31. $f(x) = 3x^4 - 5x^3; \ [0, 4]$

32. $f(x) = 0.6x^2 - 1.3x + 7.3; \ [0, 10]$

In Exercises 33–38, evaluate the definite integrals.

33. $\displaystyle\int_3^7 6x \, dx$

34. $\displaystyle\int_0^3 (3x - 4) \, dx$

35. $\displaystyle\int_{-2}^2 (10 - x^2) \, dx$

36. $\displaystyle\int_{-2}^8 (x^5 + 9x^2 - 2) \, dx$

37. $\displaystyle\int_0^5 (2.7x^2 - 1.4x + 0.8) \, dx$

38. $\displaystyle\int_1^{64} x(\sqrt{x} - \sqrt[3]{x}) \, dx$

In Exercises 39–41, determine the area of the shaded region.

39.

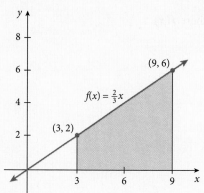

40.

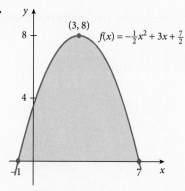

41.

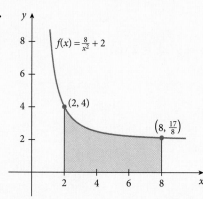

42. Consider the area under the graphs of f and g.

(a) Determine the area under the shaded region where $f(x) = 5\sqrt[3]{x}$.

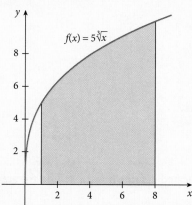

(b) Determine the area under the shaded region where $g(x) = 2\sqrt[3]{x}$.

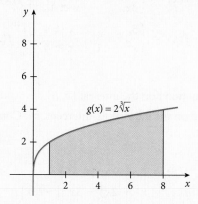

43. Consider the area under the graphs of f and g.

(a) Determine the area under the shaded region where $f(x) = \frac{x^2}{4}$.

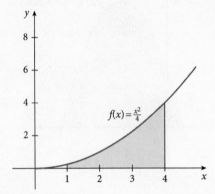

(b) Determine the area under the shaded region where $g(x) = 2\sqrt{x}$.

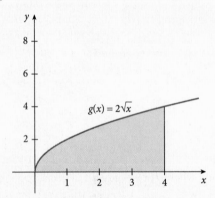

44. Consider the absolute value function $f(x) = |x| - 3$.

(a) Graph f in the standard viewing window.

(b) Rewrite f as a piecewise-defined function.

(c) Determine the x-intercepts of the graph of f.

(d) Evaluate $\int_{-2}^{3} f(x)\,dx$ and compare to $\int_{3}^{8} f(x)\,dx$.

For Exercises 45–48, complete the following.

(a) Graph the piecewise-defined function f.

(b) Use the interval addition rule to evaluate the indicated definite integral to find the area between the graph of f and the x-axis.

45. $f(x) = \begin{cases} -2x, & x < 0 \\ 3x, & x \geq 0 \end{cases}; \quad \int_{-5}^{2} f(x)\,dx$

46. $f(x) = \begin{cases} 6 - 3x, & x < 1 \\ 3x, & x \geq 1 \end{cases}; \quad \int_{0}^{6} f(x)\,dx$

47. $f(x) = \begin{cases} x^2 - 5, & x \leq -3 \\ x + 7, & x > -3 \end{cases}; \quad \int_{-5}^{2} f(x)\,dx$

48. $f(x) = \begin{cases} 5, & x < 4 \\ x^2 - 11, & x \geq 4 \end{cases}; \quad \int_{0}^{10} f(x)\,dx$

For Exercises 49–52, complete the following.

(a) Determine the x-intercepts of the graph of the function f on the interval $[a, b]$.

(b) Integrate each on the interval $[a, x\text{-intercept}]$ and on the interval $[x\text{-intercept}, b]$. Compare the results.

(c) Interpret each integral from part (b) graphically.

49. $f(x) = x - 3; \ [0, 10]$

50. $f(x) = 2 - \sqrt{x}; \ [0, 10]$

51. $f(x) = x^2 - x - 2; \ [-1, 5]$

52. $f(x) = x^3 + 2x^2 - 3x; \ [-2, 1]$

Section 6.3 Review Exercises

53. **Microeconomics—Marginal Cost:** The AddEmUp Calculator Company determines that the marginal cost function for a new calculator is given by

$$MC(x) = C'(x) = \frac{1}{20}x + 12 \qquad 0 \le x \le 300$$

where x is the number of calculators produced per day, and $MC(x)$ is the marginal cost in dollars per calculator.

(a) Evaluate $C'(160)$ and interpret.

(b) Evaluate $\int_0^{200} C'(x)\,dx$ and interpret.

54. **Microeconomics—Marginal Profit:** The Voll-E Ball Company determines that the marginal profit function for selling a new kind of football is

$$MP(x) = P'(x) = 0.0002x^2 - 0.01x + 2.85 \qquad 0 \le x \le 100$$

where x is the number of footballs made and sold, and $MP(x)$ is the marginal profit function in dollars per football.

(a) Knowing that the company breaks even if 20 footballs are sold (that is, the profit for making and selling 20 footballs is \$0), recover the profit function P.

(b) Evaluate $\int_{50}^{100} P'(x)\,dx$ and interpret.

55. **Patent Law—Foreign Patents:** During the 2000s, the number of U.S. patents issued to residents of foreign countries, particularly China, increased significantly. "There's been a significant influx of technology, and the Chinese are starting to make improvements. They are becoming innovators, not just copiers," said Chad Nydegger, attorney of the Changzhou Electronic Technology Company. (*Source:* General Patent Corporation Newsletter.) The rate of change in the number of patents issued to residents of foreign countries from 1990 to 2011 can be modeled by

$$f(x) = 0.03x^2 + 0.46x + 1.01 \qquad 0 \le x \le 21$$

where x represents the number of years since 1990, and $f(x)$ represents the rate of change in the number of patents issued, measured in thousands per year. (*Source:* U.S. Patent and Trademark Office.)

(a) Evaluate $f(15)$ and interpret.

(b) Determine the total increase in patents issued to foreign residents from 1990 to 2000.

(c) Compute $\int_{10}^{20} f(x)\,dx$ and compare to the solution of part (b).

56. **Physics—Object Displacement:** The velocity function for an object, in feet per second, is

$$v(t) = 6t - 9$$

(a) Evaluate $v(t)$ at $t = 2$.

(b) Evaluate and interpret $\int_{50}^{100} v(t)\,dt$.

Section 6.4 Review Exercises

In Exercises 57–60, use the u-substitution method to evaluate the indefinite integrals. Check by differentiating the solution.

57. $\displaystyle\int 5(5x - 7)^3\,dx$

58. $\displaystyle\int 3x^2(x^3 + 5)^4\,dx$

59. $\displaystyle\int (2x - 5)(x^2 - 5x + 3)\,dx$

60. $\displaystyle\int \frac{2x}{\sqrt{x^2 - 5}}\,dx$

In Exercises 61–64, use the u-substitution method to evaluate the indefinite integrals.

61. $\displaystyle\int (3x + 6)^8 dx$

62. $\displaystyle\int x^4 (2x^5 - 7)^3 dx$

63. $\displaystyle\int \frac{x + 2}{x^2 + 4x + 6} dx$

64. $\displaystyle\int (8x - 28)(x^2 - 7x + 9)^3 dx$

In Exercises 65–68, determine the integrals by solving the u-expression to complete the substitution.

65. $\displaystyle\int 3x(x + 4)^2 dx$

66. $\displaystyle\int x(x - 1)^6 dx$

67. $\displaystyle\int \frac{x}{\sqrt{x + 3}} dx$

68. $\displaystyle\int x\sqrt{x - 5}\, dx$

For Exercises 69–72, evaluate the definite integrals, and evaluate the limits of integration in terms of x.

69. $\displaystyle\int_{10}^{17} 2x\sqrt{x^2 - 64}\, dx$

70. $\displaystyle\int_{0}^{12} (x - 4)^3 dx$

71. $\displaystyle\int_{2}^{11} \frac{3x^2}{\sqrt{x^3 - 4}} dx$

72. $\displaystyle\int_{5}^{20} \sqrt[4]{21 - x}\, dx$

In Exercises 73–76, evaluate the definite integrals. Change the limits of integration to u-limits.

73. $\displaystyle\int_{1}^{4} \frac{2x - 5}{x^2 - 5x + 1} dx$

74. $\displaystyle\int_{0}^{2} (3x^2 - 4)(x^3 - 4x + 7)^4 dx$

75. $\displaystyle\int_{4}^{8} 3x^2\sqrt{x^3 + 17}\, dx$

76. $\displaystyle\int_{5}^{9} \frac{2x}{\sqrt{x^2 + 144}} dx$

77. Determine the exact area of the shaded region, where $f(x) = \dfrac{26x}{\sqrt{x^2 + 144}}$.

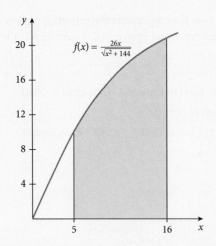

$f(x) = \frac{26x}{\sqrt{x^2 + 144}}$

78. Microeconomics—Marginal Profit: The marginal profit function for hard disks produced by the Elephant Media Company is given by

$$MP(x) = \frac{80x}{\sqrt{x^2 + 576}} \qquad 0 \le x \le 100$$

where x is the number of hard disks produced in a work shift, and $MP(x)$ is the marginal profit.

(a) Evaluate and interpret $MP(7)$.

(b) Evaluate $\int_{7}^{45} MP(x)\, dx$ and interpret.

(c) Knowing that the Elephant Media Company breaks even when 32 hard disks are sold (that is, $P(x) = 0$ when $x = 32$), recover the profit function P.

79. **Physics—Object Displacement:** Suppose that the velocity of an object is given by

$$v(t) = (t^3 - 5t)^3(3t^2 - 5)$$

where t is the time in seconds, and $v(t)$ is the velocity in meters per second.

(a) Determine the total distance moved between $t = 3$ and $t = 6$ seconds.

(b) Knowing that when $t = 3$ seconds, $s(t) = 736$ meters, determine the position function s.

Section 6.5 Review Exercises

In Exercises 80–87, evaluate the indefinite and definite integrals.

80. $\int \dfrac{3}{x} dx$

81. $\int \dfrac{3}{4} x^{-1} dx$

82. $\int 3e^x dx$

83. $\int (5e^x - 3x) dx$

84. $\int \left(\dfrac{1}{2}\right)^x dx$

85. $\int_3^5 \dfrac{6}{x} dx$

86. $\int_0^5 \dfrac{e^x - 5}{3} dx$

87. $\int_0^2 5^x dx$

In Exercises 88–95, evaluate the integrals using u-substitution.

88. $\int \dfrac{2x + 5}{x^2 + 5x - 3} dx$

89. $\int \dfrac{5}{\sqrt{x}(\sqrt{x} + 4)} dx$

90. $\int e^{3x-5} dx$

91. $\int e^x (5 - e^x)^3 dx$

92. $\int 3x^2 \cdot 5^{x^3} dx$

93. $\int_0^3 \dfrac{2x}{x^2 + 16} dx$

94. $\int_3^7 e^{2x-6} dx$

95. $\int_{-2}^2 2^{3x} dx$

96. **Marketing—Sales Rates:** The PaperCut Discount Stationery Store is having a sale on pen and pencil sets. The rate of sales is given by

$$S(t) = \dfrac{60}{t} \qquad 1 \le t \le 10$$

where t represents the number of days that the sale has been running, and $S(t)$ represents the sales rate, measured in pen and pencil sets per day.

(a) Evaluate $S(4)$ and interpret.

(b) Use an integral to determine the total increase in the number of pen and pencil sets sold between days 4 and 8.

97. **Public Economics—U.S. Book Value:** Many are concerned that the United States cannot cover its own domestic debt. "This is not true," said Joe Weisenthal of Business Insider. "Even if there were no foreign buyers of U.S. debt, the U.S. has plenty of domestic savings." (*Source:* Business Insider.) The rate of change in the amount of total net worth of households and non-profits in the United States, called the private sector's "book value," from 1950 to 2011 can be modeled by

$$f(x) = 0.114 e^{0.055x} \qquad 0 \le x \le 61$$

where x represent the number of years since 1950, and $f(x)$ represents the rate of change in book value, measured in trillions of dollars per year. (*Source:* Board of Directors of the Federal Reserve System.)

(a) Evaluate $f(30)$ and $f(60)$, and compare the results.

(b) Evaluate $\int_0^{30} f(x)\,dx$ and interpret.

(c) Evaluate $\int_{30}^{60} f(x)\,dx$ and compare with the result of part (b).

Applications of Integral Calculus

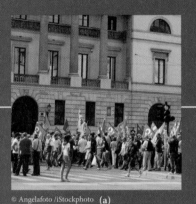

© Angelafoto /iStockphoto **(a)**

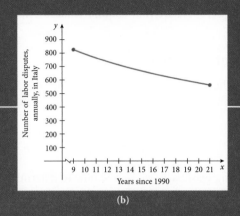

(b)

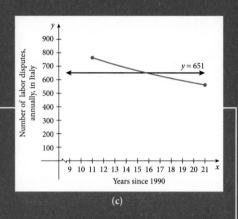

(c)

The banking crisis of 2008 caused problems in many sectors around the world. Many European countries, including Italy, were forced to austerity programs to curtail their government debts. Some of this austerity included making adjustments to pension and retirement benefits. These adjustments caused concern among some union leaders. Figure (b) shows the number of labor disputes in Italy from 1999 to 2011. In Figure (c), the horizontal line gives the average number of labor disputes, annually, from 2001 to 2011.

What We Know

In Chapter 6 we saw how the definite integral can be used to give a total accumulation. We also saw how integration and differentiation were related through the Fundamental Theorem of Calculus.

Where Do We Go

In this chapter we will see even more applications of integration. Applications will include average value, perpetuities, the Gini coefficient, consumers' and producers' surplus, and many others. We continue to strengthen the concepts set forth in Chapter 6, especially an accumulation.

SECTION OBJECTIVES

1. Determine an average value.

2. Apply an average value to a bank account.

3. Determine total income from an income stream.

4. Determine the future value of an annuity.

7.1   Average Value of a Function and the Definite Integral in Finance

In this section we examine the **average value** of a function. Compound interest will be revisited as we examine what is known as an **income stream**. The process of computing an average value for a function over an interval will be exploited throughout Chapter 7, and the concept of an income stream will also reappear in later sections of this chapter. We encourage you to work to understand these concepts as they arise, to ensure your continued success in these later sections.

Average Value of a Function

Before we analyze how to determine an average value of a function, let's first recall that the average (or mean) of n numbers, $y_1,\ y_2,\ y_3,\ \dots,\ y_n,$ is simply the sum of the numbers divided by n. That is,

$$\bar{y} = \frac{1}{n}\sum_{i=i}^{n} y_i = \frac{y_1 + y_2 + y_3 + \cdots + y_n}{n}$$

In the following Flashback, we review how to determine the average of a set of numbers. The Flashback also provides the necessary foundation for the definition of the **average value of a function.**

Flashback: Domestic Shrimp Catch Revisited

In Section 5.5 we analyzed the total domestic shrimp catch in the United States, measured in millions of pounds, from 2000 to 2008. The data are given in **Table 7.1.1**. We can represent the data graphically by plotting the data points and then making a bar graph as shown in **Figure 7.1.1**.

Table 7.1.1

Year	Shrimp caught
2000	332.5
2001	259.6
2002	227.2
2003	224.1
2004	239.1
2005	260.9
2006	278.9
2007	280.9
2008	256.6

Source: U.S. National Oceanic and Atmospheric Administration

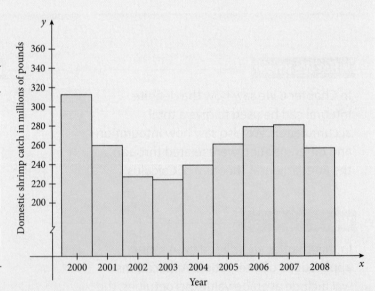

Figure 7.1.1

a. Using the data values, determine the average value of the domestic shrimp catch over the 9-year period.

b. Determine the total area of the nine rectangles in **Figure 7.1.1**.

Perform the Mathematics

a. The average value of the domestic shrimp catch over the 9-year period is given by

$$\bar{y} = \frac{1}{n} \sum_{i=i}^{n} y_i$$

$$= \frac{332.5 + 259.6 + 227.2 + 224.1 + 239.1 + 260.9 + 278.9 + 280.9 + 256.6}{9}$$

$$= \frac{2359.8}{9} = 262.2$$

So over the 9-year period from 2000 to 2008, the average value of the domestic shrimp catch was about 262.2 million pounds.

b. The base of the first rectangle is 1, and the height is the shrimp catch, 332.5. So the area of this rectangle is $1 \cdot 332.5 = 332.5$, which is just the shrimp catch for that year. Repeating this process for the other rectangles, we see that the total area of the nine rectangles is simply the sum of the values of the shrimp catch, 2359.8, which we determined in part (a). ∎

If we divide the area of the nine rectangles by the number of rectangles, 9, we get

$$\frac{2359.8}{9} = 262.2$$

which is the average value we found in the Flashback part (a). Thus, we can interpret this result to mean

$$\frac{\text{area under the bar graph}}{\text{width of the bar graph}} = \text{average value}$$

Now, if a continuous function f models the data, we can find the average value of the function over the interval $[a, b]$. Using the preceding result for bar graphs, it can be shown that

$$\frac{\text{area under the curve}}{\text{length of the interval}} = \text{average value}$$

Since we know that the area under a curve is given by $\int_a^b f(x)\,dx$, these results suggest the definition for the **average value of a continuous function f on the interval $[a, b]$**.

> **DEFINITION**
>
> **Average Value of a Continuous Function on $[a, b]$**
>
> The **average value** of a continuous function, f, on the interval $[a, b]$ is given by
>
> $$\frac{1}{b-a} \int_a^b f(x)\,dx$$

Example 1: Determining an Average Value

For the data given in the Flashback, the domestic shrimp catch in the United States can be modeled by

$$f(x) = -1.86x^3 + 31.39x^2 - 154.03x + 456.99 \qquad 1 \le x \le 9$$

where $x = 1$ corresponds to January 1, 2000, and $f(x)$ is the domestic shrimp catch in millions of pounds. Use the model to calculate the average value of the shrimp catch for the period January 1, 2000, to January 1, 2008. (*Source:* U.S. National Oceanic and Atmospheric Administration.)

> **OBJECTIVE 1**
>
> Determine an average value.

Perform the Mathematics

We first observe that $x = 1$ corresponds to January 1, 2000, and $x = 9$ corresponds to January 1, 2008. Thus, we need to determine the average value of $f(x) = -1.86x^3 + 31.39x^2 - 154.03x + 456.99$

on the interval $[1, 9]$. So we compute the average value to be

$$\frac{1}{9 - 1} \int_1^9 (-1.86x^3 + 31.39x^2 - 154.03x + 456.99)\,dx$$

$$= \frac{1}{8}\left[\frac{-1.86}{4}x^4 + \frac{31.39}{3}x^3 - \frac{154.03}{2}x^2 + 456.99x\right]\Bigg|_1^9$$

$$\approx 257.7 \text{ (rounded to the nearest tenth)}$$

This means that from January 1, 2000, to January 1, 2008, the average annual domestic shrimp catch in the United States was approximately 257.7 million pounds.   ∎

Notice that the average value determined in Example 1 was fairly close to what we found in the Flashback. Some discrepancy is expected, since the function in Example 1 is a model that *approximates* the data.

Example 2: Determining an Average Value

Determine the average value of $f(x) = x^3 - 2x^2 - 5x + 11$ on the interval $[-2, 4]$.

Perform the Mathematics

We compute the average value to be

$$\frac{1}{b - a} \int_a^b f(x)\,dx = \frac{1}{4 - (-2)} \int_{-2}^4 (x^3 - 2x^2 - 5x + 11)\,dx$$

$$= \frac{1}{6}\left[\frac{1}{4}x^4 - \frac{2}{3}x^3 - \frac{5}{2}x^2 + 11x\right]\Bigg|_{-2}^4$$

$$= \frac{1}{6}\left[\left(64 - \frac{128}{3} - 40 + 44\right) - \left(4 + \frac{16}{3} - 10 - 22\right)\right]$$

$$= \frac{1}{6}\left[\frac{76}{3} - \left(-\frac{68}{3}\right)\right] = \frac{1}{6}(48) = 8$$

So the average value of $f(x) = x^3 - 2x^2 - 5x + 11$ on $[-2, 4]$ is 8.   ∎

▶ *Try It Yourself*

Some related Exercises are 9 and 11.

The average value can be thought of as giving the typical function value for a function on the closed interval $[a, b]$. At this time, we offer the following analogy and comparison between a discrete average value and a continuous average value.

Types of values	Discrete	Continuous
Sum of values	$\sum_{i=1}^n y_i$	$\int_a^b f(x)\,dx$
Number of "points"	n	$b - a$
Average	$\frac{1}{n}\sum_{i=1}^n y_i$	$\frac{1}{b - a}\int_a^b f(x)\,dx$

Example 3: Determining an Average Value

Some Americans have become more conscious about their consumption of fat, especially saturated fat. As a result, more Americans have stopped drinking whole milk and replaced it with skim milk. The annual U.S. per capita consumption of whole milk can be modeled by

$$f(x) = 15.85e^{-0.035x} \qquad 1 \le x \le 29$$

where x is measured in years ($x = 1$ corresponds to January 1, 1980), and $f(x)$ is the annual per capita consumption of whole milk in gallons. Use the model to determine the average per capita consumption of whole milk, rounded to the nearest tenth, during the 1990s. (*Source:* U.S. Department of Agriculture.)

© Jones & Bartlett Learning

Understand the Situation

When we say "during the 1990s," we mean from January 1, 1990, to January 1, 2000. This means that we are to determine the average value of f on the interval $[11, 21]$.

Perform the Mathematics

Computing the average value gives us

$$\frac{1}{b-a}\int_a^b f(x)\,dx = \frac{1}{21-11}\int_{11}^{21} 15.85e^{-0.035x}\,dx = \frac{1}{10}\left[\frac{15.85}{-0.035}e^{-0.035x}\right]\Bigg|_{11}^{21}$$

$$= \frac{1}{10}\left[\frac{15.85}{-0.035}e^{-0.035(21)}\right] - \frac{1}{10}\left[\frac{15.85}{-0.035}e^{-0.035(11)}\right]$$

$$\approx 9.1$$

Interpret the Result

This means that during the 1990s the annual average whole milk consumption per capita was about 9.1 gallons.

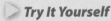 *Try It Yourself*

Some related Exercises are 25 and 57. ∎

Compound Interest

So far we have examined how the average value of a function is used in the life and social sciences. Now we turn our attention to economics, finance, and the managerial sciences. To do this, we need to recall a formula first encountered in Chapter 4 that is reviewed in the following Toolbox.

From Your Toolbox: Continuous Compound Interest Formula

If a principal of P dollars is invested into an account earning annual interest rate r (in decimal form) compounded continuously, then the amount A in the account at the end of t years is given by

$$A(t) = Pe^{rt}$$

Example 4: Determining an Average Balance and Computing a Bonus

a. Wilma deposits $5000 into a money market account earning 3.5% interest compounded continuously. Determine Wilma's average balance, to the nearest cent, over one year.

b. Suppose that the institution where Wilma has deposited her money also pays a bonus at the end of the year of 1.25% of the average balance in the account during one year. Compute the size of Wilma's bonus.

OBJECTIVE 2

Apply an average value to a bank account.

Perform the Mathematics

a.   From the formula for compound interest, we have

$$A(t) = 5000e^{0.035t}$$

We are asked to compute the average value of $A(t) = 5000e^{0.035t}$ on $[0, 1]$. So, we have

$$\frac{1}{b-a}\int_a^b A(t)\,dt = \frac{1}{1-0}\int_0^1 5000e^{0.035t}\,dt$$

$$= \frac{5000}{0.035}e^{0.035t}\Big|_0^1$$

$$= \frac{5000}{0.035}e^{0.035(1)} - \frac{5000}{0.035}e^{0.035(0)}$$

$$\approx 5088.53$$

Therefore, Wilma's average balance over one year was $5088.53.

b.   Computing 1.25% of the $5088.53 average balance from part (a) yields a bonus of

$$0.0125(5088.53) \approx 63.61$$

So Wilma's bonus was $63.61.   ∎

In Example 5, we review how to determine a total amount when given a rate of change function, as well as computing an average value.

Example 5: Determining an Account Change and an Average Balance

Stella deposits $7000 into an account where the rate of change of the amount in the account t years after the initial deposit is given by $A'(t) = 490e^{0.07t}$.

a.   Determine by how much the account changes from the end of the third year to the end of the fifth year. Round to the nearest cent.

b.   Determine the average balance, to the nearest cent, in the account over the first five years.

Perform the Mathematics

a.   Notice that we are given a **rate function**, $A'(t) = 490e^{0.07t}$. Recall from Chapter 6 that the total change in the account from the end of the third year to the end of the fifth year can be found by the definite integral. That is,

$$\int_3^5 490e^{0.07t}\,dt = \frac{490}{0.07}e^{0.07t}\Big|_3^5$$

$$= 7000e^{0.07(5)} - 7000e^{0.07(3)} \approx 1297.73$$

This means that Stella's account changed by about $1297.73 from the end of the third year to the end of the fifth year.

b.   Here we need to determine the average value of A on $[0, 5]$. The first order of business is to determine $A(t)$. To do this, we integrate

$$A(t) = \int A'(t)\,dt = \int 490e^{0.07t}\,dt$$

$$= \frac{490}{0.07}e^{0.07t} + C = 7000e^{0.07t} + C$$

So $A(t) = 7000e^{0.07t} + C$. To determine the constant C, we note that, at $t = 0$, $A(t) = 7000$. (An initial deposit corresponds to $t = 0$.) Substituting gives

$$A(t) = 7000e^{0.07t} + C$$
$$A(0) = 7000e^{0.07(0)} + C = 7000$$
$$7000 + C = 7000$$
$$C = 0$$

Since $A(t) = 7000e^{0.07t}$, we compute the average value of $A(t)$ on $[0, 5]$ as

$$\frac{1}{b-a}\int_a^b A(t)\,dt = \frac{1}{5-0}\int_0^5 7000e^{0.07t}\,dt$$
$$= \frac{1}{5}\left[\frac{7000}{0.07}e^{0.07t}\right]\Big|_0^5$$
$$= \frac{1}{5}\left[\frac{7000}{0.07}e^{0.07(5)}\right] - \frac{1}{5}\left[\frac{7000}{0.07}e^{0.07(0)}\right] \approx 8381.35$$

The average amount in Stella's account during the first five years was about \$8381.35. ∎

Continuous Stream of Income

The prior two examples reviewed the concept of continuous compounding. Continuous compounding does not mean that the bank or other institution where the money is invested is continuously placing money (the interest earned) into our account. However, we compute what is in our account as if it were. We can think of our account as being a **continuous income stream**. In fact, this is very similar to the electric company, the natural gas company, or the water company, which has a meter on our house that continuously monitors and totals the amount of electricity, natural gas, or water that we are consuming. Example 6 describes a continuous stream of income.

Example 6: Determining Total Income from an Income Stream

OBJECTIVE 3

Determine total income from an income stream.

Suppose that the rate of change of income, in thousands of dollars per year, for an oil field is projected to be $f(t) = 350e^{-0.09t}$, where t is the number of years from when oil extraction began. See **Figure 7.1.2**. Determine the total amount of money generated from this oil field during the first two years of operation.

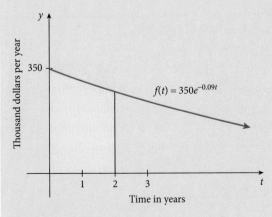

Figure 7.1.2 The area of the shaded region can be interpreted as giving the total money generated in the first 2 years.

Understand the Situation

Once again, we recognize that we have a rate function. To determine the total amount of money for the first two years, we need to determine the area under the curve on the interval [0, 2]. We use a definite integral to do this.

Perform the Mathematics

Integrating yields

$$\int_0^2 350e^{-0.09t}\,dt = \left[\frac{350}{-0.09}e^{-0.09t}\right]\Bigg|_0^2$$

$$= \frac{350}{-0.09}e^{-0.09(2)} - \frac{350}{-0.09}e^{-0.09(0)}$$

$$\approx 640.61584$$

Interpret the Result

So the first two years of operation will generate about $640,615.84 in income.

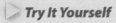 **Try It Yourself**

Some related Exercises are 41 and 43.

In reality, income from the oil field in Example 6 is not received in one lump-sum payment at the end of 2 years. The income is probably collected on a regular basis, possibly every month or quarter. In situations like Example 6, however, it is convenient to assume that the income is actually received in a **continuous stream**, that is, we assume that the income is a continuous function of time. The rate of change of income, $f(t) = 350e^{-0.09t}$ in Example 6, is called a **rate of flow function**.

DEFINITION

Total Income for a Continuous Income Stream

If f is the **rate of flow function** of a continuous income stream, then the **total income** produced from time $t = a$ to $t = b$ is found by

$$\int_a^b f(t)\,dt$$

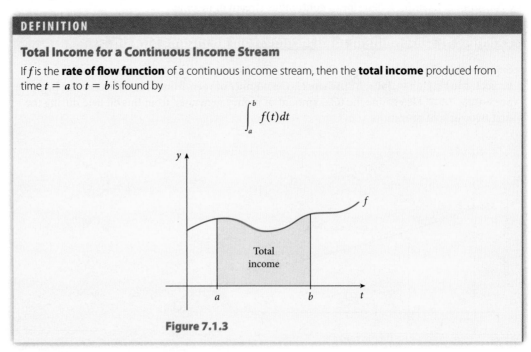

Figure 7.1.3

Annuities

We conclude this section with a look at what is known as an annuity. An **annuity** is defined as a sequence of equal payments made at equal time intervals. For example, if money is borrowed to finance the purchase of a home, a car, or even an education, the loan usually is repaid in equal monthly payments for a specified length of time. This sequence of equal payments forms an annuity. Also, payments that one receives at retirement from a pension plan, a 401(k), a 403(b), or an IRA are other examples of annuities.

Assume Maria deposits $2000 at the beginning of each year into an IRA and that her investment earns an annual interest rate of 6% compounded continuously. To compute the amount in her account after four years requires that we compute what happens to the initial deposit, the second deposit, the third deposit, and so on, for the four years. For example, using $A(t) = Pe^{rt}$, we have

- The initial deposit ($t = 0$) of $2000 is in the account for four years and grows to $2000e^{(0.06)(4)}$.
- The second deposit ($t = 1$) of $2000 is in the account for three years ($4 - 1$) and grows to $2000e^{(0.006)(4-1)} = 2000e^{0.06(3)}$.

Continuing in this manner, the total amount S in the account at the end of 4 years can be represented by

$$S = \left(\begin{array}{c}\text{value of initial payment} \\ t = 0\end{array}\right) + \left(\begin{array}{c}\text{value of second payment} \\ t = 1\end{array}\right)$$

$$+ \left(\begin{array}{c}\text{value of third payment} \\ t = 2\end{array}\right) + \left(\begin{array}{c}\text{value of final payment} \\ t = 3\end{array}\right)$$

$$S = 2000e^{0.06(4)} + 2000e^{0.06(4-1)} + 2000e^{0.06(4-2)} + 2000e^{0.06(4-3)}$$

$$S = 2000e^{0.06(4)} + 2000e^{0.06(3)} + 2000e^{0.06(2)} + 2000e^{0.06(1)}$$

$$S \approx 2542.50 + 2394.43 + 2254.99 + 2123.67 = \$9315.59$$

So the calculated total amount in the account at the end of four years is $9315.59. To see how we can approximate the sum, S, with a definite integral, we plot each term of the sum as shown in **Figure 7.1.4**.

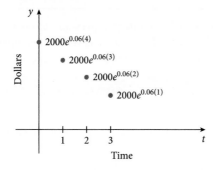

Figure 7.1.4

The sum of the amount in the IRA can be thought of as being equal to the sum of the area of the four rectangles shown in **Figure 7.1.5**, since the base (width) of each rectangle is 1. **Figure 7.1.6** shows the graph of $A(t) = 2000e^{0.06(4-t)}$ with the four rectangles. What we are about to do should look quite familiar by now. The area under A on the interval $[0, 4]$ is approximately equal to the area of the four rectangles. Hence, the area under the curve can be used to approximate the sum in the IRA at the end of four years. Example 7 illustrates the process.

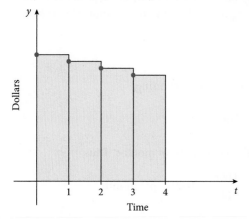

Figure 7.1.5

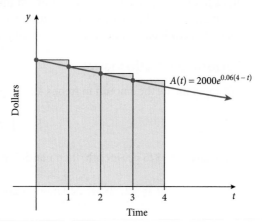

Figure 7.1.6

Example 7: Applying the Definite Integral

Write and evaluate a definite integral to find the area under the graph of A in **Figure 7.1.6** that approximates the sum in the IRA at the end of 4 years.

Perform the Mathematics

The definite integral that determines the area under the graph of A is

$$\int_0^4 2000e^{0.06(4-t)}dt$$

We evaluate the definite integral using u-substitution as follows:

Let $u = 4 - t$. Then $du = -dt$ or $-du = dt$.

By u-substitution, the integral becomes

$$-\int 2000e^{0.06u}du = -\left(\frac{2000}{0.06}e^{0.06u}\right) = -\left(\frac{2000}{0.06}e^{0.06(4-t)}\right)\Big|_0^4$$

$$= -\left(\frac{2000}{0.06}e^{0.06(4-4)} - \frac{2000}{0.06}e^{0.06(4-0)}\right)$$

$$\approx -(33{,}333.333 - 42{,}374.97) = 9041.64$$

So the approximate sum in the IRA at the end of four years is about $9041.64. ∎

Example 7 illustrates that the definite integral gives an *approximation* to the actual sum. With this in mind, the explanation preceding Example 7, as well as Example 7, outlines a procedure that we can now generalize to give an approximation for S, the **future value of an annuity**.

DEFINITION

Future Value of an Annuity

If n equal payments of P dollars are deposited into an annuity at an annual interest r (in decimal form), then the **future value**, S, for a duration of the annuity T is given by

$$S \approx \int_0^T Pe^{r(T-t)}dt$$

NOTE: It is very important to keep the units of time for T, r, and P consistent when using the formula to approximate the future value of an annuity. Example 8 illustrates.

OBJECTIVE 4

Determine the future value of an annuity.

Example 8: Determining a Future Value

Irene has just started a new job teaching. At the beginning of each month she deposits $25 into her 403(b) plan, which earns 6.6% annual interest compounded continuously. Approximate how much Irene will have in her account after 20 years.

Perform the Mathematics

We approximate the amount in Irene's 403(b) plan after 20 years by using

$$S \approx \int_0^T Pe^{r(T-t)}dt$$

If we have $P = \$25$ per month, then r and T must also be in terms of months. Thus,

$$T = 12(20) = 240 \text{ months and } r = \frac{0.066}{12} = 0.0055 \text{ per month}$$

So the approximation for S is

$$S \approx \int_0^T Pe^{r(T-t)}dt = \int_0^{240} 25e^{0.0055(240-t)}dt$$

Using a u-substitution yields $u = 240 - t$ and $du = -dt$ or $-du = dt$, and thus

$$\int_0^{240} 25e^{0.0055(240-t)}dt = -\int 25e^{0.0055u}du$$

$$= -\left(\frac{25}{0.0055}e^u\right) = -\left(\frac{25}{0.0055}e^{(240-t)}\right)\Big|_0^{240}$$

$$= -\left(\frac{25}{0.0055}e^{(240-240)} - \frac{25}{0.0055}e^{(240-0)}\right)$$

$$\approx -(4545.45 - 17{,}015.55)$$

$$= 12{,}470.10$$

So the total amount in Irene's 403(b) plan after 20 years is approximately \$12,470.10.   ∎

▶ *Try It Yourself*

Some related Exercises are 45 and 47.

Summary

In this introductory section on applications of the integral, we examined how recognizing a definite integral as a limit of a sum and recognizing the area under a curve as a definite integral were very important. The former was used to determine a formula for the average value, and the latter was used in determining a formula to approximate the future value of an annuity. We also saw how integrating a rate function gives a total accumulation. In this section, it was seen again in the context of **income streams**.

- The **average value** of a continuous function, f, on the interval $[a, b]$ is given by

$$\frac{1}{b-a}\int_a^b f(x)dx$$

- If f is the **rate of flow function** of a continuous income stream, then the **total income** produced from time $t = a$ to $t = b$ is found by

$$\int_a^b f(t)dt$$

- If n equal payments of P dollars are deposited into an annuity at an annual interest r (in decimal form), then the **future value**, S, for a duration of the annuity T is given by

$$S \approx \int_0^T Pe^{r(T-t)}dt$$

Section 7.1 Exercises

Vocabulary Exercises

1. The area under the curve of a function divided by the width of a closed interval gives the _____ _____ of the function on the interval.

2. The expression $\sum_{i=1}^n y$ gives the sum of values if the values y are _____.

3. The expression $\int_a^b f(x)\,dx$ gives the sum of values if the values $f(x)$ are _____.

4. The expression $\dfrac{1}{b-a}\int_a^b A(t)\,dt$ gives the average _____ of a continuous income account.

5. A function f that gives the rate of change of a continuous income stream is called a _____ of _____ function.

6. The expression $\int_0^T P e^{r(T-t)}\,dt$ gives the _____ value of an annuity for time T and interest rate r.

Skill Exercises

In Exercises 7–16, determine the average value of the given function on the stated interval.

7. $f(x) = 2x + 5$; $[0, 7]$

8. $f(x) = 3x + 1$; $[1, 4]$

9. $f(x) = 3x^2 - 2x$; $[-1, 2]$

10. $f(x) = 4x - 2x^2$; $[0, 3]$

11. $f(x) = x^3 + 2x^2 - x + 1$; $[0, 2]$

12. $f(x) = -2x^3 + x + 2$; $[-2, 1]$

13. $f(x) = x^{2/3}$; $[1, 8]$

14. $f(x) = \sqrt{x + 1}$; $[3, 7]$

15. $f(x) = 4e^{0.2x}$; $[0, 10]$

16. $f(x) = 2e^{-0.15x}$; $[0, 5]$

For Exercises 17–20, complete the following.

(a) Use the RTSUM program with $n = 100$ to approximate $\int_a^b f(x)\,dx$.

(b) Use the result from part (a) to approximate the average value of f on $[a, b]$.

17. $f(x) = \dfrac{2x + 1}{x^2 + 1}$; $[-1, 1]$

18. $f(x) = \dfrac{x}{x + 1}$; $[0, 4]$

19. $f(x) = \ln(1 + x^2)$; $[0, 2]$

20. $f(x) = \dfrac{1}{x^2 + 4}$; $[0, 5]$

Application Exercises

21. **Finance—Cash Reserves:** The 2013 prospectus of a certain mutual fund states that during 2012, the amount of money that was kept in cash reserves was approximately linear and can be represented by

$$CR(x) = 0.8x + 3$$

where x is the number of months after the first of the year, and $CR(x)$ is the cash reserves in millions of dollars.

(a) Compute the average cash reserves on the interval $[0, 3]$, the first quarter of 2012.

(b) Compute the average cash reserves on the interval $[9, 12]$, the last quarter of 2012.

22. **Finance—Cash Reserves:** The 2013 prospectus of a certain mutual fund states that during 2012, the amount of money that was kept in cash reserves was approximately linear and can be represented by

$$CR(x) = -0.9x + 12.5$$

where x is the number of months after the first of the year, and $CR(x)$ is the cash reserves in millions of dollars.

(a) Compute the average cash reserves on the interval $[0, 3]$, the first quarter of 2012.

(b) Compute the average cash reserves on the interval $[9, 12]$, the last quarter of 2012.

23. **Economics—Price/Demand:** The price-demand function for Linguini's Pizza Palace all-you-can-eat pizza buffet is given by

$$p(x) = -0.02x + 8.3$$

where $p(x)$ is the price in dollars, and x is the quantity demanded. Determine the average price on the demand interval $[120, 200]$.

24. **Economics—Price/Demand:** The price-demand function for Linguini's Pizza Palace specialty pizzas is given by

$$p(x) = 15.22e^{-0.015x}$$

where $p(x)$ is the price in dollars, and x is the quantity demanded. Determine the average price on the demand interval $[40, 80]$.

25. **Pharmacology—Drug Concentration:** Researchers have determined through experimentation that the percent concentration of a certain medication during the first 20 hours after it has been administered can be approximated by

$$p(t) = \frac{200t}{2t^2 + 5} \quad 0 \le t \le 20$$

where t is the time in hours after the medication is taken, and $p(t)$ is the percent concentration. Determine the average percent concentration on the interval $[0, 20]$.

26. **Pharmacology—Drug Concentration:** Researchers have determined through experimentation that the percent concentration of a certain medication during the first 20 hours after it has been administered can be approximated by

$$p(t) = \frac{300t}{6t^2 + 5} \quad 0 \le t \le 20$$

where t is the time in hours after the medication is taken, and $p(t)$ is the percent concentration.

   (a) Determine the average percent concentration on the interval $[0, 10]$.

   (b) Determine the average percent concentration on the interval $[10, 20]$.

   (c) Is the average percent concentration greater during the first 10 hours after the drug has been taken or during the second 10 hours after it has been taken? Explain.

27. **Economics—Average Cost:** Dixco Engines has determined that the cost for producing x diesel engines is given by

$$C(x) = 60{,}000 + 300x$$

where $C(x)$ is the cost in dollars.

   (a) Determine the average value of the cost function on the interval $[0, 500]$.

   (b) Determine the average cost function AC and evaluate $AC(500)$. Recall that $AC(x) = \frac{C(x)}{x}$.

   (c) Explain the differences between parts (a) and (b).

28. **Economics—Average Cost:** Digital Pet has determined that the cost for producing x virtual pets in one day is given by

$$C(x) = 150 + 3x + \frac{2x^2}{30}$$

where $C(x)$ is the cost in dollars.

   (a) Determine the average value of the cost function on the interval $[0, 200]$.

   (b) Determine the average cost function AC and evaluate $AC(200)$. Recall that $AC(x) = \frac{C(x)}{x}$.

   (c) Explain the differences between parts (a) and (b).

29. **Banking—Continuous Interest:** Jan deposits $3000 into a money market account earning 5.85% interest compounded continuously. Determine Jan's average balance, to the nearest cent, over five years.

30. **Banking—Continuous Interest:** Eugene deposits $12,000 into a money market account earning 6.5% interest compounded continuously. Determine Eugene's average balance, to the nearest cent, over three years.

31. **Banking—Continuous Interest:** Lisa deposits $6000 into a bank account that earns 4.5% interest compounded continuously. Determine Lisa's average balance, to the nearest cent, over one year.

32. **Banking—Average Balance:** Wilhelm deposits $9000 into a money market account earning 7% interest compounded continuously. The institution where Wilhelm has deposited his money also pays a bonus at the end of the year of 1% of the average balance in the account during the year.

    (a) Compute Wilhelm's average balance, to the nearest cent, over one year.

    (b) Compute the size of Wilhelm's bonus.

33. **Banking—Average Balance:** Elmer deposits $3500 into a bank account that earns 3.8% interest compounded continuously. The bank also pays a bonus at the end of the year of 0.75% of the average balance in the account during the year.

    (a) Compute Elmer's average balance, to the nearest cent, over one year.

    (b) Compute the size of Elmer's bonus.

34. **Banking—Account Analysis:** Rusty has $4000 to invest in a money market account. At Cold Cash Company, Rusty is offered an account that earns 5% interest compounded continuously with a 1% bonus at the end of the year on his average balance in the account during the year. At Money Time Company, Rusty is offered an account that earns 4.9% interest compounded continuously with a 1.2% bonus at the end of the year on his average balance in the account during the year. If Rusty's goal is to achieve the largest bonus at the end of the year, into which account should he place his money? Explain.

35. **Banking—Account Analysis:** Shannon deposits $8000 into an account where the rate of change of the amount in the account is given by

$$\frac{dA}{dt} = 480e^{0.06t}$$

    where t represents the years after the initial deposit, and A represents the amount accrued.

    (a) Determine by how much the account changes from the end of the second year to the end of the fifth year. Round to the nearest cent.

    (b) Determine the average balance, to the nearest cent, in the account over the first five years.

36. **Banking—Account Analysis:** Kevan deposits $2000 into an account where the rate of change of the amount in the account is given by

$$\frac{dA}{dt} = 110e^{0.055t}$$

    where t represents the years after the initial deposit, and A represents the amount accrued.

    (a) Determine by how much the account changed from the end of the first year to the end of the third year. Round to the nearest cent.

    (b) Determine the average balance, to the nearest cent, in the account over the first three years.

37. **Economics—Marginal Cost:** Muggy is a company that produces coffee mugs. The research department at Muggy determined the marginal cost function is given by

$$MC(x) = 1 + \frac{1}{20}x$$

    where $MC(x)$ is in dollars per mug, and x is the number of coffee mugs produced per day.

    (a) Compute the increase in cost going from a production level of 25 mugs per day to 75 mugs per day.

    (b) If daily fixed costs are $50, compute the average value of $C(x)$ over the interval $[25, 75]$ and interpret.

38. **Economics—Marginal Cost:** Binky Inc. is a company that produces pacifiers. The research department at Binky Inc. has determined the marginal cost function is given by

$$MC(x) = 0.06 + \frac{1}{300}x$$

where $MC(x)$ is in dollars per pacifier, and x is the number of pacifiers produced per day.

(a) Compute the increase in cost going from a production level of 200 pacifiers per day to 300 pacifiers per day.

(b) If daily fixed costs are \$150, compute the average value of $C(x)$ over the interval [200, 300] and interpret.

39. **Geology—Oil Production:** Geologists estimate that an oil field will produce oil at a rate given by

$$f(t) = 600e^{-0.1t}$$

where t represents the number of months since production started, and $f(t)$ represents the rate of change in oil production, measured in thousand barrels per month.

(a) Write a definite integral to estimate the total production for the first year of production.

(b) Evaluate the integral from part (a) to estimate the total production for the first year of operation. Round to the nearest whole number.

40. **Management—Oil Income:** The rate of change of income, in thousands of dollars per year, for the oil field in Exercise 39 is projected to be

$$f(t) = 8400e^{-0.1t}$$

where t is the number of months into production.

(a) Write a definite integral to estimate the total amount of money generated from this oil field for the first year of operation.

(b) Evaluate the integral from part (a) to estimate the total amount of money generated from this oil field for the first year of operation. Round to the nearest cent.

© AbleStock

41. **Business Management—Income:** Smart Ones Car Wash generates income in the first three years of operation at a rate given by

$$f(t) = 2 \qquad 0 \le t \le 10$$

where t represents the number of years in operation and $f(t)$ the rate of change in income, measured in millions of dollars per year. Determine the total income produced in the first three years of operation.

42. **Business Management—Income:** The rate of change of income produced by a vending machine located in a college dorm is given by

$$f(t) = 2500e^{0.05t} \qquad 0 \le t \le 10$$

where t represents the number of years since the installation of the vending machine, and $f(t)$ represents the rate of change in income, measured in dollars per year. Determine the total income generated by the vending machine during the first three years of operation.

43. **Business Management—Income:** The rate of change of income produced by a vending machine located in a busy airport is given by

$$f(t) = 6000e^{0.08t} \qquad 0 \le t \le 10$$

where t represents the number of years in operation and $f(t)$ the rate of change in income, measured in millions of dollars per year. Determine the total income generated by the vending machine during the first three years of operation.

44. **Business Management—Income:** Rich's Lawn Service has determined that his new landscaping company has a projected rate of change of income given by

$$f(t) = 30e^{0.15t} \qquad 0 \le t \le 10$$

where t represents the number of years since starting the business, and $f(t)$ represents the projected rate of change of income in thousands of dollars per year. Determine the total income generated by Rich's Lawn Service during the first two years of operation.

45. **Banking—Future Value of an Annuity:** At the beginning of each month, Antonio deposits \$250 into his 403(b) plan, which earns 6.5% annual interest compounded continuously. Determine how much Antonio will have in his account after 25 years. Round to the nearest cent.

46. **Banking—Future Value of an Annuity:** At the beginning of each month, Susan deposits $300 into her 403(b) plan, which earns 7.25% annual interest compounded continuously. Determine how much Susan will have in her account after 30 years. Round to the nearest cent.

47. **Banking—Future Value of an Annuity:** At the age of 22, Vicki opened an IRA. At the beginning of each year, she deposits $1800 into her IRA, which earns 8% annual interest compounded continuously. Determine how much Vicki has in her account in 43 years when she retires at age 65.

48. **Banking—Future Value of an Annuity** *(continuation of Exercise 47)*: Determine how much of Vicki's final amount is interest.

49. **Financial Planning—Retirement Income:** At the age of 25, Jeff opened an IRA. At the beginning of each year, he deposits $2000 into his IRA, which earns 8% annual interest compounded continuously. Determine how much Jeff has in his account in 40 years when he retires at age 65.

50. **Financial Planning—Retirement Income** *(continuation of Exercise 49)*: Determine how much of Jeff's final amount is interest.

51. **Financial Planning—Retirement Income:** Compare the results of Exercises 47 and 49 to answer the following.

    (a) How much did Vicki contribute to her IRA?

    (b) How much did Jeff contribute to his IRA?

    (c) By how much did Vicki's final amount surpass Jeff's final amount? Explain how Vicki's final amount could surpass Jeff's final amount if Jeff contributed more to his IRA than did Vicki.

52. **Financial Planning—College Savings:** On the birth of their new son Dylan, Bill and Lisa opened a college savings account. They deposit $50 at the beginning of each month into an account earning 5.5% annual interest compounded continuously. Determine how much is in the account in 18 years when Dylan starts college.

53. **Financial Planning—College Savings:** On the birth of their new son Randon, Donald and Melissa opened a college savings account. They deposit $75 at the beginning of each month into an account earning 5.25% annual interest compounded continuously. Determine how much is in the account in 18 years when Randon starts college.

54. **Financial Planning—College Savings:** On the birth of their new daughter Hannah, Steve and Felicia opened a college savings account. They deposit $100 at the beginning of each month into an account earning 5% annual interest compounded continuously. Determine how much is in the account in 18 years when Hannah starts college.

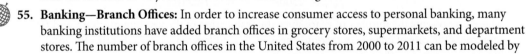 55. **Banking—Branch Offices:** In order to increase consumer access to personal banking, many banking institutions have added branch offices in grocery stores, supermarkets, and department stores. The number of branch offices in the United States from 2000 to 2011 can be modeled by

$$f(x) = 3.31x + 60.3 \qquad 0 \le x \le 11$$

where x is measured in years ($x = 1$ corresponds to January 1, 2000), and $f(x)$ represents the number of branch offices, measured in thousands. (*Source:* U.S. Federal Deposit Insurance Corporation.) Determine the average number of branch offices added per year in the decade of the 2000s.

56. **Banking—Main Offices:** While banks were increasing their branch offices, they have consolidated, and consequently decreased, their number of main bank offices. The number of main offices in the United States from 2000 to 2011 can be modeled by

$$f(x) = -0.15x + 8.28 \qquad 0 \le x \le 11$$

where x is measured in years ($x = 1$ corresponds to January 1, 2000), and $f(x)$ represents the number of main banking offices, measured in thousands. (*Source:* U.S. Federal Deposit Insurance Corporation.) Determine the average number of main banking offices closed per year in the decade of the 2000s.

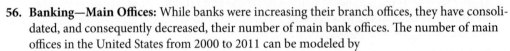 57. **Medicine—Prescription Drugs:** Many citizens have turned to alternatives such as mail-order prescriptions and preventative care medicine to counter rising prescription drug costs. The total spent in the United States on prescription drugs annually can be modeled by

$$f(x) = 0.2x^2 + 6.8x + 38.85 \qquad 1 \le x \le 22$$

where x is measured in years ($x = 1$ corresponds to January 1, 2000), and $f(x)$ represents the annual total prescription drug costs, measured in billions of dollars. (*Source:* U.S. Centers for Medicare & Medicaid Services.) Determine the average annual cost of prescription drugs in the decade of the 2000s.

58. **Family Finance—Nursing Care:** According to the AARP, in 2009, families paid $450 billion out of pocket for elderly and disabled family members. (*Source:* Kansas City Business Journal.) The annual amount spent on nursing home care can be modeled by

$$f(x) = 0.06x^2 + 3.61x + 52.67 \qquad 1 \le x \le 21$$

where x is measured in years ($x = 1$ corresponds to January 1, 2000), and $f(x)$ represents the amount spent on nursing home care, measured in billions of dollars. (*Source:* U.S. Centers for Medicare & Medicaid Services.) Determine the average annual cost of amount spent on nursing home care in the decade of the 2000s.

For Exercises 59 and 60, reviewing the integration of general exponential functions in Section 6.6 may be helpful.

59. **Ecology—Clean Energy:** Increasing fossil fuel costs, combined with concern over climate change, have resulted in an emerging green industry. The total annual revenue of the clean energy industry can be modeled by

$$f(x) = 7.02(1.09^x) \qquad 1 \le x \le 11$$

where x is measured in years ($x = 1$ corresponds to January 1, 2000), and $f(x)$ represents annual revenue of the clean energy industry, measured in billions of dollars. (*Source:* Environmental Business International.) Determine the average annual revenue of the clean energy industry in the decade of the 2000s.

60. **Ecology—Clean Energy:** In Exercise 59, we saw that revenues of the clean energy industry are increasing. The number of workers employed by those industries is increasing as well. The number of employees in clean energy industries can be modeled by

$$f(x) = 3.42 \cdot (1.08^x) \qquad 1 \le x \le 11$$

where x is measured in years ($x = 1$ corresponds to January 1, 2000), and $f(x)$ represents the number employed annually in the clean energy industry. (*Source:* Environmental Business International.) Determine the average annual number employed in the clean energy industry in the decade of the 2000s.

Concept and Writing Exercises

61. Describe in nontechnical terms how the average value of a function over a closed interval is determined.

62. In the text following Example 1, the final sentence reads, "Some discrepancy is expected, since the function in Example 1 is a model that *approximates* the data." Explain why we get an approximation when computing the average value.

63. Let $f(x) = mx + d$ represent a linear function. Determine a formula for the average value of f on the interval $[a, b]$.

64. Let $f(x) = ax^2 + bx + c$ represent a quadratic function where a, b, and c are real numbers. Determine a formula for the average value of f on the interval $[x_1, x_2]$.

65. Let $f(x) = 2x + 6$. Determine the value of b if we know that the average value of f on the interval $[0, b]$ is 10.

66. Let $f(x) = 2x^2 + 8$. Determine the value of a if we know that the average value of f on the interval $[a, 5]$ is 34.

Section Project

Maria has decided to open an IRA for retirement purposes. She has found an account that earns 7% annual interest compounded continuously. She has decided to invest $1800 each year into the IRA.

She can invest this $1800 in one of two ways: either $1800 at the beginning of each year or $150 at the beginning of each month. Maria plans on investing for 30 years, at which time she will retire. Help Maria decide between the two investment options.

(a) Determine how much is in Maria's account in 30 years if she invests $1800 at the beginning of each year.

(b) Determine how much is in Maria's account in 30 years if she invests $150 at the beginning of each month.

(c) Based on parts (a) and (b), which plan would you advise Maria to take? Explain.

7.2   Area Between Curves and Applications

In Chapter 6, as well as in Section 7.1, we saw that the area under a curve on a closed bounded interval can be determined by the definite integral. In this section we extend this concept to compute the area between two curves on a closed interval. We will also see how to determine the closed interval over which the area trapped between two curves can be calculated. The latter will require us to find points of intersection of two curves.

Area Between Curves on a Closed Interval [a, b]

We begin this section utilizing what was learned in Chapter 6, specifically, that the definite integral gives the area under a curve. Our first example reviews this important concept.

Example 1: Representing Area by a Definite Integral

a. Write a definite integral to represent the shaded area in **Figure 7.2.1**.

b. Write a definite integral to represent the area in **Figure 7.2.2**.

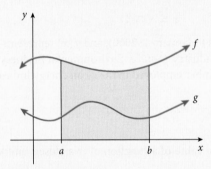

Figure 7.2.1

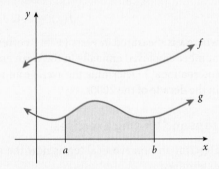

Figure 7.2.2

c. Use the results of parts (a) and (b) to write a definite integral to represent the shaded area in **Figure 7.2.3**.

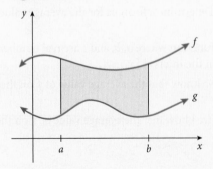

Figure 7.2.3

Perform the Mathematics

a. Since the shaded area is below the graph of f on the interval $[a, b]$, the definite integral that represents the shaded area is $\int_a^b f(x)\,dx$.

b. Since the shaded area is below the graph of g on the interval $[a, b]$, the definite integral that represents the shaded area is $\int_a^b g(x)\,dx$.

c. Here it appears that the shaded area is simply

Area under the graph of f minus area under the graph of g

$$\int_a^b f(x)\,dx - \int_a^b g(x)\,dx$$

We can use a property from Section 6.1 and write these integrals as

$$\int_a^b [f(x) - g(x)]\,dx$$ ∎

In Example 1c, what we really found was the area between the graphs of f and g on the interval $[a, b]$. In fact, the process outlined in Example 1 is exactly how we find the **area between two curves**.

Area Between Two Curves

On a closed interval $[a, b]$, the area between two continuous functions f and g is given by:
1. If $f(x) \geq g(x)$, then the area between the graphs of f and g is $\int_a^b [f(x) - g(x)]\,dx$. See **Figure 7.2.4(a)**.
2. If $g(x) \geq f(x)$, then the area between the graphs of f and g is $\int_a^b [g(x) - f(x)]\,dx$. See Figure 7.2.4(b).

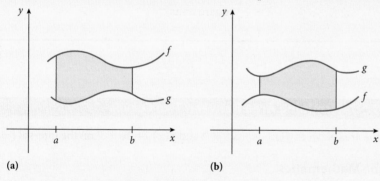

(a)   **(b)**

Figure 7.2.4 **(a)** Area of shaded region is given by $\int_a^b [f(x) - g(x)]\,dx$.
(b) Area of shaded region is given by $\int_a^b [g(x) - f(x)]\,dx$.

NOTE: An easy way to handle both cases listed is to remember the following:
To determine the area between two curves on $[a, b]$, integrate

$$\int_a^b [(\text{top curve} - \text{bottom curve})]\,dx$$

Example 2: Determining the Area Between Two Curves

Determine the area between the x-axis and $f(x) = x^2 - x - 6$ on the interval $[1, 3]$.

Perform the Mathematics

We first graph the function and shade the area as an aid in setting up the correct integral. See **Figure 7.2.5**.

OBJECTIVE 1

Determine the area between two curves.

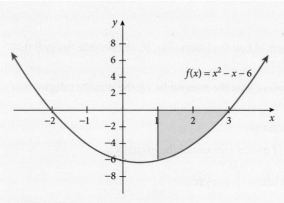

Figure 7.2.5

If we recall that the x-axis can be represented by the function $y = 0$, we can apply the techniques of this section. Thus,

$$\int_1^3 (\text{top curve} - \text{bottom curve}) \, dx = \int_1^3 [0 - (x^2 - x - 6)] \, dx$$

$$= \int_1^3 (-x^2 + x + 6) \, dx$$

$$= \left(-\frac{1}{3}x^3 + \frac{1}{2}x^2 + 6x \right) \Big|_1^3$$

$$= \left[-\frac{1}{3}(3)^3 + \frac{1}{2}(3)^2 + 6(3) \right] - \left[-\frac{1}{3}(1)^3 + \frac{1}{2}(1)^2 + 6(1) \right]$$

$$= 7\frac{1}{3} \text{ square units} \qquad \blacksquare$$

Example 3: Determining Area Between Two Curves

Determine the area between $f(x) = -x^2 + 4$ and $g(x) = 2x + 7$ on the interval $[-3, 2]$.

Perform the Mathematics

As in Example 2, we begin with a graph of f and g and shade the area. See **Figure 7.2.6**.

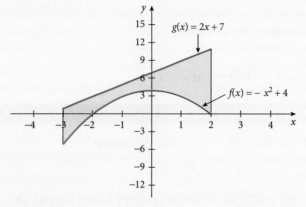

Figure 7.2.6

Since $g(x) \geq f(x)$ for all x on the interval $[-3, 2]$, the graph of g is the top curve and the graph of f is the bottom curve. Proceeding, we have

$$
\begin{aligned}
\int_{-3}^{2} [\text{top curve} - \text{bottom curve}]\, dx &= \int_{-3}^{2} [(2x + 7) - (-x^2 + 4)]\, dx \\
&= \int_{-3}^{2} (x^2 + 2x + 3)\, dx \\
&= \left(\frac{1}{3}x^3 + x^2 + 3x \right)\Bigg|_{-3}^{2} \\
&= \left(\frac{8}{3} + 4 + 6 \right) - (-9 + 9 - 9) \\
&= \frac{65}{3} \text{ square units} \qquad \blacksquare
\end{aligned}
$$

▶ **Try It Yourself**

Some related Exercises are 15 and 23.

In Example 4, we extend the concept set forth in Chapter 6 that integrating a rate of change function gives an accumulation.

Example 4: Determining Plant Growth

OBJECTIVE 2

Apply the area between two curves to tree growth.

From past records, a botanist knows that a certain species of tree grows at a rate of $\frac{3}{2}x^{-1/2}$ feet per year, where x is the age of the tree in years. However, if a special nitrogen-rich nutrient is given to the tree, it grows at a rate of $2x^{-1/2}$ feet per year. On the interval $[1, 4]$, how many more feet in growth would result from the special nitrogen-rich nutrient?

Understand the Situation

In **Figure 7.2.7** we have graphed the two curves representing the two different growth rates. To determine how many more feet of growth results from the special nitrogen rich nutrient, we need to integrate the difference of the rates from $x = 1$ to $x = 4$. In other words, we need to determine the area between the curves.

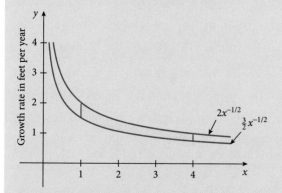

Figure 7.2.7 Additional growth is given by the shaded region.

Perform the Mathematics

The area between the curves is found by computing

$$\int_1^4 \left(2x^{-1/2} - \frac{3}{2}x^{-1/2}\right)dx = \int_1^4 \frac{1}{2}x^{-1/2}dx$$

$$= x^{1/2}\Big|_1^4 = (4)^{1/2} - (1)^{1/2} = 1$$

Interpret the Result

So on the interval [1, 4], that is, from year 1 to year 4, the nitrogen-rich nutrient would result in an additional 1 foot of growth.

▷ *Try It Yourself*

Some related Exercises are 41 and 45.   ■

Area Bounded by Two Curves

So far in this section we have presented examples in which the interval of integration has been given. We now consider how to determine the area of a region that is trapped or enclosed by two curves and when no interval is given. In these situations, where the curves completely enclose an area, we need to determine where the curves intersect.

OBJECTIVE 3

Determine the area bounded by two curves.

Example 5: Determining the Area Enclosed by Two Curves

Determine the area bounded by the curves $y = x^2 + x - 5$ and $y = 2x + 1$.

Perform the Mathematics

As always, our first step is to graph the curves so that we can see the region whose area we want. See **Figure 7.2.8**. For this region, the top curve is $y = 2x + 1$ and the bottom curve is $y = x^2 + x - 5$.

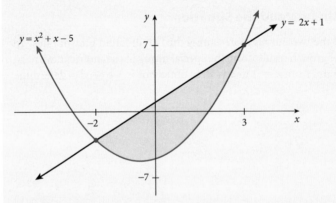

Figure 7.2.8

To determine the limits of intersection, we must find the points where the curves intersect. Algebraically, this is handled by setting the two equations equal to each other and solving for x:

$$x^2 + x - 5 = 2x + 1$$
$$x^2 - x - 6 = 0$$
$$(x - 3)(x + 2) = 0$$
$$x = 3 \quad \text{or} \quad x = -2$$

So the curves intersect at the x-values of $x = 3$ and $x = -2$. The area enclosed by the two curves is then found by integrating

$$\int_{-2}^{3} [(2x + 1) - (x^2 + x - 5)]dx = \int_{-2}^{3} (-x^2 + x + 6)dx$$

$$= \left(-\frac{1}{3}x^3 + \frac{1}{2}x^2 + 6x\right)\bigg|_{-2}^{3}$$

$$= 13.5 - \left(-7\frac{1}{3}\right)$$

$$= 20\frac{5}{6} \approx 20.83 \text{ square units} \qquad \blacksquare$$

In our next example, we revisit marginal revenue and marginal cost.

Example 6: Computing a Total Profit

OBJECTIVE 4

Apply the area bounded by two curves to total profit.

The FrigAir Company knows that its marginal cost to produce x refrigerators is given by $C'(x) = 4x + 23$ and the marginal revenue is given by $R'(x) = -0.6x + 460$, where both marginals are in dollars per unit. Compute the total profit from $x = 0$ to $x = 95$, the production level where profit is maximized.

Understand the Situation

Figure 7.2.9 has a graph of R' and C'. In Chapter 6 we saw that to determine total profit, we needed to integrate P', the marginal profit function. But, since $P(x) = R(x) - C(x)$, we have $P'(x) = R'(x) - C'(x)$. Hence, to find total profit, we need to integrate

$$\int_{0}^{95} P'(x)dx = \int_{0}^{95} [R'(x) - C'(x)]dx$$

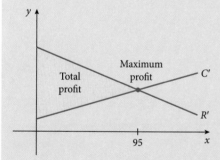

Figure 7.2.9

Perform the Mathematics

From Figure 7.2.9, we notice that the total profit is simply the area between the graphs of R' and C'. Continuing, we have

$$\int_{0}^{95} [R'(x) - C'(x)]dx = \int_{0}^{95} [(-0.6x + 460) - (4x + 23)]dx$$

$$= \int_{0}^{95} (-4.6x + 437)dx = \left(\frac{-4.6}{2}x^2 + 437x\right)\bigg|_{0}^{95}$$

$$= \left[\left(\frac{-4.6}{2}(95)^2 + 437(95)\right)\right] - \left[\frac{-4.6}{2}(0)^2 + 437(0)\right]$$

$$= 20{,}757.5$$

In Example 6 we stated that the profit is maximized at $x = 95$. As shown in Figure 7.2.9, that is where $R'(x) = C'(x)$. In general, **profit is maximized when $R'(x) = C'(x)$.**

Area Between Curves That Cross on $[a, b]$

The final scenario we consider in this section is what happens if two curves cross each other on a given interval. When this occurs, what was a top curve may become a bottom curve, as we illustrate in Example 7.

OBJECTIVE 5

Determine the area between two curves that cross on an interval.

Example 7: Determining the Area Between Two Curves That Cross on an Interval

Determine the area between the curves $y = x + 1$ and $y = \frac{2}{x}$ on the interval $\left[\frac{1}{2}, 3\right]$.

Perform the Mathematics

The first thing we do is graph the two curves on the stated interval to see which is the top curve and which is the bottom curve. We will also be able to see if the two curves cross each other in the interval. **Figure 7.2.10** shows the graph, and we have shaded the area we want to determine.

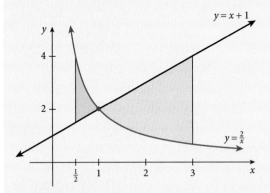

Figure 7.2.10

To the left of the point of intersection, $y = \frac{2}{x}$ is the top curve, whereas to the right it is the bottom curve. At this time we need to determine where the curves cross, or intersect if you prefer. Setting the functions equal to each other and solving for x gives us

$$x + 1 = \frac{2}{x}$$

Multiply each side by x, provided $x \neq 0$ $x^2 + x = 2$

Subtract 2 from each side $x^2 + x - 2 = 0$

Factor $(x + 2)(x - 1) = 0$

$$x = -2 \quad \text{or} \quad x = 1$$

Since $x = -2$ is not in the interval $[\frac{1}{2}, 3]$, we can ignore it. Since the curves cross each other at $x = 1$, we integrate over the intervals $[\frac{1}{2}, 1]$ and $[1, 3]$ separately and then add the results to determine the area between the curves. This gives

$$\int_{1/2}^{1}\left[\frac{2}{x} - (x + 1)\right]dx + \int_{1}^{3}\left[(x + 1) - \frac{2}{x}\right]dx$$

$$= \int_{1/2}^{1}\left(\frac{2}{x} - x - 1\right)dx + \int_{1}^{3}\left(x + 1 - \frac{2}{x}\right)dx$$

$$= \left(2\ln|x| - \frac{1}{2}x^2 - x\right)\Bigg|_{1/2}^{1} + \left(\frac{1}{2}x^2 + x - 2\ln|x|\right)\Bigg|_{1}^{3}$$

$$\approx 4.314$$

So the area between the curves $y = x + 1$ and $y = \frac{2}{x}$ on the interval $[\frac{1}{2}, 3]$ is about 4.31 square units. ∎

Try It Yourself

Some related Exercises are 33 and 35.

Technology Option

We could also use the `intersect` command on the calculator to determine the point of intersection for the two curves in Example 7. **Figure 7.2.11** shows the result of using the `intersect` command.

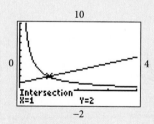

Figure 7.2.11

Summary

In this section we saw how to compute the area between two curves and how to apply this concept. To determine the area between two curves, we offer the following guidelines:

1. Graph the two curves to see the region whose area we want.

2. If an interval is not given, or if the curves cross each other on a given interval, determine the appropriate limits of integration by determining where the curves intersect each other.

3. Integrate top curve minus bottom curve on each interval.

Section 7.2 Exercises

Vocabulary Exercises

1. The area between the graphs of f and g is $\int_a^b [g(x) - f(x)]\, dx$ if the graph of f is _____ the graph of g on the interval $[a, b]$.

2. The area between the graphs of f and g is $\int_a^b [f(x) - g(x)]\, dx$ if the graph of f is _____ the graph of g on the interval $[a, b]$.

3. If we seek the area between two functions f and g where the graphs cross on the interval $[a, b]$, we must compute the point of _____ of the graphs.

4. In general, marginal profit is _____ when marginal revenue is greater than marginal _____.

5. In general, marginal profit is _____ when marginal revenue is less than marginal _____.

6. In general, profit is _____ when marginal revenue is equal to marginal cost.

Skill Exercises

For Exercises 7–14, write a definite integral to represent the area of the shaded region.

7.
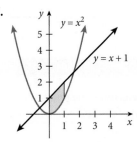

8.
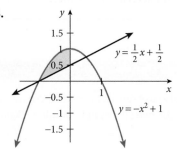

9.
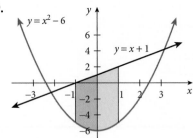

10.
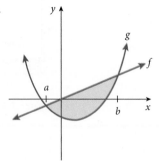

11.
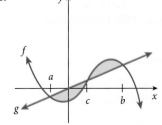

12.
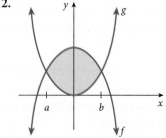

13.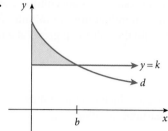

14.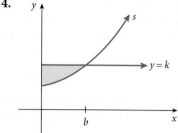

In Exercises 15–32, determine the area of the region bounded by the given conditions.

15. $f(x) = 3x + 1$ and the x-axis on $[0, 4]$.

16. $f(x) = 2x + 3$ and the x-axis on $[1, 3]$.

17. $f(x) = 2x^2 + x + 1$ and the x-axis on $[-2, 1]$.

18. $f(x) = x^2 + 2x + 3$ and the x-axis on $[-3, 2]$.

19. $f(x) = -x^2 + x + 2$ and the x-axis on $[1, 4]$.

20. $f(x) = -2x^2 - x + 1$ and the x-axis on $[-2, 3]$.

21. $f(x) = 8 - x^2$ and $g(x) = 4$ on $[-2, 2]$.

22. $f(x) = x^2 - 2$ and $g(x) = 2$ on $[-2, 2]$.

23. $f(x) = -x + 4$ and $g(x) = -x^2 + 1$ on $[-2, 2]$.

24. $f(x) = 2x + 3$ and $g(x) = -x^2 + 1$ on $[-1, 2]$.

25. $f(x) = x - 3$ and $g(x) = x^2 + x + 1$ on $[-2, 1]$.

26. $f(x) = \dfrac{1}{2}x - 1$ and $g(x) = x^2 + 2x - 1$ on $[1, 5]$.

27. $f(x) = 16 - \sqrt{x}$ and $g(x) = 8 - \sqrt[3]{x}$ on $[1, 5]$.

28. $f(x) = x^2$ and $g(x) = \sqrt{x}$ on $[0, 1]$.

29. $f(x) = 5e^{-0.15x}$ and $g(x) = 10e^{-0.08x}$ on $[0, 5]$.

30. $f(x) = 7.2e^{-0.07x}$ and $g(x) = 2.34e^{-0.11x}$ on $[1, 4]$.

31. $f(x) = \dfrac{3}{x}$ and $g(x) = \dfrac{4}{x^2}$ on $[2, 8]$.

32. $f(x) = \dfrac{3}{x}$ and $g(x) = \dfrac{4}{x^2}$ on $[1, 8]$.

In Exercises 33–40, determine the area bounded by the given curves.

33. $f(x) = 4 - x^2$ and $g(x) = -x + 2$.

34. $f(x) = -x - 3$ and $g(x) = 9 - x^2$.

35. $f(x) = -2x^2 - 5x + 3$ and $g(x) = 2x^2 + 3x - 2$.

36. $f(x) = x^2 - 6$ and $g(x) = 12 - x^2$.

37. $f(x) = 3x^2 - 2x$ and $g(x) = x^3$.

38. $f(x) = x^2$ and $g(x) = x^3$.

39. $f(x) = x^3$ and $g(x) = x$.

40. $f(x) = 2x^3$ and $g(x) = -x^3 + x^2 + 2x$.

Application Exercises

41. Bacteriology—Bacteria Growth: Suppose a bacteriologist determines that the size of a certain bacterial culture grows at a rate given by the function

$$f(t) = 4t^{5/2} \qquad t \geq 0$$

where t represents the time in minutes since the bacteria were introduced to the culture, and $f(t)$ represents the rate of change in bacterial growth, measured in milligrams per minute. The bacteriologist finds that when a special nutrient is introduced into the culture, it grows at a rate given by

$$g(t) = 4t^{18/5} \qquad t \geq 0$$

where t represents the time in minutes since the bacteria were introduced to the culture, and $f(t)$ represents the rate of change in bacterial growth, measured in milligrams per minute.

(a) Set up an integral to determine the area between the two curves on $[0, 1]$.

(b) Evaluate the integral from part (a) and interpret.

42. **Bacteriology—Bacteria Growth** *(continuation of Exercise 41)*: Suppose a bacteriologist determines that the size of a certain bacterial culture grows at a rate given by the function

$$f(t) = 4t^{5/2} \qquad t \geq 0$$

where t represents the time in minutes since the bacteria were introduced to the culture, and $f(t)$ represents the rate of change in bacterial growth, measured in milligrams per minute. The bacteriologist finds that when a special nutrient is introduced into the culture, it grows at a rate given by

$$g(t) = 4t^{18/5} \qquad t \geq 0$$

where t represents the time in minutes since the bacteria were introduced to the culture, and $f(t)$ represents the rate of change in bacterial growth, measured in milligrams per minute.

(a) Set up an integral to determine the area between the two curves on $[1, 2]$.

(b) Evaluate the integral from part (a) and interpret.

(c) On the interval $[0, 2]$, how many more bacteria result from introducing the special nutrient?

43. **Marketing—Sales Rates:** Double D cola is planning to release a new Double Dose Caffeine Cola on the market. In the past, Double D cola ran a 20-week ad campaign for any new product that it released on the market. For example, when Double D ran a 20-week ad campaign for its new caffeine-free cola, weekly sales were modeled by

$$s(t) = -0.51t^2 + 20.20t + 49.49 \qquad 0 \leq t \leq 20$$

where t represents the number of weeks after the campaign began, and $s(t)$ represents the sales in thousands of dollars per week. An agent for a TV personality has approached Double D cola and claims that, if Double D uses his client in the 20-week ad campaign, they can expect the weekly sales to be modeled by the function

$$w(t) = -0.71t^2 + 28.14t + 57.22 \qquad 0 \leq t \leq 20$$

where t represents the number of weeks after the campaign began, and $w(t)$ represents the sales in thousands of dollars per week.

(a) On the interval $[0, 20]$, determine the area between s and w and interpret.

(b) If the endorsement of the TV personality costs Double D cola \$1,000,000, will this endorsement be paid off in the first 20 weeks of the campaign? Explain.

44. **Marketing—Sales Rates:** A market analyst for Chocolate Time estimates that with no promotion, the annual sales of its candy bar Chocoloco Dream can be modeled by

$$s_1(t) = 0.76t + 8.42$$

where t represents the number of years since the product was introduced, and $s_1(t)$ represents the sales in millions of sales per year. This same analyst estimates that, with a modest promotional campaign, annual sales can be modeled by

$$s_2(t) = 11.4e^{0.05t}$$

where t represents the number of years since the product was introduced, and $s_2(t)$ represents the sales in millions of sales per year. During the first five years, what will the total increase in sales be in response to the promotion?

45. **Marketing—Sales Rates:** A market analyst for Chocolate Time estimates that with no promotion, the annual sales of its candy bar Chocoholic Chocoholic Dream can be modeled by

$$s_1(t) = 0.76t + 8.42 \qquad t \geq 0$$

where t represents the number of years since the product was introduced, and $s_1(t)$ represents the sales in millions of sales per year. This same analyst estimates that, with a modest promotional campaign, annual sales can be modeled by

$$s_2(t) = 12.15e^{0.04t} \qquad t \geq 0$$

where t represents the number of years since the product was introduced, and $s_2(t)$ represents the sales in millions of sales per year. During the first five years, what will the total increase in sales be in response to the promotion?

46. **Psychology—Memory:** A psychologist has determined that people can memorize digits at the rate modeled by

$$f(x) = 4.4e^{-0.2x}$$

where x is the time in minutes that the individuals have been memorizing, and $f(x)$ represents the rate of memorization in words per minute. She discovered that if a scent of peppermint oil is present while people are attempting to memorize and when asked to recall the digits memorized, the rate of learning is found to be modeled by

$$f(x) = 5.3e^{-0.15x}$$

where x is the time in minutes that the individuals have been memorizing, and $f(x)$ represents the rate of memorization in words per minute.

(a) Graph both rate functions in the viewing window $[0, 5]$ by $[0, 6]$.

(b) On the interval $[1, 5]$, determine the area between the two curves and interpret the result.

47. **Psychology—Memory:** A psychologist has determined that the rate at which a child learns to recognize new words can be modeled by

$$f(t) = 25t + 50\sqrt{t} \qquad t \geq 0$$

where t represents the child's age, and $f(t)$ represents the number of new words learned per year. If the child listens to special vocabulary-building audio recordings the rate at which the child recognizes new words is given by

$$g(t) = 35t + 70\sqrt{t} \qquad t \geq 0$$

where t represents the child's age, and $f(t)$ represents the number of new words learned per year.

(a) Graph both rates in the viewing window $[0, 10]$ by $[0, 600]$.

(b) On the interval $[4, 8]$, determine the area between the two curves and interpret the result.

48. **Management—Operating Costs:** Dynatronics has determined that its operating costs, in millions of dollars, have been increasing, mainly because of inflation, at a rate given by the function

$$f(x) = 1.25e^{0.12x} \qquad x \geq 0$$

where x is time measured in years since its new telephone equipment began being produced, and $f(x)$ represents the rate of change in cost in millions of dollars per year. At the beginning of the third year of production, a breakthrough in the production process resulted in costs increasing at a rate given by

$$f(x) = 0.85\sqrt{x} \qquad x \geq 0$$

where x is time measured in years since its new production process began being utilized, and $f(x)$ represents the rate of change in cost in millions of dollars per year. Compute the area between the two curves on the interval [3, 6] to determine the total savings in costs due to the breakthrough.

49. **Management—Operating Costs:** Quality Chips has determined that its costs, in millions of dollars, have been increasing, mainly because of inflation, at a rate that can be modeled by

$$f(x) = 2.67e^{0.09x}$$

where x represents the number of years since its snack food started in production, and $f(x)$ represents the rate of change in costs, measured in millions of dollars per year. At the beginning of the third year, a breakthrough in the production process resulted in costs increasing at a rate given by

$$f(x) = 0.68x^{2/3}$$

where x represents the number of years since the new production process started in production, and $f(x)$ represents the rate of change in costs, measured in millions of dollars per year. Determine the total savings in costs on the interval [3, 8] due to the breakthrough.

50. **Economics—Marginal Analysis:** The financial office at See-It Inc. determines that the marginal revenue to produce x graphing calculators is given by

$$R'(x) = -0.3x + 100$$

while the marginal cost is given by

$$C'(x) = 0.5x + \frac{750}{x + 1}$$

where x represents the number of graphing calculators produced and sold, and $R'(x)$ and $C'(x)$ are measured dollars per unit. Determine the total profit as production and sales increase from $x = 10$ to $x = 100$.

51. **Economics—Marginal Analysis:** The accounting department of Monkey Works Inc. determines that the marginal revenue to produce x squeegees is given by

$$R'(x) = -0.2x + 40$$

while the marginal cost is given by

$$C'(x) = \frac{300}{x + 1}$$

where both marginal functions are measured in dollars per unit. Determine the total profit from $x = 10$ to $x = 100$.

52. **Economics—Marginal Analysis:** The CFO of the Music Time Company determines that the marginal revenue to produce x integrated stereo amplifiers is given by

$$R'(x) = 216 - 0.24x^2$$

while marginal cost is given by

$$C'(x) = 20$$

where both marginal functions are measured in in dollars per unit. Determine the total profit from $x = 10$ to $x = 30$.

53. **Economics—Marginal Analysis:** Financial officers for Java Buddy determine that the marginal revenue to produce x coffee mugs is given by

$$R'(x) = 5 - 0.05x$$

while marginal cost is given by

$$C'(x) = 1 + \frac{40}{x}$$

where both marginal business functions are measured in dollars per unit. Determine the total profit from $x = 5$ to $x = 40$.

54. **Manufacturing—Assembly Rates:** A factory worker can assemble units at the rate given by the function

$$f(x) = -3x^2 + 13x + 11$$

where x represents the number of hours on the job, and $f(x)$ represents the assembly rate in units per hour. Determine how many units can be assembled by this worker during the first four hours.

55. **Manufacturing—Assembly Rates** (*continuation of Exercise 54*): If the factory worker in Exercise 54 has breakfast along with two cups of coffee, then she can assemble units at the rate of

$$f(x) = -2x^2 + 11x + 13$$

where x represents the number of hours on the job, and $f(x)$ represents the assembly rate in units per hour.

(a) Under these circumstances, how many units can she assemble during the first 4 hours?

(b) How many more units can she assemble in the first four hours if she has breakfast and two cups of coffee?

56. **Botany—Tree Growth:** A botanist knows from past data records that a certain species of oak tree grows at a rate given by the function

$$f(x) = \frac{4x^2 + 16x + 9}{2x + 4}$$

where x represents the age of the tree in years, and $f(x)$ represents the rate of growth of the tree in feet per year. Determine how much growth takes place on the interval [3, 8].

57. **Botany—Tree Growth** (*continuation of Exercise 56*): The botanist in Exercise 56 knows that restricting the amount of light that the oak tree receives inhibits its growth. When the light is restricted, the oak tree grows at a rate given by the function

$$f(x) = \frac{2x^2 + 16x + 9}{2x + 4}$$

where x represents the age of the tree in years, and $f(x)$ represents the rate of growth of the tree in feet per year.

(a) Under these circumstances, determine how much growth takes place on the interval [3, 8].

(b) Comparing the answer to the solution in Exercise 56, how many fewer feet of growth will result from restricting the amount of light that the tree receives?

58. **Demographics—Median Income:** About one-fourth of families with children had only one parent present in 1992—compared with about 11% of households in 1970. (*Source:* The Brookings Institution.) The rate of change in median income of a single-parent household with a *wife not present* can be modeled by the rate function

$$f(x) = 0.85e^{0.03x} \qquad 0 \le x \le 21$$

where x represents the number of years since 1990, and $f(x)$ represents the rate of change in median income, measured in thousands of dollars per year. During the same time period, the rate of change in median income of a single-parent household with a *husband not present* can be modeled by the rate function

$$g(x) = 0.75 \qquad 0 \le x \le 21$$

where x represents the number of years since 1990, and $g(x)$ represents the rate of change in median income, measured in thousands of dollars per year. (*Source:* U.S. Census Bureau.) Evaluate and interpret $\int_0^{20} [f(x) - g(x)]\, dx$.

59. Climatology—Tornado Deaths: One area to which climatologists and urban planners have given increased attention is the increasing number of deaths and injuries resulting from tornados. "With increasing urbanization, more people are affected when storms do hit, putting tornadoes in the spotlight," said Greg Carbin, the warning coordination meteorologist with the Storm Prediction Center. (*Source: Christian Science Monitor.*) The rate of change in the number of *injuries* resulting from tornados from 1995 to 2011 can be modeled by

$$f(x) = 13.59x^2 - 302x + 1551 \qquad 5 \le x \le 21$$

where x represents the number of years since 1990, and $f(x)$ represents the rate of change in the number of injuries, measured in injuries per year. During the same time period, the rate of change in the number of *lives lost* during tornados can be modeled by the function

$$g(x) = 0.47e^{0.04x} \qquad 5 \le x \le 21$$

where x represents the number of years since 1990, and $g(x)$ represents the rate of change in the number of lives lost, measured in fatalities per year. (*Source: U.S. National Weather Service.*)

(a) Compute $\int_5^{10}[f(x) - g(x)]\,dx$ and interpret. Round your answer to the nearest whole number.

(b) Now evaluate $\int_{15}^{20}[f(x) - g(x)]\,dx$ and interpret this result.

(c) Compare the results from parts (a) and (b). Is Dr. Carbin's suggestion that households should purchase a weather radio a good idea?

60. Military Science—Weapon Imports: In addition to contracting U.S. companies to procure weapons and supplies, the U.S. military exports and imports those weapons and supplies from other countries. From 2000 to 2011, the rate of change in the amount spent on military weapons and supplies *imports* can be modeled by

$$f(x) = 0.16x^3 - 1.89x^2 + 6x - 0.32 \qquad 0 \le x \le 11$$

where x represents the number of years since 2000, and $f(x)$ represents the rate of change in the amount spent on military weapons and supplies imports, measured in billions of dollars per year. During the same period of time, the rate of change in the amount spent on military weapons and supplies *exports* can be modeled by

$$g(x) = 0.28x^3 - 3.96x^2 + 16.72x - 15.54 \qquad 0 \le x \le 11$$

where x represents the number of years since 2000, and $g(x)$ represents the rate of change in the amount spent on military weapons and supplies exports, measured in billions of dollars per year. (*Source: U.S. Census Bureau.*)

(a) Use the models to determine the trade balance of U.S. military weapons and supplies from 2000 to 2005.

(b) Contrast the answer to part (a) with the trade balance of U.S. military weapons and supplies from 2005 to 2010.

61. Economics—Exports: Many economists believe that increasing exports is part of economic growth. "Increasing U.S. exports is a critical part of our economic recovery," said Fred P. Hochberg, chairman and president of the Export-Import Bank of the United States. "We are on track to double U.S. exports by the end of 2014." (*Source: BNO News.*) The rate of change in total U.S. *exports* of goods and services from 2000 to 2011 can be modeled by

$$f(x) = -7.08x^3 + 81.6x^2 - 216.6x + 122.5 \qquad 0 \le x \le 11$$

where x represent the number of years since 2000, and $f(x)$ represents the rate of change in total exports of goods and services, measured in billions of dollars per year. During the same time period, the rate of change in total U.S. *imports* of goods and services from 2000 to 2011 can be modeled by

$$g(x) = -9.32x^3 + 103.26x^2 - 296x + 333 \qquad 0 \le x \le 11$$

where x represent the number of years since 2000, and $g(x)$ represents the rate of change in total imports of goods and services, measured in billions of dollars per year. (*Source:* U.S. Census Bureau.)

(a) Use the models to determine the U.S. trade balance from 2003 to 2008.

(b) Compute $\int_8^{11}[f(x) - g(x)]\,dx$ and interpret.

Concept and Writing Exercises

62. If $f(x) \geq g(x)$ on an interval $[a, b]$, explain how $\int_a^b[f(x) - g(x)]\,dx$ determines the area between the curves.

63. If $f(x) \geq g(x)$ on an interval $[a, b]$, what do we know about the value of $\int_a^b[g(x) - f(x)]\,dx$? Explain your answer.

64. Let $f(x) = m_1x + b_1$, where m_1 and b_1 represent positive real numbers, and let $g(x) = k$, where k is a real number constant. If $f(x) \geq g(x)$ on an interval $[a, b]$, write an expression for the area between the two curves $[a, b]$.

65. Let $f(x) = k$, where k is a real number constant, and let $g(x) = m_1x + b_1$, where m_1 and b_1 represent positive real numbers. If $f(x) \geq g(x)$ on an interval $[a, b]$, write an expression for the area between the two curves on $[a, b]$.

66. Suppose we have a linear function $f(x) = 2x + 4$ and let $g(x) = k$, where k is a real number constant. If $f(x) \geq g(x)$ on an interval $[0, 5]$ and the area between the curves on that interval is 30, determine the value of k.

67. Let $f(x) = x + 3$ and $g(x) = 2x + 1$. If the area between the functions f and g is 6 on the interval $[0, b]$, find the value of b.

Section Project

The **entropy index** is a formula used for measuring inequality in data. The primary use is for income inequality. (*Source:* Urban Affairs Review.) The rate of change in per capita income for four ethnic groups is shown.

Asian and Pacific Islander:	$y_1(x) = 0.06x + 0.39$	$0 \leq x \leq 11$
White:	$y_2(x) = 0.76$	$0 \leq x \leq 11$
Black:	$y_3(x) = 0.55$	$0 \leq x \leq 11$
Hispanic:	$y_4(x) = 0.44$	$0 \leq x \leq 11$

In each case, x represents the number of years since 2000, and the function values are measured in thousands of dollars per year. (*Source:* U.S. Census Bureau.)

(a) Without computing anything, which ethnic group's per capita income was increasing the fastest? Explain your answer.

(b) Compute $\int_5^{11}[y_1(x) - y_2(x)]\,dx$ and interpret.

(c) Compute $\int_5^{11}[y_3(x) - y_4(x)]\,dx$ and interpret.

(d) Compare the results of part (a) and part (b). In which case were the differences the smallest?

(e) Suppose we want to compute the integral $\int_0^{10}[y_2(x) - y_3(x)]\,dx$. Inspect the two functions and the limits. What two numbers could we multiply to evaluate this integral?

(f) Verify your answer to part (e) by evaluating the integral $\int_0^{10}[y_2(x) - y_3(x)]\,dx$.

7.3   Economic Applications of Area Between Two Curves

In Section 7.2 we introduced how to determine the area between two curves and considered many applications of this concept. In this section we focus mainly on two specific applications of area between two curves: **consumers' and producers' surplus** and **income distributions.** Our discussion of consumers' and producers' surplus will also include a look at **market price, total consumer expenditure, revenue, market demand,** and **equilibrium point.** Our discussion on income distribution will include a look at **Lorenz curves** and the **Gini Index.**

Price-Demand Function Revisited

In Chapter 1, we were first introduced to what is known as the price-demand function. We review this function in the following Toolbox.

From Your Toolbox: Price-Demand Function

$p(x)$ gives the price at which people buy exactly x units of a product.

A common way to represent the price-demand function in economics is with the letter d. Quite simply, $d(x)$ yields the price at which exactly x units of a product will be sold. We should note that this section contains many applications that are also taught in many economics courses. Since many of you will see the letter d for the price-demand function in your economics course, we will use this notation in this section. Again, we will use d to represent the price-demand function, which we will simply call the **demand function.**

As shown in **Figure 7.3.1**, the demand function is a decreasing function. Intuitively, this seems reasonable. As the price decreases, more people will buy the product. Economists refer to the curve in Figure 7.3.1 simply as a **demand curve.** This is because it expresses the price per unit as a function of the quantity, x, in demand.

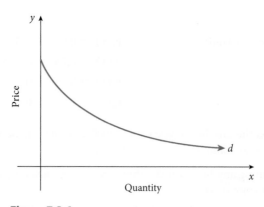

Figure 7.3.1

Example 1: Evaluating a Demand Function

Skinner Bikes, a bicycle retailer, has determined that a demand function for a certain brand of mountain bike is given by $d(x) = -0.3x + 330$, where x is the quantity demanded each month, and $d(x)$ is the price per bicycle in dollars. Compute $d(200)$, $d(250)$, and $d(300)$ and interpret each.

Perform the Mathematics

Since $d(x) = -0.3x + 330$, we have

$$d(200) = -0.3(200) + 330 = \$270 \text{ per bicycle}$$
$$d(250) = -0.3(250) + 330 = \$255 \text{ per bicycle}$$
$$d(300) = -0.3(300) + 330 = \$240 \text{ per bicycle}$$

The interpretation of these results are, respectively,

200 bicycles will be sold each month at a price of \$270 per bicycle.
250 bicycles will be sold each month at a price of \$255 per bicycle.
300 bicycles will be sold each month at a price of \$240 per bicycle. ■

Consumers' Surplus

For the sake of argument, let's assume that you really like the bicycle discussed in Example 1 and were willing to pay \$275 for it. If it currently costs, say, \$200, then in a way you "saved" \$75; the \$275 you were willing to pay minus the \$200 price. Now consider **any** price above \$200 that some consumer is willing to pay minus the actual price of \$200. The sum of these savings over all possible prices above \$200 is known as the **consumers' surplus** for this bicycle. In general, consumers' surplus measures the benefit that consumers get from an economy where competition keeps prices down.

The actual price at which an item is sold, called the **market price**, is influenced by many factors. However, when a market price is determined, we can compute the quantity that consumers demand using the demand function. Returning to our bicycle retailer in Example 1, we note that at a price of \$255 per bicycle, 250 bicycles are demanded each month.

In **Figure 7.3.2** is a shaded rectangle whose area represents what economists call the **actual consumer expenditure**. In general, the actual consumer expenditure is the amount of money that consumers spend for x bicycles at price $d(x)$ dollars per bicycle. In this case, the actual consumer expenditure is computed to be

$$(\$255)(250) = \$63{,}750$$

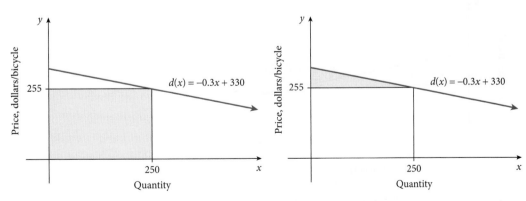

Figure 7.3.2 Actual consumer expenditure.

Figure 7.3.3 Shaded area represents consumers' surplus.

The area of the rectangle in Figure 7.3.2 is the **actual consumer expenditure,** whereas the area below the demand curve $d(x) = -0.3x + 330$ and above the rectangle gives a *total savings* that consumers realize by buying at \$255 per bicycle. See **Figure 7.3.3**. This total savings is what is known as **consumers' surplus.**

DEFINITION

Consumers' Surplus

The **consumers' surplus** is the total amount saved by consumers who are willing to pay more than market price, d_{mp}, for a product, yet are able to purchase the product for d_{mp}.

Example 2: Using a Definite Integral to Determine Consumers' Surplus

a.   Set up a definite integral to determine the area of the shaded region in Figure 7.3.3, which represents the consumers' surplus for the bicycle from Example 1.

b.   Evaluate the integral from part (a) to determine the area of the shaded region, as well as determining the consumers' surplus.

Perform the Mathematics

a.   Here we need to determine the area between two curves. We have top curve $d(x) = -0.3x + 330$ and bottom curve is $y = 255$. From our work in Section 7.2, we represent the area of the shaded region as

$$\int_0^{250} [(-0.3x + 330) - 255]\, dx = \int_0^{250} (-0.3x + 75)\, dx$$

b.   We evaluate the integral in part (b) to be

$$\int_0^{250} (-0.3x + 75)\, dx = \left(-\frac{0.3}{2}x^2 + 75x\right)\Big|_0^{250} = 9375$$

So the consumers' surplus is $9375 each month.   ■

Example 2 nicely illustrates how we can **compute consumers' surplus** for any situation.

Computing Consumers' Surplus

For a demand function d and a point (x_m, d_{mp}) on the graph of d, where x_m is called the **market demand** and d_{mp} is called the **market price**, we determine the consumers' surplus by integrating

$$\int_0^{x_m} [d(x) - d_{mp}]\, dx$$

OBJECTIVE 1

Determine consumers' surplus.

Example 3: Computing Consumers' Surplus

Assume that the market price of the bicycle in Example 1 is $210 per bicycle. Determine the consumers' surplus.

Perform the Mathematics

Recall that $d(x) = -0.3x + 330$; its graph is shown in **Figure 7.3.4**.

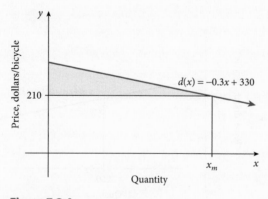

Figure 7.3.4

We know that the market price is $210, but we do not have the market demand. Since we need the market demand in our definite integral (it is the upper limit of integration), we must solve

$$210 = -0.3x + 330$$
$$-120 = -0.3x$$
$$400 = x$$

To find the consumers' surplus, we need to integrate

$$\int_0^{400} [(-0.3x + 330) - 210]\,dx = \int_0^{400} (-0.3x + 120)\,dx$$

$$= \left(-\frac{0.3}{2}x^2 + 120x \right) \Big|_0^{400}$$

$$= 24{,}000$$

So the consumers' surplus is $24,000 each month. ■

▶ **Try It Yourself**

Some related Exercises are 19 and 21.

Supply Function

Now let's take a look at things from the producers' perspective. We know that when prices go up, consumers usually demand less. However, manufacturers tend to respond to higher prices by *supplying more*. So a typical supply curve, denoted s, that relates the price per unit $s(x)$ as a function of quantity, x, supplied should be an increasing function, as shown in **Figure 7.3.5**. Notice that as quantity x increases, so does price.

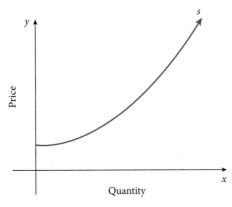

Figure 7.3.5

DEFINITION

Supply Function

The **supply function**, s, for any product gives the price $s(x)$ at which exactly x units of the product will be supplied.

The **supply function** is defined in the box to the right.

Example 4: Evaluating a Supply Function

Suppose that the supplier of the bicycle that we have discussed in this section has a supply function given by $s(x) = 0.2x + 75$, where x is the quantity supplied each month, and $s(x)$ is the price per bicycle in dollars. Compute $s(100)$, $s(200)$, and $s(300)$ and interpret each.

Perform the Mathematics

Since $s(x) = 0.2x + 75$, we have

$$s(100) = 0.2(100) + 75 = \$95 \text{ per bicycle}$$
$$s(200) = 0.2(200) + 75 = \$115 \text{ per bicycle}$$
$$s(300) = 0.2(300) + 75 = \$135 \text{ per bicycle}$$

The interpretations of these results are, respectively,

> The supplier will supply 100 bicycles each month at a price of $95 per bicycle.
>
> The supplier will supply 200 bicycles each month at a price of $115 per bicycle.
>
> The supplier will supply 300 bicycles each month at a price of $135 per bicycle.   ■

Producers' Surplus

Let's return to the scenario where the market price for a bicycle is $200 per bicycle. Suppose that the bicycle supplier is willing to stay in business if the price per bicycle dropped to $150. The fact that bicycles sell for $200 gives a "gain" of $50 per bicycle to the supplier. Now consider any price below $200 at which some supplier is willing to supply these bicycles, subtracted from the actual price of $200. The sum of these gains over all possible prices below $200 is known as the **producers' surplus**.

However, once a market price, d_{mp}, is determined for an item, we can compute the quantity that producers will supply by the supply function. Our bicycle supplier in Example 4 supplies 100 bicycles each month when the price is $95 per bicycle. The shaded region in **Figure 7.3.6** gives the actual amount that the supplier received or, if you prefer, the **revenue**. (From the consumers' perspective, this amount is called *consumer expenditure*.) We can quickly compute this amount to be $95(100) = $9500.

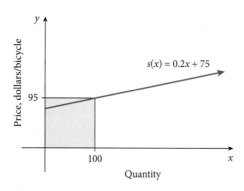

Figure 7.3.6 Shaded region represents revenue.

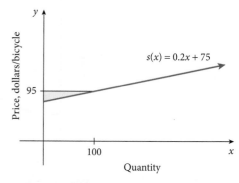

Figure 7.3.7 Shaded region represents producer's surplus.

Figure 7.3.6 shows the actual amount (or revenue) received, while **Figure 7.3.7** shows the area above the supply curve, $s(x) = 0.2x + 75$, and below the line $y = 95$. This area in Figure 7.3.7 is the extra amount that suppliers receive over what they are willing to receive. This extra amount is called the **producers' surplus**.

To determine the area of the shaded region in Figure 7.3.7, and therefore the producers' surplus, we need to determine the area between the two curves $y = 95$ and $s(x) = 0.2x + 75$. This is found by integrating

$$\int_0^{100} [95 - (0.2x + 75)]\, dx = \int_0^{100} (-0.2x + 20)\, dx$$

Notice that, just like computing consumers' surplus, **computing producers' surplus** is analogous to determining the area between two curves.

Example 5: Determining Producers' Surplus

Assume the supply function from Example 4 for the bicycle to be $s(x) = 0.2x + 75$, where $s(x)$ is the price per bicycle in dollars, and x is the number of bicycles. Determine the producers' surplus if the market price is $210 per bicycle.

Perform the Mathematics

Since $s(x) = 0.2x + 75$ and the market price is $210 per bicycle, to determine producers' surplus we must first determine the quantity supplied, or the market demand. To determine the market demand, we need to solve

$$210 = 0.2x + 75$$
$$135 = 0.2x$$
$$x = 675$$

To find producers' surplus, we integrate

$$\int_0^{675} [210 - (0.2x + 75)]\,dx = \int_0^{675} (135 - 0.2x)\,dx$$
$$= \left(135x - \frac{0.2}{2}x^2 \right)\Bigg|_0^{675}$$
$$= 45{,}562.50$$

So the producers' surplus is $45,562.50 each month. ∎

▶ **Try It Yourself**

Some related Exercises are 33 and 35.

Equilibrium Point

If we graphed a demand function, d, and a supply function, s, together on the same set of axes, we have the situation shown in **Figure 7.3.8**. The demand, x, at which the demand and supply curves intersect is called the **equilibrium demand**, while the price is called the **equilibrium price**. The point of intersection for d and s is called the **equilibrium point**. Also notice in Figure 7.3.8 that both consumers' surplus and producers' surplus can be shown together.

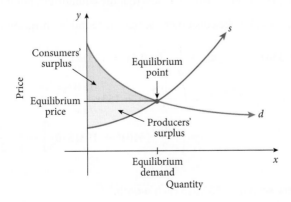

Figure 7.3.8

OBJECTIVE 3

Determine consumers' and producers' surplus at equilibrium.

Example 6: Locating an Equilibrium Point

Continuing with our bicycle scenario, if the demand function is given by $d(x) = -0.3x + 330$ and the supply function is given by $s(x) = 0.2x + 75$, complete the following.

a. Determine the equilibrium point.

b. Determine the consumers' surplus at the equilibrium demand.

c. Determine the producers' surplus at the equilibrium demand.

Perform the Mathematics

a. **Figure 7.3.9** has a graph of d and s.

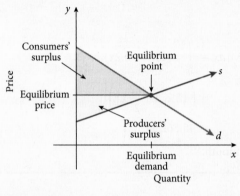

Figure 7.3.9

We determine the equilibrium point algebraically by setting $d(x) = s(x)$ and solving for x. Once we have a value for x, the equilibrium demand, we can substitute it into either $d(x)$ or $s(x)$ to determine the equilibrium price. Setting $d(x) = s(x)$ and solving for x gives us

$$-0.3x + 330 = 0.2x + 75$$

$$255 = 0.5x$$

$$510 = x$$

We have the equilibrium demand to be 510 bicycles each month. The equilibrium price is found by using either $d(x)$ or $s(x)$. Using $s(x)$, we get

$$s(x) = 0.2x + 75$$

$$s(510) = 0.2(510) + 75 = 177$$

The equilibrium price is \$177, which means that the equilibrium point is (510, \$177).

b. As shown in Figure 7.3.9, the consumers' surplus is found by integrating

$$\int_0^{510} [(-0.3x + 330) - 177]\,dx = \int_0^{510} (-0.3x + 153)\,dx$$

$$= \left(\frac{-0.3}{2}x^2 + 153x \right)\Big|_0^{510}$$

$$= \left[\frac{-0.3}{2}(510)^2 + 153(510) \right] - \left[\frac{-0.3}{2}(0)^2 + 153(0) \right]$$

$$= 39{,}015$$

So the consumers' surplus is \$39,015 each month.

c. As shown in Figure 7.3.9, the producers' surplus is determined by integrating

$$\int_0^{510} [177 - (0.2 + 75)]\,dx = \int_0^{510} (102 - 0.2x)\,dx$$

$$= \left(102x - \frac{0.2}{2}x^2 \right)\Big|_0^{510}$$

$$= \left[102(510) - \frac{0.2}{2}(510)^2 \right] - \left[102(0) - \frac{0.2}{2}(0)^2 \right]$$

$$= 26{,}010$$

So the producers' surplus is $26,010 each month.

Try It Yourself

Some related Exercises are 43 and 45.

Lorenz Curves and Gini Index

It is a fact that in our society some individuals make more money than others. Economists measure the gap between the rich and the poor by examining the proportion of the total income that is earned by the lowest 20% of the population, and then the proportion earned by the lowest 40% of the population, and so on. For example, data for income distribution in the United States in 2008 are shown in **Table 7.3.1**.

Table 7.3.1

Proportion of population	Proportion of income
0	0
0.20	0.034
0.40	0.12
0.60	0.267
0.80	0.50
1.00	1.00

Source: U.S. Census Bureau

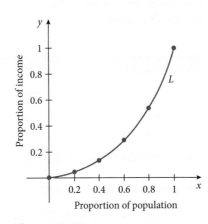

Figure 7.3.10

The table tells us that in 2008 the lowest 40% of the population earned only $0.12 = 12\%$ of the total income. The graph of $y = L(x)$ through the data points in **Figure 7.3.10** is called a **Lorenz curve**. A Lorenz curve gives us the proportion of total income earned by the lowest proportion, x, of the population. The domain for L is $[0, 1]$, and the range is $[0, 1]$.

There are two extreme cases of income distribution, which means that there are two extreme cases for Lorenz curves. They are:

1. Absolute equality of income distribution.
2. Absolute inequality of income distribution.

Absolute equality of income distribution means that everyone earns the same income. The lowest 20% earns 20% of the income, the lowest 40% earns 40% of the income, and so on. The Lorenz curve for absolute equality is $L(x) = x$, as shown in **Figure 7.3.11**. Absolute inequality of income distribution means that no one earns any income except one person who earns it all. This would yield the Lorenz curve shown in **Figure 7.3.12**.

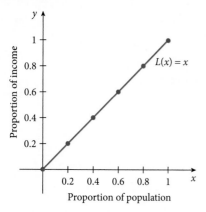

Figure 7.3.11 Lorenz curve for absolute equality.

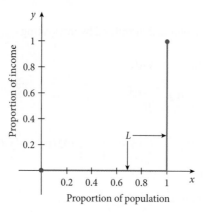

Figure 7.3.12 Lorenz curve for absolute inequality.

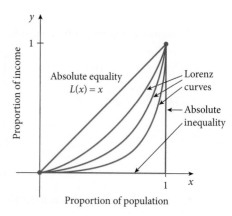

Figure 7.3.13

All Lorenz curves that we will analyze lie between these two extremes. See **Figure 7.3.13**. To measure how the actual income distribution differs from absolute equality, we will calculate the area between $L(x) = x$ (absolute equality) and the Lorenz curve modeling our income distribution.

Since the domain and range of any Lorenz curve are $[0, 1]$, the area between absolute equality and absolute inequality is $\frac{1}{2}$. (The region between these two extremes is $\frac{1}{2}$ of a square whose area is 1 square unit.) It follows that the area between any Lorenz curve and absolute equality is at most $\frac{1}{2}$. However, economists multiply this area by 2 to get a number between 0 and 1, with 0 being absolute equality and 1 being absolute inequality.

The area between a Lorenz curve and absolute equality gives a measure that economists call the **Gini Index** or the **Gini coefficient**. A larger Gini Index means more area between the Lorenz curve and absolute equality, which then means a greater inequality in income distribution.

OBJECTIVE 4

Determine a Gini Index.

Example 7: Computing a Gini Index

In Utopia Land, the Lorenz curve for income distribution is modeled by $L(x) = x^2$.

a. Graph $L(x) = x$, absolute equality, and $L(x) = x^2$, the Lorenz curve for Utopia Land on the same set of axes.

b. Compute the Gini Index for Utopia Land.

Perform the Mathematics

a. **Figure 7.3.14** gives the required graphs.

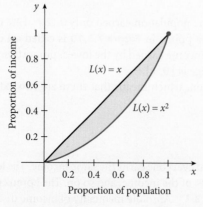

Figure 7.3.14

b. The shaded region shown in Figure 7.3.14 is the area between two curves. We compute it to be

$$\int_0^1 (x - x^2)\,dx = \left(\frac{1}{2}x^2 - \frac{1}{3}x^3\right)\Big|_0^1 = \left[\frac{1}{2}(1)^2 - \frac{1}{3}(1)^3\right] - \left[\frac{1}{2}(0)^2 - \frac{1}{3}(0)^3\right] = \frac{1}{6}$$

We now multiply by 2 to get the Gini Index of

$$2\left(\frac{1}{6}\right) = \frac{1}{3}$$

So the Gini Index for Utopia Land is $\frac{1}{3}$ or approximately 0.33. ∎

> **Computing a Gini Index**
>
> For a Lorenz curve, L, the **Gini Index** is found by
>
> $$2 \cdot \int_0^1 [x - L(x)]\,dx$$

We can now generalize the work done in Example 7 for **computing a Gini Index**.

Example 8: Computing a Gini Index

The Lorenz curve for the distribution of income in the United States in 2008 can be modeled by

$$L(x) = 2.6x^4 - 3.81x^3 + 2.38x^2 - 0.175x \qquad 0 \le x \le 1$$

Compute the Gini Index for the United States in 2008. (*Source:* U.S. Census Bureau.)

Perform the Mathematics

The Gini Index is found by

$$2 \cdot \int_0^1 [x - L(x)]\,dx = 2 \cdot \int_0^1 [x - (2.6x^4 - 3.81x^3 + 2.38x^2 - 0.175x)]\,dx$$

$$= 2 \cdot \int_0^1 (-2.6x^4 + 3.81x^3 - 2.38x^2 + 1.175x)\,dx$$

$$= 2 \cdot \left(\frac{-2.6}{5}x^5 + \frac{3.81}{4}x^4 - \frac{2.38}{3}x^3 + \frac{1.175}{2}x^2\right)\Big|_0^1$$

$$\approx 2 \cdot (0.2267)$$

$$= 0.4534$$

So the Gini Index in 2008 was about 0.4534. ∎

▶ *Try It Yourself*

Some related Exercises are 69 and 71.

Summary

In this section we studied applications for area between two curves. **Consumers' surplus** and **producers' surplus** are simply areas between two curves, and we stress that they should be viewed that way, as opposed to simply memorizing a formula. We also discussed where to locate an **equilibrium point** and the relationship between **consumer expenditure** and **revenue**. The final topic in this section dealt with income distributions and a measure of income distribution called the Gini Index. Again, this is an application of area between two curves.

- **Computing consumers' surplus:** For a demand function d and a point (x_m, d_{mp}) on the graph of d, where x_m is called the **market demand** and d_{mp} is called the **market price**, we determine the consumers' surplus by integrating $\int_0^{x_m} [d(x) - d_{mp}]\,dx$.

- **Computing producers' surplus:** For a supply function s and a point (x_m, d_{mp}) on the graph of s, where x_m is called **market demand** and d_{mp} is called **market price**, we determine the producers' surplus by integrating $\int_0^{x_m} [d_{mp} - s(x)]\,dx$.

- **Computing a Gini Index:** For a Lorenz curve, L, the **Gini Index** is found by integrating $2 \cdot \int_0^1 [x - L(x)]\,dx$.

Section 7.3 Exercises

Vocabulary Exercises

1. A function that gives the price of a product based on its quantity is called a _____ function.

2. The total amount saved by consumers who are willing to pay more than market price is called the _____ _____.

3. The total amount gained by producers who are willing to receive less than the market price is called the _____ _____.

4. The point at which a supply and a demand function intersect is called the _____ point.

5. A function that gives us the proportion of total income earned by the lowest x percent of the population is called the _____ curve.

6. A measurement of the income distribution of a country's residents is called the _____ _____.

Skill Exercises

*For Exercises 7–10, refer to **Figure 7.3.15**.*

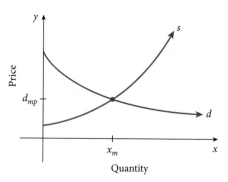

Figure 7.3.15

7. Shade the region that corresponds to the consumers' surplus. Write an integral to determine consumers' surplus.

8. Shade the region that corresponds to the producers' surplus. Write an integral to determine producers' surplus.

9. Shade the region that corresponds to the actual consumer expenditure. Determine the actual consumer expenditure.

10. Shade the region that corresponds to the revenue. Determine the actual revenue.

*For Exercises 11 and 12, refer to **Figure 7.3.16**.*

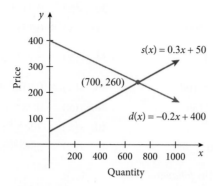

Figure 7.3.16

11. Shade the region that corresponds to the producers' surplus. Write an integral to determine producers' surplus.

12. Shade the region that corresponds to the consumers' surplus. Write an integral to determine consumers' surplus.

*For Exercises 13 and 14, refer to **Figure 7.3.17**.*

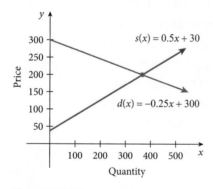

Figure 7.3.17

13. Shade the region that corresponds to the producers' surplus. Write an integral to determine producers' surplus.

14. Shade the region that corresponds to the consumers' surplus. Write an integral to determine consumers' surplus.

For Exercises 15–28, determine the consumer's surplus for the given demand function d and demand level x.

15. $d(x) = 3500 - 3x$; $x = 200$

16. $d(x) = 3500 - 3x$; $x = 500$

17. $d(x) = 220 - \dfrac{1}{3}x$; $x = 100$

18. $d(x) = 330 - \dfrac{1}{5}x$; $x = 200$

19. $d(x) = 600 - 0.6x$; $x = 150$

20. $d(x) = 540 - 0.4x$; $x = 300$

21. $d(x) = 1610 - 0.08x^2$; $x = 100$

22. $d(x) = 2720 - 0.05x^2$; $x = 200$

23. $d(x) = 400 - \dfrac{1}{3}x^2$; $x = 30$

24. $d(x) = 300 - \dfrac{1}{4}x^2$; $x = 30$

25. $d(x) = 600e^{-0.01x}$; $x = 300$

26. $d(x) = 550e^{-0.02x}$; $x = 150$

27. $d(x) = 230e^{-0.02x}$; $x = 175$

28. $d(x) = 320e^{-0.01x}$; $x = 100$

For Exercises 29–42, determine the producer's surplus for the given supply function and demand level x.

29. $s(x) = 0.2x + 50$; $x = 75$

30. $s(x) = 0.3x + 66$; $x = 90$

31. $s(x) = 0.15x + 80$; $x = 100$

32. $s(x) = 0.27x + 60$; $x = 80$

33. $s(x) = \dfrac{1}{3}x + 55$; $x = 60$

34. $s(x) = \dfrac{1}{2}x + 40$; $x = 70$

35. $s(x) = 0.05x^2 + 20$; $x = 100$

36. $s(x) = 0.01x^2 + 25$; $x = 200$

37. $s(x) = 0.1x^2 + 10$; $x = 150$

38. $s(x) = 0.2x^2 + 15$; $x = 110$

39. $s(x) = 10e^{0.02x}$; $x = 90$

40. $s(x) = 20e^{0.01x}$; $x = 120$

41. $s(x) = 13e^{0.01x}$; $x = 110$

42. $s(x) = 25e^{0.03x}$; $x = 55$

For Exercises 43–52, complete the following.

(a) Determine the equilibrium point. In Exercises 51 and 52, use the `Intersect` command on your calculator and round equilibrium demand and equilibrium price to the nearest whole number.

(b) Determine the consumers' surplus at equilibrium demand.

(c) Determine the producers' surplus at equilibrium demand.

43. $d(x) = 3553 - 13x$; $s(x) = 5.7x$

44. $d(x) = 329 - \dfrac{1}{5}x$; $s(x) = 0.27x$

45. $d(x) = 398 - 0.4x^2$; $s(x) = 0.5x + 11$

46. $d(x) = 89 - 0.25x$; $s(x) = 0.85x + 19.7$

47. $d(x) = 2743 - 0.04x^2$; $s(x) = 0.06x^2 + 20.5$

48. $d(x) = 2528 - 0.01x^2$; $s(x) = 0.02x^2 + 5$

49. $d(x) = 500 - 0.02x^2$; $s(x) = 0.2x^2 + 10$

50. $d(x) = 1426 - 0.02x^2$; $s(x) = 0.01x^2 + 13.33$

 51. $d(x) = 600e^{-0.01x}$; $s(x) = 1.125e^{0.02x}$

 52. $d(x) = 550e^{-0.02x}$; $s(x) = 20e^{0.01x}$

Application Exercises

53. Economics—Consumers' Surplus: Lindsay's Department Store has determined that the demand function for a new type of nonstick frying pan is given by

$$d(x) = -1.4x + 25$$

where x represents the number of pans demanded each day, and $d(x)$ is the price per pan, measured in dollars.

(a) Assuming that the equilibrium price is $11 per pan, determine the equilibrium demand.

(b) Determine the consumers' surplus at equilibrium demand.

54. Economics—Consumers' Surplus: The demand function for a new slow cooker is given by

$$d(x) = 98 - 7\sqrt{x}$$

where x is the number of slow cookers demanded each day, and $d(x)$ is the price per slow cooker in dollars.

(a) Assuming that the equilibrium price is $28 per slow cooker, determine the equilibrium demand.

(b) Determine the consumers' surplus at equilibrium demand.

55. Economics—Equilibrium: Frickel's Department Store has determined that the demand function for an extra-wide toaster is given by

$$d(x) = 27.3e^{-0.09x}$$

where x is the number of toasters demanded each day, and $d(x)$ is the price per toaster in dollars. Assuming that the equilibrium price is \$12.14, determine the consumers' surplus at equilibrium demand. Round equilibrium demand to the nearest whole number.

56. **Economics—Consumers' Surplus:** The Hacker's Delight has determined that the demand function for the new Doppler Don Driving Iron is given by

$$d(x) = -x^2 - x + 650$$

where x is the number of driving irons demanded each month, and $d(x)$ is the price per driving iron in dollars. Assuming that the equilibrium price is \$230, determine the consumers' surplus.

57. **Economics—Producers' Surplus:** Balata Inc., a producer of golf balls, has determined that the supply function for the new Xtrah golf ball is given by

$$s(x) = 0.24x + 3.70$$

where x is the number of dozens supplied each month, and $s(x)$ is the price per dozen.

(a) If the equilibrium price is \$19.30 per dozen, determine the equilibrium demand.

(b) Determine the producers' surplus at equilibrium demand.

58. **Economics—Producers' Surplus:** Linguini's Pizza Palace has determined that the supply function for a large pizza is given by

$$s(x) = 0.02x^2 + 0.78$$

where x is the number of pizzas supplied each day, and $s(x)$ is the price per pizza.

(a) If the equilibrium price of a pizza is \$8 per pizza, determine the equilibrium demand.

(b) Determine producers' surplus at equilibrium demand.

59. **Economics—Producers' Surplus:** Balata Inc., a producer of golf balls, has determined that the supply function for the new Equalizer golf ball is given by

$$s(x) = 6.7e^{0.02x}$$

where x is the number of dozens supplied each month, and $d(x)$ is the price per dozen. Assuming that the equilibrium price is \$19.75 per dozen, determine the producers' surplus. Round the equilibrium demand to the nearest whole number.

60. **Economics—Producers' Surplus:** Baker's Bake Shoppe, a supplier of specialty baking pans, has determined that the supply function for a certain birthday cake baking pan is given by

$$s(x) = 4e^{0.05x}$$

where x is the number of these baking pans supplied each month, and $s(x)$ is the price per pan in dollars. Assuming that the equilibrium price is \$8 per pan, determine the producers' surplus. Round equilibrium demand to the nearest whole number.

61. **Economics—Consumers'/Producers' Surplus:** BuildIt, a local home improvement store, has determined that the demand function for a 30-gallon trash can is given by

$$d(x) = 13 - 0.01x^2$$

while the related producer supply function is given by $s(x) = 0.1x + 1$, where x is the daily quantity, and $d(x)$ and $s(x)$ are in dollars per can.

(a) Determine the equilibrium point.

(b) Determine the consumers' surplus at equilibrium.

(c) Determine the producers' surplus at equilibrium.

62. **Economics—Consumers'/Producers' Surplus:** Fresh Paint, a local paint supply and paint accessories store, has determined that the demand function for a step ladder is given by

$$d(x) = -0.2x^2 + 100$$

while the related producer supply function is given by $s(x) = x + 1$, where x is the weekly quantity, and $d(x)$ and $s(x)$ are in dollars per ladder.

(a) Determine the equilibrium point. Round the equilibrium demand to the nearest whole number and round the equilibrium price to the nearest dollar.

(b) Determine the consumers' surplus at equilibrium.

(c) Determine the producers' surplus at equilibrium.

63. **Economics—Consumers'/Producers' Surplus:** Office House, an office supply store, has determined that the demand function for a floppy disk storage case is given by

$$d(x) = 30 - x$$

while the related producer supply function is given by

$$s(x) = \sqrt{x}$$

where x is the weekly quantity, and $d(x)$ and $s(x)$ are in dollars per storage case.

(a) Determine the equilibrium point.

(b) Determine the consumers' surplus at equilibrium.

(c) Determine the producers' surplus at equilibrium.

64. **Economics—Consumers'/Producers' Surplus:** Just Smell It, a scented candle shop, has determined that the demand function for a large vanilla-scented candle is given by

$$d(x) = 20 - \frac{1}{2}x$$

while the related producer supply function is given by

$$s(x) = \frac{1}{2}\sqrt{x}$$

where x is the monthly quantity, and $d(x)$ and $s(x)$ are in dollars per candle.

(a) Determine the equilibrium point. Round the equilibrium demand to the nearest whole number and round the equilibrium price to the nearest dollar.

(b) Determine the consumers' surplus at equilibrium.

(c) Determine the producers' surplus at equilibrium.

Macroeconomics—Gini Index: In Exercises 65–72, the techniques of regression were used on data from the Census Bureau to obtain Lorenz curves for selected years. Recall that x represents the proportion of the population, and L(x) represents the proportion of income. Determine the Gini Index for each year. Where appropriate, round to the nearest ten-thousandth.

65. In 1990, the Lorenz curve was given by the function $L(x) = 2.03x^4 - 3.15x^3 + 2.22x^2 - 0.01x$.

66. In 2000, the Lorenz curve was given by the function $L(x) = 2.29x^4 - 3.4x^3 + 2.23x^2 - 0.13x$.

67. In 2004, the Lorenz curve was given by the function $L(x) = 2.24x^4 - 3.29x^3 + 2.2x^2 - 0.14x$.

68. In 2005, the Lorenz curve was given by the function $L(x) = 2.29x^4 - 3.38x^3 + 2.23x^2 - 0.15x$.

69. In 2006, the Lorenz curve was given by the function $L(x) = 2.3x^4 - 3.37x^3 + 2.2x^2 - 0.14x$.

70. In 2007, the Lorenz curve was given by the function $L(x) = 2.08x^4 - 3.04x^3 + 2.08x^2 - 0.12x$.

71. In 2008, the Lorenz curve was given by the function $L(x) = 2.19x^4 - 3.21x^3 + 2.16x^2 - 0.14x$.

72. In 2009, the Lorenz curve was given by the function $L(x) = 2.17x^4 - 3.15x^3 + 2.11x^2 - 0.13x$.

Concept and Writing Exercises

73. Explain in a complete sentence the meaning of *consumers' surplus*.

74. Explain in a complete sentence what the Gini Index indicates.

75. A report from the United Nations shows that in 2010, the Scandinavian country of Denmark had a Gini Index of 0.247 while the African nation of Namibia had a Gini Index of 0.743.

(*Source:* UN Human Development Report.) Based on these indices, in which one of these countries would you prefer to start a family? Explain your answer.

76. Suppose we have a demand function of the form $d(x) = b - ax$, where a and b are constants, and let (x_m, d_{mp}) represent a point on the function d; x_m is the market demand. Compute an expression of the consumer' surplus.

77. Let $L(x) = b^2 - bx + x$ represent a Lorenz curve where b is a proper fraction. Determine an expression for the Gini Index.

78. The Gini Index is similar to a snapshot in time. The index does not depend on time. However, a collection of indices over some period of time produces a "moving picture" of income distribution. Exercises 66–72 are a collection over a period of time from 2000 to 2009. Analyze the Gini Indices over this period of time. Does the Gini Index appear to be increasing, decreasing, or staying the same over this time period? What does this mean about income distribution in the United States over this time period?

 Section Project

The Casual Day office supply store has collected the following data on the demand for a certain brand of briefcase.

Number of briefcases demanded per week	Price ($ per briefcase)
2	79.99
5	61.99
7	51.99
9	43.99
12	33.99
15	25.99

Meanwhile, the supplier of the briefcases has collected the supply/price data.

Number supplied each week	Price ($ per briefcase)
1	23.99
3	29.99
6	39.99
8	47.49
11	61.99
13	70.99
16	89.99

(a) Let x represent the quantity demanded per week. Enter the data into your calculator, plot the data, and determine a quadratic regression model for d. Round coefficients to the nearest thousandth.

(b) Now let x represent the quantity supplied each week. Enter the data into your calculator, plot the data, and determine a quadratic regression model for s. Round coefficients to the nearest thousandth.

(c) Graph d and s in the viewing window [0, 20] by [0, 70].

(d) Determine the equilibrium point by using the `Intersect` command. Round the equilibrium demand to the nearest whole number and round the equilibrium price to the nearest dollar.

(e) Determine the consumers' surplus at equilibrium.

(f) Determine the producers' surplus at equilibrium.

7.4   Integration by Parts

In this section we present a technique of integration called **integration by parts**. Integration by parts is somewhat analogous to the Product Rule for differentiating. Applications involving this new technique will include total production from oil fields and a return to continuous income streams, first presented in Section 7.1. Here we will analyze the **present value** of a continuous income stream.

Integration by Parts

In many real-life situations, managers encounter integrals that cannot be evaluated with the integration techniques that we have learned so far. Other integration techniques are needed to solve these problems. For example, the manager of a taxicab fleet knows that the variable cost per mile for maintaining a cab that has been in service for x years since 2010 is modeled by

$$C(x) = 8.3 + 0.88 \ln x$$

To find the average variable cost per mile for a cab that has been in service for 2 years since 2010, the manager would need to evaluate the integral

$$\frac{1}{2} \int_0^2 (8.3 + 0.88 \ln x) \, dx$$

We cannot evaluate this integral because we have not yet determined $\int \ln x \, dx$. A new method of integration, called **integration by parts**, can help solve this problem.

To see where the formula for integration by parts comes from, let's assume that we have two differentiable functions, u and v. We need to recall the definition of the differential from Section 3.4, as shown in the following Toolbox.

From Your Toolbox: Differential in y, dy

The differential of y, dy, is defined to be $dy = f'(x) \, dx$.

So, in differential notation, for our two functions u and v, we have

$$du = u' dx \quad \text{and} \quad dv = v' dx$$

The derivative of $u \cdot v$ is determined by the Product Rule to be

$$\frac{d}{dx}(u \cdot v) = u'v + uv'$$

Now, if we integrate both sides of this equation with respect to x, we get

$$\int \left[\frac{d}{dx}(u \cdot v) \right] dx = \int (u'v + uv') \, dx$$

$$u \cdot v = \int u'v \, dx + \int uv' \, dx$$

Since $du = u' dx$ and $dv = v' dx$, we rewrite this as

$$u \cdot v = \int v \, du + \int u \, dv$$

We now solve this equation for $\int u \, dv$ and get

$$\int u \, dv = u \cdot v - \int v \, du$$

This last equation is the formula for the integration technique known as **integration by parts**.

Integration by Parts

For two differentiable functions u and v,

$$\int u \, dv = u \cdot v - \int v \, du$$

When applying the integration by parts formula, we are really performing a *double substitution*. Example 1 shows how this works. After Example 1, we will supply some general guidelines for using integration by parts.

Example 1: Using the Integration by Parts Formula

OBJECTIVE 1

Use integration by parts to determine an indefinite integral.

Determine $\int xe^x dx$.

Perform the Mathematics

The integration by parts formula has four parts that must be determined: u, du, v, and dv. Our goal is to turn our original integral, $\int xe^x dx$, into an integral of the form $\int u\,dv$. This is where the double substitution comes in. We will let u equal some part of $xe^x dx$, and then dv will equal the rest of $xe^x dx$. We decide to select $u = x$, since $du = 1 \cdot dx$, and $dv = e^x dx$, since $\int e^x dx = e^x$. So we then have

$$u = x \qquad\qquad dv = e^x dx$$
$$du = 1 \cdot dx = dx \qquad v = \int e^x dx = e^x$$

Applying the integration by parts formula yields

$$\int u\,dv = u \cdot v - \int v\,du$$
$$\int xe^x dx = xe^x - \int e^x dx$$
$$= xe^x - e^x + C \qquad\blacksquare$$

Notice that the differentials du and dv include the dx. Also, note that when we integrate dv to get v, we can omit the constant of integration C. One constant C at the very end is enough.

The key step in using integration by parts is selecting the *two* substitutions u and dv. We rewrite the original integral in the form $\int u\,dv$ by substituting u for part of the original integrand and dv for the rest. We choose u and dv so that the resulting expression is simpler than the original. In Example 1 we chose u and dv so that the resulting expression has $\int e^x dx$, which we can easily integrate. We offer the **guidelines for choosing u and dv** in the box to the right.

Example 2: Using the Integration by Parts Formula

Determine $\int x \ln x\,dx$.

Perform the Mathematics

In the integrand $x \ln x$, we can easily integrate x, but not $\ln x$. With this in mind, we select $u = \ln x$ and $dv = x\,dx$, which means that the four required pieces are

$$u = \ln x \qquad\qquad dv = x\,dx$$
$$du = \frac{1}{x}dx \qquad\qquad v = \int x\,dx = \frac{1}{2}x^2$$

Applying the integration by parts formula yields

$$\int u\,dv = uv - \int v\,du$$
$$\int x \ln x\,dx = (\ln x) \cdot \frac{1}{2}x^2 - \int \frac{1}{2}x^2 \cdot \frac{1}{x}dx$$
$$= \frac{1}{2}x^2 \ln x - \frac{1}{2}\int x\,dx$$
$$= \frac{1}{2}x^2 \ln x - \frac{1}{2} \cdot \frac{1}{2}x^2 + C = \frac{1}{2}x^2 \ln x - \frac{1}{4}x^2 + C \qquad\blacksquare$$

Guidelines for the Selection of u and dv

1. Either select dv to be the most complicated part of the integrand that is easily integrated. Then u is the rest.

2. Or, select u so that its derivative, du, is a simpler function than u. Then dv is the rest.

▶ *Try It Yourself*

Some related Exercises are 13 and 15.

Example 3: Using the Integration by Parts Formula

Determine $\int \ln x \, dx$.

Perform the Mathematics

This one looks a little unusual in that there does not appear to be anything multiplied by $\ln x$. If we imagine $\ln x$ to be $1 \cdot \ln x$, then we can employ our guidelines. We urge caution here. It may be tempting to let $u = 1$, but this would force $dv = \ln x \, dx$. We do not know the integral of $\ln x$—that is what we are trying to determine. Basically, we have no choice but to select $u = \ln x$ and $dv = 1 \cdot dx$. This gives us

$$u = \ln x \qquad dv = 1 \cdot dx$$
$$du = \frac{1}{x} dx \qquad v = \int 1 \, dx = x$$

Applying the integration by parts formula yields

$$\int u \, dv = uv - \int v \, du$$
$$\int \ln x \, dx = x \ln x - \int x \cdot \frac{1}{x} dx$$
$$= x \ln x - \int 1 \, dx$$
$$= x \ln x - x + C \qquad \blacksquare$$

Example 4: Using the Integration by Parts Formula

Determine $\int x e^{0.2x} dx$.

Perform the Mathematics

From the guidelines, we select $u = x$ and $dv = e^{0.2x} dx$. This gives

$$u = x \qquad dv = e^{0.2x} dx$$
$$du = dx \qquad v = \int e^{0.2x} dx = \frac{1}{0.2} e^{0.2x} = 5 e^{0.2x}$$

So, via the integration by parts formula, we have

$$\int u \, dv = uv - \int v \, du$$
$$\int x e^{0.2x} dx = 5x e^{0.2x} - \int 5 e^{0.2x} dx = 5x e^{0.2x} - 5 \int e^{0.2x} dx$$
$$= 5x e^{0.2x} - 5 \left(\frac{1}{0.2} e^{0.2x} \right) + C = 5x e^{0.2x} - 25 e^{0.2x} + C \qquad \blacksquare$$

▶ *Try It Yourself*

Some related Exercises are 21 and 23.

Many times, when evaluating an integral using integration by parts and part of the integrand involves e^{ax}, choose $dv = e^{ax}dx$. We then immediately have $v = \frac{1}{a}e^{ax}$.

Example 5: Evaluating an Indefinite Integral

Determine $\int 2xe^{x^2}dx$.

Perform the Mathematics

Do not be fooled! This is nothing more than a u-substitution first encountered in Section 6.4. To integrate $\int 2xe^{x^2}dx$, we let $u = x^2$, then $du = 2xdx$, and by u-substitution we have

$$\int 2xe^{x^2}dx = \int e^u du = e^u + C = e^{x^2} + C$$

∎

Example 5 serves as a reminder that, even though we are currently learning a new integration technique in integration by parts, we should not forget our previously learned techniques. Our final example, before we consider some applications, requires two uses of the integration by parts formula.

Example 6: Utilizing the Integration by Parts Formula Twice

Determine $\int x^2 e^{2x}dx$.

Perform the Mathematics

From the guidelines and the comment following Example 4, we select $u = x^2$ and $dv = e^{2x}dx$. This gives

$$u = x^2 \qquad dv = e^{2x}dx$$

$$du = 2xdx \qquad v = \frac{1}{2}e^{2x}$$

The integration by parts formula gives

$$\int x^2 e^{2x}dx = \frac{1}{2}x^2 e^{2x} - \int \frac{1}{2}e^{2x} \cdot 2xdx = \frac{1}{2}x^2 e^{2x} - \int xe^{2x}dx$$

This time, the integration by parts formula did not produce an integral that we could quickly evaluate. However, the new integral is simpler than the original, and it looks similar to the one that we did in Example 4. So let's use integration by parts again. For the integral, $\int xe^{2x}dx$, we select $u = x$ and $dv = e^{2x}dx$. This gives

$$u = x \qquad dv = e^{2x}dx$$

$$du = dx \qquad v = \frac{1}{2}e^{2x}$$

Integration by parts yields

$$\int xe^{2x}dx = \frac{1}{2}xe^{2x} - \int \frac{1}{2}e^{2x}dx = \frac{1}{2}xe^{2x} - \frac{1}{2}\int e^{2x}dx$$

$$= \frac{1}{2}xe^{2x} - \frac{1}{2}\left(\frac{1}{2}e^{2x}\right) + C$$

$$= \frac{1}{2}xe^{2x} - \frac{1}{4}e^{2x} + C$$

OBJECTIVE 2

Use integration by parts twice to determine an indefinite integral.

Putting it all together results in

$$\int x^2 e^{2x} dx = \frac{1}{2} x^2 e^{2x} - \int x e^{2x} dx$$

$$= \frac{1}{2} x^2 e^{2x} - \left(\frac{1}{2} x e^{2x} - \frac{1}{4} e^{2x} + C \right)$$

$$= \frac{1}{2} x^2 e^{2x} - \frac{1}{2} x e^{2x} + \frac{1}{4} e^{2x} + C \qquad \blacksquare$$

▶ *Try It Yourself*

Some related Exercises are 41 and 43.

Applications

Most of the applications that we consider in this section require us to evaluate a definite integral. We will disregard whatever limits of integration we have until after we have found an indefinite integral, using whatever technique is required.

Example 7: Estimating Oil Field Production

OBJECTIVE 3

Apply integration by parts to integrating a rate function.

Geologists have estimated that an oil field will produce oil at a rate given by

$$B(t) = 5te^{-0.2t}$$

where $B(t)$ is thousands of barrels per month, t months from the start of operation. Estimate the total production in the first year of operation.

Understand the Situation

Since B is a rate function, we know that the total production in the first year of operation is given by a definite integral. So we need to evaluate

$$\int_0^{12} 5te^{-0.2t} dt$$

Perform the Mathematics

As mentioned before Example 7, we will ignore the limits of integration until after we have found an antiderivative. Using the guidelines for integration by parts, we select $u = 5t$ and $dv = e^{-0.2t} dt$. This gives us

$$u = 5t \qquad\qquad dv = e^{-0.2t} dt$$

$$du = 5dt \qquad\qquad v = \frac{1}{-0.2} e^{-0.2t} = -5e^{-0.2t}$$

So, by integration by parts, we have

$$\int 5te^{-0.2t} dt = 5t(-5e^{-0.2t}) - \int (-5e^{-0.2t}) \cdot 5dt$$

$$= -25te^{-0.2t} + 25 \int e^{-0.2t} dt$$

$$= -25te^{-0.2t} + 25 \left(\frac{1}{-0.2} e^{-0.2t} \right) + C$$

$$= -25te^{-0.2t} - 125e^{-0.2t} + C$$

Now to evaluate the *definite* integral and answer the question, we evaluate from $t = 0$ to $t = 12$ and get

$$(-25te^{-0.2t} - 125e^{-0.2t}) \Big|_0^{12}$$

$$= [-25(12)e^{-0.2(12)} - 125e^{-0.2(12)}] - [-25(0)e^{-0.2(0)} - 125e^{-0.2(0)}]$$

$$\approx 86.44$$

Interpret the Result

So, in the first year of operation, the total production is estimated to be 86.44 thousand barrels.

▷ *Try It Yourself*

Some related Exercises are 53 and 54.   ■

Example 7 reviewed the concept of integrating a rate function gives a total accumulation. In the following Flashback, we review the concept of a continuous income stream.

Flashback: Continuous Income Stream Revisited

In Section 7.1 we introduced the concept of a continuous income stream. Recall that a continuous income stream is very similar to the electric company, natural gas company, or water company that has a meter on our house that continuously monitors and totals the amount of electricity, natural gas, or water that we are consuming. Suppose that the rate of change of income in thousands of dollars per year for the oil field in Example 7 is projected to be

$$f(t) = 500e^{-0.09t}$$

where t is the number of years from when oil extraction began. See **Figure 7.4.1**. Find the total income from this field during the first year of operation.

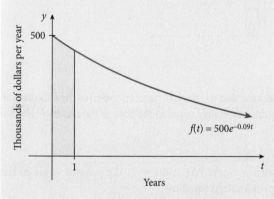

Figure 7.4.1 Shaded area gives total income.

Perform the Mathematics

Again, since we have a rate function, determining the total income generated during the first year of operation is equivalent to determining the area under the curve from $t = 0$ to $t = 1$. Integrating yields

$$\int_0^1 500e^{-0.09t}dt = \left(\frac{500}{-0.09}e^{-0.09t}\right)\Big|_0^1$$

$$= \frac{500}{-0.09}e^{-0.09(1)} - \frac{500}{-0.09}e^{-0.09(0)}$$

$$\approx 478.1600818$$

So in the first year of operation this oil field will generate approximately $478,160.08 in income.   ■

Recall from Section 7.1 that the rate of change of income in the Flashback is called the **rate of flow function**. Also in Section 7.1, we stated that in general the *total income* produced from time $t = a$ to $t = b$ is found by $\int_a^b f(t)dt$.

Now that we have properly reviewed continuous income streams, we turn our attention to the **present value of a continuous income stream**.

Present Value of a Continuous Income Stream

Suppose that we have a continuous income stream from an oil field and we know the rate of flow function. The ability to determine its **present value** is very important if we wish to sell (or buy) this oil field. We need to recall that, if P dollars are invested at an annual interest rate r compounded continuously, then the amount A after t years is given by $A = Pe^{rt}$. Solving this equation for P yields

$$P = Ae^{-rt}$$

which we say is the **present value** of A. This means that P is the amount that needs to be invested *now* at interest rate r compounded continuously so that we have A dollars t years from now.

We now generalize the present value concept to continuous income streams by revisiting rectangles and the definite integral. First, we assume that the present money can be invested at the given rate, r, compounded continuously over the given time interval. Even though this assumption may sound unrealistic, managers make these assumptions regularly based on historical data.

Suppose that f is the rate of flow function for a continuous income stream. Divide the interval $[0, T]$ into n equal subintervals of length Δt, as shown in **Figure 7.4.2**.

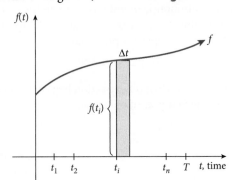

Figure 7.4.2

Present Value of a Continuous Income Stream

If f is the rate of flow function for a continuous income stream, then the **present value**, P, at annual interest rate, r, compounded continuously for T years is given by

$$P = \int_0^T f(t)e^{-rt}dt$$

In a typical subinterval, we can pick any point t_i. Let's say that it is the left-hand endpoint of the subinterval. *The total income produced over this subinterval is approximately equal to the area of the rectangle shaded in Figure 7.4.2.* This area is given by

$$f(t_i)\Delta t$$

Using the present value formula, $P = Ae^{-rt}$, with $A = f(t_i)\Delta t$ and $t = t_i$, the present value of the income received over this subinterval, P_i, is approximately equal to

$$P_i \approx f(t_i)\Delta t \cdot e^{-rt_i} = f(t_i)e^{-rt_i}\Delta t$$

To get the total **present value on the interval** $[0, T]$ requires us to sum *all* these present values. We know from our work in Chapter 6 that this is given by the definite integral $\int_0^T f(t)e^{-rt}dt$.

OBJECTIVE 4

Determine present value of a continuous income stream.

Example 8: Determining the Present Value of a Continuous Income Stream

A window washing business generates income at the rate of $3t$ thousand dollars per year, where t is the number of years from now.

a.   Determine the present value of this continuous income stream for the next seven years at 8% compounded continuously.

b.   Determine the total amount (income plus interest) produced by the window washing business over this seven-year period.

Perform the Mathematics

a. We know that $f(t) = 3t$, so the present value for the next seven years at 8% compounded continuously is

$$P = \int_0^T f(t)e^{-rt}dt = \int_0^7 3te^{-0.08t}dt$$

Since this definite integral requires integration by parts, we will ignore the limits of integration until we have determined an antiderivative. Here we select $u = 3t$ and $dv = e^{-0.08t}dt$. This yields

$$u = 3t \qquad dv = e^{-0.08t}dt$$

$$du = 3dt \qquad v = \frac{1}{-0.08}e^{-0.08t} = -12.5e^{-0.08t}$$

Using integration by parts, we have

$$\int 3te^{-0.08t}dt = 3t(-12.5e^{-0.08t}) - \int -12.5e^{-0.08t} \cdot 3dt$$

$$= -37.5te^{-0.08t} + 37.5\int e^{-0.08t}dt$$

$$= -37.5te^{-0.08t} + 37.5\left(\frac{1}{-0.08}e^{-0.08t}\right) + C$$

$$= -37.5te^{-0.08t} - 468.75e^{-0.08t} + C$$

We now evaluate this from $t = 0$ to $t = 7$ and get

$$\left. (-37.5te^{-0.08t} - 468.75e^{-0.08t}) \right|_0^7 \approx 51.053372$$

So the present value of this continuous income stream over the next 7 years is about $51,053.37.

b. The total amount (income plus interest) earned over the next seven years is equivalent to the amount earned by investing the present value $51,053.37 in an account earning 8% annual interest compounded continuously for seven years. So we use the compound interest formula and get

$$A = Pe^{rt} = 51,053.37e^{0.08(7)} \approx 89,377.73$$

So the total amount (income plus interest) produced by this window washing business over this seven-year period is $89,377.73. ∎

▶ *Try It Yourself*

Some related Exercises are 61 and 63.

Example 8 illustrates how to determine the present value of a continuous income stream and also how to determine the total amount (income plus interest) generated by the continuous income stream. If we wish to compute just the total income generated by the continuous income stream, we would evaluate $\int_0^7 3t\,dt$.

Summary

In this section we presented an integration technique known as **integration by parts**. We suggest becoming familiar with the guidelines for selecting u and dv, as well as when we need to use integration by parts. The section concluded with some applications involving integration by parts. We also determined a formula for the present value of a continuous income stream.

- **Integration by Parts Formula:** $\int u\,dv = u \cdot v - \int v\,du$
- **Guidelines for the Selection of *u* and *dv*:** (1) Select *dv* to be the most complicated part of the integrand that is easily integrated. Then *u* is the rest. (2) Or, select *u* so that its derivative, *du*, is a simpler function than *u*. Then *dv* is the rest.
- **Present Value of a Continuous Income Stream Formula:** $P = \int_0^T f(t)e^{-rt}dt$

Section 7.4 Exercises

Vocabulary Exercises

1. The integration by parts technique is often used when the integrand is a _____.
2. Given an expression for *u*, we must _____ to get the expression *du*.
3. Given an expression for *dv*, we must _____ to get an expression for *v*.
4. To integrate $\int u\,dv$, the associated functions *u* and *v* must be _____.
5. The definite integral $\int_0^T f(t)e^{-rt}dt$ gives the _____ _____ of a continuous income stream.
6. A function that gives the rate of change in income is called a rate of _____ function.

Evaluate the given indefinite integrals in Exercises 7–24. If integration by parts is required, consult the guidelines for selecting u and dv given in this section.

7. $\int 2xe^{3x}\,dx$

8. $\int 8xe^{5x}\,dx$

9. $\int xe^{4x}\,dx$

10. $\int xe^{6x}\,dx$

11. $\int xe^{-x}\,dx$

12. $\int xe^{-3x}\,dx$

13. $\int xe^{-0.03x}\,dx$

14. $\int xe^{-0.07x}\,dx$

15. $\int x^2\ln x\,dx$

16. $\int x^3\ln x\,dx$

17. $\int \ln 2x\,dx$

18. $\int \ln 3x\,dx$

19. $\int (x+3)e^x\,dx$ (Hint: Let $u = x + 3$.)

20. $\int (x-5)e^x\,dx$ (Hint: Let $u = x - 5$.)

21. $\int 6te^{-0.1t}\,dt$

22. $\int 5te^{-0.06t}\,dt$

23. $\int 1.2te^{-0.08t}\,dt$

24. $\int 2.3te^{-0.05t}\,dt$

For Exercises 25–32, evaluate the definite integrals.

25. $\int_0^2 xe^{-x}\,dx$

26. $\int_0^1 5xe^{-2x}\,dx$

27. $\int_0^5 xe^{-0.04x}\,dx$

28. $\int_0^6 xe^{-0.03x}\,dx$

29. $\int_1^e x\ln x\,dx$

30. $\int_1^3 x\ln x\,dx$

31. $\displaystyle\int_0^5 5te^{-0.06t}\,dt$

32. $\displaystyle\int_0^3 6te^{-0.1t}\,dt$

In Exercises 33–36, evaluate the integral of the form $\int(x+c)(x+d)^n\,dx$ by letting $u=x+c$ and $dv=(x+d)^n$ and applying the integration by parts formula.

33. $\displaystyle\int(x+2)(x+1)^5\,dx$

34. $\displaystyle\int(x+2)(x-3)^6\,dx$

35. $\displaystyle\int(x-2)(x+3)^6\,dx$

36. $\displaystyle\int(x-1)(x-2)^4\,dx$

For Exercises 37–46, evaluate the following by using two or more applications of integration by parts.

37. $\displaystyle\int x^2 e^{3x}\,dx$

38. $\displaystyle\int x^2 e^x\,dx$

39. $\displaystyle\int x^2 e^{-x}\,dx$

40. $\displaystyle\int x^2 e^{4x}\,dx$

41. $\displaystyle\int x^2 e^{5x}\,dx$

42. $\displaystyle\int(x+3)^2 e^x\,dx$

43. $\displaystyle\int(\ln x)^2\,dx$

44. $\displaystyle\int x(\ln x)^2\,dx$

45. $\displaystyle\int x(\ln x)^3\,dx$

46. $\displaystyle\int(\ln x)^3\,dx$

In Exercises 47–54, determine the method needed to evaluate the given integral. If the method selected is integration by parts, state the choice for u and dv. If the method selected is u-substitution, simply state the choice for u.

47. $\displaystyle\int xe^{-x}\,dx$

48. $\displaystyle\int xe^{-x^2}\,dx$

49. $\displaystyle\int\frac{1}{x\ln x}\,dx$

50. $\displaystyle\int\frac{\ln x}{x}\,dx$

51. $\displaystyle\int(x+6)e^x\,dx$

52. $\displaystyle\int(x-5)e^x\,dx$

53. $\displaystyle\int 3xe^{2x}\,dx$

54. $\displaystyle\int 3x^2 e^{x^3}\,dx$

Application Exercises

55. Geology—Oil Production: It is estimated that an oil field will produce oil at a rate given by $B(t)=6te^{-0.15t}$ thousand barrels per month, t months into production.

(a) Write a definite integral to estimate the total production for the first year of operation.

(b) Evaluate the integral from part (a) to estimate the total production for the first year of operation.

56. Geology—Oil Production: It is estimated that an oil field will produce oil at a rate given by $B(t)=7.5te^{-0.13t}$ thousand barrels per month, t months into production.

(a) Write a definite integral to estimate the total production for the first year of operation.

(b) Evaluate the integral from part (a) to estimate the total production for the first year of operation.

57. Business Management—Profit Function: The BrenKev restaurant supply company has determined that its marginal profit function is given by

$$P'(t)=2te^{0.2t}$$

where t is time in years, and $P'(t)$ is in millions of dollars per year. Knowing that $P(0)=0$, recover the profit function $P(t)$.

58. **Business Management—Profit Function:** The VivaMix Corporation has determined that its marginal profit function for their Viva blenders is given by

$$P'(t) = 3te^{0.15t}$$

where t is time in years, and $P'(t)$ is in millions of dollars per year. Knowing that $P(0) = 0$, recover the profit function $P(t)$.

59. **Business Management—Cost Function:** The BrenKev Corporation has determined that its marginal cost function is given by

$$C'(t) = te^{-0.15t}$$

where t is time in years, and $C'(t)$ is in millions of dollars per year. Assuming that $C(0) = 2$, recover the cost function $C(t)$.

60. **Business Management—Profit Function:** The VivaMix Corporation has determined that its marginal cost function is given by

$$C'(t) = 0.3te^{-0.1t}$$

where t is time in years, and $C'(t)$ is in millions of dollars per year. Assuming that $C(0) = 1.5$, recover the cost function $C(t)$.

61. **Pharmacology—Drug Absorption:** Medical researchers have determined that the rate of absorption of a certain medication is given by the rate function

$$f(t) = 4te^{-0.51t} \quad t \geq 0$$

where t represents the number of hours since the medication was taken, and $f(t)$ represents the rate of absorption, measured in milligrams per hour. Evaluate $\int_0^6 f(t)\,dt$ and interpret.

62. **Pharmacology—Drug Absorption:** Medical researchers have determined that the rate of absorption of a certain medication is given by the rate function

$$f(t) = 6te^{-0.42t} \quad t \geq 0$$

where t is the number of hours since the medication was taken, and $f(t)$ represents the rate of absorption, measured in milligrams per hour. Evaluate $\int_0^8 f(t)\,dt$ and interpret.

63. **Management—Income Rate:** Dena's Car Wash generates income at the rate of

$$f(t) = 2t$$

where t is the number of years that the car wash has been in business, and $f(t)$ represents the rate of change in income, measured in million dollars per year. Dena is interested in selling the car wash at some point in the future and wants to know how much it would be worth. Determine the present value of this continuous income stream over the first five years at a continuous compound interest rate of 6%.

64. **Management—Income Rate:** SterilizeIt is a medical instrument company that generates income at the rate given by

$$f(t) = 500,000 + 30,000t$$

where t represents the number of years that the company has been in operation, and $f(t)$ represents the rate of change in income, measured in dollars per year. Suppose an investment company is considering purchasing SterilizeIt in the future. Determine the present value of this continuous income stream over the first seven years at a continuous compound interest rate of 6.5%.

65. **Geology—Oil Production:** Suppose that an oil well for the K-28 Petroleum Corporation produces income at the rate of

$$f(t) = 200e^{-0.1t}$$

thousand dollars per year for t years. Suppose that the Cyntro Oil Company is considering the acquisition of this well from K-28 Petroleum. Determine the present value of this continuous income stream over the first four years of operation assuming a continuous compound interest rate of 5.75%.

66. **Geology—Oil Production:** Suppose that an oil well owned by the Cyntro Oil Company produces income at the rate of

$$f(t) = 150e^{-0.15t}$$

thousand dollars per year for t years. Suppose that K-28 Petroleum wants to purchase this well in order to test new equipment. Determine the present value of this continuous income stream over the first three years of operation assuming a continuous compound interest rate of 5.5%.

67. **Personal Finance—Investment Analysis:** Reggie has a choice of two investments. Each choice requires the same initial investment, and each produces a continuous income stream at an interest rate of 8% compounded continuously. The first investment is an oil well that generates income at a rate of $50t$ thousand dollars per year, where t is the number of years from now. The second investment is a natural gas well that generates income at a rate of $30 + 40t$ thousand dollars per year, where t is the number of years from now. Compare the present value of each investment to determine which the better choice is over the next five years.

68. **Personal Finance—Investment Analysis:** Repeat Exercise 67, except determine which is the better investment over the next 10 years.

69. **Labor Management—Labor Disputes:** Many European countries, including Italy, have been forced to austerity programs to curtail their government debts. Some of this austerity includes making adjustments to pensions and retirement benefits. These adjustments have caused concern among union leaders. "Obviously if there's no correction (to pension rollbacks) there'll be a protest," said Luigi Angeletti, head of the Italian labor union, UIL. (*Source: The Wall Street Journal.*) The number of labor disputes in Italy from 1999 to 2011 can be modeled by

$$f(x) = 1511 - 312\ln x \qquad 9 \le x \le 21$$

where x represents the number of years since 1990, and $f(x)$ represents the number of labor disputes annually. (*Source:* EuroStat.) Determine the average value of f on the interval $[11, 21]$ and interpret.

70. **Labor Management—Labor Disputes:** About 80% of Finnish workers are members of a trade union. (*Source:* Expat.com.) The number of labor disputes in Finland from 2000 to 2011 can be modeled by

$$f(x) = -72 + 73.6\ln x \qquad 10 \le x \le 21$$

where x represents the number of years since 1990, and $f(x)$ represents the number of labor disputes annually. (*Source:* EuroStat.) Determine the average value of f on the interval $[11, 21]$ and interpret.

71. **Nutrition—Organic Livestock:** Many consumers are turning to organic products coming from organic farms for their consumption . The number of livestock organically raised from 2000 to 2011 can be modeled by the function

$$f(x) = xe^{0.18x} \qquad 10 \le x \le 21$$

where x represents the number of years since 1990, and $f(x)$ represents the number of livestock organically raised, measured in thousands. (*Source:* U.S. Department of Agriculture.) Determine the average value of f on the interval $[11, 21]$ and interpret.

© Noel Hendrickson/Thinkstock

72. **International Trade—Military Purchases:** A country that has increased its military purchases from the United States is Australia. Concerning the purchase of C-17 military planes, Defense Minister Stephen Smith said, "We are seeking cost and availability information to enable consideration to be given to the acquisition of C-17 aircraft." (*Source:* military.com.) The amount of U.S. military sales deliveries to Australia from 1995 to 2011 can be modeled by

$$f(x) = 1.11(x - 15.5)e^{0.3x} + 38.53 \qquad 5 \le x \le 21$$

where x represents the number of years since 1990, and $f(x)$ represents the annual amount in U.S. military sales deliveries to Australia, measured in millions of USD. (*Source:* U.S. Department of Defense.) Evaluate and interpret $\dfrac{1}{21 - 11}\int_{11}^{21} f(x)\,dx$.

Concept and Writing Exercises

73. Explain the circumstances in which you would use the integration by parts technique.

74. Write a sentence on how integration by parts and the Product Rule for differentiation are related.

For Exercises 75 and 76, consider the integral $\int x(x + 1)^4 dx$.

75. Show how the integral can be evaluated using the integration by parts technique.

76. Show how the integral can be evaluated using the u-substitution technique.

77. Suppose we wish to integrate $\int (4x - 1)e^{2x^2 - x}dx$. Which technique would you use, u-substitution or integration by parts? Explain your answer.

78. Suppose we wish to integrate $\int (2x^2 - x)e^{4x - 1}dx$. Which technique would you use, u-substitution or integration by parts? Explain your answer.

 Section Project

An industry that has been challenged by recent economic events is the cable television industry. "Although it is tempting to point to over-the-top video as a potential culprit, we believe economic factors such as low housing formation and a high unemployment rate contributed to subscriber declines," said SNL Kagan media analyst Mariam Rondeli. (*Source:* Ars Technica). The number of cable TV subscribers from 2000 to 2009 is shown in the table.

Year	Number of subscribers, measured in millions
2000	66.2
2001	66.7
2003	66.1
2004	65.7
2005	65.3
2006	65.1
2007	65.1
2008	64.3
2009	62.8

(a) Use the regression capabilities of your graphing calculator to determine a linear regression model in the form

$$f(x) = ax + b \qquad 0 \le x \le 9$$

where x represents the number of years since 2000, and $f(x)$ represents the number of cable TV subscribers in millions of households annually.

(b) Inspect the model from part (a). On average, how many subscribers did the cable TV industry lose annually from 2000 to 2009?

(c) Evaluate $f(9) - f(0)$ and interpret the meaning of this value.

From 2000 to 2009, the rate of change in the annual price per year of cable TV can be modeled by the exponential function

$$g(x) = 15.76e^{0.041x} \qquad 0 \le x \le 9$$

where x represents the number of years since 2009, and $g(x)$ represents the annual cost of cable TV, measured in dollars per year.

(d) Evaluate $g(9)$ and interpret.

(e) Use the integration by parts technique to evaluate the integral $\int_0^9 f(x) \cdot g(x)\, dx$ to determine the total revenue of the cable TV industry from 2000 to 2009.

7.5   Numerical Integration

In Chapters 6 and 7 we presented several techniques of integration. In spite of all the varied techniques, we still encounter integrals that cannot be antidifferentiated by any of our methods. For example, $\int e^{-x^2}dx$ has no elementary antiderivative and cannot be evaluated using the Fundamental Theorem of Calculus. (Incidentally, $\int e^{-x^2}dx$ is related to the bell-shaped curve seen in probability and statistics courses.)

So how do we evaluate $\int_0^1 e^{-x^2}dx$ if we cannot compute an antiderivative? We approximate it numerically by interpreting it as an area under a curve. We have already seen how to approximate a definite integral using rectangles. Hence, we are familiar with some numerical methods. In this section, we explore the Trapezoidal Rule as another numerical method. The main use, and beauty, of the Trapezoidal Rule is that it is ideally suited for discrete data.

Trapezoidal Rule

To gain an understanding of the Trapezoidal Rule, we need to review the process behind rectangular approximations.

Example 1: Reviewing Rectangular Approximations

Consider $f(x) = \frac{1}{3}x^3 - 2.5x^2 + 4x + 4$. Use four left endpoint rectangles to approximate $\int_2^6 (\frac{1}{3}x^3 - 2.5x^2 + 4x + 4)dx$, that is, the area under the curve from $x = 2$ to $x = 6$.

Perform the Mathematics

Figure 7.5.1 shows a graph of f along with the four rectangles.

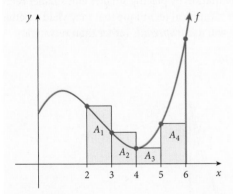

Figure 7.5.1

First, let's recall that:

- a is the left endpoint of the interval.
- b is the right endpoint of the interval.
- n is the number of rectangles.
- Δx is the width, or base, of each rectangle.

So here we have $a = 2$, $b = 6$, and $n = 4$. Then Δx is

$$\Delta x = \frac{b - a}{n} = \frac{6 - 2}{4} = 1$$

Each rectangle has a base of $\Delta x = 1$, and the height is given by $f(\text{left endpoint})$. With this in mind, we find the area of rectangle A_1 to be

$$A_1 = \text{base} \cdot \text{height}$$
$$= \Delta x \cdot f(2)$$
$$= 1 \cdot \frac{14}{3} = \frac{14}{3}$$

For rectangle A_2 we get

$$A_2 = \text{base} \cdot \text{height}$$
$$= \Delta x \cdot f(3)$$
$$= 1 \cdot \frac{5}{2} = \frac{5}{2}$$

In a similar fashion we compute the area of rectangles A_3 and A_4 to be

$$A_3 = \frac{4}{3} \quad \text{and} \quad A_4 = \frac{19}{6}$$

The sum of the four rectangles gives the rectangular approximation. This yields

$$\int_2^6 \left(\frac{1}{3}x^3 - 2.5x^2 + 4x + 4 \right) dx \approx A_1 + A_2 + A_3 + A_4$$

$$= \frac{14}{3} + \frac{5}{2} + \frac{4}{3} + \frac{19}{6} = \frac{35}{3} \quad \blacksquare$$

Recall that in Chapter 6 we improved on this approximation by placing smaller and smaller rectangles under the curve. Now we will investigate another numerical technique that may yield a better approximation with the same value for n. To do this we will use *trapezoids* rather than rectangles.

Let's look at **Figure 7.5.2**.

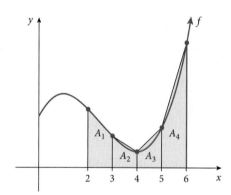

Figure 7.5.2

Here we have redrawn $f(x) = \frac{1}{3}x^3 - 2.5x^2 + 4x + 4$, but instead of four rectangles, we have drawn four trapezoids. It appears that the four trapezoids will yield a better approximation than the four rectangles did. All we need is to recall the formula for the area of a trapezoid, which is

$$A_{\text{trap}} = \frac{1}{2}(h_1 + h_2)b$$

where h_1 and h_2 represent the two heights of a trapezoid, and b represents the base. *Remember that h_1 is parallel to h_2.* Let's rework the function in Example 1 with four trapezoids.

Example 2: Approximating an Integral with Trapezoids

Consider $f(x) = \frac{1}{3}x^3 - 2.5x^2 + 4x + 4$ from Example 1. Use four trapezoids to approximate $\int_2^6 f(x)\,dx$, that is, the area under the graph of f from $x = 2$ to $x = 6$.

Perform the Mathematics

Just as with rectangles, we have $a = 2$, $b = 6$, $n = 4$, and $\Delta x = \frac{b-a}{n} = \frac{6-2}{4} = 1$. So each trapezoid has a base of $\Delta x = 1$. Notice that h_1 and h_2 are found by evaluating the function at the endpoints of the subintervals. The areas of the four trapezoids are

$$
\begin{aligned}
A_1 &= \frac{1}{2}(h_1 + h_2)b \\
&= \frac{1}{2}[f(2) + f(3)] \cdot \Delta x \\
&= \frac{1}{2}\left[\frac{14}{3} + \frac{5}{2}\right] \cdot 1 \\
&= \frac{43}{12}
\end{aligned}
$$

$$
\begin{aligned}
A_2 &= \frac{1}{2}(h_1 + h_2)b \\
&= \frac{1}{2}[f(3) + f(4)] \cdot \Delta x \\
&= \frac{1}{2}\left[\frac{5}{2} + \frac{4}{3}\right] \cdot 1 \\
&= \frac{23}{12}
\end{aligned}
$$

$$
\begin{aligned}
A_3 &= \frac{1}{2}(h_1 + h_2)b \\
&= \frac{1}{2}[f(4) + f(5)] \cdot \Delta x \\
&= \frac{1}{2}\left[\frac{4}{3} + \frac{19}{6}\right] \cdot 1 \\
&= \frac{9}{4}
\end{aligned}
$$

$$
\begin{aligned}
A_4 &= \frac{1}{2}(h_1 + h_2)b \\
&= \frac{1}{2}[f(5) + f(6)] \cdot \Delta x \\
&= \frac{1}{2}\left[\frac{19}{6} + 10\right] \cdot 1 \\
&= \frac{79}{12}
\end{aligned}
$$

The sum of the four trapezoids gives the approximation

$$
\int_2^6 \left(\frac{1}{3}x^3 - 2.5x^2 + 4x + 4\right)dx \approx A_1 + A_2 + A_3 + A_4
$$

$$
= \frac{43}{12} + \frac{23}{12} + \frac{9}{4} + \frac{79}{12} = \frac{43}{3} \qquad \blacksquare
$$

Using the Fundamental Theorem of Calculus, we can determine that the exact answer to $\int_2^6(\frac{1}{3}x^3 - 2.5x^2 + 4x + 4)dx$ is $\frac{40}{3}$. Notice that Example 2 illustrates that using four trapezoids is more accurate than using four rectangles. Also notice that, just as we did with rectangular approximations, our subintervals are of equal length when using trapezoids.

We want to look back at our work in Example 2 and refer to **Figure 7.5.3** in order to generalize the trapezoid process. To find the area of the trapezoids, we calculated

$$A_1 = \frac{1}{2}[f(a) + f(x_1)] \cdot \Delta x$$

$$A_2 = \frac{1}{2}[f(x_1) + f(x_2)] \cdot \Delta x$$

$$A_3 = \frac{1}{2}[f(x_2) + f(x_3)] \cdot \Delta x$$

$$A_4 = \frac{1}{2}[f(x_3) + f(b)] \cdot \Delta x$$

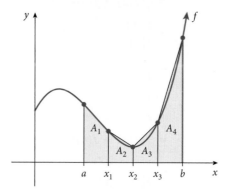

Figure 7.5.3

The total area of the four trapezoids is the sum

$$A_1 + A_2 + A_3 + A_4 = \frac{1}{2}[f(a) + f(x_1)] \cdot \Delta x + \frac{1}{2}[f(x_1) + f(x_2)] \cdot \Delta x$$
$$+ \frac{1}{2}[f(x_2) + f(x_3)] \cdot \Delta x + \frac{1}{2}[f(x_3) + f(b)] \cdot \Delta x$$

Factoring out a $\frac{1}{2}\Delta x$ yields

$$A_1 + A_2 + A_3 + A_4 = \frac{\Delta x}{2}[f(a) + f(x_1) + f(x_1) + f(x_2) + f(x_2) + f(x_3) + f(x_3) + f(b)]$$
$$= \frac{\Delta x}{2}[f(a) + 2f(x_1) + 2f(x_2) + 2f(x_3) + f(b)]$$

The last equation is the Trapezoidal Rule. In general, if n trapezoids are used, then the approximate value of the definite integral is given by the **Trapezoidal Rule**, as follows:

Trapezoidal Rule

If f is continuous on $[a, b]$, then

$$\int_a^b f(x)\,dx \approx \frac{\Delta x}{2}[f(a) + 2f(x_1) + 2f(x_2) + \cdots + 2f(x_{n-1}) + f(b)]$$

where a is the left endpoint, b is the right endpoint, n is the number of trapezoids, and $\Delta x = \frac{b-a}{n}$. Notice that $x_1 = a + \Delta x, x_2 = x_1 + \Delta x,$ and so on.

NOTES: 1. As with rectangular approximations, the larger the value of n, the more trapezoids we will use and the smaller Δx will be. Also, the result of using larger values for n is a better approximation to the definite integral.

2. There is some error involved in the trapezoidal approximation. The formula to compute the error is beyond the scope of this text. However, it is worth noting that doubling the number of trapezoids reduces the maximum error by a factor of 4.

Example 3: Using the Trapezoidal Rule

OBJECTIVE 1

Use trapezoids to approximate a definite integral.

Use the Trapezoidal Rule with four trapezoids to approximate $\int_0^2 \frac{1}{x^2 + 16} dx$.

Perform the Mathematics

We have $a = 0$, $b = 2$, $n = 4$, and $\Delta x = \frac{2 - 0}{4} = \frac{1}{2}$. So by the Trapezoidal Rule we have

$$\int_0^2 \frac{1}{x^2 + 16} dx \approx \frac{\frac{1}{2}}{2}\left[f(0) + 2f\left(\frac{1}{2}\right) + 2f(1) + 2f\left(\frac{3}{2}\right) + f(2) \right]$$

$$\approx \frac{1}{4}[0.0625 + 0.1230 + 0.1176 + 0.1096 + 0.05] \approx 0.1157$$

See **Figure 7.5.4**.

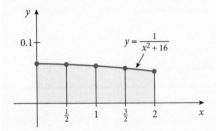

$y = \frac{1}{x^2 + 16}$

Figure 7.5.4

▶ *Try It Yourself*

Some related Exercises are 9 and 11.

◀▶ Technology Option

Because of the structure of the Trapezoidal Rule, it can be easily programmed into a calculator or computer. We supply the following program for the TI calculator family. If you do not have a TI calculator, your code may be different. For this reason, we also supply a line-by-line explanation of what the code is performing.

Code	Explanation
`Disp"ENTER FUNCTION IN Y₁"`	Stores function in Y_1.
`Prompt A, B, N`	Asks for A, B, and N, the left endpoint, right endpoint, and number of trapezoids.
`(B - A)/N → D`	Computes the base of each trapezoid.
`Y₁(A) + Y₁(B) → S`	Evaluates function in Y_1 at A and B; adds the results, and stores sum in S.
`A + D → X`	Left endpoint plus base stored in X.
`Lbl 1`	Beginning of loop.
`2*Y₁(X) + S → S`	Two times function evaluated at X; result added to sum in S; result stored in S.
`X + D → X`	Current X increased by base; result in X.
`If X < B`	If current X < right endpoint, execute next line.

`Then`	Prior line true; execute next line.
`Goto 1`	Return to beginning of loop.
`Else`	If X ≥ B, execute next line.
`S*(D/2) → S`	Current sum in S multiplied by D/2; result in S.
`Disp"TRAP APPROX IS",S`	Display trapezoid approximation.

To use the Trapezoidal Rule program to approximate $\int_{-1}^{1} e^{-x^2} dx$ using $n = 10$, 50, and 100 trapezoids, we first enter $y_1 = e^{-x^2}$ into our calculator. Next we execute the program. Results are listed in **Table 7.5.1**. Note that $a = -1$ and $b = 1$. We conclude that $\int_{-1}^{1} e^{-x^2} dx \approx 1.49$.

Table 7.5.1

n	Trapezoidal approximation to $\int_{-1}^{1} e^{-x^2} dx$
10	1.48873668
50	1.493452053
100	1.493599214

Trapezoidal Rule for Discrete Data

We have now seen how to use the Trapezoidal Rule to approximate $\int_{a}^{b} f(x) dx$. Many times, however, we have a set of data and we cannot determine a model for the data. In other words, we have no $f(x)$. In these circumstances, the Trapezoidal Rule is powerful, as shown in Example 4.

OBJECTIVE 2

Use trapezoids to approximate revenue given discrete data.

Example 4: Using the Trapezoidal Rule for Discrete Data

Table 7.5.2 shows the rate of change of revenue in millions of dollars per year for a bowling alley, where x represents the number of years since 2004.

a. Let $x = 1$ correspond to January 1, 2005, and let y represent the rate of change of revenue in millions of dollars per year. Plot the data.

b. Approximate the bowling alley's revenue from January 1, 2005, to January 1, 2011, by using the Trapezoidal Rule with the given data.

Perform the Mathematics

a. **Figure 7.5.5** gives a plot of the data.

Table 7.5.2

x	Rate of change of revenue
1	3.3
2	2
3	2.5
4	2.8
5	2.9
6	2.2
7	2

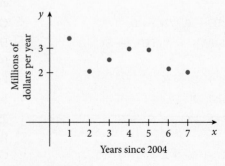

Figure 7.5.5

b. **Figure 7.5.6** is a plot of the data with the trapezoids sketched in as well.

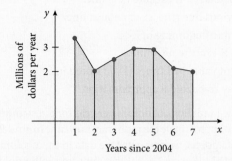

Figure 7.5.6

Here is where the power of the Trapezoidal Rule comes through. Without a model for the data, we can use the data as is and approximate the revenue. To use the Trapezoidal Rule, note that $a = 1, b = 7, n = 6,$ and $\Delta x = 1$. So by the Trapezoidal Rule we have

$$\frac{1}{2}[f(1) + 2f(2) + 2f(3) + 2f(4) + 2f(5) + 2f(6) + f(7)]$$

Now, instead of a formula for our function, we use the functional values from the data in our table. For example, from the table we know that $f(1) = 3.3, f(2) = 2,$ and so on. Proceeding, we get the approximation to be

$$\frac{1}{2}[3.3 + 2(2) + 2(2.5) + 2(2.8) + 2(2.9) + 2(2.2) + 2] = 15.05$$

So the bowling alley's revenue from January 1, 2005, to January 1, 2011, was approximately 15.05 million dollars. ∎

▶ **Try It Yourself**

Some related Exercises are 29 and 31.

◀▶ **Technology Option**

Applying the Trapezoidal Rule to a set of data can be a bit tedious, especially if there are many data points. We supply the following program that performs the Trapezoidal Rule on a set of data. Again, the code is for the TI family of calculators, and the program makes use of Lists. In L_1 (List 1) we place the values of the independent variable, whereas in L_2 (List 2) we place the values of the dependent variable.

`Disp "NUMBER DATA POINTS"`	Asks for the number of data points.
`Prompt P`	Enter number of data points.
`(L₁(P) - L₁(1))/(P - 1) → D`	Computes the base of each trapezoid; stores in D.
`L₂(1) + L₂(P) → S`	Sums first and last dependent variable values; stores sum in S.
`2 → I`	2 stored in I.
`Lbl 1`	Beginning of loop.
`2*L₂(I) + S → S`	Two times dependent variable value in current I added to value in S; result stored in S.

```
I + 1 → I                    Increment size of I.
If I < P                     If updated I < P, execute next line.
Then                         If previous line true, execute next line.
Goto 1                       Return to beginning of loop.
Else                         If I ≥ P, execute next line.
S*(D/2) → S                  Current value in S times (D/2) stored in S.
Disp"TRAP APPROX IS",S       Display trapezoidal approximation.
```

Table 7.5.3 shows the rate of change of revenue in millions of dollars per year for a restaurant, where x represents the number of years since 2004. To approximate the restaurant's revenue from 2005 to 2011 using the Trapezoidal Rule program for data points, we first enter the data in the x column into L_1 and the rate of change of revenue column into L_2. **Figure 7.5.7** shows our screen after doing this.

Table 7.5.3

x	Rate of change of revenue
1	2.5
2	2.8
3	1.2
4	0.9
5	2.7
6	2.1
7	3.1

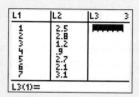

Figure 7.5.7

We now run our Trapezoidal Rule program for data points. We are first asked how many data points, and we enter a 7. The program returns 12.5. So we conclude that from 2005 to 2011, the restaurant's revenue was approximately 12.5 million dollars.

Please note that both Trapezoidal Rule programs are designed for subintervals of equal length. When using the Trapezoidal Rule program for data points, the independent variable values should be equally spaced.

OBJECTIVE 3

Apply trapezoids to analyze sulfur dioxide emissions.

Example 5: Using the Trapezoidal Rule to Analyze Sulfur Dioxide Emissions

Sulfur dioxide is a major air pollutant and can have significant impacts on the environment. In the 1970s, sulfur dioxide was singled out as being a precursor to acid rain as well as atmospheric particulates. **Table 7.5.4** shows the U.S. sulfur dioxide emissions in thousands of tons from 1950 to 2000. **Figure 7.5.8** gives a plot of the data, where $x = 1$ corresponds to the beginning of 1950 and y represents the sulfur dioxide emissions in thousands of tons. (*Source:* U.S. Environmental Protection Agency.)

Table 7.5.4

x	Sulfur dioxide emissions
1	22,358
11	22,227
21	31,161
31	25,905
41	22.433
51	16,348

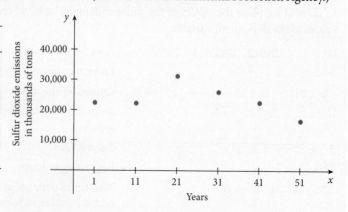

Figure 7.5.8

a. Approximate the total amount of sulfur dioxide emissions by the United States from 1950 to 2000.

b. Approximate the average amount of sulfur dioxide emissions on a yearly basis from 1950 to 2000.

Perform the Mathematics

a. By looking at **Figure 7.5.8**, we see that in order to use the Trapezoidal Rule, we have $a = 1$, $b = 51$, $n = 5$, and $\Delta x = 10$. So, by the Trapezoidal Rule, we have

$$\frac{1}{2}[f(1) + 2f(11) + 2f(21) + 2f(31) + 2f(41) + f(51)] \cdot \Delta x$$

As in Example 4, we use the functional values from **Table 7.5.4** to determine that $f(1) = 22{,}358$, $f(11) = 22{,}227$, and so on. Proceeding, we get the approximation to be

$$\frac{1}{2}[22{,}358 + 2(22{,}227) + 2(31{,}161) + 2(25{,}905) + 2(22{,}433) + 16{,}348] \cdot 10$$

$$= 1{,}210{,}790$$

This means that from 1950 to 2000, the amount of U.S. sulfur dioxide emissions was approximately 1,210,790 thousand tons or 1,210,790,000 tons.

b. To compute the average value, we take the result from part (a) and divide by $b - a$, just as we learned in Section 7.1. This yields

$$\frac{1{,}210{,}790{,}000}{51 - 1} = \frac{1{,}210{,}790{,}000}{50}$$

$$= 24{,}215{,}800$$

We conclude that from 1950 to 2000 the annual average value of the U.S. sulfur dioxide emissions was approximately 24,215,800 tons. ∎

Try It Yourself

Some related Exercises are 37 and 39.

Simpson's Rule

Another popular numerical integration technique is called **Simpson's Rule**. Simpson's Rule utilizes adjacent parabolic segments over a closed interval $[a, b]$. The sum of the areas below these parabolic segments approximates $\int_a^b f(x)\,dx$. Approximations using Simpson's Rule are very accurate, but it requires an even number n of subintervals. This could be limiting with a data set if the data do not have an even number of data points. For the curious, Simpson's Rule can be found in almost any engineering calculus textbook. We will not pursue this numerical technique in this text.

Summary

In this section we introduced our final numerical technique for integration, the **Trapezoidal Rule**. At this time the numerical techniques that we have discussed include:

- Rectangular approximations
- Built-in numerical integration on the calculator
- Trapezoidal approximations

We supplied two programs for the Trapezoidal Rule: one if we are given a function, and another if we do not have a function, or cannot determine one, for a set of data. Applications using the Trapezoidal Rule for data points were discussed. We want to stress that if some data are not easily modeled, the Trapezoidal Rule is a powerful tool to analyze data. The Trapezoidal Rule assumes that the independent variable values of the data points are equally spaced.

Section 7.5 Exercises

Vocabulary Exercises

1. The first method of numerical integration that we studied in Section 6.2 was _____ approximation.

2. A trapezoid is a four-sided figure with exactly one pair of _____ sides.

3. When performing numerical approximation of integrals, Δx represents the _____ of each rectangle or trapezoid.

4. _____ data that cannot be modeled can still be integrated using the Trapezoidal Rule.

5. We generally use numerical integration if the integrand of a definite integral has no _____.

6. A numerical integration method that uses the area under parabolic segments associated with an integrand is called _____ Rule.

Skill Exercises

In Exercises 7–16, use the Trapezoidal Rule to approximate the definite integral using $n = 4$ trapezoids.

7. $\int_{-1}^{3} x^3\,dx$

8. $\int_{0}^{2} x^4\,dx$

9. $\int_{1}^{3} \frac{5x}{5 + 2x^2}\,dx$

10. $\int_{0}^{2} \frac{1}{(1 + 2x)^2}\,dx$

11. $\int_{0}^{1} x^2 e^x\,dx$

12. $\int_{0}^{1} x^5 e^x\,dx$

13. $\int_{3}^{5} (\ln x)^2\,dx$

14. $\int_{2}^{5} (\ln x)^3\,dx$

15. $\int_{0}^{2} x^4\,dx$

16. $\int_{-1}^{3} x^3\,dx$

In Exercises 17–28, use the calculator program for the Trapezoidal Rule when given $f(x)$ to approximate the given definite integral using $n = 10$, $n = 50$, and $n = 100$ trapezoids.

17. $\int_{0}^{2} \frac{1}{(1 + 2x)^2}\,dx$

18. $\int_{1}^{3} \frac{5x}{5 + 2x^2}\,dx$

19. $\int_{1}^{4} \frac{e^{2x}}{1 + e^{4x}}\,dx$

20. $\int_{1}^{2} \frac{e^x - e^{-x}}{2}\,dx$

21. $\int_{1}^{2} \frac{e^x + e^{-x}}{2}\,dx$

22. $\int_{0}^{1} x^2 e^x\,dx$

23. $\int_{0}^{1} x^5 e^x\,dx$

24. $\int_{3}^{5} (\ln x)^2\,dx$

25. $\displaystyle\int_2^5 (\ln x)^3\, dx$

26. $\displaystyle\int_0^2 \frac{2x + 3}{\sqrt{x^2 + 2x + 5}}\, dx$

27. $\displaystyle\int_1^3 \frac{x + 2}{\sqrt[3]{6x - x^2}}\, dx$

28. $\displaystyle\int_0^3 e^{x^2}\, dx$

Application Exercises

For Exercises 29–36, use the Trapezoidal Rule to determine the solution.

29. **Business Management—Revenue Approximation:** The following data give the rate of change of revenue in thousands of dollars per week for Sparkle Time Car Wash. Here x represents the number of weeks after April 1. Approximate Sparkle Time's revenue for this period of time.

x	Rate of change of revenue
1	2.35
2	3.75
3	1.25
4	2.5
5	1.75

30. **Business Management—Sales Approximation:** The following data give the rate of change of sales, in millions of dollars per month, for Spring Brook Mall. Here x corresponds to the month of the year. Approximate Spring Brook Mall's total sales for this period of time.

x	Rate of change of sales
1	3.35
2	2.69
3	1.72
4	1.11
5	2.35
6	2.51

31. **Business Management—Cost Analysis:** The following data give the marginal cost for different levels of production at Muggy Inc. Here x represents the number of coffee mugs produced, and $C'(x)$ is in dollars per mug. Approximate the total cost in going from a production level of 10 mugs to 50 mugs.

x	10	20	30	40	50
$C'(x)$	1.5	2	2.4	3.1	3.5

32. **Business Management—Cost Analysis:** The following data give the marginal cost for different levels of production at Binky Inc. Here x represents the number of pacifiers produced, and $C'(x)$ is in dollars per pacifier. Approximate the total cost in going from a production level of 220 pacifiers to 300 pacifiers.

x	220	240	260	280	300
$C'(x)$	1.18	1.25	1.36	1.48	1.55

33. **Business Management—Profit Analysis:** The Beef Place keeps a monthly record of the rate of increase in profit $P'(x)$ during the first five months of the year. The following are the data,

where x is the month of the year, and $P'(x)$ is measured in thousands of dollars per month. Approximate the total profit during this period of time.

x	1	2	3	4	5
$P'(x)$	0.62	0.77	0.78	0.85	1.03

34. **Business Management—Profit Analysis:** Anthony has been keeping a monthly record of a company's rate of increase in profits $P'(x)$ over a one-year period. The following are the data that Anthony collected, where x is the month of the year, and $P'(x)$ is in thousands of dollars per month. Approximate the total profit for this time period.

x	$P'(x)$	x	$P'(x)$
1	6.2	7	10.7
2	7.7	8	11.1
3	7.8	9	12.1
4	8.5	10	12.3
5	10.3	11	12.1
6	10.2	12	12.5

35. **Business Management—Revenue Approximation:** The following data give the rate of change of revenue, in thousands of dollars per week, for Timber!, a local tree trimming company. Here x represents the number of weeks after May 1. Approximate the company's revenue for this time period.

x	Rate of change of revenue	x	Rate of change of revenue
1	2.25	6	2.12
2	3.75	7	3.11
3	1.35	8	2.11
4	2.55	9	1.50
5	1.85	10	3.12

36. **Business Management—Sales Analysis:** Vanessa is the leader of a sales management team for a company, and her team has been asked to prepare a year-end report. Vanessa's team gathered data showing the rate of change of sales $s'(t)$ in millions of dollars per month, for the prior year, where t is measured in months. The following are the data that were collected. Approximate the company's sales for this time period.

t	$S'(t)$	t	$S'(t)$
0	0.8	7	1.51
1	0.7	8	2.13
2	1.13	9	1.78
3	1.44	10	1.42
4	0.68	11	0.77
5	0.22	12	0.96
6	1.01		

For Exercises 37–44, use the calculator program for the Trapezoidal Rule with discrete data points, where appropriate.

37. Human Resource Management—Women in the Workforce: The data in the table give the number of females, single and married, in the U.S. labor force, in millions, from 2001 to 2009. Here x is in years where $x = 1$ corresponds to the beginning of 2001.

Year	x-value	Single women in the U.S. labor force, measured in millions	Married women in the U.S. labor force, measured in millions
2001	1	18.0	35.2
2003	3	18.4	36.0
2005	5	19.2	35.9
2007	7	19.7	36.9
2009	9	20.2	37.3

Source: U.S. Bureau of Labor Statistics

(a) Approximate the average number of single females, on a yearly basis, in the U.S. labor force from the beginning of 2001 to the beginning of 2009.

(b) Approximate the average number of married females, on a yearly basis, in the U.S. labor force from the beginning of 2001 to the beginning of 2009.

Atmospheric Science—Air Pollution: In Exercises 38, 39, and 40, use the following data for U.S. air pollution emissions for selected years from 1990 to 2010 measured in millions of tons per year. Here x is measured in years, where x = 1 corresponds to the beginning of 1990.

Year	x-value	Nitrogen oxides	Carbon monoxides	Sulfur dioxides
1990	1	25.5	154.2	23.1
1995	6	25.0	126.8	18.6
2000	11	22.6	114.5	16.3
2005	16	19.1	93.0	14.8
2010	21	16.4	77.7	11.5

Source: U.S. Environmental Protection Agency

38. Complete the following.

(a) Approximate the total amount of nitrogen oxide emissions in the United States from the beginning of 1990 to the beginning of 2010.

(b) Approximate the average amount of nitrogen oxide emissions on a yearly basis from the beginning of 1990 to the beginning of 2010.

39. Cap and trade is a method for regulating and reducing the amount of pollution emitted into the atmosphere. It is viewed as a more equitable solution to regulating pollution than a carbon tax, as it creates a commodity out of the right to emit carbon and allows the commodity to be traded on the free market.

(a) Approximate the total amount of carbon monoxide emissions in the United States from the beginning of 1990 to the beginning of 2010.

(b) Approximate the average amount of carbon monoxide emissions on a yearly basis from the beginning of 1990 to the beginning of 2010.

40. Complete the following.

(a) Approximate the total amount of sulfur dioxide emissions in the United States from the beginning of 1990 to the beginning of 2010.

(b) Approximate the average amount of sulfur dioxide emissions on a yearly basis from the beginning of 1990 to the beginning of 2010.

Labor Management—Work Stoppages: In Exercises 41 and 42, use the following data on work stoppages in the United States for selected years from 1999 to 2009. Here x is in years, where x = 1 corresponds to the beginning of 1999, and the number of workers involved is in thousands.

Year	x-value	Number of work stoppages	Number of workers involved, in thousands
1999	1	17	73
2001	3	29	99
2003	5	14	129
2005	7	22	100
2007	9	21	189
2009	11	5	13

Source: U.S. Bureau of Labor Statistics

 41. Complete the following.

(a) Approximate the total number of work stoppages in the United States from the beginning of 1999 to the beginning of 2009.

(b) Approximate the average number of work stoppages, on a yearly basis, in the United States from the beginning of 1999 to the beginning of 2009.

 42. Complete the following.

(a) Approximate the total number of workers involved in work stoppages in the United States from the beginning of 1999 to the beginning of 2009.

(b) Approximate the average number of workers involved in work stoppages, on a yearly basis, in the United States from the beginning of 1999 to the beginning of 2009.

Education—Degrees Conferred: In Exercises 43 and 44, use the following data for the number of college graduates, earning associate's degrees and bachelor's degrees, in thousands from 2000 to 2008. Here, x is in years, where x = 1 corresponds to the beginning of 2000.

Year	x-value	Associate's degrees conferred	Bachelor's degrees conferred
2000	1	565	1238
2002	3	595	1292
2004	5	665	1399
2006	7	713	1486
2008	9	756	1599

Source: U.S. National Center for Education Statistics

 43. Complete the following.

(a) Approximate the total number attaining associate's degrees from the beginning of 2000 to the beginning of 2008.

(b) Approximate the average number attaining associate's degrees, on a yearly basis, from the beginning of 2000 to the beginning of 2008.

 44. Complete the following.

(a) Approximate the total number attaining bachelor's degrees from the beginning of 2000 to the beginning of 2008.

(b) Approximate the average number attaining bachelor's degrees, on a yearly basis, from the beginning of 2000 to the beginning of 2008.

In Exercises 45–48, use the Trapezoidal Rule to answer the following.

45. Media Science—Cable TV: A business that has come upon some challenges recently is the cable TV business. "Although it is tempting to point to over-the-top video as a potential culprit, we believe economic factors such as low housing formation and a high unemployment rate contributed to subscriber declines," said SNL Kagan media analyst Mariam Rondeli. (*Source:* Ars Technica.) The rate of change in the number of cable TV subscribers in the United States from 1990 to 2011 can be modeled by

$$f(x) = 2.17e^{\frac{-x^2}{576} + \frac{x}{24}} - 0.18xe^{\frac{-x^2}{576} + \frac{x}{24}} \qquad 0 \le x \le 21$$

where x represents the number of years since 1990, and $f(x)$ represents the rate of change in the number of cable TV subscribers, measured in millions of subscribers per year. (*Source:* National Cable & Telecommunications Association.) Use the Trapezoidal Rule program, with $n = 50$, to approximate and interpret $\int_0^{21} f(x)\,dx$.

46. Media Science—Cable TV: Combining cable, Internet, and telephone services, cable TV distributors have used a bundling technique to enhance their income. The rate of change in total revenue of the cable TV industry from 1996 to 2011 can be modeled by

$$f(x) = \frac{429 \cdot (1.13^x)}{(1.13^x + 15.68)^2} \qquad 6 \le x \le 21$$

where x represents the number of years since 1990, and $f(x)$ represents the rate of change in revenue in billions of dollars per year. (*Source:* SNL Kagan Media Consultants.) Use the Trapezoidal Rule program, with $n = 50$, to approximate and interpret $\int_6^{21} f(x)\,dx$.

47. Consumerism—Domestic Car Prices: A combination of union concessions and demand for smaller, more efficient cars has resulted in a drop in average domestic new car prices in the past few years. The rate of change in the typical cost in new domestic car prices from 1990 to 2011 can be modeled by

$$f(x) = \frac{-4534x^{0.81}}{(x^{1.81} + 64.4)^2} \qquad 0 \le x \le 21$$

where x represents the number of years since 1990, and $f(x)$ represents the rate of change in new domestic car prices, measured in thousands of dollars per year. (*Source:* U.S. Bureau of Economic Analysis.)

(a) Use the model to determine how rapidly new domestic car prices were decreasing in 2009.

(b) Use the model and the Trapezoidal Rule program with $n = 50$ to determine the total reduction in new domestic car prices from 2009 to 2011.

48. Consumerism—Import Car Prices: In addition to decreases in new domestic car prices, the price of new import cars was falling as well. The rate of change in the typical cost in new import car prices from 1990 to 2011 can be modeled by

$$f(x) = \frac{-475(x^2 + 32.65x - 663.26)}{(x^2 + 10x + 500)^2} \qquad 0 \le x \le 21$$

where x represents the number of years since 1990, and $f(x)$ represents the rate of change new import car prices, measured in thousands of dollars per year. (*Source:* U.S. Bureau of Economic Analysis.)

(a) Use the model to determine how rapidly new import car prices were decreasing in 2009.

(b) Use the model and the Trapezoidal Rule program with $n = 50$ to determine the total reduction in new import car prices from 2009 to 2011.

Concept and Writing Exercises

49. Explain in your own words how the Trapezoidal Rule approximates the area under a curve.

50. Generally speaking, given the choice between using rectangular approximation (from Section 6.2) or the Trapezoidal Rule, which method would result in a better approximation? Explain your answer.

For Exercises 51 and 52, consider the function $f(x) = 6$.

51. Integrate $\int_2^5 f(x)\,dx$ by hand.

52. Use the Trapezoidal Rule to determine the area under the graph of $f(x)$ on the interval $[2, 5]$. Explain the result as compared to Exercise 51.

For Exercises 53 and 54, consider the function $f(x) = 1 + |2x - 4|$.

53. Integrate $\int_0^4 f(x)\,dx$ by hand. You will find it useful to rewrite f as a piecewise-defined function.

54. Use the Trapezoidal Rule to determine the area under the graph of $f(x)$ on the interval $[0, 4]$. Explain the result as compared to Exercise 53.

 Section Project

The cost of college and university tuition has always been a concern, with some saying that we are approaching a "tuition bubble," meaning that the cost of college exceeds the graduate's ability to pay back student loans. The following data give the average tuition and fees as well as room and board charges, in dollars, from 1990 to 2010 for public and private two- and four-year colleges. Here x is in years, where $x = 1$ corresponds to the beginning of 1990.

	Public institutions			
	Tuition and fees		**Room and board**	
***x*-value**	**Two-year colleges**	**Four-year universities**	**Two-year colleges**	**Four-year universities**
1	756	2035	1581	1728
6	1192	2977	1712	2108
11	1338	3768	1834	2628
16	1849	5939	2353	3222
21	2137	7630	2777	3911

Source: U.S. Center for Education Statistics

	Private institutions			
	Tuition and fees		**Room and board**	
***x*-value**	**Two-year colleges**	**Four-year universities**	**Two-year colleges**	**Four-year universities**
1	5196	10,348	1811	2339
6	6914	14,537	2023	3035
11	8235	19,307	2922	2157
16	12,122	25,643	3728	3855
21	11,789	30,251	3272	4376

Source: U.S. Center for Education Statistics

(a) Approximate the average tuition and fees on a yearly basis from 1990 to 2010 at public two-year colleges.

(b) Approximate the average tuition and fees on a yearly basis from 1990 to 2010 at private two-year colleges.

(c) Approximate the average room and board on a yearly basis from 1990 to 2010 at public two-year colleges.

(d) Approximate the average room and board on a yearly basis from 1990 to 2010 at private two-year colleges.

(e) Approximate the average tuition and fees on a yearly basis from 1990 to 2010 at public four-year colleges.

(f) Approximate the average tuition and fees on a yearly basis from 1990 to 2010 at private four-year colleges.

(g) Approximate the average room and board on a yearly basis from 1990 to 2010 at public four-year colleges.

(h) Approximate the average room and board on a yearly basis from 1990 to 2010 at private four-year colleges.

(i) Assume the average yearly amount determined in parts (e) and (f) for a four-year degree. How much more does it cost in tuition and fees for four years at a private four-year college than at a public four-year college?

(j) Assume the average yearly amount determined in parts (a) and (b) for a two-year degree. How much more does it cost in tuition and fees for two years at a private two-year college than at a public two-year college?

7.6 Improper Integrals

SECTION OBJECTIVES

1. Evaluate an improper integral of the form $\int_a^\infty f(x)\,dx$.

2. Evaluate an improper integral of the form $\int_{-\infty}^b f(x)\,dx$.

3. Evaluate an improper integral of the form $\int_{-\infty}^\infty f(x)\,dx$.

4. Apply an improper integral to estimate total oil well production.

5. Apply an improper integral to fund an endowment.

6. Apply an improper integral to determine a probability given a probability density function.

So far all of the definite integrals we have considered have been on a closed bounded interval $[a, b]$. In this section, we relax this condition and define integrals over an unbounded interval such as $(-\infty, b], [a, \infty)$, or even $(-\infty, \infty)$. Integrals defined on unbounded intervals are called **improper integrals** and have many applications. To understand what is happening with improper integrals, we need a brief review of limits at infinity.

Limits at Infinity

In Chapter 2 we discussed the limit concept. We saw that the definitions of the derivative and the definite integral demonstrate how central the limit concept is to calculus. Let's revisit limits at infinity and recall some concepts from our work in Chapter 2. Consult the following Toolbox.

From Your Toolbox: Limits at Infinity

1. For $f(x) = b^x$, b a real number where $b > 1$,
   (a) $\lim\limits_{x \to \infty} b^x = \infty$ and $\lim\limits_{x \to -\infty} b^x = 0$
   (b) $\lim\limits_{x \to \infty} b^{-x} = 0$ and $\lim\limits_{x \to -\infty} b^{-x} = \infty$

2. $\lim\limits_{x \to \infty} \ln x = \infty$

3. $\lim\limits_{x \to \infty} \dfrac{k}{x^n} = 0$ and $\lim\limits_{x \to -\infty} \dfrac{k}{x^n} = 0$, where k is any real constant and n is a positive real number.

Let's quickly review some limits. Remembering that the limit of a constant is that constant yields

$$\lim_{b \to \infty}\left(3 - \frac{1}{b^2}\right) = \lim_{b \to \infty} 3 - \lim_{b \to \infty} \frac{1}{b^2} = 3 - 0 = 3$$

Using other items reviewed in the Toolbox gives us

$$\lim_{b \to \infty} (e^{-2b} - 3) = \lim_{b \to \infty} e^{-2b} - \lim_{b \to \infty} 3 = 0 - 3 = -3$$

Improper Integrals

At this time one may be wondering, "Why are we discussing limits at this point in the book?" The answer becomes apparent as we consider Examples 1 and 2.

Example 1: Evaluating Definite Integrals Using the Fundamental Theorem of Calculus

Consider $f(x) = \frac{1}{x^3}$. Determine the following.

a. $\displaystyle\int_1^5 \frac{1}{x^3}\,dx$ **b.** $\displaystyle\int_1^{10} \frac{1}{x^3}\,dx$ **c.** $\displaystyle\int_1^b \frac{1}{x^3}\,dx$

Perform the Mathematics

a.   Evaluation of $\int_1^5 \frac{1}{x^3}\,dx$ will give us the area of the shaded region shown in **Figure 7.6.1**.

$$\int_1^5 \frac{1}{x^3}\,dx = \int_1^5 x^{-3}\,dx = \left(-\frac{1}{2}x^{-2}\right)\Big|_1^5 = \left(-\frac{1}{2x^2}\right)\Big|_1^5$$

$$= -\frac{1}{2(5)^2} - -\frac{1}{2(1)^2} = \frac{24}{50}$$

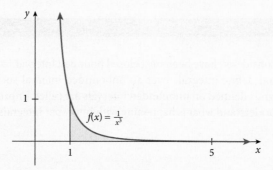

Figure 7.6.1

b.   Evaluation of $\int_1^{10} \frac{1}{x^3}\,dx$ will give us the area shown shaded in **Figure 7.6.2**.

$$\int_1^{10} \frac{1}{x^3}\,dx = \int_1^{10} x^{-3}\,dx = \left(-\frac{1}{2}x^{-2}\right)\Big|_1^{10} = \left(-\frac{1}{2x^2}\right)\Big|_1^{10}$$

$$= -\frac{1}{2(10)^2} - -\frac{1}{2(1)^2} = \frac{99}{200}$$

c.   Here we are asked to determine the area under $f(x) = \frac{1}{x^3}$ from $x = 1$ to some arbitrary point $x = b$, as shown in **Figure 7.6.3**.

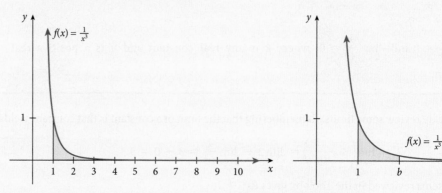

Figure 7.6.2   **Figure 7.6.3**

The process employed in parts (a) and (b) does not change.

$$\int_1^b \frac{1}{x^3}dx = \int_1^b x^{-3}dx = \left(-\frac{1}{2}x^{-2}\right)\Bigg|_1^b = \left(-\frac{1}{2x^2}\right)\Bigg|_1^b$$

$$= -\frac{1}{2(b)^2} - -\frac{1}{2(1)^2} = -\frac{1}{2b^2} + \frac{1}{2} \qquad \blacksquare$$

We are now ready to make sense out of integrating over an unbounded region.

Example 2: Evaluating an Improper Integral

Consider the graph of $f(x) = \frac{1}{x^3}$ and the shaded region shown in **Figure 7.6.4**.

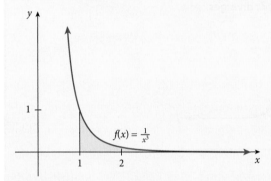

Figure 7.6.4

The shaded region is unbounded, that is, it continues indefinitely to the right. We represent the area of this unbounded region by

$$\int_1^\infty \frac{1}{x^3}dx$$

which is an example of an **improper integral**. Evaluate this integral.

Perform the Mathematics

To do this, we employ a strategy almost identical to Example 1c. That is, we pick an arbitrary point to the right of $x = 1$, call it b, and integrate $\int_1^b \frac{1}{x^3}dx$. We know from Example 1c that

$$\int_1^b \frac{1}{x^3}dx = -\frac{1}{2b^2} + \frac{1}{2}$$

Now we take the limit as $b \to \infty$ of this result and get

$$\lim_{b\to\infty}\left(-\frac{1}{2b^2} + \frac{1}{2}\right) = \lim_{b\to\infty}\left(-\frac{1}{2b^2}\right) + \lim_{b\to\infty}\frac{1}{2}$$

$$= 0 + \frac{1}{2} = \frac{1}{2} \qquad \blacksquare$$

What we have really done here in Example 2 is compute the following:

$$\lim_{b\to\infty}\left[\int_1^b \frac{1}{x^3}dx\right]$$

Notice that this evaluation took place in two stages: first, we integrated using the Fundamental Theorem of Calculus; then we did the limit.

It may contradict our intuition that an unbounded region, as in Figure 7.6.4, has a finite area of one-half square unit, but our intuition is somewhat clouded when we encounter the concept of infinity. Do not despair. Some of the greatest minds in history have struggled with the concept of infinity.

At this time we generalize what we did in Example 2, which was evaluate an **improper integral**.

Improper Integral $\int_a^\infty f(x)\,dx$

Let f be continuous and nonnegative for any $x \geq a$. Then

$$\int_a^\infty f(x)\,dx = \lim_{b \to \infty} \int_a^b f(x)\,dx$$

provided the limit exists. If the limit exists, we say the improper integral $\int_a^\infty f(x)\,dx$ **converges**. If the limit does not exist, we say the improper integral $\int_a^\infty f(x)\,dx$ **diverges**.

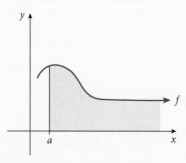

Figure 7.6.5

OBJECTIVE 1

Evaluate an improper integral of the form $\int_a^\infty f(x)\,dx$.

Example 3: Evaluating an Improper Integral

Evaluate

a. $\displaystyle\int_1^\infty \frac{1}{x^2}\,dx$

b. $\displaystyle\int_1^\infty \frac{1}{x}\,dx$

Perform the Mathematics

a.   Evaluating $\int_1^\infty \frac{1}{x^2}\,dx$ will give us the area of the shaded region in **Figure 7.6.6**.

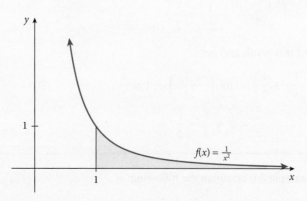

Figure 7.6.6

We begin by rewriting the integral as

$$\int_1^\infty \frac{1}{x^2}dx = \lim_{b\to\infty}\left[\int_1^b \frac{1}{x^2}dx\right]$$

We now perform the integration as follows:

$$\int_1^b \frac{1}{x^2}dx = \int_1^b x^{-2}dx = \left(-x^{-1}\right)\Big|_1^b$$

$$= -\frac{1}{x}\Big|_1^b = -\frac{1}{b}+1$$

Next we evaluate the limit to get

$$\lim_{b\to\infty}\left(-\frac{1}{b}+1\right) = \lim_{b\to\infty}\left(-\frac{1}{b}\right) + \lim_{b\to\infty}1 = 0 + 1 = 1$$

So the improper integral, $\int_1^\infty \frac{1}{x^2}dx$, converges and has a value of 1.

b.   Evaluating $\int_1^\infty \frac{1}{x}dx$ will give us the area of the shaded region shown in **Figure 7.6.7**.

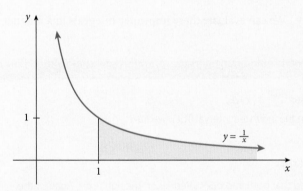

Figure 7.6.7

Proceeding as in part (a) gives us

$$\int_1^\infty \frac{1}{x}dx = \lim_{b\to\infty}\left[\int_1^b \frac{1}{x}dx\right]$$

Integrating yields

$$\int_1^b \frac{1}{x}dx = \ln x\Big|_1^b = \ln b - \ln 1 = \ln b$$

Evaluating the limit of this result gives us

$$\lim_{b\to\infty}(\ln b) = \infty$$

This means that the improper integral, $\int_1^\infty \frac{1}{x}dx$, diverges.   ∎

▶ **Try It Yourself**

Some related Exercises are 7 and 9.

We will discuss two other types of improper integrals in this section. Consider the shaded regions in Figures 7.6.8 and 7.6.9.

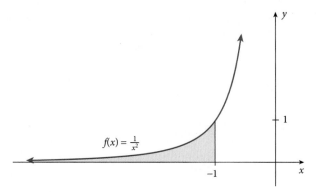

Figure 7.6.8

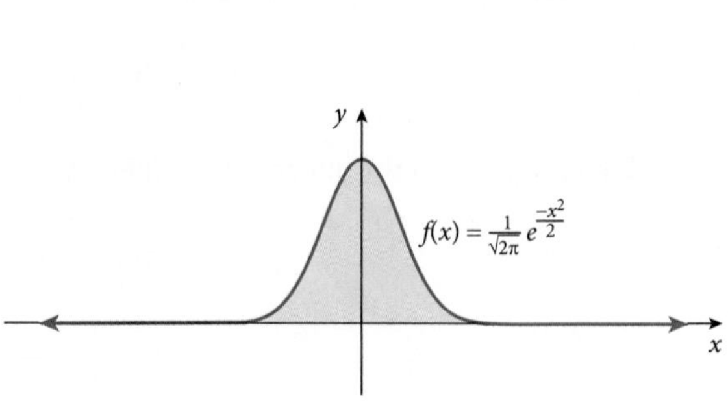

$f(x) = \frac{1}{x^2}$

Figure 7.6.9

$f(x) = \frac{1}{\sqrt{2\pi}} e^{\frac{-x^2}{2}}$

To determine the area under $f(x) = \frac{1}{x^2}$ on the interval $(-\infty, -1]$, as shown in **Figure 7.6.8**, requires that we evaluate the **improper integral** $\int_{-\infty}^{-1} \frac{1}{x^2} dx$. To determine the area under $f(x) = \frac{1}{\sqrt{2\pi}} e^{-(x^2/2)}$ on the interval $(-\infty, \infty)$, as shown in **Figure 7.6.9**, requires that we evaluate the **improper integral** $\int_{-\infty}^{\infty} \frac{1}{\sqrt{2\pi}} e^{-(x^2/2)} dx$. (The graph in Figure 7.6.9 is called the *normal curve*, which is seen frequently in statistics and probability.) We can evaluate these **improper integrals** in a manner that is similar to evaluating $\int_a^{\infty} f(x) dx$.

Improper Integrals $\int_{-\infty}^{b} f(x) dx$ and $\int_{-\infty}^{\infty} f(x) dx$

Let f be continuous and nonnegative on the indicated interval. Then we have:

1. $\int_{-\infty}^{b} f(x) dx = \lim_{a \to -\infty} [\int_a^b f(x) dx]$

2. $\int_{-\infty}^{\infty} f(x) dx = \int_{-\infty}^{c} f(x) dx + \int_c^{\infty} f(x) dx$

In 2, c is any point in $(-\infty, \infty)$, provided that both improper integrals on the right exist. Again, if the limits exist we say that the improper integral **converges.** If the limit does not exist, we say the improper integral **diverges.**

NOTE: If either integral on the right in the second formula in the box diverges, then the improper integral diverges.

OBJECTIVE 2

Evaluate an improper integral of the form $\int_{-\infty}^{b} f(x) dx$.

Example 4: Evaluating an Improper Integral

Evaluate $\int_{-\infty}^{0} e^x dx$.

Perform the Mathematics

First we rewrite the improper integral as

$$\int_{-\infty}^{0} e^x dx = \lim_{a \to -\infty} \left[\int_a^0 e^x dx \right]$$

Just as we did in the previous example, we will integrate first.

$$\int_a^0 e^x dx = e^x \Big|_a^0 = e^0 - e^a = 1 - e^a$$

Next we take the limit of this result and get

$$\lim_{a \to -\infty} (1 - e^a) = \lim_{a \to -\infty} 1 - \lim_{a \to -\infty} e^a = 1 - 0 = 1$$

So the improper integral $\int_{-\infty}^{0} e^x dx$ converges and has a value of 1. ∎

▶ **Try It Yourself**

Some related Exercises are 23 and 25.

Example 5: Evaluating an Improper Integral

OBJECTIVE 3

Evaluate an improper integral of the form $\int_{-\infty}^{\infty} f(x)\,dx$.

Evaluate $\displaystyle\int_{-\infty}^{\infty} \frac{4x^3}{(1 + x^4)^2}\,dx$.

Perform the Mathematics

We begin by rewriting the improper integral as

$$\int_{-\infty}^{\infty} \frac{4x^3}{(1 + x^4)^2}\,dx = \int_{-\infty}^{0} \frac{4x^3}{(1 + x^4)^2}\,dx + \int_{0}^{\infty} \frac{4x^3}{(1 + x^4)^2}\,dx$$

$$= \lim_{a \to -\infty}\left[\int_{a}^{0} \frac{4x^3}{(1 + x^4)^2}\,dx\right] + \lim_{b \to \infty}\left[\int_{0}^{b} \frac{4x^3}{(1 + x^4)^2}\,dx\right]$$

To integrate $\displaystyle\int_{a}^{0} \frac{4x^3}{(1 + x^4)^2}\,dx$, we use u-substitution. Let $u = 1 + x^4$. Then $du = 4x^3\,dx$. We opt to leave off the limits of integration until we have an antiderivative. Proceeding, we have

$$\int \frac{4x^3}{(1 + x^4)^2}\,dx = \int \frac{1}{u^2}\,du = \int u^{-2}\,du = -u^{-1} = -\frac{1}{u}$$

$$= -\frac{1}{(1 + x^4)}\bigg|_{a}^{0} = -1 + \frac{1}{1 + a^4}$$

We now take the limit and get

$$\lim_{a \to -\infty}\left(-1 + \frac{1}{1 + a^4}\right) = \lim_{a \to -\infty}(-1) + \lim_{a \to -\infty}\left(\frac{1}{1 + a^4}\right)$$

$$= -1 + 0 = -1$$

Integrating $\displaystyle\int_{0}^{b} \frac{4x^3}{(1 + x^4)^2}\,dx$ will result in the same antiderivative as just computed, since it has the same integrand, just different limits. Hence, we can immediately write

$$\int_{0}^{b} \frac{4x^3}{(1 + x^4)^2}\,dx = -\frac{1}{(1 + x^4)}\bigg|_{0}^{b} = -\frac{1}{1 + b^4} + 1$$

Taking the limit of this result yields

$$\lim_{b \to \infty}\left(-\frac{1}{1 + b^4} + 1\right) = \lim_{b \to \infty}\left(-\frac{1}{1 + b^4}\right) + \lim_{b \to \infty}(1)$$

$$= 0 + 1 = 1$$

Putting all of this together results in

$$\int_{-\infty}^{\infty} \frac{4x^3}{(1 + x^4)^2}\,dx = \int_{-\infty}^{0} \frac{4x^3}{(1 + x^4)^2}\,dx + \int_{0}^{\infty} \frac{4x^3}{(1 + x^4)^2}\,dx$$

$$= \lim_{a \to -\infty}\left[\int_{a}^{0} \frac{4x^3}{(1 + x^4)^2}\,dx\right] + \lim_{b \to \infty}\left[\int_{0}^{b} \frac{4x^3}{(1 + x^4)^2}\,dx\right]$$

$$= -1 + 1 = 0$$

So the improper integral converges and has a value of 0. ∎

> **Try It Yourself**
>
> Some related Exercises are 37 and 39.

Applications

One family of applications of improper integrals is in estimating the total amount of oil or natural gas that will be produced by a well, given its production rate. Example 6 illustrates.

OBJECTIVE 4

Apply an improper integral to estimate total oil well production.

Example 6: Analyzing Total Oil Production

Petroleum engineers estimate that an oil field will produce oil at a rate given by $f(t) = 600e^{-0.1t}$ thousand barrels per month, t months into production. Estimate the total amount of oil produced by this well.

Understand the Situation

Since f is a rate function, integration gives us the total amount of oil produced. The total amount of oil produced by the oil well in T months of production is given by

$$\int_0^T 600e^{-0.1t}dt$$

Since we want to know the potential output of the well, we assume that the well will operate indefinitely. Thus, the total amount of oil produced is given by

$$\int_0^\infty 600e^{-0.1t}dt = \lim_{T\to\infty}\left[\int_0^T 600e^{-0.1t}dt\right]$$

Perform the Mathematics

As we have seen in this section, we integrate first and then apply the limit concept.

$$\int_0^T 600e^{-0.1t}dt = \frac{600}{-0.1}e^{-0.1t}\Big|_0^T = -6000e^{-0.1t}\Big|_0^T$$
$$= -6000e^{-0.1T} + 6000e^{-0.1(0)}$$
$$= -6000e^{-0.1T} + 6000$$

Applying the limit results in

$$\lim_{T\to\infty}(-6000e^{-0.1T} + 6000) = 0 + 6000 = 6000$$

Interpret the Result

So the total production of the oil well is estimated to be 6000 thousand, or 6,000,000 barrels.

 **Try It Yourself**

Some related Exercises are 43 and 45.   ∎

Capital Value

We begin our study of capital value by reconsidering the idea of the present value of a continuous income stream.

In Section 7.4 we analyzed the present value of a continuous income stream. Recall that if $f(t)$ represents the rate of flow of a continuous income stream, then the present value, P, at annual interest rate r compounded continuously for T years is given by

$$P = \int_0^T f(t)e^{-rt}dt$$

The **capital value** of a continuous income stream is simply the present value on the interval $[0, \infty)$, that is,

$$\text{Capital value} = \int_0^\infty f(t)e^{-rt}dt$$

In other words, capital value gives the worth of an investment that generates income forever. B. K. O'Neill has just discovered that some land he inherited has a huge oil deposit on it. He decides to lease the oil rights to Shannon Oil for an indefinite annual payment of $50,000. Determine the capital value of this lease at an annual interest rate of 4% compounded continuously.

Perform the Mathematics

We first note that the annual payments create a continuous income stream given by

$$f(t) = 50{,}000$$

that lasts indefinitely, or forever. So the capital value is given by

$$\int_0^\infty f(t)e^{-rt}dt = \int_0^\infty 50{,}000e^{-0.04t}dt$$

$$= \lim_{T\to\infty}\left[\int_0^T 50{,}000e^{-0.04t}dt\right]$$

First, the integration gives

$$\int_0^T 50{,}000e^{-0.04t}dt = \frac{50{,}000}{-0.04}e^{-0.04t}\Big|_0^T = -1{,}250{,}000e^{-0.04t}\Big|_0^T$$

$$= -1{,}250{,}000e^{-0.04T} + 1{,}250{,}000$$

Evaluating the limit of this result gives

$$\lim_{T\to\infty}(-1{,}250{,}000e^{-0.04T} + 1{,}250{,}000) = 1{,}250{,}000$$

So the capital value is $1,250,000. ∎

Endowments and Perpetuities

A fund that creates a steady income forever (think of a continuous income stream) may also be called a **permanent endowment** or a **perpetuity**. Some scholarship funds, such as the one in Example 7, and endowed chairs at universities are examples of perpetuities.

Example 7: Funding a Scholarship

Maria Lopez, a wealthy alumna of Old State University, wants to establish a nursing student scholarship in her name at her alma mater. If the annual proceeds of the scholarship are to be $10,000 and the annual rate of interest is 6% compounded continuously, determine the amount, to the nearest cent, that Maria needs to fund this scholarship.

OBJECTIVE 5

Apply an improper integral to fund an endowment.

Perform the Mathematics

We first note that the annual proceeds determine a continuous income stream given by

$$f(t) = 10{,}000$$

that is to last forever. So the amount necessary to fund this scholarship is given by

$$\int_0^\infty 10{,}000e^{-0.06t}dt = \lim_{T \to \infty} \int_0^T 10{,}000e^{-0.06t}dt \, d$$

Integrating first produces

$$\int_0^T 10{,}000e^{-0.06t}dt = \frac{10{,}000}{-0.06}e^{-0.06t}\Big|_0^T = -166{,}666.67e^{-0.06T} + 166{,}666.67$$

Evaluating the limit yields

$$\lim_{T \to \infty}(-166{,}666.67e^{-0.06T} + 166{,}666.67) = 166{,}666.67$$

We conclude that the amount Maria needs to fund this scholarship indefinitely is $166,666.67. ∎

▶ *Try It Yourself*

Some related Exercises are 59 and 61.

Probability Density Functions

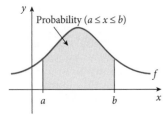

Figure 7.6.10

We conclude this section with a brief, intuitive, and informal look at how to use improper integrals with **probability density functions**. Consider an experiment that is designed so that any real number x in the interval $[a, b]$ is a possible outcome. An example of such an experiment might be to list the height or weight of a person selected at random.

Under some circumstances it is possible to determine a function f that will determine the probability that x assumes a value on a stated subinterval of $(-\infty, \infty)$. Functions that do this are called **probability density functions**. A probability density function must satisfy three conditions. See **Figure 7.6.10**.

1. For all x in $(-\infty, \infty)$, $f(x) \geq 0$.
2. The area under the graph of f is 1. That is, $\int_{-\infty}^{\infty} f(x)dx = 1$.
3. If $[a, b]$ is a subinterval of $(-\infty, \infty)$ then

$$\text{Probability } (a \leq x \leq b) = \int_a^b f(x)dx$$

OBJECTIVE 6

Apply an improper integral to determine a probability given a probability density function.

Example 8: Calculating a Probability

The length of time in minutes spent on a computer kiosk in a college dormitory has a probability density function of

$$f(x) = \begin{cases} 0.5e^{-0.5x}, & x \geq 0 \\ 0, & x < 0 \end{cases}$$

Calculate the probability that a user selected at random will spend between 1 and 2 minutes using the kiosk.

Perform the Mathematics

The first two conditions of a probability density function are left to you to verify. The probability that a user selected at random will spend between 1 and 2 minutes using the kiosk is

$$\text{Probability}(1 \le x \le 2) = \int_1^2 0.5e^{-0.5x}dx = \frac{0.5}{-0.5}e^{-0.5x}\Big|_1^2$$

$$= -e^{-0.5x}\Big|_1^2 = -e^{(-0.5)(2)} + e^{(-0.5)(1)} \approx 0.2387 \qquad ■$$

Example 9: Calculating a Probability

For the probability density function in Example 8, determine the probability that a user selected at random spends 4 or more minutes using the kiosk.

Perform the Mathematics

The probability that a user selected at random spends 4 or more minutes using the kiosk is determined by integrating

$$\text{Probability}(x \ge 4) = \int_4^\infty 0.5e^{-0.5x}dx = \lim_{b\to\infty}\left[\int_4^b 0.5e^{-0.5x}dx\right]$$

As always, we do the integration first. This yields

$$\int_4^b 0.5e^{-0.5x}dx = -e^{-0.5x}\Big|_4^b = -e^{(-0.5)(b)} + e^{-2}$$

Now we determine the limit as follows:

$$\lim_{b\to\infty}(-e^{(-0.5)(b)} + e^{-2}) = e^{-2} \approx 0.1353$$

So we have Probability$(x \ge 4) \approx 0.1353$. ■

Try It Yourself

Some related Exercises are 67 and 69.

Summary

In this section we saw that an integral in which a limit of integration was ∞ (or $-\infty$) is called an **improper integral**. We saw how to use the limit concept to evaluate improper integrals. In particular, we determined that if f is continuous and nonnegative, we have:

- $\int_a^\infty f(x)dx = \lim_{b\to\infty}\int_a^b f(x)dx$
- $\int_{-\infty}^b f(x)dx = \lim_{a\to-\infty}\left[\int_a^b f(x)dx\right]$
- $\int_{-\infty}^\infty f(x)dx = \int_{-\infty}^c f(x)dx + \int_c^\infty f(x)dx$

If the limit exists, we said that the improper integral **converges**, and if the limit does not exist, we said that the improper integral **diverges**. We applied improper integrals to determine the total output of an oil well, to assess a capital value, to determine the size of a **perpetuity**, and to briefly explore **probability density functions**.

Section 7.6 Exercises

Vocabulary Exercises

1. We call an interval of the form $(-\infty, b]$ or $[a, \infty)$ an _____ interval.

2. We call an integral that has $-\infty$ or ∞ as at least one of its limits of integration an _____ integral.

3. When we compute an integral of the form $\int_a^\infty f(x)\,dx$, we replace ∞ with b and take the _____ as b approaches infinity.

4. If the result of an integral of the form $\int_a^\infty f(x)\,dx$ is a definite number, then we say the integral _____.

5. If the result of an integral of the form $\int_a^\infty f(x)\,dx$ is either $-\infty$ or ∞ or is not defined, then we say the integral _____.

6. A fund that creates a steady income forever is called a _____.

Skill Exercises

For Exercises 7–22, evaluate each improper integral or state that it diverges.

7. $\displaystyle\int_2^\infty \frac{1}{x^4}\,dx$

8. $\displaystyle\int_4^\infty \frac{1}{x^3}\,dx$

9. $\displaystyle\int_3^\infty \frac{1}{\sqrt{x}}\,dx$

10. $\displaystyle\int_5^\infty \frac{3}{\sqrt{x}}\,dx$

11. $\displaystyle\int_2^\infty \frac{1}{x^{1.5}}\,dx$

12. $\displaystyle\int_3^\infty \frac{1}{x^{2.5}}\,dx$

13. $\displaystyle\int_1^\infty \frac{1}{x^{2/3}}\,dx$

14. $\displaystyle\int_1^\infty \frac{1}{x^{3/4}}\,dx$

15. $\displaystyle\int_1^\infty e^{-x}\,dx$

16. $\displaystyle\int_0^\infty e^{-x}\,dx$

17. $\displaystyle\int_1^\infty e^{-0.03x}\,dx$

18. $\displaystyle\int_1^\infty e^{0.03x}\,dx$

19. $\displaystyle\int_1^\infty \frac{\ln x}{x}\,dx$

20. $\displaystyle\int_{10}^\infty \ln x\,dx$

21. $\displaystyle\int_1^\infty \frac{2x}{x^2 + 3}\,dx$

22. $\displaystyle\int_0^\infty \frac{x^4}{(x^5 + 2)^2}\,dx$

For Exercises 23–34, evaluate each improper integral or state that it diverges.

23. $\displaystyle\int_{-\infty}^0 e^{2x}\,dx$

24. $\displaystyle\int_{-\infty}^0 e^{-2x}\,dx$

25. $\displaystyle\int_{-\infty}^0 e^{-x}\,dx$

26. $\displaystyle\int_{-\infty}^0 e^{2.3x}\,dx$

27. $\displaystyle\int_{-\infty}^{0} e^{0.2x}\,dx$ **28.** $\displaystyle\int_{-\infty}^{0} e^{x+1}\,dx$

29. $\displaystyle\int_{-\infty}^{-4} x^{-2}\,dx$ **30.** $\displaystyle\int_{-\infty}^{-2} x^{-4}\,dx$

31. $\displaystyle\int_{-\infty}^{-2} \frac{x^3}{(x^4-1)^2}\,dx$ **32.** $\displaystyle\int_{-\infty}^{0} \frac{3x}{(x^2+5)^4}\,dx$

33. $\displaystyle\int_{0}^{\infty} 2xe^{-x^2}\,dx$ **34.** $\displaystyle\int_{-\infty}^{0} 2xe^{-x^2}\,dx$

For Exercises 35–42, evaluate each improper integral or state that it diverges.

35. $\displaystyle\int_{-\infty}^{\infty} \frac{4x^3}{(x^4+3)^2}\,dx$ **36.** $\displaystyle\int_{-\infty}^{\infty} \frac{4x^3}{(x^4+1)^2}\,dx$

37. $\displaystyle\int_{-\infty}^{\infty} 2xe^{-x^2}\,dx$ **38.** $\displaystyle\int_{-\infty}^{\infty} 2xe^{x^2}\,dx$

39. $\displaystyle\int_{-\infty}^{\infty} e^{-x}\,dx$ **40.** $\displaystyle\int_{-\infty}^{\infty} e^{x}\,dx$

41. $\displaystyle\int_{-\infty}^{\infty} \frac{2x}{x^2+1}\,dx$ **42.** $\displaystyle\int_{-\infty}^{-\infty} \frac{1}{\sqrt{x^2+1}}\,dx$

Application Exercises

43. Geology—Oil Production: Petrohas Engineering estimates that an oil well produces oil at a rate given by

$$B(t) = 60e^{-0.05t} - 60e^{-0.1t}$$

where t represents the number of months from now, and $B(t)$ represents the oil well production, measured in thousands of barrels per month. Estimate the total amount of oil produced by this well.

44. Geology—Oil Production: A geologist estimates that an oil well produces oil at a rate of

$$B(t) = 85e^{-0.02t} - 85e^{-0.1t}$$

where t represents the number of months from now, and $B(t)$ represents the oil well production, measured in thousands of barrels per month. Estimate the total amount of oil produced by this well.

45. Geology—Natural Gas Production: GeoSurveys Inc. estimates that a natural gas well has a rate of production given by

$$CF(t) = te^{-0.25t}$$

where t represents the number of months from now, and $CF(t)$ represents the rate of production, measured in millions of cubic feet per month. Estimate the total amount of natural gas produced.

46. Geology—Natural Gas Production: EarthWorks Inc. estimates that a natural gas well has a rate of production given by

$$CF(t) = 2te^{-0.37t}$$

where t represents the number of months from now, and $CF(t)$ represents the rate of production, measured in millions of cubic feet per month. Estimate the total amount of natural gas produced.

47. **Ecology—Water Pollution:** The rate at which the hazardous chemical perchloroethylene, or PCE, is being released into Lake Stagg from an abandoned dump is given by

$$f(t) = 350e^{-0.3t}$$

where t represents the number of years from now, and $f(t)$ represents the rate at which PCE is being released into Lake Stagg, measured in tons per year. Determine the total amount of the hazardous chemical that will be released into the lake if the leak continues indefinitely.

48. **Ecology—Water Pollution:** The rate at which the toxic chemical alkylphenol is being released into the Avon River from an abandoned dump is given by

$$f(t) = 300e^{-0.25t}$$

where t represents the number of years from now, and $f(t)$ represents the rate at which alkylphenol is being introduced, measured in tons per year. Determine the total amount of alkylphenol that will be released into the Avon River if the leak continues indefinitely.

49. **Internal Medicine—Medication Dissipation:** It is a fact that when people take medication, the human body does not absorb all of it. Suppose that researchers at a pharmaceutical company know that one way to determine the amount of medication absorbed by the body is to determine the rate at which the body eliminates the medication. For a certain arthritis medication, researchers have determined that the rate at which the body eliminates the medication is given by

$$f(t) = 5e^{-0.02t} - 5e^{-0.035t}$$

where t represents the number of minutes after the medication was administered, and $f(t)$ represents the rate of absorption, measured in milliliters per minute. Determine the amount of the medication that was eliminated from the body.

50. **Internal Medicine—Medication Dissipation:** For a certain asthma medication, suppose that the rate at which the body eliminates the medication is given by

$$f(t) = 6.5e^{-0.025t} - 6.5e^{-0.05t}$$

where t represents the number of minutes after the medication was administered, and $f(t)$ represents the rate of absorption, measured in milliliters per minute. Determine the amount of asthma medication that was eliminated from the body.

51. **Real Estate—Rental Income:** Elle owns a rental property that generates an indefinite annual rent of $10,000. Determine the capital value of this property at an annual interest rate of 6% compounded continuously.

52. **Real Estate—Rental Income:** Ginger owns a rental property that generates an indefinite annual rent of $6000. Determine the capital value of this property at an annual interest rate of 8.5% compounded continuously.

53. **Entrepreneurship—Lease Income:** Austin has created a new computer game. He decides to lease the rights to his computer game to Macrohard for an indefinite annual payment of $30,000. Determine the capital value of this lease at an annual interest rate of 7.25% compounded continuously.

54. **Entrepreneurship—Lease Income:** Nikita has created a new computer game. She decides to lease the rights to her computer game to Star Systems for an indefinite annual payment of $15,000. Determine the capital value of this lease at an annual interest rate of 9.75% compounded continuously.

55. **Personal Finance—Gold Leasing:** Dylan has inherited some land that has a huge gold deposit on it. He decides to lease the rights to the gold to Hoofer Mining Company for an indefinite annual payment of $70,000. Determine the capital value of this lease at an annual interest rate of 6.5% compounded continuously.

56. **Personal Finance—Silver Leasing:** Emily owns a property that has a huge silver deposit on it. She decides to lease the rights to the silver to Hoofer Mining Company for an indefinite annual payment of $25,000. Determine the capital value of this lease at an annual interest rate of 8.8% compounded continuously.

57. **Property Management—Land Appreciation:** Lisa owns a rental property that generates an indefinite annual rent of $12,000. Determine the capital value of this property at an annual interest rate of 5.5% compounded continuously.

58. **Property Management—Rental Income:** Gene owns a rental property that generates an indefinite annual rent of $2000. Determine the capital value of this property at an annual interest rate of 7% compounded continuously.

59. **College Finance—Scholarships:** If the annual proceeds from the Emma Lou Smith scholarship fund will be $8000 indefinitely, and the annual interest rate is 7.5% compounded continuously, how much should be invested to fund this scholarship?

60. **College Finance—Scholarships:** If the annual proceeds from the Count Moncheech de Squeeg cinematography scholarship fund will be $5000 indefinitely, and the annual interest rate is 7% compounded continuously, how much should be invested to fund this scholarship?

61. **College Finance—Scholarships:** The Shannon Oil Company has decided to provide an annual scholarship in the amount of $3000 to individuals studying environmental science. If the annual rate of interest is 6.6% compounded continuously, how much should be invested to fund this scholarship perpetually?

62. **College Finance—Scholarships:** The Beagle Works Company has decided to provide an annual scholarship in the amount of $10,000 to individuals studying veterinary medicine. If the annual rate of interest is 7.75% compounded continuously, how much should be invested to fund this scholarship perpetually?

63. **College Finance—Scholarships:** You wish to leave a legacy at your alma mater in the form of a scholarship that bears your name. You want the scholarship to be an annual award of $5000. Assume that the annual interest rate is 6.25% compounded continuously.

   (a) How much should you invest to fund this annual $5000 scholarship for the next 75 years?

   (b) How much should you invest to fund this annual $5000 scholarship indefinitely?

64. **College Finance—Scholarships:** You wish to leave a legacy at your alma mater in the form of a scholarship that bears your name. You want the scholarship to be an annual award of $8000. Assume that the annual interest rate is 6.25% compounded continuously.

   (a) How much should you invest to fund this annual $8000 scholarship for the next 75 years?

   (b) How much should you invest to fund this annual $8000 scholarship indefinitely?

65. **Telephony—Pay Internet Access:** The length of Internet activity in minutes from a pay connection in a college dormitory has a probability density function given by

$$f(x) = \begin{cases} 0.3e^{-0.3x}, & x \geq 0 \\ 0, & x < 0 \end{cases}$$

   (a) Verify that f is a probability density function by showing that $\int_{-\infty}^{\infty} f(x)\,dx = 1$.

   (b) Calculate the probability that a connection selected at random lasts between 3 and 4 minutes.

   (c) Calculate the probability that a connection selected at random lasts 3 minutes or longer.

66. **Telephony—Pay Internet Access:** The length of Internet activity in minutes from a pay connection in a college cafeteria has a probability density function given by

$$f(x) = \begin{cases} 0.6e^{-0.6x}, & x \geq 0 \\ 0, & x < 0 \end{cases}$$

   (a) Verify that f is a probability density function by showing that $\int_{-\infty}^{\infty} f(x)\,dx = 1$.

   (b) Calculate the probability that a connection selected at random lasts between 1 and 2 minutes.

   (c) Calculate the probability that a connection selected at random lasts 3 minutes or longer.

67. **Library Science—Book Reserves:** The length of time for which a calculus book on a 4-hour reserve at a college library is checked out has a probability density function given by

$$f(x) = \begin{cases} \dfrac{1}{8}x, & 0 \leq x \leq 4 \\ 0, & \text{otherwise} \end{cases}$$

   (a) Verify that f is a probability density function by showing that $\int_{-\infty}^{\infty} f(x)\,dx = 1$.

   (b) Compute the probability that a student selected at random has the book checked out from 1 to 2 hours.

   (c) Compute the probability that a student selected at random has the book checked out from 30 minutes to 1 hour.

68. **Library Science—Book Reserves:** The length of time for which a solutions manual for a certain calculus book on a 6-hour reserve at a college library is checked out has a probability density function given by

$$f(x) = \begin{cases} \dfrac{1}{18}x, & 0 \leq x \leq 6 \\ 0, & \text{otherwise} \end{cases}$$

   (a) Verify that f is a probability density function by showing that $\int_{-\infty}^{\infty} f(x)\,dx = 1$.

   (b) Compute the probability that a student selected at random has the solutions manual checked out from 2 to 4 hours.

   (c) Compute the probability that a student selected at random has the solutions manual checked out from 1 to 2 hours.

69. **Criminal Science—Prison Terms:** The length of a prison term for convicted felons has a probability density function given by

$$f(x) = \begin{cases} 0.1e^{-0.1x}, & x \geq 0 \\ 0, & \text{otherwise} \end{cases}$$

   (a) Verify that f is a probability density function by showing that $\int_{-\infty}^{\infty} f(x)\,dx = 1$.

   (b) Determine the probability that a randomly selected felon has a prison term of 5 to 10 years.

   (c) Determine the probability that a randomly selected felon has a prison term of 1 to 5 years.

   (d) Determine the probability that a randomly selected felon has a prison term of 8 years or more.

70. **Consumerism—Part Failures:** The number of months until failure for a certain part on a copy machine has a probability density function given by

$$f(x) = \begin{cases} 0.02e^{-0.02x}, & x \geq 0 \\ 0, & \text{otherwise} \end{cases}$$

   (a) Verify that f is a probability density function by showing that $\int_{-\infty}^{\infty} f(x)\,dx = 1$.

   (b) Compute the probability that the part will fail in the first year.

   (c) Compute the probability that the part will fail in the second year.

71. **Consumerism—Part Failures:** The number of months until failure for a certain computer chip has a probability density function of

$$f(x) = \begin{cases} 0.03e^{-0.03x}, & x \geq 0 \\ 0, & \text{otherwise} \end{cases}$$

   (a) Verify that f is a probability density function by showing that $\int_{-\infty}^{\infty} f(x)\,dx = 1$.

   (b) Compute the probability that the part will fail in the first year.

   (c) Compute the probability that the part will fail in the second year.

72. **Customer Service—Hold Times:** At Robotics Inc. the number of minutes that a randomly selected telephone customer is on hold has a probability density function given by

$$f(x) = \begin{cases} 0.06e^{-0.06x}, & x \geq 0 \\ 0, & \text{otherwise} \end{cases}$$

(a) Verify that f is a probability density function by showing that $\int_{-\infty}^{\infty} f(x)\,dx = 1$.

(b) Compute the probability that a customer is on hold from 1 to 4 minutes.

(c) Compute the probability that a customer is on hold for 2 minutes or longer.

Concept and Writing Exercises

73. Explain in your own words the difference between an improper integral and other integrals.

74. Could an improper integral be evaluated without the use of the limit? Explain why or why not.

75. Suppose we know that $\int_{a}^{\infty} \frac{1}{x^4}\,dx = \frac{1}{81}$. Determine the value of a.

76. Suppose we know that $\int_{a}^{\infty} \frac{1}{e^x}\,dx = \frac{1}{e^4}$. Determine the value of a.

77. Given the improper integral $\int_{1}^{\infty} \frac{1}{x^k}\,dx$, for what values of k does the integral diverge?

78. Given the improper integral $\int_{0}^{\infty} \frac{1}{e^{kx}}\,dx$, for what values of k does the integral diverge?

Section Project

Peggy never finishes her calculus lecture before the end of the period, but always finishes her lectures within 5 minutes of the end of the period. The length of time that elapses between the end of the period and the end of her lecture has a probability density function given by

$$f(x) = \begin{cases} \dfrac{3}{125}x^2, & 0 \leq x \leq 5 \\ 0, & \text{otherwise} \end{cases}$$

(a) Verify that f is a probability density function by showing that $\int_{-\infty}^{\infty} f(x)\,dx = 1$.

(b) Determine the probability that the lecture continues for 1 to 2 minutes beyond the end of the period.

(c) Compute the probability that the lecture ends within 1 minute of the end of the period.

Donald never starts his calculus lecture at the beginning of a period, but always starts his lecture within 2 to 9 minutes from the beginning of the period. The length of time that elapses from 2 minutes after the beginning of the period and the beginning of his lecture is given by the probability density function given by

$$f(x) = \begin{cases} \dfrac{3}{721}x^2, & 2 \leq x \leq 9 \\ 0, & \text{otherwise} \end{cases}$$

(d) Verify that f is a probability density function by showing that $\int_{-\infty}^{\infty} f(x)\,dx = 1$.

(e) Compute the probability that his lecture begins within 3 to 5 minutes from the beginning of the period.

(f) Compute the probability that his lecture begins within 7 to 9 minutes from the beginning of the period.

7.7   Differential Equations: Separation of Variables

So far in this chapter, we have seen equations of the form

$$P'(x) = \frac{x}{15} + 20 \quad \text{and} \quad \frac{dy}{dx} = 4x^2 - 3x$$

These are examples of differential equations. Quite simply, an equation that involves a function and one (or more) of its derivatives is called a **differential equation**. Entire courses and textbooks are dedicated to the study of these equations. In this section and the next, we will study **first-order differential equations**. These are differential equations that involve a first derivative but no higher derivatives. The technique we will use to solve this kind of differential equation is called **separation of variables**.

The Differential Equation $y = f'(x)$

Since the beginning of Chapter 6, we have really been solving differential equations that have the form $y = f'(x)$. When we see

$$\frac{dy}{dx} = 3x^2$$

we are saying that the derivative of some function is $3x^2$. This differential equation is solved by integrating

$$y = \int 3x^2 dx = x^3 + C$$

OBJECTIVE 1

Determine the solution to a differential equation.

Example 1: Solving a Differential Equation

Solve the differential equation $y' = x^4$.

Perform the Mathematics

If we replace y' with $\frac{dy}{dx}$, we have

$$\frac{dy}{dx} = x^4$$

Since $\frac{dy}{dx} = x^4$, we know that y is an antiderivative of x^4. As a result,

$$y = \int x^4 dx = \frac{1}{5}x^5 + C$$

So the solution to the given differential equation is $y = \frac{1}{5}x^5 + C$.   ∎

Example 1 indicates that, in general, a differential equation of the form $y = f'(x)$ or $\frac{dy}{dx} = f(x)$ is solved by integrating $y = \int f(x)dx$.

In Example 1 we saw that the solution to $y' = x^4$ is $y = \frac{1}{5}x^5 + C$. We call $y = \frac{1}{5}x^5 + C$ the **general solution** of the differential equation $y' = x^4$ because it represents all possible solutions, one for each choice of the constant C. When the constant C is replaced with a particular value, we then have a **particular solution**. A few solutions to the differential equation $y' = x^4$ are

$$y = \frac{1}{5}x^5 + 1 \quad C = 1$$

$$y = \frac{1}{5}x^5 \qquad C = 0$$

$$y = \frac{1}{5}x^5 - 1 \quad C = -1$$

Notice in **Figure 7.7.1** that each curve has the same shape.

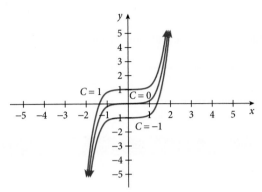

Figure 7.7.1 Graphs of $y = \frac{1}{5}x^5 + C$ for $C = -1$, $C = 0$, and $C = 1$.

This is because each curve has the same derivative $y' = x^4$. For this reason, we say that the different values for C in a general solution of a differential equation produce a *family of curves*, where the general solution $y = f(x) + C$ represents the entire family.

Separation of Variables

One important type of differential equation is called a separable differential equation. We say that a differential equation is *separable* if we can move everything involving x's to one side of the equation and everything involving y's to the other side. Example 2 illustrates the process.

| **Example 2: Finding a General Solution by Separation of Variables** | **OBJECTIVE 2** |

Determine the general solution to $\frac{dy}{dx} = \frac{x^2}{y^3}$.

Perform the Mathematics

The method known as separation of variables requires us to separate the x's from the y's algebraically.

$$\frac{dy}{dx} = \frac{x^2}{y^3}$$

Multiply both sides by dx $\qquad\qquad dy = \frac{x^2}{y^3}dx$

Multiply both sides by y^3 $\qquad\qquad y^3 dy = x^2 dx$

We now integrate both sides of the equation and get

Integrate both sides $\qquad\qquad \int y^3 dy = \int x^2 dx$

Use Power Rule for Integration $\qquad \frac{1}{4}y^4 + C_0 = \frac{1}{3}x^3 + C_1$

Combine constants $\qquad\qquad\qquad \frac{1}{4}y^4 = \frac{1}{3}x^3 + C$

Use separation of variables to determine the general solution to a differential equation.

Solving for y yields

$$y^4 = \frac{4}{3}x^3 + 4C$$

Four times constant is constant
$$y^4 = \frac{4}{3}x^3 + C$$

General solution using principal fourth root
$$y = \sqrt[4]{\frac{4}{3}x^3 + C}$$ ∎

Example 2 outlines the following process for **separation of variables**.

DEFINITION

Separation of Variables

A separable first-order differential equation has the form

$$\frac{dy}{dx} = f(x) \cdot g(y) \quad \text{or} \quad \frac{dy}{dx} = \frac{f(x)}{g(y)}, \quad g(y) \neq 0$$

It is solved by separating the variables and integrating

$$\int g(y)\,dy = \int f(x)\,dx$$

In Example 2 we found a general solution. Many times, we are given some additional information that allows us to determine the **particular solution**. We call the additional information an **initial condition**. Example 3 illustrates how to determine a particular solution.

OBJECTIVE 3

Use separation of variables to determine the particular solution to a differential equation.

Example 3: Determining a Particular Solution

Solve the differential equation $\dfrac{dy}{dx} = \dfrac{x^2}{y^3}$ with the condition that $x = 0, y = 1$.

Perform the Mathematics

From Example 2, we know that the general solution is

$$y = \sqrt[4]{\frac{4}{3}x^3 + C}$$

Given that $y = 1$, when $x = 0$, we can substitute this into the general solution and determine C.

$$y = \sqrt[4]{\frac{4}{3}x^3 + C}$$

$$1 = \sqrt[4]{\frac{4}{3}(0)^3 + C}$$

$$1 = \sqrt[4]{C}$$

$$1 = C$$

So the particular solution is $y = \sqrt[4]{\dfrac{4}{3}x^3 + 1}$. See **Figure 7.7.2**. ∎

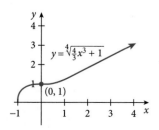

$y = \sqrt[4]{\frac{4}{3}x^3 + 1}$

$(0, 1)$

Figure 7.7.2

▶ **Try It Yourself**

Some related Exercises are 39 and 49.

Example 3 illustrates **how to use separation of variable to solve a differential equation with an initial condition**.

Several solutions to $\frac{dy}{dx} = \frac{x^2}{y^3}$ are shown in **Figure 7.7.3**, with our particular solution from Example 3 shown in blue.

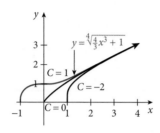

Figure 7.7.3 Graphs of $y = \sqrt[4]{\frac{4}{3}x^3 + C}$ of the solutions to $\frac{dy}{dx} = \frac{x^2}{y^3}$.

Using Separation of Variables to Solve a Differential Equation with an Initial Condition

1. Separate the variable.
2. Determine the general solution.
3. Use the initial condition to determine *C*.

◀▶ **Technology Option**

The computing capabilities of a graphing calculator can help us to visualize the solutions of a differential equation. For example, the differential equation $\frac{dy}{dx} = \frac{x^2}{y^3}$ in Example 3 tells us that the slope at any point (x, y) on a graph is given by the expression $\frac{x^2}{y^3}$. A graph of these slopes is called a **slope field**. Calculator programs can draw slope fields for us. A slope field program is included in Appendix B.

Example 4: Finding a Particular Solution When a Logarithm Is Involved

OBJECTIVE 4

Solve a differential equation involving logarithms.

Determine the particular solution for $\frac{dy}{dx} = 3xy$ with the condition $x = 0$, $y = 3$.

Perform the Mathematics

We start by separating the variables.

$$\frac{dy}{dx} = 3xy$$

$$\frac{1}{y}\frac{dy}{dx} = 3x$$

$$\frac{1}{y}dy = 3x\,dx$$

Integrating both sides yields

$$\int \frac{1}{y}dy = \int 3x\,dx$$

$$\ln y = \frac{3}{2}x^2 + C \quad y > 0$$

Recall the definition of a logarithm, $y = \ln x$ if and only if $e^y = x$. This allows us to rewrite this equation as

$$y = e^{(3/2)x^2 + C}$$
$$y = e^{(3/2)x^2} \cdot e^C$$

Since e^C represents an arbitrary constant, we can simply write C for e^C and represent the general solution as

$$y = C \cdot e^{(3/2)x^2}$$

Now, applying the initial condition $x = 0$, $y = 3$,

$$3 = C \cdot e^{(3/2)(0)^2}$$
$$3 = C \cdot e^0 = C \cdot 1 = C$$

So the particular solution is $y = 3e^{(3/2)x^2}$. ∎

▶ *Try It Yourself*

Some related Exercises are 63 and 67.

Sometimes we will find that there can be more than one general solution to a differential equation. This is illustrated in Example 5.

Example 5: Finding All General Solutions to a Differential Equation

Find the general solution to $y \cdot \frac{dy}{dx} = 2x$.

Perform the Mathematics

Since x's and y's are already separated, we can simply multiply by dx and then integrate both sides.

$$y \cdot \frac{dy}{dx} = 2x$$
$$y \, dy = 2x \, dx$$
$$\int y \, dy = \int 2x \, dx$$
$$\frac{1}{2}y^2 = x^2 + C$$
$$y^2 = 2x^2 + 2C$$
$$y^2 = 2x^2 + C$$
$$y = \pm\sqrt{2x^2 + C}$$

So we have two square roots, one positive and the other negative. Thus, these two solutions

$$y = \sqrt{2x^2 + C} \quad \text{and} \quad y = -\sqrt{2x^2 + C}$$

together are the general solution to the differential equation. ∎

Example 6: Solving a Differential Equation by *u*-Substitution

OBJECTIVE 5

Solve a differential equation using *u*-substitution.

Determine the general solution for $\frac{dy}{dx} = xy - 3x$.

Perform the Mathematics

To separate the variables for this differential equation, we need to factor on the right side.

$$\frac{dy}{dx} = xy - 3x$$

Factor right side

$$\frac{dy}{dx} = x(y - 3)$$

Multiply by dx and divide by $y - 3$

$$\frac{dy}{y - 3} = x\,dx$$

Integrate both sides

$$\int \frac{1}{y - 3}\,dy = \int x\,dx$$

Let $u = y - 3$, then $du = dy$

$$\int \frac{1}{u}\,du = \int x\,dx$$

Integrate and assume $u > 0$

$$\ln u = \frac{1}{2}x^2 + C$$

Resubstitute $y - 3$ for u

$$\ln(y - 3) = \frac{1}{2}x^2 + C$$

Definition of logarithm

$$y - 3 = e^{(1/2)x^2 + C}$$

Rule of exponents

$$y - 3 = e^{(1/2)x^2} \cdot e^C$$

Replace constant e^C by C

$$y - 3 = Ce^{(1/2)x^2}$$

Add 3 to both sides

$$y = Ce^{(1/2)x^2} + 3 \qquad \blacksquare$$

Applications

Many of the applications of separable differential equations involve differential rate equations, where a rate of change function is given. Usually, we use the separation of variables technique to recover an original function.

Example 7: Application of Separation of Variables

OBJECTIVE 6

Apply separation of variables.

The annual rate of increase in employees between 1970 and 2005 at the NewMedia Company can be modeled by the differential rate equation

$$\frac{dy}{dt} = 0.3t^{0.6} \qquad 0 \leq t \leq 35$$

where t represents the number of years since 1970, and y represents the number of employees in hundreds.

a. Knowing that the company had 1800 employees in 1985, find the employee growth function y for the company.

b. Use the result from part (a) to estimate the number of employees at the company in 2003.

Perform the Mathematics

a. We start by determining the general solution to the differential rate equation.

$$\frac{dy}{dt} = 0.3t^{0.6}$$

$$dy = 0.3t^{0.6}dt$$

$$\int dy = \int 0.3t^{0.6}dt$$

$$y = \frac{0.3}{1.6}t^{1.6} + C$$

$$y = \frac{3}{16}t^{1.6} + C$$

Since the company had 1800 employees in 1985, we know that $y = 18$ when $t = 15$. This allows us to solve for C.

$$y = \frac{3}{16}t^{1.6} + C$$

$$18 = \frac{3}{16}(15)^{1.6} + C$$

$$18 - \frac{3}{16}(15)^{1.6} = C$$

$$3.72 \approx C$$

So the employee growth of the company can be modeled by

$$y = \frac{3}{16}t^{1.6} + 3.72$$

b. To estimate the number of employees at the company in 2003, we need to use the model found in part (a) and evaluate when $t = 33$.

$$y = \frac{3}{16}(33)^{1.6} + 3.72$$

$$y \approx 54.14$$

So, according to the model, the NewMedia Company had about 54.14 hundred, or 5414, employees in 2003.   ∎

Summary

- A differential equation of the form $y = f'(x)$ or $\frac{dy}{dx} = f(x)$ is solved by integrating $y = \int f(x)dx$.

- A separable first-order differential equation $\frac{dy}{dx} = f(x) \cdot g(y)$, $g(y) \neq 0$, is solved by separating the variables and integrating $\int g(y)dy = \int f(x)dx$.

Section 7.7 Exercises

Vocabulary Exercises

1. An equation that involves a function and one or more of its derivatives is called a _____ equation.

2. The technique that involves placing the x-expression on one side of an equation and the y-expression on the other side is called _____ of _____.

3. The solution to a differential equation is called a _____ _____ if no initial condition is specified.

4. The solution to a differential equation is called a _____ _____ if an initial condition is provided.

5. A graph of the possible solutions of a differential equation for various values of C is called a _____ _____ _____.

6. We can check the answer to a differential equation by computing the _____ of the solution.

Skill Exercises

For Exercises 7–20, determine if the given function f is the solution to the given differential equation by determining the derivative f'.

7. $f(x) = 5x + 3$; $y' = 5$

8. $f(x) = x^2$; $y' = 2x - 7$

9. $f(x) = x^5 - 4$; $y' = x^4$

10. $f(x) = x^6 + x - 4$; $y' = 6x^5 - 1$

11. $f(x) = \dfrac{1}{x}$; $y' = -\dfrac{1}{x^2}$

12. $f(x) = \dfrac{1}{x}$; $xy' = -y$

13. $f(x) = 2x^3 + 10$; $y' = 3x^2$

14. $f(x) = x^2 + 1$; $3xy' = 6y - 6$

15. $f(x) = \ln x$; $xy' = 1$

16. $f(x) = \dfrac{1}{2}\ln x$; $2xy' = 1$

17. $f(x) = \dfrac{3}{e^{4x}}$; $y' + 4y = 0$

18. $f(x) = xe^x$; $xy' = y(x + 1)$

19. $f(x) = e^{3x} + 5$; $y' + 15 = 3y$

20. $f(x) = \dfrac{e^{2x} + x^2}{2}$; $y' - 3 = e^{2x} + x$

In Exercises 21–30, determine the general solution to the given differential equation.

21. $\dfrac{dy}{dx} = -2$

22. $\dfrac{dy}{dx} = e$

23. $\dfrac{dy}{dx} = 1 - 3x$

24. $\dfrac{dy}{dx} = 2x$

25. $\dfrac{dy}{dx} = 5x^3 + x - 2$

26. $\dfrac{dy}{dx} = 10x^4 - x^2 + 2$

27. $\dfrac{dy}{dx} = \sqrt{x} + 2$

28. $\dfrac{dy}{dx} = \sqrt[4]{x} + x$

29. $\dfrac{dy}{dx} = \dfrac{5}{x^2}$

30. $\dfrac{dy}{dx} = x^2 - \dfrac{1}{x^2}$

For Exercises 31–36, graph a family of curves for the general solutions of the given differential equations using the values $C = 0$, $C = -2$, and $C = 3$.

31. $\dfrac{dy}{dx} = 2$

32. $\dfrac{dy}{dx} = -5$

33. $\dfrac{dy}{dx} = 3x + 1$

34. $\dfrac{dy}{dx} = 1 - 2x$

35. $\dfrac{dy}{dx} = \sqrt{x} + 3$

36. $\dfrac{dy}{dx} = \sqrt{x} + 2$

For Exercises 37–50, determine the particular solution to the differential equation using the given conditions.

37. $\dfrac{dy}{dx} = 12x; \quad x = 1, y = 8$

38. $\dfrac{dy}{dx} = 12x; \quad x = 1, y = -1$

39. $\dfrac{dy}{dx} = 3x^2 + 4x; \quad x = 1, y = 6$

40. $\dfrac{dy}{dx} = xe^{x^2}; \quad x = 0, y = 2$

41. $\dfrac{dy}{dx} = 7 - 4x; \quad x = 0, y = 3$

42. $\dfrac{dy}{dx} = 7 - 4x; \quad x = -1, y = 3$

43. $\dfrac{dy}{dx} = x^3 - 2x; \quad x = 0, y = -2$

44. $\dfrac{dy}{dx} = -x^2 - x; \quad x = 1, y = 1$

45. $\dfrac{dy}{dx} = x^2 + 2x + 3; \quad x = 0, y = 4$

46. $\dfrac{dy}{dx} = \sqrt[3]{x^2} - x; \quad x = 0, y = 6$

47. $\dfrac{dy}{dx} = 2x^{0.3}; \quad x = 0, y = 4$

48. $\dfrac{dy}{dx} = 4x^{1.7}; \quad x = 0, y = 1$

49. $\dfrac{dy}{dx} = 1 - \dfrac{2}{x}; \quad x = 1, y = 6$

50. $\dfrac{dy}{dx} = 3 - \dfrac{1}{x}; \quad x = 1, y = 8$

For Exercises 51–60, determine the general solution to the separable differential equation.

51. $\dfrac{dy}{dx} = \dfrac{x}{y}$

52. $\dfrac{dy}{dx} = \dfrac{3y}{x}$

53. $\dfrac{dy}{dx} = 3x^2y$

54. $\dfrac{dy}{dx} = 4x^3y$

55. $\dfrac{dy}{dx} = e^{x+y}$ (Hint: Begin by rewriting the right side as two factors.)

56. $\dfrac{dy}{dx} = e^{x-y}$ (Hint: Begin by rewriting the right side as two factors.)

57. $\dfrac{dy}{dx} = (2x + 2)y$

58. $\dfrac{dy}{dx} = 4xy + 7y$

59. $\dfrac{dy}{dx} = \sqrt{xy}$ (Hint: Begin by rewriting the right side as two factors.)

60. $\dfrac{dy}{dx} = \left(\dfrac{y}{x}\right)^2$

For Exercises 61–70, determine the particular solution to the separable differential equation using the given initial condition.

61. $\dfrac{dy}{dx} = \dfrac{x}{y}; \quad x = 0, y = -1$

62. $\dfrac{dy}{dx} = 2xy; \quad x = 0, y = 1$

63. $\dfrac{dy}{dx} = \dfrac{y}{x}; \quad x = 1, y = 3$

64. $\dfrac{dy}{dx} = 2xy^4; \quad x = 0, y = 1$

65. $\dfrac{dy}{dx} = \dfrac{x + 1}{xy}; \quad x = 1, y = 2$

66. $\dfrac{dy}{dx} = \dfrac{1 + x^2}{xy}; \quad x = 2, y = 4$

67. $x \cdot \dfrac{dy}{dx} = xy; \quad x = 0, y = 1$

68. $x \cdot \dfrac{dy}{dx} = x^4y; \quad x = 3, y = 10$

69. $\dfrac{dy}{dx} = e^{x-y}; \quad x = 0, y = 0$

70. $\dfrac{dy}{dx} = xye^{x^2}; \quad x = 0, y = e$

Application Exercises

71. Economics—Marginal Cost: The Strike Now Company determines that the marginal cost for their MaxiPak of lighters is given by the differential rate equation

$$\frac{dC}{dx} = -0.02x + 6 \qquad 0 \leq x \leq 300$$

where x represents the number of MaxiPaks produced, and C represents the cost in dollars.

(a) Determine the general solution of the cost function C.

(b) Find the particular cost function C, knowing that the cost of producing 10 MaxiPaks of lighters is $400.

72. Economics—Marginal Revenue: The Sleep-4-Ever Company determines that the marginal revenue for their queen-sized mattress is given by the differential rate equation

$$\frac{dR}{dx} = 157$$

where x is the number of queen-sized mattresses sold, and R is the revenue in dollars.

(a) Determine the general solution of the revenue function R.

(b) Find the particular revenue function R, knowing that there is zero revenue for zero sales.

73. Marketing—Consumer Acceptance: Suppose that the ChocoCola Company has changed their formula for their Chocolate Cherry Cola. The public is slow in accepting this new formula, but the company is winning over Chocolate Cherry Cola drinkers over time. The company's research department determines that the rate of acceptance of the new cola can be modeled by the differential equation

$$\frac{dA}{dt} = t + 1.1 \qquad 0 \leq t \leq 12$$

where t represents the number of months since the new formula for Chocolate Cherry Cola arrived on the market, and A represents the percentage of the Chocolate Cherry Cola drinkers who have accepted the new formula.

(a) Evaluate $\frac{dA}{dt}$ when $t = 10$ and interpret.

(b) Determine the general solution for the differential rate equation and interpret.

(c) Find the particular solution, knowing that 5% of the drinkers had accepted the new formula as soon as it was released. In other word, the initial condition is $t = 0$, $A = 5$.

74. Conservationism—Frog Population: Suppose a conservationist determines that the rate of population growth of the rare blue-spotted frog on a wildlife reserve during a yearlong study can be modeled by the differential rate equation

$$\frac{dp}{dt} = \frac{15}{t^{0.6}} \qquad 1 \leq t \leq 12$$

where t is the number of months since the start of the study, and p represents the number of frogs.

(a) Evaluate $\frac{dp}{dt}$ when $t = 5$ and interpret.

(b) Determine the particular solution $p(t)$ of the differential equation, knowing that the number of frogs observed after $t = 9$ months of the study was 90.

(c) Evaluate $p(5)$ and interpret.

 75. Microeconomics—Park Revenues: A recreational activity that has been seemingly immune to economic fluctuations is attendance at amusement parks. "[Because of] a greater willingness to take vacations, along with the incentive of enticing price promotions offered by many leading theme parks, we anticipate an increase in theme park sales in the coming years," said Fiona

© Edwin Verin/ShutterStock, Inc.

O'Donnell, senior analyst at Mintel Market Research. (*Source:* Theme Park Post.) From 2000 to 2011, the increase in revenues at amusement parks can be modeled by the differential equation

$$\frac{dR}{dt} = 1.36 \qquad 0 \le t \le 11$$

where t represents the number of years since 2000, and R represents the annual revenue at amusement parks, measured in billions of dollars. (*Source:* U.S. Bureau of Economic Analysis.)

(a) Determine the general solution to the differential rate equation.

(b) Knowing that 31.1 billion dollars were spent in amusement parks in 2000, determine the particular solution to the differential equation.

(c) Evaluate $R(10)$ and interpret the result.

 76. Sociology—Gambling: New avenues such as online gambling and state sanctioning of casino gambling have resulted in a steady increase in money spent on games of chance. From 2000 to 2011, the increase in the amount Americans spent on gambling can be modeled by the differential equation

$$\frac{dy}{dx} = 3.21 \qquad 0 \le x \le 11$$

© LiquidLibrary

where x represents the number of years since 2000, and y represents the amount Americans spent on gambling, measured in billions of dollars. (*Source:* U.S. Bureau of Economic Analysis.)

(a) Determine the general solution to the differential rate equation.

(b) Knowing that Americans spent 81 billion dollars on gambling in 2000, determine the particular solution to the differential equation.

(c) Evaluate $y(11)$ and interpret the result.

 77. Macroeconomics—Home Loans: During the decade of the 2000s, the United States experienced a "housing bubble" in which many Americans purchased homes that they could barely afford through subprime mortgages. The rate of change in the number of conventional bank loans for housing financing from 2000 to 2009 can be modeled by the differential equation

$$\frac{dy}{dx} = -60.24x + 225.59 \qquad 0 \le x \le 9$$

where x represents the number of years since 2000, and y represents the annual number of conventional housing loans, measured in thousands. (*Source:* U.S. Department of Housing and Urban Development.)

(a) Determine the general solution to the differential rate equation.

(b) Determine the particular solution to the differential equation using the initial condition $(x, y) = (0, 695)$.

(c) Use the solution from part (b) to complete the table. According to the table, during which year was home financing via conventional loans at its greatest?

Year	Number of conventional loans, in thousands
2001	
2003	
2005	
2007	
2009	

 78. Macroeconomics—Home Loans: Federal Housing Authority (FHA)-insured loans are a type of federal assistance and have historically allowed lower-income Americans to borrow money for the purchase of a home that they may not be able to otherwise afford. The rate of change in the number of FHA loans awarded from 2000 to 2009 can be modeled by

$$\frac{dy}{dx} = 4.14x - 24.89 \qquad 0 \le x \le 9$$

where x represents the number of years since 2000, and y represents the number of FHA loans awarded annually, measured in thousands. (*Source:* U.S. Department of Housing and Urban Development.)

(a) Determine the general solution to the differential rate equation.

(b) Find the particular solution to the differential equation given the initial condition $(x, y) = (0, 138)$.

(c) Use the solution from part (b) to complete the table. According to the table, during which year was home financing via FHA loans at its lowest?

Year	Number of conventional loans, in thousands
2001	
2003	
2005	
2007	
2009	

 79. Personal Finance—Healthcare Premiums: With an aging population, many Americans have become concerned over the increasing cost of their medical insurance. The rate of change in healthcare insurance premiums from 2001 to 2011 can be modeled by the differential rate equation

$$\frac{dy}{dx} = \frac{117.52}{x^{0.74}} \qquad 1 \le x \le 11$$

where x represents the number of years since 2000, and y represents the annual expenditures on healthcare insurance premiums, measured in billions of dollars. (*Source:* Office of the Actuary, National Health Expenditure Group.) Find the particular solution to this differential equation given the condition that when $x = 5$, $y = 691$.

 80. Public Health—Medicare: Enacted in 2006, the Medicare Part D plan, a federal program, is used to subsidize the costs of prescription drugs for Medicare beneficiaries in the United States. This unfunded program has resulted in increased federal expenditures in the Medicare program. From 2001 to 2011, the rate of increase in annual public expenditures in the federal Medicare program can be modeled by the differential equation

$$\frac{dy}{dx} = \frac{72.85}{x^{0.65}} \qquad 1 \le x \le 11$$

where x represents the number of years since 2000, and y represents the public expenditure in the Medicare program, measured in billions of dollars annually. (*Source:* U.S. Centers for Medicare and Medicaid Services.) Find the particular solution to this differential equation given the condition that when $x = 4$, $y = 311$.

Concept and Writing Exercises

81. Explain in your own words what is a differential equation.

82. Is $x^2 - x = 20$ a differential equation? Explain why or why not.

83. Explain the difference between the general solution and a particular solution to a differential equation.

84. To get a particular solution to a first-order differential equation, we need an initial condition. How many conditions do you think we need to get the particular solution to a second-order differential equation $y = f''(x)$? Explain your reasoning.

85. Let a and b represent positive real numbers. Determine the general solution to the differential equation $\frac{dy}{dx} = ax^b$.

86. If a represents any real number, determine the general solution to the differential equation $\frac{dy}{dx} = \frac{e}{a}$.

 Section Project

The New Skin Company tracks the marginal revenue (in hundreds of dollars), at various weekly sales levels, of its designer face cream. The data are summarized in **Table 7.7.1**.

Table 7.7.1

Sales level, x	Marginal revenue, $\dfrac{dR}{dx}$
1	3.97
5	3.91
10	3.77
15	3.68
20	3.60
25	3.48
30	3.43

(a) Use the regression capabilities of your calculator to find a linear regression model for the marginal revenue in the form

$$\frac{dR}{dx} = ax + b \qquad 1 \le x \le 30$$

where x represents the sales level, and R represents the revenue in hundreds of dollars.

(b) Knowing that there is no revenue when sales are zero, determine the revenue function R.

(c) Find the sales level x that will maximize the revenue.

(d) Determine the price-demand function p.

7.8 Differential Equations: Growth and Decay

Quite often in the business, life, and social sciences, we wish to study the behavior of a quantity as it increases or decreases over time. In this section, we will study **exponential growth** and **exponential decay** phenomena that are modeled by a specific type of differential equation. These growth and decay models have the properties of exponential functions that we have studied so far. We will also examine how exponential growth and decay models affect both the business and life sciences. We will also look at **limited** and **logistic models** in which the growth is limited by some y-value. In earlier chapters, we called these y-values **horizontal asymptotes**.

Unlimited Growth

There are many applications in which *the rate of change increases proportionally to the amount present*. With compound interest, for example, we know that as interest in the account accumulates, the account balance grows. In the social sciences, we know that as a population increases, more individuals are added to the population, and the population increases faster. These are all examples of **unlimited growth models**. If we call y the quantity at any given time t, the **exponential growth model** satisfies the equation

(rate of change in quantity) = (is proportional to)(the amount present)

Now we can write this as a differential equation with constant k as

$$y' = ky$$

To solve this differential equation, we rely on our separation of variables technique. Since the independent variable for this kind of model is usually time t, we can write

$$y' = ky$$

Replace y' with $\frac{dy}{dt}$ $$\frac{dy}{dt} = ky$$

Divide by y and integrate both sides $$\int \frac{1}{y}dy = \int k\,dt$$

Integrate and assume $y > 0$ $$\ln y = kt + C$$

Definition of logarithm $$y = e^{kt+C}$$

Rule of exponents $$y = e^{kt} \cdot e^{C}$$

Replace constant e^{C} with C $$y = Ce^{kt}$$

DEFINITION

Unlimited Growth Model

The solution to the differential equation $y' = ky$ is the **unlimited growth model**

$$y = Ce^{kt}$$

with initial value when $t = 0$ of $y = C$. See **Figure 7.8.1**.

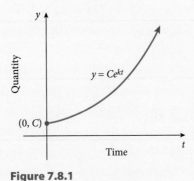

Figure 7.8.1

NOTES:

1. When the rate of growth is proportional to the present size, we have a condition called **unlimited growth.** The term *unlimited* means that y grows without limit.
2. Whenever we encounter this situation, a differential equation of the form $y' = ky$, we can immediately write the solution as $y = Ce^{kt}$. Remember that C is the initial value.
3. k is called the **rate of growth.**

Example 1: Writing an Exponential Growth Model with a Given Rate

OBJECTIVE 1

Write an exponential growth model with a given rate.

A biological researcher initially starts a culture of 200 bacteria and knows from past experience that the number of bacteria grows proportionally to the amount present. Past studies have shown that the rate of growth is 11% each hour.

a.   Write an exponential growth model for the number of bacteria present after t hours.

b.   Evaluate y when $t = 6$ and interpret.

c.   Evaluate y' when $t = 6$ and interpret.

Perform the Mathematics

a.   We know that when the time starts, the number of bacteria is 200. Thus, when $t = 0$, $y = 200$, the initial amount is $C = 200$. This gives us the exponential growth model

$$y = 200e^{kt}$$

Since past studies have shown the rate of growth to be 11% each hour, we have $k = 0.11$. So the function for the number of bacteria present after t hours is given by

$$y = 200e^{0.11t}$$

b.   Evaluating y at $t = 6$ gives

$$y = 200e^{0.11(6)} = 200e^{0.66} \approx 386.96$$

This means that, according to the model, there will be about 387 bacteria present after 6 hours.

c.   The derivative of $y = 200e^{0.11t}$ is

$$y' = \frac{d}{dt}[200e^{0.11t}] = 200e^{0.11t}(0.11) = 22e^{0.11t}$$

Evaluating y' at $t = 6$ yields

$$y' = 22e^{0.11(6)} = 22e^{0.66} \approx 42.57$$

Thus, after 6 hours the number of bacteria in the culture is growing at the rate of about 42.57 bacteria per hour.   ∎

▶ *Try It Yourself*

Some related Exercises are 29 and 31.

In practice, we are usually not given the growth rate for a differential equation as we were in Example 1. Instead, we are usually given an initial amount C and then given some other ordered pair. We can use this ordered pair to determine the rate of growth. (This technique will be demonstrated in Example 2.)

Exponential Decay

DEFINITION

Exponential Decay

The general exponential model $y = Ce^{-kt}$ is the solution to the differential equation $y' = -ky$. C is called the **initial amount,** and k is the **decay rate.**

Another family of applications relating to exponential models is called **exponential decay**. This type of model is used to exhibit behavior in which the rate of change decreases proportionally to the amount present. In this case, the differential equation for such behavior is

(rate of change in quantity) = (is proportional to)(the amount present)
$$y' = -ky$$

The **exponential decay model** associated with this differential equation is $y = Ce^{-kt}$.

Exponential decay models are solved in the same manner as the growth models. One common application of exponential decay models is the *half-life of a radioactive substance*. Half-life is the amount of time over which a radioactive substance will be reduced by half to an inert substance.

Example 2: Determining Radioactive Decay

A scientist at the Los Altos Laboratory has a 100-milligram sample of the radioactive substance cobalt-60. The half-life of cobalt-60 is known to be 5.3 years.

a. Determine an exponential decay model for the number of milligrams of cobalt-60 present after t years.

b. How many years will it take for the sample of cobalt-60 to decay to 10 milligrams?

OBJECTIVE 2

Write an exponential decay model for cobalt-60.

Perform the Mathematics

a. We know that the initial amount is 100 milligrams, so this gives us

$$y = Ce^{-kt}$$
$$y = 100e^{-kt}$$

where t represents the number of years, and y is the amount of cobalt-60 present. Knowing the half-life of cobalt-60 is 5.3 years tells us that, when $t = 5.3$, $y = \frac{1}{2}(100) = 50$ milligrams. Using this information, we can determine k to be

$$50 = 100e^{-k(5.3)}$$

$$\frac{1}{2} = e^{-k(5.3)}$$

$$\ln\left(\frac{1}{2}\right) = \ln e^{-k(5.3)}$$

$$\ln\left(\frac{1}{2}\right) = -k(5.3)$$

$$-\frac{\ln\left(\frac{1}{2}\right)}{5.3} = k$$

$$0.13 \approx k$$

So the exponential decay model for the cobalt-60 sample is

$$y = 100e^{-0.13t}$$

b. Here we need the t-value so that $y = 10$. So we need to solve the equation $10 = 100e^{-0.13t}$ for t. Solving for t gives us

$$10 = 100e^{-0.13t}$$

$$\frac{1}{10} = e^{-0.13t}$$

$$\ln\left(\frac{1}{10}\right) = \ln e^{-0.13t}$$

$$\ln\left(\frac{1}{10}\right) = -0.13t$$

$$-\frac{\ln\left(\frac{1}{10}\right)}{0.13} = t$$

$$17.71 \approx t$$

This means that it will take about 17.71 years for the 100-milligram cobalt-60 sample to decay to 10 milligrams. ∎

Limited Growth Models

There are some pitfalls to the unlimited growth model. For example, a town of 4000 that is currently growing at a rate of 18% per year has an unlimited growth model of

$$y = 4000e^{0.18t}$$

If we use the model to project the population 50 years from now, we get

$$y = 4000e^{0.18(50)}$$
$$= 4000e^9 \approx 32{,}412{,}336 \text{ people}$$

It seems rather unrealistic to believe that a town with a population of 4000 will swell to over 32 million in 50 years. There must be a reasonable limit, or upper bound, for almost every such quantity. For these kind of situations, we use the **limited growth model**. Here the rate of change is directly proportional to the distance from a limiting point, or upper bound. If we denote L as the upper bound and y as the value of the function at time t, then this distance from the upper bound is given by the difference $L - y$. So the differential equation for this situation is expressed as

$$(\text{Rate of growth}) = (\text{constant of variation})(\text{distance from upper bound } L)$$

or, mathematically,

$$y' = k(L - y)$$

For these applications, we assume that the initial condition is $y = 0$ when $t = 0$. Let's now use the separation of variables technique to solve this differential equation. Since time t is usually our independent variable, we can write

$$\frac{dy}{dt} = k(L - y)$$

Now, separating the variables t and y, we get

$$\frac{1}{L - y}dy = kdt$$

$$\int \frac{1}{L - y}dy = \int kdt$$

We need to use a u-substitution on the left-hand side. Let $u = L - y$ so that $du = -dy$ to get

$$\int \frac{1}{u}(-du) = \int kdt$$

$$-\int \frac{1}{u}du = \int kdt$$

$$-\ln u = kt + C$$

$$-\ln (L - y) = kt + C$$

$$\ln (L - y) = -kt - C$$

Now, using the definition of a logarithm gives us

$$L - y = e^{-kt - C}$$
$$L - y = e^{-kt} \cdot e^{-C}$$
$$-y = -L + e^{-kt} \cdot e^{-C}$$
$$y = L - e^{-kt} \cdot e^{-C}$$

Using the initial condition that $t = 0$ and $y = 0$, we can find the value of the constant e^{-C}.

$$0 = L - e^{-k(0)} \cdot e^{-C}$$
$$0 = L - e^{-C}$$
$$e^{-C} = L$$

Thus, we get the solution

$$y = L - Le^{-kt}$$

Factoring L from the right side, we get the **limited growth model** as a function of t to be

$$y = L(1 - e^{-kt})$$

DEFINITION

Limited Growth Model

The solution to the differential equation $y' = k(L - y)$ is the **limited growth model** given by the function

$$y = L(1 - e^{-kt})$$

where L is the **limit of growth** and k is the **rate of growth**.

NOTES:

1. The limit of growth is L. This is also known as the upper bound.
2. y is increasing, and the graph is concave down for all $t \geq 0$.

Applications of the limited growth model include learning curves, company growth, depreciation of equipment, and diffusion of mass media information.

Example 3: Deriving a Limited Growth Model

OBJECTIVE 3

Derive a limited growth model.

In a study of the memorization of unrelated shapes, a person can memorize no more than 150 of them in three hours. In a psychology class study, it is found that, on average, a subject can memorize 23 of the shapes after a 10-minute period.

a. Write a limited growth model where t is the time used to memorize the shapes, and y is the number of shapes memorized after t minutes.

b. Estimate the number of shapes that could be memorized in 45 minutes.

Perform the Mathematics

a. To determine the particular solution, we need to find the value of k. Given that $L = 150$ (the upper bound) and $y = 23$ when $t = 10$, we can get k by solving

$$y = L(1 - e^{-kt})$$

$$23 = 150(1 - e^{-k \cdot 10})$$

$$\frac{23}{150} = 1 - e^{-k \cdot 10}$$

$$e^{-k \cdot 10} = 1 - \frac{23}{150}$$

$$e^{-k \cdot 10} = \frac{127}{150}$$

$$\ln(e^{-k \cdot 10}) = \ln\left(\frac{127}{150}\right)$$

$$-k \cdot 10 = \ln\left(\frac{127}{150}\right)$$

$$k = -\frac{1}{10}\ln\left(\frac{127}{150}\right) \approx 0.017$$

This gives us the learning curve model $y = 150(1 - e^{-0.017t})$. The graph is shown in **Figure 7.8.2**.

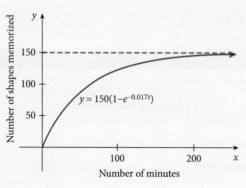

Figure 7.8.2

b. To estimate the number of shapes memorized in 45 minutes, we need to evaluate y when $t = 45$ to get

$$y = 150(1 - e^{-0.017(45)}) = 150(1 - e^{-0.765}) \approx 80.20$$

So according to the model, about 80 shapes can be memorized in a 45 minute period. ∎

▶ **Try It Yourself**

Some related Exercises are 37 and 39.

Logistic Growth Model

Many phenomena in the business, life, and social sciences follow a combination of initial rapid growth, only to flatten out for large values of t. The population of many developing countries follows this pattern of a rapid increase in population early on and then a leveling off as the country develops. See **Figure 7.8.3**.

This results in a differential equation that exhibits a combination of both unlimited growth and limited growth. In this type of differential equation, the rate of change y' is proportional to the amount currently present times the distance from an upper bound L. That is,

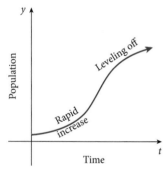

Figure 7.8.3

$$\begin{pmatrix} \text{Rate of} \\ \text{growth} \end{pmatrix} = \begin{pmatrix} \text{constant of} \\ \text{variation} \end{pmatrix} \cdot \begin{pmatrix} \text{amount currently} \\ \text{present} \end{pmatrix} \cdot \begin{pmatrix} \text{distance from} \\ \text{upper bound } L \end{pmatrix}$$

Mathematically, this is

$$y' = ky(L - y)$$

By using separation of variables with independent variable t, we get

$$\frac{dy}{dt} = y' = ky(L - y)$$

$$\int \frac{1}{y(L - y)} dy = \int k dt$$

Computing the general solution to this differential equation is done in the Exercises. The result is the **logistic growth model**.

DEFINITION

Logistic Growth Model

The solution to the differential equation $y' = ky(L - y)$ is the **logistic growth model** given by the function

$$y = \frac{L}{1 + ae^{-kLt}}$$

where L is the limit of growth and a and k are real number constants.

NOTE: In the logistic growth model, notice that when $t = 0$, $y = \frac{L}{1+a}$. This means that $(0, \frac{L}{1+a})$ is the initial value.

Applications of the logistic growth model include new product sales, dispersion of mass media information and rumors, spread of disease, and long-term population growth. One common application of logistic growth models occurs in the *harvest models* of trees. The model applies in this situation because, as the population of trees in a forest or tree farm increases, the room for roots to spread becomes limited, the amount of sunlight hitting the trees can become restricted, and the number of seedlings that have room and light to grow in becomes less and less. This results in a tapering of the number of trees that can thrive in a fixed area. Consequently, conservationists use a logistic model for the population size for trees.

Example 4: Deriving a Logistic Growth Model

OBJECTIVE 4

Derive a logistic growth model.

Suppose that commercial tree growers estimate that the maximum number of pine trees that can be supported at a large nursery is 15,000. The nursery was started with 1200 harvestable trees on the land. After 14 years, they find that 6000 of the trees are ready to be harvested. Write a differential equation for the rate of growth in population of the nursery trees, then determine a logistic model for the number of trees that are harvestable after t years.

Perform the Mathematics

In the differential equation $y' = ky(L - y)$, we know that the upper bound is $L = 15{,}000$, so we get $y' = ky(15{,}000 - y)$. To get the harvest function, we know its general solution has the form

$$y = \frac{15{,}000}{1 + ae^{-k \cdot 15{,}000t}}$$

To simplify our computations, we replace $k \cdot 15{,}000$ with another constant b.

$$y = \frac{15{,}000}{1 + ae^{-bt}}$$

Since the nursery started with 1200 trees, we know that $y = 1200$ when $t = 0$.

$$y = \frac{15{,}000}{1 + ae^{-bt}}$$

$$1200 = \frac{15{,}000}{1 + ae^{-b(0)}}$$

$$1200 = \frac{15{,}000}{1 + a}$$

$$1 + a = \frac{15{,}000}{1200}$$

$$a = \frac{15{,}000}{1200} - 1 = 11.5$$

This gives us the general solution

$$y = \frac{15{,}000}{1 + 11.5e^{-bt}}$$

To get the proper value of b, we use the condition that $y = 6000$ when $t = 14$. This gives us

$$y = \frac{15{,}000}{1 + 11.5e^{-bt}}$$

$$6000 = \frac{15{,}000}{1 + 11.5e^{-b(14)}}$$

$$1 + 11.5e^{-b(14)} = \frac{15{,}000}{6000}$$

$$11.5e^{-b(14)} = \frac{15{,}000}{6000} - 1$$

$$e^{-b(14)} = \frac{1}{11.5}\left(\frac{15{,}000}{6000} - 1\right)$$

$$-b(14) = \ln\left[\frac{1}{11.5}\left(\frac{15{,}000}{6000} - 1\right)\right]$$

$$b = -\frac{1}{14}\left\{\ln\left[\frac{1}{11.5}\left(\frac{15{,}000}{6000} - 1\right)\right]\right\}$$

$$b \approx 0.15$$

Thus, the logistic model for the number of trees that are harvestable after t years is

$$y = \frac{15{,}000}{1 + 11.5e^{-0.15t}}$$

The graph of model for the first five decades is displayed in **Figure 7.8.4**.

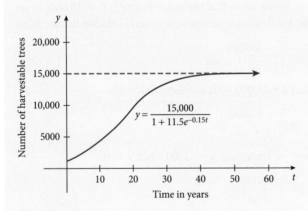

Figure 7.8.4

Many times, we are not given information such as the upper bound or growth rate for an application. As a matter of fact, these are pieces of information that are expected to be discovered. In practice, logistic model applications come from collected data and are generated by regression techniques.

Summary

In this section, we focused on four different types of growth and decay models. We found that each differential equation from which the model came could be found by separation of variables.

- The **exponential growth model** $y = Ce^{kt}$ is the solution to the differential equation $y' = ky$.

- The **exponential decay model** $y = Ce^{-kt}$ is the solution to the differential equation $y' = -ky$.

- The **limited growth model** $y = L(1 - e^{-kt})$ is the solution to the differential equation $y' = k(L - y)$.

- The **logistic growth model** $y = \dfrac{L}{1 + ae^{-kLt}}$ is the solution to the differential equation $y' = ky(L - y)$.

Section 7.8 Exercises

Vocabulary Exercises

1. The solution to the differential $y' = ky$ yields an _____ growth model.
2. For the differential equation $y' = -k \cdot f(t)$, we call k the rate of _____.
3. For the unlimited growth model $y = ce^{kt}$, the value of c represents the _____ value.
4. We call a model of the form $y = L(1 - e^{-kt})$ a _____ growth model.
5. For the growth model $y = L(1 - e^{-kt})$, we call L the _____ bound.
6. A growth model of the form $y = \dfrac{L}{1 + ae^{-kLt}}$ is referred to as a _____ growth model.

Skill Exercises

In Exercises 7–18, complete the following.

(a) Use the definition of the limited growth model to write the corresponding differential equation in the form $y' = k(L - y)$.

(b) Identify the limit of growth L and evaluate y when $t = 0$ to determine the initial value.

7. $y = 10(1 - e^{-t})$ 8. $y = 5(1 - e^{-t})$
9. $y = 8(1 - e^{-2t})$ 10. $y = 8(1 - e^{-3t})$
11. $y = 20(1 - e^{-0.5t})$ 12. $y = 16(1 - e^{-0.3t})$
13. $y = 12 - 12e^{-0.1t}$ 14. $y = 24 - 24e^{-0.2t}$
15. $y = 1.5(1 - e^{-0.2t})$ 16. $y = 0.01(1 - e^{-0.3t})$
17. $y = 15.5 - 15.5e^{-0.1t}$ 18. $y = 62 - 62e^{-0.02t}$

For Exercises 19–28, complete the following.

(a) Use the definition of the logistic growth model to write the corresponding differential equation in the form $y' = ky(L - y)$.

(b) Evaluate y when $t = 0$ to determine the initial value and identify the limit of growth L.

19. $y = \dfrac{30}{1 + e^{-t}}$ 20. $y = \dfrac{15}{1 + e^{-t}}$

21. $y = \dfrac{50}{1 + 2e^{-t}}$ **22.** $y = \dfrac{10}{1 + 5e^{-t}}$

23. $y = \dfrac{150}{1 + e^{-2t}}$ **24.** $y = \dfrac{90}{1 + e^{-2t}}$

25. $y = \dfrac{400}{1 + 2e^{-2t}}$ **26.** $y = \dfrac{2500}{1 + 5e^{-4t}}$

27. $y = \dfrac{1000}{1 + 4.5e^{-0.9t}}$ **28.** $y = \dfrac{4000}{1 + 6e^{-1.2t}}$

Application Exercises

For Exercises 29–32, use an unlimited growth model.

29. Demographics—Population Growth: The population of Sensenville is increasing at a rate of 5% per year. The current population of the town is 50,000. Assume that the rate of change in the population is increasing proportionally to the population size.

   (a) Write a differential equation and an initial condition for the population of Sensenville.

   (b) Write an exponential growth model for the population of Sensenville in *t* years.

   (c) Use the model to estimate the population in 10 years.

30. Demographics—Population Growth: The population of Burgentown is increasing at a rate of 2% per year. The current population of the city is 250,000. Assume that the rate of change in the population is increasing proportionally to the population size.

   (a) Write a differential equation and an initial condition for the population of Burgentown.

   (b) Write an exponential growth model for the population of Burgentown in *t* years.

   (c) Use the model to estimate the population in 10 years.

31. Biology—Bacteria Growth: The number of bacteria in a culture starts at 5000 and increases at a rate of 12% per hour. Assume that the rate of change in the number of bacteria is increasing proportionally to the number of bacteria present.

   (a) Write a differential equation for the number of bacteria present after *t* hours.

   (b) Write an exponential growth model of the bacteria present after *t* hours.

   (c) Find the number of bacteria after $t = 3$, 12, and 24 hours.

32. Biology—Bacteria Growth: The number of bacteria in a culture starts at 500 and increases at a rate of 7% per hour. Assume that the rate of change in the number of bacteria is increasing proportionally to the number of bacteria present.

   (a) Write a differential equation for the number of bacteria present after *t* hours.

   (b) Write an exponential growth model of the bacteria present after *t* hours.

   (c) Find the number of bacteria after $t = 0$, 3, 6, and 12 hours.

For Exercises 33–36, use an exponential decay model.

33. Physics—Radioactive Half-life: Suppose a scientist has a 3000-milligram sample of the radioactive substance phosphorus-32. The half-life of phosphorus-32 is known to be 14.2 days.

   (a) Determine an exponential decay model for the number of milligrams of phosphorus-32 present after *t* days.

   (b) How many days will it take for the sample of phosphorus-32 to decay to 2000 milligrams?

34. Physics—Radioactive Half-life: Suppose that a scientist has a 300-gram sample of the radioactive substance radon. The half-life of radon is known to be 3.82 days.

   (a) Determine an exponential decay model for the number of grams of radon present after *t* days.

   (b) How many days will it take for the sample of radon to decay to 50 grams?

35. **Demographics—Population Decrease:** The population of Devonburg was 920,000 in 2005. Because of changing economic conditions in the community, the population had decreased to 700,000 in 2010. Assume that the rate of change in the population is decreasing proportionally to the population size.

   (a) Write a differential equation and an initial condition for the population of Devonburg.

   (b) Write an exponential decay model for the population of Devonburg in t years.

   (c) Use the model to estimate the population in 2016.

36. **Demographics—Population Decrease:** The population of Grandland was 604,000 in 2006. Because of urban flight, the population decreased to 500,000 in 2011. Assume that the rate of change in the population is decreasing proportionally to the population size.

   (a) Write a differential equation and an initial condition for the population of Grandland.

   (b) Write an exponential decay model for the population of Grandland in t years.

   (c) Use the model to estimate the population in 2017.

For Exercises 37–40, use a limited growth model.

37. **Management—Employee Training:** Suppose that the Wesson and Selverstone Investment Group has introduced a new computer system for its 3000 employees. The developer of the software has determined that the employees will learn the new system according to the model

$$y = 3000(1 - e^{-0.3t}) \qquad t \geq 0$$

   where t represents the number of weeks since the system was installed, and y represents the number of employees who have learned the new system.

   (a) Evaluate y and y' when $t = 2$ and interpret.

   (b) According to the model, how many weeks will it take for at least 2500 employees to learn the new system?

38. **Epidemiology—Flu Epidemic:** Consider a college dormitory of 600 students where a flu epidemic is infecting students according to the model

$$y = 600(1 - e^{-0.22t}) \qquad t \geq 0$$

   where t represents the number of days since the flu was first diagnosed, and y represents the number of dorm residents infected.

   (a) Evaluate y and y' when $t = 4$ and interpret.

   (b) According to the model, how many days will it take until at least 400 students have been infected by the flu?

39. **Management—Employee Training:** The MidWest Fabricating Company installs a new metal press for its 300 employees. The company that makes the press has determined through experience that employees learn to use the new press in a way that follows the limited growth model

$$y = L(1 - e^{-kt}) \qquad t \geq 0$$

   where t is the number of weeks since the new press was installed, and y represents the number of employees who have learned to use the new equipment. They find that after the first week, 40 employees have learned to use the new metal press.

   (a) Identify the limit of growth L.

   (b) Solve the equation $40 = 300(1 - e^{-k(1)})$ to determine the rate of growth k.

   (c) Graph the model for the first half-year (that is, 26 weeks) after installation.

   (d) If the company has set a goal of having half of the employees skilled at using the machine in the first month (that is, in the first 4 weeks), will the company reach its goal?

40. **Conservation—Wildlife Management:** A municipal park is being overrun by 200 deer, so park officials decide to open the park to hunting for a short time in an attempt to curb the deer

population. The officials believe that the number of deer killed by the hunters will follow the limited growth model

$$y = L(1 - e^{-kt}) \qquad t \geq 0$$

where t is the number of days since the hunting season began, and y represents the number of deer killed. After the first two days of hunting, 24 deer were killed.

(a) Identify the limit of growth L.

(b) Solve the equation $24 = 200(1 - e^{-k(2)})$ to determine the rate of growth k.

(c) Graph y to display the number of deer killed if the hunting season lasted 30 days.

(d) If the officials plan to call an end to the hunting season when the deer population is cut in half, how many days will this take?

For Exercises 41–44, use a logistic growth model.

41. **Consumer Technology—Digital Video Recorders:** According to estimates from the Whitehead Research Group, the percentage of households that own digital video recorders in a certain county is given by the logistic model

$$f(t) = \frac{90}{1 + 22e^{-0.7t}} \qquad 0 \leq t \leq 10$$

where t represents the number of years since 2000, and $f(t)$ represents the percentage of households that own digital video recorders.

(a) According to the model, what percentage of the households in the county owned digital video recorders in 2009?

(b) Evaluate $f'(t)$ when $t = 6$ and interpret.

(c) According to the model, what is the maximum number of households in the county that are expected to own a digital video recorder?

42. **Sociology—Group Behavior:** On a certain university campus, the 12,000 students who attend home basketball games are quickly swept by the new fad of tossing toilet paper onto the basketball court after the first basket is made for the home team. Research students in the sociology department think that the fad started with about 100 fans and that the number of students at the home games who throw toilet paper follow the logistic model

$$f(x) = \frac{12,000}{1 + 119e^{-0.8x}} \qquad 0 \leq x \leq 12$$

where x is the number of home games completed, and $f(x)$ represents the number of students expected to toss toilet paper onto the basketball court after the first basket is made for the home team.

(a) Evaluate y when $x = 8$ and interpret.

(b) If the university's athletic director threatens to search all students coming to the game for toilet paper when at least half of the student fans are engaging in the activity, how many home games will this take?

43. **Aquaculture—Fish Harvest:** The Top Cove Fishery established a harvesting area that is designed to hold a maximum of 7500 catfish. The fishery was initially stocked with nearly 1000 catfish. After two years, the number of catfish in the fishery had grown to nearly 2500.

(a) What is the limit of growth L?

(b) Solve the equation $2500 = \dfrac{7500}{1 + 6.5e^{-b(2)}}$ to determine the rate of growth where $b = kL$.

(c) Write the logistic function in the form $f(t) = \dfrac{L}{1 + ae^{-bt}}$ for the first 5 years since the harvesting area was established, where $b = kL$.

(d) Conservationists define the *maximum sustained yield* (MSY) as the largest number of the population that can be removed while sustaining the population. For logistic functions, the MSY is found at the inflection point of the curve. That is the point $(\frac{\ln a}{b}, \frac{L}{2})$ where $b = kL$. Determine the MSY for the harvesting area.

44. Sociology—Dispersion of News: Consider a town with a population of 30,000 whose city council president has just resigned. He announces the resignation at a closed session of the 10-member city council. It takes only six hours for 27,300 citizens to learn about the resignation.

(a) What is the limit of growth L?

(b) Solve the equation $27{,}300 = \dfrac{30{,}000}{1 + 3000e^{-b(6)}}$ to determine the rate of growth, where $b = kL$.

(c) Write the logistic function in the form $f(t) = \dfrac{L}{1 + ae^{-bt}}$ for the first seven hours after the president's announcement.

45. E-Commerce—Company Growth: A company that has experienced rapid growth is online merchandiser Amazon.com. "The company's management team has a proven track record of succeeding through investment cycles (2004 to 2006), and Amazon is building a platform from which to generate substantial free-cash-flow growth for many, many years," said Mark S. Mahaney, an analyst with Citigroup. (*Source: Barron's Magazine.*) From 2000 to 2011, the net revenue of Amazon.com has been growing at an annual rate of 28% per year. (*Source: Amazon company filings.*)

(a) Write a differential equation in the form $\frac{dy}{dt} = ky$ for Amazon's revenue growth.

(b) Solve the differential equation from part (a) to get an unlimited growth model of the form

$$f(t) = Ce^{kt} \qquad 0 \le t \le 11.$$

(c) Knowing that Amazon's net revenue in 2000 was 1.28 billion dollars, write the particular solution of the general solution from part (b).

(d) Evaluate $f(14)$ and interpret the result.

46. Agribusiness—Grain-Export: Led by demand for wheat and soybeans, the amount of exports from the United States to China has been increasing rapidly over the past two decades. From 1990 to 2011, exports from the United States to China have been increasing at a rate of 16.5% per year. (*Source: U.S. Department of Commerce.*)

(a) Write a differential equation in the form $\frac{dy}{dt} = ky$ for the increase in U.S. exports to China.

(b) Solve the differential equation from part (a) to get an unlimited growth model of the form

$$f(t) = Ce^{kt} \qquad 0 \le t \le 11$$

where t represents the number of years since 1990, and $f(t)$ represents the annual exports from the United States to China, measured in millions of dollars.

(c) Knowing that in 1990 there were 300 million dollars of exports from the United States to China, write the particular solution of the general solution from part (b).

(d) Evaluate $f(13)$ and interpret the result.

47. Public Funding—Veteran's Care: A consequence to the recent wars in Iraq and Afghanistan is the increased costs in medical care for veterans. "This presents a big challenge—and one that we have no choice but to step up to meet if we are going to avoid many of the same mistakes we saw with the Vietnam generation," said Patty Murray, member of the Senate Veterans' Affairs Committee. (*Source: Seattle Times.*)

(a) Knowing that the increase in cost can be expressed by the differential equation $\frac{dy}{dt} = 0.01(160.95 - y)$, write a limited growth function of the form

$$f(t) = L(1 - e^{-kt}) \qquad 0 \le t \le 21$$

where t represents the number of years since 1990, and $f(t)$ represents the annual cost of medical and hospital care for veterans in billions of dollars.

(b) Solve the equation $f(t) = 37$ for t. Round t to the nearest whole number. Interpret the meaning of the values of t and $f(t)$.

(c) Evaluate the limit $\lim\limits_{t\to\infty} f(t)$ and interpret its meaning.

 48. Economics—ACSI: The American Customer Satisfaction Index (ACSI) is an economic indicator that measures the satisfaction of consumers across the U.S. economy. The 2011 ACSI shows that five of the top 10 most hated companies are airlines. (*Source:* ACSI.)

(a) Knowing that the increase in customer service complaints (unhelpful employees, inadequate meals and cabin service, treatment of delayed passengers, etc.) can be expressed by the differential equation $\frac{dy}{dt} = 0.12(1864 - y)$, write a limited growth function of the form

$$f(t) = L(1 - e^{-kt}) \quad 1 \leq t \leq 11$$

where t represents the number of years since 2000, and $f(t)$ represents the annual number of customer service complaints.

(b) Solve the equation $f(t) = 950$ for t. Round t to the nearest whole number. Interpret the meaning of the values of t and $f(t)$.

(c) Evaluate the limit $\lim\limits_{t\to\infty} f(t)$ and interpret its meaning.

 49. Investment Banking—Mutual Funds: Mutual funds are a managed type of collective investment that pools money from many investors to buy stocks, bonds, and short-term money market investments. Many investors use mutual funds to generate a stream of income while they are in their retirement years. The total in accounts in mutual funds in the United States from 1990 to 2011 can be modeled by the logistic growth model

$$f(t) = \frac{201{,}337}{1 + 399e^{-0.10t}} \quad 0 \leq t \leq 21$$

where t represents the number of years since 1990, and $f(t)$ represents the annual total in accounts in mutual funds, measured in billions of dollars. (*Source:* Board of Governors of the Federal Reserve System.)

(a) Evaluate and interpret $f(10) - f(0)$.

 (b) Graph f in the viewing window $[0, 25]$ by $[500, 6000]$, then use the `Intersect` command to determine the t-value when $f(t) = 3000$. Round the t-value to the nearest whole number and interpret each coordinate.

 50. Environmental Science—Carbon Emissions: A country that has been increasingly concerned about its carbon emissions is Australia. "On average, each person in Australia now emits more than five tons of carbon a year, while in China the figure is only one ton per year," says Michael Raupach of the Commonwealth Scientific and Industrial Research Organization. (*Source: The Sydney Morning Herald.*) The amount of carbon monoxide emissions in Australia from 1990 to 2011 can be modeled by the logistic growth curve

$$f(t) = \frac{555.9}{1 + 1.5e^{-0.1t}} \quad 0 \leq t \leq 21$$

where t represents the number of years since 1990, and $f(t)$ represents the amount of carbon monoxide emissions, measured in billions of metric tons. (*Source:* The International Energy Statistics Database.)

(a) Evaluate and interpret $f(20) - f(10)$.

 (b) Graph f in the viewing window $[0, 25]$ by $[200, 500]$, then use the `Intersect` command to determine the t-value when $f(t) = 305$. Round the t-value to the nearest whole number and interpret each coordinate.

Concept and Writing Exercises

51. State in your own words the difference between a limited growth model and a logistic growth model.

For Exercises 52–56, consider the differential equation corresponding to the logistic growth model, $y' = ky(L - y)$.

52. Use the separation of variables technique to write an equation with expressions involving y on the left side and the constant k on the right.

53. Verify that the derivative of the expression $\frac{1}{L}\ln\left(\frac{y}{L - y}\right)$ is $\frac{1}{y(L - y)}$ using the differentiation rules. Keep in mind that L is a constant.

54. Show that when we solve the equation $\frac{1}{L}\ln\left(\frac{y}{L - y}\right) = kt + C$ for y, we get $y = \frac{L}{1 + ae^{-kLt}}$, where $a = e^{-CL}$. Do this by isolating y and dividing the numerator and denominator of the resulting fraction by ae^{kLt}.

55. Verify that $y = \frac{L}{1 + ae^{-kLt}}$ satisfies the differential equation $y' = ky(L - y)$ and explain why $y' > 0$ for any nonnegative value of t. Note that this shows that the logistic function is increasing over its domain.

56. Assuming $kLt > 0$, use the properties of limits and exponential functions to explain why $\lim_{t \to \infty}\left(\frac{L}{1 + ae^{-kLt}}\right) = L$.

 Section Project

The Belgian Pierre-François Verhulst developed the logistic growth model in 1838 to explain the demographic transition where the rate of reproduction is proportional to both the existing population and the amount of available resources, all else being equal. To test its accuracy of this model, consider the U.S. population from 1790 to 1980 given in **Table 7.8.1**.

Table 7.8.1

Year	Population (in millions)	Year	Population (in millions)
1790	3.929	1890	62.980
1800	5.308	1900	76.212
1810	7.240	1910	92.228
1820	9.638	1920	106.022
1830	12.861	1930	123.203
1840	17.063	1940	132.165
1850	23.192	1950	151.326
1860	31.443	1960	179.324
1870	50.189	1970	203.302
1880	50.189	1980	226.549

Source: U.S. Census Bureau

(a) Make a scatterplot of the data, where t is the number of years past 1800, and $-10 \leq t \leq 180$.

(b) Use your calculator to get a logistic regression model y that represents the U.S. population where t is the number of years past 1800. Now graph the scatterplot along with the logistic model y for $-10 \leq t \leq 250$ and then for $0 \leq t \leq 370$.

(c) Determine the year in which the rate of growth in the U.S. population started to decrease. (*Hint:* Find the inflection point.) What do you notice about this answer and the answer to part (b)?

(d) According to the model, what is the maximum size that the U.S. population can reach?

(e) The current U.S. population can be found using a search engine. Compare the current year's population to the prediction from the model. How do you account for the difference?

Section 7.1 Review Exercises

In Exercises 1–4, determine the average value of the given function on the stated interval.

1. $f(x) = 3x + 4;$ $[0, 6]$

2. $f(x) = x^2 - 2x;$ $[1, 5]$

3. $f(x) = 4x^3 - 3x^2 + 2;$ $[0, 5]$

4. $f(x) = 3e^{-0.2x};$ $[0, 2]$

 5. For $f(x) = \ln(1 + x)$, complete the following.

   (a) Use your RTSUM program to approximate $\int_a^b f(x)\,dx$ on $[0, 8]$. Use $n = 100$.

   (b) Use your approximation from part (a) to determine an (approximate) average value for $f(x)$ on $[0, 8]$.

6. **Economics—Price-Demand:** The price-demand function for the five-minute special lunch at the Reggie's Veggies Restaurant is given by

$$p(x) = 9.10 - 0.07x$$

where $p(x)$ is the price in dollars, and x is the quantity demanded. Determine the average price in the demand interval $[24, 60]$.

7. **Economics—Cost Analysis:** The Tinnitus Phone Factory has determined that the cost for producing x telephones is given by

$$C(x) = 1300 + 18x$$

where $C(x)$ is the cost in dollars.

   (a) Determine the average value of the cost function over the interval $[0, 400]$.

   (b) Determine the average cost function $AC(x) = \frac{C(x)}{x}$ and evaluate $AC(400)$.

   (c) Explain the differences between parts (a) and (b).

8. **Banking—Money Markets:** Frederick deposits $1400 into a money market account earning 5.25% interest compounded continuously. Determine Frederick's average balance, to the nearest cent, over one year.

9. **Banking—Account Balances:** Penultimate Savings and Loan offers a money market account that earns 4.7% interest compounded continuously. At the end of the year, the account also pays a bonus of 0.85% of the average balance in the account during the year. Carlene deposits $2400 into a Penultimate money market account.

   (a) Compute Carlene's average balance, to the nearest cent, over one year.

   (b) Compute the size of Carlene's bonus.

10. **Banking—Account Balances:** Sylvia deposits $800 into an account where the rate of change of the amount in the account is given by $A'(t) = 56e^{0.07t}$ t years after the initial deposit.

   (a) Determine by how much the account will change from the end of the first year to the end of the fourth year. Round to the nearest cent.

   (b) Determine the average balance, to the nearest cent, in the account over the first four years.

11. **Small Business—Game Income:** The rate of change of the total income produced by a pinball machine located in a college dorm is given by

$$f(t) = 4000e^{0.09t}$$

where t is the time in years since the installation of the pinball machine, and $f(t)$ is in dollars per year. Determine the total income generated by the pinball machine during its first five years of operation.

12. **Personal Finance—IRAs:** Calvin contributes $2000 to his IRA account at the beginning of each year for 30 years. The account earns 6.2% annual interest compounded continuously. Determine how much money Calvin has in his account after 30 years.

13. **Personal Finance—College Savings:** Upon adopting their daughter Felicia, Chris and Pat opened a college savings account. They deposit $65 at the beginning of each month into an account earning 4.8% annual interest, compounded continuously. Determine how much is in the account in 16 years, when Felicia starts college.

 14. **Industrial Economics—Trucking Revenue:** A part of the industrial sector that has remained strong over the past decade is truck transportation of goods. "We have much better net revenue growth momentum going into 2011 than we had the last few years," said John P. Wiehoff, President of C.H. Robinson Trucking. (*Source: The Journal of Commerce.*) The annual revenue from truck transportation from 2003 to 2010 can be modeled by

$$f(x) = -2.2x^2 + 37.54x + 73.7 \quad 1 \le x \le 10$$

where $x = 1$ corresponds to January 1, 2000, and $f(x)$ represents the of revenue from truck transportation, measured in billions of dollars. Use the model to calculate the average annual revenue in the trucking industry for the period from January 1, 2000, to January 1, 2009.

Section 7.2 Review Exercises

For Exercises 15 and 16, write a definite integral to represent the area of the shaded region.

15.

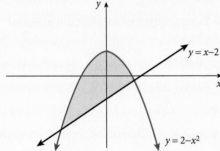

16.

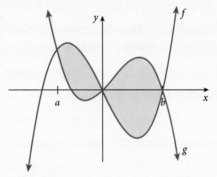

In Exercises 17–24, determine the area of the region bounded by the given conditions.

17. $f(x) = 9 - x^2$ and the x-axis on $[-3, 3]$
18. $f(x) = x^3 - 5x^2 - 2x + 3$ and the x-axis on $[1, 5]$
19. $f(x) = x^2 - 4$ and $g(x) = x + 2$ on $[-2, 3]$
20. $f(x) = x^3 + 2x^2 - 20$ and $g(x) = -3x^2 + 4x$ on $[-5, 2]$
21. $f(x) = 4x + 6$ and $g(x) = x^2 - 6$ on $[-2, 6]$
22. $f(x) = 4 + \sqrt{x}$ and $g(x) = \sqrt[3]{x}$ on $[0, 6]$
23. $f(x) = 5e^x$ and $g(x) = 3.7^x$ on $[-2, 5]$
24. $f(x) = \frac{8}{x^3}$ and $g(x) = \frac{2}{x}$ on $[1, 4]$

In Exercises 25–28, determine the area of the region bounded by the given curves.

25. $f(x) = x^2$ and $g(x) = x + 6$
26. $f(x) = x^3 + 2x^2 - 5x - 6$ and the x-axis
27. $f(x) = x^3$ and $g(x) = 2x^2 + 8x$
28. $f(x) = \sqrt[3]{x}$ and $g(x) = \frac{1}{4}x$

29. **Small Business—Monthly Sales:** The Krazy Dog Company is planning to sell a new tricycle for dogs. It estimates that its monthly sales, $s(t)$, in units per month, t months after the product is released, will be given by

$$s(t) = 1.7t^2 - 16.1t + 160$$

An agent for Lucky, the star of the famous movie *Rover's Revenge*, claims that if Lucky is featured in an ad campaign the monthly sales, $m(t)$, in units per month, t months after the product is released, will be

$$m(t) = 164e^{0.09t}$$

On the interval $[0, 15]$, determine the area between the graphs of s and m then interpret.

30. **Bacteriology—Bacteria Growth:** The size of bacteria culture A grows at a rate of $20x^{1.5}$ bacteria per minute, and the size of bacteria culture B grows at a rate of $10x^{2.5}$ bacteria per minute.

    (a) Set up an integral to determine the area between the two curves on $[0, 2]$.

    (b) Evaluate the integral from part (a) and interpret.

    (c) Set up an integral to determine the area between the two curves on $[2, 4]$.

    (d) Evaluate the integral from part (c) and interpret.

    (e) On the interval $[0, 4]$, determine which culture has more bacterial growth and by how many bacteria.

31. **Operations Research—Process Innovation:** The Loose Nail Furniture Company has determined that its costs, in thousands of dollars, have been increasing mainly due to inflation at a rate given by $215e^{0.14x}$, where x is time measured in years since a certain item started being produced. At the beginning of the fourth year, a breakthrough in the production process resulted in costs increasing at a rate given by $80x^{3/4}$. Determine the total savings in the interval $[4, 8]$ due to the breakthrough.

32. **Economics—Deriving Profit:** The Lost Sole Shoe Company determines that the marginal revenue to produce x pairs of shoes is given by

    $$MR(x) = 85 - 0.005x^2$$

    while the marginal cost is given by

    $$MC(x) = 32$$

    where both marginal functions are in dollars per pair of shoes. Determine the total profit from $x = 45$ to $x = 84$.

33. **Operations Research—Assembly Rates:** Fred and Ginger are assembly workers at the Clobini Factory. Fred assembles units at the rate of

    $$f(x) = -5x^2 + 23x + 31$$

    units per hour and Ginger assembles units at the rate of

    $$g(x) = -4x^2 + 25x + 32$$

    where x is the number of hours on the job, during the first four hours on the job.

    (a) How many units does Fred assemble during his first four hours?

    (b) How many more units does Ginger assemble than Fred during the same time period?

 34. **International Finance—Trade with Hong Kong:** The sizable trade surplus that the United States has with Hong Kong has raised concerns among some international financiers. "The bulging surplus will intensify trade friction and protectionist initiatives," said Wang Qian, an economist at JPMorgan Chase & Co. in Hong Kong. "It will also strengthen the pressure for further appreciation of the yuan." (*Source:* Bloomberg News.) The imports and exports between the United States and Hong Kong from 1990 to 2011 can be modeled by the following functions.

$$\text{Imports: } I(x) = -1.76x^3 + 15x^2 + 115.8x + 9361 \qquad 0 \le x \le 21$$
$$\text{Exports: } E(x) = 8.76x^3 - 240x^2 + 2294x + 6148 \qquad 0 \le x \le 21$$

where x represents the number of years since 1990, and $I(x)$ and $E(x)$ represent the respective annual amount of imports and exports for the United States and Hong Kong, measured in millions of dollars. (*Source:* U.S. Census Bureau.)

(a) $\displaystyle\int_0^2 (I(x) - E(x))\,dx$ and interpret the result.

(b) $\displaystyle\int_{15}^{20} (I(x) - E(x))\,dx$ and interpret the result.

Section 7.3 Review Exercises

Economics—Consumers/Producers Surplus: *For Exercises 35–38, refer to the figure.*

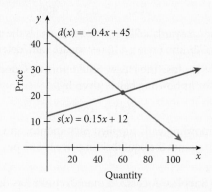

35. Shade the region that corresponds to the consumers' surplus. Write an integral to determine the consumers' surplus.

36. Shade the region that corresponds to the producers' surplus. Write an integral to determine the producers' surplus.

37. Shade the region that corresponds to the actual consumer expenditure. Determine the actual consumer expenditure.

38. Shade the region that corresponds to the actual revenue. Determine the actual revenue.

In Exercises 39–42, for each demand function given and demand level x, determine the consumers' surplus.

39. $d(x) = 450 - 0.8x; \quad x = 100$

40. $d(x) = 3600 - 0.02x^2; \quad x = 300$

41. $d(x) = 100 - \sqrt{x}; \quad x = 36$

42. $d(x) = 225e^{-0.01x}; \quad x = 250$

In Exercises 43–46, for each supply function given and demand level x, determine the producers' surplus.

43. $s(x) = 0.5x + 1200; \quad x = 400$

44. $s(x) = \dfrac{1}{15}x + 10; \quad x = 150$

45. $s(x) = 0.02x^2 + 25; \quad x = 50$

46. $s(x) = 84e^{0.02x}; \quad x = 200$

Economics: Consumers'/Producers' Surplus: *For Exercises 47–50, complete the following.*

(a) Determine the equilibrium point. Round equilibrium demand and equilibrium price to the nearest whole number.

(b) Determine the consumers' surplus at equilibrium demand.

(c) Determine the producers' surplus at equilibrium demand.

47. $d(x) = 250 = 0.3x; \quad s(x) = 60 + 0.2x$

48. $d(x) = 1280 - 0.8x; \quad s(x) = 272 + 1.6x$

49. $d(x) = 4500 - 0.04x^2; \quad s(x) = 0.03x^2 + 1700$

50. $d(x) = 700e^{-0.05x}; \quad s(x) = 50e^{0.06x}$

51. **Economics—Consumers' Surplus:** Knutsen Bolts Hardware has determined that the demand function for a new type of electric drill is given by

$$d(x) = 22 - 2\sqrt{x}$$

where x is the number of drills demanded each day, and $d(x)$ is the price per drill in dollars.

(a) Assuming that the equilibrium price is $14 per drill, determine the equilibrium demand. Round to the nearest whole number.

(b) Determine the consumers' surplus at equilibrium demand.

52. **Economics—Consumers' Surplus:** Carpets Galore has determined that the demand function for a certain kind of carpet is given by

$$d(x) = -0.5x + 400$$

where x is the number of square yards sold per week, and $d(x)$ is the price per square yard in dollars. Assuming that the equilibrium price is $18 per square yard, determine the consumers' surplus.

53. **Economics—Producers' Surplus:** The Heavy Stuff Sporting Goods company has determined that the supply function for its bowling ball is given by

$$s(x) = 18e^{0.007x}$$

where x is the number of bowling balls supplied each month, and $s(x)$ is the price per bowling ball in dollars. Assuming that the equilibrium price is $47 per bowling ball, determine the producers' surplus.

54. **Economics—Producers' Surplus:** A cushion manufacturer has determined that the supply function for a certain kind of cushion is given by

$$s(x) = 0.04x^2 + 11$$

where x is the number of these cushions supplied each week, and $s(x)$ is the price per cushion in dollars. Assuming that the equilibrium price is $23.96, determine the producers' surplus.

55. **Economics—Consumers'/Producers' Surplus:** Springs 'n' Things has determined that the demand function for its deluxe clipboard is given by

$$d(x) = 12.48 - 0.005x^2$$

while the related producer supply function is given by

$$s(x) = \sqrt{x}$$

where x is the weekly quantity, and $d(x)$ and $s(x)$ are in dollars per clipboard.

(a) Determine the equilibrium point.

(b) Determine the consumers' surplus at equilibrium.

(c) Determine the producers' surplus at equilibrium.

 56. **Macroeconomics—Income Distribution:** The following Lorenz curve can be used to model the income distribution in the United States in 1992. Recall that x represents the proportion of the population, and $L(x)$ represents the proportion of income. (*Source:* U.S. Census Bureau.)

$$L(x) = \frac{0.39x - 0.04}{1 - 0.69x + 0.37x^2}$$

Determine the Gini Index for 1992. Round to the nearest ten-thousandth.

Section 7.4 Review Exercises

For Exercises 57–62, evaluate the given integrals. If integration by parts is required, consult the guidelines for selecting u and dv given in Section 7.4.

57. $\displaystyle\int 4xe^{2x}\,dx$

58. $\displaystyle\int xe^{-0.05x}\,dx$

59. $\int x^4 \ln x \, dx$

60. $\int (x - 2)e^x \, dx$

61. $\int 8te^{-0.2t} \, dt$

62. $\int (x + 5)(x - 2)^7 \, dx$

Exercises 63–66 require two (or more) applications of integration by parts.

63. $\int x^2 e^{-x} \, dx$

64. $\int (x + 1)^2 e^x \, dx$

65. $\int (\ln x)^4 \, dx$

66. $\int x(\ln x)^4 \, dx$

In Exercises 67–70, determine the method needed to evaluate the given integral. If the method selected is integration by parts, state the choice for u and dv. If the method selected is u-substitution, simply state the choice for u.

67. $\int x^2 e^{-x^3} \, dx$

68. $\int 2x \ln x \, dx$

69. $\int xe^{5x} \, dx$

70. $\int \dfrac{(\ln x)^5}{x} \, dx$

71. Geology—Oil Production: It is estimated that an oil field will produce oil at a rate given by

$$B(t) = 8te^{-0.17t}$$

thousand barrels per month, *t* months into production.

(a) Write a definite integral to estimate the total production for the first year of operation.

(b) Evaluate the integral from part (a) to estimate the total production for the first year of operation.

72. Economics—Profit Function: Hoppin Pepper Inc. has determined that its marginal profit function is given by

$$P'(t) = 42te^{0.12t}$$

where *t* is time in years, and $P'(t)$ is in thousands of dollars per year. Assuming that $P(0) = 0$, determine $P(t)$.

73. Banking—Continuous Interest: Kevin's business generates income at the rate of $90t$ thousand dollars per year, where *t* is the number of years since the business started. Determine the present value of this continuous income stream over the first four years at a continuous compound interest rate of 7%.

74. Banking—Continuous Interest: Suppose that a gold mine produces income at the rate of $140e^{-0.08t}$ thousand dollars per year, where *t* is the number of years since mining began. Determine the present value of this continuous income stream over the first six years of operation, assuming a continuous compound interest rate of 5%.

 75. Macroeconomics—Greek Fuel Prices: Some say that if a global recession occurs, Greece would have served as a warning sign. "A negative feedback loop now appears to be in the making in both the United States and Europe. Both economies are dangerously close to a recession," said Joachim Fels of Morgan Stanley. (*Source: St. Louis Post-Dispatch.*) The result has been a rise in consumer costs to Greeks, centered on fuel prices. The average cost of gas in Greece from 2008 to 2012 can be modeled by

$$f(x) = 0.75 + 0.66 \ln x \qquad 1 \le x \le 5$$

where $x = 1$ corresponds to January 1, 2008, and $f(x)$ represents the average fuel price, measured in euros per liter. (*Source: Europe's Energy Portal.*) Determine the average value of *f* on [1, 5] and interpret.

76. Ecology—Australian Water Usage: Australians have realized the need to conserve their resources for future generations, including the use of water. "Just like Australians have done with water consumption, we need to make a change in the way we think about energy use," said Mark Dreyfus of the Federal Labor Party. (*Source: The Sydney Morning Herald.*) The annual amount of water usage in Australia from 2005 and projected through 2060 can be modeled by

$$f(x) = 6.47 + 8.66 \ln x \qquad 5 \le x \le 60$$

where x represents the number of years since 2000, and $f(x)$ represents the annual water usage measured in cubic kilometers. (*Source:* Pardee Center for International Futures.) Determine the average value of f on [5, 55] and interpret.

Section 7.5 Review Exercises

In Exercises 77–80, use the Trapezoidal Rule to approximate the given definite integral using $n = 4$ trapezoids.

77. $\displaystyle\int_{-3}^{1} x^5 \, dx$

78. $\displaystyle\int_{4}^{8} (\ln x)^4 \, dx$

79. $\displaystyle\int_{0}^{1} (x^2 - 2x) \, dx$

80. $\displaystyle\int_{0}^{8} \frac{1}{x+1} \, dx$

In Exercises 81–84, use the Trapezoidal Rule program to approximate the given definite integral using $n = 10$, $n = 50$, and $n = 100$ trapezoids.

81. $\displaystyle\int_{1}^{11} \sqrt{x^2 + 3x + 2} \, dx$

82. $\displaystyle\int_{3}^{8} x^2 e^{0.5x} \, dx$

83. $\displaystyle\int_{1}^{5} \frac{(\ln x)^3}{x} \, dx$

84. $\displaystyle\int_{0}^{10} \frac{3x + 4}{x^2 + 2} \, dx$

85. Economics—Revenue Analysis: The following data give the rate of change of the year-to-date revenues in dollars per month for Jack's Handcrafted Birdhouses. Here x represents the number of months after May 1.

x	Rate of change of year-to-date revenue
1	2047
2	3124
3	2387
4	1937
5	2183
6	2706

Use the Trapezoidal Rule to approximate the revenues for Jack's Handcrafted Birdhouses for this period of time.

86. Economics—Revenue Analysis: The data in the table to the left give the marginal revenue for different levels of production at David's DVD Division. Here x represents the number of DVDs produced, and $R'(x)$ is in dollars per DVD.

x	$R'(x)$
100	30
200	20
300	15
400	10
500	8

Use the Trapezoidal Rule to approximate the total revenue increase that would be obtained by going from a production level of 100 to 500 DVDs.

 87. Economics—Cost Analysis: The GHI company has determined the marginal cost for different levels of production. The following data have been collected, where x represents the number of items produced, and $C'(x)$ is in thousands of dollars per item.

x	$C'(x)$		x	$C'(x)$
0	3.8		14	5.8
2	3.6		16	6.9
4	3.2		18	7.8
6	3.4		20	9.4
8	3.9		22	12.7
10	4.2		24	18.5
12	4.7			

Assume that $C(0) = 0$ and use the Trapezoidal Rule program to approximate the total cost of producing 24 items.

Section 7.6 Review Exercises

In Exercises 88–97, evaluate each improper integral or state that it is divergent.

88. $\displaystyle\int_5^\infty \frac{1}{x^3}dx$

89. $\displaystyle\int_1^\infty \frac{4}{\sqrt[3]{x}}dx$

90. $\displaystyle\int_{10}^\infty \frac{3}{x}dx$

91. $\displaystyle\int_0^\infty (8x - 2)\,dx$

92. $\displaystyle\int_0^\infty e^{0.1x}dx$

93. $\displaystyle\int_5^\infty \frac{4x}{2x^2 + 3}dx$

94. $\displaystyle\int_{-\infty}^0 e^{3x}dx$

95. $\displaystyle\int_{-\infty}^0 xe^{-0.2x^2}dx$

96. $\displaystyle\int_{-\infty}^\infty e^{0.05x}dx$

97. $\displaystyle\int_{-\infty}^\infty \frac{2x}{\sqrt{x^2 + 5}}dx$

98. Geology—Oil Production: An oil well has been estimated to produce oil at a rate given by

$$B(t) = 43e^{-0.03t} - 43e^{-0.08t}$$

thousand barrels per month, where t is the number of months from now. Estimate the total amount of oil that will be produced by this well.

99. Banking—Continuous Interest: Kevin owns a rental property that generates an indefinite annual rent of $14,000. Determine the capital value of this property at an annual interest rate of 5.25% compounded continuously.

100. Banking—Continuous Interest: A scholarship fund is to provide an annual scholarship in the amount of $8000. If the annual rate of interest is 9%, compounded continuously, how much should be invested to fund this scholarship perpetually?

101. Ecology—Ocean Pollution: The rate at which a certain toxic chemical is being released into an ocean from an abandoned dump is given by

$$f(t) = 260e^{-0.2t}$$

tons per year t years from now. Determine the total amount of the toxic chemical that will be released into the ocean if the leak continues indefinitely.

102. Library Science—Book Circulation: The number of days for which a book is checked out of the Springfield Public Library has a probability density function given by

$$f(x) = \begin{cases} \dfrac{2}{441}x, & 0 \leq x \leq 21 \\ 0, & \text{otherwise} \end{cases}$$

(a) Verify that $\int_{-\infty}^{\infty} f(x)\,dx = 1$.

(b) Compute the probability that the length of time for which a book is checked out is from 7 to 14 days.

(c) Compute the probability that the length of time for which a book is checked out is from 14 to 21 days.

103. Management—Letter Delivery: In a certain country, the number of days for a mailed letter to reach its destination has a probability density function given by

$$f(x) = \begin{cases} 0.08e^{-0.08x}, & x \geq 0 \\ 0, & \text{otherwise} \end{cases}$$

(a) Verify that $\int_{-\infty}^{\infty} f(x)\,dx = 1$.

(b) Compute the probability that a letter takes two to five days to reach its destination.

(c) Compute the probability that a letter takes 10 or more days to reach its destination.

Section 7.7 Review Exercises

For Exercises 104–109, determine the general solution to the differential equation.

104. $\dfrac{dy}{dx} = \sqrt{y}$

105. $\dfrac{dy}{dx} = \dfrac{y^2}{x^2}$

106. $\dfrac{dy}{dx} = 3x^2$

107. $\dfrac{dy}{dx} = 4y$

108. $\dfrac{dy}{dx} = \dfrac{y - 2}{4}$

109. $\dfrac{dy}{dx} = 10 - 5y$

For Exercises 110–115, determine the particular solution to the differential equation.

110. $\dfrac{dy}{dx} = \dfrac{2x}{y}, \quad x = 1, \quad y = 4$

111. $\dfrac{dy}{dx} = 4y^2, \quad x = 0, \quad y = -1$

112. $\dfrac{dy}{dx} = 3y, \quad x = 0, \quad y = -1$

113. $\dfrac{dy}{dx} = x(4 - x), \quad x = 0, \quad y = \dfrac{1}{2}$

114. $\dfrac{dy}{dx} = x^2 - 4x, \quad x = 0, \quad y = 8$

115. $\dfrac{dy}{dx} = 2x - x^2, \quad x = 0, \quad y = 25$

116. Economics—Cost Analysis: The Swing-Thing Company has found that the marginal cost of producing their spiral slide add-on is given by the differential rate equation

$$\frac{dC}{dx} = 91$$

where x represents the number of spiral slides produced, and C represents the cost in dollars.

(a) Determine the general solution for the cost function C.

(b) Find the cost function for the spiral slides, knowing that it costs $1250 to produce five spiral slides.

117. Public Safety—Traffic Accidents: Changes in the minimum drinking age, national maximum speed limits, and technical driver aids such as antilock brakes have resulted in a reduction in traffic accidents over the past few decades. The rate of change in traffic accidents from 1980 to 2011 can be modeled by the differential equation

$$\frac{df}{dx} = -0.22 \qquad 0 \le x \le 31$$

where x represents the number of years since 1980, and $f(x)$ represents the annual number of traffic accidents, in thousands. (*Source:* U.S. Department of Transportation.)

(a) Determine the general solution to the differential equation.

(b) Knowing that there were 13.4 thousand traffic accidents in 2000, find the particular solution to the differential equation.

(c) If the number of traffic accidents continues to decline at this rate, estimate the number of traffic accidents in 2018.

Section 7.8 Review Exercises

In Exercises 118–121, write an exponential growth model in the form $f(t) = a \cdot e^{kt}$ that satisfies the given differential equation and initial condition.

118. $f'(t) = 0.7 \cdot f(t);$ $f(0) = 300$

119. $f'(t) = 1.37 \cdot f(t);$ $f(0) = 8.2$

120. $f'(t) = -0.07 \cdot f(t);$ $f(0) = 5$

121. $f'(t) = -0.87 \cdot f(t);$ $f(0) = 384$

122. Demographics—Population Growth: The population of Popuville is increasing at a rate of 11% per year. The current population of the town is 42,000. Assume that the rate of change in the population is increasing proportionally to the population size.

(a) Write a differential equation and an initial condition for the population of Popuville.

(b) Write an exponential growth model for the population of Popuville after t years.

(c) Use the model to estimate the population after 15 years.

123. Bacteriology—Bacteria Growth: The number of bacteria in a culture starts at 300 and increases at a rate of 8% per hour. Assume that the rate of change in the number of bacteria is increasing proportionally to the number of bacteria present.

(a) Write a differential equation for the number of bacteria present after t hours.

(b) Write an exponential growth model for the number of bacteria present after t hours.

(c) Complete the table.

Number of hours, t	Number of bacteria, p(t)
0	
2	
4	
10	
20	

(d) Graph the exponential growth model for the values $0 \le t \le 20$.

Calculus of Several Variables

© LiquidLibrary   **(a)**

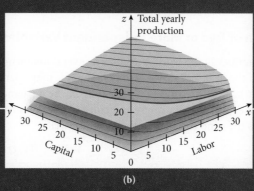

(b)

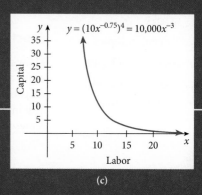

(c)

When manufacturing a product such as a golf cart, there are many inputs used in the production process. Economists tend to reduce all of the different inputs to two, labor and capital. This allows economists to determine a production function that relates the quantity produced as a function of labor and capital. The graph of a production function is a surface in three dimensions. Figure (b) shows the graph of a production function for the manufacturing of golf carts. The surface of the graph of the production function has been sliced with a horizontal plane. The curve that results from where the plane and the graph of the production function intersect is shown in Figure (c). Economists call such a curve an isoquant. An isoquant gives all combinations of labor and capital that yield a fixed quantity to be produced. Knowing all combinations of labor and capital that yield a fixed quantity to be produced is important for any business manager.

What We Know

In the first seven chapters of this text, we saw how the derivative and the integral can be used in a variety of settings. In all of these settings the functions we applied calculus to were functions of a single independent variable.

Where Do We Go

In this chapter we extend our calculus knowledge to functions that involve two independent variables. We will still see that rates of change are central to the derivative concept in the many varied applied settings.

Chapter Sections

8.1 Functions of Several Variables

8.2 Level Curves, Contour Maps, and Cross-Sectional Analysis

8.3 Partial Derivatives and Second-Order Partial Derivatives

8.4 Maxima and Minima

8.5 Lagrange Multipliers

8.6 Double Integrals

Chapter Review Exercises

SECTION OBJECTIVES

1. Evaluate a function of two independent variables.
2. Determine a cost function of two independent variables.
3. Plot points in 3-space.
4. Graph planes in 3-space.
5. Analyze a Cobb-Douglas production function.

8.1   Functions of Several Variables

In this section we introduce the concept of a function of two or more independent variables and the coordinate system used to analyze the graphs of such functions. Some calculators can graph a function of two independent variables. At this time we suggest that you consult the owner's manual of your calculator to determine if your calculator does this.

Function of Two or More Independent Variables

Many quantities in the world depend on several independent variables. For example, the wind chill index and the heat index, two numbers that we may hear on the local weather report, each depend on two variables. The wind chill depends on the air temperature and the wind speed, whereas the heat index depends on the air temperature and the relative humidity. In mathematical terms, if WCI represents the wind chill index, then we may write

$$WCI = f(v, t)$$

where v is the wind speed in miles per hour, and t is the air temperature in degrees Fahrenheit. We can evaluate a function that has more than one independent variable. Before we show how to do this, we first present a definition for a **function of two independent variables**.

DEFINITION

Function of Two Independent Variables

An equation of the form $z = f(x, y)$ represents a **function of two independent variables** if, for each ordered pair of real number (x, y), the equation determines a unique real number z.

NOTE: The variables x and y are the **independent variables** and z is the **dependent variable**.

The **domain** of a function of two variables is the set of all ordered pairs (x, y) for which z is defined. Example 1 illustrates how to evaluate a function of two independent variables.

OBJECTIVE 1

Evaluate a function of two independent variables.

Example 1: Evaluating a Function of Two Independent Variables

Consider $f(x, y) = 2x^2 - 3y^3$. Determine the following.

a. $f(2, 1)$ **b.** $f(-1, 2)$ **c.** $f(1, -2)$

Perform the Mathematics

a. Evaluating a function of two independent variables is just like evaluating a function with only one independent variable. That is, we substitute 2 in for x and 1 in for y. This produces

$$f(x, y) = 2x^2 - 3y^3$$
$$f(2, 1) = 2(2)^2 - 3(1)^3 = 8 - 3 = 5$$

So we have $f(2, 1) = 5$.

b. Evaluating as we did in part (a) yields

$$f(x, y) = 2x^2 - 3y^3$$
$$f(-1, 2) = 2(-1)^2 - 3(2)^3 = 2 - 24 = -22$$

So we have $f(-1, 2) = -22$.

c. Evaluating as we did in parts (a) and (b) gives

$$f(x, y) = 2x^2 - 3y^3$$
$$f(1, -2) = 2(1)^2 - 3(-2)^3 = 2 + 24 = 26$$

So we have $f(1, -2) = 26$. ■

▶ *Try It Yourself*

Some related Exercises are 11 and 13.

In Example 2, we illustrate how a function of two variables can be constructed to determine a cost function.

Example 2: Determining a Cost Function

OBJECTIVE 2

Determine a cost function of two independent variables.

DynaBall Corporation manufactures two types of golf balls. One has a balata cover favored by professionals and other low-handicap players. The other has a durable cover favored by high-handicap players. The cost to make a balata ball is $0.97, whereas the cost to make a durable-cover ball is $0.89. Determine a cost function, C, as a function of x and y, where x is the number of balata-covered balls produced, and y is the number of durable-cover balls produced, for the production of the balata- and durable-covered balls.

Perform the Mathematics

The cost function for the balata balls is simply

$$C(x) = (\text{unit cost}) \cdot (\text{quantity}) = 0.97x$$

For the durable-covered balls, the cost function is

$$C(y) = (\text{unit cost}) \cdot (\text{quantity}) = 0.89y$$

In the language of functions, we are asked to determine cost as a function of x **and** y. Intuitively, this should be the sum of the cost functions $C(x)$ and $C(y)$. This gives us

$$C(x, y) = 0.97x + 0.89y$$ ■

The cost function in Example 2, $C(x, y) = 0.97x + 0.89y$, gives us the cost of producing x balata balls and y durable-cover balls. This function is an example of a **function of two independent variables x and y.** If this company decides to expand its golf ball line by producing and selling a special low-compression ball for senior citizens, we would introduce a third independent variable and rewrite our cost function as a *function of three independent variables*. For the most part, we will be discussing functions of two independent variables in this chapter. For many of the problems that we encounter, we may have to place restrictions on the domain to determine the realistic domain. Example 3 illustrates this idea.

Example 3: Determining a Domain and Evaluating a Function of Two Variables

In Example 2 we determined that the cost function to produce x balata balls and y durable-cover balls is given by $C(x, y) = 0.97x + 0.89y$.

a. Determine the reasonable domain for $C(x, y)$.

b. Evaluate $C(250, 700)$ and interpret.

Perform the Mathematics

a.   At first glance, we notice that $C(x, y)$ is defined for all values of x and y. In other words, any ordered pair (x, y) is in the domain. But since $C(x, y)$ represents a *cost function*, we claim that it is unrealistic for x or y to be negative. (How does one produce -3 golf balls?) It is possible for x or y to be 0, so we could write that the domain is all ordered pairs (x, y) with real numbers x and y such that $x \geq 0$ and $y \geq 0$.

b.   We evaluate $C(250, 700)$ by simply substituting as follows.

$$C(x, y) = 0.97x + 0.89y$$
$$C(250, 700) = 0.97(250) + 0.89(700) = 865.50$$

This means that it costs \$865.50 to produce 250 balata balls and 700 durable-cover balls. ■

Cartesian Coordinates in 3-Space

So far in this text we have considered graphs of functions where the function has one independent variable and one dependent variable. The graph of this type of function is determined by points in the xy-plane that we called **ordered pairs**. To graph a function of two independent variables and one dependent variable, we simply extend this concept. Since we now have a total of three variables, two independent and one dependent, we need three coordinate axes, and we plot points determined by **ordered triples** (x, y, z). In **Figure 8.1.1** we have three coordinate axes that meet, at right angles to each other, at a point called the **origin** $(0, 0, 0)$. Think of our familiar xy-plane as being horizontal, like the floor of a room, while the z-axis extends vertically above and below the plane. See **Figure 8.1.2**.

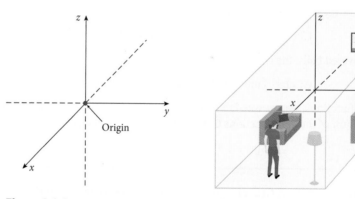

Figure 8.1.1   **Figure 8.1.2**

In Figure 8.1.1 we have labeled the x-, y-, and z-axes. Consider the solid portion of each axis to be the positive portion of the axis, whereas the dashed part corresponds to the negative portion of each axis. These three axes may be used to determine the location of any point in three-dimensional space.

Any point in three-dimensional space, or **3-space**, can be represented with an ordered triple (x, y, z). When considering points in space, imagine the first two coordinates of an ordered triple as indicating where to go in the xy-plane. The third coordinate, the z-coordinate, tells us whether to go up or down or to stay put. Example 4 shows how to plot points in space.

OBJECTIVE 3

Plot points in 3-space.

Example 4: Plotting Points in Space

Plot the following points.

a. $(2, 3, 1)$ **b.** $(-2, -4, -2)$

Perform the Mathematics

a. As shown in **Figure 8.1.3(a)**, we first locate (2, 3) in the *xy*-plane. We start at the origin and move 2 units along the positive *x*-axis and then 3 units to the right (in the positive direction), taking care to remain parallel to the *y*-axis. We use a " *X* " to locate (2, 3) in the *xy*-plane. Since our *z*-coordinate is a *positive* 1, we move up 1 unit directly above this location, taking care to remain parallel to the *z*-axis, and plot the point (2, 3, 1) as shown in Figure 8.1.3(b).

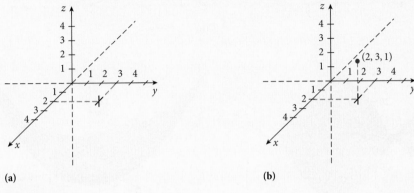

(a)   (b)

Figure 8.1.3

b. In a manner similar to part (a), we first locate (−2, −4) in the *xy*-plane. See **Figure 8.1.4(a)**. Here our *z*-coordinate is a *negative* 2, so we move down 2 units from this location and plot the point (−2, −4, −2) as shown in Figure 8.1.4(b).

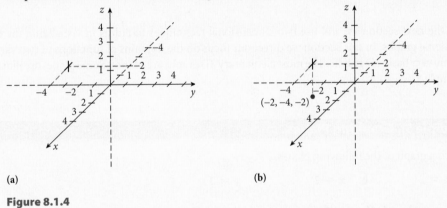

(a)   (b)

Figure 8.1.4   ■

▶ *Try It Yourself*

Some related Exercises are 31 and 37.

Graphing in 3-Space

Recall that for a function *f* of one independent variable, the graph of $y = f(x)$ is a curve that is above or below the *x*-axis. This curve is a collection of all points (x, y) such that $y = f(x)$. Also note that the curve lies in the *xy*-plane.

For a function of two independent variables, say $z = f(x, y)$, the graph of $f(x, y)$ is a collection of all points (x, y, z) such that $z = f(x, y)$. The graph of $z = f(x, y)$ is called a **surface** and is either above or below the *xy*-plane. Recall that an equation of the form $z = f(x, y)$ represents a function of two independent variables if for each ordered pair of real numbers (x, y), the equation determines a unique real number *z*. We can think of the domain of $z = f(x, y)$ as being any ordered pair (x, y) for which *z* is defined.

As we can imagine, graphing functions of two variables is difficult since it involves drawing three-dimensional (or 3D) graphs. Computers and some calculators can generate 3D graphs fairly quickly. **Figure 8.1.5** shows several graphs of surfaces.

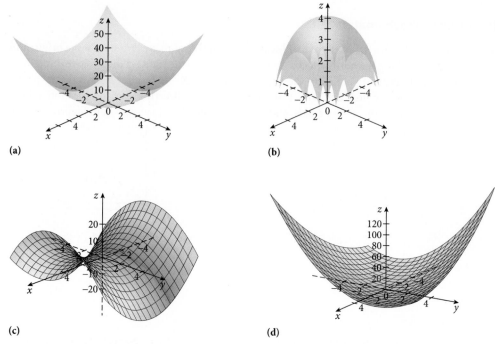

Figure 8.1.5 (a) Graph of $z = x^2 + y^2$. **(b)** Graph of $z = \sqrt{16 - x^2 - y^2}$. **(c)** Graph of $z = x^2 - y^2$. **(d)** Graph of $z = x^2 - 2xy + 2y^2$.

In the next section we will use two-dimensional techniques to aid us in visualizing the three-dimensional graphs. In this section we direct our focus on the domains of functions of two variables and some very basic 3D graphs. The most elementary 3D graphs are **planes**. Some planes are illustrated in Example 5.

<table>
<tr><td>**OBJECTIVE 4**</td></tr>
<tr><td>Graph planes in 3-space.</td></tr>
</table>

Example 5: Graphing Planes

Sketch a graph of the following planes.

a. $z = 2$ **b.** $x = 2$ **c.** $y = 3$

Perform the Mathematics

a.   The equation $z = 2$ is satisfied by all ordered triples having the form $(x, y, 2)$. To find a point on this graph, we locate any point (x, y) in the xy-plane and then move up 2 units. The result is a plane 2 units above the xy-plane and parallel to the xy-plane. The surface is shown in **Figure 8.1.6**.

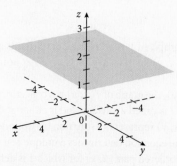

Figure 8.1.6 Graph of $z = 2$.

b. The equation $x = 2$ is satisfied by all ordered triples of the form $(2, y, z)$. To find any point on this graph, we locate any point of the form $(2, y)$ in the xy-plane or, equivalently, any point on the line $x = 2$ in the xy-plane and plot all points above, on, or below this line. The graph of $x = 2$ is a plane parallel to the yz-plane, as shown in **Figure 8.1.7**.

c. The equation $y = 3$ is satisfied by all ordered triples of the form $(x, 3, z)$. To find any point on this graph, we locate any point of the form $(x, 3)$ in the xy-plane or, equivalently, any point on the line $y = 3$ in the xy-plane and plot all points above, on, or below this line. The graph of $y = 3$ is a plane parallel to the xz-plane, as shown in **Figure 8.1.8**.

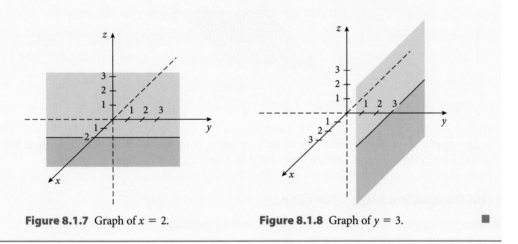

Figure 8.1.7 Graph of $x = 2$.　　　　**Figure 8.1.8** Graph of $y = 3$.　■

In Example 5, all the planes that we graphed had only one variable in the equation. We notice that when this occurs the resulting graph is a plane that is parallel to one of the **coordinate planes**. The three coordinate axes determine three coordinate planes, which are the xy-, the xz-, and the yz-planes. See **Figure 8.1.9**.

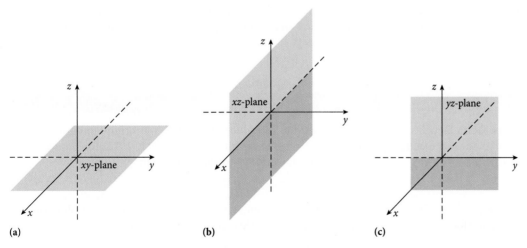

Figure 8.1.9 **(a)** Graph of the xy-plane. **(b)** Graph of the xz-plane. **(c)** Graph of the yz-plane.

Planes parallel to the coordinate planes are used in the next section when we study *level curves* and *cross sections* as part of our two-dimensional analysis of a 3D surface. Since this will be used in future work, we offer a summary of **graphing planes** in the box to the right.

Graphing Planes Parallel to Coordinate Axes Planes

For any real constant c, we have

1. The graph of $z = c$ is parallel to the xy-plane.
2. The graph of $x = c$ is parallel to the yz-plane.
3. The graph of $y = c$ is parallel to the xz-plane.

Most applications that we encounter in this chapter will have the domain restricted to quadrant I in the xy-plane, that is, $x \geq 0$ and $y \geq 0$. **Figure 8.1.10** shows the domain for most applications that we will examine.

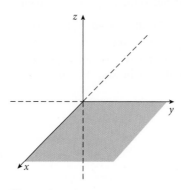

Figure 8.1.10

The surface of a 3D graph is entirely above, entirely below, or has portions above and portions below this quadrant of the xy-plane. The last part of this section focuses on one such application.

Cobb-Douglas Production Function

When producing any item, many inputs are used in the production process. Mathematically, if we let Q represent the output, then we have

$$Q = f(x_1, x_2, x_3, \ldots, x_k)$$

where each $x_1, x_2, x_3, \ldots$ is a different input. Clearly, Q is a function of several variables. However, economists tend to reduce all of these different inputs in this **production function** to two, L and K, and use

$$Q = f(L, K)$$

where Q is the number of units of output, L is the number of units of **labor**, and K is the number of units of **capital**. Capital includes many items, such as buildings, equipment, and insurance. This production function was developed and made popular by the mathematician Charles Cobb and the economist Paul Douglas and is called the **Cobb-Douglas production function**. The most general form of a **Cobb-Douglas production function** is given in the box to the left.

OBJECTIVE 5

Analyze a Cobb-Douglas production function.

Example 6: Analyzing a Cobb-Douglas Production Function

The Cobb-Douglas production function was introduced in 1928. It was originally constructed for all the manufacturing output in the United States for the years 1899 to 1922. The production function for all U.S. manufacturing output from 1899 to 1922 is given by

$$Q = 1.01 L^{0.75} K^{0.25}$$

where Q is the total yearly production, K is the capital investment, and L is the total labor force. Using x for labor and y for capital, we can rewrite this production function as

$$Q = f(x, y) = 1.01 x^{0.75} y^{0.25}$$

a. Determine the reasonable domain of $Q = f(x, y)$.

b. Determine whether the graph of Q is above or below the xy-plane.

c. Evaluate Q when $x = 50$ and $y = 60$ and interpret.

Perform the Mathematics

a. Since x represents the total labor force and y represents the total capital investment, we conclude that realistically, neither can be negative. So the reasonable domain is all ordered pairs (x, y) such that $x \geq 0$ and $y \geq 0$. Graphically, it is the region shaded in **Figure 8.1.11**.

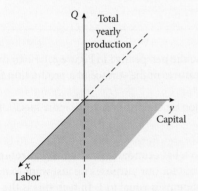

Figure 8.1.11

b. From part (a) we know that $x \geq 0$ and $y \geq 0$. If $x = 0$ or if $y = 0$, we have $Q = 0$. If $x \neq 0$ and $y \neq 0$, then $Q > 0$. So we conclude that the graph of Q, the surface, lies above the xy-plane. (It intersects the x-axis and the y-axis.) **Figure 8.1.12** gives two different perspectives of the graph of Q. In Figure 8.1.12(a), the axes are in a standard position. In Figure 8.1.12(b), the axes are rotated 180°.

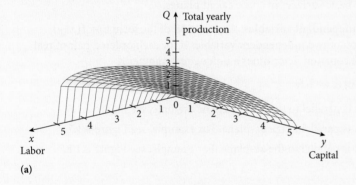

(a)

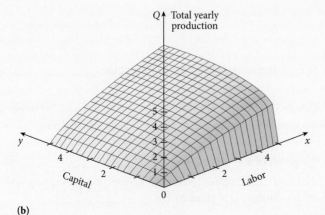

(b)

Figure 8.1.12 Graph of $Q = 1.01x^{0.75}y^{0.25}$.

c. To evaluate Q when $x = 50$ and $y = 60$, we simply substitute these values for x and y into the production function, as follows:

$$Q = f(x, y) = 1.01x^{0.75}y^{0.25}$$
$$Q = f(50, 60) = 1.01(50)^{0.75}(60)^{0.25} \approx 52.86$$

Thus, when 50 units of labor and 60 units of capital are used, the number of units produced is approximately 52.86. ■

For future graphs of production functions, we will use the perspective in Figure 8.1.12(b) unless stated otherwise. This perspective shows the important features of the surface of a production function quite nicely.

When Cobb and Douglas first introduced their production function, they gave the more specific form

$$Q = aL^bK^{1-b}$$

They used this form because they restricted the results to what economists call **constant return to scales**. Return to scales is discussed in economics theory, but for our purposes we just want to notice that, to have constant return to scales, the sum of the exponents is equal to 1. In fact, this is the only difference between Cobb and Douglas's specific form and the general form presented before Example 6.

Summary

In this section we introduced the concept of a function of two independent variables. We saw that the domain of a function of two variables is in the xy-plane and the graph is in 3-space. We plotted points in 3-space, looked at some surfaces in space, and graphed some very basic surfaces in space called **planes**.

- **Function of two independent variables**: An equation of the form $z = f(x, y)$ represents a function of two independent variables if for each ordered pair of real numbers (x, y), the equation determines a unique real number z.

- For any real constant c, we have:

  The graph of $z = c$ is parallel to the xy-plane. For example, see Figure 8.1.6.

  The graph of $x = c$ is parallel to the yz-plane. For example, see Figure 8.1.7.

  The graph of $y = c$ is parallel to the xz-plane. For example, see Figure 8.1.8.

Section 8.1 Exercises

Vocabulary Exercises

1. In a function of the form $z = f(x, y)$, the variables x and y represent the _____ variables.

2. In a function of the form $z = f(x, y)$ the variable z is the _____ variable.

3. A point (x, y, z) resulting from equation of function $z = f(x, y)$ is called an ordered _____ .

4. For a real number constant c, the graph of $x = c$ is parallel to the _____ plane.

5. If L and K represent units of labor and capital respectively with a, b and c representing real number constants, then the total production Q is given by the _____ _____ function $Q = aL^bK^c$.

6. To graph of function of the form $z = f(x, y)$ we graph in _____ with an x-, y-, and z-axis.

Skill Exercises

In Exercises 7–10, evaluate the function $f(x, y) = x^2 - 2xy + 3y^2$ at the indicated values.

7. $f(2, 3)$ **8.** $f(-1, 4)$

9. $f(-2, -1)$ **10.** $f(3, -2)$

In Exercises 11–14, evaluate the function $f(x, y) = \dfrac{x^2 + y^2}{x + y}$ for the indicated values.

11. $f(3, 1)$ **12.** $f(-1, 2)$

13. $f(-2, -3)$ **14.** $f(1, -2)$

In Exercises 15–18, evaluate the function $f(x, y) = \dfrac{3x - y + 1}{x^2 - y^2}$ for the indicated values.

15. $f(1, 4)$ **16.** $f(-2, -5)$

17. $f(-2, 2)$ **18.** $f(2, 0)$

In Exercises 19–22, evaluate the function $f(x, y) = y + x\ln x + xe^y$ for the indicated values.

19. $f(1, 2)$ **20.** $f(1, 0)$

21. $f(e, 0)$ **22.** $f(e^2, 1)$

For Exercises 23–28, determine the domain of the given function. Recall that the domain in this case is the set of all ordered pairs (x, y) in the xy-plane.

23. $f(x, y) = x + 2y$ **24.** $f(x, y) = 2x - y$

25. $f(x, y) = \dfrac{1}{x + y}$ **26.** $f(x, y) = \dfrac{3x}{x - y}$

27. $f(x, y) = \dfrac{3}{x - 4y}$ **28.** $f(x, y) = e^y + x\ln x$

For Exercises 29–42, plot the given points.

29. $(4, 4, 2)$ **30.** $(-3, 2, -1)$

31. $(-2, 1, 2)$ **32.** $(-3, 4, 1)$

33. $(-3, -1, 1)$ **34.** $(2, -2, 3)$

35. $(1, -3, -2)$ **36.** $(2, -3, -2)$

37. $(4, 2, -1)$ **38.** $(2, 5, -3)$

39. $(-1, -3, -2)$ **40.** $(-2, -4, -1)$

41. $(2, -4, 0)$ **42.** $(-3, 1, 0)$

For Exercises 43–50, determine if the given point is above or below the xy-plane. Explain your answer without plotting points.

43. $(2, 2, 3)$ **44.** $(3, 1, 2)$

45. $(-3, -2, 5)$ **46.** $(-1, -2, 4)$

47. $(4, -2, -1)$ **48.** $(2, 0, -5)$

49. $(0, 3, -0.5)$ **50.** $(1, -1, 0.1)$

In Exercises 51–56, sketch the graph of the given plane in 3-space.

51. $x = 3$ **52.** $x = -2$

53. $z = 3$ **54.** $z = -2$

55. $y = -1$ **56.** $y = 2$

Application Exercises

57. **Management—Weekly Sales:** Seamount Boats spends x thousand dollars each week on newspaper advertising and y thousand dollars each week on radio advertising. The company has weekly sales, in tens of thousands of dollars, given by

$$S(x, y) = 2x^2 + y$$

   (a) Determine $S(5, 3)$ and interpret.

   (b) Determine $S(3, 5)$ and interpret.

58. **Management—Weekly Sales:** Lakeway Boats spends x thousand dollars each week on radio advertising and y thousand dollars each week on television advertising. The company has weekly sales, in tens of thousands of dollars, given by

$$S(x, y) = 3x + 2y^3$$

   (a) Determine $S(2, 6)$ and interpret.

   (b) Determine $S(6, 2)$ and interpret.

59. **Advertising—Ticket Sales:** Tube Town, a recently opened water park, spends x thousand dollars on radio advertising and y thousand dollars on television advertising. The park has weekly ticket sales, in tens of thousands of dollars, given by

$$TS(x, y) = 1.5x^2 + 3.2y^3$$

   (a) Determine $TS(1, 0.5)$ and interpret.

   (b) Determine $TS(0.5, 1)$ and interpret.

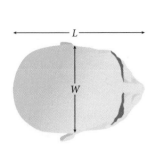

60. **Anthropology—Cephalic Index:** In their study of human groupings, anthropologists often use an index that indicates the shape of the head, the *cephalic index*. The cephalic index is given by

$$C(W, L) = 100 \cdot \frac{W}{L}$$

   where W is the width and L is the length of an individual's head. Both measurements are made across the top of the head and are in inches. Determine $C(6, 8.2)$ and $C(8.8, 9.7)$.

61. **Hematology—Poiseuille's Law:** Poiseuille's law states that the resistance, R, for blood flowing in a blood vessel is given by

$$R(L, r) = K \cdot \frac{L}{r^4}$$

   where K is a constant, L is the length of the blood vessel, and r is the radius of the blood vessel. Determine $R(36, 1)$ and $R(36, 2)$.

 62. **External Medicine—Body Surface Area:** An individual's body surface area is approximated by the DuBois and DuBois Formula and is given by the function

$$BSA(w, h) = 0.007184w^{0.425} \cdot h^{0.725}$$

   where BSA is in square meters, w is weight in kilograms, and h is height in centimeters. (*Source:* Archives of Internal Medicine.) Evaluate $BSA(70, 160)$ and interpret.

 63. **Meteorology—Wind Chill:** The wind chill index is modeled by

$$f(v, T) = 35.74 + 0.6215T - 35.75v^{0.16} + 0.4275Tv^{0.16}$$

   where v represents the velocity of the wind in miles per hour, T is the actual air temperature in degrees Fahrenheit, and $f(v, T)$ is the wind chill index in degrees Fahrenheit. (*Source:* U.S. National Weather Service.) Wind chill is the temperature that it actually feels like. Evaluate $f(25, 10)$ and interpret. Round your answer to the nearest integer.

 64. **Forest Science—Doyle's Law:** The *Doyle log rule* is one method used to determine the yield of a log, measured in board-feet. In English, the rule states

Deduct 4 inches from the diameter of the log as an allowance for slab; square one-quarter of the remainder and multiply the result by the length of the log in feet. (Source: U.S. Forest Service.)

Mathematically, this is translated as

$$f(d, L) = \left(\frac{d - 4}{4}\right)^2 \cdot L$$

where d is the diameter in inches, L is the length in feet, and $f(d, L)$ is the number of board-feet. Determine $f(30, 12)$ and interpret.

65. **Microeconomics—Cost Analysis:** The Leaf Eater Company manufactures and sells leaf blowers and a special 10-foot blower attachment to clean gutters. The monthly cost function, in dollars, for the company is given by

$$C(x, y) = 1000 + 35x + 1.5y$$

where x is the number of leaf blowers produced each month, and y is the number of 10-foot blower attachments produced each month. Determine $C(50, 30)$ and interpret.

66. **Microeconomics—Revenue Analysis:** The Leaf Eater Company from Exercise 65 has determined the following price functions for leaf blows and attachments:

$p = 120 - 0.8x - 0.1y$ the price in dollars for a leaf blower
$q = 52.3 - 0.15x + 0.015y$ the price in dollars for a 10-foot attachment

In each, x represents the number of leaf blowers sold each month, and y represents the number of 10-foot blower attachments sold each month.

(a) Determine the revenue function $R(x, y)$.

(b) Determine $R(50, 30)$ and interpret.

67. **Microeconomics—Profit Analysis:** Refer to the functions given in Exercises 65 and 66.

(a) Using the cost function from Exercise 65 and the revenue function from Exercise 66, determine the profit function $P(x, y)$.

(b) Determine $P(50, 30)$ and interpret.

68. **Microeconomics—Profit Analysis:** The Chalet Bicycle Company manufactures 21-speed racing bicycles and 21-speed mountain bicycles. Let x represent the weekly demand for a 21-speed racing bicycle and y represent the weekly demand for a 21-speed mountain bicycle. The weekly price-demand equations are given by

$p = 350 - 4x + y$ the price in dollars for a 21-speed racing bicycle
$q = 450 + 2x - 3y$ the price in dollars for a 21-speed mountain bicycle

The cost function is given by

$$C(x, y) = 390 + 95x + 100y$$

(a) Determine the weekly revenue function $R(x, y)$.

(b) Evaluate $R(15, 20)$ and interpret.

(c) Determine the weekly profit function $P(x, y)$.

(d) Evaluate $P(15, 20)$ and interpret.

69. **Microeconomics—Profit Analysis:** A T-shirt maker produces two types of tie-dyed T-shirts. The full-rainbow T-shirt costs $5 each to produce, and the partial-rainbow T-shirt costs $4 each to produce. She sells the full-rainbow T-shirts for $12 each and the partial-rainbow T-shirts for $9.50 each.

(a) Determine the cost function $C(x, y)$ for making x full-rainbow T-shirts and y partial-rainbow T-shirts. Assume that the fixed costs are $50.

(b) Determine the revenue function $R(x, y)$.

(c) Determine the profit function $P(x, y)$.

70. **Banking—Interest Rates:** If $5000 is invested at an annual interest rate of r (in decimal form) compounded quarterly for t years, the total amount accumulated is given by

$$f(r, t) = 5000\left(1 + \frac{r}{4}\right)^{4t}$$

Determine $f(0.075, 15)$ and interpret.

71. **Banking—Interest Rates:** If P dollars is invested at an annual interest rate of 6.125% compounded monthly for t years, the total amount accumulated is given by

$$f(P, t) = P\left(1 + \frac{0.06125}{12}\right)^{12t}$$

Determine $f(2000, 20)$ and interpret.

72. **Banking—Interest Rates:** If P dollars is invested at an annual interest rate of 6% compounded continuously for t years, the total amount accumulated is given by

$$f(P, t) = Pe^{0.06t}$$

Determine $f(3000, 10)$ and interpret.

73. **Banking—Interest Rates:** If $2000 is invested at an annual interest rate of r (in decimal form) compounded continuously for t years, the total amount accumulated is given by

$$f(r, t) = 2000e^{rt}$$

Determine $f(0.0725, 25)$ and interpret.

74. **Economics—Labor Analysis:** A golf club manufacturer has a Cobb-Douglas production function given by

$$Q = f(x, y) = 21x^{0.3}y^{0.75}$$

where x is utilization of labor and y is utilization of capital. Determine the number of units of golf clubs produced when 200 units of labor and 75 units of capital are used.

75. **Economics—Labor Analysis:** A sports shoe manufacturer has a Cobb-Douglas production function given by

$$Q = f(x, y) = 42x^{0.37}y^{0.66}$$

where x is utilization of labor and y is utilization of capital. Determine the number of units of sports shoe produced when 300 units of labor and 100 units of capital are used.

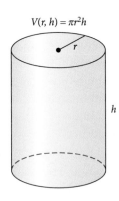

$V(r, h) = \pi r^2 h$

76. **Geometry—Cylinder Volume:** The volume of a cylinder, such as a soup can, is given by

$$V(r, h) = \pi r^2 h$$

where r is the radius and h is the height.

(a) Determine $V(2, 6)$ and interpret. (Each is measured in inches.)

(b) If the radius and height are equal, we can express the volume of the cylinder as a function of one variable, either r or h. Determine $V(r)$ and $V(h)$.

77. **Cognitive Study—IQ Values:** An individual's IQ is defined to be

$$IQ = f(a, m) = 100 \cdot \frac{m}{a}$$

where m is the individual's mental age (as determined by a test) in years, and a is the individual's actual age in years.

(a) Evaluate $f(9, 12)$ and interpret.

(b) Evaluate $f(12, 9)$ and interpret.

(c) Evaluate $f(12, 12)$ and interpret.

(d) Determine IQ if $a = m$. In your own words, what does this tell us about an IQ of 100?

78. **Economics—Capital Analysis**: A bicycle seat manufacturer has a Cobb–Douglas production function given by

$$Q = f(x, y) = 22x^{0.75}y^{0.25}$$

where x is utilization of labor, and y is utilization of capital. Determine the number of units of bicycle seats produced when 100 units of labor and 50 units of capital are used.

79. **Economics—Capital Analysis** (*continuation of Exercise 78*): Suppose that we wish to know what combinations of labor and capital would result in 2000 units of bicycle seats being produced. In other words, we want to know what values of x and y satisfy

$$2000 = 22x^{0.75}y^{0.25}$$

If we solve this equation for y, we get

Divide each side by $22x^{0.75}$	$\dfrac{2000}{22x^{0.75}} = y^{0.25}$
Raise each side to the fourth power	$\left(\dfrac{2000}{22x^{0.75}}\right)^4 = y$
Simplify the left side of the equation	$(90.91x^{-0.75})^4 = y$

(a) In the viewing window $[0, 200]$ by $[0, 1000]$, graph $y = (90.91x^{-0.75})^4$. The curve that you see is called an *isoquant*. We will discuss isoquants in Section 8.2.

(b) Use the `Trace` command to approximate the number of units of capital needed if 70 units of labor are used to produce 2000 units of bicycle seats.

(c) Use the `Trace` command to approximate the number of units of capital needed if 100 units of labor are used to produce 2000 units of bicycle seats.

(d) Determine three combinations of labor and capital that will produce 2000 units of bicycle seats. (Answers may vary.)

80. **Economics—Capital Analysis**: The O'Neill Corporation has 10 soft drink bottling plants located in the United States. In a recent year, the data for each plant gave the number of labor hours (in thousands), capital (total net assets, in millions), and the total quantity produced (in thousands of gallons). The data are shown in **Table 8.1.1**. The plants all use the same technology, so a production function can be determined. A Cobb-Douglas production function modeling these data is given by

$$Q = f(L, K) = 1.64L^{0.623}K^{0.357}$$

Table 8.1.1

Labor	Capital	Quantity
100	11	68
100	13	72
110	14	79
125	16	89
133	17	95
140	20	105
151	23	114
152	23	115
160	24	120
166	26	127

(a) Determine $f(125, 18)$ and interpret.

(b) Determine $f(70, 30)$ and interpret.

Concept and Writing Exercises

81. Explain the difference between a function of one independent variable and a function of two independent variables.

82. Can we graph a function of three independent variables? Explain why or why not.

83. Given the function $f(x, y) = x^2 + y$, determine x knowing $f(x, 5) = 21$.

84. Given the function $f(x, y) = \frac{x}{2 + y}$, determine y knowing $f(5, y) = \frac{1}{2}$.

For Exercises 85 and 86, complete the numerical table for the given values of x and y.

85. For the function $f(x, y) = 3x + y^2$.

	$x = -3$	$x = -2$	$x = -1$	$x = 0$	$x = 1$	$x = 2$	$x = 3$
$y = -3$							
$y = -2$							
$y = -1$							
$y = 0$							
$y = 1$							
$y = 2$							
$y = 3$							

86. For the function $f(x, y) = xy - 2x$.

	$x = -3$	$x = -2$	$x = -1$	$x = 0$	$x = 1$	$x = 2$	$x = 3$
$y = -3$							
$y = -2$							
$y = -1$							
$y = 0$							
$y = 1$							
$y = 2$							
$y = 3$							

 Section Project

The program in Appendix B that performs multiple regression is needed to do this section project. Suppose the Basich Company has eight plants in the United States producing handheld calculators. In a recent year, the data for each plant gave the number of labor hours (in thousands), capital (in millions), and total quantity produced. The data are shown in the **Table 8.1.2**. The plants all use the same technology, so a production function can be determined.

Table 8.1.2

Labor	Capital	Quantity
96	15	12,234
103	20	13,907
104	21	14,187
109	22	14,903
111	27	15,915
121	30	17,507
122	31	17,768
127	36	19,047

(a) Enter the data into your calculator and execute the program to determine a Cobb-Douglas production function to model the data. Round all values to the nearest hundredth.

(b) Suppose that Richard, CEO of the Basich Company, wants to know all combinations of labor and capital that produce 15,000 calculators. To do this, use the production function from part (a) and rename L with an x and K with a y. Substitute 15,000 for Q and solve the resulting equation for y.

(c) Graph the equation found in part (b) in an appropriate viewing window. Use the `Trace` command to determine three realistic combinations of labor and capital that will produce 15,000 calculators.

8.2   Level Curves, Contour Maps, and Cross-Sectional Analysis

SECTION OBJECTIVES

1. Determine a level curve.
2. Construct a contour map.
3. Construct vertical cross sections.
4. Apply cross-sectional analysis to a production function.

In Section 8.1 we were introduced to functions of two independent variables and saw that the graphs of these functions are surfaces in 3-space. **Figure 8.2.1** shows a graph of a production function.

As mentioned in Section 8.1, graphing surfaces in 3-space is very difficult without the aid of a computer. For this reason **cross-sectional analysis** can aid in understanding the behavior of a surface by analyzing the surface using graphs in two dimensions. For example, consider **Figure 8.2.2**, which shows the surface in Figure 8.2.1 sliced with a horizontal plane $z = 4$.

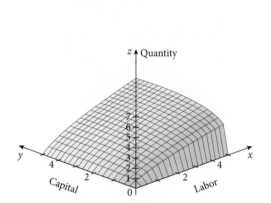

Figure 8.2.1 Graph of $z = 1.3x^{0.75}y^{0.3}$.

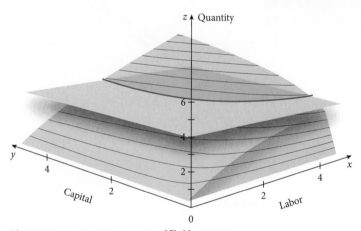

Figure 8.2.2 Graph of $z = 1.3x^{0.75}y^{0.3}$ and the horizontal plane $z = 4$.

The horizontal plane $z = 4$ intersects the surface to form a curve that is parallel to the xy-plane, which means that we can graph the curve in the xy-plane. Slicing of surfaces with horizontal planes and graphing the resulting curves in a single xy-plane is where we begin our analysis of surfaces in 3-space.

Horizontal Cross Sections, Level Curves, and Isoquants

To have an idea of what we will be doing, let's consider the **topographical map** shown in **Figure 8.2.3**.

A topographical map is simply a two-dimensional graph of a three-dimensional surface, such as a mountain. The lines we see on the topographical map connect points with the same elevation. In mathematics, we call lines that connect points of equal elevation **contour lines** or **level curves**. Also, we call a collection of contour lines a **contour map**. Assuming that the elevation between the contour lines on the topographical map changes by a constant amount, we see that the more closely packed the lines are, the steeper the terrain. The more spread apart the contour lines are, the flatter the terrain.

A topographical map gives a good overall picture of the terrain, indicating where hills and flat areas are located. In Example 1 we employ the concept illustrated in the topographical map to a surface in 3-space.

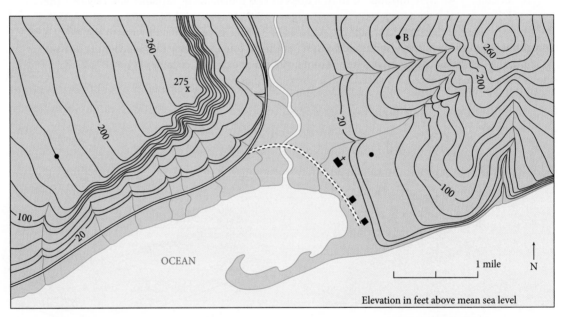

Figure 8.2.3

OBJECTIVE 1

Determine a level curve.

Example 1: Determining a Level Curve

An economist for the Linger Golf Cart Corporation has computed a production function for the manufacture of their golf carts to be

$$f(x, y) = Q = 1.3x^{0.75}y^{0.25}$$

where Q is the number of golf carts produced each week, x is the number of labor hours each day, and y is the daily usage of capital investment. What combinations of labor hours each day and daily usage of capital investment will result in 13 golf carts being produced each week?

Perform the Mathematics

We are being asked to determine *all* possible combinations of labor hours, x, and daily usage of capital investment, y, such that $Q = 13$. In other words, we want to know for what values of x and y does

$$13 = 1.3x^{0.75}y^{0.25}$$

Solving this equation for y gives

$$13 = 1.3x^{0.75}y^{0.25}$$
$$10 = x^{0.75}y^{0.25}$$
$$10x^{-0.75} = y^{0.25}$$
$$y = (10x^{-0.75})^4 = 10,000x^{-3}$$

To determine all combinations of x and y, we can simply graph $y = (10x^{-0.75})^4 = 10,000x^{-3}$ in the xy-plane. **Figure 8.2.4** shows a graph of $f(x, y) = Q = 1.3x^{0.75}y^{0.25}$, and **Figure 8.2.5** shows a graph of $y = 10,000x^{-3}$. The graph in Figure 8.2.5 answers the question.

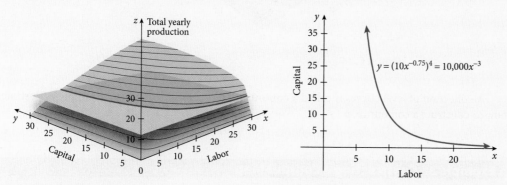

Figure 8.2.4 Graph of $f(x, y) = Q = 1.3x^{0.75}y^{0.25}$ and the horizontal plane $Q = 13$.   **Figure 8.2.5**

Economists call the graph in Figure 8.2.5 an **isoquant** (*iso* means "same," and *quant* is short for quantity). It gives all combinations of x and y that yield a production of 13 golf carts each week. Graphically, it is the result of slicing the surface given by $f(x, y) = Q = 1.3x^{0.75}y^{0.25}$ with the horizontal plane $Q = 13$. (We can think of the variable Q as behaving like the variable z in our xyz-coordinate system.)

Technology Option

To see the isoquant shown in Figure 8.2.5 on your calculator, simply graph $y = 10,000x^{-3}$ in the viewing window $[0, 20]$ by $[0, 30]$. See **Figure 8.2.6**.

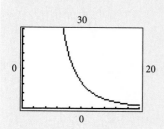

Figure 8.2.6

If we take the curve in Figure 8.2.5 and lift it 13 units above the xy-plane, we get a picture of the behavior of the *surface* 13 units above the xy-plane. **Figure 8.2.7** illustrates this.

As we saw with the topographical map, the general mathematical word for an isoquant is **level curve**. It has this name since every point on the curve has the same dependent variable value. Also, recall that another name for level curve is a **contour line**.

DEFINITION

Level Curve or Contour Line

A **level curve**, or **contour line**, is obtained from a surface $z = f(x, y)$ by slicing it with a horizontal plane $z = c$. The equation for the level curve at height c is given by

$$c = f(x, y)$$

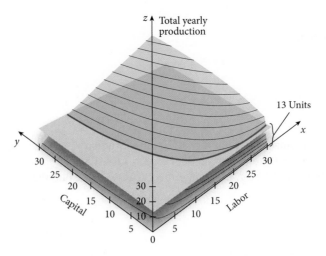

Figure 8.2.7

A collection of level curves is called a **contour map** or **contour diagram**. Example 2 illustrates how to construct a contour map.

OBJECTIVE **2**

Construct a contour map.

Example 2: Constructing a Contour Map

Let $z = f(x, y) = x^2 + y^2$. Construct a contour map using $c = 2, 4,$ and 6. Compare the contour map to the graph of the surface.

Perform the Mathematics

The level curve at any height c is given by

$$c = x^2 + y^2$$

This is simply the equation of a circle, centered at the origin, with a radius of $\sqrt{c}$. So the level curve at $c = 2$ is found by graphing

$$2 = x^2 + y^2$$

For $c = 4$, we graph $4 = x^2 + y^2$, and for $c = 6$, we graph $6 = x^2 + y^2$. The contour map is shown in **Figure 8.2.8**, and the graph of $z = x^2 + y^2$ is shown in **Figure 8.2.9**.

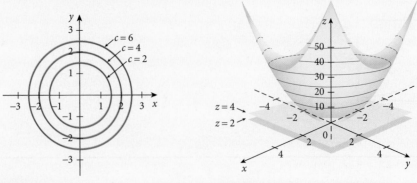

Figure 8.2.8   **Figure 8.2.9**

Notice that the graph of $z = x^2 + y^2$ gets steeper as we move farther away from the origin. This can be observed on the contour map in Figure 8.2.8, since the level curves become more tightly packed together as we move away from the origin.

▶ *Try It Yourself*

Some related Exercises are 9 and 11.

Example 3: Constructing a Contour Map in an Application

Recall that the golf cart manufacturer in Example 1 has a production function of $Q = f(x, y) = 1.3x^{0.75}y^{0.25}$, where Q is the number of golf carts produced each week, x is the number of labor hours each day, and y is the daily usage of capital investment. Construct a contour map (here, a collection of isoquants) using $c = 10, 20, 30,$ and 40.

Perform the Mathematics

Recall that we can think of Q as being the same as z. The isoquant (level curve) at any production level (height) is given by

$$c = 1.3x^{0.75}y^{0.25}$$

Solving this equation for y gives

$$c = 1.3x^{0.75}y^{0.25}$$

$$\frac{c}{1.3} = x^{0.75}y^{0.25}$$

$$\frac{c}{1.3}x^{-0.75} = y^{0.25}$$

$$y = \left(\frac{c}{1.3}x^{-0.75}\right)^4 = \frac{c^4}{2.8561}x^{-3}$$

We now substitute the given values of c into this equation and graph the resulting equations, as shown in **Figure 8.2.10**. The graph of the surface is in **Figure 8.2.11**.

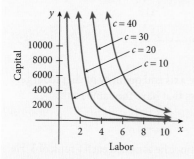

Figure 8.2.10

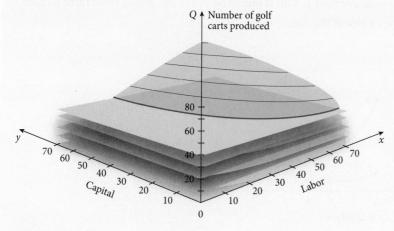

Figure 8.2.11

Notice in Examples 2 and 3 that the contour map is constructed on one xy-plane. Also notice that the values of c in both examples are equally spaced. Having equally spaced c-values and constructing a contour map on a single xy-plane are necessary if we want the contour map to give us an idea of the behavior of the surface.

Indifference Curves

So far we have seen that the level curves of a production function are called isoquants. We now look at another level curve from economics that is called an **indifference curve**. First we need a brief discussion on what economists call a **utility function**. Utility functions and indifference curves are the basis of the modern theory of consumer behavior.

We say that an individual derives *satisfaction* or *utility* from commodities consumed during a given time period. In the time period, the individual will consume a large variety of different commodities. Economists refer to this collection of different commodities as a **commodity bundle**. For different commodity bundles, economists assume that each individual compares alternative commodity bundles and states a preference. Example 4 illustrates this idea.

Example 4: Analyzing Commodity Bundles

Tom enjoys eating cheeseburgers and drinking colas. He was asked to rank the following commodity bundles (of cheeseburgers and colas) for a typical week, with a preferred bundle assigned a higher number. His ranking are given in **Table 8.2.1**.

Table 8.2.1

Bundle	Colas (x)	Cheeseburgers (y)	Rank
A	3	5	3
B	4	3	3
C	5	2	3
D	1	4	1
E	2	2	1
F	3	1	1

a. Which bundle(s) is(are) most preferred by Tom?

b. Make a plot of the data, placing quantity of colas on the x-axis and quantity of cheeseburgers on the y-axis. Connect with a smooth curve those points that have the same rank.

Perform the Mathematics

a. It appears that Tom prefers bundles A, B, and C the most since he assigned each a rank of 3. He is said to be *indifferent* among these three bundles. Tom prefers bundles D, E, and F the least since he assigned each a rank of 1. Tom is said to be *indifferent* among these three bundles.

b. **Figure 8.2.12** gives a plot of the data.

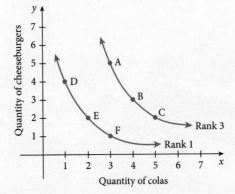

Figure 8.2.12

The two curves seen in Figure 8.2.12 are called **indifference curves**, since the consumer in question, Tom, is indifferent to any bundle on each curve. In other words, on the rank 3 indifference curve, Tom would receive the same utility from bundle A as from bundle B or C.

Since the indifference curves in Figure 8.2.12 are labeled with rank 1 and rank 3, we can easily imagine that the *indifference curves are level curves* of some function. The function in question is called a **utility function** and is denoted

$$U(x, y) = f(x, y)$$

The utility function assigns a numerical value (or utility level) to commodity bundles for goods x and y. Without going too far into economics theory, all that we require of the utility function is that it reflect the same rankings that a consumer assigns to alternative commodity bundles. So if a consumer prefers bundle A to bundle D, the utility function has to assign a *larger* numerical value to bundle A than to bundle D. Example 5 shows how this is done.

Example 5: Sketching an Indifference Map

Suppose that the utility from consuming x colas and y cheeseburgers is given by $U(x, y) = \sqrt{xy}$. Draw a contour map for $c = 1, 2, 3,$ and 4. Here we will have four indifference curves. A collection of indifference curves is also known as an **indifference map**.

Perform the Mathematics

As we have seen so far in this section, we can algebraically set this up as

$$U(x, y) = \sqrt{xy}$$
$$c = \sqrt{xy}$$
$$c^2 = xy$$
$$y = \frac{c^2}{x}$$

Now we substitute 1, 2, 3, and 4 in for c and graph the following in the xy-plane.

$$y = \frac{1}{x}, \quad y = \frac{4}{x}, \quad y = \frac{9}{x}, \quad \text{and} \quad y = \frac{16}{x}$$

Figure 8.2.13 is the contour map, and **Figure 8.2.14** is a graph of the utility function. Again, notice that the curves in Figure 8.2.13 are the result of slicing the surface in Figure 8.2.14 with horizontal planes 1, 2, 3, and 4 units, respectively, above the xy-plane.

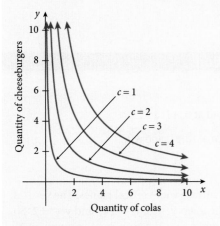

Figure 8.2.13

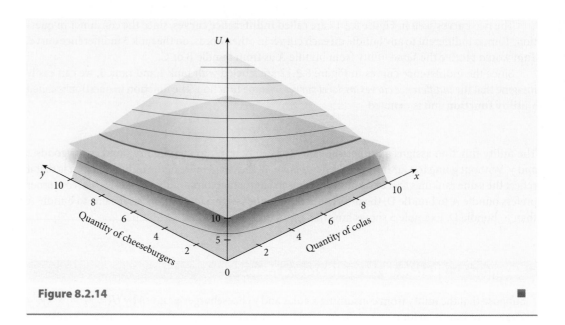

Figure 8.2.14   ■

▶ **Try It Yourself**

Some related Exercises are 21 and 23.

Vertical Cross Sections

So far we have seen how slicing a surface with a horizontal plane produces a level curve that can be drawn in the xy-plane. A collection of several slices at different heights produces a collection of level curves that gives us an idea of the behavior of the surface in 3-space. However, if we look at **vertical cross sections** of the form $x = c$ and $y = c$, we can sometimes improve on our 3-space visualization.

In Example 2, we sliced the surface with planes of the form $z = c$. (Look back at Figure 8.2.9.) Notice that these planes are perpendicular to the z-axis and that the collection of level curves, that is, the contour map, was graphed on the xy-plane. The procedure for vertical cross sections is basically the same. If we slice a surface with the planes of the form $x = c$ (these will be vertical planes), the planes are perpendicular to the x-axis, and the resulting cross sections are graphed on the yz-plane. Similarly, if we slice a surface with planes of the form $y = c$ (again, vertical planes), the planes are perpendicular to the y-axis, and the resulting cross sections are graphed on the xz-plane. We demonstrate this process in Example 6.

OBJECTIVE 3

Construct vertical cross sections.

Example 6: Constructing Vertical Cross Sections

Consider $z = y^2 - x^2$.

a.   Sketch the cross sections on the yz-plane with y fixed at 0, ±1, and ±2.

b.   Sketch the cross section on the yz-plane with x fixed at 0, ±1, and ±2.

Perform the Mathematics

a.   Here y is fixed, which means that we are slicing the surface with vertical planes. The vertical planes are perpendicular to the y-axis and are of the form $y = 0$, $y = \pm1$, and $y = \pm2$. So we need to graph the following curves in the xz-plane.

$$z = -x^2 \qquad (y = 0)$$
$$z = 1 - x^2 \qquad (y = \pm 1)$$
$$z = 4 - x^2 \qquad (y = \pm 2)$$

To graph these curves, we simply need to treat z as the dependent variable and x as the independent variable. The graphs of these curves are given in **Figure 8.2.15**, and the graph of the surface with a vertical cross section is shown in **Figure 8.2.16**.

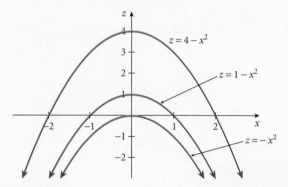

Figure 8.2.15

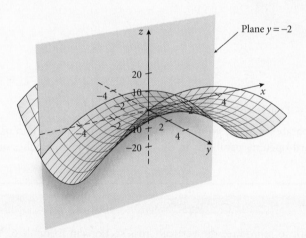

Figure 8.2.16 Vertical plane $y = -2$ intersects the surface in a curve.

b. Here x is fixed, which means that we are slicing the surface with vertical planes. The vertical planes are perpendicular to the x-axis and are of the form $x = 0$, $x = \pm 1$, and $x = \pm 2$. Hence, we need to graph the following curves in the yz-plane:

$$z = y^2 \qquad (x = 0)$$
$$z = y^2 - 1 \qquad (x = \pm 1)$$
$$z = y^2 - 4 \qquad (x = \pm 2)$$

Notice that here z is the dependent variable and y is the independent variable. The graphs of these cross sections are given in **Figure 8.2.17**, and the graph of the surface with a vertical cross section is shown in **Figure 8.2.18**.

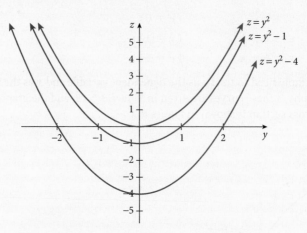

Figure 8.2.17

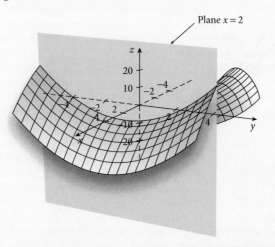

Figure 8.2.18 Vertical plane $x = 2$ intersects the surface in a curve.

Notice the upward-opening parabolas in the y-direction, which matches our work in part (b). Also, notice the downward-opening parabolas in the x-direction, which matches our work in part (a).   ∎

 Technology Option

We can use our graphing calculator to reproduce the vertical cross sections in Example 6. To get the cross section shown in Figure 8.2.15, since z is the dependent variable, it acts like y for the calculator purposes. Here, x is still the independent variable. So we enter and graph $y_1 = 4 - x^2$, $y_2 = 1 - x^2$, and $y_3 = -x^2$ to reproduce Figure 8.2.15. See **Figure 8.2.19**.

To get the cross section in Figure 8.2.17, since z is the dependent variable, it acts like y for the calculator purposes. Here, since y is the independent variable, it acts like x for the calculator. So we enter and graph $y_1 = x^2$, $y_2 = x^2 - 1$, and $y_3 = x^2 - 4$ to reproduce Figure 8.2.17. See **Figure 8.2.20**.

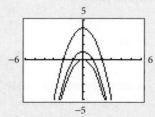

Figure 8.2.19

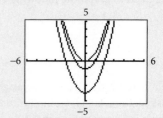

Figure 8.2.20

Before we look at our concluding example, we want to mention that our work with vertical cross sections will be very important in Section 8.3 when we tackle derivatives of functions of two variables. These derivatives, called *partial derivatives*, are related to vertical cross sections.

Example 6: Applying Cross-Sectional Analysis to a Production Function

OBJECTIVE 4

Apply cross-sectional analysis to a production function.

Consider the production function for the Linger Golf Cart Corporation in Example 1. It is

$$f(x, y) = Q = 1.3x^{0.75}y^{0.25}$$

where Q is the number of golf carts produced per week, x is the number of labor hours each day, and y is the daily usage of capital investment. Construct a cross section on an xy-plane with y fixed at $y = 20$. Approximate how many labor hours are required each day to produce 10 golf carts per week.

Perform the Mathematics

We begin by replacing the dependent variable Q with z so that we have

$$z = 1.3x^{0.75}y^{0.25}$$

Again, since y is fixed, we are slicing the surface with a vertical plane perpendicular to the y-axis of the form $y = 20$. So we need to graph in an xz-plane the following:

$$z = 1.3x^{0.75}y^{0.25} = 1.3x^{0.75}(20)^{0.25} \approx 2.75x^{0.75}$$

The graph of the cross section is given in **Figure 8.2.21**, and the graph of the surface is given in **Figure 8.2.22**.

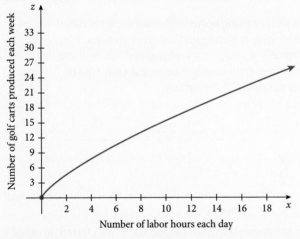

Figure 8.2.21 Graph of $z = 2.75x^{0.75}$ representing the vertical cross section when y is fixed at $y = 20$.

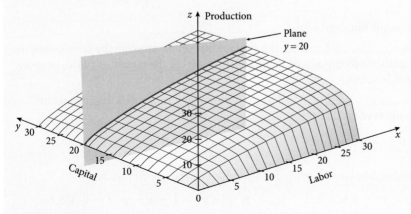

Figure 8.2.22

Now, to approximate how many labor hours are required each day to produce 10 golf carts per week requires us to substitute $z = 10$ into $z = 2.75x^{0.75}$ and solve for x.

$$z = 2.75x^{0.75}$$

Substitute $z = 10$ $10 = 2.75x^{0.75}$

Divide both sides by 2.75 $\dfrac{10}{2.75} = x^{0.75}$

Raise both sides to the $\frac{4}{3}$ power $\left(\dfrac{10}{2.75}\right)^{4/3} = (x^{0.75})^{4/3}$

Simplify $5.59 \approx x$

This means that approximately 5.6 labor hours each day will produce a quantity of 10 golf carts per week when the capital investment is fixed at $y = 20$ units each day.   ■

▶ *Try It Yourself*

Some related Exercises are 43 and 45.

Summary

We began this section by analyzing a surface using horizontal cross sections, called **level curves**. We then analyzed surfaces in space using vertical cross sections. We sliced a surface with planes of the form $x = c$ and $y = c$ and graphed the resulting cross sections on the yz-plane and xz-plane, respectively. Notice that each type of cross section tells us something about the shape of the surface.

Section 8.2 Exercises

Vocabulary Exercises

1. The study of slicing the graphs of functions of two independent variables with horizontal and vertical planes is called _____ analysis.

2. A two-dimensional graph of a three-dimensional surface such as the Earth's terrain is called a _____ map.

3. Lines that connect points of equal elevation are called contour lines or _____ _____.

4. A collection of contour lines make up a _____ _____.

5. In economics, an _____ is a contour line drawn through the set of points at which the same quantity of output is produced while changing the quantities of two or more independent variable values.

6. An _____ _____ shows how consumers would react to different combinations of products.

Skill Exercises

In Exercises 7–14, sketch a contour map of the given function for the specific c-values.

7. $f(x, y) = x + y + 1$ $c = 0, 1, 2, 3$ 8. $f(x, y) = 1 - x - y$ $c = 0, 3, 6, 9$

9. $f(x, y) = 2x + 3y + 6$ $c = 0, 2, 4, 6$ 10. $f(x, y) = 2y - 3x + 1$ $c = 0, 2, 4, 6$

11. $f(x, y) = \sqrt{16 - x^2 - y^2}$ $c = 0, 2, 4$ **12.** $f(x, y) = \sqrt{25 - x^2 - y^2}$ $c = 1, 3, 5$

13. $f(x, y) = e^x + y$ $c = 0, 2, 4, 6$ **14.** $f(x, y) = \ln x + y - 1$ $c = 0, 2, 4, 6$

For Exercises 15–20, make a contour map of the given production functions for the specified c-values.

15. $Q = 24x^{0.5}y^{0.5}$ $c = 10, 20, 30, 40$ **16.** $Q = 36.5x^{0.5}y^{0.5}$ $c = 100, 200, 300, 400$

17. $Q = 5.6x^{0.65}y^{0.41}$ $c = 10, 20, 30, 40$ **18.** $Q = 6.3x^{0.7}y^{0.35}$ $c = 20, 40, 60, 80$

19. $Q = 180x^{0.7}y^{0.3}$ $c = 100, 200, 300, 400$ **20.** $Q = 120x^{0.75}y^{0.25}$ $c = 100, 200, 300, 400$

In Exercises 21–26, make a contour map of the given utility function for the specified c-values.

21. $U(x, y) = \sqrt{xy}$ $c = 2, 4, 6, 8$ **22.** $U(x, y) = x^{0.25}y^{0.75}$ $c = 1, 2, 3, 4$

23. $U(x, y) = x^{0.75}y^{0.25}$ $c = 1, 2, 3, 4$ **24.** $U(x, y) = x^{0.65}y^{0.35}$ $c = 1, 2, 3, 4$

25. $U(x, y) = x^{0.55}y^{0.45}$ $c = 1, 2, 3, 4$ **26.** $U(x, y) = x^{0.55}y^{0.45}$ $c = 2, 4, 6, 8$

For Exercises 27–34, sketch on the appropriate plane, either the xz-plane or the yz-plane, cross sections for the given function and specified vertical plane values.

27. $z = 15x^{0.5}y^{0.5}$ $x = 2, 4, 6$ **28.** $z = 15x^{0.5}y^{0.5}$ $x = 3, 6, 9$

29. $z = 15x^{0.5}y^{0.5}$ $y = 2, 4, 6$ **30.** $z = 15x^{0.5}y^{0.5}$ $y = 3, 6, 9$

31. $z = 120x^{0.75}y^{0.25}$ $x = 10, 20, 30$ **32.** $z = 120x^{0.75}y^{0.25}$ $x = 20, 40, 60$

33. $z = 120x^{0.75}y^{0.25}$ $y = 20, 40, 60$ **34.** $z = 120x^{0.75}y^{0.25}$ $y = 10, 20, 30$

Application Exercises

35. Economics—Labor/Capital Analysis: The Cranky Corporation has three crankshaft producing plants in the United States and a Cobb-Douglas production function of

$$Q = f(x, y) = 1.2x^{0.7}y^{0.3}$$

where x represents the number of labor hours (in thousands), y represents the capital (total net assets in dollars), and Q represents the quantity produced (in thousands). Sketch isoquants for $Q = 20, 40,$ and 60.

36. Economics—Labor/Capital Analysis: Ling Incorporated has four pretzel-making plants in the United States and a Cobb-Douglas production function of

$$Q = f(x, y) = 2.3x^{0.2}y^{0.8}$$

where x represents the number of labor hours (in hundreds), y represents the capital (total net assets in dollars), and Q represents the quantity produced (in hundreds of pounds). Sketch isoquants for $Q = 30, 60,$ and 90.

37. Economics—Utility Analysis: Suppose that the manager at the Pizza Casa determines that the utility from consuming colas and pizza in a typical week is given by

$$U(x, y) = x^{2/3}y^{1/3}$$

where x represents the number of colas consumed, y represents the number of slices of pizzas consumed, and U represents the resulting utility level. Sketch an indifference map for $c = 1, 2,$ and 3.

38. Economics—Utility Analysis: Suppose that the owner of the Sock Hop Ole Time Restaurant finds that the utility from consuming milkshakes and cheeseburgers in a typical week is given by

$$U(x, y) = x^{1/3}y^{2/3}$$

where x represents the number of milkshakes consumed, y represents the number of cheese-burgers consumed, and U represents the resulting utility level. Sketch an indifference map for $c = 1, 2,$ and 3.

39. **Anthropology—Cephalic Index:** In their study of human groupings, anthropologists often use an index called the **cephalic index**. This index expresses the ratio of the maximum breadth of a skull to its maximum length. The cephalic index is given by

$$z = C(x, y) = 100 \cdot \frac{x}{y}$$

where x is the width and y is the length of an individual's head. Both measurements are made across the top of the head, and both are in inches. Construct a contour map for $c = 75, 80, 85,$ and 90.

 40. **Forestry Science—Doyle's Rule:** The **Doyle log rule** is one method used to determine the yield of a log, measured in board feet. Mathematically, this rule is given by

$$f(x, y) = \left(\frac{x - 4}{4} \right)^2 \cdot y$$

where x is the diameter of the log in inches, y is the length of the log in feet, and $f(x, y)$ is the number of board-feet. (*Source:* U.S. Forest Service.)

(a) Construct a contour map for $c = 400, 500, 600,$ and 700.

(b) Describe what the level curve for $c = 500$ represents using a complete sentence.

41. **Economics—Labor/Capital Analysis:** The O'Neill Corporation has 10 soft-drink bottling plants in the United States. The research office determines that the Cobb-Douglas production function associated with the total output of the plants is given by

$$Q = f(x, y) = 1.64x^{0.6}y^{0.4}$$

where x is the number of labor hours (in thousands), y is the capital (total net assets in millions), and Q is the quantity produced (in thousands of gallons). Sketch isoquants for $Q = 80,$ 100, 120, and 140.

42. **Economics—Labor/Capital Analysis** (*continuation of Exercise 41*): James, the CEO of the O'Neill Corporation, wants to keep the number of labor hours at each plant fixed at $x = 110$.

(a) By fixing labor at $x = 110$, is James looking at a vertical or horizontal cross section?

(b) Graph the cross section when $x = 110$.

(c) Approximate how much capital is required to produce 100.

(d) Describe what the graph in part (b) represents using a complete sentence.

43. **Economics—Labor/Capital Analysis** (*continuation of Exercise 41*): James, the CEO of the O'Neill Corporation, wants to keep capital at each plant fixed at $y = 18$.

(a) By fixing capital at $y = 18$, is James looking at a vertical or horizontal cross section?

(b) Graph the cross section when $y = 18$.

(c) Approximate how many labor hours would be required to produce 100.

(d) Describe what the graph in part (b) represents using a complete sentence.

44. **Economics—Labor/Capital Analysis:** The Tressel Golf Club Company has completed its research and has determined that its output is modeled by a Cobb-Douglas production function given by

$$Q = f(x, y) = 21x^{0.3}y^{0.75}$$

where x is the utilization of labor, y is the utilization of capital, and Q is the number of units of golf clubs produced. Sketch the isoquants for $Q = 500, 1000, 1500,$ and 2000.

45. **Economics—Labor/Capital Analysis** (*continuation of Exercise 44*): The CEO of the Tressel Golf Club Company wants to keep labor fixed at $x = 100$ units.

(a) By fixing labor at $x = 100$ units, is the CEO looking at a vertical or horizontal cross section?

(b) Graph the cross section when $x = 100$.

(c) Approximate how many units of capital would be required to produce 1500 units of golf clubs.

(d) Describe what the graph in part (b) represents using a complete sentence.

46. **Economics—Labor/Capital Analysis** (*continuation of Exercise 44*): The CEO of the Tressel Golf Club Company wants to keep capital fixed at $y = 100$ units.

(a) By fixing capital at $y = 100$ units, is the CEO looking at a vertical or horizontal cross section?

(b) Graph the cross section when $y = 100$.

(c) Approximate how many units of labor would be required to produce 1500 units of golf clubs.

(d) Describe what the graph in part (b) represents using a complete sentence.

47. **Economics—Labor/Capital Analysis:** The Brody Corporation has 11 plants worldwide and a Cobb-Douglas production function given by

$$Q = f(x, y) = 1.7x^{0.8}y^{0.2}$$

where x is the number of units of labor (in thousands), y is the number of units of capital (in millions), and Q is the number of units of quantity produced. Construct isoquants for $c = 300$, 400, 500, and 600.

48. **Economics—Labor/Capital Analysis** (*continuation of Exercise 47*): Cheryl, the CEO of the Brody Corporation, wants to keep labor fixed at $x = 350$ units.

(a) Graph the vertical cross section when $x = 350$.

(b) Approximate how much capital is required to produce 560 units.

49. **Economics—Labor/Capital Analysis** (*continuation of Exercise 47*): Cheryl, the CEO of the Brody Corporation, wants to keep capital fixed at $y = 50$ units.

(a) Graph the vertical cross section when $y = 50$.

(b) Approximate how many units of labor are required to produce 560 units.

50. **Geometry—Cylinder Volume:** The volume of a cylinder with radius r and height h is given by $V(r, h) = \pi r^2 h$. Substituting x for r and y for h gives $V(x, y) = \pi x^2 y$. Make a contour map for $c = 10$, 20, 30, and 40.

51. **Geometry—Cylinder Volume:** Suppose that the height of the cylinder in Exercise 50 is fixed at 4 inches, that is, $y = 4$.

(a) Graph the vertical cross section when $y = 4$.

(b) Approximate the radius necessary to produce a cylinder with a volume of approximately 113 cubic inches.

(c) Use the graph to approximate the volume if the radius is 6 inches.

(d) In your own words, what does the graph in part (a) represent?

52. **Cognitive Science—IQ Score:** The intelligence quotient is defined to be a number representing a person's reasoning ability as compared to an average for their age, taken as 100. This IQ value can be computed using the function

$$IQ = f(x, y) = \frac{y}{x} \cdot 100$$

where y is the individual's mental age in years, and x is the individual's actual age in years.

(a) Make a contour map for $c = 90$, 100, 110, and 120. Keep in mind that $x > 0$ and $y > 0$.

(b) Write three combinations of actual age and mental age that give an IQ of 120.

53. **Cognitive Science—IQ Score:** For the IQ function in Exercise 52, consider a vertical cross section by keeping mental age fixed at 20, that is, $y = 20$. Then the IQ function becomes

$$IQ = f(x, 20) = \frac{20}{x} \cdot 100 = \frac{2000}{x}$$

where x represents the chronological age.

(a) To accommodate our calculator, substitute y for IQ. This yields the function $y = \frac{2000}{x}$. Graph this function in the viewing window $[0, 60]$ by $[0, 300]$.

(b) The point $(15, 133\frac{1}{3})$ is on the graph. Interpret the meaning of this point.

(c) The point $(16, 125)$ is on the graph. Interpret the meaning of this point.

 54. Cognitive Science—IQ Score: For the IQ function in Exercise 52, consider a vertical cross section by keeping chronological age fixed at 20, that is, $x = 20$. Our function then becomes

$$IQ = f(20, y) = \frac{y}{20} \cdot 100 = 5y$$

(a) To accommodate our calculator, substitute y for IQ and x for y. This gives $y = 5x$. Graph this function in the viewing window $[0, 60]$ by $[0, 300]$.

(b) The point $(30, 150)$ is on the graph. Interpret the meaning of this point.

(c) The point $(31, 155)$ is on the graph. Interpret the meaning of this point.

(d) The point $(29, 145)$ is on the graph. Interpret the meaning of this point.

(e) Fill in the blanks. *When actual age is fixed at 20, each 1-year increase in mental age* _____ *(increases/decreases) IQ by* _____ *units.*

Concept and Writing Exercises

55. Other than meteorology, name a profession that takes advantage of contour maps.

56. Describe how meteorologists use level curves and contour maps.

57. Suppose we have a function of two independent variables $z = f(x, y) = x^2 + 2y$ and construct a level curve for $c = 5$. If $x = -3$, determine the value of y.

58. Suppose we have a function of two independent variables $z = f(x, y) = \frac{x + y}{y}$ and construct a level curve for $c = 2$. If $y = 4$, determine the value of x.

59. Suppose that we have a contour map of a large rounded hill. What function could be used to represent this contour map?

60. For a function of two independent variable and a real number constant a, is $f(ax, ay) = az$? If true, explain why. If false, provide a counterexample.

 ## Section Project

The program in Appendix B that performs multiple regression is necessary to do this section project. The Riblet Engine Company has six plants around the United States to produce small engines. In a recent year the data for each plant gave the number of labor hours (in thousands), capital (in millions), and total quantity produced, is shown in the table.

Labor	Capital	Quantity
42	5	27,825
44	6	31,058
47	7	34,633
49	9	39,762
50	10	42,201
52	13	48,621

(a) The plants all operate with the same technology, so a production function can be determined. Enter the data into your calculator and execute the program to determine a Cobb-Douglas production function to model the data. Round all values to the nearest hundredth.

(b) Graph isoquants for $q = 20,000, 30,000, 40,000,$ and $50,000$.

(c) The CEO of the company, Gene, wants to keep labor fixed at 48. Is Gene asking for a vertical or horizontal cross section?

(d) Graph the cross section when labor is fixed at 48.

(e) Use the graph to approximate how much capital is required to produce a quantity of 35,000.

8.3   Partial Derivatives and Second-Order Partial Derivatives

SECTION OBJECTIVES

1. Compute a partial derivative.

2. Evaluate and interpret a partial derivative.

3. Apply a partial derivative to determine marginal revenue.

4. Compute marginal productivity of labor and capital.

5. Determine second-order partial derivatives.

In this section we discuss how to measure the rate of change of a function of two independent variables. We will observe how the vertical cross sections discussed in Section 8.2 are central to the study of this rate of change, which is known as **partial derivatives**. Partial derivatives simply measure the rate of change of a function of two independent variables, and they will be compared to the derivative that was discussed in Section 2.4. We also will discuss **second-order** partial derivatives. Second-order partial derivatives are analogous to second derivatives, which we discussed in Section 5.4.

Partial Derivatives

We have encountered many interpretations and applications of the derivative of a function of one independent variable, f. Naturally, we would like to extend this rate of change concept to a function of two variables, $f(x, y)$. To do this, we consider how the function values change as x changes, that is, when y is kept constant, and we consider how the function values change as y changes, that is, when x is kept constant. Geometrically, we are talking about the **vertical cross sections** introduced in Section 8.2.

Flashback: O'Neill Corporation Revisited

In Section 8.2 we were introduced to the O'Neill Corporation. Recall that the O'Neill Corporation has 10 soft-drink bottling plants and a production function of

$$Q = f(x, y) = z = 1.64x^{0.6}y^{0.4}$$

where x is the number of labor hours (in thousands), y is the capital (total net assets in millions), and Q is the quantity produced (in thousands of gallons). James, the CEO of the O'Neill Corporation, wants to keep the number of labor hours at each plant fixed at $x = 110$. Graph the cross section when $x = 110$ and describe what the graph represents.

Perform the Mathematics

Figure 8.3.1 gives the graph of the surface and the plane $x = 110$, and **Figure 8.3.2** is a graph of the cross section. Notice that the cross section is graphed on the yz-plane.

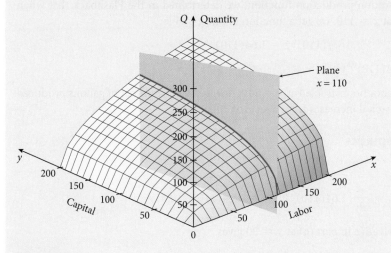

Figure 8.3.1

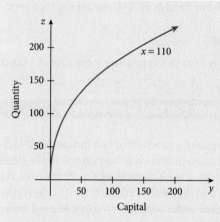

Figure 8.3.2

The graph of the cross section shows us that quantity is a function of capital when labor is held constant at $x = 110$. Specifically, we have

$$Q = f(x, y) = 1.64x^{0.6}y^{0.4}$$
$$= f(110, y) = 1.64(110)^{0.6}y^{0.4}$$

So by holding labor constant at $x = 110$, we now have quantity as a function of one variable, capital or y.   ■

Let's take another look at Figure 8.3.2. Notice that when labor is fixed at $x = 110$ the vertical cross section is a function of *one* variable. (All along the curve in Figure 8.3.2, x is held constant, $x = 110$.) As we saw in the Flashback, this function of one variable has the form

$$f(110, y) = 1.64(110)^{0.6}y^{0.4}$$

From our work with derivatives, we know that we can find the rate at which $f(110, y) = 1.64(110)^{0.6}y^{0.4}$ changes, as y changes, by taking the derivative. What exactly does all this tell us? Example 1 provides the answer.

Example 1: Determining a Rate of Change

For the O'Neill Corporation production function, we determined in the Flashback that when labor is held constant at $x = 110$, we get a function of one variable:

$$f(110, y) = 1.64(110)^{0.6}y^{0.4}$$

a.   Determine $\frac{d}{dy}[1.64(110)^{0.6}y^{0.4}]$.

b.   If the number of labor hours is fixed at $x = 110$, how fast is the number of gallons produced increasing as the capital increases from a level of 20?

Perform the Mathematics

a.   The derivative is

$$\frac{d}{dy}[1.64(110)^{0.6}y^{0.4}] = 1.64(110)^{0.6}(0.4)y^{-0.6}$$

b.   Evaluating the derivative in part (a) at $y = 20$ gives

$$1.64(110)^{0.6}(0.4)(20)^{-0.6} \approx 1.8244$$

Evaluating the derivative at $y = 20$ gives the *slope of the tangent line to the curve at $y = 20$*. See **Figure 8.3.3**. The curve in Figure 8.3.3 is the graph of the cross section when $x = 110$, so it is the graph of quantity as a function of capital when labor is fixed at 110. This tells us that, when the number of labor hours is held constant at $x = 110$ and capital is at 20 units, as capital increases 1 unit, production is increasing at a rate of about $1.8 \frac{\text{thousand gallons}}{\text{unit of capital}}$.

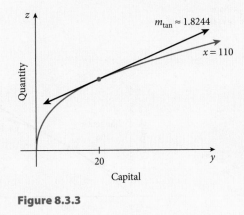

Figure 8.3.3

Let's summarize what was done in the Flashback and Example 1.

- When x is held constant at $x = 110$, we produce a **vertical cross section** of the surface.
- The graph of this cross section in the yz-plane shows production as a function of *one* variable.
- We can differentiate the function of one variable to give us a rate of change.

Definition of Partial Derivatives

The two-dimensional analysis that we did in the Flashback and in Example 1 tells us something remarkable about the three-dimensional surface. Since x is held constant at $x = 110$ and we looked specifically at the scenario when $y = 20$, this means that we were at the point $(110, 20, 91.22)$ on the graph of the surface (z-coordinate rounded to the nearest hundredth) of our function. If we are at the point $(110, 20, 91.22)$ on the surface and move *in the y-direction*, that is, parallel to the y-axis, the surface is changing at a rate of approximately $1.8 \frac{\text{units of } z}{\text{unit of } y}$. This is exactly the same as the slope of the tangent line to the curve that resulted from slicing the surface with the plane $x = 110$. See **Figure 8.3.4**.

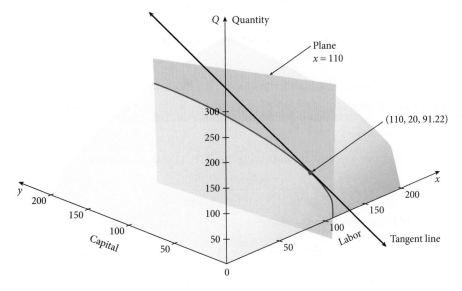

Figure 8.3.4

The process of holding x constant at any value $x = a$, moving in the y-direction (that is, parallel to the y-axis), and determining the rate of change of the surface is one interpretation of the **partial derivative of $f(x, y)$ with respect to y**. In a completely analogous manner, the process of holding y constant at any value $y = b$, moving in the x-direction (that is, parallel to the x-axis), and determining the rate of change of the surface is one interpretation of the **partial derivative of $f(x, y)$ with respect to x**. See **Figure 8.3.5(a)–(c)**.

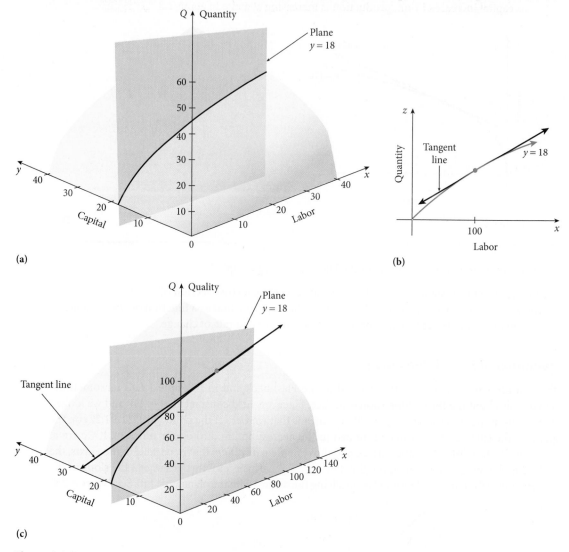

(a)

(b)

(c)

Figure 8.3.5

The previous explanation serves as a basis for the formal definition of a **partial derivative**.

DEFINITION

Partial Derivative

Let $f(x, y)$ be a function of two variables. The **partial derivative of $f(x, y)$ with respect to x** is

$$f_x(x, y) = \lim_{h \to 0} \frac{f(x + h, y) - f(x, y)}{h}$$

The **partial derivative of $f(x, y)$ with respect to y** is

$$f_y(x, y) = \lim_{h \to 0} \frac{f(x, y + h) - f(x, y)}{h}$$

Of course, each of these partial derivatives exists only if the appropriate limit exists.

The important practical point on how to **compute partial derivatives** is given in the following box.

Computing Partial Derivatives

1. The **partial derivative of $f(x, y)$ with respect to x** is found by treating y as a constant and performing our ordinary differentiation techniques. The notation for this particular derivative is $f_x(x, y)$ or $\frac{\partial f}{\partial x}$. The units of this partial derivative are units of f per unit of x.

2. The **partial derivative of $f(x, y)$ with respect to y** is found by treating x as a constant and performing our ordinary differentiation techniques. The notation for this particular derivative is $f_y(x, y)$ or $\frac{\partial f}{\partial y}$. The units of this partial derivative are units of f per unit of y.

Example 2: Computing a Partial Derivative

OBJECTIVE 1

Compute a partial derivative.

Consider $f(x, y) = x^2y - y^2$.

a.   Determine $f_x(x, y)$.   b.   Determine $f_y(x, y)$.

Perform the Mathematics

a.   To determine $f_x(x, y)$, we treat y as a constant and take the derivative with respect to x. This gives

$$f(x, y) = x^2y - y^2$$

$$f_x(x, y) = \frac{\partial}{\partial x}[x^2y] - \frac{\partial}{\partial x}[y^2]$$

$$f_x(x, y) = 2x \cdot y - 0 = 2xy$$

Observe that $\frac{\partial}{\partial x}[y^2] = 0$ because y is treated as a constant.

b.   To compute $f_y(x, y)$, we treat x as a constant and take the derivative with respect to y. This gives

$$f(x, y) = x^2y - y^2$$

$$f_y(x, y) = \frac{\partial}{\partial y}[x^2y] - \frac{\partial}{\partial y}[y^2] = 0$$

$$f_y(x, y) = x^2 \cdot 1 - 2y = x^2 - 2y$$ ∎

▶ *Try It Yourself*

Some related Exercises are 9 and 11.

Example 3: Computing Partial Derivatives

Consider $f(x, y) = x^2y + y^3x - 2xy + y$.

a.   Determine $f_x(x, y)$.   b.   Determine $f_y(x, y)$.

Perform the Mathematics

a.   To determine $f_x(x, y)$, we treat y as a constant and differentiate with respect to x.

$$f(x, y) = x^2y + y^3x - 2xy + y$$

$$f_x(x, y) = (2 \cdot x)y + y^3 \cdot 1 - 2 \cdot 1 \cdot y + 0$$

$$= 2xy + y^3 - 2y$$

b.   To determine $f_y(x, y)$, we treat x as a constant and differentiate with respect to y.

$$f(x, y) = 2x^3y^4 + \frac{1}{3}y^3 + e^{3x} - e^{2y}$$

$$f_y(x, y) = x^2 \cdot 1 + (3y^2) \cdot x - 2x \cdot 1 + 1$$

$$= x^2 + 3y^2x - 2x + 1 \qquad \blacksquare$$

▶ *Try It Yourself*

Some related Exercises are 13 and 15.

OBJECTIVE 2

Evaluate and interpret a partial derivative.

Example 4: Evaluating and Interpreting a Partial Derivative

Let $f(x, y) = 2x^3y^4 + \frac{1}{3}y^3 + e^{3x} - e^{2y}$.

a.   Determine $f_x(x, y)$.

b.   Evaluate $f_x(x, y)$ at $(1, \frac{3}{2})$ and interpret.

Perform the Mathematics

a.   To determine $f_x(x, y)$, we treat y as a constant and differentiate with respect to x.

$$f(x, y) = 2x^3y^4 + \frac{1}{3}y^3 + e^{3x} - e^{2y}$$

$$f_x(x, y) = (6x^2)y^4 + 0 + 3e^{3x} - 0 = 6x^2y^4 + 3e^{3x}$$

b.   Evaluating $f_x(x, y)$ at $(1, \frac{3}{2})$ yields

$$f_x\left(1, \frac{3}{2}\right) = 6(1)^2\left(\frac{3}{2}\right)^4 + 3e^{3(1)}$$

$$= 30.375 + 3e^3$$

$$\approx 90.63$$

This means that if we are at the point $(1, \frac{3}{2}, 11.25)$ on the surface and move along the surface in the x-direction, parallel to the x-axis, the function is changing at a rate of approximately $90.63 \frac{\text{units of } f}{\text{unit of } x}$. Equivalently, we say that at the point $(1, \frac{3}{2}, 11.25)$ the slope of the surface in the x-direction is approximately 90.63. See **Figure 8.3.6**.

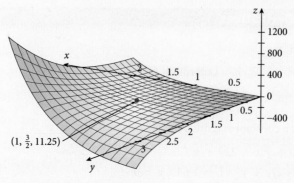

Figure 8.3.6 $\qquad\qquad\qquad\qquad\qquad\qquad\qquad\qquad\qquad\qquad\qquad\blacksquare$

> ▶ *Try It Yourself*
>
> Some related Exercises are 37 and 39.

Applications

In the next example, we analyze the relationship between a partial derivative and **marginal analysis**.

| **Example 5: Analyzing Marginal Revenue** | **OBJECTIVE 3** |

OBJECTIVE 3

Apply a partial derivative to determine marginal revenue.

A company sells leaf blowers and a special 10-foot attachment to the leaf blower to clean gutters. Let x and y be the number of leaf blowers sold per month and the number of 10-foot attachments sold per month, respectively. Suppose that

$$p = 1200 - 8x - y, \qquad \text{the price in dollars for a leaf blower}$$
$$q = 523 - 1.5x - 0.15y, \qquad \text{the price in dollars for an attachment}$$

a. Determine the revenue function $R(x, y)$.

b. Determine $R_x(x, y)$ and $R_y(x, y)$. Interpret each.

c. Evaluate and interpret $R_x(50, 25)$.

Perform the Mathematics

a. Since revenue equals price times quantity, we have

$$R(x, y) = x(1200 - 8x - y) + y(523 - 1.5x - 0.15y)$$
$$= 1200x - 8x^2 - xy + 523y - 1.5xy - 0.15y^2$$
$$= 1200x - 8x^2 - 2.5xy + 523y - 0.15y^2$$

b. To determine $R_x(x, y)$, we treat y as a constant and differentiate with respect to x.

$$R(x, y) = 1200x - 8x^2 - 2.5xy + 523y - 0.15y^2$$
$$R_x(x, y) = 1200 \cdot 1 - 8 \cdot 2x - 2.5(1)(y) + 0 - 0$$
$$= 1200 - 16x - 2.5y$$

$R_x(x, y)$ gives the amount that each additional leaf blower adds to the total revenue. In other words, it gives the **marginal revenue** for leaf blowers.

To determine $R_y(x, y)$, we treat x as a constant and differentiate with respect to y.

$$R(x, y) = 1200x - 8x^2 - 2.5xy + 523y - 0.15y^2$$
$$R_y(x, y) = 0 - 0 - 2.5(x)(1) + 523(1) - 0.15(2y)$$
$$= -2.5x + 523 - 0.3y$$

$R_y(x, y)$ gives the amount that each additional 10-foot attachment adds to the total revenue. In other words, it is the **marginal revenue** for the 10-foot attachments.

c. Since $x = 50$ and $y = 25$, we have

$$R_x(50, 25) = 1200 - 16(50) - 2.5(25)$$
$$= 337.5$$

This means that when 50 leaf blowers and 25 of the 10-foot attachments have been sold, the company receives $337.50 for each additional leaf blower sold. ∎

Example 6: Computing Partial Derivatives of a Volume Formula

Recall that the formula for the volume of a cylinder is given by $V(r, h) = \pi r^2 h$, where r is the radius and h is the height. Evaluate and interpret $V_r(2, 3)$ and $V_h(2, 3)$.

Understand the Situation

In the formula for volume of a cylinder, $V(r, h) = \pi r^2 h$, r and h are the independent variables. Thus, $V_r(r, h)$ is found by keeping h constant and differentiating with respect to r, whereas $V_h(r, h)$ is found by keeping r constant and differentiating with respect to h.

Perform the Mathematics

Keeping h constant, we compute $V_r(r, h)$ as

$$V_r(r, h) = 2\pi r h$$

To determine $V_h(r, h)$, we keep r constant, which gives

$$V_h(r, h) = \pi r^2$$

Evaluating $V_r(2, 3)$ gives us

$$V_r(2, 3) = 2\pi(2)(3) = 12\pi$$

Evaluating $V_h(2, 3)$ yields

$$V_h(2, 3) = \pi(2)^2 = 4\pi$$

Interpret the Results

$V_r(2, 3)$ tells us that when $r = 2$ cm and $h = 3$ cm, and the height is held constant at $h = 3$ cm, the volume is increasing at a rate of about 12π cm^3 per cm increase of the radius. $V_h(2, 3)$ tells us that when $r = 2$ cm and $h = 3$ cm, and the radius is held constant at $r = 2$ cm, the volume is increasing at a rate of about 4π cm^3 per cm increase of the height.

 Try It Yourself

Some related Exercises are 69 and 71.   ∎

DEFINITION

Marginal Productivity of Labor and Capital

For any production function of the form $Q = f(x, y) = ax^m y^n$, where a, m, and n are positive constants and x represents units of labor and y represents units of capital, we have the following:

- $f_x(x, y)$ gives the approximate change in productivity per unit change in labor and is called **marginal productivity of labor**.
- $f_y(x, y)$ gives the approximate change in productivity per unit change in capital and is called **marginal productivity of capital**.

Marginal Productivity of Labor and Capital

Example 1 foreshadowed the concept known as **marginal productivity of labor** and **marginal productivity of capital**. These two partial derivatives are very important in economics.

OBJECTIVE 4

Compute marginal productivity of labor and capital.

Example 7: Computing Marginal Productivity of Labor and Capital

Recall that the production function for the O'Neill Corporation is given by

$$Q = f(x, y) = 1.64x^{0.6}y^{0.4}$$

where x is the number of labor hours in thousands, y is the capital in millions, and Q is the quantity in thousands of gallons.

a.   Compute the marginal productivity of labor and the marginal productivity of capital.

b.   Evaluate the marginal productivity of labor at $x = 100$ and $y = 18$ and interpret.

c.   Evaluate the marginal productivity of capital at $x = 110$ and $y = 20$ and interpret.

Perform the Mathematics

a.   Since marginal productivity of labor is $f_x(x, y)$, we have

$$f(x, y) = 1.64x^{0.6}y^{0.4}$$
$$f_x(x, y) = 1.64(0.6)x^{-0.4}y^{0.4}$$
$$= 0.984x^{-0.4}y^{0.4}$$

Marginal productivity of capital is $f_y(x, y)$, which is

$$f(x, y) = 1.64x^{0.6}y^{0.4}$$
$$f_y(x, y) = 1.64x^{0.6}(0.4)y^{-0.6}$$
$$= 0.656x^{0.6}y^{-0.6}$$

b.   Since $x = 100$ and $y = 18$, we need to evaluate $f_x(100, 18)$, which gives

$$f_x(100, 18) = 0.984(100)^{-0.4}(18)^{0.4} \approx 0.5$$

This means that when the O'Neill Corporation uses 100 units of labor and 18 units of capital and keeps capital fixed at 18 units, production is increasing at a rate of approximately 0.5 $\frac{\text{thousand gallons}}{\text{unit of labor}}$.

c.   Since $x = 110$ and $y = 20$, we need to evaluate $f_y(110, 20)$, which yields

$$f_y(110, 20) = 0.656(110)^{0.6}(20)^{-0.6} \approx 1.8$$

This means that when the O'Neill Corporation uses 110 units of labor and 20 units of capital and keeps labor fixed at 110 units, production is increasing at a rate of approximately 1.8 $\frac{\text{thousand gallons}}{\text{unit of labor}}$.   ∎

For some amounts currently invested in labor and in capital, a natural question to ask is whether production would increase more rapidly if additional resources were invested in labor or in capital. We can **approximate the change in productivity** as shown in the box to the right.

Approximating Change in Productivity

1.   The change in productivity with respect to labor is the product of

$$f_x(x, y) \cdot (\text{change in } x)$$

2.   The change in productivity with respect to capital is the product of

$$f_y(x, y) \cdot (\text{change in } y)$$

Example 8: Allocating Additional Resources

Suppose that a production function is given by

$$f(x, y) = 4.23x^{0.37}y^{0.66}$$

where x represents dollars in millions spent on labor, and y represents dollars in millions spent on capital equipment. Currently, $x = 5$ and $y = 1$. Would production increase more by spending an additional $1 million on labor or $500,000 on capital equipment?

Understand the Situation

We need to compute and compare $f_x(5, 1) \cdot 1$ and $f_y(5, 1) \cdot (0.5)$. Notice that since capital, y, is in millions, $500,000 is the same as $0.5 million.

Perform the Mathematics

Proceeding with the partial derivatives, we have

$$f_x(x, y) = 4.23(0.37)x^{-0.63}y^{0.66} = 1.5651x^{-0.63}y^{0.66}$$
$$f_x(5, 1) = 1.5651(5)^{-0.63}(1)^{0.66} \approx 0.568$$

So the change in productivity with respect to labor is $f_x(5, 1) \cdot 1 \approx 0.568$. Continuing, we get for capital

$$f_y(x, y) = 4.23(0.66)x^{0.37}y^{-0.34} = 2.7918x^{0.37}y^{-0.34}$$
$$f_y(5, 1) = 2.7918(5)^{0.37}(1)^{-0.34} \approx 5.06$$

The change in productivity with respect to capital is $f_y(5, 1) \cdot (0.5) \approx 2.53$.

Interpret the Result

Since the change in productivity with respect to capital is greater than the change in productivity with respect to labor, production increases more by spending $500,000 on capital equipment than by spending $1 million on labor.

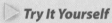 **Try It Yourself**

Some related Exercises are 79 and 80.   ■

Second-Order Partial Derivatives

Just like the second derivatives for a function of one variable, there are **second-order partial derivatives**. These second-order partial derivatives are very important for Section 8.4, when we discuss locating relative extrema on the surface of $z = f(x, y)$.

DEFINITION

Second-Order Partial Derivatives

If $z = f(x, y)$, then the four possible second-order partial derivatives are

$$f_{xx}(x, y) = \frac{\partial}{\partial x}\left(\frac{\partial f}{\partial x}\right) \qquad f_{yy}(x, y) = \frac{\partial}{\partial y}\left(\frac{\partial f}{\partial y}\right)$$

$$f_{xy}(x, y) = \frac{\partial}{\partial y}\left(\frac{\partial f}{\partial x}\right) \qquad f_{yx}(x, y) = \frac{\partial}{\partial x}\left(\frac{\partial f}{\partial y}\right)$$

NOTE: In subscript notation, we differentiate from left to right. That is, to compute f_{yx}, we first compute f_y and then differentiate this result with respect to x.

OBJECTIVE 5

Determine second-order partial derivatives.

Example 9: Computing a Second-Order Partial Derivative

Determine all four second-order partial derivatives of

$$f(x, y) = x^2y + y^3x - 2xy + y$$

Perform the Mathematics

First we calculate $f_x(x, y)$. It is given by

$$f_x(x, y) = 2xy + y^3 - 2y$$

We can now compute $f_{xx}(x, y)$ from this result by finding the derivative of $f_x(x, y)$ with respect to x.

$$f_{xx}(x, y) = \frac{\partial}{\partial x}[f_x(x, y)]$$

$$= \frac{\partial}{\partial x}[2xy + y^3 - 2y]$$

$$= 2y + 0 - 0 = 2y$$

Also, we can compute $f_{xy}(x, y)$ from $f_x(x, y)$ by finding the derivative of $f_x(x, y)$ with respect to y.

$$f_{xy}(x, y) = \frac{\partial}{\partial y}[f_x(x, y)]$$

$$= \frac{\partial}{\partial y}[2xy + y^3 - 2y]$$

$$= 2x + 3y^2 - 2$$

To determine the other two second-order partial derivatives, we return to our original function, $f(x, y) = x^2y + y^3x - 2xy + y$, and compute $f_y(x, y)$. This gives

$$f_y(x, y) = x^2 + 3y^2x - 2x + 1$$

From this result we have

$$f_{yy}(x, y) = \frac{\partial}{\partial y}[f_y(x, y)] = 6yx$$

$$f_{yx}(x, y) = \frac{\partial}{\partial x}[f_y(x, y)] = 2x + 3y^2 - 2$$

Notice that $f_{xy}(x, y) = f_{yx}(x, y)$. This is true for many functions that we encounter, but not all functions. ∎

▶ **Try It Yourself**

Some related Exercises are 49 and 51.

Summary

In this section we saw how the vertical cross sections introduced in Section 8.2 are key to our understanding of a partial derivative. We supplied the following on computing partial derivatives:

- The **partial derivative of $f(x, y)$ with respect to x** is found by treating y as a constant and performing our ordinary differentiation techniques. The notation for this particular derivative is $f_x(x, y)$ or $\frac{\partial f}{\partial x}$.

- The **partial derivative of $f(x, y)$ with respect to y** is found by treating x as a constant and performing our ordinary differentiation techniques. The notation for this particular derivative is $f_y(x, y)$ or $\frac{\partial f}{\partial y}$.

We encountered several applications of partial derivatives, including marginal productivity of labor and marginal productivity of capital. Given a production function of the form $Q = f(x, y) = ax^m y^n$, these were defined as

- $f_x(x, y)$ gives the approximate change in productivity per unit change in labor and is called **marginal productivity of labor**.

- $f_y(x, y)$ gives the approximate change in productivity per unit change in capital and is called **marginal productivity of capital**.

The section concluded with a look at second-order partial derivatives. If $z = f(x, y)$, then the four possible **second-order partial derivatives** are

$$f_{xx}(x, y) = \frac{\partial}{\partial x}\left(\frac{\partial f}{\partial x}\right) \qquad f_{yy}(x, y) = \frac{\partial}{\partial y}\left(\frac{\partial f}{\partial y}\right)$$

$$f_{xy}(x, y) = \frac{\partial}{\partial y}\left(\frac{\partial f}{\partial x}\right) \qquad f_{yx}(x, y) = \frac{\partial}{\partial x}\left(\frac{\partial f}{\partial y}\right)$$

Section 8.3 Exercises

Vocabulary Exercises

1. The partial derivative of $f(x, y)$ with respect to x is found by treating _____ as a constant.

2. The partial derivative of $f(x, y)$ with respect to y is found by treating _____ as a constant.

3. For a production function $Q = f(x, y) = ax^m y^n$, where x and y represent units of labor and capital, respectively, $f_x(x, y)$ represents the marginal productivity of _____.

4. For a production function $Q = f(x, y) = ax^m y^n$, where x and y represent units of labor and capital, respectively, $f_y(x, y)$ represents the marginal productivity of _____.

5. For $z = f(x, y)$, a second-order partial derivative $\frac{\partial}{\partial x}\left(\frac{\partial f}{\partial x}\right)$ can be written as _____.

6. For $z = f(x, y)$, a second-order partial derivative $\frac{\partial}{\partial x}\left(\frac{\partial f}{\partial y}\right)$ can be written as _____.

Skill Exercises

For Exercises 7–36, determine the partial derivatives $f_x(x, y)$ and $f_y(x, y)$.

7. $f(x, y) = 5x^2 - 6y^3$

8. $f(x, y) = 3x + 2y + 10$

9. $f(x, y) = 2xy - y^2 + 1$

10. $f(x, y) = 3x^4 + 3x + y^2$

11. $f(x, y) = x^3 y^2$

12. $f(x, y) = x^2 y^4$

13. $f(x, y) = x^2 + 3x^2 y^3 - 2y^3 - xy$

14. $f(x, y) = 2x^4 + x^2 y^2 - 3y^2 - y$

15. $f(x, y) = 2x^3 - y^2 + 2x - 3$

16. $f(x, y) = 3x^4 - 2y^3 + 3x^2 - 5xy$

17. $f(x, y) = (2x + 3y)^3$

18. $f(x, y) = (3x - 5y)^4$

19. $f(x, y) = 37.21x^{0.15} y^{0.87}$

20. $f(x, y) = 2.41x^{0.27} y^{0.71}$

21. $f(x, y) = e^x \ln y$

22. $f(x, y) = e^y \ln x$

23. $f(x, y) = \dfrac{y}{x}$

24. $f(x, y) = \dfrac{x}{y}$

25. $f(x, y) = 3x^2 - 2x^3 y^4 + 7$

26. $f(x, y) = 4y^3 - 5x^2 y + 2$

27. $f(x, y) = (x^2 + y^3)^2$

28. $f(x, y) = (x^3 - y^2)^4$

29. $f(x, y) = \ln(x^2 + y^3)$

30. $f(x, y) = \ln(y^2 - x)$

31. $f(x, y) = \dfrac{x^2 y}{y + x}$

32. $f(x, y) = \dfrac{xy^2}{y - x}$

33. $f(x, y) = 3x^2 y^3 + e^{x+y}$

34. $f(x, y) = -2x^3 y - e^{x-y}$

35. $f(x, y) = e^{xy} + \ln y$

36. $f(x, y) = y^2 e^{xy} + \ln x$

For the functions given in Exercises 37–46, evaluate the partial derivative at the given values and interpret.

37. $f(x, y) = 4y^3 - 5x^2 y + 2$; $f_x(1, 2)$

38. $f(x, y) = 3x^4 - 2y^3 + 3x^2 - 5xy$; $f_x(2, 1)$

39. $f(x, y) = 2x^4 + x^2 y^2 - 3y^2 - y$; $f_y(2, 3)$

40. $f(x, y) = \ln(y^2 - x)$; $f_y(2, 5)$

41. $f(x, y) = \ln(y^2 - x)$; $f_x(2, 5)$

42. $f(x, y) = 37.21x^{0.15} y^{0.87}$; $f_y(5, 2)$

43. $f(x, y) = 37.21x^{0.15} y^{0.87}$; $f_x(5, 2)$

44. $f(x, y) = 3x^2 y^3 + e^{x+y}$; $f_y(1, 2)$

45. $f(x, y) = -2x^3 y - e^{x-y}$; $f_x(2, 4)$

46. $f(x, y) = \dfrac{\ln x}{x^2 - y^2}$; $f_x(4, 3)$

For Exercises 47–62, determine the partial derivatives $f_{xx}(x, y)$, $f_{xy}(x, y)$, $f_{yx}(x, y)$, and $f_{yy}(x, y)$.

47. $f(x, y) = 4y^3 - 5x^2 y + 2$

48. $f(x, y) = 2x^3 - x^2 y^3 + 2y^4$

49. $f(x, y) = 3x^2 - 2x^3 y^2 + 2x^3$

50. $f(x, y) = 3x^4 + 2x^3 y^2 - y$

51. $f(x, y) = 5xy - 6x^3 + 7y$

52. $f(x, y) = 5xy^3 - 7x^3 y$

53. $f(x, y) = y^5 - 2x^3 y^2 + 7x^2$

54. $f(x, y) = y^3 + 3x^2 y - 8x^3$

55. $f(x, y) = y^2 e^{xy} + \ln x$

56. $f(x, y) = x^3 e^{xy} + \ln y$

57. $f(x, y) = 5.2x^{0.65}y^{0.4}$

58. $f(x, y) = 2.41x^{0.27}y^{0.71}$

59. $f(x, y) = \dfrac{2y}{x}$

60. $f(x, y) = \dfrac{y^2}{x}$

61. $f(x, y) = ye^{xy}$

62. $f(x, y) = y^2 e^{xy}$

Application Exercises

63. Managerial Science—Sales Analysis: Seamego Boats spends x thousand dollars each week on newspaper advertising and y thousand dollars each week on radio advertising. The company has weekly sales, in tens of thousands of dollars, given by

$$S(x, y) = 2x^2 + y$$

(a) Determine $S_x(x, y)$ and $S_y(x, y)$.

(b) Determine $S_x(3, 5)$ and $S_y(3, 5)$ and interpret each.

64. Managerial Science—Sales Analysis: Hilltop Boats spends x thousand dollars each week on radio advertising and y thousand dollars each week on television advertising. The company has weekly sales, in tens of thousands of dollars, given by

$$S(x, y) = 3x + 2y^3$$

(a) Determine $S_x(x, y)$ and $S_y(x, y)$.

(b) Determine $S_x(2, 6)$ and $S_y(2, 6)$ and interpret each.

65. Advertising—Ticket Sales: Tube Town, a recently opened water park, spends x thousand dollars on radio advertising and y thousand dollars on television advertising. The park has weekly ticket sales, in tens of thousands of dollars, of

$$TS(x, y) = 1.5x^2 + 3.2y^2$$

(a) Determine $TS_x(x, y)$ and $TS_y(x, y)$.

(b) Determine $TS_x(1, 0.5)$ and $TS_y(1, 0.5)$ and interpret each.

66. Meteorology—Wind Chill: In Section 8.1 we saw that the wind chill index can be modeled by the function

$$f(v, T) = 35.74 + 0.6215T - 35.75v^{0.16} + 0.4275Tv^{0.16}$$

where v is the wind speed in miles per hour, and T is the actual air temperature in degrees Fahrenheit. (*Source:* National Oceanic and Atmospheric Administration.)

(a) Determine $f_T(v, T)$ and $f_v(v, T)$.

(b) Evaluate $f(25, 5)$, $f_T(25, 5)$, and $f_v(25, 5)$ and interpret each.

67. Banking—Compound Interest: If an amount P dollars is invested at an annual interest rate of 6% compounded continuously for t years, the total amount accumulated is given by

$$f(P, t) = Pe^{0.06t}$$

Determine $f_P(P, t)$ and $f_t(P, t)$ and interpret each.

68. Cytology—Cell Study: In human cells, sodium ions are transported from inside the cell to outside the cell; at the same time, potassium ions are transported from outside the cell to inside. When this takes place across a nerve fiber in a nerve cell, a slight negative electric charge is left behind in the nerve fiber. The amount of voltage that will develop across a membrane is given by the Nernst equation,

$$f(x, y) = -61\log\frac{x}{y}$$

where x is the concentration of potassium ions inside the nerve fiber, y is the concentration of potassium ions outside the nerve fiber, and $f(x, y)$ is measured in millivolts. Determine $f_x(x, y)$ and $f_y(x, y)$ and interpret each.

69. **Geometry—Conical Volume:** The formula for the volume of a cone is given by

$$V(r, h) = \frac{1}{3}\pi r^2 h$$

where r is the radius and h is the height.

(a) Determine $V_r(r, h)$ and $V_h(r, h)$.

(b) Compute $V(3, 5)$, $V_r(3, 5)$, and $V_h(3, 5)$ and interpret each.

70. **Hematology—Poiseuille's Law:** Poiseuille's law states that the resistance, R, for blood flowing in a blood vessel is given by the function

$$R(L, r) = k \cdot \frac{L}{r^4}$$

where k is a constant, L is the length of the blood vessel, and r is the radius of the blood vessel.

(a) Determine $R_L(L, r)$ and $R_r(L, r)$.

(b) Evaluate $R_L(6, 0.3)$ and $R_r(6, 0.3)$ and interpret each.

71. **Anthropology—Cephalic Index:** In their study of human groupings, anthropologists often use an index called the cephalic index. The cephalic index is given by

$$C(W, L) = 100 \cdot \frac{W}{L}$$

where W is the width and L is the length of an individual's head. Both measurements are made across the top of the head and are in inches. Determine $C_W(6, 8.2)$ and $C_L(6, 8.2)$ and interpret each.

 72. **Dermatology—Body Surface Area:** An individual's body surface area is approximated by the function of two independent variables

$$BSA(w, h) = 0.007184w^{0.425}h^{0.725}$$

where BSA is in square meters, w is weight in kilograms, and h is height in centimeters. This formula is known as the DuBois and DuBois formula. (*Source: Archives of Internal Medicine.*)

(a) Determine $BSA_w(w, h)$ and $BSA_h(w, h)$.

(b) Evaluate $BSA(70, 160)$, $BSA_w(70, 160)$, and $BSA_h(70, 160)$ and interpret each.

73. **Economics—Revenue Analysis:** The Chalet Bicycle Company manufactures 21-speed racing bicycles and 21-speed mountain bicycles. Let x represent the weekly demand for a 21-speed racing bicycle and y represent the weekly demand for a 21-speed mountain bicycle. The weekly price-demand equations are given by

$$p = 350 - 4x + y, \qquad \text{the price in dollars for a 21-speed racing bicycle}$$
$$q = 450 + 2x - 3y, \qquad \text{the price in dollars for a 21-speed mountain bicycle}$$

(a) Determine the revenue function $R(x, y)$.

(b) Determine $R_x(x, y)$ and $R_y(x, y)$. Interpret each.

(c) Evaluate $R(15, 20)$ and $R_x(15, 20)$ and interpret each.

74. **Economics—Revenue Analysis** (*continuation of Exercise 73*): The cost function for Chalet Bicycle Company is given by

$$C(x, y) = 390 + 95x + 100y$$

(a) Determine the weekly profit function $P(x, y)$.

(b) Determine $P_x(x, y)$ and $P_y(x, y)$ and interpret each.

(c) Evaluate $P(15, 20)$ and $P_x(15, 20)$ and interpret each.

75. **Economics—Marginal Revenue:** The Kerr Company produces two types of graphing calculators, x units of a 2D graphing calculator and y units of a 3D graphing calculator, each month. The monthly revenue, in dollars, is given by

$$R(x, y) = 70x + 95y + 0.5xy - 0.04x^2 - 0.04y^2$$

(a) Determine the marginal revenue function for the 2D graphing calculator.

(b) Determine the marginal revenue function for the 3D graphing calculator.

76. **Economics—Marginal Revenue** *(continuation of Exercise 75):* The Kerr Company knows that the cost function for producing the two types of calculators is

$$C(x, y) = 4x + 5y + 3200$$

(a) Determine the marginal cost function for the 2D graphing calculator.

(b) Determine the marginal cost function for the 3D graphing calculator.

77. **Economics—Profit Analysis:** The Smolki Engine Company produces two types of small engines, x thousand two-stroke engines and y thousand four-stroke engines. The vice president of the company, Jamie, has determined that the revenue and cost functions for a year are, in thousands of dollars,

$$R(x, y) = 42x + 39.5y$$
$$C(x, y) = 0.5x^2 + 0.8xy + y^2 + 5x + 4y$$

(a) Determine the profit function $P(x, y)$.

(b) Determine $P_x(x, y)$ and $P_y(x, y)$ and interpret each.

(c) Compute $P(18, 10)$ and $P_x(18, 10)$ and interpret each.

(d) Compute $P_y(18, 10)$ and interpret.

78. **Economics—Labor/Capital Analysis:** A golf club manufacturer has a Cobb-Douglas production function given by

$$Q = f(x, y) = 21x^{0.3} y^{0.75}$$

where x is the utilization of labor (in millions), y is the utilization of capital (in millions), and Q is the number of units of golf clubs produced.

(a) Compute $f_x(x, y)$ and $f_y(x, y)$.

(b) If the golf club manufacturer is currently using 150 units of labor and 100 units of capital, determine the marginal productivity of labor and the marginal productivity of capital.

(c) Would production increase more by spending an additional \$1 million on labor or \$500,000 on capital? Explain.

79. **Economics—Labor/Capital Analysis:** The Kevshan Company has a production function of the form

$$Q = f(x, y) = 161x^{0.8} y^{0.25}$$

where x is the number of labor hours in thousands, y is the capital equipment in millions, and Q is the quantity of calculators produced.

(a) Compute $f_x(x, y)$ and $f_y(x, y)$.

(b) If the Kevshan Company is now using 111 units of labor and 25 units of capital, determine the marginal productivity of labor and the marginal productivity of capital and interpret each.

80. **Economics—Labor/Capital Analysis:** The Baker Compact Disc Company has eight plants around the United States that manufacture compact disc players. The Cobb-Douglas production function for the eight plants is given by

$$Q = f(x, y) = 3.14x^{0.6} y^{0.45}$$

where x is the dollars (in millions) spent on labor, y is the dollars (in millions) spent on capital, and Q is the quantity produced (in thousands).

(a) Compute $f_x(x, y)$ and $f_y(x, y)$.

(b) If the Baker Compact Disc Company is currently using 1.7 units of labor and 5 units of capital, determine the marginal productivity of labor and the marginal productivity of capital and interpret each.

(c) Would production increase more by spending an additional \$1 million on labor or \$500,000 on capital? Explain.

81. **Economics—Labor/Capital Analysis:** The Toth Engine Company has a Cobb-Douglas production function given by

$$Q = f(x, y) = 1434.33x^{0.6}\, y^{0.45}$$

where x is the number of labor hours in thousands, y is the capital in millions, and Q is the total quantity produced.

(a) Compute $f_x(x, y)$ and $f_y(x, y)$.

(b) If the Toth Engine Company is currently using 47 units of labor and 8 units of capital, determine the marginal productivity of labor and the marginal productivity of capital and interpret each.

82. **Economics—Labor/Capital Analysis:** Anastasia's, a manufacturer of skis, has a monthly production function given by

$$Q = f(x, y) = 210x^{0.6}\, y^{0.4}$$

where Q is the number of pairs of skis produced, x is the amount of labor used, and y is the amount of capital used.

(a) Compute $f_x(x, y)$ and $f_y(x, y)$.

(b) If Anastasia's is currently using 15 units of labor and 23 units of capital, determine the marginal productivity of labor and the marginal productivity of capital and interpret each.

83. **Economics—Labor/Capital Analysis:** The Jendrag Motorcycle Company has a production function given by

$$Q = f(x, y) = 60x^{0.55}\, y^{0.5}$$

where x is the dollars (in millions) spent on labor, y is the dollars (in millions) spent on capital, and Q is the quantity produced (in thousands).

(a) Compute $f_x(x, y)$ and $f_y(x, y)$.

(b) If the Jendrag Company is currently using 220 units of labor and 140 units of capital, determine the marginal productivity of labor and the marginal productivity of capital and interpret each.

(c) Would production increase more by spending an additional $1 million on labor or $1.5 million on capital? Explain.

Concept and Writing Exercises

84. Describe in a sentence the geometric meaning of the partial derivative of $f(x, y)$ with respect to x.

85. Describe in a sentence the geometric meaning of the partial derivative of $f(x, y)$ with respect to y.

86. For $f(x, y) = x^2 + y^2 - xy + 3y$, determine values for x and y such that $f_x(x, y) = 0$ and $f_y(x, y) = 0$ simultaneously.

87. For $f(x, y) = x^2 + y^2 - 8x + 2y + 7$, determine values for x and y such that $f_x(x, y) = 0$ and $f_y(x, y) = 0$ simultaneously.

88. A function $f(x, y) = ax^2 + bxy + cy^2$ has partial derivative values $f_x(1, 2) = 4$, $f_y(1, 2) = -12$, and $f_{xy}(1, 2) = 0$. Solve a system of equations to determine a, b, and c.

89. A function $f(x, y) = ax^2 + bxy + cy^2$ has partial derivative values $f_{xx}(x, y) = 4$, $f_{yy}(x, y) = -8$, and $f_{xy}(x, y) = 3$. Solve a system of equations to determine a, b, and c.

Section Project

The program in Appendix B that performs multiple regression is needed to do the section project. The IronWorks Company has six plants in the U.S. Midwest that make fireplace accessories. Data for each plant give the dollars (in millions) spent on labor, the dollars (in millions) spent on capital, and the total quantity produced (in thousands) for 2009. The plants all operate at the same level of technology, so a production function can be determined.

Labor	Capital	Quantity
2.2	1.7	10.7
2.3	1.6	10.8
2.5	1.75	11.8
2.9	1.8	13.3
3	1.85	13.7
3.2	2	14.7

(a) Enter the data into your calculator and execute the program to determine a Cobb-Douglas production function. Round all values to the nearest hundredth.

(b) On average, IronWorks is currently investing 2.7 units of labor and 1.8 units of capital. Find the marginal productivity of labor and marginal productivity of capital at these levels of labor and capital investment and interpret each.

(c) At its current levels of capital and labor investment, would production increase more by spending, on average, an additional \$800,000 on labor or \$400,000 on capital?

8.4 Maxima and Minima

SECTION OBJECTIVES

1. Determine critical points.
2. Determine relative extrema.
3. Determine maximum revenue.

We are now ready for a brief analysis of extreme values for a function of two variables. If we can locate the high points (**relative maxima**) and the low points (**relative minima**) on a surface, then we can optimize the quantity represented by the surface. For example, we could determine a maximum profit or a minimum cost. This should sound familiar, because we optimized functions of one variable in Sections 5.2 and 5.4. There we learned how the **Second Derivative Test** aided us in locating the relative extrema. We shall see in this section that there is a Second Derivative Test for relative extrema of a function of two variables. Just as we did with a function of one variable, we must first discuss the concept of **critical points**.

Relative Extrema, Critical Points, and the Second Derivative Test

We begin our analysis by making an assumption. Given $z = f(x, y)$, we assume that all second-order partial derivatives exist for $f(x, y)$ in some circular region in the xy-plane. This assumption guarantees that we are dealing with what is known as a **smooth surface**. A smooth surface is one that has no edges (like a shoe box), sharp points (like the point of a nail), or breaks (like the San Andreas fault line). These surfaces are illustrated in **Figure 8.4.1**.

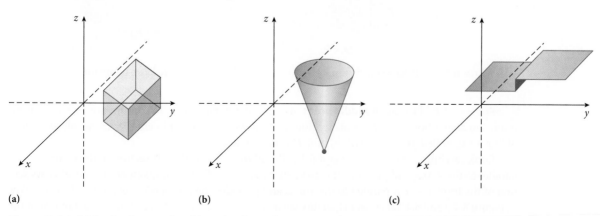

(a)   **(b)**   **(c)**

Figure 8.4.1 **(a)** Surface has an edge. **(b)** Surface has a sharp point. **(c)** Surface has a break.

Finally, we are not going to concern ourselves with boundary points. Thus, we are not going to concern ourselves with absolute extrema.

In Section 5.1 we defined relative extrema for a function of one variable as follows.

From Your Toolbox: Definition of Relative Extrema

For some open interval containing c
1. f has a **relative maximum** at c if $f(c) \geq f(x)$ for all x in the interval.
2. f has a **relative minimum** at c if $f(c) \leq f(x)$ for all x in the interval.

We extend this definition to functions of two variables to get a **definition of relative extrema in 3-space.**

DEFINITION

Definition of Relative Extrema in 3-Space

Let $f(x, y)$ be a function of two variables. The value $f(a, b)$ is
1. A **relative maximum** if $f(a, b) \geq f(x, y)$ for all points (x, y) in some circular region in the xy-plane around (a, b), where (a, b) is the center of the circular region.
2. A **relative minimum** if $f(a, b) \leq f(x, y)$ for all points (x, y) in some circular region in the xy-plane around (a, b), where (a, b) is the center of the circular region.

NOTE: The circular region described in these definitions should be small. If it is too large, it may contain some values of $f(x, y)$ that are larger or smaller than the relative extrema.

Figure 8.4.2 illustrates a relative maximum, and **Figure 8.4.3** illustrates a relative minimum.

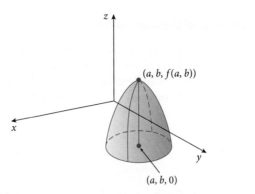

Figure 8.4.2 Relative maximum.

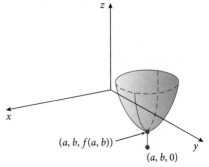

Figure 8.4.3 Relative minimum.

Figure 8.4.4 illustrates a **saddle point**, which is neither a relative maximum nor a relative minimum. No matter how small a circular region we draw with (a, b) as the center, there are always values of $f(x, y)$ such that $f(x, y) > f(a, b)$ and $f(x, y) < f(a, b)$.

Look carefully at Figures 8.4.2 and 8.4.3. Recall that a vertical cross section of the form $x = c$ is parallel to the y-axis and provides the geometric meaning of $f_y(x, y)$. Likewise, a vertical cross section of the form $y = c$ is parallel to the x-axis and provides the geometric meaning of $f_x(x, y)$. Now in Figure 8.4.2 and 8.4.3 it appears that **any** vertical cross section will have a horizontal tangent at the relative extreme, in particular, both $f_x(x, y) = 0$ and $f_y(x, y) = 0$.

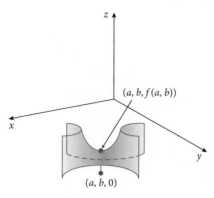

Figure 8.4.4 Saddle point.

Recall our definition of a **critical value** of a function of one variable given in Section 5.1.

From Your Toolbox: Definition of Critical Value

A **critical value** for f is an x-value in the domain of f for which $f'(x) = 0$ or $f'(x)$ is undefined.

The possibilities for a function of two variables are beyond the scope of this text. We shall restrict our discussion to **critical points** for which $f_x(x, y) = 0$ and $f_y(x, y) = 0$.

DEFINITION

Critical Point
The point (a, b) is a **critical point** for $f(x, y)$ if $f_x(a, b) = 0$ **and** $f_y(a, b) = 0$.

NOTE: It is called a critical *point* because the domain of $f(x, y)$ is a region in the *xy*-plane.

Example 1: Determining Critical Points

OBJECTIVE 1

Determine critical points.

Determine the critical points for $f(x, y) = x^2 + y^2 - 8x + 2y + 7$.

Perform the Mathematics

Critical points are determined by setting the partial derivatives $f_x(x, y) = 0$ and $f_y(x, y) = 0$ and solving the system of equations for x and y. The partial derivatives are

$$f_x(x, y) = 2x - 8 \quad \text{and} \quad f_y(x, y) = 2y + 2$$

Setting each equal to zero and solving yields

$$0 = 2x - 8 \quad \text{and} \quad 0 = 2y + 2$$
$$4 = x \quad \text{and} \quad -1 = y$$

So the only critical point is the point $(4, -1)$. Notice that this point is in the *xy*-plane. If we wanted to know the point on the surface, we would find the z-coordinate by evaluating $f(4, -1)$. This yields a z-coordinate of -10. So the point on the surface would be $(4, -1, -10)$. See **Figure 8.4.5**.

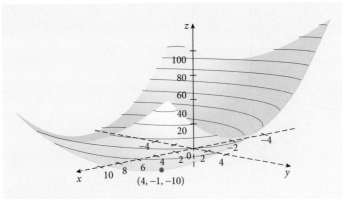

Figure 8.4.5

▶ *Try It Yourself*

Some related Exercises are 7 and 9.

Example 2: Determining Critical Points

Determine the critical points for $f(x, y) = x^2 + y^2 - xy + 3y$.

Perform the Mathematics

The partial derivatives are

$$f_x(x, y) = 2x - y \quad \text{and} \quad f_y(x, y) = 2y - x + 3$$

Setting each to zero gives

$$0 = 2x - y \quad \text{and} \quad 0 = 2y - x + 3$$

Here we have a system of two equations. We need to determine values of x and y that satisfy *both* equations. There are several methods that we could use to solve this system. We opt to use an algebraic technique known as *substitution*. A review of the substitution method for solving a system of equations is given in Appendix A. This method yields

$$0 = 2x - y \quad \text{and} \quad 0 = 2y - x + 3$$
$$y = 2x$$

We substitute $y = 2x$ into the second equation for y and get

$$0 = 2(2x) - x + 3$$
$$0 = 4x - x + 3$$
$$0 = 3x + 3$$
$$-1 = x$$

Now, we simply substitute $x = -1$ into either of the original equations and solve for y. We choose the first equation, and this gives

$$0 = 2x - y$$
$$0 = 2(-1) - y$$
$$-2 = y$$

So $x = -1$ and $y = -2$ are the solutions to both equations. Thus, our critical point is $(-1, -2)$.

▶ *Try It Yourself*
Some related Exercises are 13 and 15.

Now that we know how to find critical points, our next task is to determine whether we have a relative extremum at a given critical point. Just as not all critical values of a function of one variable give relative extrema, there is no guarantee that we will find a relative extremum at critical points of a function of two variables. For example, the saddle point in **Figure 8.4.4** is a critical point that does not produce a relative extremum. The following **Second Derivative Test** enables us to determine the shape of the surface in the vicinity of critical points, which in turn tells us if we have a relative extremum or not.

Second Derivative Test

Let $z = f(x, y)$ be a function of two variables such that $f_{xx}(x, y)$, $f_{yy}(x, y)$, and $f_{xy}(x, y)$ all exist. If $f_x(a, b) = 0$ and $f_y(a, b) = 0$, that is (a, b) is a critical point, then we define a number D to be

$$D = f_{xx}(a, b) \cdot f_{yy}(a, b) - [f_{xy}(a, b)]^2$$

The Second Derivative Test has the following form:
1. If $D > 0$ and $f_{xx}(a, b) < 0$, then $f(a, b)$ is a **relative maximum.**
2. If $D > 0$ and $f_{xx}(a, b) > 0$, then $f(a, b)$ is a **relative minimum**.
3. If $D < 0$, then $f(a, b)$ is a **saddle point**.
4. If $D = 0$, the test gives no information about $f(a, b)$.

Although this test appears complicated, it is fairly easy to apply, as Example 3 illustrates.

Example 3: Determining Relative Extrema

Determine any relative extrema for $f(x, y) = x^2 + y^2 - xy + 3y$.

OBJECTIVE 2

Determine relative extrema.

Perform the Mathematics

This is the function from Example 2. There we determined the partial derivatives to be

$$f_x(x, y) = 2x - y \quad \text{and} \quad f_y(x, y) = 2y - x + 3$$

And the only critical point occurred at $x = -1$ and $y = -2$, that is, at $(-1, -2)$. To use the Second Derivative Test, we need the following second-order partial derivatives

$$f_{xx}(x, y) = 2, \quad f_{yy}(x, y) = 2, \quad \text{and} \quad f_{xy}(x, y) = -1$$

We now evaluate each second-order partial derivative at our critical point $(-1, -2)$.

$$f_{xx}(-1, -2), \quad f_{yy}(-1, -2) = 2, \quad \text{and} \quad f_{xy}(-1, -2) = -1$$

Now we define D to be

$$D = f_{xx}(-1, -2) \cdot f_{yy}(-1, -2) - [f_{xy}(-1, -2)]^2$$
$$= 2 \cdot 2 - [-1]^2 = 4 - 1 = 3$$

Since $D > 0$ and $f_{xx}(-1, -2) > 0$, the Second Derivative Test asserts that there is a relative minimum at $(-1, -2)$. Specifically, on the *surface* there is a relative minimum at $(-1, -2, -3)$. ∎

We offer guidelines on **applying the Second Derivative Test** in the box to the left.
Before we address some applications, let's do Example 4 to ensure that the process is understood.

Applying the Second Derivative Test

1. Determine all critical points (a, b).
2. Determine $f_{xx}(x, y), f_{yy}(x, y),$ and $f_{xy}(x, y)$.
3. Evaluate the second-order partial derivatives at each critical point.
4. Determine D.
5. Apply the Second Derivative Test.

Example 4: Determining Relative Extrema

Determine the relative extrema for $f(x, y) = x^3 + 3x^2 - y^2 + 2y + 4$.

Perform the Mathematics

1. We begin by locating the critical points. First, the partial derivatives are

$$f_x(x, y) = 3x^2 + 6x \quad \text{and} \quad f_y(x, y) = -2y + 2$$

Setting each equal to zero and solving yields

$$0 = 3x^2 + 6x \quad \text{and} \quad 0 = -2y + 2$$
$$0 = 3x(x + 2) \qquad\qquad 1 = y$$
$$x = 0, x = -2$$

So here we have two critical points, when $x = 0$ and $y = 1$ as well as when $x = -2$ and $y = 1$. As ordered pairs, this gives the critical points as $(0, 1)$ and $(-2, 1)$.

2. Next we determine the second-order partial derivatives to be

$$f_{xx}(x, y) = 6x + 6, \quad f_{yy}(x, y) = -2, \quad \text{and} \quad f_{xy}(x, y) = 0$$

3. Evaluating the second-order partial derivatives at the critical point $(0, 1)$ gives

$$f_{xx}(0, 1) = 6, \quad f_{yy}(0, 1) = -2, \quad \text{and} \quad f_{xy}(0, 1) = 0$$

Evaluating the second-order partial derivatives at the critical point $(-2, 1)$ gives

$$f_{xx}(-2, 1) = -6, \quad f_{yy}(-2, 1) = -2, \quad \text{and} \quad f_{xy}(-2, 1) = 0$$

4. For the critical point $(0, 1)$, we determine D to be

$$D = f_{xx}(0, 1) \cdot f_{yy}(0, 1) - [f_{xy}(0, 1)]^2$$
$$= 6(-2) - [0]^2 = -12$$

For the critical point $(-2, 1)$, we determine D to be

$$D = f_{xx}(-2, 1) \cdot f_{yy}(-2, 1) - [f_{xy}(-2, 1)]^2$$
$$= (-6)(-2) - [0]^2 = 12$$

5. For the critical point $(0, 1)$, since $D < 0$, the Second Derivative Test asserts that we have a **saddle point** at $(0, 1)$. Specifically, we have a saddle point at $(0, 1, 5)$. For the critical point $(-2, 1)$, since $D > 0$ and $f_{xx}(-2, 1) < 0$, the Second Derivative Test tells us that we have a **relative maximum** at $(-2, 1)$. Specifically, we have a relative maximum at $(-2, 1, 9)$. **Figure 8.4.6** shows the surface with the saddle point and the relative maximum.

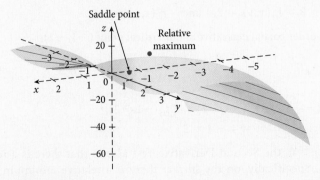

Figure 8.4.6

> **Try It Yourself**
> Some related Exercises are 25 and 27.

Applications

The final two examples of this section illustrate how, through applications, the Second Derivative Test is a very powerful method in locating relative extrema.

Example 5: Maximizing Revenue

OBJECTIVE 3

Determine maximum revenue.

The All Clear Company sells two types of car windshield wiper fluid: *regular*, which can be used at temperatures above 10°F, and a *no-freeze*, which can be used at temperatures above −30°F. Let x represent the number of gallons of no-freeze sold each year, in millions, and let y represent the number of gallons of regular sold each year, also in millions. Suppose that

$$p = 3.9 - 0.5x - 0.1y, \quad \text{the price in dollars of a gallon of no-freeze}$$
$$q = 3.9 - 0.1x - 0.8y, \quad \text{the price in dollars of a gallon of regular}$$

a. Determine the revenue function $R(x, y)$.

b. Determine x and y such that revenue is maximized and find the maximum revenue.

Perform the Mathematics

a. As always, revenue equals price times quantity, which gives

$$R(x, y) = x(3.9 - 0.5x - 0.1y) + y(3.9 - 0.1x - 0.8y)$$
$$= 3.9x - 0.5x^2 - 0.1xy + 3.9y - 0.1xy - 0.8y^2$$
$$= -0.5x^2 - 0.8y^2 + 3.9x + 3.9y - 0.2xy$$

b. Since we want to maximize revenue, we first locate any critical points of $R(x, y)$. Computing $R_x(x, y)$ and $R_y(x, y)$ yields

$$R_x(x, y) = -x + 3.9 - 0.2y \quad \text{and} \quad R_y(x, y) = -1.6y + 3.9 - 0.2x$$

Setting each equal to zero and solving gives

$$0 = -x + 3.9 - 0.2y \quad \text{and} \quad 0 = -1.6y + 3.9 - 0.2x$$

We opt to solve by using the substitution method, as we did in Example 2. From the first equation we have

$$x = 3.9 - 0.2y$$

Substituting this into the second equation yields

$$0 = -1.6y + 3.9 - 0.2(3.9 - 0.2y)$$
$$0 = -1.6y + 3.9 - 0.78 + 0.04y$$
$$0 = -1.56y + 3.12$$
$$y = 2$$

Substituting $y = 2$ into the first equation gives us

$$0 = -x + 3.9 - 0.2(2)$$
$$x = 3.5$$

So we have a critical point at $(3.5, 2)$. We now proceed to determine if the critical point yields a relative extremum. First we compute the second-order partial derivatives to be

$$R_{xx}(x, y) = -1, \quad R_{yy}(x, y) = -1.6, \quad \text{and} \quad R_{xy}(x, y) = -0.2$$

Evaluating our critical point at these second-order partial derivatives gives

$$R_{xx}(3.5, 2) = -1, \quad R_{yy}(3.5, 2) = -1.6, \quad \text{and} \quad R_{xy}(3.5, 2) = -0.2$$

Now, for the critical point $(3.5, 2)$ we define D as

$$D = R_{xx}(3.5, 2) \cdot R_{yy}(3.5, 2) - [R_{xy}(3.5, 2)]^2$$
$$= (-1)(-1.6) - [-0.2]^2 = 1.56$$

Since $D > 0$ for the critical point $(3.5, 2)$ and $R_{xx}(3.5, 2) < 0$, we have a relative maximum at $x = 3.5$ and $y = 2$. So All Clear needs to sell 3.5 million gallons of the no-freeze and 2 million gallons of the regular to maximize revenue for the year. The maximum revenue is

$$R(3.5, 2) = -0.5(3.5)^2 - 0.8(2)^2 + 3.9(3.5) + 3.9(2) - 0.2(3.5)(2)$$
$$= 10.725$$

So the maximum revenue is \$10.725 million dollars. ∎

Example 6: Maximizing Volume

North-South Airlines allows each passenger to carry up to three suitcases as long as the sum of the width, length, and height of each suitcase is less than or equal to 60 inches. Determine the dimensions of a suitcase of maximum volume that a passenger may carry with this restriction.

Understand the Situation

First we sketch a picture of a generic suitcase. See **Figure 8.4.7**. We represent length with an x, width by y, and height by z. The airline's restriction is that

$$x + y + z \le 60$$

The volume of our suitcase is given by

$$V = x \cdot y \cdot z$$

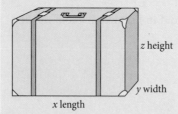

Figure 8.4.7

Notice that the volume is a function of three independent variables. We need to somehow rewrite this as a function of two variables so that we may use the techniques learned in this section. Since we believe that the largest volume for the suitcase results when $x + y + z = 60$, we solve this equation for z and get

$$z = 60 - x - y$$

We substitute this into our equation for volume and get a function of two variables

$$V(x, y) = xy(60 - x - y) = 60xy - x^2y - xy^2$$

Perform the Mathematics

To maximize volume, we proceed by determining the critical points and applying the Second Derivative Test. First, we determine $V_x(x, y)$ and $V_y(x, y)$.

$$V_x(x, y) = 60y - 2xy - y^2 \quad \text{and} \quad V_y(x, y) = 60x - x^2 - 2xy$$

Setting each equal to zero yields

$$0 = 60y - 2xy - y^2 \quad \text{and} \quad 0 = 60x - x^2 - 2xy$$
$$0 = y(60 - 2x - y) \quad \text{and} \quad 0 = x(60 - x - 2y)$$

Thus,

$$0 = y \quad \text{or} \quad 0 = 60 - 2x - y \quad \text{and} \quad 0 = x \quad \text{or} \quad 0 = 60 - x - 2y$$

We can immediately reject the situation when $x = 0$ and $y = 0$, since this would produce a suitcase with a volume of 0. We are then left with solving

$$0 = 60 - 2x - y \quad \text{and} \quad 0 = 60 - x - 2y$$

Using the substitution method, we solve the first equation for y and get

$$y = 60 - 2x$$

Substituting this into our second equation gives

$$0 = 60 - x - 2(60 - 2x)$$
$$0 = 60 - x - 120 + 4x$$
$$0 = -60 + 3x$$
$$x = 20$$

Substituting $x = 20$ into the first equation to determine y gives

$$y = 60 - 2(20) = 20$$

So our critical point is $x = 20$ and $y = 20$. Next we need the second-order partial derivatives

$$V_{xx}(x, y) = -2y, \quad V_{yy}(x, y) = -2x, \quad \text{and} \quad V_{xy}(x, y) = 60 - 2x - 2y$$

Evaluating each second-order partial derivative at the critical point $(20, 20)$ yields

$$V_{xx}(20, 20) = -40$$
$$V_{yy}(20, 20) = -40$$
$$V_{xy}(20, 20) = 60 - 2(20) - 2(20) = 60 - 40 - 40 = -20$$

For the critical point $(20, 20)$, we determine D to be

$$D = V_{xx}(20, 20) \cdot V_{yy}(20, 20) - [V_{xy}(20, 20)]^2$$
$$= (-40)(-40) - [-20]^2 = 1200$$

Since $D > 0$ for the critical point $(20, 20)$ and $V_{xx}(20, 20) < 0$, we have a relative maximum at $x = 20, y = 20$.

Interpret the Result

So the dimensions that would maximize volume would be 20 by 20 by 20 inches.

▷ *Try It Yourself*

Some related Exercises are 41 and 43.   ■

Summary

In this section we defined the **relative extrema** of a function of two variables.

Let $f(x, y)$ be a function of two variables. The value $f(a, b)$ is

- A **relative maximum** if $f(a, b) \geq f(x, y)$ for all points (x, y) in some circular region in the xy-plane around (a, b), where (a, b) is the center of the circular region.

- A **relative minimum** if $f(a, b) \leq f(x, y)$ for all points (x, y) in some circular region in the xy-plane around (a, b), where (a, b) is the center of the circular region.

We also defined critical points for a function of two variables as follows:

- The point (a, b) is a **critical point** for $f(x, y)$ if $f_x(a, b) = 0$ **and** $f_y(a, b) = 0$

We presented the **Second Derivative Test,** which allows us to determine if the critical point produces a relative maximum, a relative minimum, a saddle point, or none of these.

- If $f_x(a, b) = 0$ and $f_y(a, b) = 0$, that is, (a, b) is a critical point, then we define a number D to be

$$D = f_{xx}(a, b) \cdot f_{yy}(a, b) - [f_{xy}(a, b)]^2$$

The Second Derivative Test has the following form:

1. If $D > 0$ and $f_{xx}(a, b) < 0$, then $f(a, b)$ is a **relative maximum.**

2. If $D > 0$ and $f_{xx}(a, b) > 0$, then $f(a, b)$ is a **relative minimum**.

3. If $D < 0$, then $f(a, b)$ is a **saddle point**.

4. If $D = 0$, the test gives no information about $f(a, b)$.

Section 8.4 Exercises

Vocabulary Exercises

1. An ordered triple $(a, b, f(a, b))$ is a relative _____ if $f(a, b) \leq f(x, y)$ for all points in a circular region in the xy-plane around (a, b).

2. An ordered triple $(a, b, f(a, b))$ is a relative _____ if $f(a, b) \geq f(x, y)$ for all points in a circular region in the xy-plane around (a, b).

3. A smooth surface for the graph of a function of two independent variables is one whose _____ _____ partial derivatives exist in some circular region in the xy-plane.

4. A point that is neither a relative maximum nor a relative minimum is called a _____ point.

5. If (a, b) is a point such that $f_x(a, b) = 0$ and $f_y(a, b) = 0$, then (a, b) is a _____ point.

6. An optimization technique in which we compute $D = f_{xx}(a, b) \cdot f_{yy}(a, b) - [f_{xy}(a, b)]^2$ for a critical point (a, b) is called the _____ _____ Test.

Skill Exercises

For Exercises 7–16, determine the critical points for the given functions.

7. $f(x, y) = x^2 + y^2 - 4x + 6y - 4$

8. $f(x, y) = x^2 + y^2 - 6x - 4y + 3$

9. $f(x, y) = x^2 + y^2 + 2x - 4y - 3$

10. $f(x, y) = x^2 + y^2 - 8x + 2y - 2$

11. $f(x, y) = 2x^2 + 3y^2 - 8x + 6y + 5$

12. $f(x, y) = x^2 + y^2 + xy - 6y + 1$

13. $f(x, y) = x^2 + y^2 - xy - 3y + 5$

14. $f(x, y) = x^2 + y^2 - xy - 3x - 3y + 2$

15. $f(x, y) = x^2 - y^2 + xy + 2x - 9y - 5$

16. $f(x, y) = 2x^2 + 3y^2 + 2xy + 4x - 8y + 3$

For Exercises 17–36, use the Second Derivative Test to locate any relative extrema or saddle points.

17. $f(x, y) = x^2 + y^2 - 6x - 4y + 3$

18. $f(x, y) = x^2 + y^2 - 4x + 6y - 4$

19. $f(x, y) = x^2 + y^2 - 6x + 4y + 2$

20. $f(x, y) = 3x^2 - y^2 - 4x + 6y + 1$

21. $f(x, y) = -x^2 - 3y^2 - 4x - 4y - 1$

22. $f(x, y) = x^2 + y^2 - xy - 3y + 5$

23. $f(x, y) = x^2 + y^2 + xy - 6y + 1$

24. $f(x, y) = x^2 - y^2 + xy + 2x - 9y - 5$

25. $f(x, y) = x^2 + y^2 - xy - 3x - 3y + 2$

26. $f(x, y) = 3x^2 + y^2 + xy + 3y + 4$

27. $f(x, y) = \dfrac{4}{3}x^3 - y^2 - 4x^2 + 2y - 1$

28. $f(x, y) = 4x^3 - 6x^2 - 24x + 2y^2 - 4y + 6$

29. $f(x, y) = 2y^3 - 6y^2 - 18y + 2x^2 - 4x + 12$

30. $f(x, y) = x^3 + 3xy - y^3$

31. $f(x, y) = x^3 - 3xy + y^3$

32. $f(x, y) = 2x^3y - 2x + 16y - 5$

33. $f(x, y) = e^{-x^2 - y^2}$

34. $f(x, y) = e^{xy}$

35. $f(x, y) = 6xy + \dfrac{12}{x} - \dfrac{3}{y}$

36. $f(x, y) = 4xy + \dfrac{8}{x} - \dfrac{2}{y}$

Application Exercises

37. **Economics—Robotics Cost:** The annual cost for labor and specialized robotics equipment for an automobile manufacturer, in millions of dollars, is given by

$$C(x, y) = 5x^2 + 5xy + 7.5y^2 - 40x - 45y + 135$$

where x is the amount, in millions of dollars, spent each year on labor, and y is the amount, in millions of dollars, spent each year on robotics equipment.

(a) Determine how much should be spent on each, per year, to minimize cost.

(b) Determine the minimum cost.

38. **Economics—Jean Profits:** New Jeans, a specialty blue jeans manufacturer, produces two types of blue jeans each day, x pairs of straight leg and y pairs of wide leg. The daily profit function, in dollars, is given by

$$P(x, y) = 78x + 3xy - 3x^2 - y^2 - 2y$$

(a) How many straight-leg jeans and how many wide-leg jeans should be produced and sold each day to maximize profit?

(b) What is the maximum profit?

39. **Economics—Wok Profits:** The Wokon Company produces two types of woks each day, x number of regular sized and y number of jumbo sized. The daily profit from the sale of x number of regular-sized woks and y number of jumbo-sized woks is given by

$$P(x, y) = -0.3y^3 - 0.2x^2 + 6xy - 2$$

where $P(x, y)$ is measured in hundreds of dollars.

(a) How many regular-sized woks and how many jumbo-sized woks should be sold each day to maximize profit?

(b) What is the maximum profit?

© iStockphoto/Thinkstock

40. **Economics—Engine Profits:** The BigBang Engine Company produces two types of engines, x thousand two-stroke engines and y thousand four-stroke engines. Vice President Kathy determined the revenue and cost functions for a year to be (in millions of dollars)

$$R(x, y) = 42x + 39.5y$$
$$C(x, y) = 0.5x^2 + 0.8xy + y^2 + 5x + 4y$$

(a) Determine how many two-stroke engines and how many four-stroke engines should be produced each year to maximize profit.

(b) What is the maximum profit?

41. **Economics—Potato Chip Profits:** Crispy Chips produces two types of potato chips each year, x 18-ounce bags (in millions) of sour cream and chives and y 18-ounce bags (in millions) of barbecue. The revenue and cost functions for the year, in millions of dollars, are

$$R(x, y) = 1.5x + 2y$$
$$C(x, y) = x^2 - xy + 2y^2 + 2.5x - 9y + 2$$

(a) Determine the profit function $P(x, y)$.

(b) How many 18-ounce bags of each type of chip should be produced each year to maximize profit?

(c) What is the maximum profit?

42. **Economics—Calculator Profits:** The Palton Company produces two types of graphing calculators each month, x units of a 2D graphing calculator and y units of a 3D graphing calculator. The monthly revenue and cost functions, in dollars, are

$$R(x, y) = 75x + 101y + 0.05xy - 0.04x^2 - 0.04y^2$$
$$C(x, y) = 13.98x + 15.02y + 32{,}000$$

(a) Determine how many 2D graphing calculators and how many 3D graphing calculators should be sold each month to maximize profit.

(b) What is the maximum profit?

43. **Economics—Bicycle Revenue:** Sammy's Cycles manufactures 21-speed touring bicycles and 21-speed mountain bicycles. Let x represent the weekly demand for a 21-speed touring bicycle, and let y represent the weekly demand for a 21-speed mountain bicycle. The weekly price-demand equations are given by

$$p = 349 - 4x + y, \qquad \text{the price in dollars for a 21-speed touring bicycle}$$
$$q = 446 + 2x - 3y, \qquad \text{the price in dollars for a 21-speed mountain bicycle}$$

(a) Determine the revenue function $R(x, y)$.

(b) How many of each type of bicycle should be produced each week to maximize revenue?

(c) What is the maximum revenue?

44. **Economics—Bicycle Profits** (*continuation of Exercise 43*): Sammy's Cycles has a cost function given by

$$C(x, y) = 390 + 95x + 96y$$

(a) Determine the profit function $P(x, y)$.

(b) How many of each type of bicycle should be produced to maximize profit?

(c) What is the maximum profit?

45. **Economics—Leaf Blower Revenue:** The Leaf Chewer Company manufactures and sells leaf blowers and a special 8-foot blower attachment to clean gutters. Let x represent the number of leaf blowers produced and sold each month, and let y represent the number of special 8-foot attachments produced and sold each month. The monthly price-demand equations are given by

$$p = 228 - 8x - y, \qquad \text{the price in dollars for a leaf blower}$$
$$q = 31 - x - 0.15y, \qquad \text{the price in dollars for a 8-foot attachment}$$

(a) Determine the revenue function $R(x, y)$.

(b) Determine how many leaf blowers and how many 8-foot attachments should be produced and sold each month to maximize revenue.

46. Economics—Leaf Blower Revenue *(continuation of Exercise 45):* The monthly cost function for the Leaf Chewer Company is given by

$$C(x, y) = 1000 + 28x + 5y$$

(a) Determine the monthly profit function $P(x, y)$.

(b) Determine how many leaf blowers and how many 8-foot attachments should be sold each month to maximize profit.

(c) What is the maximum profit?

Concept and Writing Exercises

47. Let $f(x, y) = ax^2 + by^2 + c$. Show that for any real numbers a, b, and c that the critical value is $(0, 0)$.

48. You have been challenged to design a rectangular box with no top and two parallel partitions that holds 125 cubic inches. Determine the dimensions that will require the least amount of material to do this.

49. In the Second Derivative Test, we define $D = f_{xx}(a, b) \cdot f_{yy}(a, b) - [f_{xy}(a, b)]^2$, where (a, b) is a critical point. What condition for D is necessary for $f(x, y)$ to have a relative extremum?

50. For a function $f(x, y)$ and corresponding critical value (a, b), suppose that $f_{xy}(a, b) = 6$ and $f_{xx}(a, b) = 3$. What value of $f_{yy}(a, b)$ is necessary for f to have a saddle point at $(a, b, f(a, b))$?

 51. For Example 5, determine the critical point by setting each partial derivative to zero, solving for y, graphing, and using the `Intersect` command.

52. For Example 6, how is the height of the suitcases computed to be 20 inches?

Section Project

Federal Express states that the maximum length plus girth for its FedEx Overnight Freight and its FedEx 2Day Freight shipments is 300 inches. (*Source:* Federal Express.) See **Figure 8.4.8**.

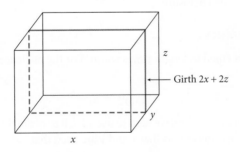

Girth $2x + 2z$

Figure 8.4.8

(a) Determine the dimensions of the largest-volume package that can be sent by FedEx Overnight Freight and FedEx 2Day Freight. Assume that the length is y.

(b) To avoid U.S. Domestic Freight Service charges, any package sent by Federal Express must have a length plus girth not exceeding 165 inches. Determine the dimensions of the largest-volume package that can be sent by Federal Express without incurring Domestic Freight Service charges. Assume that the length is y.

SECTION OBJECTIVES

1. Use the method of Lagrange.

2. Apply the method of Lagrange to a production function.

3. Apply the method of Lagrange to determine optimal increase in production.

8.5   Lagrange Multipliers

Many optimization problems that we encounter are actually **constrained** by some external circumstances. For example, North-South Airlines in Example 6 from Section 8.4 had a **constraint** that length plus width plus height of a suitcase cannot exceed 60 inches. We maximized the volume of a suitcase, $V = x \cdot y \cdot z$, subject to the constraint $x + y + z \le 60$. Other examples include maximizing an area of an enclosure given a finite amount of fence, or maximizing production given a budget constraint. Each of these problems may be solved by using the **Method of Lagrange Multipliers**.

Method of Lagrange Multipliers

The method that we are about to study is for solving maximum or minimum problems when dealing with restrictions on the independent variables. The method is named after the French mathematician Joseph Louis Lagrange (1736–1813), who discovered the method at the age of 19. We introduce the method through a Flashback.

Flashback: Beloved Beagle Revisited

In Section 5.4, we saw that one of the authors needs to build an enclosure for his beloved beagle. He has 200 feet of fence and wants to make a rectangular enclosure along a straight wall of his house. See **Figure 8.5.1**.

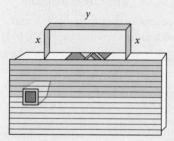

Figure 8.5.1

a. Determine a function of two independent variables for the area of the enclosure.

b. Determine any restrictions on x and y, assuming that we want to use all 200 feet of fence to maximize the area of the enclosure.

Perform the Mathematics

a. We know that area is equal to length times width. For the enclosure shown in Figure 8.5.1, this gives

$$A(x, y) = xy$$

b. We only have 200 feet of fence and we wish to use all of it to make this enclosure. From Figure 8.5.1 we determine that this means that x and y are such that

$$2x + y = 200$$

This equation represents the restrictions placed on the independent variables x and y. These restrictions placed on x and y are called **constraints**. ∎

In short, what we are asked to do in situations like the Flashback is to

Maximize $A(x, y) = xy$

Subject to $2x + y = 200$ or $2x + y - 200 = 0$

This is a specific example of the more **general form of maximum/minimum problems** of the following form.

DEFINITION

General Form of Maximum/Minimum Problems

Maximize/Minimize $\qquad z = f(x, y)$

Subject to $\qquad\qquad g(x, y) = 0$

In Section 5.4 we determined the dimensions that maximized the area of the enclosure in the Flashback by solving the *constraint equation* for y and substituted the result into the area equation. This gave us a function of one independent variable, which we solved using methods developed in Section 5.4. So why do we need a new method to solve this problem? There are several reasons.

1. Sometimes our constraint equation is complicated and cannot be solved for one variable.
2. The **Method of Lagrange Multipliers** generalizes to functions of several variables subject to one or more constraints.

Theorem of Lagrange

The relative extrema of the function $z = f(x, y)$ subject to the constraint equation $g(x, y) = 0$ occur among those points (x, y) for which there exists a value of λ (*lambda*),

$$F(x, y, \lambda) = f(x, y) + \lambda \cdot g(x, y)$$

where we have the following:

$$F_x(x, y, \lambda) = 0, \quad F_y(x, y, \lambda) = 0, \quad \text{and} \quad F_\lambda(x, y, \lambda) = 0$$

provided all partial derivatives exist.

NOTE: The λ (Greek letter *lambda*) in the definition of $F(x, y, \lambda)$ is called the *Lagrange Multiplier*. To organize our work when **using the method of Lagrange**, we use the following steps.

How to Use the Method of Lagrange

1. Write the situation in the form

   Maximize (or minimize) $\qquad z = f(x, y)$

   Subject to $\qquad\qquad\qquad g(x, y) = 0$

2. Define $F(x, y, \lambda) = f(x, y) + \lambda \cdot g(x, y)$
3. Determine the partial derivatives

   $$F_x(x, y, \lambda), \quad F_y(x, y, \lambda), \quad \text{and} \quad F_\lambda(x, y, \lambda)$$

4. Solve the system of equations

   $$F_x(x, y, \lambda) = 0, \quad F_y(x, y, \lambda) = 0, \quad F_\lambda(x, y, \lambda) = 0$$

5. The maximum (or minimum) of $f(x, y)$ is among the values in step 4. Simply evaluate $z = f(x, y)$ at each.

OBJECTIVE 1

Use the method of Lagrange.

Example 1: Utilizing the Method of Lagrange

Use the Method of Lagrange to determine the maximum area of an enclosure for the beloved beagle in the Flashback.

Perform the Mathematics

We will break down the Method of Lagrange into the five steps just given.

Step 1: Write the situation in the appropriate form.

$$\text{Maximize} \qquad A(x, y) = xy$$
$$\text{Subject to} \qquad 2x - y - 200 = 0$$

Step 2: Define function F using the Lagrange multiplier λ.

$$F(x, y, \lambda) = A(x, y) + \lambda \cdot g(x, y)$$
$$= xy + \lambda(2x - y - 200)$$

Step 3: Determine the partial derivatives.

$$F_x(x, y, \lambda) = y + \lambda(2 + 0 - 0) = y + 2\lambda$$
$$F_y(x, y, \lambda) = x + \lambda(0 + 1 - 0) = x + \lambda$$
$$F_\lambda(x, y, \lambda) = 0 + 1 \cdot (2x + y - 200) = 2x + y - 200$$

Step 4: Solve the system. The solutions to the system are critical points of F.

$$y + 2\lambda = 0$$
$$x + \lambda = 0$$
$$2x + y - 200 = 0$$

To solve this system, notice from the first two equations that we have

$$y = -2\lambda \quad \text{and} \quad x = -\lambda$$

If we substitute these values for x and y into the third equation and solve for λ, we have

$$2(-\lambda) + (-2\lambda) - 200 = 0$$
$$-4\lambda - 200 = 0$$
$$-4\lambda = 200$$
$$\lambda = -50$$

Substituting $\lambda = -50$ into the first two equations yields

$$y = -2\lambda \qquad \text{and} \quad x = -\lambda$$
$$y = -2(-50) \quad \text{and} \quad x = -(-50)$$
$$y = 100 \qquad \text{and} \quad x = 50$$

Step 5: Here it is important to realize exactly what the Theorem of Lagrange states. Quite simply, the Method of Lagrange finds only critical points. It does not tell whether the function is maximized, minimized, or neither at the critical point. In this problem, as in each problem that we do, we must know that the maximum (or minimum) does exist. It then follows that the maximum (or minimum) must occur at a critical point as determined by Lagrange. For this situation, there is certainly an optimal area enclosure for the beagle. Since there is only one possibility from Step 4, it follows that the dimensions are $x = 50$ feet by $y = 100$ feet, and the maximum enclosure is $A(50, 100) = 50(100) = 5000$ square feet. ∎

Example 2: Utilizing the Method of Lagrange

$$\text{Minimize} \quad f(x, y) = x^2 + 3y^2$$
$$\text{Subject to} \quad x + y = 2$$

Perform the Mathematics

Step 1: Write in an appropriate form by writing the constraint equation as $g(x, y) = 0$.

$$\text{Minimize} \quad f(x, y) = x^2 + 3y^2$$
$$\text{Subject to} \quad g(x, y) = x + y - 2 = 0$$

Step 2: Define function F using the Lagrange multiplier λ.

$$F(x, y, \lambda) = f(x, y) + \lambda \cdot g(x, y)$$
$$= x^2 + 3y^2 + \lambda(x + y - 2)$$

Step 3: Determine the partial derivatives

$$F_x(x, y, \lambda) = 2x + \lambda$$
$$F_y(x, y, \lambda) = 6y + \lambda$$
$$F_\lambda(x, y, \lambda) = x + y - 2$$

Step 4: Solve the system

$$2x + \lambda = 0$$
$$6y + \lambda = 0$$
$$x + y - 2 = 0$$

From the first two equations, we have

$$2x = -\lambda \quad \text{and} \quad 6y = -\lambda$$

Setting these equal to each other yields

$$2x = 6y$$
$$y = \frac{1}{3}x$$

Since we have y in terms of x, we can return to our constraint equation, $x + y - 2 = 0$, and substitute for y, which yields

$$x + \frac{1}{3}x - 2 = 0$$
$$\frac{4}{3}x = 2$$
$$x = \frac{3}{2}$$

From $y = \frac{1}{3}x$, we also have

$$y = \frac{1}{3}\left(\frac{3}{2}\right) = \frac{1}{2}$$

Step 5: There is only one solution from Step 4, $x = \frac{3}{2}$ and $y = \frac{1}{2}$. So the function is minimized at $f(\frac{3}{2}, \frac{1}{2})$. This gives

$$f\left(\frac{3}{2}, \frac{1}{2}\right) = \left(\frac{3}{2}\right)^2 + 3\left(\frac{1}{2}\right)^2 = 3$$

The minimum value of $f(x, y) = x^2 + 3y^2$ subject to $x + y = 2$ is 3. ∎

▶ *Try It Yourself*

Some related Exercises are 13 and 15.

Geometric Analysis

Now let's look at what is happening geometrically. If we add a *constraint*, $g(x, y) = 0$, in the *xy*-plane, it gives the sketch in **Figure 8.5.2**. Now we look at only that part of the surface that is directly above our constraint equation. This yields our *constrained maximum*, (x_0, y_0, z_0), which we notice in Figure 8.5.2 is different from our *unconstrained (or free) maximum*, (x, y, z).

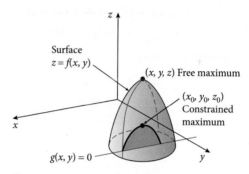

Figure 8.5.2

To gain another perspective, let's return to our two-dimensional setting and look at the level curves that we discussed in Section 8.2. Recall that a level curve is the result of slicing the surface with a horizontal plane.

Example 3: Analyzing Level Curves and a Constraint Equation

In Example 1 we

$$\text{Maximized} \quad A(x, y) = xy$$
$$\text{Subject to} \quad g(x, y) = 2x + y - 200 = 0$$

Sketch level curves for $A(x, y)$ for c-values of 3000, 5000, 7000 and graph the constraint equation in the *xy*-plane.

Perform the Mathematics

Recall that to sketch level curves for $A(x, y)$, we graph equations of the form

$$c = xy \quad \text{or} \quad y = \frac{c}{x}$$

So for $c = 3000$, $c = 5000$, and $c = 7000$, we need to graph

$$y = \frac{3000}{x}, \quad y = \frac{5000}{x}, \quad \text{and} \quad y = \frac{7000}{x}$$

We also need to graph the constraint equation

$$g(x, y) = 2x + y - 200$$

When $g(x, y) = 0$, the graph is a line in the *xy*-plane. Thus we need to graph

$$2x + y - 200 = 0 \quad \text{or} \quad y = 200 - 2x$$

In **Figure 8.5.3** we show the graphs of $y = \frac{3000}{x}$, $y = \frac{5000}{x}$, $y = \frac{7000}{x}$, and $y = 200 - 2x$.

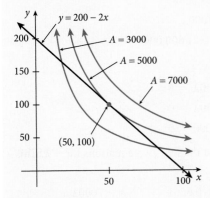

Figure 8.5.3

From Figure 8.5.3, we can make several observations.

1. The constraint line shows all combinations of x and y that we could use to exhaust our 200 feet of fence.
2. The level curve corresponding to $A = 7000$ is not possible since it does not intersect the graph of our constraint equation.
3. The maximum value for A on the constraint equation occurs where the constraint is tangent to the level curve $A = 5000$.

Example 3 demonstrates that the critical points of our function $F(x, y, \lambda)$ in the Lagrange multipliers method occur at points where a level curve of $f(x, y)$ is tangent to the constraint curve $g(x, y)$.

Example 4: Maximizing a Production Function

OBJECTIVE 2

Apply the method of Lagrange to a production function.

Recall that the O'Neill Corporation has a production function of

$$Q = f(x, y) = 1.64x^{0.6}y^{0.4}$$

where x is labor hours, y is capital, and $f(x, y)$ is thousands of gallons produced. Each unit of labor is \$15,000 and each unit of capital is \$7000, and the total expense for both (a budget constraint) is limited to \$1.8 million. (These figure are for each of the O'Neill Corporation's 10 plants.) Determine the number of units of labor and capital needed to maximize production and give the maximum production.

Perform the Mathematics

This appears to be a maximization problem requiring the Method of Lagrange multipliers.

Step 1: We want to maximize $f(x, y) = 1.64x^{0.6}y^{0.4}$. Since each unit of labor is \$15,000 and each unit of capital is \$7000, we have the constraint equation $15,000x + 7000y = 1,800,000$. So our problem has the form

Maximize $f(x, y) = 1.64x^{0.6}y^{0.4}$

Subject to $g(x, y) = 15,000x + 7000y - 1,800,000 = 0$

Step 2: Define $F(x, y, \lambda) = f(x, y) + \lambda \cdot g(x, y)$ using Lagrange multiplier λ.

$$F(x, y, \lambda) = 1.64x^{0.6}y^{0.4} + \lambda(15,000x + 7000y - 1,800,000)$$

Step 3: Determine the partial derivatives

$$F_x(x, y, \lambda) = 0.984x^{-0.4}y^{0.4} + 15{,}000\lambda$$

$$F_y(x, y, \lambda) = 0.656x^{0.6}y^{-0.6} + 7000\lambda$$

$$F_\lambda(x, y, \lambda) = 15{,}000x + 7000y - 1{,}800{,}000$$

Step 4: Solve the system

$$0.984x^{-0.4}y^{0.4} + 15{,}000\lambda = 0$$

$$0.656x^{0.6}y^{-0.6} + 7000\lambda = 0$$

$$15{,}000x + 7000y - 1{,}800{,}000 = 0$$

To solve this nasty-looking system, let's solve the first two equations for $-\lambda$. This gives

$$0.984x^{-0.4}y^{0.4} = -15{,}000\lambda \quad \text{and} \quad 0.656x^{0.6}y^{-0.6} = -7000\lambda$$

$$\frac{0.984x^{-0.4}y^{0.4}}{15{,}000} = -\lambda \quad \text{and} \quad \frac{0.656x^{0.6}y^{-0.6}}{7000} = -\lambda$$

Now we can set these equal to each other to give

$$\frac{0.984x^{-0.4}y^{0.4}}{15{,}000} = \frac{0.656x^{0.6}y^{-0.6}}{7000}$$

Now for the "trick." Multiply both sides by $x^{0.4}y^{0.6}$.

$$x^{0.4}y^{0.6}\left(\frac{0.984x^{-0.4}y^{0.4}}{15{,}000}\right) = x^{0.4}y^{0.6}\left(\frac{0.656x^{0.6}y^{-0.6}}{7000}\right)$$

$$\frac{0.984x^{-0.4+0.4}y^{0.4+0.6}}{15{,}000} = \frac{0.656x^{0.6+0.4}y^{-0.6+0.6}}{7000}$$

$$\frac{0.984y}{15{,}000} = \frac{0.656x}{7000}$$

$$y = \frac{15{,}000}{0.984}\left(\frac{0.656x}{7000}\right) \approx 1.43x$$

Substituting this back into the budget constraint equation for y gives

$$15{,}000x + 7000(1.43x) - 1{,}800{,}000 = 0$$

$$25{,}010x = 1{,}800{,}000$$

$$x \approx 71.97$$

From $y \approx 1.43x$, we also determine that

$$y \approx 1.43(71.97) \approx 102.92$$

Step 5: Since there is only one solution from Step 4, $x \approx 71.97$ and $y \approx 102.92$, we conclude that $x \approx 71.97$ units of labor and $y \approx 102.92$ units of capital maximize production. The maximum production is approximately

$$f(71.97, 102.92) = 1.64(71.97)^{0.6}(102.92)^{0.4}$$

$$\approx 136.19 \text{ thousand gallons} \qquad \blacksquare$$

▶ **Try It Yourself**

Some related Exercises are 31 and 32.

Marginal Productivity of Money

In Example 4 we did not determine the value of $-\lambda$. In economics, $-\lambda$ has a practical meaning. To determine $-\lambda$ in Example 4, we could use either

$$\frac{0.984x^{-0.4}y^{0.4}}{15,000} = -\lambda \quad \text{or} \quad \frac{0.656x^{0.6}y^{-0.6}}{7000} = -\lambda$$

from step 4 and substitute in our values for x and y to get

$$-\lambda \approx 7.57 \times 10^{-5} = 0.0000757$$

The first thing we observe is that $-\lambda$ is positive. Now let's investigate the meaning of $-\lambda$ by looking at what happens when the budget for the O'Neill Corporation increases from $1.8 million to $1.9 million. How does this change x, y, and Q? Example 5 shows us.

Example 5: Maximizing a Production Function

If the budget for the O'Neill Corporation has been increased from $1.8 million to $1.9 million, what combinations of x and y maximize Q?

Perform the Mathematics

Here, we are to

$$\text{Maximize} \quad Q = f(x, y) = 1.64x^{0.6}y^{0.4}$$
$$\text{Subject to} \quad g(x, y) = 15,000x + 7000y - 1,900,000 = 0$$

The process outlined in Steps 2 through 4 of Example 4 is the same here, except 1,800,000 is replaced with 1,900,000. We save the details for you to work out. The solution is $x \approx 75.97$ and $y \approx 108.64$. The maximum production with $x \approx 75.97$ units of labor and $y \approx 108.64$ units of capital is about $Q \approx 143.76$ thousand gallons. ∎

Notice that our value for Q in Example 5 is approximately equal to our value for Q from Example 4 plus $10^5 \cdot (-\lambda)$. In other words,

$$Q \text{ from Example 5} \approx Q \text{ from Example 4} + 10^5 \cdot (-\lambda)$$
$$\approx 136.19 + 10^5(7.57 \times 10^{-5}) \approx 143.76$$

Thus, in a production function setting, it appears that:

- $-\lambda$ gives the extra production achieved by increasing the budget by *one* unit. Since the budget was increased by $100,000 = 10^5$ in Example 5, we multiplied $-\lambda$ by 10^5.
- The value of $-\lambda$ approximates the increase in the optimal value of Q when the budget is increased by *one* unit.

In the language of calculus, this means that $-\lambda$ **gives the rate of change of the optimal value of Q as the budget increases** or in other words, the **marginal productivity of money**.

DEFINITION

Marginal Productivity of Money

Let $Q = f(x, y)$ be a production function and $g(x, y) = 0$ be a budget constraint. For

$$F(x, y, \lambda) = f(x, y) + \lambda \cdot g(x, y)$$

$-\lambda$ is the **marginal productivity of money** at (x, y), which is the extra production achieved by increasing the budget one unit.

NOTE: $-\lambda$ in this setting is always a positive number.

Now from our work in Example 5 and the discussion immediately following Example 5, we can observe where the following **formula for optimal increase in production** arises.

DEFINITION

Formula for Optimal Increase in Production

Assume that $-\lambda$ is the marginal productivity of money and P additional dollars is available to the budget. The **optimal increase in production** is given by $-\lambda P$.

Example 6 ties together elements from Examples 4 and 5 and the marginal productivity of money concept.

OBJECTIVE 3

Apply the method of Lagrange to determine optimal increase in production.

Example 6: Maximizing Production and Determining Optimal Increase

RollyOn Tires has a Cobb-Douglas production function of

$$Q = f(x, y) = 250x^{0.25}y^{0.75}$$

where x represents units of labor and y represents units of capital. Suppose that each unit of labor is \$100 and each unit of capital is \$250. Assume the budget is limited to \$100,000.

a.   Determine x and y that maximize production and determine the maximum production.

b.   Determine the marginal productivity of money for the division of labor and capital determined in part (a) and interpret. Find the optimal increase in production if an additional \$15,000 is budgeted.

Perform the Mathematics

a.   **Step 1:**   We want to

$$\text{Maximize} \quad f(x, y) = 250x^{0.25}y^{0.75}$$
$$\text{Subject to} \quad g(x, y) = 100x + 250y - 100{,}000 = 0$$

Step 2: Define $F(x, y, \lambda) = f(x, y) + \lambda g(x, y)$ using the Lagrange multiplier λ.

$$F(x, y, \lambda) = 250x^{0.25}y^{0.75} + \lambda(100x + 250y - 100{,}000)$$

Step 3: Determine the partial derivatives.

$$F_x(x, y, \lambda) = 62.5x^{-0.75}y^{0.75} + 100\lambda$$
$$F_y(x, y, \lambda) = 187.5x^{0.25}y^{-0.25} + 250\lambda$$
$$F_\lambda(x, y, \lambda) = 100x + 250y - 100{,}000$$

Step 4: As we did in Example 4, we solve the first two equations for $-\lambda$.

$$62.5x^{-0.75}y^{0.75} = -100\lambda \quad \text{and} \quad 187.5x^{0.25}y^{-0.25} = -250\lambda$$

$$\frac{62.5x^{-0.75}y^{0.75}}{100} = -\lambda \quad \text{and} \quad \frac{187.5x^{0.25}y^{-0.25}}{250} = -\lambda$$

Setting these equal to each other and multiplying both sides of the equation by $x^{0.75}y^{0.25}$ gives

$$x^{0.75}y^{0.25}\left(\frac{62.5x^{-0.75}y^{0.75}}{100}\right) = x^{0.75}y^{0.25}\left(\frac{187.5x^{0.25}y^{-0.25}}{250}\right)$$

$$\frac{62.5y}{100} = \frac{187.5x}{250}$$

$$y = \frac{100}{62.5}\left(\frac{187.5x}{250}\right) = 1.2x$$

Substituting this into the budget constraint equation for y gives

$$100x + 250(1.2x) - 100,000 = 0$$
$$400x = 100,000$$
$$x = 250$$

From $y = 1.2x$, we also determine that $y = 300$.

Step 5: So $x = 250$ units of labor and $y = 300$ units of capital maximize production at $f(250, 300) = 71,658.2$ units. We round this to 71,658 units.

b. The marginal productivity of money at $x = 250$ and $y = 300$ is

$$-\lambda = \frac{62.5x^{-0.75}y^{0.75}}{100} = \frac{62.5(250)^{-0.75}(300)^{0.75}}{100} \approx 0.717$$

Thus, for each additional dollar available to the budget, production increases by about 0.717 units. The optimal increase in production is $-\lambda P$. Since $P = \$15,000$, $-\lambda P \approx 0.717(15,000) = 10,755$. Hence, we now have

$$\begin{array}{l} 71,658 \text{ (maximum production with a budget of \$100,000)} \\ \underline{+10,755 \text{ (additional production from extra \$15,000 added to the budget)}} \\ 82,413 \end{array}$$

This 82,413 is the optimal production possible with $115,000. ∎

▶ *Try It Yourself*

Some related Exercises are 33 and 35.

Summary

In this section we have seen how to maximize or minimize problems that arise when we have **constraints** on our independent variables. The **Method of Lagrange Multipliers** was used to solve these problems.

How to Use the Method of Lagrange

1. Write the situation in the form

   Maximize (or minimize) $z = f(x, y)$

   Subject to $g(x, y) = 0$

2. Define $F(x, y, \lambda) = f(x, y) + \lambda \cdot g(x, y)$

3. Determine the partial derivatives

   $$F_x(x, y, \lambda), \quad F_y(x, y, \lambda), \quad \text{and} \quad F_\lambda(x, y, \lambda)$$

4. Solve the system of equations

   $$F_x(x, y, \lambda) = 0, \quad F_y(x, y, \lambda) = 0, \quad F_\lambda(x, y, \lambda) = 0$$

5. The maximum (or minimum) of $f(x, y)$ is among the values in step 4. Simply evaluate $z = f(x, y)$ at each.

The section concluded with problems from the world of economics, specifically marginal productivity of money.

- $-\lambda$ is the **marginal productivity of money** at (x, y), which is the extra production achieved by increasing the budget one unit.

- Assume that $-\lambda$ is the marginal productivity of money and P additional dollars is available to the budget. The **optimal increase in production** is given by $-\lambda P$.

Section 8.5 Exercises

Vocabulary Exercises

1. The method used to solve relative extrema problem when constraint equations are involved is that of _____ multipliers.

2. The Greek letter λ is pronounced _____.

3. When we cite a constraint equation, we often refer to the equation by using the phrase "_____ to."

4. After determining the partial derivative of the Lagrangian function $F(x, y, \lambda)$, we form a _____ of equations.

5. For a Cobb-Douglas production function $Q = f(x, y)$, the value of $-\lambda$ is the _____ _____ of money at labor and capital level (x, y).

6. For a Cobb-Douglas production function $Q = f(x, y)$, the value $-\lambda P$ is the _____ _____ in production for additional dollars P that is available to the budget.

Skill Exercises

For Exercises 7–24, use the Method of Lagrange Multipliers to solve each problem.

7. Maximize $f(x, y) = -3xy$ subject to $x + y = 1$.

8. Maximize $f(x, y) = 2xy$ subject to $2x + y = 20$.

9. Maximize $f(x, y) = xy$ subject to $2x + y = 12$.

10. Maximize $f(x, y) = 2xy - 5$ subject to $x + y = 12$.

11. Minimize $f(x, y) = xy$ subject to $x - y = -8$.

12. Maximize $f(x, y) = 3xy + x$ subject to $x + y = 1$.

13. Minimize $f(x, y) = x^2 + 2y^2$ subject to $2x + y = 8$

14. Minimize $f(x, y) = x^2 + y^2 - 7$ subject to $x + 2y = 10$.

15. Maximize $f(x, y) = 2xy - 4x$ subject to $x + y = 12$.

16. Maximize $f(x, y) = 25 - x^2 - y^2$ subject to $2x + y = 10$.

17. Maximize $f(x, y) = 1.64x^{0.6}y^{0.4}$ subject to $12{,}000x + 5000y = 1{,}100{,}000$.

18. Maximize $f(x, y) = 125x^{0.75}y^{0.25}$ subject to $100x + 250y = 100{,}000$.

19. Minimize $f(x, y) = x^2 + y^2 - xy - 4$ subject to $x + y - 6 = 0$.

20. Maximize $f(x, y) = 16 - x^2 - y^2$ subject to $x + 2y - 6 = 0$.

21. Maximize $f(x, y) = xy$ subject to $x^2 + y^2 = 8$.

22. Minimize $f(x, y) = xy$ subject to $x^2 + y^2 = 8$.

23. Maximize $f(x, y) = e^{xy}$ subject to $x^2 + y^2 = 8$.

24. Minimize $f(x, y) = e^{x^2 + y^2}$ subject to $x + y = 5$.

Application Exercises

25. **Macroeconomics—Minimizing Cost:** On-the-Go Music produces two types of personal music players each day, x units of its regular model and y units of its joggers' model. The cost, in dollars, is given by

$$C(x, y) = 0.1x^2 + 0.2y^2$$

Because of limitations on the supply of parts, it is necessary that

$$x + y = 180$$

   (a) Determine how many of each type of personal music player should be produced to minimize cost.

   (b) Determine the minimum cost.

26. **Macroeconomics—Minimizing Cost:** DeeDee's produces two types of diapers each week, x packages of its regular absorbent and y packages of its extra absorbent. The cost, in dollars, is given by

$$C(x, y) = 0.1x^2 + 0.5y^2$$

Because of limited materials, it is necessary that

$$x + y = 1200$$

   (a) Determine how many of each type of diaper should be produced to minimize cost.

   (b) Determine the minimum cost.

27. **Macroeconomics—Minimizing Cost:** SalPal produces two types of personal security devices each day, x units of its flashlight and pepper spray device and y units of its flashlight and siren device. The cost, in dollars, is given by

$$C(x, y) = 0.1x^2 + 0.2y^2$$

Because of shipping restrictions, it is necessary that

$$x + y = 120$$

   (a) Determine how many of each type of security device should be produced to minimize cost.

   (b) Determine the minimum cost.

28. **Macroeconomics—Minimizing Cost:** Rikki's Bicycle Company makes two models of bicycles, 10-speeds and 21-speeds. Each week, the cost to make x 10-speeds and y 21-speeds is given by

$$C(x, y) = 20{,}000 + 100x + 140y - xy$$

where $C(x, y)$ is in dollars. Determine the number of 10-speeds and the number of 21-speeds that should be produced each week to minimize costs if the total number produced each week is 200 bicycles.

29. **Macroeconomics—Minimizing Production Cost:** The Riblet Engine Company has installed a newer, more efficient production line, which is alongside an older production line that is still in use. Let x represent the number of units produced on the older line, and let y represent the number of units produced on the new line. The cost of a production run is given by

$$C(x, y) = 2x^2 - xy + y^2 + 250$$

where $C(x, y)$ is in dollars. A production run has been scheduled for which both lines will be used to complete an order of 104 units.

   (a) Determine the number of units produced by each line in order to minimize the cost.

   (b) Determine the minimum production cost.

30. **Macroeconomics—Maximizing Production:** Sumption Sonic Incorporated has a production function for the manufacture of solar cells of

$$Q = f(x, y) = 2.5x^{0.7}y^{0.3}$$

where Q is the number of solar cells produced per week, x is the number of units of labor each week, and y is the number of units of capital. A unit of labor costs \$400, a unit of capital costs \$500, and the total budget for both is \$180,000.

(a) Determine the number of units of labor and the number of units of capital required to maximize production.

(b) Determine the maximum production.

31. **Macroeconomics—Maximizing Production:** The Linger Golf Cart Corporation has a production function for the manufacture of golf carts given by

$$Q = f(x, y) = 13x^{0.75}y^{0.3}$$

where Q is the number of golf carts produced per week, x is the number of labor hours per week, and y is the weekly capital investment. A unit of labor is \$15 and a unit of capital is \$50, and the total budget for both is limited to \$18,000.

(a) Determine the number of units of labor and the number of units of capital required to maximize production.

(b) Determine the maximum production.

32. **Macroeconomics—Maximizing Production:** The $(RB)^2$ Company has a production function of

$$Q = f(x, y) = 1610x^{0.8}y^{0.25}$$

where Q is the quantity of handheld calculators produced per year, x is the number of labor hours in thousands, and y is the capital equipment. A unit of labor is \$15,000 and a unit of capital is \$4000, and the total budget for both is limited to \$1.5 million.

(a) Determine the number of units of labor and the number of units of capital required to maximize production.

(b) Determine the maximum production.

33. **Macroeconomics—Optimal Production:** The Arnold Engine Company has a production function of

$$Q = f(x, y) = x^{0.6}y^{0.45}$$

where x is the number of units of labor, y is the number of units of capital, and Q is the total quantity produced. Each unit of labor is \$16,000 and each unit of capital is \$4000, and the total budget for both is limited to \$1.7 million.

(a) Determine the number of units of labor and capital needed to maximize production.

(b) Determine the marginal productivity of money for the labor and capital found in part (a) and interpret.

(c) Determine the optimal increase in production if an additional \$200,000 is budgeted.

34. **Macroeconomics—Optimal Production:** Hogrefe's Incorporated has a production function given by

$$Q = f(x, y) = x^{0.7}y^{0.3}$$

where x is the number of units of labor, y is the number of units of capital, and Q is the total quantity produced. Each unit of labor is \$1200, each unit of capital is \$800, and the total budget for both is limited to \$480,000.

(a) Determine the number of units of labor and capital needed to maximize production.

(b) Determine the marginal productivity of money for the labor and capital found in part (a) and interpret.

(c) Determine the optimal increase in production if an additional \$20,000 is budgeted.

35. **Macroeconomics—Optimal Production:** Hoffman Laboratories has a production function given by

$$Q = f(x, y) = 5.6x^{0.65}y^{0.41}$$

where Q is the quantity of pain-killing tablets produced (in thousands) per week, x is the weekly units of labor, and y is the weekly units of capital. Each unit of labor is $2000, each unit of capital is $1500, and the budget for both is limited to $800,000.

(a) Determine the number of units of labor and capital needed to maximize production.

(b) Determine the marginal productivity of money for the labor and capital found in part (a) and interpret.

(c) Determine the optimal increase in production if an additional $50,000 is budgeted.

36. **Macroeconomics—Optimal Production:** Blair's Inc. has a production function given by

$$Q = f(x, y) = 25x^{0.8}y^{0.2}$$

where Q is the total quantity produced, x is the number of units of labor used, and y is the number of units of capital used. Each unit of labor costs $30, each unit of capital costs $60, and the budget for both is $300,000.

(a) Determine the number of units of labor and capital needed to maximize production.

(b) Determine the marginal productivity of money for the labor and capital found in part (a) and interpret.

(c) Determine the optimal increase in production if an additional $40,000 is budgeted.

37. **Macroeconomics—Optimal Production:** New River Outfitters has a production function given by

$$Q = f(x, y) = 100x^{0.6}y^{0.4}$$

where x is the number of units of labor, y is the number of units of capital, and Q is the quantity produced. Each unit of labor costs $100, each unit of capital is $150, and the budget for both is $60,000.

(a) Determine the number of units of labor and capital needed to maximize production.

(b) Determine the marginal productivity of money for the labor and capital found in part (a) and interpret.

(c) Determine the optimal increase in production if an additional $10,000 is budgeted.

38. **Macroeconomics—Optimal Production:** The Bowen Company has a production function given by

$$Q = f(x, y) = 25x^{0.65}y^{0.35}$$

where x is the number of units of labor, y is the number of units of capital, and Q is the quantity produced. Each unit of labor costs $200, each unit of capital costs $250, and the budget for both is $130,000.

(a) Determine the number of units of labor and capital needed to maximize production.

(b) Determine the marginal productivity of money for the labor and capital found in part (a) and interpret.

(c) Determine the optimal increase in production if an additional $20,000 is budgeted.

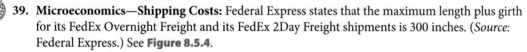

39. **Microeconomics—Shipping Costs:** Federal Express states that the maximum length plus girth for its FedEx Overnight Freight and its FedEx 2Day Freight shipments is 300 inches. (*Source: Federal Express.*) See **Figure 8.5.4**.

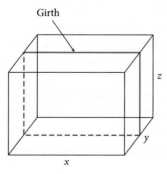

Figure 8.5.4

Use the Method of Lagrange Multipliers to determine the dimensions of the largest-volume package that can be sent by FedEx Overnight Freight and FedEx 2Day Freight. Assume that the length is y.

 40. **Microeconomics—Shipping Costs:** To avoid U.S. Domestic Freight Services charges, any packages sent by Federal Express must have the length plus girth not exceed 165 inches. (*Source:* Federal Express.) See the preceding figure. Use the method of Lagrange multipliers to determine the dimensions of the largest-volume package that can be sent by Federal Express without incurring Domestic Freight Service charges. Assume that the length is y.

Concept and Writing Exercises

41. When using the Method of Lagrange Multipliers, we refer to the idea of a constraining equation. Explain the meaning of this constraint equation using a complete sentence.

42. Research Joseph Louis Lagrange to find in which country he was born.

43. A company wishes to enclose a rectangular parking lot using an existing building as one side of the boundary and adding fencing for the other boundaries. If 625 feet of fencing is available, determine the dimensions of the largest parking lot that can be enclosed.

44. Determine the dimensions of a rectangular garden with an area of 5000 square feet that minimizes the cost of fencing if one side costs three times as much as the other three sides.

45. Explain a situation when we would not use the Method of Lagrange Multipliers.

 ### Section Project

This section project synthesizes information studied in Chapter 8 and uses the following data. Also, the program in Appendix B that performs multiple regression is needed to do the section project.

The Gantzler Corporation has six plants around the United States that produce plastic tubing. In a recent year the following data for each plant gave the number of labor hours (in thousands), capital (in millions), and quantity produced.

Labor	Capital	Quantity
38	3.1	3440
39	3.2	3533
41	3.7	3800
42	3.8	3894
42	3.81	3896
43	3.9	3987

(a) The plants have the same technology, so a production function can be determined. Enter the data into your calculator and execute the program to determine a Cobb-Douglas production function to model the data. Use x for labor and y for capital, and round all values to the nearest hundredth.

(b) Compute $f_x(x, y)$ and $f_y(x, y)$.

(c) If, on average, the Gantzler Corporation is now using 41 units of labor and 3.6 units of capital, determine the marginal productivity of labor and the marginal productivity of capital and interpret each.

(d) Suppose that each unit of labor costs $12,000 and each unit of capital costs $2000. The budget for both is $1.1 million. Determine the number of units of labor and capital needed to maximize production.

(e) Determine the marginal productivity of money for the division of labor and capital found in part (d).

(f) Determine the optimal increase in production if an additional $100,000 is budgeted.

In this chapter we have seen how to generalize the derivative concept to functions of two variables. A natural question to ask is whether we can do the same with integration. The answer is yes, and we give a brief description of the process in this section. To fully understand the process, we begin the section by looking at **partial antidifferentiation**.

SECTION OBJECTIVES

1. Antidifferentiate a function of two variables.

2. Evaluate a partial antiderivative.

3. Evaluate iterated integrals.

4. Write and evaluate a double integral.

5. Compute an average value.

6. Compute a volume.

Partial Antidifferentiation and Iterated Integrals

In Chapter 6 we saw that antidifferentiation is the reverse operation of differentiation. In other words, antidifferentiation can "undo" differentiation. Remember that at that time we were dealing with functions of a single variable. To antidifferentiate a function of two variables with respect to one variable, we simply treat all other variables as if they are constants. This means that when we see

$$\int f(x, y)\,dx \quad \text{(Notice the } dx \text{ in this integral)}$$

we antidifferentiate $f(x, y)$ with respect to x and treat y as a constant. Likewise, when we see

$$\int f(x, y)\,dy \quad \text{(Notice the } dy \text{ in this integral)}$$

we antidifferentiate $f(x, y)$ with respect to y and treat x as a constant. Example 1 illustrates the process.

Example 1: Antidifferentiating Functions of Two Variables

Determine the following indefinite integrals.

a. $\int (x^2 y + 3y^2 x)\,dx$ **b.** $\int (x^2 y + 3y^2 x)\,dy$

OBJECTIVE 1

Antidifferentiate a function of two variables.

Perform the Mathematics

a. Here we are antidifferentiating with respect to x, so we treat y as a constant. We simply apply some properties of antidifferentiation from Chapter 6. Consult the following Toolbox.

From Your Toolbox: Properties of Indefinite Integrals

1. $\int [f(x) + g(x)]\,dx = \int f(x)\,dx + \int g(x)\,dx.$
2. $\int k \cdot f(x)\,dx = k \cdot \int f(x)\,dx$, where k is some constant.

Using these properties, we integrate as follows.

$$\int (x^2 y + 3y^2 x)\,dx = \int x^2 y\,dx + \int 3y^2 x\,dx$$

$$= y \int x^2\,dx + 3y^2 \int x\,dx$$

$$= y\left(\frac{1}{3}x^3\right) + 3y^2\left(\frac{1}{2}x^2\right) + C(y)$$

$$= \frac{1}{3}x^3 y + \frac{3}{2}x^2 y^2 + C(y)$$

Notice that the **constant** is not just a C. It can be *any function of y*, since for any function of y we know

$$\frac{\partial}{\partial x}[C(y)] = 0$$

b. Here we antidifferentiate with respect to y, so we treat x as a constant. This gives

$$\int (x^2y + 3y^2x)\,dy = \int x^2y\,dy + \int 3y^2x\,dy$$

$$= x^2\int y\,dy + x\int 3y^2\,dy$$

$$= x^2\left(\frac{1}{2}y^2\right) + x(y^3) + C(x)$$

$$= \frac{1}{2}x^2y^2 + xy^3 + C(x)$$

Notice here that our antiderivative has its arbitrary constant $C(x)$.   ■

In Chapter 6 we encouraged checking our antidifferentiation by differentiating the answer. We still encourage this type of checking. In Example 1 our result can be checked by partial differentiation. To check the result in Example 1a, we compute the partial derivative

$$\frac{\partial}{\partial x}\left[\frac{1}{3}x^3y + \frac{3}{2}x^2y^2 + C(y)\right] = \left(\frac{1}{3}\cdot 3x^2\right)y + \left(\frac{3}{2}\cdot 2x\right)y^2 + 0$$

$$= x^2y + 3xy^2$$

Example 1 shows us that we can easily antidifferentiate function of two variables as long as we remember which variable is being treated as a constant. In Example 2 we extend the process to evaluating definite integrals.

<table>
<tr><td>**OBJECTIVE 2**</td></tr>
<tr><td>Evaluate a partial antiderivative.</td></tr>
</table>

Example 2: Evaluating Partial Antiderivatives

Evaluate the following definite integrals.

a. $\displaystyle\int_0^1 (x^2y + 3y^2x)\,dx$ **b.** $\displaystyle\int_0^1 (x^2y + 3y^2x)\,dy$

Perform the Mathematics

a. Using the result from Example 1a, we have

$$\int_0^1 (x^2y + 3y^2x)\,dx = \left(\frac{1}{3}x^3y + \frac{3}{2}x^2y^2\right)\Bigg|_{x=0}^{x=1}$$

$$= \left(\frac{1}{3}(1)^3y + \frac{3}{2}(1)^2y^2\right) - \left(\frac{1}{3}(0)^3y + \frac{3}{2}(0)^2y^2\right)$$

$$= \frac{1}{3}y + \frac{3}{2}y^2$$

Notice that this definite integral produces a function of y.

b. Using the result from Example 1b, we have

$$\int_0^1 (x^2y + 3y^2x)\,dy = \left(\frac{1}{2}x^2y^2 + xy^3\right)\Big|_{y=0}^{y=1}$$

$$= \left(\frac{1}{2}x^2(1)^2 + x(1)^3\right) - \left(\frac{1}{2}x^2(0)^2 + x(0)^3\right)$$

$$= \frac{1}{2}x^2 + x$$

Notice that this definite integral produces a function of x. ∎

▶ *Try It Yourself*

Some related Exercises are 15 and 17.

In Example 2, we noticed the following:

1. $\displaystyle\int_0^1 (x^2y + 3y^2x)\,dx = \frac{1}{3}y + \frac{3}{2}y^2 = f(y)$

2. $\displaystyle\int_0^1 (x^2y + 3y^2x)\,dy = \frac{1}{2}x^2 + x = f(x)$

So it appears that when we integrate and evaluate a definite integral of the form

$$\int_a^b f(x, y)\,dx$$

the result is a function of a single variable, y. (It could also produce a constant.) Likewise, when we integrate and evaluate a definite integral of the form

$$\int_c^d f(x, y)\,dy$$

the result is a function of a single variable, x. (It could also produce a constant.) Since each of these produces a function of a single variable, they in turn could be an integrand of a second integral. In Example 3, we evaluate what are called **iterated integrals**. The word *iterated* simply means "repeated."

Example 3: Evaluating Iterated Integrals

Evaluate the following.

a. $\displaystyle\int_0^1\left[\int_0^1 (x^2y + 3y^2x)\,dx\right]dy$ **b.** $\displaystyle\int_0^1\left[\int_0^1 (x^2y + 3y^2x)\,dy\right]dx$

OBJECTIVE 3

Evaluate iterated integrals.

Perform the Mathematics

a. The brackets indicate the order in which the definite integrals are to be evaluated. Hence, we first evaluate the inside integral. From Example 2a, we have

$$\int_0^1 (x^2y + 3y^2x)\,dx = \frac{1}{3}y + \frac{3}{2}y^2$$

This result is now the integrand for the outer integral. This gives

$$\int_0^1 \left[\int_0^1 (x^2 y + 3y^2 x) dx \right] dy = \int_0^1 \left(\frac{1}{3} y + \frac{3}{2} y^2 \right) dy$$

$$= \left(\frac{1}{6} y^2 + \frac{1}{2} y^3 \right) \Big|_0^1$$

$$= \left(\frac{1}{6} + \frac{1}{2} \right) - (0 + 0) = \frac{2}{3}$$

b.   From Example 2b, the inside integral is

$$\int_0^1 (x^2 y + 3y^2 x) dy = \frac{1}{2} x^2 + x$$

This result is now the integrand for the outer integral. This gives

$$\int_0^1 \left[\int_0^1 (x^2 y + 3y^2 x) dy \right] dx = \int_0^1 \left(\frac{1}{2} x^2 + x \right) dx$$

$$= \left(\frac{1}{6} x^3 + \frac{1}{2} x^2 \right) \Big|_0^1$$

$$= \left(\frac{1}{6} + \frac{1}{2} \right) - (0 + 0) = \frac{2}{3}$$ ∎

Double Integrals

It is no accident that the results in Examples 3a and 3b are identical. Examples 1 through 3 suggest the following definition of a **double integral**.

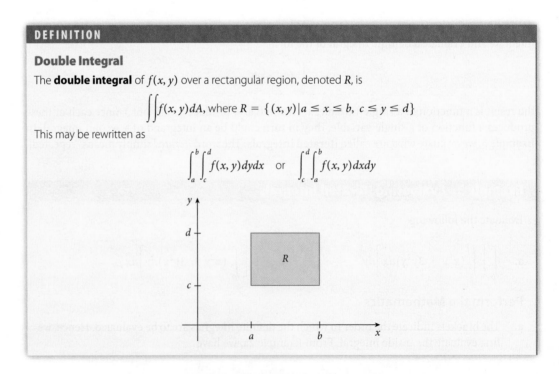

> ### DEFINITION
>
> **Double Integral**
> The **double integral** of $f(x, y)$ over a rectangular region, denoted R, is
> $$\iint f(x, y) dA, \text{ where } R = \{ (x, y) \mid a \le x \le b, \ c \le y \le d \}$$
> This may be rewritten as
> $$\int_a^b \int_c^d f(x, y) dy dx \quad \text{or} \quad \int_c^d \int_a^b f(x, y) dx dy$$

Notice that in the double integral, $\iint f(x, y) dA$, R is called the **region of integration**. This region is in the xy-plane. Also, dA indicates that either order of integration, $dy dx$ or $dx dy$, may be used.

We need to mention that a more general definition of double integrals exists. The one that we supply is more applicable to the functions that we will study.

Example 4: Writing and Evaluating a Double Integral

For $\iint (2x + y)\,dA$, where $R = \{(x, y)\,|\,1 \le x \le 3,\, 0 \le y \le 2\}$:

a.   Sketch the region of integration R in the xy-plane.

b.   Write and evaluate a double integral with $dy\,dx$ order of integration.

c.   Write and evaluate a double integral with $dx\,dy$ order of integration.

Perform the Mathematics

a.   The region of integration is shown in **Figure 8.6.1**.

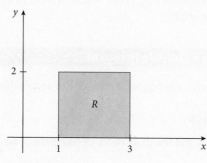

Figure 8.6.1

b.   A $dy\,dx$ order of integration means that the innermost integral has limits determined by the bounds for y and the outermost integral has limits determined by the bounds for x. This gives

$$\int_1^3 \int_0^2 (2x + y)\,dy\,dx$$

We evaluate by performing the innermost integration first. This gives

$$\int_1^3 \left[\int_0^2 (2x + y)\,dy \right] dx = \int_1^3 \left[\left(2xy + \frac{1}{2}y^2 \right) \Big|_{y=0}^{y=2} \right] dx$$

$$= \int_1^3 (4x + 2)\,dx$$

$$= (2x^2 + 2x) \Big|_1^3$$

$$= (18 + 6) - (2 + 2) = 20$$

c.   A $dx\,dy$ order of integration means that the innermost integral has limits determined by the bounds for x and the outermost integral has limits determined by the bounds for y. This gives

$$\int_0^2 \int_1^3 (2x + y)\,dx\,dy$$

Evaluating, starting with the innermost integral, yields

$$\int_0^2\left[\int_1^3(2x+y)dx\right]dy = \int_0^2\left[(x^2+xy)\Big|_{x=1}^{x=3}\right]dy$$

$$= \int_0^2(8+2y)dy = (8y+y^2)\Big|_0^2$$

$$= (16+4)-(0+0) = 20 \qquad \blacksquare$$

▶ *Try It Yourself*

Some related Exercises are 19 and 21.

Again Example 4 illustrates that, as long as our function is continuous, either order of integration (*dydx* or *dxdy*) yields the same result.

Example 5: Evaluating a Double Integral

Given $\int_{-1}^1\int_0^2 x^2 e^{-y}dxdy$, sketch the region of integration and evaluate the double integral.

Perform the Mathematics

From the limits of integration and the *dxdy* order of integration, we have

$$-1 \le y \le 1 \quad\text{and}\quad 0 \le x \le 2$$

The region is shown in **Figure 8.6.2**.

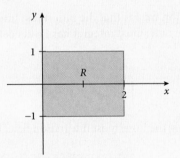

Figure 8.6.2

Evaluating, starting with the innermost integral, gives

$$\int_{-1}^1\left[\int_0^2 x^2 e^{-y}dx\right]dy = \int_{-1}^1\left[\left(\frac{1}{3}x^3 e^{-y}\right)\Big|_{x=0}^{x=2}\right]dy = \int_{-1}^1\frac{8}{3}e^{-y}dy$$

$$= \left(-\frac{8}{3}e^{-y}\right)\Big|_{-1}^1 = -\frac{8}{3}e^{-1}+\frac{8}{3}e \qquad \blacksquare$$

▶ *Try It Yourself*

Some related Exercises are 31 and 33.

Average Value over Rectangular Regions

In Section 7.1 we defined the average value of a function, f, over an interval $[a, b]$. Consult the following Toolbox.

From Your Toolbox: Average Value of a Function

The *average value* of a continuous function, f, on the interval $[a, b]$ is given by

$$\frac{1}{b - a}\int_a^b f(x)\,dx$$

We can extend the average value concept to functions of two variables.

DEFINITION

Average Value over Rectangular Regions

The **average value** of $f(x, y)$ over rectangular region R is defined to be

$$\frac{1}{(b - a)(c - d)}\iint f(x, y)\,dA$$

or

$$\frac{1}{\text{area of } R}\iint f(x, y)\,dA$$

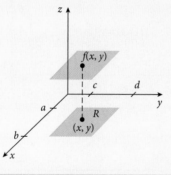

NOTE: Observe that $(b - a)(c - d)$ is simply the area of rectangular region R.

Example 6: Computing an Average Value

Determine the average value of $f(x, y) = x + y$ over the region R shown in **Figure 8.6.3**.

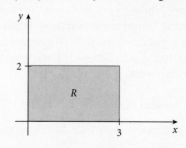

Figure 8.6.3

OBJECTIVE 5

Compute an average value.

Perform the Mathematics

From Figure 8.6.3 we see that R is simply

$$0 \leq x \leq 3 \quad \text{and} \quad 0 \leq y \leq 2$$

Hence, the area of R is 6. Using a $dydx$ order of integration, we compute the average value to be

$$
\begin{aligned}
\frac{1}{6}\int_0^3\int_0^2 (x+y)\,dydx &= \frac{1}{6}\int_0^3\left[\int_0^2 (x+y)\,dy\right]dx \\
&= \frac{1}{6}\int_0^3\left[\left(xy+\frac{1}{2}y^2\right)\Big|_{y=0}^{y=2}\right]dx \\
&= \frac{1}{6}\int_0^3 (2x+2)\,dx \\
&= \frac{1}{6}(x^2+2x)\Big|_0^3 \\
&= \frac{1}{6}[(9+6)-(0+0)] = \frac{1}{6}(15) = 2.5
\end{aligned}
$$

This means that 2.5 is simply the average of the z-values for $z = f(x,y) = x + y$ over the region R. See **Figure 8.6.4**.

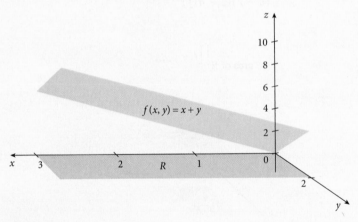

Figure 8.6.4

Example 7: Computing an Average Value of a Production Function

The Boroff Engine Company has a production function given by

$$Q = f(x,y) = x^{0.6}y^{0.45}$$

where Q is the number of engines produced per week, x is the number of employees at the company, and y is the weekly operating budget in thousands of dollars. Because the company uses a temporary labor agency, it uses anywhere from 40 to 50 employees each week, and its operating budget is anywhere from $10,000 to $15,000 each week. Determine the average number of engines that the company can produce each week.

Understand the Situation

Since the Boroff Engine Company uses anywhere from 40 to 50 employees each week, we know that

$$40 \leq x \leq 50$$

Also, since its operating budget is anywhere from \$10,000 to \$15,000 each week, we can represent y in thousands by

$$10 \leq y \leq 15$$

Thus, we need to determine the average value of $Q = f(x, y) = x^{0.6}y^{0.45}$ over the rectangular region shown in **Figure 8.6.5**.

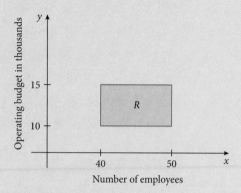

Figure 8.6.5

Perform the Mathematics

The area of the region is found to be $(50 - 40)(15 - 10) = 50$. So the average value is

$$\frac{1}{50}\int_{40}^{50}\int_{10}^{15} x^{0.6}y^{0.45}\,dy\,dx = \frac{1}{50}\int_{40}^{50}\left[\int_{10}^{15} x^{0.6}y^{0.45}\,dy\right]dx$$

$$= \frac{1}{50}\int_{40}^{50}\left[\left(\frac{1}{1.45}x^{0.6}y^{1.45}\right)\bigg|_{10}^{15}\right]dx$$

$$= \frac{1}{50}\int_{40}^{50}\left(\frac{1}{1.45}x^{0.6}(15)^{1.45} - \frac{1}{1.45}x^{0.6}(10)^{1.45}\right)dx$$

$$\approx \frac{1}{50}\int_{40}^{50}(34.99x^{0.6} - 19.44x^{0.6})\,dx$$

$$= \frac{1}{50}\int_{40}^{50} 15.55x^{0.6}\,dx$$

$$= \frac{15.55}{50}\int_{40}^{50} x^{0.6}\,dx = \frac{15.55}{50}\left[\left(\frac{1}{1.6}x^{1.6}\right)\bigg|_{40}^{50}\right]$$

$$= \frac{15.55}{50}\left[\frac{1}{1.6}(50)^{1.6} - \frac{1}{1.6}(40)^{1.6}\right] \approx 30.51$$

Interpret the Result

This means that, to the nearest engine, the Boroff Engine Company can produce an average of 31 engines each week. **Figure 8.6.6** gives a graph of $Q = f(x, y) = x^{0.6}y^{0.45}$ over the region R. Again, the result of 31 is simply the average of the Q-values over R.

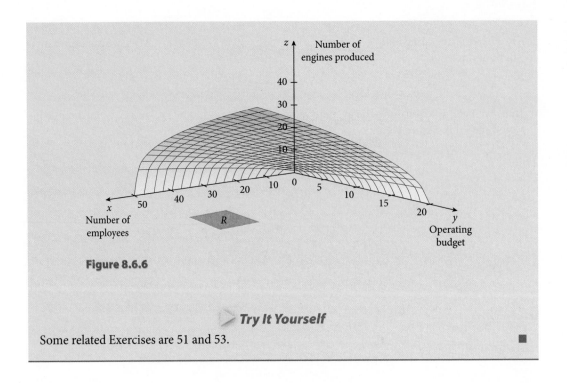

Figure 8.6.6

▷ **Try It Yourself**

Some related Exercises are 51 and 53.   ∎

Volume

We conclude this section with a discussion on volume. In Chapter 6 we saw that the area of the region under a curve given by $y = f(x)$ on the interval $[a, b]$ is determined by $\int_a^b f(x)\,dx$, as long as $f(x)$ is nonnegative on $[a, b]$. See **Figure 8.6.7**. Here, we can determine the *volume* of the *solid* under a **surface** given by $f(x, y)$ over rectangular region R using double integrals. See **Figure 8.6.8**.

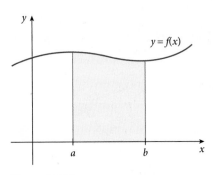

Figure 8.6.7

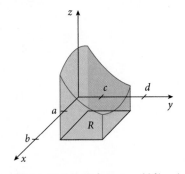

Figure 8.6.8 Volume $= \iint f(x, y)\,dA$.

DEFINITION

Volume under a Surface

If $z = f(x, y)$ is nonnegative and continuous ($f(x, y) \geq 0$) over rectangular region R,
$R = \{(x, y) \,|\, a \leq x \leq b, c \leq y \leq d\}$, then the volume of the solid under $f(x, y)$ over R is given by

$$V = \iint f(x, y)\,dA$$

Example 8: Computing a Volume

Determine the volume under $f(x, y) = 4 - x^2 - y^2$ above the region

$$R = \{(x, y) \mid 0 \le x \le 1, 0 \le y \le 1\}$$

Perform the Mathematics

Figure 8.6.9 shows the region, and **Figure 8.6.10** shows the volume that we are trying to determine.

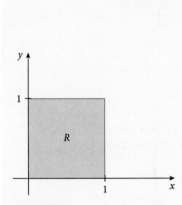

Figure 8.6.9

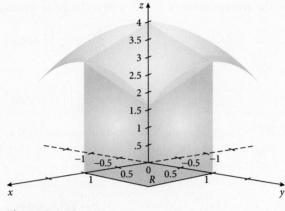

Figure 8.6.10

The volume is given by

$$V = \iint (4 - x^2 - y^2)\,dA$$

We select a *dydx* order of integration, which gives

$$V = \int_0^1 \int_0^1 (4 - x^2 - y^2)\,dy\,dx = \int_0^1 \left[\left(4y - x^2 y - \frac{1}{3}y^3 \right) \Big|_{y=0}^{y=1} \right] dx$$

$$= \int_0^1 \left(4 - x^2 - \frac{1}{3} \right) dx = \int_0^1 \left(\frac{11}{3} - x^2 \right) dx$$

$$= \left(\frac{11}{3}x - \frac{1}{3}x^3 \right) \Big|_0^1 = \left(\frac{11}{3} - \frac{1}{3} \right) - (0 - 0)$$

$$= \frac{10}{3} \text{ cubic units} \qquad \blacksquare$$

▶ *Try It Yourself*

Some related Exercises are 43 and 45.

Summary

In this section we have seen how to extend the concept of antidifferentiation to functions of two variables. Partial antidifferentiation led to iterated integrals, which in turn led us to double integrals. We then extended the concept of average value to functions of two variables. We concluded the section with a brief look at how a double integral can be used to determine the volume of a solid under a surface over a rectangular region.

- The **double integral** of $f(x, y)$ over a rectangular region, denoted R, is

$$\iint f(x, y)\, dA, \text{ where } R = \{(x, y)\,|\,a \le x \le b,\ c \le y \le d\}$$

This may be rewritten as

$$\int_a^b \int_c^d f(x, y)\, dy\, dx \quad \text{or} \quad \int_c^d \int_a^b f(x, y)\, dx\, dy$$

- The **average value** of $f(x, y)$ over rectangular region R is defined to be

$$\frac{1}{(b - a)(c - d)}\iint f(x, y)\, dA$$

or

$$\frac{1}{\text{area of } R}\iint f(x, y)\, dA$$

- If $z = f(x, y)$ is nonnegative and continuous ($f(x, y) \ge 0$) over rectangular region R, $R = \{(x, y)\,|\,a \le x \le b, c \le y \le d\}$, then the volume of the solid under $f(x, y)$ over R is given by

$$V = \iint f(x, y)\, dA$$

Section 8.6 Exercises

Vocabulary Exercises

1. The expression $\int f(x, y)\, dy$ is an example of _____ antidifferentiation.

2. An integral of the form $\int_c^d \int_a^b f(x, y)\, dx\, dy$ is called an _____ integral.

3. An integral of the form $\iint f(x, y)\, dA, R = \{(a, b)\,|\,a \le x \le b,\ c \le y \le d\}$, is called a _____ integral.

4. In the integral $\iint f(x, y)\, dA, R = \{(a, b)\,|\,a \le x \le b,\ c \le y \le d\}$, we call R the _____ of integration.

5. If we compute $\dfrac{1}{\text{area of } R}\iint f(x, y)\, dA$, we are computing the _____ value over a rectangular region.

6. If $f(x, y) \ge 0$ and continuous over $R = \{(a, b)\,|\,a \le x \le b,\ c \le y \le d\}$, then the _____ of the solid under $f(x, y)$ and over R is given by $\iint f(x, y)\, dA$.

Skill Exercises

Evaluate the integral in Exercises 7–18.

7. $\displaystyle\int_1^3 3x^2 y^2\, dy$

8. $\displaystyle\int_1^4 2x^3 y^3\, dy$

9. $\displaystyle\int_1^3 3x^2 y^2\, dx$

10. $\displaystyle\int_1^4 2x^3 y^3\, dx$

11. $\displaystyle\int_0^2 (2x + 3y)\,dy$

12. $\displaystyle\int_0^3 (3x - 2y)\,dy$

13. $\displaystyle\int_0^2 (2x + 3y)\,dx$

14. $\displaystyle\int_0^3 (3x - 2y)\,dx$

15. $\displaystyle\int_1^2 (x^3y^2 - 2xy)\,dy$

16. $\displaystyle\int_2^4 (x^2y^3 - 3xy)\,dy$

17. $\displaystyle\int_1^2 (x^3y^2 - 2xy)\,dx$

18. $\displaystyle\int_2^4 (x^2y^3 - 3xy)\,dx$

For Exercises 19–28, complete the following.

(a) Sketch the region of integration.

(b) Write and evaluate a double integral with a $dy\,dx$ order of integration.

(c) Write and evaluate a double integral with a $dx\,dy$ order of integration.

19. $\displaystyle\iint 2xy\,dA;\ R = \{(x, y)\,|\,0 \le x \le 2,\ 0 \le y \le 3\}$

20. $\displaystyle\iint 3xy\,dA;\ R = \{(x, y)\,|\,0 \le x \le 2,\ 0 \le y \le 3\}$

21. $\displaystyle\iint (3x + y)\,dA;\ R = \{(x, y)\,|\,0 \le x \le 1,\ 0 \le y \le 2\}$

22. $\displaystyle\iint (4x - y)\,dA;\ R = \{(x, y)\,|\,2 \le x \le 4,\ 1 \le y \le 2\}$

23. $\displaystyle\iint \sqrt{xy}\,dA;\ R = \{(x, y)\,|\,1 \le x \le 16,\ 1 \le y \le 4\}$

24. $\displaystyle\iint \sqrt{xy}\,dA;\ R = \{(x, y)\,|\,1 \le x \le 4,\ 1 \le y \le 16\}$

25. $\displaystyle\iint x^2y^2\,dA;\ R = \{(x, y)\,|\,1 \le x \le 2,\ 0 \le y \le 1\}$

26. $\displaystyle\iint x^2y^2\,dA;\ R = \{(x, y)\,|\,0 \le x \le 1,\ 1 \le y \le 2\}$

27. $\displaystyle\iint \left(2 - \frac{1}{2}x^2 + y^2\right)dA;\ R = \{(x, y)\,|\,0 \le x \le 1,\ 0 \le y \le 1\}$

28. $\displaystyle\iint \left(3 + x^2 - \frac{1}{2}y^2\right)dA;\ R = \{(x, y)\,|\,0 \le x \le 1,\ 0 \le y \le 1\}$

In Exercises 29–34, sketch the region of integration and then evaluate the double integral.

29. $\displaystyle\int_0^1 \int_0^2 xy\,dy\,dx$

30. $\displaystyle\int_1^3 \int_1^2 5xy\,dy\,dx$

31. $\displaystyle\int_1^2 \int_1^3 (x^2y + y)\,dx\,dy$

32. $\displaystyle\int_1^3 \int_1^2 (x^2y - y)\,dx\,dy$

33. $\displaystyle\int_{-1}^1 \int_0^2 x^2e^y\,dx\,dy$

34. $\displaystyle\int_{-1}^1 \int_0^2 x^2e^y\,dy\,dx$

In Exercises 35–42, determine the average value of the function over the region R.

35. $f(x, y) = x^2 + y^2;\ R = \{(x, y)\,|\,0 \le x \le 1,\ 0 \le y \le 1\}$

36. $f(x, y) = x^2 + y^2;\ R = \{(x, y)\,|\,0 \le x \le 2,\ 0 \le y \le 2\}$

37. $f(x, y) = 9 - x^2 - y^2;\ R = \{(x, y)\,|\,0 \le x \le 1,\ 0 \le y \le 1\}$

38. $f(x, y) = 9 - x^2 - y^2$; $R = \{(x, y) | 0 \le x \le 2, 0 \le y \le 2\}$

39. $f(x, y) = 2xy$; $R = \{(x, y) | 0 \le x \le 2, 0 \le y \le 2\}$

40. $f(x, y) = 2xy$; $R = \{(x, y) | 0 \le x \le 1, 0 \le y \le 1\}$

41. $f(x, y) = x^2 e^y$; $R = \{(x, y) | 0 \le x \le 2, 0 \le y \le 1\}$

42. $f(x, y) = x^2 e^y$; $R = \{(x, y) | 0 \le x \le 1, 0 \le y \le 2\}$

In Exercises 43–50, determine the volume under $f(x, y)$ above the given region R.

43. $f(x, y) = 9 - x^2 - y^2$; $R = \{(x, y) | 0 \le x \le 1, 0 \le y \le 2\}$

44. $f(x, y) = 12 - x^2 - y^2$; $R = \{(x, y) | 0 \le x \le 1, 0 \le y \le 2\}$

45. $f(x, y) = x^2 + 2y^2$; $R = \{(x, y) | 0 \le x \le 1, 0 \le y \le 2\}$

46. $f(x, y) = 2x^2 + y^2$; $R = \{(x, y) | 0 \le x \le 2, 0 \le y \le 2\}$

47. $f(x, y) = 3xy$; $R = \{(x, y) | 0 \le x \le 2, 0 \le y \le 1\}$

48. $f(x, y) = 2xy$; $R = \{(x, y) | 0 \le x \le 1, 0 \le y \le 1\}$

49. $f(x, y) = x^2 e^y$; $R = \{(x, y) | 0 \le x \le 2, 0 \le y \le 1\}$

50. $f(x, y) = x^2 e^y$; $R = \{(x, y) | 0 \le x \le 1, 0 \le y \le 2\}$

Application Exercises

51. Manufacturing—Average Production: Sumption Sonic Inc. has a production function given by

$$Q = f(x, y) = 250x^{0.7}y^{0.3}$$

where Q is the number of solar cells produced each week, x is the number of employees at the company, and y is the weekly operating budget in thousands. Because the company uses a temporary labor agency, it uses anywhere from 30 to 40 employees each week, and its operating budget is anywhere from $150,000 to $180,000 each week. Determine the average number of solar cells that the company can produce each week.

52. Manufacturing—Average Production: The Linger Golf Cart Corporation has a production function given by

$$Q = f(x, y) = 1.3x^{0.75}y^{0.3}$$

where Q is the number of golf carts produced each week, x is the number of employees at the company, and y is the weekly operating budget in thousands. Because the company uses a temporary labor agency, it uses anywhere from 10 to 15 employees each week, and its operating budget is anywhere from $12,000 to $16,000 each week. Determine the average number of golf carts that the company can produce each week.

53. Manufacturing—Average Production: The Basich Company has a production function given by

$$Q = f(x, y) = 161x^{0.8}y^{0.25}$$

where Q is the number of handheld calculators produced each week, x is the number of employees at the company, and y is the weekly operating budget in thousands. Because the company uses a temporary labor agency, it uses anywhere from 70 to 80 employees each week, and its operating budget is anywhere from $20,000 to $30,000 each week. Determine the average number of handheld calculators that the company can produce each week.

54. Manufacturing—Average Production: Hoffman Laboratories has a production function given by

$$Q = f(x, y) = 5.6x^{0.65}y^{0.41}$$

where Q is the number of pain-killing tablets produced (in thousands) each week, x is the number of employees at the company, and y is the weekly operating budget in thousands. Because the company uses a temporary labor agency, it uses anywhere from 25 to 35 employees each week, and its operating budget is anywhere from $40,000 to $50,000 each week. Determine the average number of pain-killing tablets that the company can produce each week.

55. **Meteorology—Average Weather:** The temperature, in degrees Fahrenheit, x miles east and y miles north of a reporting weather station is given by

$$f(x, y) = 36 + 2x - 3y$$

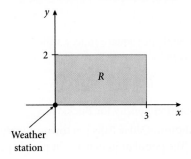

Determine the average temperature over the region shown in the figure.

56. **Meteorology—Average Weather:** The temperature, in degrees Fahrenheit, x miles east and y miles north of a reporting weather station is given by

$$f(x, y) = 45 + x - 2y$$

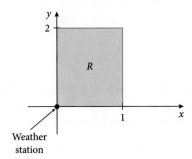

Determine the average temperature over the region shown in the figure.

57. **Demography—Average Population:** Cedar Falls is a town roughly shaped like a square, as shown in the figure.

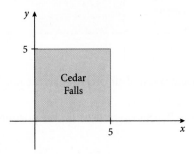

The **population density**, in people per square mile, x miles east and y miles north of the southwest corner of the town is given by

$$h(x, y) = 30{,}000e^{-y}$$

Determine the average population density for Cedar Falls.

58. **Demography—Average Population:** Bucksville is a town situated along a fairly straight river and is fairly rectangular, as shown in the figure.

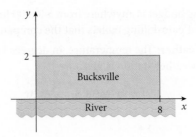

The population density, in people per square mile, x miles east and y miles north of the southwest corner of the town is given by

$$h(x, y) = 15,000(5 - y^2)$$

Determine the average population density for Bucksville.

59. **Demography—Total Population:** Cedar Falls is a town roughly shaped like a square, as shown in the figure in Exercise 57. The population density, in people per square mile, x miles east and y miles north of the southwest corner of the town is given by

$$h(x, y) = 30,000e^{-y}$$

Determine the *total population* of Cedar Falls. (To do this, integrate the population density over region R. This illustrates that a double integral can determine a *continuous sum*.)

60. **Demography—Total Population:** Bucksville is a town situated along a fairly straight river and is fairly rectangular, as shown in the figure in Exercise 58. The population density, in people per square mile, x miles east and y miles north of the southwest corner of the town is given by

$$h(x, y) = 15,000(5 - y^2)$$

Determine the total population of Bucksville.

61. **Climatology—Air Pollution:** A heavy industrial complex in Bakerstown emits pollution into the atmosphere. Because of the prevailing winds, the air pollution (in parts per million) x miles east and y miles north of the complex is given by

$$f(x, y) = 125 - 10x^2 - 10y^2$$

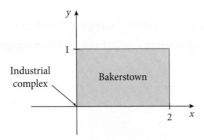

Determine the average concentration of air pollution in the northeast part of Bakerstown. See the figure.

62. **Climatology—Air Pollution:** Repeat Exercise 61 for the portion of Bakerstown shown in the figure.

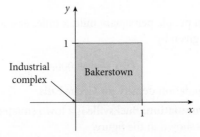

63. Cognitive Science—Average IQ: An individual's IQ is defined to be

$$IQ = f(a.m) = \frac{m}{a} \cdot 100$$

where m is the individual's mental age in years, and a is the individual's actual age in years. In a calculus class, the mental age varies from 23 to 32 years, and the actual age varies from 19 to 25 years. Determine the average IQ for this calculus class.

64. Cognitive Science—Average IQ: Repeat Exercise 63, except here the mental age varies from 23 to 45 years and the actual age varies from 22 to 32 years.

Concept and Writing Exercises

65. The double integral can be used to compute the volume of a solid in 3-space. Explain in a sentence or two how this is done.

66. Can double integrals be used to compute the area between two curves in the xy-plane? Explain how or why it cannot be done.

67. Let a, b, c, and d represent real number constants. Either show that the following statement is true or explain why it is false. If $\int_a^b \int_c^d f(x) \cdot g(y) dx dy$, then $\int_c^d f(x) dx \cdot \int_a^b g(y) dy$.

68. Can a *triple* integral be computed? Does it have a geometric interpretation? Explain your answers.

69. Suppose $f(x, y) = k$, where k represents a real number constant. Compute
$f(x, y) = k$; $R = \{(x, y) | a \le x \le b, c \le y \le d\}$.

Section Project

In their study of human groupings, anthropologists often use an index called the *cephalic index*. The cephalic index is given by

$$C(W, L) = \frac{W}{L} \cdot 100$$

where W is the width and L is the length of an individual's head. Both measurements are made across the top of the head and are in inches. In a calculus class, the width of heads varies from 6 to 8.5 inches and the length of heads varies from 8 to 10 inches.

(a) Determine the average value of the cephalic index for this class.

(b) Repeat part (a), except here the width of heads varies from 6.5 to 8.2 inches and the length of heads varies from 7.6 to 9.8 inches.

Section 8.1 Review Exercises

In Exercises 1–4, evaluate $f(x, y) = \dfrac{x^2 + 3}{x - y^2}$ for the indicated values.

1. $f(3, 5)$

2. $f(-2, 4)$

3. $f(10, 3)$

4. $f(20, -4)$

In Exercises 5 and 6, determine the domain of the given function. Recall that here the domain is the set of all ordered pairs (x, y) in the xy-plane.

5. $f(x, y) = \sqrt{x + y}$

6. $f(x, y) = \dfrac{5x}{x - 2y}$

In Exercises 7–10, plot the given points.

7. $(5, 0, 0)$

8. $(3, 8, -5)$

9. $(0, -4, 3)$

10. $(-2, -6, 5)$

In Exercises 11–14, without plotting the point, determine if the given point is above or below the xy-plane. Explain.

11. $(-3, -4, 2)$

12. $(4, 3, -8)$

13. $(6, -2, -1)$

14. $(-5, 9, 4)$

In Exercises 15 and 16, sketch the graph of the given plane in 3-space.

15. $x = -5$

16. $z = 2$

17. Economics—Profit Analysis: The Moondoe Coffee Company has determined that its monthly profit, in dollars, is given by

$$P(x, y) = 0.8x + 1.2y - 6000$$

where x is the number of regular cups of coffee sold in a month, and y is the number of cups of cappuccino sold in a month. Determine $P(4000, 3000)$ and interpret.

18. Management—Sales Analysis: The Stellar Stereo Store spends x thousand dollars each month on television advertising and y thousand dollars each month on direct mail advertising. The company has monthly sales given by

$$S(x, y) = x^2 + 3y$$

(a) Determine $S(4, 7)$ and interpret.

(b) Determine $S(7, 4)$ and interpret.

19. Management—Profit Analysis: The WriteNow Company sells a regular calligraphy set for $12 and a deluxe calligraphy set for $21. The cost to produce these items is $7 for the regular set and $13 for the deluxe set. The company also has fixed monthly expenses totaling $5000.

(a) Determine the cost $C(x, y)$ of producing x regular sets and y deluxe sets in a month.

(b) Determine the revenue function $R(x, y)$.

(c) Determine the profit function $P(x, y)$.

20. Management—Profit Analysis: The Digital Calculator Company manufactures scientific calculators and graphing calculators. Let x represent the weekly demand for scientific calculators, and let y represent the demand for graphing calculators. The weekly price-demand equations are given by

$$p = 60 - 0.3x + 0.1y, \text{ the price in dollars for a scientific calculator}$$

$$q = 100 + 0.2x - 0.5y, \text{ the price in dollars for a graphing calculator}$$

The cost function is given by

$$C(x, y) = 3000 + 25x + 35y$$

(a) Determine the weekly revenue function $R(x, y)$.

(b) Evaluate $R(40, 60)$ and interpret.

(c) Determine the weekly profit function $P(x, y)$.

(d) Evaluate $P(40, 60)$ and interpret.

21. **Banking—Continuous Interest:** If an amount P dollars is invested at an annual interest rate of r (in decimal form) compounded continuously for 10 years, the total amount accumulated is given by

$$f(P, r) = Pe^{10r}$$

Determine $f(4500, 0.045)$ and interpret.

22. **Geometry—Pyramid Volume:** The volume of a pyramid with a square base is given by

$$V(x, h) = \frac{1}{3}x^2 h$$

where x is the length of one side of the square base, and h is the height of the pyramid. Assume that x and h are measured in inches.

(a) Determine $V(8, 12)$ and interpret.

(b) If the height is equal to the side length of the square base, the volume can be expressed as a function of one variable, either x or h. Determine $V(x)$ and $V(h)$.

Section 8.2 Review Exercises

For Exercises 23–26, make a contour map of the given function for the specified c-values.

23. $f(x, y) = 2x - y + 3$	$c = 0, 2, 4, 6$
24. $f(x, y) = x^2 - y$	$c = -2, 0, 2, 4$
25. $f(x, y) = 16 - x^2 - y^2$	$c = 0, 4, 8, 12$
26. $f(x, y) = x - e^y$	$c = 0, 1, 2, 3$

In Exercises 27–30, make a contour map of the given production functions for the specified c-values.

27. $Q = 10x^{0.5}y^{0.5}$	$c = 10, 20, 30, 40$
28. $Q = 2.7x^{0.3}y^{0.75}$	$c = 20, 40, 60, 80$
29. $Q = 8.9x^{0.62}y^{0.42}$	$c = 100, 200, 300, 400$
30. $Q = 80x^{0.21}y^{0.79}$	$c = 50, 100, 150, 200$

In Exercises 31–34, make a contour map of the given utility function for the specified c-values.

31. $U = \sqrt[3]{xy^2}$	$c = 2, 4, 6, 8$
32. $U = x^{0.4}y^{0.6}$	$c = 1, 2, 3, 4$
33. $U = x^{0.15}y^{0.85}$	$c = 2, 4, 6, 8$
34. $U = x^{0.3}y^{0.7}$	$c = 1, 2, 3, 4$

In Exercises 35–38, sketch on the appropriate plane, either the xz-plane or yz-plane, cross sections for the given function for the specified vertical plane values. See Example 6 in Section 8.2.

35. $z = 20x^{0.3}y^{0.7}$	$x = 2, 4, 6$
36. $z = 20x^{0.3}y^{0.7}$	$y = 2, 4, 6$
37. $z = x^2 - y$	$y = 0, 1, 2, 3$
38. $z = x^2 - y$	$x = 0, 1, 2, 3$

39. **Cross-Sectional Analysis—Capital Costs:** The McGillicutty Corporation has 12 manufacturing plants and has a Cobb-Douglas production function of

$$Q = f(x, y) = 1.87x^{0.7} y^{0.3}$$

where x is the number of labor hours (in thousands), y is the capital (total net assets in millions), and Q is the quantity produced (in thousands of units). Sketch isoquants for $Q = 60, 80, 100,$ and 120.

40. **Cross-Sectional Analysis—Capital Costs:** Refer to Exercise 39. Lucille, the CEO of the McGillicutty Corporation, wants to keep the capital at each plant fixed at $y = 4.1$.

    (a) By fixing capital at $y = 4.1$, is Lucille looking at a vertical or horizontal cross section?

    (b) Graph the cross section when $y = 4.1$.

    (c) Approximate how many labor hours would be required to produce a quantity of 150,000 units (that is, $Q = 150$)?

41. **Cross-Sectional Analysis—Capital Costs:** Refer to Exercises 39 and 40. Instead of keeping the capital fixed, suppose that Lucille, the CEO of the McGillicutty Corporation, wants to keep the number of labor hours at each plant fixed at $x = 240$.

    (a) By fixing labor at $x = 240$, is Lucille looking at a vertical or horizontal cross section?

    (b) Graph the cross section when $x = 240$.

    (c) Approximate how much capital is required to produce 150,000 units (that is, $Q = 150$).

 42. **Cross-Sectional Analysis—Capital Costs:** The MaxiSound Company has seven plants around the United States that produce speakers for computers. In a recent year, the data for each plant gave the number of labor hours (in thousands), capital (in millions), and total quantity of speakers produced as shown in the table.

Labor	Capital	Quantity
24	4.3	13,016
26	4.9	14,565
29	5.6	16,500
31	6.1	17,852
38	5.8	18,600
45	6.4	21,052
46	7.3	23,084

   (a) The plants all use the same technology, so a production function can be determined. Enter the data into your calculator and execute the multiple regression program in Appendix B to determine a Cobb-Douglas production function to model the data. Round all values to the nearest hundredth.

   (b) Construct isoquants for $Q = 12,000, 16,000, 20,000,$ and 24,000.

Section 8.3 Review Exercises

For Exercises 43–50, determine $f_x(x, y)$ and $f_y(x, y)$.

43. $f(x, y) = 5x - 3y^2$

44. $f(x, y) = 2x^3 - 5xy + y^8$

45. $f(x, y) = (8y^2 - 3x)^5$

46. $f(x, y) = 5.83x^{0.38} y^{0.61}$

47. $f(x, y) = \dfrac{x + y}{3x}$

48. $f(x, y) = \ln(5x - y^3)$

49. $f(x, y) = e^{x-y} + 5x^3 y$

50. $f(x, y) = \ln xy - e^{x/y}$

In Exercises 51–54, for the given function, evaluate the stated partial derivative and interpret.

51. $f(x, y) = 5x^2 + 3xy^4 - 2y^5$; determine $f_y(3, 5)$.

52. $f(x, y) = \ln(x^2 + 5y)$; determine $f_x(2, 7)$.

53. $f(x, y) = 12.85x^{0.38} y^{0.62}$; determine $f_y(3, 2)$.

54. $f(x, y) = \dfrac{\ln 2y}{x^2 + y}$; determine $f_x(-2, 5)$.

In Exercises 55–58, determine $f_{xx}(x, y), f_{xy}(x, y),$ and $f_{yy}(x, y)$.

55. $f(x, y) = -4x^3 + 10xy + xy^5$

56. $f(x, y) = 8x^2 y^7 - x^3 y^5 + 12x^2$

57. $f(x, y) = (x + 2y)^2$

58. $f(x, y) = \ln(3x - y)$

59. For $f(x, y) = 6x^2 - 4xy + 5y^2 + x - 3y + 17$, determine values for x and y such that $f_x(x, y) = 0$ and $f_y(x, y) = 0$ simultaneously.

60. Economics—Revenue Analysis: The Heavy Stuff Athletic Club spends x thousand dollars each year on television advertising and y thousand dollars each year on newspaper advertising. The club has weekly revenues (in thousands of dollars) given by

$$R(x, y) = 1.5x + 0.6x^2$$

(a) Determine $R_x(x, y)$ and $R_y(x, y)$ and interpret.

(b) Determine $R_x(12, 8)$ and interpret.

(c) Determine $R_y(12, 8)$ and interpret.

61. Economics—Revenue Analysis: Pixel Power, Inc., manufactures 15-inch computer monitors and 17-inch computer monitors. Let x represent the weekly demand for a 15-inch monitor, and let y represent the weekly demand for a 17-inch monitor. The weekly price-demand equations are given by

$$p = 480 - 0.6x + 0.2y, \text{ the price in dollars for a 15-inch monitor}$$
$$q = 720 + 0.3x - 0.7y, \text{ the price in dollars for a 17-inch monitor}$$

(a) Determine the weekly revenue function $R(x, y)$.

(b) Determine $R_x(x, y)$ and $R_y(x, y)$ and interpret each.

(c) Determine $R_x(640, 850)$ and $R_y(640, 850)$ and interpret each.

62. Economics—Profit Analysis *(continuation of Exercise 61):* The weekly cost function for Pixel Power, Inc., is given by

$$C(x, y) = 12{,}000 + 128x + 163y$$

(a) Determine the weekly profit function $P(x, y)$.

(b) Determine $P_x(x, y)$ and $P_y(x, y)$ and interpret each.

(c) Determine $P_x(640, 850)$ and $P_y(640, 850)$ and interpret each.

63. Economics—Profit Analysis: The WatchIt Company produces two types of watches, x hundred analog and y hundred digital, each month. The monthly revenue and cost functions are, in hundreds of dollars,

$$R(x, y) = -3.5x^2 - 3.5xy + 1.2y^2 + 175x + 215y$$
$$C(x, y) = 15xy + 65x + 85y + 600$$

(a) Determine the monthly profit function $P(x, y)$.

(b) Determine $P_x(x, y)$ and $P_y(x, y)$ and interpret each.

(c) Determine $P_x(8, 11)$ and $P_y(8, 11)$ and interpret each.

(d) Determine $P(8, 11)$ and interpret.

64. Economics—Marginal Productivity: The Superior Gizmo Company has a production function given by

$$Q = f(x, y) = 5.64x^{0.25} y^{0.8}$$

where x is the number of labor hours in thousands, y is the capital in millions, and Q is the total quantity produced in thousands.

(a) Compute $f_x(x, y)$ and $f_y(x, y)$.

(b) If the Superior Gizmo Company is currently using 38 units of labor and 1.6 units of capital, determine the marginal productivity of labor and the marginal productivity of capital and interpret each.

(c) Would production increase more by increasing the labor by 1200 labor hours or by increasing the capital by $40,000?

Section 8.4 Review Exercises

In Exercises 65–68, determine the critical points for the given function.

65. $f(x, y) = x^2 + y^2 + 8x - 6y + 8$

66. $f(x, y) = x^2 - y^2 + 4x + 10y - 3$

67. $f(x, y) = x^2 + y^2 - 5xy + 6x - 8y + 7$

68. $f(x, y) = 3x^2 + 5y^2 - 8xy + 2x + 6y + 3$

In Exercises 69–76, use the Second Derivative Test to locate any relative extrema and saddle points.

69. $f(x, y) = x^2 + y^2 - 2x + 8y - 7$

70. $f(x, y) = 2x^2 + 5y^2 + 6x - 2y + 12$

71. $f(x, y) = x^2 - 2y^2 + xy + 3x - 3y$

72. $f(x, y) = \dfrac{4}{3}y^3 + x^2 - 4x - 8y - 17$

73. $f(x, y) = 2x^3 - 2x^2 + 4xy + y^2$

74. $f(x, y) = \dfrac{1}{5}x^5 - y^2 - 16x + 6y - 8$

75. $f(x, y) = e^{x^2 - y^2 + 4y}$

76. $f(x, y) = 4xy + \dfrac{1}{x} + \dfrac{2}{y}$

77. Economics—Maximizing Revenue: Roberta's Hair Salon offers a basic haircut and a deluxe haircut, which includes shampoo and styling. Let x represent the daily demand for basic haircuts, and let y represent the daily demand for deluxe haircuts. The daily price-demand equations are given by

$$p = 12 - 0.3x + 0.1y, \text{ the price in dollars for a basic haircut}$$
$$q = 20 + 0.1x - 0.2y, \text{ the price in dollars for a deluxe haircut}$$

(a) Determine the revenue function $R(x, y)$.

(b) How many basic haircuts and how many deluxe haircuts should be given per day in order to maximize revenue?

(c) What is the maximum daily revenue?

78. Economics—Maximizing Profit *(continuation of Exercise 77):* Roberta's Hair Salon has a daily cost function given by

$$C(x, y) = 200 + 4x + 5y$$

(a) Determine the profit function $P(x, y)$.

(b) How many basic haircuts and how many deluxe haircuts should be given per day in order to maximize profit?

(c) What is the maximum daily profit?

79. Economics—Maximizing Profit: The WriteNow Company makes marking pens with either a fine tip or a large tip. Let x represent the number of thousands of fine-tip pens, and let y

represent the number of thousands of large-tip pens produced each day. The daily profit function is given by

$$P(x, y) = -3x^2 - 2y^2 + 2xy + 800x + 1200y - 400$$

where $P(x, y)$ is the profit in dollars for producing x thousand fine-tip pens and y thousand large-tip pens.

(a) How many fine-tip pens and how many large-tip pens should be sold each day to maximize profit?

(b) What is the maximum profit?

80. **Geometry—Minimizing Surface Area:** You have been challenged to design a rectangular box with no top and three parallel partitions that will hold 80 cubic inches. Determine the dimensions that will require the least amount of material to do this.

Section 8.5 Review Exercises

In Exercises 81–86, use the Method of Lagrange Multipliers to solve each problem.

81. Maximize $f(x, y) = 5xy$ Subject to $x + y = 10$

82. Minimize $f(x, y) = 3xy$ Subject to $x - 2y = 6$

83. Maximize $f(x, y) = 16 - 2x^2 + 5xy - 18y^2$ Subject to $x + 3y = 6$

84. Minimize $f(x, y) = 3 + x^2 + y^2$ Subject to $2x - y = 10$

85. Maximize $f(x, y) = 3.8x^{0.35}y^{0.6}$ Subject to $120x + 260y = 2500$

86. Minimize $f(x, y) = xy$ Subject to $x^2 + 4y^2 = 8$

87. **Economics—Maximizing Production:** The Liberty Bell Telephone Company has a production function for the manufacture of portable telephones of

$$Q = f(x, y) = 28x^{0.6}y^{0.4}$$

where Q is the number of thousands of telephones produced per week, x is the number of units of labor each week, and y is the number of units of capital. A unit of labor is $500, and a unit of capital is $800. The total expense for both is limited to $4000.

(a) Determine the number of units of labor and the number of units of capital required to maximize production.

(b) Determine the maximum production.

88. **Economics—Marginal Productivity:** The Dizzy Dog Company has a production function for the manufacture of dog toys given by

$$Q = f(x, y) = 4.3x^{0.55}y^{0.5}$$

where Q is the quantity of dog toys produced (in thousands) per week, x is the number of units of labor, and y is the number of units of capital. Each unit of labor is $600, and each unit of capital is $400. The budget for both is limited to $24,000.

(a) Determine the number of units of labor and the number of units of capital required to maximize production.

(b) Determine the marginal productivity of money for the division of labor and capital found in part (a) and interpret.

(c) Determine the optimal increase in production if an additional $2000 is allotted toward the budget.

89. **Economics—Minimizing Cost:** A family plans to plant a garden with an area of 300 square feet. Three sides of the garden will have fencing, and the fourth side will have a wall that costs five times as much (per linear foot) as the fencing. Use the Method of Lagrange Multipliers to determine the dimensions of the garden that will minimize the cost.

90. Management—Shipping Dimensions: Packages weighing less than 15 pounds sent by Priority Mail are charged at the standard rate (based on the weight of the package), provided that the length plus girth is no more than 84 inches. Use the Method of Lagrange Multipliers to determine the dimensions of the largest volume package that can be sent via Priority Mail at the standard rate. Assume that the length is *y*. (*Source:* United States Postal Service.)

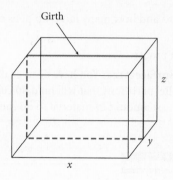

Section 8.6 Review Exercises

In Exercises 91–94, evaluate the given integral.

91. $\displaystyle\int_5^8 2x^2 y^3 \, dx$

92. $\displaystyle\int_0^5 (3x - 4y) \, dx$

93. $\displaystyle\int_2^6 (x^2 y - 4y^3) \, dx$

94. $\displaystyle\int_3^9 (x^2 y^3 + 4xy) \, dx$

For Exercises 95 and 96, complete the following.

(a) Sketch the region of integration.

(b) Write and evaluate a double integral with a *dy dx* order of integration.

(c) Write and evaluate a double integral with a *dx dy* order of integration.

95. $\displaystyle\int (3x - 4xy) \, dA$ $R = \{(x, y) \,|\, 0 \le x \le 5, \ 0 \le y \le 2\}$

96. $\displaystyle\int \sqrt[3]{xy} \, dA$ $R = \{(x, y) \,|\, 0 \le x \le 8, \ 1 \le y \le 27\}$

For Exercises 97 and 98, complete the following.

(a) Sketch the region of integration.

(b) Evaluate the given double integral.

97. $\displaystyle\int_0^4 \int_1^3 3xy^2 \, dy \, dx$

98. $\displaystyle\int_{-2}^6 \int_0^3 (2x + ye^x) \, dx \, dy$

In Exercises 99 and 100, determine the volume under $f(x, y)$ above the given region R.

99. $f(x, y) = 25 - x^2 - y^2$ $R = \{(x, y) \,|\, 0 \le x \le 3, \ 0 \le y \le 4\}$

100. $f(x, y) = \dfrac{e^x}{y}$ $R = \{(x, y) \,|\, 0 \le x \le 4, \ 1 \le y \le e\}$

In Exercises 101–104, determine the average value of the function over the region R.

101. $f(x, y) = x^2 + 3xy - y^2$ $R = \{(x, y) \,|\, 0 \le x \le 3, \ 0 \le y \le 4\}$

102. $f(x, y) = 3x + 4y^3$ $R = \{(x, y) \,|\, 0 \le x \le 5, \ 0 \le y \le 2\}$

103. $f(x, y) = x^3 y$ $\qquad$ $R = \{(x, y) | 1 \le x \le 5,\ 3 \le y \le 7\}$

104. $f(x, y) = e^{x + y/2}$ $\qquad$ $R = \{(x, y) | 0 \le x \le 1,\ 0 \le y \le 2\}$

105. Operations Research—Average Production: The Cripton Company has a production function given by

$$Q = f(x, y) = 11.8x^{0.68} y^{0.45}$$

where Q is the number of radios produced (in thousands) each week, x is the number of employees at the company, and y is the weekly operating budget in thousands. Because the company uses a temporary labor agency, it has anywhere from 18 to 30 employees each week, and its operating budget is anywhere from \$16,000 to \$24,000 each week. Determine the average number of radios that the company produces each week.

106. Meteorology—Average Temperature: The temperature, in degrees Fahrenheit, x miles east and y miles north of a reporting weather station is given by

$$f(x, y) = 68 + 2x - 3y$$

Determine the average temperature over the region shown in the figure.

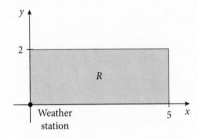

107. Meteorology—Air Pollution: Brandonville is a town whose shape is fairly rectangular, as shown. A heavy industrial complex located at the southwest corner of the town emits pollution into the atmosphere. The air pollution (in parts per million) x miles east and y miles north of the complex is given by

$$f(x, y) = 150 - x^2 - 2y^2$$

Determine the average concentration of air pollution in Brandonville.

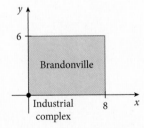

108. Demography—Average Population: Refer to Exercise 107. The population density of Brandonville, in people per square mile, x miles east and y miles north of the southwest corner of the town is given by

$$h(x, y) = 8000(3 + x + y^2)$$

Determine the average population of Brandonville.

Essentials of Algebra

Exponents and Radicals

If a is a real number and n is a positive integer, then a^n *is the product of n factors of a*. That is, $a^n = a \cdot a \cdot a \cdots \cdots a$. We can use this definition to give us the **Properties of Exponents**.

DEFINITION

Properties of Exponents

Let m and n be integers and a and b be real numbers. Then

1. $a^0 = 1, a \neq 0$

2. $a^1 = a$

3. $a^m \cdot a^n = a^{m+n}$

4. $\dfrac{a^m}{a^n} = a^{m-n}, a \neq 0$

5. $(ab)^n = a^n \cdot b^n$

6. $\left(\dfrac{a}{b}\right)^n = \dfrac{a^n}{b^n}, b \neq 0$

7. $a^{-n} = \dfrac{1}{a^n}, a \neq 0$

8. $a^{m/n} = \sqrt[n]{a^m} = (\sqrt[n]{a})^m$

Property 8 is a special type of exponent called a **rational exponent**. An expression containing a rational exponent can be written as a radical expression, and vice versa. Many times in calculus we wish to write expressions in the exponential form $ax^{m/n}$.

The properties of exponents can also be applied to write expressions given in rational exponent form in radical form.

Example 1: Writing Expressions in Radical Form

Write the following expressions in radical form.

a. $x^{3/2}$

b. $7x^{-1/4}$

Perform the Mathematics

a. Using exponent property 8, we see that $x^{3/2} = \sqrt{x^3}$.

b. We start by applying exponent property 7 to get $7x^{-1/4} = \dfrac{7}{x^{1/4}}$. Exponent property 8 then

gives us $7x^{-1/4} = \dfrac{7}{x^{1/4}} = \dfrac{7}{\sqrt[4]{x}}$. ∎

Systems of Equations

A **system of equations** is a collection of two or more equations with each having one or more variables. The following are examples of systems of equations.

$$2x + y = 6 \qquad \begin{aligned} 2x + y - z &= 2 \\ x + 3y + 2z &= 1 \\ x + y + z &= 2 \end{aligned}$$

$$x - y = 6$$

A **solution** to a system of equations is the values of the variables that make each equation in the system true. For example, $x = 4, y = -2$ is a solution to the first system above

$$\text{since} \quad \begin{aligned} 2(4) + (-2) &= 6 \qquad \text{is true} \\ 4 - (-2) &= 6 \qquad \text{is true}. \end{aligned}$$

The algebraic method used to solve many of the systems in this textbook is the **Substitution Method.**

Substitution Method

1. Choose one of the equations and solve for one of the variables.
2. Substitute the result in the remaining equation(s).
3. If one equation with one variable results, solve the equation. Otherwise, repeat step 1.
4. Find the values of the remaining variables by back-substitution.

Example 2: Solving a System of Equations by Substitution

Solve the system of equations $\begin{aligned} 3x + 2y &= 11 \\ -x + y &= 3 \end{aligned}$ by the substitution method.

Perform the Mathematics

We begin by solving the second equation for y to get $y = x + 3$. Now we substitute this result into the first equation to get

$$3x + 2(x + 3) = 11$$
$$3x + 2x + 6 = 11$$
$$5x + 6 = 11$$
$$5x = 5$$
$$x = 1$$

Now we back-substitute $x = 1$ in the second equation to determine y.

$$-1 + y = 3$$
$$y = 4$$

So the solution to the system is $x = 1, y = 4$. ∎

Many of the systems we solve have one solution, just as in Example 2. We call these systems **consistent.** If a system has no solutions, it is called **inconsistent.** Some systems have an infinite number of solutions. These systems are called **dependent.**

Another technique used to solve systems of equations is the **graphical method.** This method is based on the following observation.

For a consistent system, the coordinates of the points of intersection on the graphs of the equations are the solutions of the system.

Technology Option

In order to verify the solution of the system $\begin{array}{c} 3x + 2y = 11 \\ -x + y = 3 \end{array}$ by the graphical method, we need to graph each equation in the system. To graph the system, we need to solve each equation for y to get

$$y_1 = \frac{11 - 3x}{2}$$
$$y_2 = x + 3$$

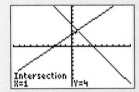

Figure A.1

The graph of the equations is shown in **Figure A.1** with the `Intersect` command used to determine the intersection point. We see that the result is the same as in Example 2.

Sometimes we must solve a system of three equations with three variables. The graphical method is of little use in this case, so we use the substitution method.

Example 3: Solving a System of Three Equations

Solve the system of equations $\begin{array}{rcl} x + z &=& 0 \\ y + 2z &=& 0 \\ 2x + y &=& 200 \end{array}$ using the substitution method.

Perform the Mathematics

We begin by solving the first equation for x and the second equation for y to get $x = -z$ and $y = -2z$, respectively. Now we substitute these expressions in the third equation to get

$$2(-z) + (-2z) = 200$$
$$-2z - 2z = 200$$
$$-4z = 200$$
$$z = -50$$

Since we know that $z = -50$, we can substitute this value into the first equation in the system to get $x - 50 = 0$; so we see that $x = 50$. Using $z = -50$ again, we back-substitute in the second equation to get

$$y + 2(-50) = 0$$
$$y - 100 = 0$$
$$y = 100$$

So the solution to this system is $x = 50$, $y = 100$, and $z = -50$. ∎

Properties of Logarithms

We can define a logarithm base b of x by saying that $y = \log_b x$ if and only if $b^y = x$, where $b > 0$, $b \neq 1$, $x > 0$. We can use this definition to prove the **Properties of Logarithms**. Some of these properties are proved in Appendix C.

DEFINITION

Properties of Logarithms

Let $b > 0$, $b \neq 1$, where m and n are positive real numbers and r is any real number. Then

1. $\log_b 1 = 0$
2. $\log_b b = 1$
3. $\log_b(m \cdot n) = \log_b m + \log_b n$
4. $\log_b\left(\dfrac{m}{n}\right) = \log_b m - \log_b n$
5. $\log_b m^r = r\log_b m$
6. $\log_b b^r = r$

Some logarithms have special bases. If $b = 10$, then $\log_{10} x = \log x$ is a **common logarithm**. Many applications in this textbook use the number e as the base. This is called a **natural logarithm** and is denoted by $\log_e x = \ln x$.

Example 4: Using the Properties of Logarithms

Use the properties of logarithms to write the following as a sum, difference, or product of logs.

a. $\log_b \dfrac{(x - 3)y}{x}$

b. $\log_b[(x + 7)y]^4$

Perform the Mathematics

a. $\log_b \dfrac{(x - 3)y}{x} = \log_b[(x - 3)y] - \log_b x$ Logarithm property 4

$\qquad\qquad\quad = \log_b(x - 3) + \log_b y - \log_b x$ Logarithm property 3

b. $\log_b[(x + 7)y]^4 = 4\log_b[(x + 7)y]$ Logarithm property 5

$\qquad\qquad\quad = 4[\log_b(x + 7) + \log_b y]$ Logarithm property 3 ■

Working with Logarithms

There are times when we need to solve **logarithmic equations.** The method used to solve these equations is based on the following property:

$$\text{If } \log_b m = \log_b n, \text{ then } m = n.$$

Here m and n are called the **arguments** of the logarithmic expressions. The next example illustrates the use of this property.

Example 5: Solving Logarithmic Equations

Solve the equation for x: $\log_b(x + 6) - \log_b(x + 2) = \log_b x$

Perform the Mathematics

We start by rewriting the left side as a single logarithmic expression to get

$$\log_b\left(\frac{x + 6}{x + 2}\right) = \log_b x$$

Now we use the property for solving logarithmic equations by setting the arguments of the logarithmic expression equal to each other.

$$\frac{x + 6}{x + 2} = x$$
$$x + 6 = x(x + 2)$$
$$x + 6 = x^2 + 2x$$
$$0 = x^2 + x - 6$$
$$0 = (x - 2)(x + 3)$$
$$x = 2 \ \text{ or } \ x = -3$$

Since we cannot take the logarithm of a negative number, we see that $x = -3$ is not a solution. So the only solution to the equation is $x = 2$. ∎

Another way we use logarithms is to solve **exponential equations.** If we are given an equation of the form $b^x = k$, we can solve for x by taking the logarithm of each side of the equation and using logarithm property 5. Since it does not matter which base of logarithm we use, we usually take the natural logarithm of each side.

Example 6: Solving Exponential Equations

Solve the equation $3 \cdot 5^x = 21$ for x.

Solution

$$\begin{array}{ll} 3 \cdot 5^x = 21 & \text{Given equation} \\ 5^x = 7 & \text{Divide by 3} \\ \ln(5^x) = \ln 7 & \text{Take the natural logarithm of both sides} \\ x \ln 5 = \ln 7 & \text{By logarithm property 5} \\ x = \dfrac{\ln 7}{\ln 5} \approx 1.21 & \text{Divide by } \ln 5 \end{array}$$ ∎

Appendix A Exercises

In Exercises 1–6, write the expression in exponential form.

1. $\dfrac{-4}{x^6}$

2. $\dfrac{17}{x^2}$

3. $1.6\sqrt{x}$

4. $7.2\sqrt[4]{x}$

5. $\dfrac{8}{\sqrt[3]{x^2}}$

6. $\dfrac{-1}{\sqrt[4]{x^5}}$

In Exercises 7–12, write the expressions in radical form.

7. $x^{3/7}$

8. $x^{1/8}$

9. $6.3x^{4/5}$

10. $2.1x^{2/5}$

11. $2x^{-2/3}$

12. $5x^{-1/2}$

In Exercises 13–18, solve the system of equations using the substitution method.

13. $y = x + 2$
$x + 2y = 16$

14. $x - y = 2$
$2x + y = 13$

15. $x + 2y = 8$
$3x - 4y = 9$

16. $2x - y = -21$
$4x + 5y = 7$

17. $4x + 3z = 4$
$2y - 6z = -1$
$8x + 4y + 3z = 9$

18. $x + y + z = 2$
$2x + y - z = 5$
$x - y + z = -2$

For Exercises 19 and 20, solve the system of equations using the graphical method.

 19. $2x + 5y = 21$
$3x - 2y = -16$

 20. $3x - 4y = 1$
$2x + 3y = 12$

In Exercises 21–26, use the properties of logarithms to write the following as a sum, difference, or product of logs.

21. $\log_b\left(\dfrac{2x}{5y}\right)^3$

22. $\log_b\left(\dfrac{7y}{3x}\right)^2$

23. $\log\left(\dfrac{xy^2}{y + 4}\right)$

24. $\log\left(\dfrac{x^5 y^3}{x - 2}\right)$

25. $\ln\sqrt{\dfrac{5x^3}{y^9}}$

26. $\ln\sqrt[3]{\dfrac{y^4}{x^2}}$

In Exercises 27–30, solve the logarithmic equation for x.

27. $\log_b(x - 2) = \log_b(x - 7) + \log_b 4$

28. $\log_b x - \log_b(x + 1) = \log_b 5$

29. $\ln(4x + 5) - \ln(x + 3) = \ln 3$

30. $\ln(4x - 2) = \ln 4 - \ln(x - 2)$

In Exercises 31–34, solve the exponential equation for x.

31. $3^x = 6$

32. $4^x = 12$

33. $2^{5x+2} = 8$

34. $10^{3x-7} = 5$

Calculator Programs

All the programs in this section are written for the Texas Instruments TI-84 family. To convert the programs for your model of calculator, consult your owner's manual.

Secant Line Slope Program

This program allows us to quickly compute the slope of secant lines over smaller and smaller intervals. The function for which you are computing the secant line slope over smaller and smaller intervals must first be entered into Y_1 in your calculator. This program then allows us to approximate the tangent line slope at a point.

```
PROGRAM:MSEC
Input"X-VALUE",X
For(I,0,3)
1/(10^I)→H
(Y₁(X+H)−Y₁(X))/H→M
Disp M
End
Disp"PRESS A KEY"
Disp"TO CONTINUE"
Pause
For(J,0,3)
−1/(10^J)→H
(Y₁(X+H)−Y₁(X))/H→M
Disp M
End
Stop
```

Quadratic Formula Program

This program approximates the real roots of the quadratic equation $ax^2 + bx + c = 0$ or reports that there are no real roots. The values of the coefficients a, b, and c are entered after the program is executed.

```
PROGRAM:QUADFORM
Prompt A,B,C
(B^2−4*A*C)→D
If D<0
Then
Disp"NO REAL SOLUTIONS"
Goto 1
Else
```

```
(-B+√(D))/(2*A)→M
(-B-√(D))/(2*A)→N
Disp"SOLUTIONS",M,N
Lbl 1
```

Example 1

Use the Quadratic Formula Program to approximate the roots of the quadratic equation $2x^2 - 4x = 3$.

Perform the Mathematics

First we need to write the equation in the standard form $2x^2 - 4x - 3 = 0$. We enter $a = 2$, $b = -4$, and $c = -3$. After the program is executed, we get the output shown in **Figure B.1**.

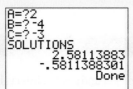

```
A=?2
B=?-4
C=?-3
SOLUTIONS
          2.58113883
         -.5811388301
                 Done
```

Figure B.1   ∎

Newton's Method Program

We can extend the ideas of tangent lines and linear approximations discussed in Section 3.4 to derive what is known as Newton's Method, named after one of the founding fathers of calculus, Isaac Newton. This method can be used to numerically determine zeros (or roots) of differentiable functions. For the method, we let f be a differentiable function and suppose that r is a real number zero. If x_n is an approximation to r, then the next approximation is given by $x_{n+1} = x_n - \dfrac{f(x_n)}{f'(x_n)}$, provided that $f'(x_n) \neq 0$, where $n = 0, 1, 2, \ldots$. Before running the program, $f(x)$ must be stored in Y_1 and its derivative $f'(x)$ stored in Y_2. On execution of the program, an initial guess $X = x_0$ is prompted.

```
PROGRAM:NEWTON
Prompt X
Lbl A
X-Y₁(X)/Y₂(X)→X
Disp X
Pause
Goto A
```

Example 2

For the function $f(x) = x^2 - 4x - 2$, compute the first three iterations of an approximation to a real root of f using the Newton's Method program. Use the initial guess $x = -2$.

Perform the Mathematics

First we enter $f(x)$ in Y_1 and $\frac{d}{dx}(x^2 - 4x - 2) = 2x - 4$ in Y_2 and then execute the program. The first three iterations are displayed in **Figure B.2**.

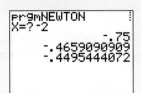

Figure B.2

The Enter key must be pressed to get the next approximation. ∎

Left/Right Sum Program

This program combines the left sum and right sum methods, which were discussed in Section 6.2. To run the program, the integrand of $\int_a^b f(x)\,dx$ must be entered in Y_1. When executed, the program prompts the user to enter values for the limits of integration a and b and the desired number of rectangles n.

```
PROGRAM:LFTRTSUM
Prompt A,B,N
(B−A)/N→D
0→T
0→Z
For (I,0,N−1)
A+I*D→X
T+D*Y₁→T
End
Disp"LEFT SUM APPROX",T
For (J,1,N)
A+J*D→X
Z+D*Y₁→Z
End
Disp"RIGHT SUM APPROX",Z
Disp"DIFFERENCE",abs(T−Z)
```

Example 3

Use the Left/Right Sum Program to approximate $\displaystyle\int_0^3 \frac{1}{1+x}\,dx$ using $n = 20$ rectangles.

Perform the Mathematics

Here we enter $\dfrac{1}{1+x}$ in Y_1. After the program is executed, we enter $a = 0$, $b = 3$, and $n = 20$. The output is shown in **Figure B.3**.

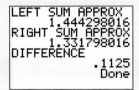

Figure B.3 ∎

Midsum Program

This program is another way to approximate the area under a curve based on the rectangular approximation methods discussed in Section 6.2. It uses the midpoint rule, which states that the area under the graph of f on the closed interval $[a, b]$ can be approximated by

$$\text{Area} \approx \sum_{i=1}^{n} f\left(\frac{x_i + x_{i-1}}{2}\right)\Delta x$$

where the height of each rectangle is determined by the middle of its base. The integrand of $\int_a^b f(x)\,dx$ must be entered in Y_1. When executed, the program prompts the user to enter values for the limits of integration a and b and the number of desired rectangles n.

```
PROGRAM:MIDSUM
Prompt A,B,N
0→T
(B−A)/N→H
For(I,1,N,1)
A+I*H→F
A+(I−1)*H→G
(F+G)/2→X
T+Y₁*H→T
End
Disp"APPROX AREA",T
```

Example 4

Use the Midsum Program to approximate $\displaystyle\int_0^3 \frac{1}{1+x}\,dx$ using $n = 20$ rectangles.

Perform the Mathematics

We begin by entering $\dfrac{1}{1+X}$ in Y_1. When we enter $a = 0$, $b = 3$, and $n = 20$ after the program is executed, we get the output shown in **Figure B.4**.

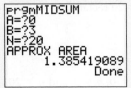

Figure B.4

Slope Field Program

This program sketches the slope field, or family of solution curves, for the differential equation $\frac{dy}{dx} = f(x, y)$. Here $\frac{dy}{dx}$ is entered in Y_1.

```
PROGRAM:SLPFILD
Clr Draw
Fnoff
7*(Xmax-Xmin)/83→H
7*(Ymax-Ymin)/55→K
6.25/H^2→A
6.25/K^2→B
Xmin+0.5*H→X
Ymin+0.5*K→V
1→L
Lbl 1
V→Y
1→J
Lbl 2
Y₁→T
1/√(A+B*T²)→M
T*M→N
Line (X-M,Y-N,X+M,Y+N)
Y+K→Y
IS>(J,8)
Goto 2
X+H→X
IS>(L,12)
Goto 1
```

Example 5

Graph the slope field for the differential equation $\frac{dy}{dx} = \frac{x}{2y}$ in the viewing window $[-3, 3]$ by $[-3, 3]$.

Perform the Mathematics

We enter the differential equation as shown in **Figure B.5**.

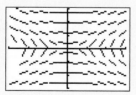

Figure B.5   **Figure B.6**

When the program is executed, we get the output shown in **Figure B.6**. ∎

Multiple Regression Program

This program models the Cobb-Douglas production function $Q(L, K) = kL^aK^b$, where the labor data (L) is entered in list L_1, the capital data in L_2, and the quantity data in L_3. On execution, the program prompts the user to enter the number of data points used.

```
PROGRAM:MULTIREG
Disp"NUMBER OF DATA POINTS"
Prompt N
log(L₁)→L₁
log(L₂)→L₂
log(L₃)→L₃
(sum(L₁))/N→P
(sum(L₂))/N→Q
(sum(L₃))/N→R
sum(L₁*L₃)−N*P*R→A
sum(L₂*L₃)−N*Q*R→B
sum((L₁)^2)−N*P^2→C
sum(L₁*L₂)−N*P*Q→D
sum((L₂)^2)−N*Q^2→E
(A*E−B*D)/(C*E−D^2)→F
(B*C−A*D)/(C*E−D^2)→G
R−F*P−G*Q→H
Disp"EXPONENT FOR L IS",F
Disp"EXPONENT FOR K IS",G
Disp"CONSTANT IS",10^H
```

Example 6

The Brandy Corporation has 11 plants worldwide. In a recent year, the data for each plant gave the number of labor hours (in thousands), capital (total net assets, in millions), and total quantity produced, as follows:

Labor	250	270	300	320	350	400	440	440	450	460	460
Capital	30	34	44	50	70	76	84	86	104	110	116
Quantity	245	240	300	320	390	440	520	520	580	600	600

The plants all use the same technology, so a production function can be determined. Enter the data and execute the program to determine a Cobb-Douglas production function to model the data.

Perform the Mathematics

We enter the labor, capital, and quantity data in lists L_1, L_2, and L_3, respectively. The output is displayed in **Figure B.7**.

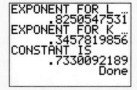

```
EXPONENT FOR L
      .8250547531
EXPONENT FOR K
      .3457819856
CONSTANT IS
      .7330092189
            Done
```

Figure B.7

So, based on the data, the Cobb-Douglas production function can be modeled by $Q(L, K) = 0.73L^{0.83}K^{0.35}$.

Selected Proofs

The solution of the quadratic equation $ax^2 + bx + c = 0$ is given by $x = \dfrac{-b \pm \sqrt{b^2 - 4ac}}{2a}$.

Proof: We will solve the quadratic equation $ax^2 + bx + c = 0$ by completing the square.

$ax^2 + bx = -c$	Subtract c
$x^2 + \dfrac{b}{a}x = -\dfrac{c}{a}$	Divide by a
$x^2 + \dfrac{b}{a}x + \left(\dfrac{b}{2a}\right)^2 = \left(\dfrac{b}{2a}\right)^2 - \dfrac{c}{a}$	Complete the square
$x^2 + \dfrac{b}{a}x + \left(\dfrac{b}{2a}\right)^2 = \dfrac{b^2}{4a^2} - \dfrac{c}{a}$	Simplify the right side
$\left(x + \dfrac{b}{2a}\right)^2 = \dfrac{b^2}{4a^2} - \dfrac{c}{a}$	Write the left side as a squared binomial
$x + \dfrac{b}{2a} = \pm\sqrt{\dfrac{b^2}{4a^2} - \dfrac{c}{a}}$	Take the square root of both sides
$x + \dfrac{b}{2a} = \pm\sqrt{\dfrac{b^2 - 4ac}{4a^2}}$	Write the radicand as a single fraction
$x + \dfrac{b}{2a} = \pm\dfrac{\sqrt{b^2 - 4ac}}{2a}$	Take the square root of the denominator on the right side
$x = -\dfrac{b}{2a} \pm \dfrac{\sqrt{b^2 - 4ac}}{2a}$	Subtract $-\dfrac{b}{2a}$
$x = \dfrac{-b \pm \sqrt{b^2 - 4ac}}{2a}$	Write the right side as a single fraction ∎

Properties of Logarithms

Here we will prove three of the properties of logarithms listed in Appendix A.

Property: $\log_b(m \cdot n) = \log_b m + \log_b n$

Proof: Let $x = \log_b m$ and $y = \log_b n$. Then, by the definition of logarithm, we may write $b^x = m$ and $b^y = n$. Multiplying these equations gives

$$b^x \cdot b^y = m \cdot n = b^{x+y}$$

Using the definition of logarithm, we may write the equation $m \cdot n = b^{x+y}$ as

$$\log_b(m \cdot n) = x + y$$

Since $x = \log_b m$ and $y = \log_b n$, we get

$$\log_b(m \cdot n) = \log_b m + \log_b n \qquad \blacksquare$$

Property: $\log_b\left(\dfrac{m}{n}\right) = \log_b m - \log_b n$

Proof: Let $x = \log_b m$ and $y = \log_b n$. Then, by the definition of logarithm, we may write $b^x = m$ and $b^y = n$. Dividing these equations gives

$$\frac{b^x}{b^y} = \frac{m}{n} = b^{x-y}$$

Using the definition of logarithm, we may write the equation $\dfrac{m}{n} = b^{x-y}$ as

$$\log_b\left(\frac{m}{n}\right) = x - y$$

Since $x = \log_b m$ and $y = \log_b n$, we get

$$\log_b\left(\frac{m}{n}\right) = \log_b m - \log_b n \qquad \blacksquare$$

Property: $\log_b m^r = r \log_b m$

Proof: Let $x = \log_b m$. Then, by the definition of logarithm, we may write the equation in exponential form as $b^x = m$. If we raise both sides of this equation to the rth power, we get

$$(b^x)^r = m^r \qquad b^{xr} = m^r$$

By the definition of logarithm, we can write the equation $b^{xr} = m^r$ as

$$xr = \log_b m^r$$

Since $x = \log_b m$, we have $r \log_b m = \log_b m^r$. $\qquad \blacksquare$

Power Rule

If $f(x) = x^n$, where n is any real number, then $f'(x) = nx^{n-1}$. Equivalently $\frac{d}{dx}[x^n] = nx^{n-1}$, and if $y = x^n$, then $\frac{dy}{dx} = nx^{n-1}$.

Proof: Here we will prove the Power Rule for n being a positive integer. By the definition of derivative, we have $\dfrac{d}{dx}[x^n] = \lim\limits_{h \to 0} \dfrac{(x + h)^n - x^n}{h}$. From the Binomial Theorem in algebra, we know that $(x + h)^n$ can be written as

$$(x + h)^n = x^n + nx^{n-1}h + \frac{n(n - 1)}{2}x^{n-2}h^2 + \cdots + nxh^{n-1} + h^n$$

Substituting this expression into the definition of derivative, we get

$$\frac{d}{dx}[x^n] = \lim_{h \to 0} \frac{\left[x^n + nx^{n-1}h + \frac{n(n-1)}{2}x^{n-2}h^2 + \cdots + nxh^{n-1} + h^n\right] - x^n}{h}$$

$$= \lim_{h \to 0} \frac{nx^{n-1}h + \frac{n(n-1)}{2}x^{n-2}h^2 + \cdots + nxh^{n-1} + h^n}{h}$$

$$= \lim_{h \to 0} \left[nx^{n-1} + \frac{n(n-1)}{2}x^{n-2}h + \cdots \left(\begin{array}{c}\text{other terms} \\ \text{with } x \text{ and} \\ h \text{ factors}\end{array}\right) \cdots + nxh^{n-2} + h^{n-1}\right]$$

Since every term in the brackets except the first contains h as a factor, every term except the first approaches zero as $h \to 0$, and we get

$$\frac{d}{dx}[x^n] = nx^{n-1} \qquad \blacksquare$$

Constant Multiple Rule

If $f(x) = k \cdot g(x)$, where k is any real number, then $f'(x) = k \cdot g'(x)$, assuming that g is differentiable. Equivalently, $\frac{d}{dx}[k \cdot g(x)] = k \cdot g'(x)$.

Proof: Recall that the definition of the derivative of a function f is $\frac{d}{dx}[f(x)] = \lim_{h \to 0} \frac{f(x+h) - f(x)}{h}$. Using this definition for $f(x) = k \cdot g(x)$, we get

$$\frac{d}{dx}[k \cdot g(x)] = \lim_{h \to 0} \frac{k \cdot g(x+h) - k \cdot g(x)}{h} \qquad \text{Using the definition of derivative}$$

$$= \lim_{h \to 0} \frac{k[g(x+h) - g(x)]}{h} \qquad \text{Factor } k \text{ from the numerator}$$

$$= k \cdot \lim_{h \to 0} \frac{g(x+h) - g(x)}{h} \qquad \text{From Limit Theorem 3 in Section 2.1}$$

$$= k \cdot g'(x) \qquad \text{By definition}$$
$$\lim_{h \to 0} \frac{g(x+h) - g(x)}{h} = g'(x) \qquad \blacksquare$$

Sum Rule

If $h(x) = f(x) + g(x)$, where f and g are differentiable functions, then $h'(x) = f'(x) + g'(x)$. Equivalently, $\frac{d}{dx}[f(x) + g(x)] = \frac{d}{dx}[f(x)] + \frac{d}{dx}[g(x)] = f'(x) + g'(x)$.

Proof: We want to show that $\frac{d}{dx}[f(x) + g(x)] = f'(x) + g'(x)$, so, by the definition of derivative,

$$\frac{d}{dx}[f(x) + g(x)] = \lim_{h \to 0} \frac{[f(x + h) + g(x + h)] - [f(x) + g(x)]}{h}$$
Using the definition of derivative

$$= \lim_{h \to 0} \frac{f(x + h) + g(x + h) - f(x) - g(x)}{h}$$
Removing the parentheses

$$= \lim_{h \to 0} \frac{f(x + h) - f(x) + g(x + h) - g(x)}{h}$$
Interchanging the two middle terms in the numerator

$$= \lim_{h \to 0} \left[\frac{f(x + h) - f(x)}{h} + \frac{g(x + h) - g(x)}{h} \right]$$
Rewriting the difference quotient as two fractions

$$= \lim_{h \to 0} \left[\frac{f(x + h) - f(x)}{h} \right] + \lim_{h \to 0} \frac{g(x + h) - g(x)}{h}$$
Limit Theorem 4 in Section 2.1

$$= f'(x) + g'(x)$$
By the definition of derivative ∎

Difference Rule

If $h(x) = f(x) - g(x)$, where f and g are differentiable functions, then $h'(x) = f'(x) - g'(x)$.

Equivalently,

$$\frac{d}{dx}[f(x) - g(x)] = \frac{d}{dx}[f(x)] - \frac{d}{dx}[g(x)] = f'(x) - g'(x)$$

Proof: We want to show that $\frac{d}{dx}[f(x) - g(x)] = f'(x) - g'(x)$, so, by the definition of derivative,

$$\frac{d}{dx}[f(x) - g(x)] = \lim_{h \to 0} \frac{[f(x + h) - g(x + h)] - [f(x) - g(x)]}{h}$$
Using the definition of derivative

$$= \lim_{h \to 0} \frac{f(x + h) - g(x + h) - f(x) + g(x)}{h}$$
Removing the parentheses

$$= \lim_{h \to 0} \frac{f(x + h) - f(x) - g(x + h) + g(x)}{h}$$
Interchanging the two middle terms in the numerator

$$= \lim_{h \to 0} \frac{[f(x + h) - f(x)] - [g(x + h) - g(x)]}{h}$$
Factor −1 from the last two terms in the numerator

$$= \lim_{h \to 0} \left[\frac{f(x + h) - f(x)}{h} - \frac{g(x + h) - g(x)}{h} \right]$$
Rewriting the difference quotient as two fractions

$$= \lim_{h \to 0} \left[\frac{f(x + h) - f(x)}{h} \right] - \lim_{h \to 0} \frac{g(x + h) - g(x)}{h}$$
Limit Theorem 4 in Section 2.1

$$= f'(x) - g'(x)$$
By the definition of derivative ∎

Product Rule

If $h(x) = f(x) \cdot g(x)$, where f and g are differentiable functions, then

$$h'(x) = f'(x) \cdot g(x) + f(x) \cdot g'(x)$$

Equivalently,

$$\frac{d}{dx}[f(x) \cdot g(x)] = \frac{d}{dx}[f(x)] \cdot g(x) + f(x) \cdot \frac{d}{dx}[g(x)] = f'(x) \cdot g(x) + f(x) \cdot g'(x)$$

Proof: We want to show that $\frac{d}{dx}[f(x) \cdot g(x)] = f'(x) \cdot g(x) + f(x) \cdot g'(x)$, so, by the definition of derivative,

$$\frac{d}{dx}[f(x) \cdot g(x)] = \lim_{h \to 0} \frac{f(x + h) \cdot g(x + h) - f(x) \cdot g(x)}{h}$$

Now we subtract and add the term $f(x) \cdot g(x + h)$ in the numerator to get

$$= \lim_{h \to 0} \frac{f(x + h) \cdot g(x + h) - f(x) \cdot g(x + h) + f(x) \cdot g(x + h) - f(x) \cdot g(x)}{h}$$

Factoring $g(x + h)$ from the first two terms and $f(x)$ from the last two terms gives

$$= \lim_{h \to 0} \frac{[f(x + h) - f(x)] \cdot g(x + h) + f(x) \cdot [g(x + h) - g(x)]}{h}$$

Writing the difference quotient as two fractions yields

$$= \lim_{h \to 0} \left[\frac{f(x + h) - f(x)}{h} \cdot g(x + h) \right] + \lim_{h \to 0} \left[f(x) \cdot \frac{g(x + h) - g(x)}{h} \right]$$

$$= f'(x) \cdot g(x) + f(x) \cdot g'(x) \qquad \blacksquare$$

Quotient Rule

If $h(x) = \dfrac{f(x)}{g(x)}$, where f and g are differentiable functions, then

$$h'(x) = \frac{f'(x) \cdot g(x) - f(x) \cdot g'(x)}{[g(x)]^2} \quad \text{where } g(x) \neq 0$$

Equivalently,

$$\frac{d}{dx}\left[\frac{f(x)}{g(x)} \right] = \frac{\frac{d}{dx}[f(x)] \cdot g(x) - f(x) \cdot \frac{d}{dx}[g(x)]}{[g(x)]^2}$$

Proof: We want to show that $\dfrac{d}{dx}\left[\dfrac{f(x)}{g(x)} \right] = \dfrac{f'(x) \cdot g(x) - f(x) \cdot g'(x)}{[g(x)]^2}$, so, by the definition of derivative,

$$\frac{d}{dx}\left[\frac{f(x)}{g(x)} \right] = \lim_{h \to 0} \frac{\dfrac{f(x + h)}{g(x + h)} - \dfrac{f(x)}{g(x)}}{h}$$

Multiplying the numerator and denominator by $g(x + h) \cdot g(x)$ gives

$$= \lim_{h \to 0} \frac{f(x + h) \cdot g(x) - f(x) \cdot g(x + h)}{h \cdot g(x + h) \cdot g(x)}$$

Subtracting and adding the term $f(x) \cdot g(x)$ in the numerator yields

$$= \lim_{h \to 0} \frac{f(x + h) \cdot g(x) - f(x) \cdot g(x) + f(x) \cdot g(x) - f(x) \cdot g(x + h)}{h \cdot g(x + h) \cdot g(x)}$$

$$= \lim_{h \to 0} \frac{[f(x + h) - f(x)]g(x) - f(x)] - g(x) + g(x + h)]}{h \cdot g(x + h) \cdot g(x)}$$

If we write the numerator in the form of two difference quotients and take the limit, we get

$$= \lim_{h \to 0} \left[\frac{\dfrac{f(x + h) - f(x)}{h} \cdot g(x) - f(x) \cdot \dfrac{g(x + h) - g(x)}{h}}{g(x + h) \cdot g(x)} \right]$$

$$= \frac{f'(x) \cdot g(x) - f(x) \cdot g'(x)}{[g(x)]^2}$$

∎

Differentiability Implies Continuity

If f is differentiable at $x = c$, then f is continuous at $x = c$.

Proof: Let f be differentiable at $x = c$. So we know that $f'(c)$ exists, and we need to show that $\lim_{x \to c} f(x) = f(c)$ or, equivalently, $\lim_{x \to c} [f(x) - f(c)] = 0$. Assuming $x \neq c$, we begin by multiplying and dividing the expression in the limit argument by $(x - c)$ to get

$$\lim_{x \to c} [f(x) - f(c)] = \lim_{x \to c} \left[(x - c) \cdot \frac{f(x) - f(c)}{x - c} \right]$$

By Limit Theorem 5 in Section 2.1, we have

$$\lim_{x \to c} [f(x) - f(c)] = \lim_{x \to c} (x - c) \cdot \lim_{x \to c} \frac{f(x) - f(c)}{x - c}$$

$$= 0 \cdot f'(c) = 0$$

Hence, $\lim_{x \to c} f(x) = f(c)$ and f is continuous at $x = c$. ∎

Fundamental Theorem of Calculus

If f is a continuous function defined on a closed interval $[a, b]$ and F is an antiderivative of f, then $\int_a^b f(x)\,dx = F(b) - F(a)$.

Proof: Let f be a continuous function with $f(x) > 0$ for all x in $[a, b]$. Now we define an area function A, which represents the area under the graph of f from a to x. See **Figure C.1**.

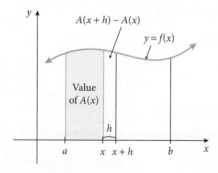

Figure C.1

We need to first show that $\left(\begin{array}{c}\text{Area under the graph}\\ \text{of } f \text{ on } [a, b]\end{array}\right) = A(b) - A(a)$. Since f is continuous on $[a, b]$, we know that $\int_a^x f(t)\,dt$ exists, and so we define $A(x) = \int_a^x f(t)\,dt$. We see that $A(a) = \int_a^a f(t)\,dt = 0$, and so $A(b) = A(b) - 0 = A(b) - A(a)$.

So we have $\left(\begin{array}{c}\text{Area under the graph}\\ \text{of } f \text{ on } [a, b]\end{array}\right) = A(b) - A(a)$. Now we need to show that A is an antiderivative of f. By the definition of derivative, we have $A'(x) = \lim\limits_{h \to 0} \dfrac{A(x + h) - A(x)}{h}$. Analyzing this limit, we see the following:

- $A(x + h)$ is the area under the graph of f between a and $x + h$.
- $A(x)$ is the area under the graph of f between a and x.
- $A(x + h) - A(x)$ is the area between x and $x + h$.

For a small value of h, x and $x + h$ are values close to one another, and since f is continuous, $f(x + h)$ is close in value to $f(x)$. In other words,

$$A(x + h) - A(x) \approx \left(\begin{array}{c}\text{Area of rectangle with height}\\ f(x) \text{ and width } h\end{array}\right) = f(x) \cdot h$$

So, for a small value h, $\dfrac{A(x + h) - A(x)}{h} \approx \dfrac{f(x) \cdot h}{h} = f(x)$. Thus, as $h \to 0$ we see

that $A'(x) = \lim\limits_{h \to 0} \dfrac{A(x + h) - A(x)}{h} = f(x)$. Hence, A is an antiderivative of f. ■

Derivation of the Least-Squares Formulas

Given a collection of data points $(x_1, y_1), (x_2, y_2), (x_3, y_3), \ldots, (x_n, y_n)$, the linear regression model for the data has the form $y = mx + b$, where

$$m = \frac{n\sum xy - \sum x \cdot \sum y}{n\sum x^2 - (\sum x)^2} \quad \text{and} \quad b = \frac{\sum y \cdot \sum x^2 - \sum x \cdot \sum xy}{n\sum x - (\sum x)^2}$$

Proof: Let $(x_1, y_1), (x_2, y_2), (x_3, y_3), \ldots, (x_n, y_n)$ be a set of data points. To determine a regression line, we need to find values m and b so that the sum of the squares of the residuals $d = y - (mx + b)$ is a minimum. If we write m and b in the form of a function of two independent variables, we see that we need to minimize

$$\begin{aligned}f(m, b) = \sum d^2 &= \sum [y - (mx + b)]^2\\ &= (y_1 - mx_1 - b)^2 + (y_2 - mx_2 - b)^2(y_3 - mx_3 - b)^2\\ &\quad + \cdots + (y_n - mx_n - b)^2\end{aligned}$$

The partial derivative of f with respect to m is

$$\begin{aligned}\frac{\partial f}{\partial m} &= 2(y_1 - mx_1 - b)(-x_1) + 2(y_2 - mx_2 - b)(-x_2)\\ &\quad + 2(y_3 - mx_3 - b)(-x_1) + \cdots + 2(y_n - mx_n - b)(-x_n)\\ &= 2[(-x_1y_1 - x_2y_2 - x_3y_3 - \cdots - x_ny_n) + m(x_1^2 + x_2^2 + x_3^2 + \cdots + x_n^2)\\ &\quad + b(x_1 + x_2 + x_3 + \cdots + x_n)\\ &= 2[-\sum xy + m\sum x^2 + b\sum x]\end{aligned}$$

The partial derivative of f with respect to b is

$$\frac{\partial f}{\partial b} = 2(y_1 - mx_1 - b)(-1) + 2(y_2 - mx_2 - b)(-1)$$

$$+ 2(y_3 - mx_3 - b)(-1) + \cdots + 2(y_n - mx_n - b)(-1)$$

$$= 2[(-y_1 - y_2 - y_3 - \cdots - y_n) + m(x_1 + x_2 + x_3 + \cdots + x_n)$$

$$+ b(1 + 1 + 1 + \cdots + 1)$$

$$= 2[-\Sigma y + m\Sigma x + bn]$$

Now, if we set $\frac{\partial f}{\partial m} = 0$ and $\frac{\partial f}{\partial b} = 0$ and solve the system of equations, we get

$$\begin{cases} -\Sigma xy + m\Sigma x^2 + b\Sigma x = 0 \\ -\Sigma y + m\Sigma x + bn = 0 \end{cases} \quad \text{or} \quad \begin{matrix} m\Sigma x^2 + b\Sigma x = \Sigma xy & \text{(I)} \\ m\Sigma x + bn = \Sigma y & \text{(II)} \end{matrix}$$

To solve for m, we multiply equation (I) by n and equation (II) by $-\Sigma x$, and then add the equations to get

$$nm\Sigma x^2 - m(\Sigma x)^2 = n\Sigma xy - \Sigma x\Sigma y$$

$$m[n\Sigma x^2 - (\Sigma x)^2] = n\Sigma xy - \Sigma x\Sigma y$$

$$m = \frac{n\Sigma xy - \Sigma x\Sigma y}{n\Sigma x^2 - (\Sigma x)^2}$$

To solve for b, we multiply equation (I) by $-\Sigma x$ and equation (II) by Σx^2, and then add the equations to get

$$bn\Sigma x - b(\Sigma x)^2 = \Sigma x^2\Sigma y - \Sigma x\Sigma xy$$

$$b[n\Sigma x - (\Sigma x)^2] = \Sigma x^2\Sigma y - \Sigma x\Sigma xy$$

$$b = \frac{\Sigma y \cdot \Sigma x^2 - \Sigma x \cdot \Sigma xy}{n\Sigma x - (\Sigma x)^2}$$

Answers to Odd-Numbered Exercises

Chapter 1

Section 1.1

1. Range **3.** Domain **5.** Evaluating

7.

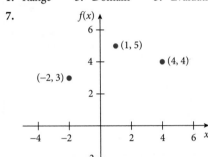

9.

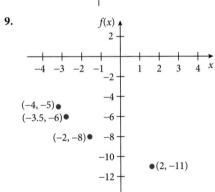

11.

x	y
2	6
4	5
6	4
8	2
10	0

13.

x	y
10	60
20	50
30	40
40	20
50	10

15.

x	y
-3	-30
-2	-10
-1	0
0	10
2	10

17. Function **19.** Not a function **21.** Function

23. Not a function **25.** Function **27.** $f(2) = 3$; $(2, 3)$

29. $f(1) = 1$; $(1, 1)$ **31.** $g(3) = 9$; $(3, 9)$ **33.** $g(0) = 0$; $(0, 0)$

35. $g(\frac{1}{2}) = \frac{1}{4}$; $(\frac{1}{2}, \frac{1}{4})$ **37.** $g(-0.25) = 0.0625$; $(-0.25, 0.0625)$

39. **(a)** t **(b)** $y = f(t)$ **(c)** $f(10) \approx 10$
 (d) $f(30) \approx 20$ **(e)** $[0, \infty)$ **(f)** $[0, \infty)$

41.

Interval Notation	Inequality Notation	Number Line
$[-\frac{1}{2}, 5)$	$-\frac{1}{2} \le x < 5$	
$[-2, \infty)$	$x \ge -2$	
$(\infty, -3)$	$x < -3$	
$(-1, 10)$	$-1 < x < 10$	
$(-\infty, 3) \cup (3, \infty)$	$x < 3$ or $x > 3$	

43. $(-\infty, \infty)$ **45.** $(-\infty, \infty)$ **47.** $(-\infty, 5) \cup (5, \infty)$

49. $(-\infty, 6]$ **51.** $(-\infty, \infty)$ **53.** $(-\infty, 0) \cup (0, 2) \cup (2, \infty)$

55. The graph $f(x)$ in viewing window $[-3, 3]$ by $[0, 2]$ is displayed below:

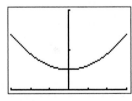

Domain $(-\infty, \infty)$; range $[\frac{1}{2}, \infty)$

57. The graph $f(x)$ in viewing window $[-10, 10]$ by $[-0.7, 0.3]$ is displayed below:

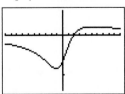

Domain $(-\infty, \infty)$; range $[-0.43, 0.1]$

59. The domain is $(-\infty, \infty)$. The range is $(-\infty, \infty)$.

61. The domain is $(-5, \infty)$. The range is $(-\infty, 7]$.

63. The domain is $(-50, 100]$. The range is $[0, 10)$.

65. (a)

x	0	5	10	15	20	25	30	35
$p(x)$	41.25	40	38.75	37.5	36.25	35	33.75	32.5

(b)

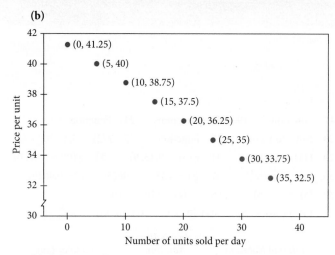

(c) $[32.5, 41.25]$

(d) $p(22) = 35.75$. If 22 units of Teddy Bear designer lingerie are sold per day, the price will be \$35.75 per unit.

67. \$36.75

69.

Year	Number of Housing Starts in the Midwestern U.S. (in thousands)
0	318
1	330
2	350
3	374
4	356
5	357
6	280
7	210
8	135
9	97

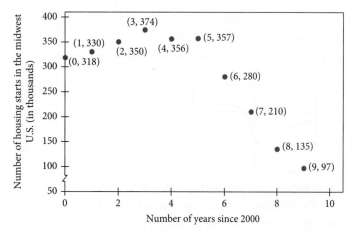

71.

Year	American Per Person Consumption of Tea (in gallons)
0	7.3
10	6.9
15	7.9
20	7.8
24	8.0
25	8.0
26	8.4
27	8.4
28	8.0

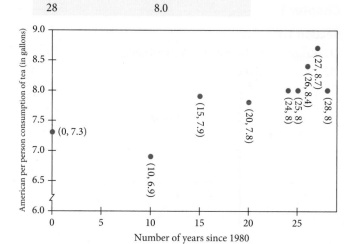

73.

Year	Total Cases on U.S. Supreme Court Docket (in thousands)
0	6.3
5	7.6
10	9.0
15	9.6
16	10.3
17	9.6
18	9.0
19	9.3

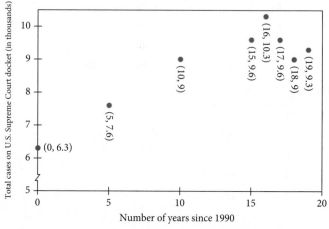

75. Yes, a nonzero number can divide into zero.

77. Only the number five would be excluded from the domain to avoid a division by zero.

79. Zero; negative

Section 1.2

1. Linear **3.** Point-slope

5. Constant

7. $m = -5$

9. $m = -\dfrac{5}{6}$

11. $m = 1.1$

13. $m = 1$

15. $m = -\dfrac{5}{3}$

17. $m = -4$

19. $m = \dfrac{17}{4}$

21. $y - 6 = -3(x - 2)$ or $y + 3 = -3(x - 5)$

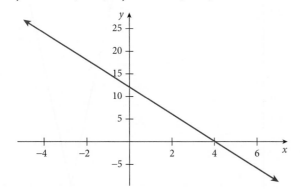

23. $y - 5 = 4(x - 3)$ or $y - 13 = 4(x - 5)$

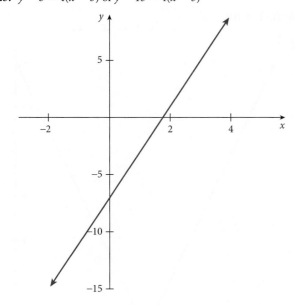

25. $x = 0$

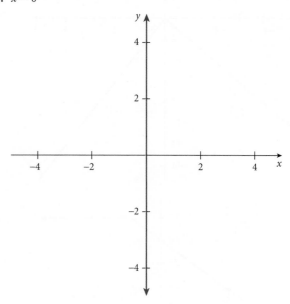

27. $x = -2$

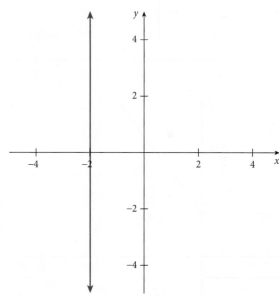

29. $y = 3(x - 8) + 4$

31. $y = 5(x - 0) - 13$ **33.** $y = 0.3(x + 4)$

35. $f(x) = -\dfrac{1}{4}(x + 1) + 2$ or $f(x) = -\dfrac{1}{4}(x - 3) + 1$

37. $y = \dfrac{1}{8}(x + 12)$ or $y = \dfrac{1}{8}x + \dfrac{3}{2}$

39. $f(x) = -\dfrac{47}{40}x + 6$ or $f(x) = -\dfrac{47}{40}(x - 4) + 1.3$

41. y-intercept is $(0, 5)$; x-intercept is $(1.25, 0)$

43. y-intercept is $(0, 120)$; x-intercept is $(400, 0)$

45. y-intercept is $(0, 8.98)$; x-intercept is $(-44.9, 0)$

47. -4.5; decreasing   **49.** 4; increasing

51. 4; increasing   **53.** $\frac{1}{3}$; increasing

55. 14; increasing

57.

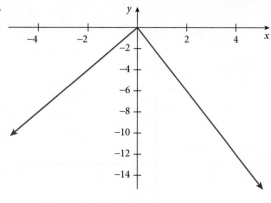

59.

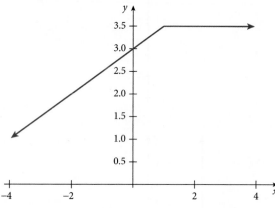

61.

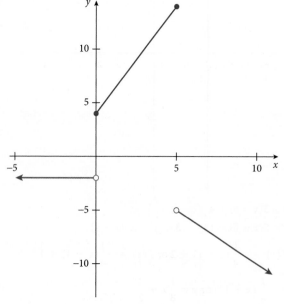

63. (a)

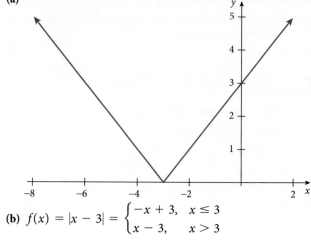

(b) $f(x) = |x - 3| = \begin{cases} -x + 3, & x \leq 3 \\ x - 3, & x > 3 \end{cases}$

65. (a)

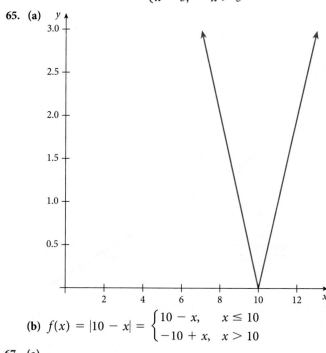

(b) $f(x) = |10 - x| = \begin{cases} 10 - x, & x \leq 10 \\ -10 + x, & x > 10 \end{cases}$

67. (a)

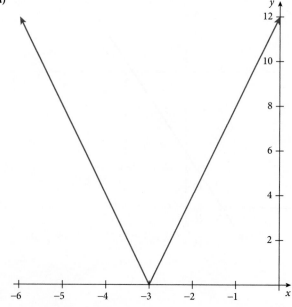

(b) $f(x) = |4x + 12| = \begin{cases} -4x - 12, & x \le -3 \\ 4x + 12, & x > -3 \end{cases}$

69. (a) $f(x) = -150x + 800$ **(b)** $5\frac{1}{3}$ years

71. (a) $f(x) = -3000x + 40{,}000$ **(b)** \$25,000

73. $C(x) = 147.75x + 2700$ **75.** $C(x) = 575x + 1250$

77. $C(x) = 80x + 650$

79. (a) Fixed costs are \$550; variable costs are \$1.25 per bottle.

    (b) $C(70) = 637.5$. It costs \$637.50 to produce 70 bottles.

    (c) 115 bottles

    (d) \$1.25 per bottle

    (e) \$1.25

81. (a) Fixed costs are \$20; variable costs are \$0.15 per mile.

    (b) $C(x) = 0.15x + 20$

    (c) $(0, \infty]$

    (d) 38; It cost \$38 to rent a truck and drive 120 miles.

    (e) \$0.15; Marginal cost

83. (a) \$50; Corresponds to 0 hours of labor

    (b) \$15; Called variable or marginal cost

    (c) $C(x) = 15x + 50$

    (d) \$117.50

85. (a) decreased; 0.15

    (b) 7.53. In the year 2005, there were 7.53 thousand main banking offices.

    (c) 2008

87. For a linear depreciation model, the slope m will be negative since that would mean that the function decreases from left to right.

89. $y_2 = -2$

91. $(6, \infty)$

Section 1.3

1. Parabola   **3.** Vertex   **5.** Discriminant

7. (a) $(-3, -4)$

    (b) Increasing on $(-3, \infty)$; decreasing on $(-\infty, -3)$

9. (a) $(3, 15)$

    (b) Increasing on $(-\infty, 3)$; decreasing on $(3, \infty)$

11. (a) $(-0.6, -4.8)$

    (b) Increasing on $(-0.6, \infty)$; decreasing on $(-\infty, -0.6)$

13. (a) $\left(\dfrac{-25}{14}, \dfrac{-209}{280} \right) \approx (-1.79, -0.75)$

    (b) Increasing on $(\frac{-25}{14}, \infty)$; decreasing on $(-\infty, \frac{-25}{14})$

15. (a) $(-1, 1.4)$

    (b) Increasing on $(-\infty, -1)$; decreasing on $(-1, \infty)$

17. $-4; 4$ **19.** $0; -2$ **21.** $2; 3$ **23.** $-2.5; \dfrac{10}{3}$

25. $\dfrac{1 + \sqrt{5}}{2}; \dfrac{1 - \sqrt{5}}{2}$ **27.** $\dfrac{7 + \sqrt{5}}{22}; \dfrac{7 - \sqrt{5}}{22}$

29. No real roots

31. The following graphs are graphed in viewing window $[-10, 10]$ by $[-30, 30]$.

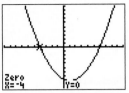 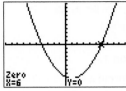

The zeros are $x = -4$ and $x = 6$.

33. The following graph is graphed in viewing window $[-10, 10]$ by $[-30, 30]$.

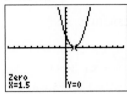

The zero is $x = 1.5$.

35. The following graph is graphed in viewing window $[-5, 5]$ by $[-10, 10]$.

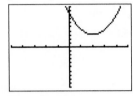 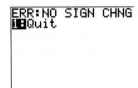

No real zeros

37. (a) $m_{sec} = -5$ **(b)** $m_{sec} = -5$ **(c)** $m_{sec} = -5$

39. (a) $m_{sec} = 5$ **(b)** $m_{sec} = 3$ **(c)** $m_{sec} = 7$

41. (a) $m_{sec} = 3$ **(b)** $m_{sec} = 2$ **(c)** $m_{sec} = 1.5$

43. 101; Over a five-year period, the number of citations was increasing at an average rate of 101 citations per year.

44. 51; Over a ten-month period, the number of sunglasses sold was increasing at an average rate of 51 sunglasses per month.

45. (a) 96; After 3 seconds, the rock is 96 feet above the ground.

    (b) 16; Between 1 and 3 seconds, the rock was rising at an average speed of 16 feet per second.

47. (a) 2033; This means that after 3 hours, there were 2,033 bacteria in the colony.

    (b) 17; From 3 to 6 hours after the colony was started, the colony was growing at an average rate of 17 bacteria per hour.

49. (a) 5,845

    (b) 157; The number of United States Supreme Court cases was increasing on average by 157 court cases per year.

51. (a) 45.16; This means that the number of nonprofit health and medical associations was increasing on average by 45.16 associations per year during 1980–1990.

    (b) 68.56; The number of nonprofit health and medical associations was increasing on average by 68.56 associations per year during 1990–2000.

    91.96; The number of nonprofit health and medical associations was increasing on average by 91.96 associations per year during 2000–2010.

    (c) 1980–1990

53. For a quadratic function, a secant line is a line that passes through two points of the graph of the function.

55. $f(x) = 2x^2 + 12x + 19$

57. $\Delta x = 3$

Section 1.4

1. Operations **3.** Price **5.** Break-even

7. (a) $(f+g)(x) = 10x + 2; (-\infty, \infty)$

(b) $(f-g)(x) = 2x + 4; (-\infty, \infty)$

(c) $(f \cdot g)(x) = 24x^2 + 6x - 3; (-\infty, \infty)$

(d) $\left(\dfrac{f}{g}\right)(x) = \dfrac{6x+3}{4x-1}; (-\infty, 0.25) \cup (0.25, \infty)$

9. (a) $(f+g)(x) = x^2 + \sqrt{x+5} + 5; [-5, \infty)$

(b) $(f-g)(x) = -x^2 + \sqrt{x+5} - 5; [-5, \infty)$

(c) $(f \cdot g)(x) = -x^2\sqrt{x+5} + 5\sqrt{x+5}; [-5, \infty)$

(d) $\left(\dfrac{f}{g}\right)(x) = \dfrac{\sqrt{x+5}}{x^2+5}; [-5, \infty)$

11. 1 **13.** 336 **15.** -11 **17.** $\dfrac{15}{13}$

19. $R(x) = 2.55x$ **21.** $R(x) = -3.1x^2$

23. $R(x) = -0.3x^2 + 20x$

25. $R(x) = -0.1x^2 + 50x$

27. (a) $\left(5\dfrac{2}{3}, 283\dfrac{1}{3}\right)$ **(b)** $283.33

29. (a) $(91.78, 188.61)$ and $(0.218, 5.44)$.

(b) $5.44 or $188.61

31. (a) Graph of the revenue and cost functions:

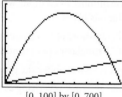

[0, 100] by [0, 700]

(b)

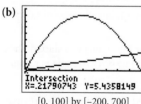 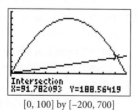

Intersection X=.21790743 Y=5.4358149

Intersection X=91.782093 Y=188.56419

[0, 100] by [-200, 700] [0, 100] by [-200, 700]

The break-even points are $(0.218, 5.44)$ and $(91.78, 188.61)$.

(c) $5.44 or $188.61

33. (a) Graph of the revenue and cost functions:

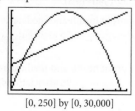

[0, 250] by [0, 30,000]

(b)

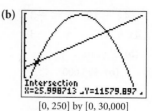 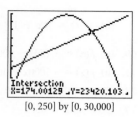

Intersection X=25.998713 Y=11579.897

Intersection X=174.00129 Y=23420.103

[0, 250] by [0, 30,000] [0, 250] by [0, 30,000]

The break-even points are $(26, 11, 580)$ and $(174, 23, 420)$.

(c) $11, 580 or $23, 420.

35. (a) $P(x) = -6x^2 + 118x - 240$

(b) $x \approx 2.30$ or $x \approx 17.36$

(c) $(9.83, 340.17)$; We have a maximum profit of approximately $340.17 when 9.83 units are sold.

37. (a) $P(x) = -5x^2 + 75.2x - 155$

(b) $x \approx 2.47$ or $x \approx 12.57$

(c) $(7.52, 127.752)$; We have a maximum profit of $127.75 at $x = 7.52$.

39. Constant **41.** Not a polynomial **43.** Quadratic

45. As $x \to -\infty, f(x) \to -\infty$ and as $x \to \infty, f(x) \to \infty$.

47. As $x \to -\infty, f(x) \to \infty$ and as $x \to \infty, f(x) \to -\infty$.

49. As $x \to -\infty, f(x) \to -\infty$ and as $x \to \infty, f(x) \to -\infty$.

51. As $x \to -\infty, f(x) \to \infty$ and as $x \to \infty, f(x) \to -\infty$.

53. As $x \to -\infty, f(x) \to -\infty$ and as $x \to \infty, f(x) \to -\infty$.

55. (a) $P(x) = -0.89x^2 + 25.7x - 61.8$

(b) $(14.44, 123.73)$; We have a maximum profit of approximately $123.73 when 14.44 units are sold.

57. (a) $P(x) = R(x) - C(x) = -x^2 + 30x - 200$

(b) $(15, 25)$; We have a maximum profit of $25 when 15 units are sold.

59. (a) $P(x) = -x^2 + 20x - 80$

(b) $(10, 20)$; We have a maximum profit of $20 when 10 units are sold.

61. (a)

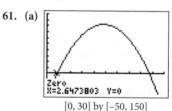

Zero X=2.6473803 Y=0

Zero X=26.229024 Y=2E-11

[0, 30] by [-50, 150] [0, 30] by [-50, 150]

(b) The zeros of the profit function, $(2.65, 0)$ and $(26.23, 0)$, correspond to the break-even points.

63. (a)

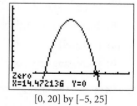

Zero X=5.527864 Y=0

Zero X=14.472136 Y=0

[0, 20] by [-5, 25] [0, 20] by [-5, 25]

(b) The zeros of the profit function, $(5.53, 0)$ and $(14.47, 0)$, correspond to the break-even points.

65. (a) 42; 53

(b) $g(f(x)) = 24x - 6$

$f(g(x)) = 24x + 5$

67. (a) 74; 67

(b) $g(f(x)) = 25x^2 - 15x + 4$

$f(g(x)) = 5x^2 + 15x + 17$

69. (a) $\dfrac{1}{8}; \dfrac{1}{8}$

(b) $g(f(x)) = \dfrac{1}{x^3}$

$f(g(x)) = \dfrac{1}{x^3}$

71. (a) 5; $\sqrt{7}$

(b) $g(f(x)) = x + 3$

$f(g(x)) = \sqrt{x^2 + 3}$

73. Let $g(x) = x + 3$ and $f(x) = x^3$.

75. Let $g(x) = \frac{1}{x+3}$ and $f(x) = x^2$.

77. Let $g(x) = x - 2$ and $f(x) = \sqrt[4]{x}$.

79. Let $g(x) = \sqrt{x}$ and $f(x) = 2 - 3x$.

89. (a) $(f - g)(20) = -245$

The Handi-Neighbor Hardware Store makes $245 less than the U-Do-It Store when they both sell 20 hammers.

(b) $(f + g)(20) = 635$

Both stores will make $635 combined if each sells 20 hammers.

91. $\frac{2225}{15} \approx 148.33$; The revenue generated by producing and selling LifeMax elliptical is increasing on average by approximately $148.33 per elliptical from the 5th to the 20th elliptical sold.

93. 3; The profit generated by making and selling the touch screen ski gloves was increasing on average by $3 per ski glove from the 35th to the 45th ski glove sold.

95. (a) $f(28) = 14.36$; The per capita consumption of ice cream in the year 2008 was 14.36 pounds per person.

$g(28) = 304.16$; In the year 2008, there were approximately 304.16 million people in the United States.

(b) $(f \cdot g)(28) = 4367.7376$; In the year 2008, approximately 4367 million pounds of ice cream were consumed in the United States.

(c) $(f \cdot g)(x) = -0.3135x^2 + 25.0244x + 3912.84$; The total number of pounds of ice cream consumed in millions where x represents the number of years since 1980.

97. (a) $(f - g)(x) = 0.09x^2 - 1.14x + 2.57$; This function gives the total amount in the Highway Trust Fund in billions of dollars where x represents the number of years since 1990.

(b) 0.92; This means that in the year 2001, there was $0.92 billion or $920 million in the Highway Trust Fund.

99. (a) $r(t) = 1.3t$

(b) $V(t) = 2.9293\pi\, t^3$; $V(t)$ is the volume of the sphere as a function of time.

(c) $V(6) = 632.736\pi \approx 1987.80$; After 6 seconds, the volume of the sphere is about 1987.80 in³.

101. Yes, the individual functions $f(x)$ and $g(x)$ are polynomials and each has a domain of all real numbers.

103. $a_2 + b_2$ **105.** Nine

Section 1.5

1. Asymptote   **3.** Radical   **5.** Power

7. (a) $(-\infty, 0) \cup (0, \infty)$

(b) Hole at $x = 0$

9. (a) $(-\infty, 2) \cup (2, \infty)$

(b) Vertical asymptote at $x = 2$

11. (a) $(-\infty, -2) \cup (-2, \infty)$

(b) Hole at $x = -2$

13. (a) $(-\infty, -3) \cup (-3, -1) \cup (-1, \infty)$

(b) Hole at $x = -1$ and a vertical asymptote at $x = -3$

15. (a) $(-\infty, -1.5) \cup (-1.5, 2) \cup (2, \infty)$

(b) Hole at $x = -1.5$ and a vertical asymptote at $x = 2$

17. No horizontal asymptote

19. $y = 3$

21. $y = 3$

23. $y = 1$

25. $y = 2$

27. y-intercept is $(0, 0)$; x-intercept is $(0, 0)$

29. y-intercept is $(0, \frac{-3}{4})$; x-intercept is $(0.5, 0)$

31. $(0, -0.5)$; x-intercept is $(-2, 0)$

33. $(0, -\frac{2}{9})$; x-intercept is $(2, 0)$

35. y-intercept is $(0, \frac{4}{3})$; x-intercepts are $(-2, 0)$ and $(2, 0)$

37. y-intercept is $(0, 5)$; no x-intercepts

39. y-intercept is $(0, -2)$; x-intercept is approximately $(0.631,0)$

41. y-intercept is $(0, 0.5)$; x-intercept is $(-0.657, 0)$

43.

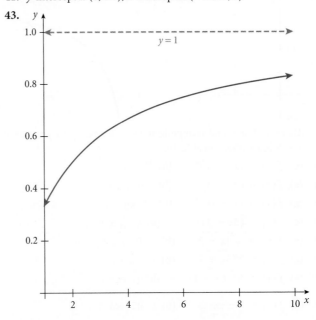

There is a horizontal asymptote at $y = 1$.

45.

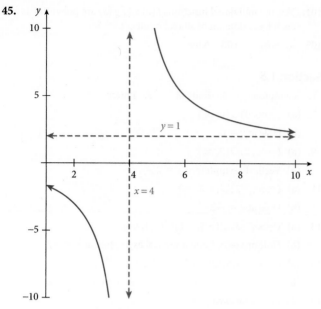

There is a horizontal asymptote at $y = 1$ and a vertical asymptote at $x = 4$.

47.

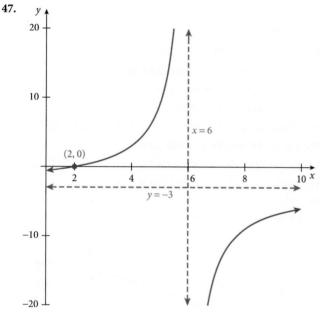

There is a horizontal asymptote at $y = -3$, a vertical asymptote at $x = 6$ and x-intercept at $(2, 0)$.

49. (a) $f(x) = (x - 1)^{1/4}$ (b) $[1, \infty)$

51. (a) $f(x) = (4 - x)^{1/3}$ (b) $(-\infty, \infty)$

53. (a) $f(x) = (x - 4)^{3/2}$ (b) $[4, \infty)$

55. (a) $f(x) = (3x + 4)^{-1/2}$ (b) $\left(-\frac{4}{3}, \infty\right)$

57. (a) $f(x) = \sqrt{4x + 3}$ (b) $[-0.75, \infty)$

59. (a) $f(x) = \sqrt[3]{3x + 1}$ (b) $(-\infty, \infty)$

61. (a) $f(x) = \sqrt{(6x - 1)^3}$ (b) $[\frac{1}{6}, \infty)$

63. (a) $f(x) = \dfrac{1}{\sqrt{6x + 3}}$ (b) $(-0.5, \infty)$

65. (a) $f(85) = 170$; It will cost \$170 million to remove 85% of the pollutants.

(b)

x	5	50	70	90	95	100
$f(x)$	≈ 1.6	30	70	270	570	undefined

(c) \$29,400   (d) $f(100)$ is undefined.

67. (a) $f(10) = 18.2$; After 10 days, the person had about 18 out of 20 items remembered.

(b) $f(100) = 19.82$; After 100 days, the person had about 20 out of 20 items remembered.

69. (a) $S(10) = \frac{6003}{101} \approx 59.44$; If the company spends \$10,000 on advertising, the income from the sales will be about \$5,944,000.

(b) $y = 60$; The most income the company can generate by advertising is \$6,000,000.

71. (a) $f(70) = \frac{7}{5}\sqrt{70} \approx 11.71$; If the tower is 70 feet tall, an observer can see about 11.71 miles into the forest.

(b) $x = \frac{21025}{49} \approx 429.08$; The tower should be about 429.08 feet high.

73. (a) $f(300) = 28.1\sqrt[3]{300} \approx 188$; There are about 188 plant species in 300 square miles of rainforest.

(b) $m_{sec} = \frac{28.1}{469} = 0.0599$; From 1728 square miles to 2197 square miles, the average increase in plant species is about 0.0599 species per square mile.

75. (a) 16.75; In the year 1990, the total cost for the food assistance programs was \$16.75 billion.

(b) $m_{sec} = 1.479$; From 1990 to 2000, the average amount spent on the food assistance programs grew by \$1.479 billion dollars per year.

(c) $m_{sec} = 1.849$; The amount for federal food assistance programs grew more rapidly during the decade from 2000 to 2010.

77. (a) $x = \frac{35}{3}$; The vertical asymptote is not in the reasonable domain for this model.

(b) $m_{sec} = 7.97$; The number of unemployed in Ireland has been increasing by an average rate of 7.97 thousand people per year from the years 2005 to 2010.

79. $(0, c)$

81. $m_{sec} = \dfrac{a(x_1 + \Delta x)^b - a(x_1)^b}{\Delta x}$

83. The x-intercept is $\left(-\frac{b}{a}, 0\right)$ assuming $x = -\frac{b}{a}$ does not make the denominator equal to zero. The y-intercept is $\frac{b}{d}$ assuming $x = 0$ is in the domain of the function.

Chapter 1 Review Exercises

Section 1.1 Review Exercises

1.

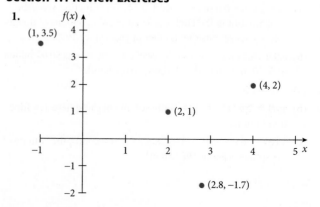

3. r is the independent variable; A is the dependent variable.

5. Function

Section 1.2 Review Exercises

7. $f(3) = 5; (3, 5)$

9. (a) x (b) y (c) $f(3) = 2$

   (d) $f(1) = 2$ (e) $[1, 4]$ (f) $[1, 5]$

11. $[2, \infty)$ **13.** $(-\infty, -2) \cup (-2, 2) \cup (2, \infty)$

15. The domain is $(-2, \infty)$; the range is $[-3, \infty)$.

17. $m_{sec} = \dfrac{2}{3}$ **19.** $y - 5 = \dfrac{5}{4}(x - 0)$ **21.** $y = \dfrac{3}{4}x - 7$

23. (a) $f(x) = -40x + 800$ (b) Five

25. (a) $C(x) = 75x + 2600$

   (b) $C(20) = 4100$; It will cost \$4,100 to produce 20 heart rate monitors.

   (c) \$75

27.

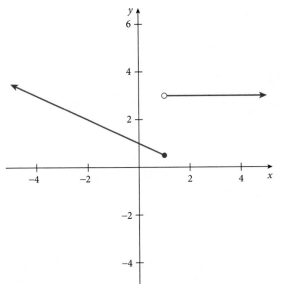

29. (a) \$2,500   (b) \$60 per fan; This is also the variable cost.

   (c) $C(x) = 60x + 2500$ (d) \$7,540

Section 1.3 Review Exercises

31. (a) $(-\frac{3}{2}, 16.5)$ (b) $(-\infty, 16.5]$

33. (a) $(1.6, -5.24)$ (b) $(1.6, \infty)$ **35.** $-4; 2.5$

37. The graphs are displayed in viewing window $[-5, 5]$ by $[-5, 5]$.

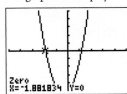

and

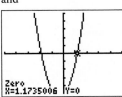

The two zeros are $x = -1.88$ and $x = 1.17$.

39. (a) $m_{sec} = -1$ (b) $m_{sec} = 1$ (c) $m_{sec} = 5$

41. (a) Concave down; the leading coefficient is negative.

   (b)

x	$f(x)$
4	56
5	82.4
6	93.1
7	88.1
8	67.4

   (c) $f(5) = 82.4$; In the year 2005, there were 82.4 thousand, or 82,400 computer system engineers employed.

   (d) 2006

   (e) The vertex is given by $(-\frac{b}{2a}, f(-\frac{b}{2a}))$ with $a = -7.85$ and $b = 97.05$. This gives us $x = -\frac{b}{2a} = -\frac{97.05}{2(-7.85)} \approx 6.18$. Rounding to 6 would conform with our answer in part d.

   (f) $m_{sec} = 10.7$; The number of computer systems engineers employed grew on average by approximately 10.7 thousand engineers per year from 2004 to 2007.

Section 1.4 Review Exercises

43. (a) $(f + g)(x) = \sqrt{x + 2} + x - 2$; the domain is $[-2, \infty)$.

   (b) $(f - g)(x) = \sqrt{x + 2} - x + 2$; the domain is $[-2, \infty)$.

   (c) $(f \cdot g)(x) = x\sqrt{x + 2} - 2\sqrt{x + 2}$; the domain is $[-2, \infty)$.

   (d) $\left(\dfrac{f}{g}\right)(x) = \dfrac{\sqrt{x + 2}}{x - 2}$; the domain is $[-2, 2) \cup (2, \infty)$.

45. $(f - g)(4) = -29$ **47.** $\left(\dfrac{f}{g}\right)(3) = 0.16$

49. $R(x) = 145x - 0.15x^2$

51. $P(x) = -0.08x^2 + 16x - 400; (100, 400)$

53. (a) Quadratic   (b) Constant   (c) Linear   (d) Quartic

55. As $x \to -\infty, f(x) \to \infty$ and as $x \to \infty, f(x) \to \infty$.

57. (a) As $x \to -\infty, f(x) \to -\infty$ and as $x \to \infty, f(x) \to \infty$.

   (b)

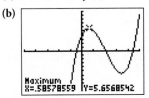 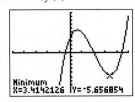

We have a "peak" at $(0.59, 5.66)$ and a "valley" at $(3.41, -5.66)$.

   (c) $f(x)$ is decreasing on $(0.59, 3.41)$ and increasing on $(-\infty, 0.59) \cup (3.41, \infty)$.

Section 1.5 Review Exercises

59. (a) $(-\infty, -1.5) \cup (-1.5, \infty)$

   (b) Vertical asymptote at $x = -1.5$

   (c) No horizontal asymptote

61. (a) $(-\infty, -3) \cup (-3, 3) \cup (3, \infty)$

   (b) Hole at $x = 3$

   (c) Horizontal asymptote at $y = 2$

63. The y-intercept is $(0, -\frac{16}{9})$; the x-intercept is $(4, 0)$.

65. The y-intercept is $(0, 2)$; $f(x)$ has no x-intercepts.

67.

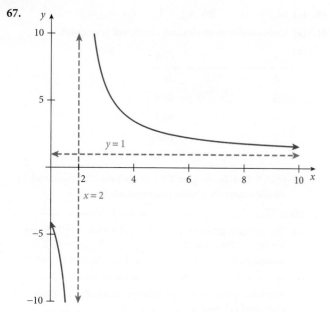

69. (a) $\sqrt[4]{(x-2)^3} = (x-2)^{3/4}$ (b) $[2, \infty)$

71. (a) $f(64) = 2$; It will take 2 seconds for an object to fall 64 feet.

   (b) $f(144) = 3$; It will take 3 seconds for an object to fall 144 feet.

Chapter 2

Section 2.1

1. Limit   **3.** Right; right-hand   **5.** Graphically

7. (a) $\lim\limits_{x \to 1^-} f(x) = -1$ (b) $\lim\limits_{x \to 1^+} f(x) = -1$

   (c) $\lim\limits_{x \to 1^+} f(x) = -1$

11. (a) $\lim\limits_{x \to -2^-} f(x) = 48$ (b) $\lim\limits_{x \to -2^+} f(x) = 48$

   (c) $\lim\limits_{x \to -2} f(x) = 48$

13. (a) $\lim\limits_{x \to 1^-} f(x) = 2$ (b) $\lim\limits_{x \to 1^+} f(x) = 2$ (c) $\lim\limits_{x \to 1} f(x) = 2$

15. (a) $\lim\limits_{x \to -2^-} f(x) = 12$ (b) $\lim\limits_{x \to -2^+} f(x) = 12$

   (c) $\lim\limits_{x \to -2} f(x) = 12$

17. (a) $\lim\limits_{x \to 0^-} f(x) = 1$ (b) $\lim\limits_{x \to 0^+} f(x) = 1$ (c) $\lim\limits_{x \to 0} f(x) = 1$

19. $\lim\limits_{x \to 1} f(x) = -6$

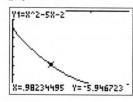 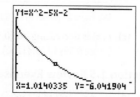

21. $\lim\limits_{x \to -1} g(x) = 3$

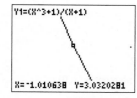 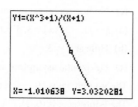

23. $\lim\limits_{x \to 3} f(x) = 4$

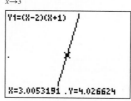 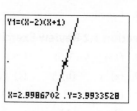

25. $\lim\limits_{x \to 4} f(x) = 0.25$

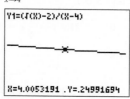 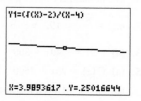

27. $\lim\limits_{x \to -2} f(3x + 1) = -5$

29. $\lim\limits_{x \to 10} \sqrt{x - 5} = \sqrt{5} \approx 2.2361$

31. $\lim\limits_{x \to 3} \dfrac{x^2 - 9}{x - 3} = 6$

33. $\lim\limits_{x \to 0} |x - 2| = 2$

35. $\lim\limits_{x \to 1} \dfrac{|x - 1|}{|x|} = \dfrac{1 - 1}{1} = 0$

37. $\lim\limits_{x \to -1} \dfrac{x^2 - 1}{x + 1} = -2$

39. $\lim\limits_{x \to 0} \sqrt{2x + 3} = \sqrt{3} \approx 1.732$

41. $\lim\limits_{x \to -5} \dfrac{x^2 - 25}{x - 5} = 0$

43. $\lim\limits_{x \to 1} \dfrac{x^2 - 1}{x + 1} = 0$

45. $\lim\limits_{x \to 2} (x + 1)^2 \cdot (3x - 1)^3 = 1125$

47. $\lim\limits_{x \to 3} \dfrac{x^2 - x - 6}{x - 3} = 5$

49. $\lim\limits_{x \to -2} \dfrac{x + 2}{x^2 + 5x + 6} = 1$

51. $\lim\limits_{x \to 4} \dfrac{\sqrt{x} - 2}{x - 4} = \dfrac{1}{4}$

53. (a) $\lim\limits_{x \to -1} f(x) = 1$ (b) $f(-1) = 1$

   (c) $\lim\limits_{x \to 3^-} f(x) = 1$ (d) $\lim\limits_{x \to 3^+} f(x) = 2$

   (e) $\lim\limits_{x \to 3} f(x)$ does not exist.   (f) $f(1) = 2$

   (g) $\lim\limits_{x \to 1} f(x) = 3$ (h) $\lim\limits_{x \to 1^-} f(x) = 3$

   (i) $\lim\limits_{x \to 1^+} f(x) = 3$

55. $\lim\limits_{h \to 0} f(2x + h) = 2x + (0) = 2x$

57. $\lim\limits_{h \to 0} \dfrac{3xh + h^2}{h} = 3x.$

59. $\lim\limits_{h \to 0} \dfrac{4x^2h + 2h^2}{h} = 4x^2$

61. $\lim\limits_{h \to 0} \dfrac{\sqrt{x + h} - \sqrt{x}}{h} = \dfrac{\sqrt{x}}{2x}$

63. (a) $f(1 + h) = 2h + 5$

(b) $\lim\limits_{x \to 0} \dfrac{f(1 + h) - f(1)}{h} = 2$

65. (a) $f(1 + h) = h^2 + 2h$ (b) $\lim\limits_{x \to 0} \dfrac{f(1 + h) - f(1)}{h} = 2$

67. (a) $f(1 + h) = h^2 + 2$ (b) $\lim\limits_{x \to 0} \dfrac{f(1 + h) - f(1)}{h} = 0$

69. (a) $f(1 + h) = |1 + h|$ (b) $\lim\limits_{x \to 0} \dfrac{f(1 + h) - f(1)}{h} = 1$

71. (a) $f(1 + h) = \sqrt{1 + h}$

(b) $\lim\limits_{x \to 0} \dfrac{f(1 + h) - f(1)}{h} = \dfrac{1}{2}$

73. f is continuous at $x = -2$ since the function and limit exists at $x = 2$ and they equal the same value.

75. f is continuous at $x = 3$ since the function and limit exists at $x = 3$ and they equal the same value.

77. f is not continuous at $x = 3$ since $f(3)$ is not defined.

79. f is not continuous at $x = 5$ since $f(5)$ is not defined.

81. f is not continuous at $x = 1$ since the limit as x approaches 1 does not exist.

83. $(-\infty, \infty)$

85. $(-\infty, -5) \cup (-5, \infty)$

87. $(-\infty, -1) \cup (-1, 3) \cup (3, \infty)$

89. $[-1.5, \infty)$

91. (a) $N(10) = 1948$; This means that when $10,000 is spent on advertising, 194,800 items are sold.

(b) $\lim\limits_{x \to 10} 2000 - \dfrac{520}{x} = 1948$; As the level of advertising expenditures approaches $10,000 the number of items sold approaches 194,800 items.

93. (a) $AC(10) = 2257.35$; The average cost of producing 10 units of a product is $2257.35.

(b) $\lim\limits_{x \to 10} \dfrac{22500 + 7.35x}{10} = 2257.35$; As the level of production approaches 10 units, the average cost approaches $2257.35 per unit.

95. (a) $c(100) = 30$; It will cost $30 to rent a medium-size car and drive it 100 miles.

(b) $\lim\limits_{x \to 100} c(m) = 30$; The limit of the cost of the rental as the miles driven approaches 100 miles is $30.

(c) $c(200) = 60$; It will cost $60 to rent a medium-size car and drive it 200 miles.

(d) $\lim\limits_{x \to 100} c(m)$ does not exist.

(e) The jump in the graph represents the end of the first day and the beginning of the second day.

97. No. For instance, for the function $f(x) = \dfrac{\sqrt{x} - 3}{x - 9}$, we saw in Exercise 26 that the limit as x approached 9 was $\frac{1}{6}$ but, $f(9) \ne \frac{1}{6}$.

99. (a)

x	0	0.9	0.99	0.999	1	1.001	1.01	1.1	2
$f(x) = x + 1$	1	1.9	1.99	1.999	?	2.001	2.01	2.1	3
$g(x) = \frac{x^2 - 1}{x - 1}$	1	1.9	1.99	1.999	?	2.001	2.01	2.1	3

(b) $f(1) = 2$. $g(1)$ does not exist. Thus, $x = 1$ is in the domain of f but not of g.

(c) From the chart, $\lim\limits_{x \to 1} f(x) = 1$ and $\lim\limits_{x \to 1} g(x) = 1$. So, the limit does exist as $x \to 1$.

101. The limit will not exist since $\lim\limits_{x \to 3} h(x) = \lim\limits_{x \to 3} \dfrac{f(x)}{g(x)} =$

$\dfrac{\lim\limits_{x \to 3} f(x)}{\lim\limits_{x \to 3} g(x)} = \dfrac{5}{0}$.

103. This means as the x values approach 3, the functional values approach 4.

105. The functional value $f(-1)$ does not exist because -1 is not in the domain of the function. Since the left-hand limit equals the right-hand limit equals 3, the limit exists.

107. $\lim\limits_{x \to 5} h(x) = 5 - 3\sqrt{5}$

Section 2.2

1. Asymptote   **3.** Vertical   **5.** Horizontal asymptote

7. (a) $\lim\limits_{x \to 0^-} f(x) = -\infty$ (b) $\lim\limits_{x \to 0^+} f(x) = \infty$

(c) $\lim\limits_{x \to 0} f(x)$ does not exist.

9. (a) $\lim\limits_{x \to 1^-} f(x) = \infty$ (b) $\lim\limits_{x \to 1^+} f(x) = \infty$

(c) $\lim\limits_{x \to 1} f(x) = \infty$

11. (a) $\lim\limits_{x \to -2^-} f(x) = -0.2$ (b) $\lim\limits_{x \to -2^+} f(x) = -0.2$

(c) $\lim\limits_{x \to -2} f(x) = -0.2$

13. (a) $\lim\limits_{x \to 0^-} f(x) = -\infty$ (b) $\lim\limits_{x \to 0^+} f(x) = \infty$

(c) $\lim\limits_{x \to 0} f(x)$ does not exist.

15. $\lim\limits_{x \to 0} \dfrac{2}{x^3}$ does not exist.   **17.** $\lim\limits_{x \to -2} \dfrac{x + 2}{x^2 - x - 6} = -\dfrac{1}{5}$

19. $\lim\limits_{x \to 3} \dfrac{x - 3}{x^2 - 9} = \dfrac{1}{6}$ **21.** Does not exist.

23. $\lim\limits_{x \to -\infty} \dfrac{2x + 5}{x - 1} = 2$ **25.** $\lim\limits_{x \to \infty} \dfrac{3x^2 - x + 2}{2x^2 + x - 5} = 1.5$

27. $\lim\limits_{x \to \infty} \dfrac{2x^2 + 2x + 1}{5x^3 + 3x - 5} = 0$

29. $\lim\limits_{x \to -\infty} \dfrac{2x^2 + 2x + 1}{5x^3 + 3x - 5} = 0$

31. $y = 2$ **33.** $y = 0$ **35.** $\lim\limits_{x \to 2^+} f(x) = -\infty$

37. $\lim\limits_{x \to 2} f(x)$ does not exist.

39. $\lim\limits_{x \to -1} f(x)$ does not exist.

41. $\lim\limits_{x \to \infty} f(x) = 0$

43. (a) $[0, 100)$

(b) $C(40) = 148\frac{2}{3} \approx 148.67$; It will cost approximately $148,670 to clean up 40% of the pollutants.

(c) $\lim\limits_{x \to 100^-} C(x) = \infty$; As the amount of pollutants removed approaches 100%, the cost grows without bound.

45. $\lim\limits_{x \to \infty} N(x) = 2000$; As the amount of money spent increases without bound, the number of items purchased approaches 2000 hundred items or 200,000 items.

47.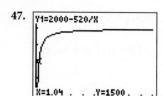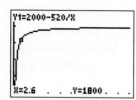

An interval for x that would result in function values between 1500 and 1800 is [1.04, 2.6].

49. $\lim\limits_{x\to\infty} AC(x) = 6.85$; As production increases without bound, the average cost approaches $6.85 per unit.

51. (a) $P(12) = \frac{940}{1.24} \approx 758$; After one year, there are approximately 758 bass in the lake.

(b) $\lim\limits_{t\to\infty} P(t) = 3500$; As time increases without bound, the bass population in the lake approaches 3,500.

53. (a) $x = \frac{35}{3}$ is a vertical asymptote for the graph of f, but it is not in the reasonable domain of f.

(b) $\lim\limits_{x\to\infty} f(x) = \frac{203}{3} \approx 67.67$; This means that as time increases without bound, the number of unemployed in Ireland approaches 67.67 thousand.

55. (a) $f(10) = \frac{1177700}{1589} \approx 741.16$; This means that in the year 2000, the number of mobile cellular subscribers internationally will be approximately 741.16 million.

(b) We have $\lim\limits_{x\to 25.89^-} f(x) = \infty$ while $\lim\limits_{x\to 25.89^+} f(x) = -\infty$. This implies that $x = 25.89$ is a vertical asymptote for the graph of f, but it is not in the reasonable domain of f.

(c) $\lim\limits_{x\to\infty} f(x) = -2089$; This value does not make sense for the function's range since you have a negative amount of cellular subscribers.

57. $\lim\limits_{x\to\infty} f(x) = 2770$; This means that as time increases without bound, the number of nonconnected PACs will approach 2770.

59. Since the left-hand limit does not equal the right-hand limit corresponding to $\lim\limits_{x\to 0}\frac{1}{x}$, we would say the limit does not exist. For $f(x) = \frac{1}{x^2}$, approaching from 0 either side will have the function values increase without bound and thus $\lim\limits_{x\to 0}\frac{1}{x^2} = \infty$.

61. If c was nonzero and n equaled 4, which is the highest degree, then:

$$\lim_{x\to\infty}\frac{5x^4+6}{cx^4+10} = \lim_{x\to\infty}\frac{5x^4+6}{cx^4+10}\cdot\frac{\frac{1}{x^4}}{\frac{1}{x^4}} = \lim_{x\to\infty}\frac{5+\frac{6}{x^4}}{c+\frac{10}{x^4}} = \frac{5+0}{c+0} = \frac{5}{c}.$$

If c was equal to zero, then the limit of the function would increase without bound.

63. $\lim\limits_{x\to 0^-}\frac{k}{x^5} = \infty$ and $\lim\limits_{x\to 0^+}\frac{k}{x^5} = -\infty$

65. This means that as x approaches zero from either direction, the function values increase without bound.

67. $a = 2\sqrt{3}$ or $a = -2\sqrt{3}$

69. The function values will approach infinity if n is a whole number less than 4.

71. The function values will approach zero if n is a whole number greater than 4.

73. (a) The graph of $f(x)$:

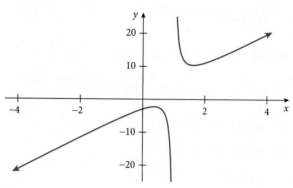

The graph of $g(x)$:

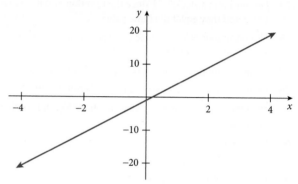

(b)
$$\begin{array}{r} 5x - 1 \\ x-1\overline{)5x^2 - 6x + 3} \\ -(5x^2 - 5x) \\ \hline -x + 3 \\ -(-x+1) \\ \hline 2 \end{array}$$

Notice that the quotient is $g(x)$.

(c)

x	$f(x)$	$g(x)$
10	≈ 49.222	49
100	≈ 499.020	499
1000	≈ 4999.002	4999

Section 2.3

1. Tangent **3.** h **5.** Instantaneous

7. Instantaneous rate of change **9.** Average rate of change

11. Instantaneous rate of change **13.** Average rate of change

15. Average rate of change **17.** Instantaneous rate of change

19. 6 **21.** 59 **23.** -1 **25.** 0.75

27. $2(20)^{-0.8} \approx 0.182$ **29.** $2.1(15)^{-1.2} \approx 0.081$ **31.** 1

33. (a) Positive between points A and B and also between points D and E; negative between points B and C; zero between points C and D

(b) Positive between points B and C; negative between points C and D; zero between points A and B

35. 3 **37.** 7 **39.** 99 **41.** 0.75 **43.** 0.17440

45. -0.11239 **47.** (a) $32.50 per coat (b) 31

49. 0.49; In the year 2006, the percentage of freshmen that classify their political orientation as liberal was increasing at a rate of around 0.49% per year.

51. -1.08355; In the year 1997, the amount of federal research grants to the NEH was decreasing by about $-\$1.08355$ million per year.

53. The average rate of change determines how the function changes for two given values for the independent variable. The instantaneous rate of change determines how the function changes for one given value for the independent variable.

55. $\dfrac{f(b) - f(a)}{b - a} = 4$

57. $b = 20$

59. The instantaneous rate of change does not exist at $x = 1$ since the left-hand limit does not equal the right-hand limit.

61. The instantaneous rate of change at $x = 2$ is a since the left-hand limit equals the right-hand limit which equals a.

Section 2.4

1. m_{tan} **3.** $f'(x)$ **5.** 2   **7.** -6 **9.** -6 **11.** $\dfrac{1}{2}$

13. 36.6   **15.** $-\dfrac{1}{9}$ **17.** $y = 2x + 1$ **19.** $y = -6x - 4$

21. $y = -6x + 16$ **23.** $y = \dfrac{1}{2}x + 2$

25. $y = -13.6x + 56.7$ **27.** $y = -\dfrac{1}{9}x + \dfrac{2}{3}$ **29.** 2

31. -2 **33.** $2x$ **35.** $2x - 2$ **37.** $4.2x + 3.2$

39. $-4x + 3$ **41.** $-4.6x + 3.1$ **43.** $3x^2 + 2x$

45. $f'(-1) = 2$ and $f'(2) = 2$ **47.** $f'(-3) = -8$ and $f'(2) = 2$

49. $f'(0) = 3.2$ and $f'(2) = 11.6$

51. $f'(3) = -9$ and $f'(6) = -21$

53. $f'(-1) = 1$ and $f'(1) = 5$

55. f is not differentiable at $x = -2$ (since it is not continuous at $x = -2$) and at $x = 2$ (since it is not continuous at $x = 2$).

57. f is not differentiable at $x = -2$ (since it has a corner at $x = -2$) and at $x = 1$ (since it has a corner at $x = 1$).

59. f is not differentiable at $x = 2$ (since it has a corner at $x = 2$).

61. f is differentiable for all values of x.

63. (a) $[0, \infty)$ (b) $f'(x) = \dfrac{1}{2}x^{-1/2} = \dfrac{1}{2\sqrt{x}}$ (c) $(-\infty, 0]$

   (d) f has a vertical tangent at $x = 0$.

65. (a) $(-\infty, \infty)$ (b) $f'(x) = \dfrac{2}{3}x^{-1/3} = \dfrac{2}{3\sqrt[3]{x}}$

   (c) $x = 0$ (d) f has a corner at $x = 0$.

67. (a) $(-\infty, -1)\cup(-1, \infty)$

   (b) $f'(x) = \dfrac{2x \cdot (x + 1) - (x^2 - 1) \cdot (1)}{(x + 1)^2} = 1$ as long as $x \neq -1$.

   (c) f' is undefined at $x = -1$.

   (d) f is not continuous at $x = -1$.

69. (a) $(-\infty, 2)\cup(2, \infty)$

   (b) $f'(x) = \dfrac{2x \cdot (x - 2) - (x^2 - 4) \cdot (1)}{(x - 2)^2} = 1$ as long as $x \neq 2$.

   (c) f' is undefined at $x = 2$.

   (d) f is not continuous at $x = 2$.

71. (a) $1000 - 4x$

   (b) $P'(200) = 200$; When 200 modems are produced and sold, the profit is increasing at a rate of \$200 per modem.

   $P'(300) = -200$; When 300 modems are produced and sold, the profit is decreasing at a rate of \$200 per modem.

73. (a) $R(50) = 7500$; The total revenue from the production and sale of 50 sets of Junior Golf Clubs is \$7500.

   $R(150) = 7500$; The total revenue from the production and sale of 150 sets of Junior Golf Clubs is \$7500.

   (b) $200 - 2x$

   (c) $R'(50) = 100$; When 50 sets of Junior Golf Clubs are produced and sold, the revenue is increasing at a rate of \$100 per set.

   $R'(150) = -100$; When 150 sets of Junior Golf Clubs are produced and sold, the revenue is decreasing at a rate of \$100 per set.

75. (a) $2x + 15$

   (b) $C'(40) = 95$; When 40 refrigerators are produced and sold, the total cost is increasing at a rate of \$95 per refrigerator.

   $C'(100) = 215$; When 100 refrigerators are produced and sold, the total cost is increasing at a rate of \$215 per refrigerator.

77. (a) $N(20) = 1974$; $N'(20) = 1.3$; When \$20,000 is spent on advertising, sales will be 197,400 items, and sales are increasing at a rate of 130 items per thousand advertising dollars.

   (b) $N(36) \approx 1985.555$; $N'(36) = \frac{520}{(36)^2} \approx 0.401$; When \$36,000 is spent on advertising, sales will be 198,556 items, and sales are increasing at a rate of 40 items per thousand advertising dollars.

   (c) Decreasing

79. (a) $f'(x) = 0.16$

   (b) $f'(5) = 0.16$; This means that in the year 1995, the annual Thai rice export is increasing by 0.16 millions of metric tons per year.

81. (a) $f'(x) = -1.1x + 22.57$

   (b) $f'(15) = 6.87$; This means that in 2005, the amount of power generated from nuclear power plants was increasing by 6.87 billion kilowatt hours per year.

83. The derivative of a function is the function that tells the slope of the tangent line at any given value of x.

85. $f(x) = x^2; c = 8$

87. $f'(x) = a$

89. $f'(x) = 3ax^2 + 2bx + c$

91. (a) $f(1) = 3$

   (b) The derivative does not exist at $x = 1$ since the limit does not exist.

Chapter 2 Review Exercises

Section 2.1 Review Exercises

1. (a) $\lim_{x \to 0^-} f(x) = 1$ (b) $\lim_{x \to 0^+} f(x) = 1$

   (c) $\lim_{x \to 0} f(x) = 1$

3. $\lim_{x \to 3.1} f(x) \approx 23.59$

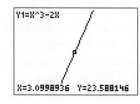

5. $\lim_{x \to 2}(7x^3 - 10x) = 36$

7. $\lim_{x \to -3} |x - 5| = 8$ 9. $\lim_{x \to 10} \dfrac{1}{x - 10}$ does not exist.

11. (a) $\lim_{x \to -4} f(x) = 2$ (b) $\lim_{x \to 0^-} f(x) = 0$

   (c) $f(0) = -\dfrac{3}{2}$ (d) $\lim_{x \to 2} f(x)$ does not exist.

   (e) $\lim_{x \to 3} f(x) = 1$ (f) $f(3) = 1$

13. $\lim_{h \to 0} \dfrac{2x^2 h - 9h}{h} = 2x^2 - 9$

15. (a) $f(2 + h) = 12 + 12h + 3h^2$

   (b) $\lim_{h \to 0} \dfrac{f(2 + h) - f(2)}{h} = 12$

17. (a) $C(400) = 38000$; It will cost \$38,000 to produce 400 scare bears with glowing eyes.

   (b) $\lim_{x \to 100} C(x) = 37000$; As the production approaches 100 scare bears, the production cost approaches \$37,000.

   (c) $AC(25) = 1460$; When 25 scare bears are produced, the average cost will be \$1460 per bear.

19. (a) The graph is displayed in viewing window [0, 40] by [0, 10]:

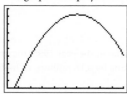

   (b) $f(20) = 8.96; f(25) = 9.26; f(30) = 8.56$

   (c) $\lim_{h \to 0} \dfrac{f(24 + h) - f(24)}{h} = 0$

   (d) 24 tablespoons

Section 2.2 Review Exercises

21. (a) $\lim_{x \to 3^-} f(x) = \infty$ (b) $\lim_{x \to 3^+} f(x) = \infty$

   (c) $\lim_{x \to 3} f(x) = \infty$

23. (a) $(-\infty, 0) \cup (0, \infty)$ (b) Does not exist.

25. $\lim_{x \to -\infty} \dfrac{x + 2}{2x - 3} = 0.5$ 27. $\lim_{x \to \infty} \dfrac{-4x^2 - 3x + 11}{8x^3 - 5} = 0$

29. $y = \dfrac{1}{3}$

31. No. The limit does not equal the functional value at $x = -4$.

33. Yes

35. (a) Yes since the limit and the functional value exist at $x = 2$ and they equal to each other.

   (b) No since $h(3)$ does not exist.

37. (a) $\lim_{x \to 40} AC(x) = 10.95$; When forty copies of a lithograph are produced, the average cost is \$10.95 per copy.

   (b) $\lim_{x \to \infty} AC(x) = 5$; As the number of copies increases without bound, the average cost approaches \$5 per copy.

Section 2.3 Review Exercises

39. $m_{sec} = 73$ 41. $m_{sec} \dfrac{-0.205811138}{1.7} \approx -0.121$

43. At $x = 4$, $\lim_{h \to 0} \dfrac{f(x + h) - f(x)}{h} = 2.5$.

45. At $x = 4$, $\lim_{h \to 0} \dfrac{f(x + h) - f(x)}{h} \approx 0.612$.

Section 2.4 Review Exercises

47. $f'(4) = 48$

49. (a) $y = 4x + 3$

   (b) The graph is displayed in viewing window $[-5, 5]$ by $[-5, 5]$.

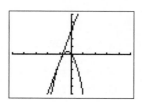

51. $f'(x) = 4x^3$ 53. $f'(x) = -\dfrac{2}{x^3}$

55. $f'(x) = 6x; f'(9) = 54$

57. (a) $q = 300$ (b) $q = 300$ (c) \$40,000

59. Since the function is discontinuous at $x = 0$, it is not differentiable at $x = 0$.

61. f is continuous at $x = 0$; f is differentiable at $x = 0$ and $f'(0) = -1$.

Chapter 3

Section 3.1

1. Differentiation   3. Zero   5. Derivative

7. Position   9. $f'(x) = 0$ 11. $f'(x) = 0$

13. $f'(x) = 6x^{6-1} = 6x^5$ 15. $f'(x) = 3 \cdot 4x^{4-1} = -12x^3$

17. $f'(x) = 2 \cdot \dfrac{2}{3}x^{2/3-1} = \dfrac{4}{3x^{1/3}} = \dfrac{4}{3\sqrt[3]{x}}$

19. $f'(x) = -3 \cdot \dfrac{-1}{3}x^{-1/3-1} = \dfrac{1}{\sqrt[3]{x^4}}$

21. $f'(x) = \dfrac{2}{3} \cdot 4x^{4-1} = \dfrac{8}{3}x^3$

23. $f'(x) = -\dfrac{2}{5} \cdot \dfrac{5}{3}x^{5/3-1} = -\dfrac{2\sqrt[3]{x^2}}{3}$ 25. $f'(x) = 6x^2 + 8x - 7$

27. $f'(x) = 6x - 2$ 29. $f'(x) = -10x - 6$

31. $f'(x) = -15x^2 + 7$

33. $f'(x) = \dfrac{3}{2}x^2 + \dfrac{6}{5}x - \dfrac{2}{3}$

35. $f'(x) = 2.62x + 2.05$

37. $f'(x) = -0.4x + 10.5x^2 - 1.6x^3$

39. $f'(x) = 3.45x^2 - 4.6x + 2.53$

41. $f'(x) = \dfrac{3}{2}x^{-1/2} + \dfrac{1}{2} - 10x = \dfrac{3}{2\sqrt{x}} + \dfrac{1}{2} - 10x$

43. $f'(x) = \dfrac{1}{3}x^{-2/3} + 2x - 9x^2 = \dfrac{1}{3\sqrt[3]{x^2}} + 2x - 9x^2$

45. $f'(x) = \dfrac{2}{3}x^{-1/3} + 2x^{-3/2} = \dfrac{2}{3\sqrt[3]{x}} + \dfrac{2}{\sqrt{x^3}}$

47. $f'(x) = 3.1725x^{0.35}$ **49.** $f'(x) = 0 - 10x^{-3} = -\dfrac{10}{x^3}$

51. $f'(x) = 4x - x^{-2} = 4x - \dfrac{1}{x^2}$

53. $f'(x) = -\dfrac{9}{2}x^{-5/2} + 2x^{-3/2} = -\dfrac{9}{2\sqrt{x^5}} + \dfrac{2}{\sqrt{x^3}}$

55. $f'(x) = -1.175x^{-3/2} + \dfrac{4.6}{3}x^{-5/3} = -\dfrac{1.175}{\sqrt{x^3}} + \dfrac{23}{15\sqrt[3]{x^5}}$

57. (a) $(-\infty, \infty)$ (b) $f(x) = \dfrac{x^3}{4} - \dfrac{3x^2}{4} + \dfrac{1}{3}$

   (c) $f'(x) = 0.75x^2 - 1.5x$ (d) $(-\infty, \infty)$

59. (a) $(-\infty, 0)\cup(0, \infty)$

   (b) $f(x) = 2x^2 + 3x - 1 + 3x^{-1}$

   (c) $f'(x) = 4x + 3 - 0 - 3x^{-2} = 4x + 3 - \dfrac{3}{x^2}$

   (d) $(-\infty, 0)\cup(0, \infty)$

61. (a) $(-\infty, 0)\cup(0, \infty)$

   (b) $f(x) = 7x^2 - 50 + x^{-1}$

   (c) $f'(x) = 14x - 0 - x^{-2} = 14x - \dfrac{1}{x^2}$

   (d) $(-\infty, 0)\cup(0, \infty)$

63. (a) $(0, \infty)$

   (b) $f(x) = 2x^{5/2} - 7x^{3/2} + 1.5x^{-1/2}$

   (c) $f'(x) = 5x^{3/2} - 10.5x^{1/2} - 0.75x^{-3/2}$

   (d) $(0, \infty)$

65. (a) $f'(x) = 3x^2$ (b) 3 (c) $y = 3x + 2$

   (d) The function and the tangent line are graphed in viewing window $[-2, 2]$ by $[-4, 4]$.

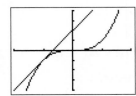

67. (a) $f'(x) = \dfrac{-1}{x^2}$ (b) $-\dfrac{1}{9}$ (c) $y = \dfrac{-1}{9}x + \dfrac{2}{3}$

(d) The function and the tangent line are graphed in viewing window $[-3, 8]$ by $[-3, 3]$.

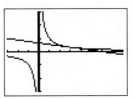

69. (a) $f'(x) = \dfrac{2}{3\sqrt[3]{x}}$

   (b) $\dfrac{1}{3}$

   (c) $y = \dfrac{1}{3}x + \dfrac{4}{3}$

   (d) The function and the tangent line are graphed in viewing window $[0, 20]$ by $[0, 8]$.

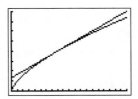

71. (a) $N'(x) = \dfrac{520}{x^2}$

   (b) $N(10) = 2448; N'(10) = 5.2$; When \$10,000 is spent on advertising, 244,800 CDs are sold and sales are increasing at a rate of 520 CDs per thousand dollars spent.

73. (a) $C'(x) = 11 - \dfrac{3.5}{\sqrt{x}} + 0.045\sqrt{x}$

   (b) $C(300) \approx 6334.64; C'(300) \approx 11.58$; When 300 coats are produced per week, the total cost is about \$6,334.64 and the costs are increasing at a rate of about \$11.58 per coat.

75. (a) $f'(x) = \dfrac{-186.42}{x^3}$

   (b) $f(1) = 93.21; f'(1) = -186.42$; At one mile downwind from the plant, the amount of sulfur dioxide in the air is 93.21 parts per million and the amount of sulfur dioxide in the air is decreasing at a rate of 186.42 parts per million per mile.

77. $s'(65) = 0.24$; When 65 dozen golf balls are supplied, the price per dozen golf balls increases by 0.24 dollars or 24 cents per dozen.

79. $s'(31) \approx 0.06$; When 31 baking pans are supplied, the price is increasing by 0.06 dollars or 6 cents per pan.

81. $f(9) = 21.42; f'(9) = 1.08$; In the year 2004, the number of Hispanic voters was 21.42 million and increasing at a rate of 1.08 million Hispanic voters per year.

83. $f'(5) = -3.06$; The unemployment rate in Bulgaria is decreasing by 3.06% per year in the year 2005.

85. Since the tangent line to any point on the graph of a constant function would just be the constant function itself and since the slope of a constant line is zero, then the derivative of a constant is zero.

87. (a) $y_1 = \dfrac{1}{6}x + \dfrac{3}{2}$ **(b)** $y_2 = -6x + 57$

    (c) The viewing window is $[-10.16, 20.16]$ by $[-10, 10]$.

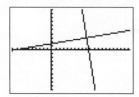

91. (a) 4   **(b)** 0   **(c)** Does not exist

Section 3.2

1. Multiplied   **3.** Square

5. $f'(x) = 2x \cdot (2x + 1) + x^2 \cdot (2) = 6x^2 + 2x$

7. $f'(x) = 3x^2 \cdot (3x^2 + 2x - 5) + x^3 \cdot (6x + 2)$
$\qquad = 15x^4 + 8x^3 - 15x^2$

9. $f'(x) = 12x^3 \cdot (2x^2 - 9x + 1) + (3x^4) \cdot (4x - 9)$
$\qquad = 36x^5 - 135x^4 + 12x^3$

11. $f'(x) = -10x \cdot (3x^3 + 5x - 7) + (-5x^2) \cdot (9x^2 + 5)$
$\qquad = -75x^4 - 75x^2 + 70x$

13. $f'(x) = 3 \cdot (2x - 1) + (3x + 4) \cdot 2 = 12x + 5$

15. $f'(x) = 5 \cdot (3x^3 + 2x^2 + 1) + (5x + 3) \cdot (9x^2 + 4x)$
$\qquad = 60x^3 + 57x^2 + 12x + 5$

17. $f'(x) = (6x - 2) \cdot (2x^2 + 5x - 7) + (3x^2 - 2x + 1) \cdot (4x + 5)$
$\qquad = 24x^3 + 33x^2 - 58x + 19$

19. $f'(x) = (x^{-1/2} + 4) \cdot (3x - 4) + (2x^{1/2} + 4x - 3) \cdot 3$

$\qquad = 24x + 9\sqrt{x} - 25 - \dfrac{4}{\sqrt{x}}$

21. $f'(x) = (3.6x^{1/5} - 5) \cdot (4x^{5/3} + 2x - 5)$
$\qquad\qquad + (3x^{6/5} - 5x) \cdot \left(\dfrac{20}{3}x^{2/3} + 2\right)$

23. $f'(x) = 1.5x^{-1/2} \cdot (2x^{1/2} - x^{-3}) + (3x^{1/2} - 5) \cdot (x^{-1/2} + 3x^{-4})$

$\qquad = 6 - \dfrac{5}{\sqrt{x}} + \dfrac{7.5}{x^3\sqrt{x}} - \dfrac{15}{x^4}$

25. $f'(x) = \left(\dfrac{2}{3}x^{-1/3} + 1\right) \cdot (x^{-1} + x^{-2})$
$\qquad\qquad + (x^{2/3} + x + 1) \cdot (-x^{-2} - 2x^{-3})$

$\qquad = -\dfrac{1}{3x\sqrt[3]{x}} - \dfrac{2}{x^2} - \dfrac{4}{3x^2\sqrt[3]{x}} - \dfrac{2}{x^3}$

27. $f'(x) = \dfrac{1 \cdot (x + 1) - (x + 2) \cdot 1}{(x + 1)^2} = \dfrac{-1}{(x + 1)^2}$

29. $y' = \dfrac{4 \cdot (2x + 1) - (4x - 3) \cdot 2}{(2x + 1)^2} = \dfrac{10}{(2x + 1)^2}$

31. $f'(x) = \dfrac{(6x - 5) \cdot (5x^2 + 3x + 2) - (3x^2 - 5x + 1) \cdot (10x + 3)}{(5x^2 + 3x + 2)^2}$

$\qquad = \dfrac{34x^2 + 2x - 13}{(5x^2 + 3x + 2)^2}$

33. $f'(x) = \dfrac{(1.5x^{-1/2}) \cdot (6x - 1) - (3x^{1/2} - 5) \cdot 6}{(6x - 1)^2}$

$\qquad = \dfrac{-9x + 30\sqrt{x} - 1.5}{\sqrt{x}(6x - 1)^2}$

35. $f'(x) = \dfrac{(3x^2 - 6x + 2) \cdot (x - 1) - [(x^2 + 2)(x - 3)] \cdot 1}{(x - 1)^2}$

$\qquad = \dfrac{2x^3 - 6x^2 + 6x + 4}{(x - 1)^2}$

37. $f'(x) = \dfrac{(30x^5 + 60x^3 + 4x) \cdot (x - 4) - [(5x^4 + 2)(x^2 + 3)] \cdot 1}{(x - 4)^2}$

$\qquad = \dfrac{25x^6 - 120x^5 + 45x^4 - 240x^3 + 2x^2 - 16x - 6}{(x - 4)^2}$

39. $f'(x) = (0) \cdot (4x^3 + 2x^2 - 3x - 5) + \dfrac{1}{2} \cdot (12x^2 + 4x - 3)$

$\qquad = 6x^2 + 2x - \dfrac{3}{2}$

41. (a) $f'(x) = 2x^3 - 10x + 2x^3 = 4x^3 - 10x$ **(b)** $y = -6x + 2$

43. (a) $f'(x) = 5x^4 + 3x^2 + 2x$ **(b)** $y = 10x - 6$

45. (a) $f'(x) = \dfrac{-3}{(x - 1)^2}$ **(b)** $y = -3x + 10$

47. (a) $f'(x) = \dfrac{-6x^2 + 18x - 6}{(-2x + 3)^2}$ **(b)** $y = -\dfrac{6}{5}x - \dfrac{1}{5}$

49. (a) $y = 20x - 36$

    (b) The graphs are displayed in the viewing window $[-3, 3]$ by $[-10, 20]$.

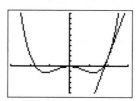

51. (a) $y = 7x + 10$

    (b) The graphs are displayed in the viewing window $[-2, 2]$ by $[-10, 20]$.

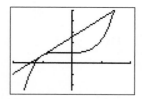

53. (a) $y = 6.92x - 8.64$

    (b) The equations are displayed in the viewing window $[-1/2, 5]$ by $[-5, 15]$.

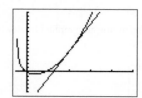

55. $f'(x) = 3x^2 + 8x - 7$ **57.** $f'(x) = 24x^3 + 42x^2 - 2x - 21$

59. $f'(x) = 21x^2\sqrt{x} - 7.5x\sqrt{x} + 6\sqrt{x} - \dfrac{1}{\sqrt{x}}$

61. (a) $q(3) = 30(3) - 0.5(3)^2 = 90 - 4.5 = 85.5$; $q'(3) = 30 - 3 = 27$; Three months after a new computer hits the market, the monthly sales are 8550 units and the monthly sales are increasing at a rate of 2700 computers per month.

(b) $p(3) = 2200 - 34(3)^2 = 2200 - 306 = 1894; p'(3) = -68(3) = -204$; After three months on the market, the price of the computer was \$1894 and the price of the computer was decreasing at a rate of \$204 per month.

(c) $R(t) = (2200 - 34t^2)(30t - 0.5t^2)$

(d) $R(3) = 161{,}937; R'(3) = 33{,}696$; After three months, the total monthly revenue from sales of a new computer was \$16,193,700 and revenue was increasing at a rate of \$3,369,600 per month.

63. (a) $q(3) = 85.5; q'(3) = 27$; Three months after a new CD-ROM drive hits the market, the monthly sales are 8,550 CD-ROM units and the monthly sales are increasing at a rate of 2700 CD-ROM drives per month.

(b) $p(3) = 211; p'(3) = -2(3) = -6$; After three months on the market, the price of the CD-ROM drive was \$211 and the price was decreasing at a rate of \$6 per month.

65. (a) $C'(x) = \dfrac{11300}{(100 - x)^2}$

(b) $C(50) = 113$; It will cost \$113,000 to clean up 50% of the pollutants.

$C'(50) = 4.52$; When 50% of the pollutants is removed, the cost of removing the pollutants is increasing at a rate of \$4520 per 1%.

67. (a) 100   **(b)** $P'(t) = \dfrac{68}{(1 + 0.02t)^2}$

(c) $P(5) = \frac{450}{1.1} \approx 409$; After five months, there are about 409 bass in the lake.

$P'(5) = \frac{68}{1.21} \approx 56.2$; After five months, the bass population was increasing at a rate of about 56.2 bass per month.

69. (a) $f'(x) = 0.24x^{1/2} + 0.9728x^{-1/2} + 2.1104$

(b) $f'(18) \approx 3.38$; In 2008, the amount spent on boats and pleasure craft was increasing by approximately 3.38 billion dollars per year.

71. (a) $f'(x) = \dfrac{0.0824}{(0.03x - 0.35)^2}$

(b) $f'(9) = 12.875$; This means that the number of unemployed in Ireland was increasing by 12.875 thousand or 12,875 people per year in 2009.

73. (a) $h(x) = \dfrac{f(x)}{g(x)} \cdot 100 = \dfrac{1.65x + 13.13}{2.45x^2 + 26.52x + 419.01} \cdot 100$

$= \dfrac{165x + 1313}{2.45x^2 + 26.52x + 419.01}$

The graph is below in a [0, 20] by [0, 4] window:

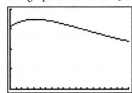

(b) $h'(x) = \dfrac{-404.25x^2 - 6433.7x + 34{,}315.89}{(2.45x^2 + 26.52x + 419.01)^2}$

(c) $h'(3) \approx 0.042$; The percentage of profits spent on taxes was increasing by 0.042% per year in 1993.

$h'(x) \approx -0.074$; The percentage of profits spent on taxes was decreasing by 0.074% per year in 2008.

75. $k'(x) = f'(x) \cdot g(x) \cdot h(x) + f(x) \cdot g'(x) \cdot h(x) + f(x) \cdot g(x) \cdot h'(x)$

Section 3.3

1. Composite   **3.** Generalized Power

5. $f'(x) = 2(x + 1)(1) = 2x + 2$

7. $f'(x) = 3(x - 5)^2(1) = 3(x - 5)^2$

9. $f'(x) = 2(2 - x)(-1) = 2x - 4$

11. $f'(x) = 3(2x + 4)^2(2) = 6(2x + 4)^2$ or $24(x + 2)^2$

13. $f'(x) = 5(5 - 2x)^4(-2) = -10(5 - 2x)^4$

15. $f'(x) = 5(3x^2 + 7)^4(6x) = 30x(3x^2 + 7)^4$

17. $f'(x) = 2(x^3 - 2x^2 + x)(3x^2 - 4x + 1) = (6x^2 - 8x + 2)(x^3 - 2x^2 + x)$ or $2x(3x^2 - 4x + 1)(x^2 - 2x + 1)$

19. $f'(x) = 9(x^3 - 4)^2(3x^2) = 27x^2(x^3 - 4)^2$

21. $f'(x) = 50(5x^2 - 3x - 1)^9(10x - 3)$

23. $f'(x) = 55(4x^2 - x - 4)^{54}(8x - 1)$

25. $f'(x) = \dfrac{1}{2}(2x - 4)^{-1/2}(2) = \dfrac{1}{\sqrt{2x - 4}}$

27. $f'(x) = \dfrac{1}{3}(x^2 + 2x)^{-2/3} \cdot (2x + 2) = \dfrac{2x + 2}{3\sqrt[3]{(x^2 + 2x)^2}}$

29. $f'(x) = -2(5x - 2)^{-3}(5) = \dfrac{-10}{(5x - 2)^3}$

31. $f'(x) = -\dfrac{1}{2}(x^2 + 2x + 4)^{-3/2}(2x + 2)$

$= \dfrac{-x - 1}{\sqrt{(x^2 + 2x + 4)^3}}$

33. $f'(x) = -\dfrac{1}{4}(3x^3 - x)^{-5/4}(9x^2 - 1)$

$= \dfrac{-9x^2 + 1}{4\sqrt[4]{(3x^3 - x)^5}}$

35. (a) $f'(x) = \dfrac{-3}{(3x + 4)^2}$ **(b)** $f'(x) = \dfrac{-3}{(3x + 4)^2}$

37. (a) $f'(x) = \dfrac{-10}{(x - 2)^3}$ **(b)** $f'(x) = \dfrac{-10}{(x - 2)^3}$

39. (a) $f'(x) = \dfrac{-4x - 4}{(x^2 + 2x + 3)^2}$ **(b)** $f'(x) = \dfrac{-4x - 4}{(x^2 + 2x + 3)^2}$

41. $y = 6x - 5$ **43.** $y = -4x + 5$

45. $y = 256x - 496$ **47.** $x = 2$

49. $f'(x) = \dfrac{1}{2}(x^2 + 5)^{-1/2}(2x) = \dfrac{x}{\sqrt{x^2 + 5}}$

51. $f'(x) = \dfrac{1}{3}(2x - 1)^{-2/3}(2) = \dfrac{2}{3\sqrt[3]{(2x - 1)^2}}$

53. $f'(x) = -\dfrac{5}{2}(2x - 8)^{-3/2}(2) = \dfrac{-5}{\sqrt{(2x - 8)^3}}$

55. $f'(x) = -\dfrac{64}{3}(5x^2 - 6x + 3)^{-4/3}(10x - 6)$

$= \dfrac{-128(5x - 3)}{\sqrt[3]{(5x^2 - 6x + 3)^4}}$

57. $f'(x) = 1(x - 4)^3 + x(3)(x - 4)^2 \cdot 1$
$= 4(x - 4)^2(x - 1)$

59. $f'(x) = 1(x^2 + 3x)^{1/2} + x\left(\dfrac{1}{2}\right)(x^2 + 3x)^{-1/2}(2x + 3)$

$\quad = \dfrac{x(4x + 9)}{2\sqrt{x}(x + 3)}$

61. $f'(x) = \dfrac{3x^2(3x - 8)^2 - 2x^3(3x - 8)(3)}{(3x - 8)^4} = \dfrac{3x^2(x - 8)}{(3x - 8)^3}$

63. $f'(x) = 3(x + 3)^2[1](2x - 1)^2 + 2(x + 3)^3(2x - 1)[2]$

$\quad = (x + 3)^2(2x - 1)(10x + 9)$

65. $f'(x) = \dfrac{1}{2}\left[\dfrac{x + 3}{x - 3}\right]^{-1/2} \cdot \dfrac{1 \cdot (x - 3) - (x + 3) \cdot 1}{(x - 3)^2}$

$\quad = \dfrac{-3}{(x - 3)\sqrt{(x - 3)(x + 3)}}$

67. $f'(x) = \dfrac{1.84x + 1.15}{(4x^2 + 5x + 6)^{0.77}}$

69. $f'(x) = 1.03(x + 3)^{0.03}$ **71.** $f'(x) = 1.7568(x + 1)^{0.22}$

73. $f'(70) \approx -36.51$; When the group is 70 years old, the number that are surviving is decreasing at a rate of about 36.51 people per year.

75. $f'(10) \approx 192.09$; In the tenth year of the study, the number of students enrolled in the Arts and Sciences was increasing by about 192.09 students per year.

77. (a) $f(20) \approx 187.08$; It will take about 187.08 minutes to learn 20 items on the list.

(b) $f'(x) = \dfrac{3.75x - 15}{\sqrt{x - 6}}$

(c) $f'(20) \approx 16.04$; After 20 items in the list have been learned, the time required to learn an item on the list is increasing at a rate of about 16.04 minutes per item.

79. (a) $f(15) \approx 129.75$; After 15 years of operation, approximately 129.75 thousands gallons of toxic material is entering the lake.

(b) $f'(x) = \dfrac{16\left(\dfrac{4}{5}x^{1/5} + 2\right)^3}{25x^{4/5}}$

(c) $f'(15) = \dfrac{16\left(\dfrac{4}{5}(15)^{1/5} + 2\right)^3}{25(15)^{4/5}} \approx 2.82$; After the company has been in operation for 15 years, the amount of toxic material entering the lake is increasing at a rate of about 2.82 thousand gallons per year.

81. $f'(8) \approx 13.66$; In 2008, the monthly minimum wage in Romania is increasing by approximately 13.66 Euros per month.

83. (a) $f'(x) = \dfrac{1.150182}{(1.57x + 17.27)^{0.34}}$

(b) $f(35) \approx 18.70$; $f'(35) \approx 0.27$; In 2005, the approximate number of unmarried women in the U.S. labor force was 18.70 million and was increasing by 0.27 million women per year.

85. $f'(x) = \dfrac{3}{x}$ **87.** $f'(x) = -\dfrac{1}{2x}$

89. First, you would need to use the Product Rule and within the Product Rule, you will need to use two Chain Rules followed by the Sum and Difference and Constant Product Rules as needed.

91. $h'(x) = \dfrac{1}{2}(x + (x + x^{1/2})^{1/2})^{-1/2} \cdot \left[1 + \dfrac{1}{2}(x + x^{1/2})^{-1/2} \cdot \left(1 + \dfrac{1}{2}x^{-1/2}\right)\right]$

93. $\dfrac{dy}{dx} = \dfrac{\sqrt{2t}}{2\sqrt{t + 1}}$

Section 3.4

1. Difference **3.** Actual **5.** Small **7.** $dy = 6\,dx$

9. $dy = (-6x + 2)\,dx$ **11.** $dy = \dfrac{-5}{(x - 1)^2}\,dx$

13. $dy = \dfrac{1}{(x + 1)^2}\,dx$ **15.** $dy = \left(\dfrac{1}{2\sqrt{x}} - \dfrac{2}{x^2}\right)dx$

17. $dy = \left(\dfrac{-1}{2x\sqrt{x}} + \dfrac{2}{3\sqrt[3]{x}}\right)dx$ **19.** $dy = \dfrac{-4x}{(x^2 - 1)^2}\,dx$

21. $dy = \left(5.1\sqrt[10]{x^7} + \dfrac{28}{5\sqrt[5]{x}}\right)dx$ **23.** $\Delta y = 0.21$; $dy = 0.2$

25. $\Delta y = -398$; $dy = -390$ **27.** $\Delta y \approx 1.58$; $dy \approx 1.67$

29. $\Delta y \approx 0.035$; $dy \approx 0.035$ **31.** $\Delta y \approx -0.0802$; $dy \approx -0.0889$

33. $\Delta y = 0.4606$; $dy = 1.2$ **35.** 5.1 **37.** 2.9630

39. 1.9938 **41.** 0.28280 **43.** 6.28

45. (a) $dy = \dfrac{55}{\sqrt{x}}\,dx$ **(b)** $dy \approx 13.91$

47. (a) $dy = (0.66x^2 - 4.7x + 14.32)\,dx$ **(b)** $dy \approx 1401.96$

49. (a) $dy = (120 - 4.8x)\,dx$ **(b)** $dy = 72$

51. (a) $dy = -0.07$; From 2009 to 2010, the average number of days an acute inpatient stays in a hospital in Japan is decreasing by about 0.07 days.

(b) $y = -0.07x + 19.13$; When $x = 14$, $y = -0.07(14) + 19.13 = 18.15$. In 2014, the number of days an acute inpatient stays in the hospital in Japan will be about 18.15 days.

53. (a) $dy = -90$; From 2008 to 2009, the number of bank-issued credit cards in circulation has decreased by about 90 million credit cards.

(b) $y = -90x + 1146.2$; When $x = 12$, $y = -90(12) + 1146.2 = 66.2$. In 2012, the number of bank-issued credit cards will be about 66.2 million credit cards.

55. Δy provides the actual change in the variable y whereas dy gives the approximate change. For small changes in x, $\Delta y \approx dy$.

57. $f(1.96) \approx -4.12$; $f(2.04) \approx -3.88$

Section 3.5

1. Marginal **3.** Marginal cost **5.** Price

7. (a) $MC(x) = 23$

(b) $MC(10) = 23$; The estimated cost of producing the 11th unit is $23.

(c) $C(11) - C(10) = 23$; This matches exactly our estimate in part b.

9. (a) $MC(x) = x + 12.7$

(b) $MC(11) = 23.7$; The estimated cost of producing the 12th unit is $23.70.

(c) $C(12) - C(11) = 24.2$; Our estimate in part b is off by $0.50.

11. (a) $MC(x) = 0.6x^2 - 6x + 50$

(b) $MC(30) = 410$; The estimated cost of producing the 31st unit is $410.

(c) $C(31) - C(30) = 425.2$; Our estimate in part b is off by $15.20.

13. (a) $R(x) = 6x$; $P(x) = x - 500$

(b) $MC(x) = 5$; $MP(x) = 1$

15. (a) $R(x) = \dfrac{-x^2}{20} + 15x$; $P(x) = \dfrac{-3x^2}{50} + 8x - 1000$

(b) $MC(x) = \dfrac{x}{50} + 7$; $MP(x) = \dfrac{-3x}{25} + 8$

17. (a) $R(x) = -0.005x^2 + 7x$; $P(x) = 0.001x^3 - 0.005x^2 + 3x - 100$

(b) $MC(x) = -0.003x^2 + 4$; $MP(x) = 0.003x^2 - 0.01x + 3$

19. $MC(x) = 40 - 0.002x$; $MC(200) = 39.6$; The estimated daily cost of producing the 201st unit is $39.60.

21. $MC(x) = 35 - 0.02x$; $MC(200) = 31$; The estimated cost to produce the 201st rain coat is $31.

23. (a) $MC(x) = 200 - 0.4x$; $MC(500) = 0$

(b) It would be recommended to increase production since there would be no increase in cost.

25. (a) $AC(x) = \frac{15000}{x} + 100 - 0.001x$; $AC(100) = 249.9$; When 100 patio swings are produced per day, the average cost per day to produce each swing is $249.90.

(b) $MAC(x) = -\frac{15000}{x^2} - 0.001$; $MAC(100) = -1.501$; When 100 patio swings are produced each day, the average cost per day to produce each swing is decreasing at a rate of $1.501 per swing.

27. (a) $MP(x) = -0.02x + 12$; $MP(700) = -2$; The estimated weekly loss of producing and selling the 701st radio is $2.

(b) Profit is decreasing when the Hanash Corporation is producing and selling 700 radios per week.

29. $MP(x) = 2 - \dfrac{1}{2\sqrt{x}}$; $MP(110) \approx 1.95233$; The estimated monthly profit of selling the 111th subscription is about $1.95.

31. (a) $AP(x) = 2 - \dfrac{1}{\sqrt{x}}$; $AP(110) \approx 1.90465$; When 110 subscriptions are sold per month, the average profit per month for each subscription is about $1.90.

(b) $MAP(x) = \dfrac{1}{2x\sqrt{x}}$; $MAP(110) \approx 0.0004334$; When 110 subscriptions are sold per month, the average profit per month for each subscription is increasing at a rate of about $0.00043 per subscription.

33. (a) $P(x) = -0.001x^2 + 16x - 5000$

(b) $P(1000) = 10,000$; The profit from producing and selling 1000 action figures is $10,000.

(c) $MP(1000) = 14$; The estimated profit of producing and selling the 1001st action figure is $14.

35. (a) $R(x) = \frac{-x^2}{30} + 200x$

(b) $P(x) = \frac{-x^2}{30} + 140x - 72000$; $x = 601$ is the smallest production level and $x = 3599$ is the largest production level.

(c) $P'(3000) = -60$; The estimated loss of producing and selling the 3001st picture frame is $60.

37. (a) $C(1001) \approx 42,530$ **(b)** $C(2001) \approx 57,100$

39. If marginal profit were to equal zero, then marginal revenue minus marginal cost would equal zero and consequently, marginal revenue would equal marginal cost.

41. We would determine the level of production by setting the marginal profit function equal to c and solving for x.

43. We would determine the level of production by setting the profit function equal to c and solving for x.

45. $MR(x) = k$ **47.** $MP(x) = 2ax + k - b$

Chapter 3 Review Exercises

Section 3.1 Review Exercises

1. $f'(x) = 8x^7$ **3.** $f'(x) = \dfrac{12}{7\sqrt[7]{x^{10}}}$

5. $f'(x) = 12x + 5$ **7.** $f'(x) = \dfrac{3}{2}x^2 - 6x + \dfrac{2}{3}$

9. $f'(x) = -21.3x^2 - 10.04x + 11.19$

11. $f'(x) = \dfrac{1}{5\sqrt[5]{x^4}} - \dfrac{1}{2\sqrt{x}} - \dfrac{1}{x^2}$

13. (a) $(-\infty, 0) \cup (0, \infty)$ **(b)** $f(x) = 3x^2 - 2x + 6 + 2x^{-1}$

(c) $f'(x) = 6x - 2 - \dfrac{2}{x^2}$

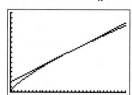

(d) $(-\infty, 0) \cup (0, \infty)$

15. (a) $f'(x) = -2x^{-3} = \dfrac{-2}{x^3}$ **(b)** $\dfrac{-1}{32}$ **(c)** $y = \dfrac{-1}{32}x + \dfrac{3}{16}$

(d) The graph is displayed in viewing window $[0, 8]$ by $[0, 0.25]$

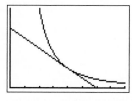

17. (a) $f'(t) = 6 + \dfrac{9}{2\sqrt{t}} + 0.03\sqrt{t}$

(b) $f'(\frac{1}{3600}) \approx 276$; After the first second, the energy usage of the appliance is increasing at a rate of 276 watt-hours per hour.

19. (a) $m_{sec} = 900$

(b)

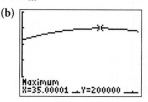

Maximum
X=35.00001 Y=200000

In the 35th week, the average price of a home was at its maximum level of $200,000.

Section 3.2 Review Exercises

21. $f'(x) = 30x^4 - 12x^3 + 24x^2$

23. $f'(x) = 24x^3 - 21x^2 - 38x + 9$

25. $f'(x) = 42x - \dfrac{141\sqrt{x}}{2} + 27 - \dfrac{2}{\sqrt{x}}$

27. $f'(x) = \dfrac{-8x^2 + 66x + 13}{(3x^2 - x + 9)^2}$

29. (a) $f'(x) = -27x^2 - 30x$ (b) $y = -168x + 204$

(c) The graph is displayed in viewing window $[0, 4]$ by $[-800, 0]$

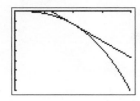

Section 3.3 Review Exercises

31. $f'(x) = 3(x + 2)^2$ **33.** $f'(x) = -3(8 - x)^2$

35. $f'(x) = 8(2x + 5)^3$ **37.** $f'(x) = 6(2x - 5)(x^2 - 5x + 3)$

39. $f'(x) = \dfrac{4x - 5}{3\sqrt[3]{(2x^2 - 5x + 7)^2}}$

41. (a) $f'(x) = \dfrac{-6}{(2x + 9)^2}$ (b) $f'(x) = \dfrac{-6}{(2x + 9)^2}$

43. (a) $f'(x) = \dfrac{-18}{(3x + 5)^4}$ (b) $f'(x) = \dfrac{-18}{(3x + 5)^4}$

45. $y = 400x + 368$ **47.** $y = 32x - 80$

49. $f'(x) = \dfrac{7}{2\sqrt{7x - 12}}$ **51.** $f'(x) = \dfrac{-6}{(\sqrt{4x + 5})^3}$

53. $f'(x) = (x^2 + 5)^2(7x^2 + 5)$

55. $f'(x) = \dfrac{15x - 1}{2(5x - 6)\sqrt{5x - 6}}$

57. $f'(x) = \dfrac{0.67(6x - 1)}{(3x^2 - x + 1)^{0.33}}$

59. $f'(x) = \dfrac{-0.7(3x^2 - 2x + 5)}{(x^3 - x^2 + 5x + 1)^{1.7}}$

61. (a) $f'(x) = 2.268\,(1.8x + 6.3)^{0.26}$

(b) $f'(8) \approx 4.986$; This means that in 1998, the net profits for petroleum and coal corporations was increasing by about 4.986 billion dollars per year.

(c) The graph is given below for the viewing window $[0, 21]$ by $[0, 7]$:

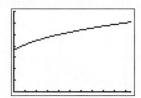

Section 3.4 Review Exercises

63. $dy = 4\,dx$ **65.** $dy = \dfrac{-8}{(x - 5)^2}\,dx$

67. $dy = \left(\dfrac{1}{2\sqrt{x}} + \dfrac{12}{x^5}\right)dx$

69. $dy = \left(4x^3 - 2 + \dfrac{2}{3\sqrt[3]{x}}\right)dx$

71. $dy = \dfrac{20x}{(x^2 + 5)^2}\,dx$

73. $dy = \left(3.4x^{0.7} - \dfrac{4}{x^{0.2}}\right)dx$

75. $\Delta y = 2.34$; $dy = 2.3$

77. $\Delta y = 0.1$; $dy = \approx 0.1051$

79. $\Delta y \approx -0.34$; $dy = -0.4$

81. 8.0625

83. 2.0094

85. (a) $f'(x) = (90 - 5.4x)dx$

(b) $dy = 68.4$

87. In Exercise 85, $\Delta y = 65.7$; Our estimate in #85 was off by $6840 - 6570 = 270$ shirts.

In Exercise 86, $\Delta y = 65.93$; Our estimate in #86 was off by $6780 - 6593 = 187$ shirts.

Section 3.5 Review Exercises

89. (a) $MC(x) = 18$

(b) $MC(12) = 18$; Thus, it will cost about \$18 to produce the 13th unit.

(c) $C(13) - C(12) = 18$; This matches our answer in part b exactly.

91. (a) $MC(x) = 26.7$

(b) $MC(8) = 26.7$; Thus, it will cost about \$26.70 to produce the 9th unit.

(c) $C(9) - C(8) = 26.7$; This matches our answer in part b exactly.

93. (a) $MC(x) = \dfrac{1}{2}x + 12$

(b) $MC(31) = 27.5$; Thus, it will cost about \$27.50 to produce the 32nd unit.

(c) $C(32) - C(31) = 27.75$; This is only \$0.25 off from our answer in part b.

95. (a) $R(x) = 11x$

(b) $P(x) = 4x - 250$

(c) $MP(x) = 4$

(d) $MC(x) = 7$; $MR(x) = 11$

(e) $MR(x) - MC(x) = 4$; This is the same as part c.

97. (a) $R(x) = \dfrac{-x^2}{15} + 50x$

(b) $P(x) = \dfrac{-x^2}{6} + 47x - 850$

(c) $MP(x) = \dfrac{-x}{3} + 47$

(d) $MC(x) = \dfrac{1}{5}x + 3$; $MR(x) = \dfrac{-2x}{15} + 50$

(e) $MR(x) - MC(x) = \dfrac{-x}{3} + 47$; This is the same as part c.

99. (a) $R(x) = -0.005x^2 + 10x$

(b) $P(x) = 0.001x^3 - 0.005x^2 + 2x - 100$

(c) $MP(x) = 0.003x^2 - 0.01x + 2$

(d) $MC(x) = -0.01x + 10$

(e) $MR(x) - MC(x) = 0.003x^2 - 0.01x + 2$; This is the same as part c.

101. (a) $C(x) = 85x + 5600$ (b) $MC(x) = 85$

103. (a) $MAC(x) = -5600x^{-2} = \dfrac{-5600}{x^2}$

(b) $MAC(250) \approx -0.09$; When 250 bicycles are produced, then average cost per bike is decreasing at a rate of about \$0.09 per bike.

105. (a) $P(x) = -0.02x^2 + 65x - 5600$

(b) Either 1500 or 1750 bicycles should be manufactured.

107. (a) Since $R(x) = 6.5x$; $P(x) = -\frac{x^2}{5500} + \frac{25}{6}x - 1500$

(b) $P(3000) \approx 9363.6364$; When 3000 watercolor sets are produced and sold, Colorama Company earns about \$9,363.64 in profits.

(c) $MP(x) = -\dfrac{x}{2750} + \dfrac{25}{6}$

(d) $MP(3000) \approx 3.0758$; The approximate profit from producing and selling the 3001st water color set is \$3.08.

(e) $P(3001) = 9366.7122$

(f) Two ten-thousandths

109. (a) $C(x) = 4.5x + 12000$; $R(x) = -0.002x^2 + 20x$

(b) $P(x) = -0.002x^2 + 15.5x - 12000$

(c) $x = 873$ is the smallest production level and $x = 6,877$ is the largest production level.

(d) $P'(x) = -0.004x + 15.5$

(e) $P'(2500) = 5.5$; When 2500 books are produced, the profit is increasing at a rate of \$5.50 per book.

111. 22,910 **113.** 38,910 **115.** 72,320

117. (a) $C(x) = 119,890x + 123,450$

(b) $R(x) = -3000x^3 - 50x^2 + 300,000x$ (Note: d is so small that we can ignore it)

(c) $MAC(x) = -123,450x^{-2} = \dfrac{-123,450}{x^2}$

(d) $MAC(1.5) \approx -54,866.67$; When 1500 printers are produced, we estimate that the average cost per 1000 printers is decreasing by about \$54,866.67 for each additional 1000 printers.

Chapter 4

Section 4.1

1. e **3.** General **5.** Decay **7.** Logistic

9. (a)

x	-5	-4	-3	-2	-1	0	1	2	3	4	5
$(\frac{1}{3})^x$	243	81	27	9	3	1	$\frac{1}{3}$	$\frac{1}{9}$	$\frac{1}{27}$	$\frac{1}{81}$	$\frac{1}{243}$

(b)

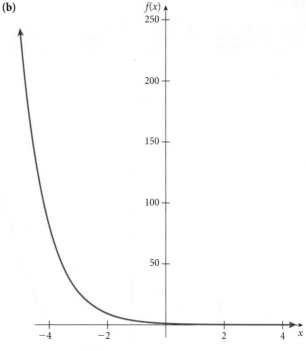

Exponential decay function

11. (a)

x	-5	-4	-3	-2	-1	0	1	2	3	4	5
4^x	$\frac{1}{1024}$	$\frac{1}{256}$	$\frac{1}{64}$	$\frac{1}{16}$	$\frac{1}{4}$	1	4	16	64	256	1024

(b)

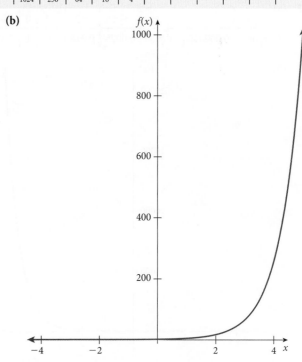

Exponential growth function

13. (a)

x	-5	-4	-3	-2	-1	0	1	2	3	4	5
$(2.3)^x$	0.016	0.036	0.082	0.19	0.43	1	2.3	5.29	12.17	27.98	64.36

(b)

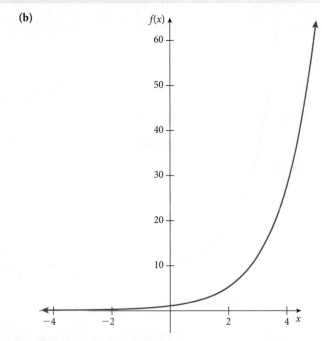

Exponential growth function

15. (a)

x	-5	-4	-3	-2	-1	0	1	2	3	4	5
$(0.7)^x$	5.95	4.16	2.92	2.04	1.43	1	0.7	0.49	0.34	0.24	0.17

(b)

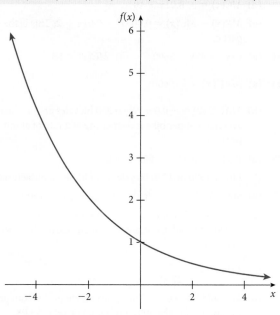

Exponential decay function

17. (a)

x	-5	-4	-3	-2	-1	0	1	2	3	4	5
e^{2x}	0.000045	0.00034	0.0025	0.018	0.14	1	7.39	54.60	403.43	2980.96	22026.47

(b)

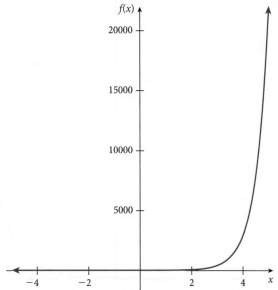

Exponential growth function

19. (a)

x	-5	-4	-3	-2	-1	0	1	2	3	4	5
$e^{0.3x}$	0.22	0.30	0.41	0.55	0.74	1	1.35	1.82	2.46	3.32	4.48

(b)

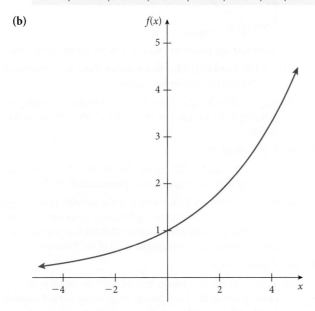

Exponential growth function

21. (a)

x	-5	-4	-3	-2	-1	0	1	2	3	4	5
$e^{-1.6x}$	2980.96	601.85	121.51	24.53	4.95	1	0.20	0.041	0.0082	0.0017	0.00034

(b)

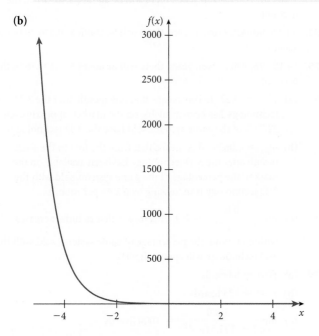

Exponential decay function

23. (a) $f(7) \approx 6.29$; That means that after seven days from when the wound occurred, the size of the wound is about 6.29 square inches.

(b) $m_{sec} \approx -2.71$; This means that from the first to the eighth day, the wound is shrinking in size by about 2.71 inches per day on average.

(c) $\lim\limits_{x \to \infty} [30(\frac{4}{5})^x] = 0$; As time continues without bound, the size of the wound approaches zero square inches. That is to say that the wound has healed.

25. (a) $f(5) \approx 5307.39$; That means that in 2005, the number of traffic fatalities in Sweden was about 5,307 fatalities.

(b) $m_{sec} \approx -416.96$; This means that from the fifth to the ninth year after the safety program began, the number of traffic fatalities in Sweden has decreased by around 416 fatalities per year.

(c) Since the base of the exponential function is less than one, this function will be a decreasing function, which implies that the number of traffic fatalities will decrease over time and therefore the program is effective.

(d)

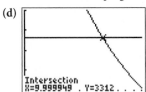

Intersection
X=9.999949 . Y=3312 . . . .

That means in around 10 years from the beginning of the program, the number of annual traffic fatalities will reach 3,312 fatalities.

27. (a) $f(1) = 45.448$; This means that in 2001, the approximate amount spent on direct marketing was about 45.448 billion dollars.

(b) $m_{sec} \approx 2.01$; This means that from 2001 to 2006, the amount spent on direct marketing was increasing on average by 2.01 billion dollars per year.

(c) Looking at the exponential model, the base is greater than one, which means that the amount spent is increasing.

29. $120   **31.** $3,573.05   **33.** $2,564.44

35. (a) $3,173.75   **(b)** $3202.06   **(c)** $3,216.87

37. (a) $37,534.15   **(b)** $38,464.55   **(c)** $38,533.00

39. (a) 7227.37; If $5500 is invested at a 5.5% interest rate compounded quarterly, there will be about $7,227.37 in the account after five years.

(b) $1,727.37

41. (a) 14353.51; If $10000 is invested at a 7.25% interest rate compounded monthly, there will be about $14,353.51 in the account after five years.

(b) $4,353.51

43. (a) $A(t) = 4000\left(1 + \dfrac{0.0575}{12}\right)^{12t}$

(b)

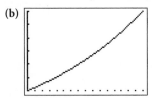

(c)

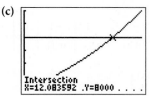

Intersection
X=12.083592 .Y=8000 . . . .

It will take approximately twelve years for the money to double.

45. (a) $f(5) \approx 3,789,831$; After five minutes, there are approximately 3,789,831 bacteria in the colony.

(b) $m_{sec} \approx 1,076,901.90$; In the first ten minutes, the colony was growing at an average rate of about 1,076,901.90 bacteria per minute.

47. (a) Exponential decay

(b) $f(2) \approx 19,392.58$; This means that in 2002, the amount of fish caught by Norwegians was approximately 19,393 tons.

(c) $m_{sec} \approx -2368.69$; This means that the amount of fish caught by Norwegians was decreasing by about 2,369 tons per year from 2002 to 2008, which implies that the fish population is not growing as fast to meet the need of the fisheries.

49. (a) Exponential growth

(b) $f(3) \approx 21.23$; This means that in 2003, the number of participants in the Food Stamp Program was approximately 21.23 million people.

(c) $m_{sec} \approx 1.49$; This means that from 2002 to 2009, the number of participants in the Food Stamp Program was increasing by approximately 1.49 million people per year.

51. 1,469.61; After seven years, there will be about $1,469.61 in the account.

53. 30,226.90; After seven years, there will be about $30,226.90 in the account.

55. 14,103.76; After seven years, there will be about $14,103.76 in the account.

57. (a) $f(1) \approx 0.2271$; This means that one month after the 3-D technology has been available on the market, approximately 22.71% of the game systems sold have the 3-D technology.

(b) $m_{sec} \approx 0.0629$; This means that from the first to the sixth month after the 3-D technology has been available on the market, the percentage of the game systems sold with the 3-D technology is increasing by 6.29% per month.

(c) $\lim\limits_{t \to \infty} \dfrac{0.9}{1 + 4e^{-0.3t}} = 0.9$; This means that as time increases without bound, the percentage of game systems sold with the 3-D technology will approach 90%.

59. (a) $f(0) \approx 8$ lizards

(b) $f(8) \approx 723$ lizards

(c) $\lim\limits_{t \to \infty} \dfrac{1000}{1 + 121.51e^{-0.72t}} = 1000$ lizards

61. (a) $f(6) \approx 13,101.39$

(b) $m_{sec} \approx 6233.94$; From 2006 to 2011, the number of AIDS cases was increasing by approximately 6,234 cases per year.

(c) $\lim\limits_{t \to \infty} \dfrac{46,820}{1 + 246e^{-0.76t}} = 46,820$; This means that as time continues without bound, the number of new AIDS cases in the United States will approach 46,820 cases.

63. The general exponential function can have any positive real number that is not equal to one as the base whereas the natural exponential function must have the natural number e as its base.

67. $(-\infty, 0)$

69. The value of b, the base, is greater than one since the graph is increasing from left to right.

71. $f(-1) = b^{-1} = \dfrac{1}{b}; f(0) = b^0 = 1; f(1) = b^1 = b$

Section 4.2

1. One-to-one **3.** Reflected **5.** 10

7. $f(g(3)) = 54; f(g(-2)) = -36$

9. $f(g(0)) = 2; g(f(0)) = 1$

11. $g(f(x)) = 24x - 6; f(g(x)) = 24x + 5$

13. $g(f(x)) = 25x^2 - 15x + 4; f(g(x)) = 5x^2 + 15x + 17$

15. $g(f(x)) = \dfrac{1}{x^3}; f(g(x)) = \dfrac{1}{x^3}$

17. $g(f(x)) = x + 3; f(g(x)) = \sqrt{x^2 + 3}$

27. $\log_2(\tfrac{3}{5}) = \log_2(3) - \log_2(5)$

29. $\log(8 \cdot 20) = \log(8) + \log(20)$

31. $\ln(\sqrt{26}) = \ln([26]^{1/2}) = \dfrac{1}{2}\ln(26)$

33. $\log_3\left(\dfrac{4\sqrt{3}}{9}\right) = \log_3(4) - 1.5$

35. $5 = \log_2(32)$ **37.** $-3 = \log_2\left(\dfrac{1}{8}\right)$ **39.** $1 = \ln(e)$

41. $x = \dfrac{\ln(6375)}{\ln(2)} \approx 2.67$ **43.** $x = \ln(31) \approx 3.43$

45. $x = \dfrac{\ln(42) - \ln(1.21)}{\ln(0.3)} \approx -2.95$

47. **(a)** $r(t) = 1.3t$.

  (b) $V(t) = 2.9293\pi \, t^3$; $V(t)$ is the volume of the sphere as a function of time.

  (c) $V(6) \approx 1987.80$; After 6 seconds, the volume of the sphere is about 1987.80 in³.

49. **(a)** $f(x) = x^{0.27}; g(x) = 1351x + 1355$

  (b) $h(11) - h(1) \approx 5.251$; In 2011, farms received 5.251 billion dollars more from direct federal subsidies than what they had received in 2001.

  (c) $m_{\text{sec}} \approx 0.5251$; That means that from 2001 to 2011, the amount of direct federal subsidies paid to farms was increasing on average by 0.5251 billion dollars per year.

51. **(a)** $f(10) \approx 103.87$; That means that in 2010, approximately 103.87 billion dollars was spent by Americans on foreign travel.

  (b) $f(6) - f(1) \approx 17.56$

  (c) $m_{\text{sec}} \approx 3.51$

53. **(a)** $x = e^{272/169} \approx 5$; That means that in 2005, the number of gross ticket sales for Broadway plays reached 828 million dollars.

  (b) $m_{\text{sec}} \approx 68.00$

55. Various answers. Graphically, if you were to draw $f(x) = b^x$ and $g(x) = \log_b x$, they would be reflections of each other over the line $y = x$ for $b > 0$ and $b \neq 1$.

57. $\log\left(\dfrac{ab}{c}\right) = \log(a) + \log(b) - \log(c)$

59. $\ln\left(\dfrac{a\sqrt{c}}{b}\right) = \log(a) + \dfrac{1}{2}\log(c) - \log(b)$

Section 4.3

1. Equivalent

3. General

5. Rewriting

7. $f'(x) = 7e^x$

9. $f'(x) = 2(4 + e^x) + 2x(e^x) = 2xe^x + 2e^x + 8$

11. $f'(x) = -10(5 - e^x)^{-2}(-e^x) = \dfrac{10e^x}{(5 - e^x)^2}$

13. $f'(x) = 8xe^x + 4x^2e^x = 4xe^x(x + 2)$

15. $f'(x) = \tfrac{1}{2}(12 - e^x)^{-1/2} \cdot (-e^x) = \dfrac{-e^x}{2\sqrt{12 - e^x}}$

17. $f'(x) = \dfrac{e^x(x^3 - 1) - (e^x - 10)3x^2}{(x^3 - 1)^2}$

$= \dfrac{x^3e^x - e^x - 3x^2e^x + 30x^2}{(x^3 - 1)^2}$

19. $f'(x) = \dfrac{e^x(e^x - 1) - (e^x + 1)e^x}{(e^x - 1)^2} = \dfrac{e^{2x} - e^x - e^{2x} - e^x}{(e^x - 1)^2}$

$= \dfrac{-2e^x}{(e^x - 1)^2}$

21. $f'(x) = 2e^x + 2xe^x - 1$

23. $f'(x) = e^{2x-1}(2) = 2e^{2x-1}$

25. $f'(x) = e^{\sqrt{x}}\dfrac{1}{2}x^{-1/2} = \dfrac{e^{\sqrt{x}}}{2\sqrt{x}}$

27. $f'(x) = 5e^{2x} + 5xe^{2x}(2) = 5e^{2x} + 10xe^{2x}$

29. $f'(x) = 0$ **31.** $f'(x) = 10^x \ln(10)$

33. $f'(x) = (\tfrac{1}{3})^x \ln(\tfrac{1}{3}) = -(\tfrac{1}{3})^x \ln(3)$

35. $f'(x) = 3x^2 \cdot 0.3^x + x^3 \cdot 0.3^x \ln(0.3) = x^2 0.3^x(3 + x \ln(0.3))$

37. $f'(x) = 10^{x+3} \ln(10) \cdot (1) = 10^{x+3} \ln(10) = 10^x \cdot 10^3 \ln(10)$

$= 10^{x+3} \ln(10)$

39. $f'(x) = 9^{1/x} \ln(9) \cdot (-1)x^{-2} = \dfrac{-9^{1/x} \ln(9)}{x^2}$

41. $f'(x) = e^x + xe^x - (5^2)^x \ln(5)(2) = e^x + xe^x - 2 \cdot 25^x \ln(5)$

43. **(a)** $f'(x) = 0.42e^{0.2x}$

  (b) $f'(3) \approx 0.77$; Three days after an animal specimen is exposed to a new pesticide, the diameter of the tumor is increasing at a rate of about 0.77 millimeters per day.

45. **(a)** $p'(t) = 12(0.8)^t \ln(0.8)$

  (b) $p'(1) \approx -2.14$; One year after the factory opened, the fish population was decreasing at a rate of about 214 fish per year.

  $p'(8) \approx -0.45$; The decay rate one year after the factory opened was more than 4.75 times the decay rate nine years after the factory opened.

47. (a) $A'(t) = 130e^{0.065t}$

(b) $A(5) \approx 2768.06$; After five years, there is about \$2,768.06 in the account.

$A'(5) \approx 179.92$; After five years, amount in the account was growing at a rate of about \$179.92 per year.

49. (a) $A'(t) = 250e^{0.05t}$

(b) $A(9) \approx 7841.56$; After nine years, there is about \$7,841.56 in the account.

$A'(9) \approx 392.08$; After nine years, the amount in the account was growing at a rate of about \$392.08 per year.

51. (a) $f'(x) = -18391.4e^{-0.2x}$

(b) $f'(2) = -12,328.12$

53. (a) $f'(x) = 1.12e^{0.8x}$

(b) $f'(9) \approx 1500.16$

55. (a) $f'(x) = 47 \cdot (1.08)^x \cdot \ln(1.08)$

(b) $f'(4) \approx 4.92$; $f'(8) \approx 6.70$; Revenues from motion pictures was growing by about 4.92 million dollars per year in 2004 while it was growing by about 6.70 million dollars per year in 2008 and thus, revenues had a higher rate of change in 2008.

59. $(-0.9127653, 1.1183256)$; The graph of $g(x) = e^x$ is bold:

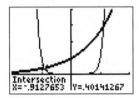 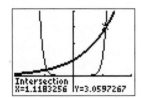

65. By letting $f(x) = e^x$, we have $\lim\limits_{x \to 0} \frac{e^x - 1}{x} = \lim\limits_{x \to 0} \frac{e^x - e^0}{x} =$

$\lim\limits_{x \to 0} \frac{f(x) - f(0)}{x} = \lim\limits_{x \to 0} \frac{f(0 + x) - f(0)}{x} = f'(0)$. Since $f(x) = e^x$,

then $f'(x) = e^x$ which implies $\lim\limits_{x \to 0} \frac{e^x - 1}{x} = f'(0) = e^0 = 1$.

Section 4.4

1. Natural   **3.** Model   **5.** General

7. $f'(x) = 5\left(\dfrac{1}{x}\right) = \dfrac{5}{x}$ **9.** $f'(x) = 6\left(\dfrac{1}{x}\right) = \dfrac{6}{x}$

11. $f'(x) = 12x^2 \cdot \ln(x) + 4x^3\left(\dfrac{1}{x}\right) = 4x^2(3\ln(x) + 1)$

13. $f'(x) = \dfrac{15x^4 \cdot (\ln(x)) - 3x^5 \cdot \dfrac{1}{x}}{(\ln(x))^2} = \dfrac{3x^4(5\ln(x) - 1)}{(\ln(x))^2}$

15. $f'(x) = -12\left(\dfrac{1}{x}\right) = \dfrac{-12}{x}$

17. $f'(x) = \dfrac{1}{x + 7}(1) = \dfrac{1}{x + 7}$

19. $f'(x) = \dfrac{1}{2x - 5}(2) = \dfrac{2}{2x - 5}$

21. $f'(x) = \dfrac{1}{x^2 + 3}(2x) = \dfrac{2x}{x^2 + 3}$

23. $f'(x) = \dfrac{1}{2} \cdot \dfrac{1}{2x + 5} \dfrac{d}{dx}(2x + 5) = \dfrac{1}{2x + 5}$

25. $f'(x) = 6(\ln(x))^5\left(\dfrac{1}{x}\right) = \dfrac{6[\ln(x)]^5}{x}$

27. $f'(x) = \dfrac{1}{2} \cdot \dfrac{1}{2}x^{-1/2} \cdot \ln(x) + \dfrac{1}{2}x^{1/2} \cdot \dfrac{1}{x} = \dfrac{\ln(x)}{4\sqrt{x}} + \dfrac{1}{2\sqrt{x}}$

29. $f'(x) = \dfrac{[2x + 2](\ln(x + 5)) - (x^2 + 2x + 3)\dfrac{1}{x + 5}}{(\ln(x + 5))^2}$

$= \dfrac{2x + 2}{\ln(x + 5)} - \dfrac{x^2 + 2x + 3}{(x + 5)(\ln(x + 5))^2}$

31. $f'(x) = \dfrac{1}{\ln(10)}\left(\dfrac{1}{x}\right) = \dfrac{1}{x\ln(10)}$

33. $f'(x) = \dfrac{6}{\ln(3)}\left(\dfrac{1}{x}\right) = \dfrac{6}{x\ln(3)}$

35. $f'(x) = 2x\log 9(x) + x^2 \cdot \dfrac{1}{\ln(9)}\left(\dfrac{1}{x}\right) = \dfrac{x(2\ln(x) + 1)}{\ln(9)}$

37. $f'(x) = \dfrac{1}{\ln(2)} \cdot \dfrac{1}{5x + 3} \cdot (5) = \dfrac{5}{(5x + 3)\ln(2)}$

39. $f'(x) = \dfrac{1}{\ln(10)} \cdot \dfrac{1}{x + 3} \cdot (1) - \dfrac{1}{\ln(10)} \cdot \dfrac{1}{x^2 + 1} \cdot (2x)$

$= \dfrac{x^2 + 1 - 2x^2 - 6x}{(x + 3)(x^2 + 1)\ln(10)} = \dfrac{-x^2 - 6x + 1}{(x + 3)(x^2 + 1)\ln(10)}$

41. $y = \dfrac{1}{2}x - 1 + \ln(2)$ **43.** $y = x - 1$

45. $y = 4x - 4$ **47.** $y = \dfrac{6}{e}x - 5$

49. (a) $f'(t) = \dfrac{12}{t}$

(b) $f(12) \approx 780$; $f'(12) = \dfrac{12}{12} = 1$

51. (a) $f'(x) = \dfrac{5}{x\ln(2)}$

(b) $f'(2) \approx 3.61$; When Vectrum has been on the market for two years, the number of people using the drug was growing at a rate of about 3,610 people per year.

$f'(2) \approx 0.721$; When Vectrum has been on the market for ten years, the number of people using the drug was growing at a rate of about 721 prescriptions per year.

53. (a) $f'(x) = \dfrac{12.22}{x}$

(b) $f'(9) = 1.35\overline{7}$; $f'(19) \approx 0.643$; The USPS revenue was growing approximately 0.714 billion dollars per year slower in 2009 as compared to 1999.

55. (a) $f'(x) = \dfrac{1}{\ln 10} \cdot \dfrac{3986}{x}$

(b) $f'(2) \approx 865.59$; $f'(9) \approx 192.34$; It appears that the Corrections Act was effective in reducing the prison population.

57. $f'(x) = \dfrac{d}{dx}[\ln(\ln(x))] = \dfrac{1}{\ln(x)} \cdot \dfrac{d}{dx}[\ln(x)] = \dfrac{1}{\ln(x)} \cdot \dfrac{1}{x}$

59. $x = 0$

63. $\lim\limits_{x \to \infty} h(x) = \infty$

Chapter 4 Review Exercises

Section 4.1 Review Exercises

1. (a) All values are rounded to the nearest hundred or two significant digits.

x	-5	-4	-3	-2	-1	0	1	2	3	4	5
$f(x)$	97.66	39.06	15.63	6.25	2.5	1	0.4	0.16	0.256	0.010	0.0041

(b)

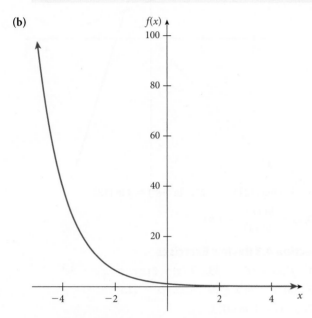

(c) Exponential decay function

3. (a) All values are rounded to the nearest hundred or two significant digits.

x	-5	-4	-3	-2	-1	0	1	2	3	4	5
$f(x)$	0.018	0.041	0.091	0.20	0.45	1	2.23	4.95	11.02	24.53	54.60

(b)

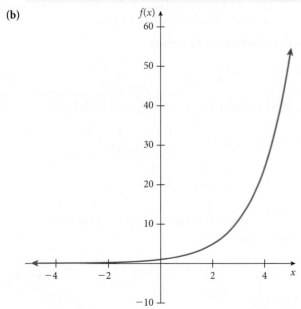

(c) Exponential growth function

5. (a) $f(1) = 8.2$; On day 1 of the sale, 8.2% of the customers responded to the sale.

    (b) $m_{\text{sec}} = 4.35$; Between the fourth and seventh day of the sale, the percentage of the customers responding to the sale was increasing by an average rate of about 4.35% per day.

    (c) $x = \dfrac{\ln{(0.1875)}}{\ln{(0.71)}} \approx 5$ days into the sale

7. (a) $47,635.59

    (b) $49,054.48

    (c) $49,160.18

9. (a) $A(t) = 5000\left(1 + \dfrac{0.062}{12}\right)^{12t}$

    (b)

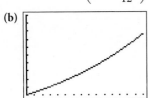

    (c)

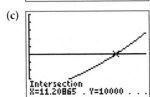

Intersection
X=11.20865 . Y=10000 . . .

It will take about 11.21 years for the money to double.

11. (a) $A(t) = 2500e^{0.057t}$

    (b) $A(4) \approx 3140.21$; After four years, about $3140.21 will be in the account.

    (c)

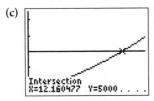

Intersection
X=12.160477 Y=5000 . . . .

It will take about 12.16 years for the money to double.

13. (a) $f(35) \approx 70$; In 1975, there were about 70 leopards.

    (b) $m_{\text{sec}} \approx 2.99$; Between 1987 and 1990, the leopard population was growing at an average rate of about 2.99 leopards per year.

Section 4.2 Review Exercises

15. $g(f(-2)) = 15$

17. (a) Domain of f is $[\frac{2}{3}, \infty)$; Domain of g $(-\infty, \infty)$

    (b) $f(g(x)) = \sqrt{9x^2 + 16}$

    (c) $g(f(x)) = 9x$

19. $f(x) = x + 5$ and $g(x) = x^2$

21. $f(x) = \dfrac{x - 4}{x + 10}$ and $g(x) = 3x$

23. (b)

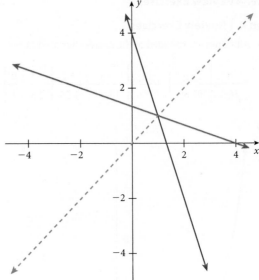

25. $5 = \log_3{(243)}$ **27.** $\ln{(18^3)} = 3\ln{(18)}$

29. $x = \dfrac{\ln{(13)}}{\ln{(4)}} \approx 1.85$

Section 4.3 Review Exercises

31. $f'(x) = 3e^x$ **33.** $f'(x) = 21x^2e^x + 7x^3e^x - 3e^x$

35. (a) $y = 3e^2x - 3e^2 \approx 22.17x - 22.17$

37. $f'(x) = 4^x \ln{(4)}$

39. $f'(x) = 4x^3 \cdot 0.7^x + x^4 \cdot 0.7^x \cdot \ln{(0.7)}$
$= x^3 \cdot 0.7^x[4 + x\ln{(0.7)}]$

41. $f'(x) = 5^{x^2-1}\ln{(5)} \cdot 2x = 2x \cdot 5^{x^2-1} \cdot \ln{(5)}$

43. (a) $f'(x) = 56.53e^{0.14x} \cdot 0.14 = 7.9142e^{0.14x}$

    (b) $f'(21) \approx 149.70$; This means that in 2011, the annual amount of agricultural exports from Peru was increasing by about 149.70 million dollars per year.

Section 4.4 Review Exercises

45. $f'(x) = -6 \cdot \dfrac{1}{x} = \dfrac{-6}{x}$

47. $f'(x) = 20x^4 \cdot \ln{(x)} + 4x^5 \cdot \dfrac{1}{x} = 4x^4[5\ln{(x)} + 1]$

49. $f'(x) = -2(\ln{(x)})^{-3} \cdot \dfrac{1}{x} = \dfrac{-2}{x[\ln{(x)}]^3}$

51. $f'(x) = 3 \cdot \dfrac{1}{x} \cdot \dfrac{1}{\ln{(5)}} = \dfrac{3}{x\ln{(5)}}$

53. $f'(x) = \dfrac{1}{(6x - 5)\ln{(10)}} \cdot 6 = \dfrac{6}{(6x - 5)\ln{(10)}}$

55. (a) $y = 4e^{-4}x + 12$

57. $f'(x) = 3e^{5x} + 3xe^{5x}(5) - \dfrac{1}{x + 4} = 3e^{5x} + 15xe^{5x} - \dfrac{1}{x + 4}$

59. (a) $f'(x) = \dfrac{10.54}{x}$

    (b) $f'(20) = 0.527$; This means that in 2010, the amount of donations to colleges and universities was increasing by 0.527 billion dollars per year.

Chapter 5

Section 5.1

1. Decreasing   **3.** Constant   **5.** Minimum

7. Increasing on $(-1, 2)$: Decreasing on $(-\infty, -1) \cup (2, \infty)$; Constant at $x = -1$ and at $x = 2$

9. Increasing on $(-\infty, -1)$; Decreasing on $(1, \infty)$; Constant on $[-1, 1]$

11. Increasing on $(-2, 0) \cup (1, \infty)$; Decreasing on $(-\infty, -2) \cup (0, 1)$; Constant at $x = -2$, at $x = 0$, and at $x = 1$

13. (a) $(-\infty, -2) \cup (0, 2)$

(b) $(-2, 0) \cup (2, \infty)$

(c) Derivative is zero at $x = -2$, $x = 0$, and $x = 2$; Undefined nowhere

15. (a) $(-\infty, -2) \cup (0, 2) \cup (2, \infty)$ (b) $(-2, 0)$

(c) Undefined at $x = -2$ and $x = 2$; derivative is zero at $x = 0$

17. (a) $(-\infty, 2) \cup (2, \infty)$ (b) None

(c) Undefined at $x = 2$; Zero nowhere

19. f has no critical values.   **21.** $x = 2.5$

23. $x = -5$ and $x = 3$ **25.** $x = -0.5$

27. f has no critical values.   **29.** f has no critical values.

31. $x = -1$ **33.** f has no critical values.

35. (a) Decreasing on $(-\infty, 2.5)$; Increasing on $(2.5, \infty)$

(b) Relative minimum at $x = 2.5$

37. (a) Increasing on $(-\infty, -5) \cup (3, \infty)$; Decreasing on $(-5, 3)$

(b) Relative minimum at $x = 3$; Relative maximum at $x = -5$

39. (a) Increasing on $(-0.5, \infty)$; Decreasing nowhere

(b) No relative extrema

41. (a) Increasing on $(0, \infty)$ (b) No relative extrema

43. (a) $(-\infty, -1) \cup (-1, \infty)$ (b) No relative extrema

45. Relative maximum at $x = 2.5$

47. Relative maximum at $x = -1$; Relative minimum at $x = 0$

49. Relative maximum at $x = -3$; Relative minimum at $x = 1$

51. Various answers. One such graph is below:

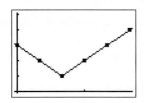

53. Various answers. One such graph is below:

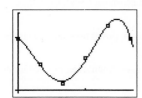

55. (a) Increasing on $[0, 150)$; Decreasing on $(150, 300]$

(b) Relative maximum at $(150, 22{,}500)$; There is a maximum revenue of \$22,500 when 150 modems are produced and sold.

57. (a) \$6.67

(b) $R(x) = 15.22xe^{-0.015x}$

(c) Increasing on $[0, 66.67)$; Decreasing on $(66.67, \infty)$

(d) Relative maximum at $(67, 373.27)$; When the demand is 67 pizzas, the maximum revenue of \$373.27 is obtained.

59. (a) $p'(t) = \dfrac{-400t^2 + 1000}{(2t^2 + 5)^2}; t = 1.58$

(b) Increasing on $[0, 1.58)$; Decreasing on $(1.58, 20]$

(c) Relative maximum at $(1.58, 27.62)$; Thus, 1.58 hours after the drug has been given, the concentration achieves a maximum value of about 27.62%.

61. (a) $AC(x) = \dfrac{150}{x} + 3 + \dfrac{2x}{30}; x \approx 47.43$

(b) Decreasing on $(0, 47.43)$; Increasing on $(47.43, 200]$

(c) Relative minimum at $(47, 9.32)$; The minimum average cost of \$9.32 per digital pet is achieved when 47 pets are produced per day.

63. (a) Increasing on $[0, 7.36)$; Decreasing on $(7.36, 10]$

(b) Relative maximum at $(7, 1804)$; Minnesotan hog production reached a maximum in 2007 with 1804 million pounds produced.

65. (a) $x \approx 0.423$

(b) Year 2000; 4,309 thousand people

67. (a) $f'(x) = 2.72x - 8.21$ (b) $x \approx 3.02$ (c) 1983

71. $a = \dfrac{2}{9}; b = \dfrac{1}{3}; c = -\dfrac{4}{3}; d = \dfrac{7}{9}$ **75.** $x = -\dfrac{b}{2a}$

Section 5.2

1. Maximum   **3.** Closed   **5.** Extreme

7. Absolute maximum of 1 at $x = -2$; Absolute minimum of -8 at $x = 1$

9. Absolute maximum of 8.21 at $x = -0.79$; Absolute minimum of -4 at $x = 2$

11. Absolute maximum of 0 at $x = 0$ and at $x = 3$; Absolute minimum of -4 at $x = -1$ and at $x = 2$

13. Absolute maximum of 29 at $x = -2$; Absolute minimum of 4.89 at $x = 0.75$

15. Absolute maximum of 2401 at $x = 2$; Absolute minimum of 0 at $x = 0.71$

17. Absolute maximum of 1 at $x = 1$; Absolute minimum of -1 at $x = -1$

19. Absolute maximum of -0.5 at $x = 0$; Absolute minimum of -1 at $x = 1$

21. Absolute maximum of $\frac{1}{2}$ at $x = 1$; Absolute minimum of $\frac{1}{17}$ at $x = 4$

23. Absolute maximum of 25.68 at $x = -0.34$; Absolute minimum of -60.42 at $x = 2.89$

25. Absolute maximum at $\left(-2, \frac{71}{3}\right)$; Absolute minimum at $\left(3, -\frac{11}{2}\right)$

27. (a) \$46   (b) \$1,936

29. (a) 7 days   (b) 15 bacteria per cubic centimeter

31. (a) $A(x) = x(200 - 2x) = -2x^2 + 200x$ (b) $x = 50$

(c) 5,000 square feet

33. (a) $A(x) = (x + 10)\left(\dfrac{1440}{x} + 16\right) = 16x + 1600 + \dfrac{14400}{x}$

(b) $x = 30; y = 48$ **(c)** 2,560 square feet

35. (a) After about 1.41 hours

(b) 0.198 milligrams per cubic centimeter

37. (a) $AC(x) = 0.03x^2 - 2.4x + 73.16 + \dfrac{102.27}{x}$

(b) $x \approx 41.01$

(c) $AC(41) \approx \$27.68$ per unit; $C(41) \approx \$1135.06$

39. (a) $P(x) = 0.11x^3 - 13.12x^2 + 488.05x - 102.27$

(b) \$5,701.58 **(c)** 29.67 **(d)** \$223.97 per unit

41. (a) $C(x) = 12x + 100$ **(b)** $R(x) = -\dfrac{1}{3}x^2 + \dfrac{80}{3}x$

(c) $x = 22$

45. From 2000 to 2010, the revenue from alarm system sales reached its lowest point of about \$2,766.28 million in sales around 2001 and its highest point of \$5,697 million in 2010.

47. (a) $f'(x) = -2.94x^2 - 37.46x + 1071; x \approx 13.75.$

(b) From 1990 to 2010, the amount spent on local radio advertising was at its lowest in 1990 with \$6,621 million dollars and at its highest around 2004 with nearly \$15,258.49 million dollars spent.

49. Two or three depending on the location of the vertex in relation to the closed interval.

51. It appears the absolute maximum occurs at the left endpoint of the interval $[0, 5]$:

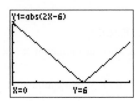

It appears that the minimum occurs between $[0, 5]$, namely at the point $(3, 0)$:

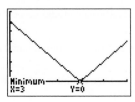

53. $[-3, 3]$

Section 5.3

1. Second **3.** Concave down **5.** Inflection

7. $f'(x) = \dfrac{d}{dx}[-4x^5 - 6x^3 + 7x] = -20x^4 - 18x^2 + 7$

$f''(x) = \dfrac{d}{dx}[-20x^4 - 18x^2 + 7] = -80x^3 - 36x$

$f'''(x) = \dfrac{d}{dx}[-80x^3 - 36x] = -240x^2 - 36$

9. $f''(x) = \dfrac{d}{dx}[7x^3 - 3x^2 + 4x + 5] = 21x^2 - 6x + 4$

$f''(x) = \dfrac{d}{dx}[21x^2 - 6x + 4] = 42x - 6$

$f'''(x) = \dfrac{d}{dx}[42x - 6] = 42$

11. $f'(x) = \dfrac{d}{dx}[e^x] = e^x$

$f''(x) = \dfrac{d}{dx}[e^x] = e^x$

$f'''(x) = \dfrac{d}{dx}[e^x] = e^x$

13. $f'(x) = \dfrac{d}{dx}[x^{1/2}] = \dfrac{1}{2}x^{-12} = \dfrac{1}{2\sqrt{x}}$

$f''(x) = \dfrac{d}{dx}\left[\dfrac{1}{2}x^{-1/2}\right] = \dfrac{1}{2} \cdot \dfrac{-1}{2}x^{-3/2} = \dfrac{-1}{4}x^{-3/2} = -\dfrac{1}{4\sqrt{x^3}}$

$f'''(x) = \dfrac{d}{dx}\left[\dfrac{-1}{4}x^{-3/2}\right] = \dfrac{-1}{4} \cdot \dfrac{-3}{2}x^{-5/2} = \dfrac{3}{8}x^{-5/2} = \dfrac{3}{8\sqrt{x^5}}$

15. $f'(x) = \dfrac{d}{dx}[\ln(x)] = \dfrac{1}{x}$

$f''(x) = \dfrac{d}{dx}\left[\dfrac{1}{x}\right] = \dfrac{d}{dx}[x^{-1}] = -x^{-2} = -\dfrac{1}{x^2}$

$f'''(x) = \dfrac{d}{dx}\left[-\dfrac{1}{x^2}\right] = \dfrac{d}{dx}[-x^{-2}] = 2x^{-3} = \dfrac{2}{x^3}$

17. (a) Concave up on $(1, \infty)$; Concave down on $(-\infty, 1)$

(b) Increasing on $(1, \infty)$; Decreasing on $(-\infty, 1)$

19. (a) Concave up on $(-\infty, \infty)$; Concave down nowhere

(b) Increasing on $(-\infty, \infty)$; Decreasing nowhere

21. (a) Concave up on $(-\infty, -1) \cup (1, \infty)$; Concave down on $(-1, 1)$

(b) Increasing on $(-\infty, -1) \cup (1, \infty)$; Decreasing on $(-1, 1)$

23. Concave up on $(-2, \infty)$; Concave down on $(-\infty, -2)$; $(-2, -25)$ is an inflection point.

25. f is concave up on $(-\infty, \frac{1}{6})$; Concave down on $(\frac{1}{6}, \infty)$; $(\frac{1}{6}, -\frac{134}{9})$ is an inflection point.

27. Concave up on $(-\infty, \infty)$; Concave down nowhere; No inflection points

29. Concave up nowhere; Concave down on $(-\infty, \infty)$; No inflection points

31. f is concave up on $(0, \infty)$; Concave down on $(-\infty, 0)$; No inflection points

33. f is concave up nowhere; Concave down on $(-\infty, 0) \cup (0, \infty)$; No inflection points

35. f is concave up on $(-\infty, \infty)$; Concave down nowhere; No inflection points

37. f is concave up on $(-\infty, \infty)$; Concave down nowhere; No inflection points

39. Concave up nowhere; Concave down on $(-1, \infty)$; No inflection points

41. (a) f is increasing on $(-0.75, \infty)$; Decreasing on $(-\infty, -0.75)$

(b) Relative minimum at $(-0.75, -0.125)$

(c) Concave up on $(-\infty, \infty)$; Concave down nowhere

(d) No inflection points

43. (a) f is increasing on $(-\infty, -5) \cup (1, \infty)$; Decreasing on $(-5, 1)$

(b) Relative minimum at $(1, -13)$

(c) Concave up on $(-2, \infty)$; Concave down on $(-\infty, -2)$

(d) $(-2, 41)$ is an inflection point.

45. (a) f is increasing on $(-0.17, 1.28)$; Decreasing on $(-\infty, -0.17) \cup (1.28, \infty)$

(b) Relative maximum at $(1.28, -4.54)$

(c) Concave up on $(-\infty, \frac{5}{9})$; Concave down on $(\frac{5}{9}, \infty)$

(d) $(\frac{5}{9}, -\frac{1667}{243})$ is an inflection point.

47. (a) Increasing on $(-\infty, -2.24) \cup (2.24, \infty)$; Decreasing on $(-2.24, 0) \cup (0, 2.24)$

(b) Relative minimum at $(2.24, 4.47)$.

(c) Concave up on $(0, \infty)$; Concave down on $(-\infty, 0)$

(d) No inflection points

49. (a) Increasing on $(1, \infty)$; Decreasing on $(-\infty, 1)$

(b) Relative minimum at $(1, 0)$

(c) Concave up nowhere; Concave down on $(-\infty, 1) \cup (1, \infty)$

(d) No inflection points

51. (a) Increasing on $(-\infty, \infty)$; Decreasing nowhere

(b) No relative extrema

(c) Concave up on $(-\infty, \infty)$; Concave down nowhere

(d) No inflection points

53. (a) Increasing on $(-3, \infty)$; Decreasing nowhere

(b) No relative extrema

(c) Concave up nowhere; Concave down on $(-3, \infty)$

(d) No inflection points

55. Various answers. One such graph is below:

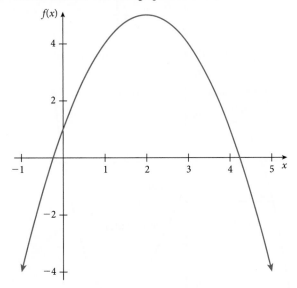

57. Various answers. One such graph is below:

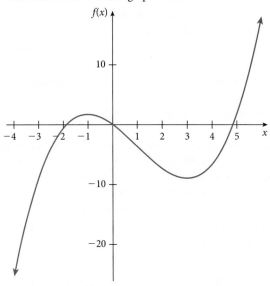

59. (a) $C(5) = 50$; $C'(5) = 28$; When 5 items are produced, the cost to produce the sixth item is about $2,800.

(b) Increasing on $(2, \infty)$; Decreasing on $[0, 2)$; Relative minimum of 1 at $x = 2$

(c) $(2, 20)$

61. (a) $P(35) = 393,812.5$; $P'(35) = 21,197.50$; When 3500 refrigerators are produced and sold, the profit gained from producing and selling the next 100 refrigerators is about $21,197.50.

(b) Increasing on $(-\infty, 64.49)$; Decreasing on $(64.49, \infty)$; Relative maximum at $(64.49, 27,199.28)$.

63. Since $P''(x) = MP'(x)$, when $MP'(x) > 0$, MP is increasing, then $P''(x) > 0$, P is concave up, and when $MP'(x) < 0$, MP is decreasing, then $P''(x) < 0$, P is concave down.

65. $(15, 5500)$; This is the point that the greatest growth in sales is occurring due to advertising.

67. (a) $f(50) \approx 1.00$; When the arterial pressure is 50 mm Hg, the blood flow is about 1 mL/min. $f'(50) \approx 0.026$; When the arterial pressure is 50 mm Hg, the blood flow is increasing at a rate of about 0.026 mL/min per mm Hg.

69. (a) $f(20) \approx 3.61$; When a person is 20 years old, the cardiac output is about $3.61 \, \frac{\text{liters per minute}}{\text{square meters}}$. $f(20) \approx -0.045$; When a person is 20 years old, the cardiac output is decreasing by about $0.045 \, \frac{\text{liters per minute}}{\text{square meters}}$ per year.

(b) Increasing nowhere; Decreasing on $[10, 80]$

71. (a) In 1980, CO emissions were decreasing at a rate of 2.82 million tons per year while in 2005, the emissions were decreasing at a rate of 4.32 million tons per year.

(b) The CO emissions are decreasing at an increasing rate since, as time goes on, the rate of emissions becomes more and more negative.

73. (a) During 2009 and 2001   (b) $x = 10.25$

(b) During 2004

83. $f'(x) \cdot g(x) + f(x) \cdot g'(x) > 0$

85. $f''(x) \cdot g(x) + 2f'(x) \cdot g'(x) + f(x) \cdot g''(x) > 0$

Section 5.4

1. Open   3. Maximum   5. Extrema

7. Relative minimum of $-\frac{10}{3}$ at $x = \frac{1}{3}$

9. Relative maximum of 1.63 at $x \approx -1.52$; Relative minimum of -40.15 at $x \approx 2.85$

11. Relative maximum of -6.5 at $x = -3$; Relative minimum of $\frac{-20}{3}$ at $x = -2$

13. Relative maximum of 7 at $x = -2$; Relative minimum of -6.5 at $x = 1$

15. Relative minimum of -12.95 at $x = -2.2$; Relative maximum of 6.04 at $x = -0.07$; Relative minimum of -2.88 at $x = 1.57$

17. Relative minimum of -32 at $x = -2$; Relative maximum of 16 at $x = 0$; Relative minimum of -32 at $x = 2$

19. Relative maximum of 1 at $x = 0$; No relative minimum

21. Relative minimum of -5 at $x = 0$; No relative maximum

23. $(\sqrt{3}, 4\sqrt{3}) \approx (1.73, 6.93)$

25. $\left(\frac{\sqrt{2}}{2}, 4\sqrt{2} - 3\right) \approx (0.71, 2.66)$

27. $(\sqrt[4]{2}, 6\sqrt{2} - 2) \approx (1.19, 6.49)$

29. $\left(\sqrt[4]{\frac{2}{3}}, 2\sqrt{6} - 1\right) \approx (0.90, 3.90)$

31. $(1.5\sqrt{2}, -6\sqrt{2}) \approx (2.12, -8.49)$

33. $\left(\frac{\sqrt{15}}{3}, -6\sqrt{10}\right) \approx (1.29, -7.75)$

35. $(\sqrt[4]{1.5}, -2\sqrt{6} + 3) \approx (1.11, -1.90)$

37. $\left(\sqrt[3]{\frac{4}{3}}, -\sqrt[3]{121.5} + 2\right) \approx (1.10, -2.95)$

39. (a) $A(x) = x\left[\frac{1}{3}(400 - 2x)\right] = -\frac{2}{3}x^2 + \frac{400}{3}x$

    (b) $x = 100$ feet; $y = \frac{200}{3}$ feet

    (c) 6,666.67 square feet

41. (a) $x = \frac{1320}{\pi} \approx 420.17$ feet; $y = 0$ feet

    (b) 138,655.79 square feet

43. (a) $x = 28.28$ feet; $y = 42.43$ feet   (b) \$1357.65

45. (a) $V(x) = 4x^3 - 44x^2 + 120$

    (b) $l = 8.38$ in; $w = 6.38$ in; $h = 1.81$ in

    (c) 96.77 cubic inches

47. (a) $x = 6.12$ in and $y = 9.16$ in   (b) \$6.73

49. (a) $AC(x) = 100x + 400 + \frac{11000}{x}$ (b) $x \approx 10.49$

    (c) Around 10.49 years

51. (a) $R(x) = -0.75x^2 + 200x$; $(0, 200]$.

    (b) \$100.25 per room; \$13,333.25

53. 63 trees   55. 60 trees

57. The population should grow to 125,000 deer and it will sustain a maximal yearly harvest of 31,250 deer.

59. The population should grow to 50,000 rabbits and it will sustain a maximal yearly harvest of 7,500 rabbits.

61. Each lot size has 50 desks and 8 orders are placed per year.

63. Each lot size has 160 cases and 3 orders are placed per year.

65. Each lot size has 100 stoves and 25 orders are placed per year.

67. Each lot size has 240 jeans and 5 orders are placed per year.

69. Each lot size has 200 shoes and 5 orders are placed per year.

71. Each lot size has 2000 games and 4 production runs are done per year.

73. If $f'(c) = 0$ implies there is a horizontal tangent line at $x = c$ to the graph of f and $f''(c) < 0$ implies f is concave down around $x = c$ and thus $(c, f(c))$ is an relative maxima for f.

75. $\left(\frac{\sqrt{ab}}{a}, 2\sqrt{ab}\right)$ 77. $\left(\frac{\sqrt{ac}}{a}, 2\sqrt{ac} - b\right)$

Section 5.5

1. Zeros (or roots)   3. Increasing; decreasing

5. Concave up; concave down

7. Positive on $(-5, -1) \cup (\frac{13}{4}, 7)$; Negative on $(-\infty, -5) \cup (-1, \frac{13}{4}) \cup (7, \infty)$

9. Positive on $(-\infty, -3) \cup (1, 5)$; Negative on $(-3, 1) \cup (5, \infty)$

11. Concave up on $(-1, 4)$; Concave down on $(-\infty, -1) \cup (4, \infty)$

13. Positive on $(-\infty, -5) \cup (-\frac{5}{3}, \infty)$; Negative on $(-5, -\frac{5}{3})$

15. Positive on $(-3, 1) \cup (5, \infty)$; Negative on $(-\infty, -3) \cup (1, 5)$

17. Concave up on $(-\infty, -1) \cup (3, \infty)$; Concave down on $(-1, 3)$

19.

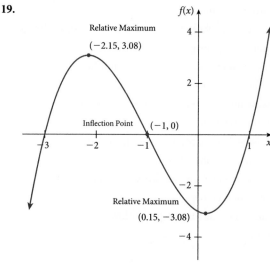

21.

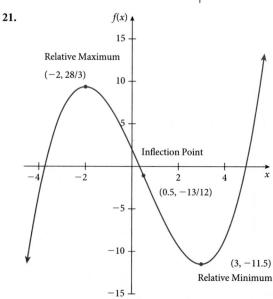

23.

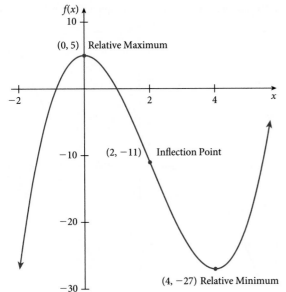

25.

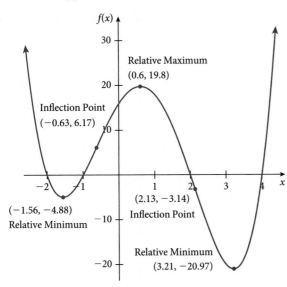

27.

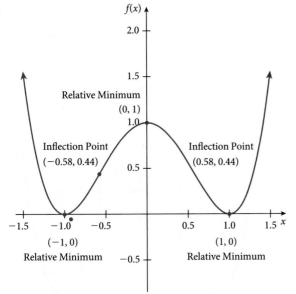

29.

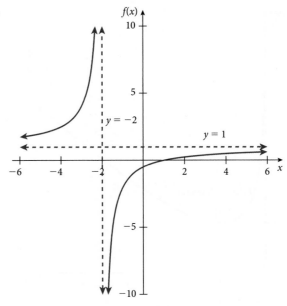

31.

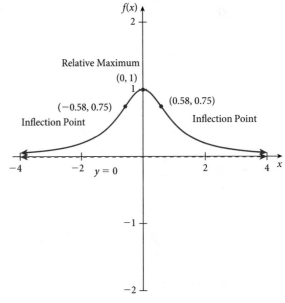

33.

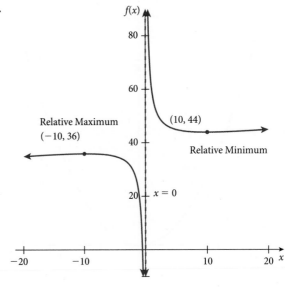

35.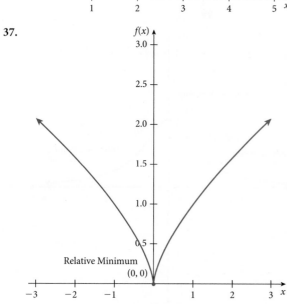

37.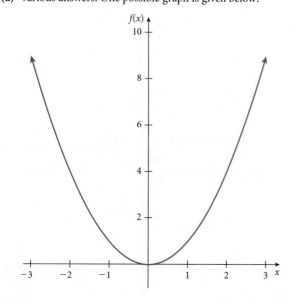

39. (a) Various answers. One possible graph is given below:

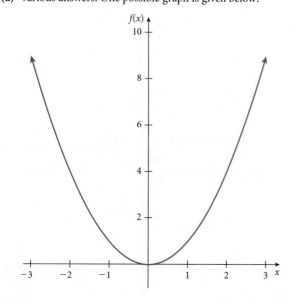

(b) Various answers. One possible graph is given below:

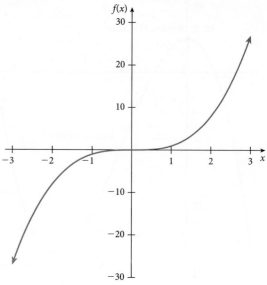

41. (a) Various answers. One possible graph is given below:

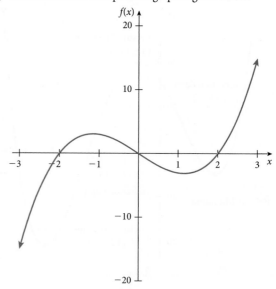

(b) Various answers. One possible graph is given below:

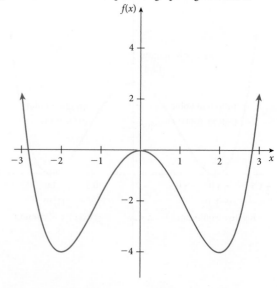

43. (a) $R'(x) = \dfrac{d}{dx}[216x - 0.08x^3] = 216 - 0.24x^2.$

Increasing on $[0, 30]$; Decreasing on $(30, 50]$

(b) Concave down on $(0, 50]$

45. (a) $AP(x) = -0.01x + 12 - \dfrac{2000}{x}$

(b)
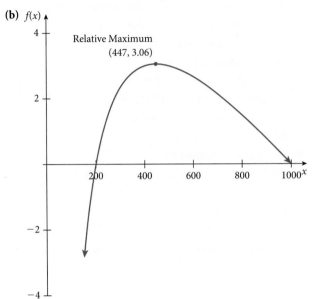

Relative Maximum
(447, 3.06)

47. The marginal profit function; The average profit function

49. (a) $C'(x) = 3 + \dfrac{2x}{15}$

(b)
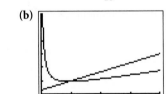

(c) $(47.43, 9.324)$

(d) The x-coordinate of the result from part c is the same as the x-coordinate of the relative minimum for AC in #48, part b.

51. (a) $AC(t) = 100t + 400 + \dfrac{11000}{t}$ **(b)** $t \approx 10.5$ years

(c) $AC(10.5) \approx 2{,}498$

55. (a) $f(20) \approx 62.83$; When the wind velocity is 20 miles per hour, the percent total heat loss by convection is about 62.83%.

$f'(20) \approx 0.79$. When the wind velocity is 20 miles per hour, the percent total heat loss by convection is increasing at a rate of about 0.79% per mile per hour.

57. Up   **59.** Up

61. The second derivative is the rate at which the rate of the function is increasing or decreasing.

Section 5.6

1. Elasticity   **3.** Arc   **5.** Point   **7.** $E_a = 1.1$

9. $E_a \approx 1.07$ **11.** $E_a \approx 0.68$ **13.** $x = 6 - 0.01p$

15. $x = \sqrt{\dfrac{300}{p - 10}}$ **17.** $x = 300 - 25p$

19. $x = \sqrt{300 - p^2}$ **21.** $x = -10\ln\left(\dfrac{P}{100}\right)$

23. $E(10) \approx 0.29$; Inelastic   **25.** $E(8) \approx 0.94$; Inelastic

27. $E(15) = 1$; Unitary   **29.** $E(30) = \dfrac{3}{4}$; Inelastic

31. $E(30) = 2$; Elastic   **33.** $E(200) = 4$; Elastic

35. $E(19) \approx 0.0350$; Inelastic

37. $E(40) = 2$; Elastic

39. (a) $d(5) = 10$; When the price is set at \$5, the number of mouse pads sold is 1000 mouse pads.

$d(6) = 8$; When the price is set at \$6, the number of mouse pads sold is 800 mouse pads.

(b) $E_a = 1$; This means that a change in price will cause a relatively equal change in demand.

41. $E_a = 1.9125$; This means that a change in price will cause a relatively large change in demand.

43. $E_a \approx 2.01$; This means that a change in price will cause a relatively large change in demand.

45. (a) Inelastic   **(b)** Raised

47. (a) Elastic   **(b)** Lowered

49. (a) $E(4) \approx 0.19$; The price should be raised to increase revenue.

(b) \$12.50 per quart

51. (a) $E(1.5) \approx 0.58$; The price should be raised to increase revenue.

(b) \$1.83 per night-light

53. (a) $E(100) \approx 0.51$; The price should be raised to increase revenue.

(b) \$116.71 per T.V.

55. When $E(p) < 1$, revenue is increasing while when $E(p) > 1$, revenue is decreasing. Thus, when $E(p) = 1$, we have $R'(p) = d(p)(1 - E(p)) = d(p)(1 - 1) = 0$, which implies $(p, R(p))$ is a maximum for the function R.

57. $E(p) = \dfrac{2p^2}{a - p^2}$

59. $E(p) = bp$

Section 5.7

1. Vertical   **3.** y **5.** Circle   **7.** $\dfrac{dy}{dx} = -2$

9. $\dfrac{dy}{dx} = -\dfrac{1}{6y}$ **11.** $\dfrac{dy}{dx} = -\dfrac{x}{y}$ **13.** $\dfrac{dy}{dx} = \dfrac{4x^3 - 15x^2}{3y^2}$

15. $\dfrac{dy}{dx} = \left(\dfrac{y}{x}\right)^{3/4}$ **17.** $\dfrac{dy}{dx} = -\dfrac{2x + y}{x}$ **19.** $\dfrac{dy}{dx} = \dfrac{x^2 + 4y}{y^2 - 4x}$

21. $\dfrac{dy}{dx} = \dfrac{6x - y}{x + y}$ **23.** $\dfrac{dy}{dx} = -\dfrac{y}{x(\ln(x) + 1)}$

25. $\dfrac{dy}{dx} = \dfrac{e^y + 2x}{2y - xe^y}$ **27.** $\dfrac{dy}{dx} = \dfrac{3x^2}{5^y \ln(5)}$

29. $\dfrac{dy}{dx} = \dfrac{1}{10^{y-2} \ln(10)}$ **31.** $\dfrac{dy}{dx} = 6x\sqrt[3]{y^2}$

33. (a) $\dfrac{dy}{dx} = -\dfrac{1}{2}$ (b) $\dfrac{dy}{dx} = -\dfrac{1}{2}$

35. (a) $\dfrac{dy}{dx} = \dfrac{4x^3 - y}{x}$ (b) $\dfrac{dy}{dx} = 3x^2 + \dfrac{4}{x^2}$

37. (a) $\dfrac{dy}{dx} = \dfrac{-2xy^2 + y}{x}$ (b) $\dfrac{dy}{dx} = \dfrac{-x^2 + 1}{(x^2 + 1)^2}$

39. $y = -1.5x + 6.5$ 41. $y = 2$ 43. $y = 2x - 1$

45. $y = 2x - 2$ 47. $2\dfrac{dx}{dt} + 3\dfrac{dy}{dt} = 0$

49. $2x\dfrac{dx}{dt} - 3\dfrac{dy}{dt} = 0$ 51. $2x\dfrac{dx}{dt} + 2y\dfrac{dy}{dt} = 5\dfrac{dx}{dt}$

53. $5y\dfrac{dx}{dt} + 5x\dfrac{dy}{dt} + 4y^3\dfrac{dy}{dt} = \dfrac{dx}{dt}$ 55. $\dfrac{dy}{dt} = -\dfrac{2}{7}$

57. $\dfrac{dx}{dt} = -4$ 59. $\dfrac{dy}{dt} = 1.5$

61. (a) $\dfrac{dp}{dx} = -2x$

   (b) $\dfrac{dp}{dx}\bigg|_{(11,\,29)} = -22$; At a demand level of 1100 mini picture frames and a price of $29, the price is decreasing at a rate of $22 per hundred picture frames.

63. (a) $\dfrac{dp}{dx} = -\dfrac{p + 2}{x}$

   (b) $\dfrac{dp}{dx}\bigg|_{(20,\,48)} = -2.5$; At a demand level of 2000 tents and a price of $48, the price is decreasing at a rate of $2.50 per hundred tents.

65. 0.955 inches per minute

67. 1.67 inches per minute

69. 1.01 square inches per minute

71. $\dfrac{32}{3}$ feet per second

73. (a) $\dfrac{dR}{dt} = 250\dfrac{dx}{dt} - \dfrac{4}{5}x\dfrac{dx}{dt}$

   (b) Increasing at a rate of $34,000 per month

75. (a) $2x\dfrac{dx}{dt} + 2y\dfrac{dy}{dt} = 0$

   (b) 6.87 feet per second

77. (a) $\dfrac{dy}{dt} = -2500(1 + x)^{-2}\dfrac{dx}{dt} = \dfrac{-2500}{(1 + x)^2}\dfrac{dx}{dt}$.

   (b) Decreasing at a rate of 160 bass per year

79. When an equation is solved for one variable, then explicit differentiation is used. If an equation is not solved for one variable, you will need to use implicit differentiation, which usually requires the Chair Rule.

81. $\dfrac{dy}{dx} = \dfrac{aby - ax^{a-1}}{by^{b-1} - abx}$

83. $a\dfrac{dx}{dt} + b\dfrac{dy}{dt} = 0$

Chapter 5 Review Exercises

Section 5.1 Review Exercises

1. (a) Increasing on $(-\infty, -2) \cup (3, \infty)$
   (b) Decreasing on $(-2, 3)$
   (c) Constant at $x = -2$ and $x = 3$

3. (a) Increasing on $(-1, 1)$ (b) Decreasing on $(1, \infty)$
   (c) Constant on $(-\infty, -1)$ and at $x = 1$

5. (a) Increasing on $(-\infty, -3) \cup (1, \infty)$
   (b) Decreasing on $(-3, 1)$
   (c) Constant at $x = -3$ and $x = 1$

7. (a) Increasing on $(-4, -1) \cup (3, \infty)$
   (b) Decreasing on $(-\infty, -4) \cup (-1, 3)$
   (c) Constant at $x = -4$, $x = 1$, and $x = 3$

9. (a) f has no critical values
   (b) Increasing on $(-\infty, \infty)$; Decreasing nowhere
   (c) No relative extrema

11. (a) $x = -4$ and $x = 2$
   (b) Increasing on $(-\infty, -4) \cup (2, \infty)$; Decreasing on $(-4, 2)$
   (c) Relative maximum at $(-4, 100)$; Relative minimum at $(2, -8)$

13. (a) $x = \dfrac{2}{3}$
   (b) Increasing on $(\frac{2}{3}, \infty)$; Decreasing on $(-\infty, \frac{2}{3})$
   (c) Relative minimum at $(\frac{2}{3}, 0)$

15. Relative maximum at $x = \dfrac{8}{3}$

17. Relative maximum at $x = -3$; Relative minimum at $x = 3$

19.

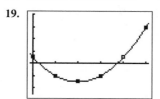

21. (a) $p(30) \approx 1.83$ (b) $R(x) = 4.5xe^{-0.03x}$
   (c) Increasing on $[0, 33.333)$; Decreasing on $(33.333, \infty)$
   (d) Relative maximum of 55.18 at $x = 33.33$; The total revenue peaks at $55,180 when 33,330 containers are produced and sold.

23. (a) $AC(x) = \dfrac{800}{x} + 15 + \dfrac{x}{18}$ (b) $x = 120$
   (c) Decreasing on $(0, 120)$; Increasing on $(120, 500)$
   (d) Relative minimum of 28.33 at $x = 120$; The average cost is minimized at $28.33 per phone when 120 phones are produced per day.

Section 5.2 Review Exercises

25. Absolute maximum of 53 at $x = 10$; Absolute minimum of -3.25 at $x = 2.5$

27. Absolute maximum of 95 at $x = 5$; Absolute minimum of -5.66 at $x = 1.41$

29. Absolute maximum of 11.18 at $x = 5$; Absolute minimum of 0 at $x = 0$

31. $x = 15$; $y = 30$

33. (a) $AC(x) = 0.001x^2 - 0.6x + 217 + \dfrac{7200}{x}$

(b) About 333 units　　(c) \$149.71 per unit

35. (a) $P(x) = -0.003x^3 + 2.1x^2 - 218.7x - 7200$

(b) \$49,394.57　　(c) About 407 units　　(d) \$277.51 per unit

Section 5.3 Review Exercises

37. $f'(x) = \dfrac{d}{dx}[x^3 - 7x^2 + 5] = 3x^2 - 14x.$

$f''(x) = \dfrac{d}{dx}[3x^2 - 14x] = 6x - 14$

$f'''(x) = \dfrac{d}{dx}[6x - 14] = 6.$

39. $f'(x) = \dfrac{d}{dx}[(x - 5)^{1/3}] = \dfrac{1}{3}(x - 5)^{-2/3}(1) = \dfrac{1}{3\sqrt[3]{(x - 5)^2}}.$

$f''(x) = \dfrac{d}{dx}\left[\dfrac{1}{3}(x - 5)^{-2/3}\right] = \dfrac{1}{3} \cdot \dfrac{-2}{3}(x - 5)^{-5/3}$

$= \dfrac{-2}{9\sqrt[3]{(x - 5)^5}}.$

$f'''(x) = \dfrac{d}{dx}\left[\dfrac{-2}{9}(x - 5)^{-5/3}\right] = \dfrac{-2}{9} \cdot \dfrac{-5}{3}(x - 5)^{-8/3}$

$= \dfrac{10}{27\sqrt[3]{(x - 5)^8}}.$

41. (a) Concave up on $(\frac{1}{3}, \infty)$; Concave down on $(-\infty, \frac{1}{3})$

(b) $(\frac{1}{3}, -5.41)$ is an inflection point.

43. (a) Concave up on $[0, \infty)$; Concave down nowhere

(b) No inflection points

45. (a) Concave up on $(-\infty, 0)$; Concave down $(0, \infty)$

(b) No inflection points

47. (a) Concave up on $(-\infty, \infty)$; Concave down nowhere

(b) No inflection points

49. (a) Increasing on $(-\infty, -3) \cup (2, \infty)$; Decreasing on $(-3, 2)$

(b) Relative minimum at $(2, -44)$

(c) Concave up on $(-0.5, \infty)$; Concave down on $(-\infty, -0.5)$

(d) $(-0.5, 18.5)$ is an inflection point.

51. (a) Decreasing on $(-\infty, 2)$; Increasing nowhere

(b) No relative extrema

(c) Concave up nowhere; Concave down on $(-\infty, 2)$

(d) No inflection points

53. (a) $P(4) = 155$; When 400 cookies are baked and sold per day, the daily profit is \$155.

$P'(4) = 68$; When 400 cookies are baked and sold per day, the profit is increasing at a rate of \$68 per 100 cookies.

(b) Increasing on $[0, 9)$; Decreasing on $(9, \infty)$; Relative maximum at $(9, 357.5)$; Profit is maximized at \$357.50 when 900 cookies are baked and sold.

(c) Increasing on $[0, 2)$; Decreasing on $(2, \infty)$; P is concave up on $[0, 2)$ and concave down on $(2, \infty)$.

(d) $(2, 11)$

55. (a)
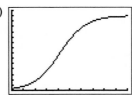

(b) $f(84) \approx 1369$; Eighty-four years after the goats were introduced, there were about 1,369 goats.

$f'(84) \approx 2.70$; Eighty-four years after the goats were introduced, the goat population was increasing at a rate of about 2.7 goats per year.

(c) $f''(t) = \dfrac{-487.62e^{-0.09t}(1 - 43e^{-0.09t})}{(1 + 43e^{-0.09t})^3}$

(d) Concave up on $[0, 41.79)$; Concave down on $(41.79, 100]$

(e) 42 years after it was introduced

Section 5.4 Review Exercises

57. (a) $AC(x) = 0.0015x + 0.6 + \dfrac{4.5}{x}$

(b) 54,770 newspapers

59. Each lot size has 25 fax machines and 32 orders are placed per year.

61. V has an absolute maximum volume of 5,488 cubic inches when $x = 14$ in and $y = 28$ in.

Section 5.5 Review Exercises

63. Increasing on $(-\infty, -3) \cup (-1, 3)$; Decreasing on $(-3, -1) \cup (3, \infty)$

65. Increasing on $(-2, 1)$; Decreasing on $(-\infty, -2) \cup (1, \infty)$

67. Relative maximum at $(-3, -2)$ and at $(3, 4)$; Relative minimum at $(-1, -4)$

69. Relative minimum of -31 at $x = -6$

71. Relative maximum of 0 at $x = 0$; Relative minimum of -2 at $x = 1$

73.
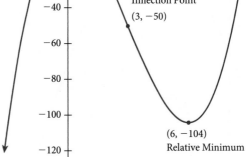

75.

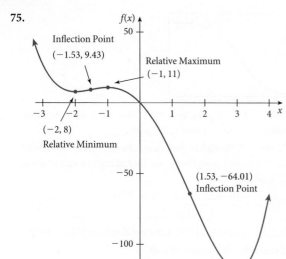

77.

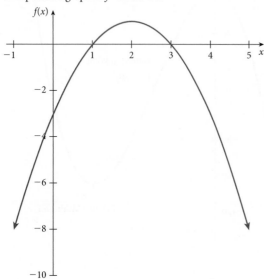

79. One possible graph of f' would be:

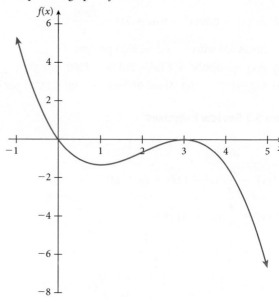

One possible graph of f'' would be:

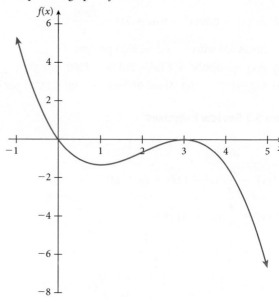

Section 5.6 Review Exercises

81. $E_a = = 0.225$ **83.** $E_a \approx 1.21$

85. **(a)** $d(8) = 334$; When the price is set at \$8, the number of FileFinder programs sold is 334,000 programs.

     $d(9) = 332$; When the price is set at \$9, the number of FileFinder programs sold is 332,000 programs.

   (b) $E_a \approx 0.048$; A change in price will cause a relatively small change in demand.

87. $x = 600 - \dfrac{100}{3}p$ **89.** $x = \sqrt{500 - p^2}$

91. **(a)** $E(p) = \dfrac{5p}{180 - 5p}$ **(b)** $E(20) = \dfrac{5}{4}$; Elastic

93. **(a)** $E(p) = \dfrac{p}{2(200 - p)}$ **(b)** $E(100) = \dfrac{1}{2}$; Inelastic

95. **(a)** $E(p) = \dfrac{1.3p}{35 - p}$ **(b)** $E(15) = 0.975$; Inelastic

   (c) Since the demand is inelastic, the prices should be raised to increase revenue.

Section 5.7 Review Exercises

97. $\dfrac{dy}{dt} = \dfrac{3}{5}$ **99.** $\dfrac{dy}{dx} = \dfrac{2x - 30x^5}{3y^2}$

101. $\dfrac{dy}{dx} = \dfrac{2xy}{3y^2 - x^2}$ **103.** $\dfrac{dy}{dx} = \dfrac{y \ln(y)}{y - x}$

105. **(a)** $\dfrac{dy}{dx} = -\dfrac{2}{3}$ **(b)** $\dfrac{dy}{dx} = -\dfrac{2}{3}$

107. **(a)** $\dfrac{dy}{dx} = \dfrac{2 - 3x^2y}{x^3 - 1}$

   (b) $\dfrac{dy}{dx} = \dfrac{-4x^3 - 6x^2 - 2}{(x^3 - 1)^2}$

109. $y = 1.25x - 2.25$ **111.** $y = \dfrac{4}{3}x - \dfrac{1}{3}$

113. (a) $\dfrac{dp}{dx} = -\dfrac{p + 10}{x}$

(b) $\dfrac{dp}{dx}\bigg|_{(15,\,30)} = -2\dfrac{2}{3}$; When the demand is 15,000 lamps, the price per lamp is decreasing at a rate of about $2.67 per thousand lamps.

115. $3x^2\dfrac{dx}{dt} + 3y^2\dfrac{dy}{dt} = 6\dfrac{dx}{dt}$ **117.** $y\dfrac{dx}{dt} + x\dfrac{dy}{dt} - 3y^2\dfrac{dy}{dt} = 2\dfrac{dx}{dt}$

119. $\dfrac{dx}{dt} = -\dfrac{1 + 4}{2 - 3} \cdot (4) = 5 \cdot 4 = 20$ **121.** $\dfrac{dy}{dt} = -11$

123. (a) $\dfrac{dS}{dt} = 8\pi r\dfrac{dr}{dt}$ (b) 140.74 square centimeters per second

Chapter 6

Section 6.1

1. Antiderivative **3.** General Antiderivative **5.** Rate

7. Yes **9.** No **11.** Yes **13.** No **15.** Yes

17. $\dfrac{x^5}{5} + C$ **19.** $\dfrac{x^{3.31}}{3.31} + C$ **21.** $-\dfrac{1}{2t^2} + C$

23. $\dfrac{4x^{9/4}}{9} + C$ **25.** $\dfrac{3}{2}x^{2/3} + C$ **27.** $\dfrac{2x^7}{35} + C$

29. $x^2 + 3x + C$ **31.** $\dfrac{x^2}{3} + 4x + C$ **33.** $t^3 + t^2 + 10t + C$

35. $x - \dfrac{2x^3}{3} + \dfrac{3x^4}{4} + C$ **37.** $2.07x^3 + 0.015x^2 - 4.01x + C$

39. $-\dfrac{1}{x} + \dfrac{3}{2x^2} + C$ **41.** $3x + \dfrac{4}{3}x^{3/2} + C$

43. $-\dfrac{3}{2x^2} + \dfrac{4x^{5/2}}{5} - 4x + C$ **45.** $\dfrac{t^3 - 2t}{6} + C$

47. $\dfrac{1}{20z^2} - \dfrac{2}{z} + \dfrac{z^4}{4} + C$ **49.** $\dfrac{200}{113}t^{1.13} + 5t + C$

51. $f(x) = -2x + 4$ **53.** $f(x) = \dfrac{5}{2}x^2$

55. $f(x) = x^2 - 3x + 4$ **57.** $f(t) = 500t - 0.025t^2 + 40$

59. $f(x) = -\dfrac{2}{x} - \dfrac{3}{2x^2} - x + \dfrac{13}{2}$ **61.** $f(t) = 3t^{5/3} + 3t^{2/3} + 1$

63. $f(t) = -\dfrac{t^4 + 1}{2t^2} + 5$

65. (a) $P(q) = 40q - 0.025q^2$ (b) 7000

67. (a) $R(x) = 0.0000115x^3 - 0.015x^2 + 3.75x$

(b) $p(x) = 0.0000115x^2 - 0.015x + 3.75$

(c) $2.40

69. (a) $AC(x) = \dfrac{100}{x} + 1.50$ (b) $C(x) = 100 + 1.50x$

(c) $C(100) = 250$; The cost of producing 100 banners is $250.

71. (a) $f(x) = 1.08x + 11.7$

(b) The function f gives the number of Hispanic voters in millions for x years since 1995.

(c) 23.58 million

73. (a) 3.92; In 2000, the number of cases of botulism was increasing by 3.92 cases per year.

(b) $f(x) = 0.11x^2 - 0.48x + 88$ (c) About 122

75. (a) $f(x) = 0.08x^3 - 1.24x^2 + 3.34x + 15.27$

(b) $f(5) = 10.97$

77. $f(x) = kx + b - k \cdot a$

79. $AC(x) = \dfrac{a}{2}x^2 + bx + \dfrac{d}{x} + C$; $C(x) = \dfrac{a}{2}x^3 + bx^2 + k + Cx$

81. $f(x) = -\dfrac{1}{14}x^2 + \dfrac{4}{7}x + \dfrac{36}{7}$

Section 6.2

1. Step size **3.** Limits **5.** Left

7. $\Delta x = \dfrac{1}{2}$ **9.** $\Delta x = \dfrac{1}{2}$ **11.** $\Delta x = \dfrac{1}{2}$

13. (a) $\dfrac{81}{4}$ square units (b) $\dfrac{99}{4}$ square units

15. (a) 28 square units (b) 60 square units

17. (a) $\dfrac{77}{8}$ square units (b) $\dfrac{65}{8}$ square units

19.

	Left	Right
10	9.24	6.84
100	8.1204	7.8804
1000	8.012004	7.988004

21.

	Left	Right
10	30.24	51.04
100	38.9664	41.0464
1000	39.89606	40.10406

23.

	Left	Right
10	33.25	39
100	35.79625	36.37125
1000	36.05459	36.11209

25. (a) $\displaystyle\int_3^9 10\,dx$ (b) 60 square units

27. (a) $\displaystyle\int_0^{20} \dfrac{1}{2}\,dx$ (b) 100 square units

29. (a) $\displaystyle\int_1^2 3\,dx + \int_2^4 (x + 1)\,dx$ (b) 14 square units

31. $\dfrac{45}{2}$ **33.** 10 **35.** 14 **37.** $\dfrac{128}{3}$ **39.** 9 **41.** $\dfrac{112}{3}$

43. $\dfrac{165}{4}$ **45.** $\dfrac{433}{12}$ **47.** $-\dfrac{85}{2}$ **49.** $\dfrac{20}{3}$

51.
```
fnInt(Y₁,X,1,4)
         9.254869524
```

53.
```
fnInt(Y₁,X,3,7)
         1.609437912
```

55.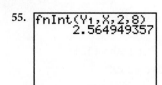

57. (a)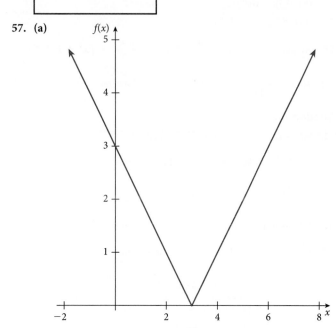

(b) $f(x) = \begin{cases} x - 3, & x \ge 3 \\ 3 - x, & x < 3 \end{cases}$ **(c)** $\dfrac{29}{2}$ square units

59. (a)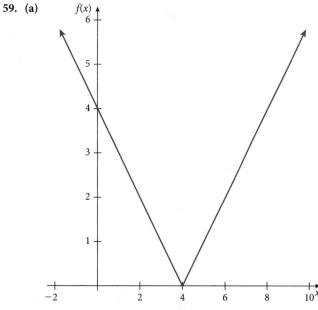

(b) $f(x) = \begin{cases} x - 4, & x \ge 4 \\ 4 - x, & x < 4 \end{cases}$ **(c)** 20 square units

61. $\dfrac{25}{2}$ **63.** $\dfrac{14}{3}$ **65.** $\dfrac{106}{3}$

67. (a) $x = -2$

(b) $-\dfrac{9}{2}; \dfrac{49}{2}$

(c) On the interval $[-5, -2]$, the area is below the x-axis. Thus, we should expect $\int_{-5}^{-2}(x + 2)\,dx$ to be negative. On the interval $[-5, 5]$, the area is above the x-axis. Thus we should expect $\int_{-2}^{5}(x + 2)\,dx$ to be positive.

69. (a) $x = 4$ **(b)** $-\dfrac{16}{3}; \dfrac{16}{3}$

(c) On the interval $[0, 4]$, the area is below the x-axis. Thus, we should expect $\int_{0}^{4}(2\sqrt{x} - 4)\,dx$ to be negative. On the interval $[4, 9]$, the area is above the x-axis. Thus, we should expect $\int_{4}^{9}(2\sqrt{x} - 4)\,dx$ to be positive.

71. (a) $x = 3$ **(b)** $-\dfrac{80}{3}; \dfrac{10}{3}$

(c) On the interval $[-1, 3]$, the area is below the x-axis. Thus, we should expect $\int_{-1}^{3}(x^2 - 9)\,dx$ to be negative. On the interval $[3, 4]$, the area is above the x-axis. Thus, we should expect $\int_{3}^{4}(x^2 - 9)\,dx$ to be positive.

73. Rectangles can be used to fill in the space between the curve and the x-axis where more rectangles mean a better approximation.

75. kb **77.** $\dfrac{k \cdot a^{(1+k)/k}}{1 + k}$

Section 6.3

1. Continuous **3.** Increase **5.** Rate

7. (a) $\displaystyle\int_{50}^{150} 5.5\,dx$

(b) 550; The total increase in cost of producing 50 to 150 bandanas is $550.

9. (a) $\displaystyle\int_{5}^{10} (-1.02x + 2.02)\,dx$

(b) 62.75; The number of weeks of the Double D cola ad campaign increases from 5 to 10 weeks, the total increase in sales is 62.75 thousand dollars.

11. (a)

(b) 1500; The cost of producing the first 50 pairs of walking shoes is $1,500.

13. **(a)** 2850; The total revenue of producing the 10th through 40th water-proof headset radios is $2,850.

(b)

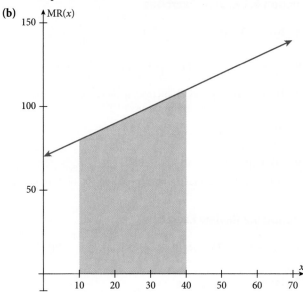

15. 270; The total revenue generated from producing and selling the first 200 tickets is $270.

17. 730; The total increase in profit from producing and selling 10 to 20 machines is $730.

19. -6.67; This means that as the number of banners increases from 10 to 30, the total decrease in average cost is $6.67.

21. **(a)** 8; After 2 seconds, the velocity is 8 feet per second.

(b) 120; From 2 to 8 seconds, the total displacement of the object was 120 feet.

23. **(a)** 9; After 1 second, the velocity is 9 feet per second.

(b) 60; From 1 to 5 seconds, the total displacement of the object is 60 feet.

25. **(a)** 50.2; After 1 second, the velocity is 50.2 feet per second.

(b) 122.4; From 1 to 5 seconds, the total displacement of the object is 122.4 feet.

27. **(a)** 56; After 1 second, the velocity is 56 feet per second.

(b) 112; From 0 to 2 seconds, the total displacement of the object is 112 feet.

29. **(a)** 159; The golf ball is moving at a rate of 159 feet per second after 3 seconds.

(b) 621; Over the first 3 seconds, the ball traveled 621 feet.

31. 0.7%   **33.** 155.7 billion kilowatt hours

35. **(a)** -2.9; This means that in 2006, the unemployment rate in Bulgaria was decreasing by 2.9% per year.

(b) -3.02; This means that from 2005 to 2006, the unemployment rate in Bulgaria decreased by 3.02%.

37. 0.125; From 2015 to 2020, the population in Michigan is expected to grow by 0.125 million or 125,000 people.

39. -3.225; The percentage of smokers in the U.S. decreased by 3.225% from 2000 to 2005.

41. 250; The total increase in cost of producing the first 75 fobs is $250.

43. 1990 to 2000: -13.01; 2000 to 2010: -6.72; The decrease in the amount of federal research grants was the greatest during 2000 to 2010.

45. kb **47.** $n = 6$ **49.** During 2005

Section 6.4

1. Composite   **3.** Power   **5.** x **7.** $\dfrac{(3x + 4)^3}{3} + C$

9. $\dfrac{(4x^2 + 1)^4}{4} + C$ **11.** $\dfrac{2(x^2 - 1)^{3/2}}{3} + C$

13. $\dfrac{(x^3 - 2x)^2}{2} + C$ **15.** $\dfrac{3(x^3 - 5)^{2/3}}{2} + C$

17. $-\dfrac{1}{2(2x^2 + 3)^2} + C$ **19.** $\dfrac{(4x + 7)^5}{20} + C$

21. $\dfrac{(x^2 - 3)^6}{12} + C$ **23.** $\dfrac{2\sqrt{x^3 - 5}}{3} + C$

25. $-\dfrac{1}{4(10x^2 - 4)} + C$ **27.** $\dfrac{2(1 - x^{-1})^{3/2}}{3} + C$

29. $\dfrac{(x - 7)^{11}}{11} + C$ **31.** $-\dfrac{1}{8(x^2 + 2x + 5)^4} + C$

33. $\dfrac{3(x^2 + 3x - 10)^2}{2} + C$

35. $\dfrac{2(x + 1)^{5/2}}{5} - \dfrac{2(x + 1)^{3/2}}{3} + C$

37. $\dfrac{(x - 1)^7}{7} + \dfrac{(x - 1)^6}{6} + C$

39. $2(x - 1)^{3/2} + 6(x - 1)^{1/2} + C$ **41.** 4   **43.** 0   **45.** 52

47. $2\sqrt{3}$ **49.** $\dfrac{1}{8}$ **51.** 4   **53.** $\dfrac{9}{20}$ **55.** $\dfrac{18}{19}$

57. $\dfrac{1640}{6561}$ **59.** $\dfrac{52}{3}$ **61.** 0   **63.** $\dfrac{24}{5}$

65. $\dfrac{1}{2}$ square units   **67.** $\dfrac{16\sqrt{2}}{3}$ square units

69. **(a)** 6; When daily production is 75 fobs, it cost about $6 to produce the 76th fob.

(b) 250; The total increase in daily costs of producing the 1st 75 fobs is $250.

(c) $C(x) = 10\sqrt{x^2 + 10,000} + 2000$

(d) $3000

71. **(a)** About $144.40

(b) 120; The total increase in profit of producing and selling from 20 to 100 flags each month is $12,000.

(c) $P(x) = \sqrt{2x^2 - 400} - 20$

73. **(a)** About 1.588 feet   **(b)** $s(t) = \sqrt{t^2 + 9} + 3$

75. We should consider using the u-substitution technique whenever our integrand is a composition of functions.

77. $\dfrac{(ax + b)^{n+1}}{n + 1} + C$

79. $\dfrac{1}{4(-n + 1)}(ax^2 + b)^{-n+1} + C$

Section 6.5

1. Natural   **3.** $[1, a]$ **5.** $e^x + C$ **7.** $2 \ln |x| + C$

9. $-\dfrac{2}{3} \ln |x| + C$ **11.** $\ln |x|(1 + e) + C$ **13.** $3 \ln 5$

15. $\dfrac{1}{2} \ln \dfrac{5}{2} \approx 0.458$ **17.** $\dfrac{1}{2} \ln |2x + 3| + C$

19. $\dfrac{1}{2} \ln (x^2 + 1) + C$ **21.** $\dfrac{1}{2} \ln |x^2 - 4x + 9| + C$

23. $\dfrac{1}{4} \ln (x^4 + 10) + C$ **25.** $\ln |\ln 2x| + C$

27. $4 \ln \dfrac{\sqrt{2} + 4}{5}$ **29.** $\dfrac{2}{3} \ln 7$ **31.** $2e^x + C$

33. $e^x - x + C$ **35.** $4e^x + \dfrac{x^2}{2} - x + C$ **37.** $e - 2$

39. $\dfrac{1}{2}(e^2 - e + 4)$ **41.** $\dfrac{1}{4}e^{4x+1} + C$ **43.** $e^{x^2+1} + C$

45. $\ln (e^x + 2) + C$ **47.** $3 \ln |1 - e^{-x}| + C$

49. $2e^{\sqrt{x}} + C$ **51.** $-\dfrac{1}{4}(e^{-12} - e^{-4})$ **53.** $2 - e^{-1}$

55. $\dfrac{1}{2}e^{x^2} + C$ **57.** $\dfrac{(e^{4x} + 4)^2}{8} + C$ **59.** $\dfrac{e^{x \ln 5}}{\ln 5} + C$

61. $\dfrac{5\left(\dfrac{1}{2}\right)^x}{\ln \dfrac{1}{2}} + C$ **63.** $\dfrac{-7^{-x}}{\ln 7} + C$

65. $\dfrac{10^6 - 10^3}{3 \ln 10} \approx 144{,}620$ **67.** $\dfrac{121}{16 \ln 2} \approx 1.894$

69. **(a)** 20; The sales rate after 2 days is 20 dresses per day.

    (b) About 28 dresses

71. **(a)** 1.56; At a production level of 20 faucets, the cost to produce the 21st faucet is about $1.56.

    (b) About $100.39

73. **(a)** -10.82; This means that in 2005, the number of mobile homes manufactured in the southern U.S. decreased by 10.82 thousand homes per year.

    (b) About 37.50 thousand mobile homes

75. **(a)** In 2008, the number of food stamp recipients was increasing by about 1.71 million people per year.

    (b) 10.88; From 2000 to 2008, the number of participants in the Food Stamp Program was about 10.88 million people.

77. **(a)** In 2003, the amount of subprime loans was increasing by about 112.20 billion dollars per year.

    (b) 603.62; From 2003 to 2006, the total amount financed in subprime loans was about 603.62 billion dollars.

79. A division by zero would result.

81. $\dfrac{a}{k + 1} \ln |x^{k+1} + b| + C$

83. $\dfrac{k^{ax+b}}{a \ln (k)} + C$

Chapter 6 Review Exercises

Section 6.1 Review Exercises

1. No   **3.** Yes   **5.** $\dfrac{x^8}{8} + C$ **7.** $-\dfrac{1}{4t^4} + C$

9. $\dfrac{x^8}{16} - 3x^2 + C$ **11.** $\dfrac{x^3}{3} + 0.2x^2 - \dfrac{2x^{7/2}}{7} + C$

13. Decreasing on $(-\infty, 2)$; Increasing on $(2, \infty)$; Concave up on $(-\infty, \infty)$; Relative minimum at $x = 2$

15. $\dfrac{3x^2}{2} + 4x + 11$ **17.** $y = \dfrac{2x^{3/2}}{3} - 23$

19. **(a)** $P(q) = 48q - 0.015q^2 - 3050$

    (b) 17, 200; The total profit from selling 500 units of product is $17,200.

Section 6.2 Review Exercises

21. $\Delta x = \dfrac{1}{3}$ **23.** $\Delta x = \dfrac{1}{8}$ **25.** **(a)** 50   **(b)** 75

27. **(a)** $\dfrac{55}{8}$ **(b)** $\dfrac{91}{8}$

29.

	Left Sum	Right Sum
10	705.6	961.6
100	816.576	842.176
1000	828.0538	830.6138

31.

	Left Sum	Right Sum
10	211.8298	391.0298
100	285.5104	303.4304
1000	293.5047	295.2967

33. 120   **35.** $\dfrac{104}{3}$ **37.** 99

39. 24 square units   **41.** 15 square units

43. **(a)** $\dfrac{16}{3}$ square units   **(b)** $\dfrac{32}{3}$ square units

45. **(a)**

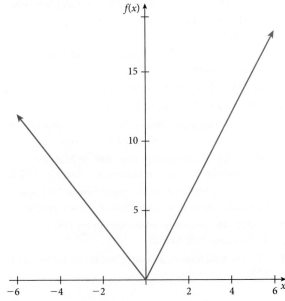

(b) 31 square units

47. (a)

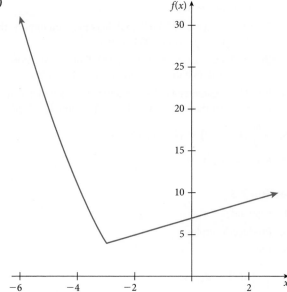

(b) About 55.17 square units

49. (a) $x = 3$ **(b)** $-\dfrac{9}{2}; \dfrac{49}{2}$

(c) On the interval $[0, 3]$, the area is below the x-axis. Thus, we should expect $\int_0^3 (x - 3)\,dx$ to be negative. On the interval $[3, 10]$, the area is above the x-axis. Thus, we should expect $\int_3^{10} (x - 3)\,dx$ to be positive.

51. (a) $x = 2$ **(b)** $-\dfrac{9}{2}; \dfrac{45}{2}$

(c) On the interval $[-1, 2]$, the area is below the x-axis. Thus, we should expect $\int_{-1}^2 (x^2 - x - 2)\,dx$ to be negative. On the interval $[2, 5]$, the area is above the x-axis. Thus, we should expect $\int_2^5 (x^2 - x - 2)\,dx$ to be positive.

Section 6.3 Review Exercises

53. (a) 20; At a production level of 160 calculators, the cost of producing the 161st calculator is about $20.

(b) 3400; The total cost of producing the first 200 calculators is $3400.

55. (a) 14.66; In 2005, the number of patents issued was increasing by 14.66 thousand patents per year.

(b) 43,100 patents

(c) 149.1; From 2000−2010, over three times the number of patents were issued as compared to 1990−2000.

Section 6.4 Review Exercises

57. $\dfrac{1}{4}(5x - 7)^4 + C$ **59.** $\dfrac{1}{2}(x^2 - 5x + 3)^2 + C$

61. $\dfrac{(3x + 6)^9}{27} + C$ **63.** $\dfrac{1}{2}\ln(x^2 + 4x + 6) + C$

65. $3\left(\dfrac{(x + 4)^4}{4} - \dfrac{4(x + 4)^3}{3} \right) + C$

67. $\dfrac{2(\sqrt{x + 3})^3}{3} - 6\sqrt{x + 3} + C$ **69.** 2106

71. About 68.856   **73.** 0   **75.** $\dfrac{22{,}876}{3}$ **77.** 182

79. (a) 299,215,620 meters   **(b)** $s(t) = \dfrac{(t^2 - 5t)^4}{4} - 4448$

Section 6.5 Review Exercises

81. $\dfrac{3}{4}\ln|x| + C$ **83.** $5e^x - \dfrac{3x^2}{2} + C$ **85.** $6\ln\dfrac{5}{3}$

87. $\dfrac{24}{\ln 5}$ **89.** $10\ln|\sqrt{x} + 4| + C$ **91.** $-\dfrac{(5 - e^x)^4}{4} + C$

93. $2\ln\dfrac{5}{4}$ **95.** $\dfrac{1365}{64\ln 2} \approx 30.77$

97. (a) $f(30) \approx 0.59$; $f(60) \approx 3.09$; In 2010, the book value is growing at a rate that is over 5 times the rate of growth in 1980.

(b) About 8.72; From 1950 to 1980, the book value grew by about 8.72 trillion dollars.

(c) 45.41; From 1980 to 2010, the book value grew by about 45.41 trillion dollars.

Chapter 7

Section 7.1

1. Average value   **3.** Continuous   **5.** Rate; flow

7. 12   **9.** 2   **11.** $\dfrac{14}{3}$ **13.** $\dfrac{93}{35}$

15. $2e^2 - 2 \approx 12.78$ **17. (a)** About 1.59   **(b)** About 0.795

19. (a) About 1.45   **(b)** 0.725

21. (a) $4.2 million   **(b)** $11.4 million

23. $5.10   **25.** About 12.70%

27. (a) $135,000   **(b)** $AC(x) = \dfrac{60{,}000}{x} + 300$; $420

(c) Part (a) is the average total cost of building 500 engines, while part (b) is the average cost of producing each engine if 500 are produced.

29. About $3,484.85   **31.** About $6,137.05

33. (a) About $3,567.35   **(b)** $26.76

35. (a) $1,778.90   **(b)** About $9329.57

37. (a) About $175   **(b)** About $167.71

39. (a) $\displaystyle\int_0^{12} 600e^{-0.1t}\,dt$ **(b)** 4,193 thousand barrels

41. $6 million   **43.** $20,343.69   **45.** $188,234.72

47. $679,206.56   **49.** $588,313.25

51. (a) $77,400   **(b)** $80,000

(c) $90,893.31; This is because Vicki's contributions had 3 years longer to earn interest.

53. $26,962.51   **55.** About 80 offices

57. About $83.65 billion   **59.** 12.14 billion dollars

61. To find the average value of a function, find the total area and divide by the width of that area.

63. $\dfrac{m}{2}(b + a) + d$ **65.** $b = 4$

Section 7.2

1. Below   **3.** Intersection   **5.** Negative; cost

7. $\displaystyle\int_0^1 (x + 1 - x^2)\, dx$ **9.** $\displaystyle\int_{-1}^1 (x + 7 - x^2)\, dx$

11. $\displaystyle\int_a^c [g(x) - f(x)]\, dx + \int_c^b [f(x) - g(x)]\, dx$

13. $\displaystyle\int_0^b [(d)x - k]\, dx$ **15.** 28 square units

17. $\dfrac{15}{2}$ square units   **19.** $\dfrac{59}{6}$ square units

21. $\dfrac{32}{3}$ square units   **23.** $\dfrac{52}{3}$ square units

25. 15 square units   **27.** About 30.88 square units

29. About 23.62 square units   **31.** About 2.66 square units

33. $\dfrac{9}{2}$ square units   **35.** 18 square units   **37.** $\dfrac{1}{2}$ square units

39. $\dfrac{1}{2}$ square units

41. (a) $\displaystyle\int_0^1 (4t^{5/2} - 4t^{18/5})\, dt$

   (b) About 0.27; During the first minute, the bacteria without nutrient added exceeds the bacteria with nutrient added by about 0.27 milligrams.

43. (a) 1209.27; If Double D cola uses the TV personality in its ads, sales are expected to increase by 1.2 million dollars during the first twenty weeks.

   (b) Double D will still have an increase of about $209,270 after paying the fee for the TV personality.

45. About $15.7 million

47. (a)

   (b) About 435.032; With the special tapes, a child can learn about 435 more new words between the ages of 4 and 8.

49. About $11.6 million   **51.** About $1,944.83

53. About $80.94

55. (a) About 97 units   (b) About 13 more units

57. (a) Around 52.30 feet   (b) About 34.7 fewer feet

59. (a) About 390.57; From 1995 to 2000, there were around 391 more injured people from tornadoes than there were lives lost.

   (b) About 2276.51; From 2005 to 2010, there were around 2277 more injured people from tornadoes than there were lives lost.

(c) It would be advisable for households to purchase a weather radio since the number of injuries has greatly increased in more recent times.

61. (a) The U.S. spent 122.3 billion dollars more on imports than it did on exports from 2003 to 2008.

   (b) 1623.42; The U.S. spent 1,623.42 billion dollars more on exports than it did on imports from 2008 to 2011.

63. It will be negative since the area under $f(x)$, denoted $\int_a^b f(x)\, dx$, will be larger than the area under $g(x)$, denoted $\int_a^b g(x)\, dx$.

65. $-\dfrac{m_1}{2}(b^2 - a^2) + (k - b_1)(b - a)$

67. $b = 2 + 2\sqrt{2}$

Section 7.3

1. Demand

3. Producer's surplus

5. Lorenz

7. $\displaystyle\int_0^{x_m} [d(x) - d_{mp}]\, dx$

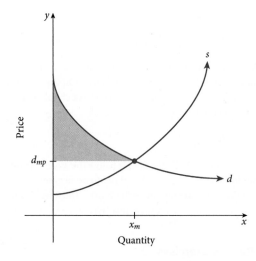

9. $(d_{mp})\,(x_m)$

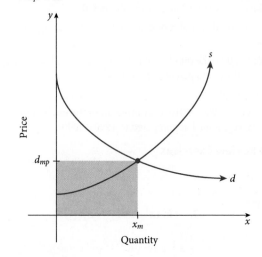

11. $\displaystyle\int_{0}^{700} (210 - 0.3x)\, dx$

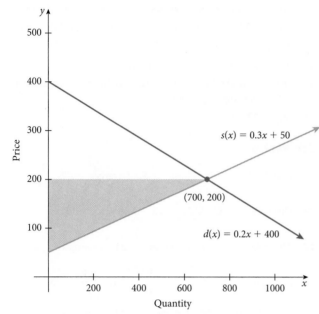

13. $\displaystyle\int_{0}^{360} (90 - 0.25x)\, dx$

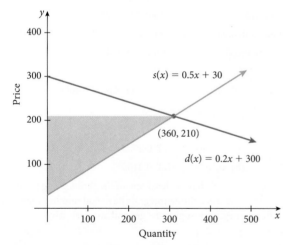

15. $60,000   **17.** $1,666.67   **19.** $6,750

21. $53,333.33   **23.** $6,000   **25.** $48,051.10

27. $9,937.29   **29.** $562.50   **31.** $750   **33.** $600

35. $33,333.33   **37.** $225,000   **39.** $2,919.86

41. $1,690.54

43. (a) $(190, 1{,}083)$ (b) $234,650   (c) $102,885

45. (a) $(430, 226)$ (b) $36,980   (c) $46,225

47. (a) $(165, 1{,}654)$ (b) $119,790   (c) $179,685

49. (a) $(35,255)$ (b) $5,716.67   (c) $5,716.67

51. (a) $(209, 74)$ (b) $37,112.77   (c) $11,845.42

53. (a) 10 frying pans   (b) $70   **55.** About $59

57. (a) 65 golf balls   (b) $507   **59.** About $415

61. (a) $(30, 4)$ (b) $180   (c) $45

63. (a) $(25, 5)$ (b) $312.5   (c) $41.67   **65.** 0.293

67. 0.4223   **69.** 0.4383   **71.** 0.429

73. Consumers' surplus is the amount "saved" by consumers by purchasing an item they were planning to buy at a higher price for a lower price.

75. It would be preferential to start a family in Denmark since the income is more evenly distributed amongst the population whereas in Namibia has a small population of people that controls most of the wealth.

77. $bx^2 - 2b^2x$

Section 7.4

1. Product   **3.** Integrate   **5.** Present value

7. $e^{3x}\left(\dfrac{2}{3}x - \dfrac{2}{9}\right) + C$ **9.** $e^{4x}\left(\dfrac{1}{4}x - \dfrac{1}{16}\right) + C$

11. $-e^{-x}(x + 1) + C$ **13.** $e^{-0.03x}\left(\dfrac{-100}{3}x - \dfrac{10{,}000}{9}\right) + C$

15. $\dfrac{1}{3}x^3\left(\ln x - \dfrac{1}{3}\right) + C$ **17.** $x(\ln 2x - 1) + C$

19. $e^x(x + 2) + C$ **21.** $-60e^{-0.1t}(t + 10) + C$

23. $-15e^{-0.08t}(t + 12.5) + C$ **25.** About 0.594

27. About 10.95   **29.** About 2.097   **31.** About 51.300

33. $(x + 1)^6\left(\dfrac{6x + 13}{42}\right) + C$ **35.** $(x + 3)^7\left(\dfrac{7x - 19}{56}\right) + C$

37. $\dfrac{e^{3x}}{27}(9x^2 - 6x + 2) + C$ **39.** $-e^{-x}(x^2 + 2x + 2) + C$

41. $\dfrac{e^{5x}}{125}(25x^2 - 10x + 2) + C$

43. $x(\ln x)^2 - 2x \ln x + 2x + C$

45. $\dfrac{x^2}{8}\left(4(\ln x)^3 - 6(\ln x)^2 + 6\ln x - 3\right) + C$

47. By parts: $u = x; dv = e^{-x}\,dx$ **49.** u-substitution: $u = \ln x$

51. By parts: $u = x + 6; dv = e^x\,dx$

53. By parts: $u = 3x; dv = e^{2x}\,dx$

55. (a) $\displaystyle\int_{0}^{12} 6te^{-0.15t}dt$ (b) About 143.243 thousand barrels

57. $P(t) = 10e^{0.2t}(t - 5) + 50$

59. $C(t) = -\dfrac{20}{9}e^{-0.15t}(3t + 20) + \dfrac{418}{9}$

61. About 12.451; After six hours, about 12.451 mg was absorbed.

63. About $20,520,000   **65.** About $593,534

67. Since the present value of the oil well is $408,874 and the present value of the gas well is $508,330, the natural gas well is a better investment.

69. About 651.187; In Italy from 2001 to 2011, there were about 651 labor disputes on average annually.

71. About 354.045; From 2001 to 2011, there was an average number of about 354 thousand livestock organically raised each year.

73. Typically, when the integrand is written as a product and a u-substitution will not work, you would want to consider the integration by parts method.

77. A u-substitution would be best since the derivative of the exponent $2x^2 - x$ is a part of the integrand. Thus, we should pick $u = 2x^2 - x$.

Section 7.5

1. Rectangular   **3.** Width

5. Elementary antiderivative   **7.** 22   **9.** 1.48

11. 0.76   **13.** 3.83   **15.** 70,625

17. $a = 0, b = 2$

n trapezoids	Trap Approximation
10	0.4128406644
50	0.4005283873
100	0.4001322241

19. $a = 1, b = 4$

n trapezoids	Trap Approximation
10	0.0690011825
50	0.0671679267
100	0.0671104334

21. $a = 1, b = 2$

n trapezoids	Trap Approximation
10	2.453701923
50	2.451740936
100	2.451679645

23. $a = 0, b = 1$

n trapezoids	Trap Approximation
10	0.4091397464
50	0.396143122
100	0.3957354568

25. $a = 2, b = 5$

n trapezoids	Trap Approximation
10	6.178488696
50	6.172491939
100	6.172304409

27. $a = 1, b = 3$

n trapezoids	Trap Approximation
10	4.051132552
50	4.049970734
100	4.049934358

29. $9550   **31.** $100   **33.** $3225   **35.** $21,025

37. (a) About 19.1 million   **(b)** About 36.3 million

39. (a) About 2,251.25 million tons
(b) About 150.08 million tons

41. (a) About 194 work stoppages   **(b)** About 19.4 per year

43. (a) About 5,267 thousand   **(b)** About 658.375 thousand

45. About 6.24 million

47. (a) About $670 per year   **(b)** About $1,220

49. The Trapezoidal Rule finds the sums of the areas that fit under the curve using trapezoids.

51. 18   **53.** 12

Section 7.6

1. Unbounded   **3.** Limit   **5.** Diverges   **7.** $\dfrac{1}{24}$

9. Diverges   **11.** $\sqrt{2}$ **13.** Diverges   **15.** $\dfrac{1}{e}$

17. About 32.35   **19.** Diverges   **21.** Diverges   **23.** $\dfrac{1}{2}$

25. Diverges   **27.** 5   **29.** $\dfrac{1}{4}$ **31.** $-\dfrac{1}{60}$ **33.** 1

35. 0   **37.** 0   **39.** Diverges   **41.** Diverges

43. 600 thousand barrels   **45.** 16 million cubic feet

47. About 1,166.67 tons   **49.** About 107 ml

51. About $166,667   **53.** About $413,793

55. About $1,076,923   **57.** About $218,182

59. About $106,667   **61.** About $45,454

63. (a) $79,263   **(b)** About $80,000

65. (a) 1   **(b)** About 0.1054   **(c)** About 0.4066

67. (a) 1   **(b)** 0.1875   **(c)** $\dfrac{3}{64}$

69. (a) 1   **(b)** 0.2387   **(c)** 0.2983

71. (a) 1   **(b)** 0.3023   **(c)** 0.2109

73. Improper integrals have at least one of the limits going to either $-\infty$ or ∞ whereas definite integrals have real number limits and indefinite integrals just give the general antiderivative.

75. $a = 3$ **77.** $k \le 1$

Section 7.7

1. Differential   **2.** Separation; variables

3. General solution   **4.** Particular solution

5. Family of curves   **7.** Yes   **9.** No   **11.** Yes

13. No   **15.** Yes   **17.** Yes   **19.** Yes

21. $y = -2x + C$

23. $y = x - \dfrac{3x^2}{2} + C$

25. $y = \dfrac{5x^4}{4} + \dfrac{x^2}{2} - 2x + C$

27. $y = \dfrac{2x^{3/2}}{3} + 2x + C$ **29.** $y = -\dfrac{5}{x} + C$

31.

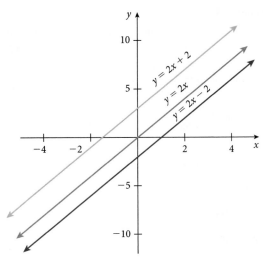

33.

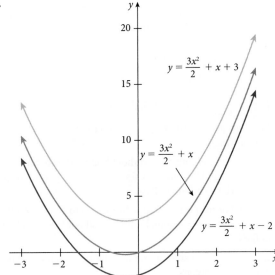

35.

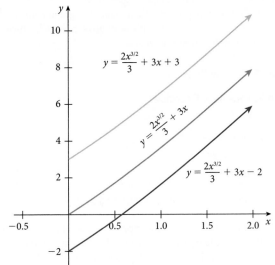

37. $y = 6x^2 + 2$ **39.** $y = x^3 + 2x^2 + 3$

41. $y = 7t - 2t^2 + 3$ **43.** $y = \dfrac{x^4}{4} - x^2 - 2$

45. $y = \dfrac{t^3}{3} + t^2 + 3t + 4$ **47.** $y = \dfrac{2(x)^{1.3}}{1.3} + 4$

49. $y = x - 2\ln|x| + 5$ **51.** $y^2 = x^2 + C$

53. $y = Ce^{x^3}$ **55.** $e^{-y} = C - e^x$ **57.** $y = Ce^{x(x+2)}$

59. $y^{1/2} = \dfrac{x^{3/2}}{3} + C$ **61.** $y = -e^{1/2x^2}$

63. $y = 3x$ **65.** $y = \sqrt{2x + 2\ln|x| + 2}$

67. $y = e^x$ **69.** $y = x$

71. (a) $C(x) = -0.01x^2 + 6x + C$

  (b) $C(x) = -0.01x^2 + 6x + 341$

73. (a) 11.1; Ten months after its introduction, the new percentage of drinkers accepting the new formula is increasing at a rate of 11.1% per month.

  (b) The percentage of drinkers accepting the new formula after t months is given by $A(t) = \dfrac{t^2}{2} + 1.1t + C$ where C is determined by initial conditions.

  (c) $A(t) = \dfrac{t^2}{2} + 1.1t + 5$

75. (a) $R(t) = 1.36t + C$

  (b) $R(t) = 1.36t + 31.1$

  (c) 44.7; In 2010, the annual revenue at amusement parks will be 44.7 billion dollars.

77. (a) $y = -30.12x^2 + 225.59x + C$

  (b) $y = -30.12x^2 + 225.59x + 695$

  (c)

Year	Number of Conventional Loans, in thousands
2001	890.47
2003	1100.69
2005	1069.95
2007	798.25
2009	285.59

During 2003, the number of conventional loans for home financing was at its greatest.

79. $y = 452x^{0.26} + 4.14$

81. A differential equation is an equation that contains functions and the derivatives of those functions.

83. A general solution has the arbitrary constant C whereas a particular solution gives the answer for one scenario and has a specific value for C under a certain restriction.

85. $y = \dfrac{a}{b + 1}x^{b+1} + C$

Section 7.8

1. Unlimited   **3.** Initial   **5.** Upper

7. (a) $y' = (10 - y)$ (b) $f(0) = 0; L = 10$

9. (a) $y' = 2(8 - y)$ (b) $f(0) = 0; L = 8$

11. (a) $y' = 0.5(20 - y)$ (b) $f(0) = 0; L = 20$

13. (a) $y' = 0.1(12 - y)$ (b) $f(0) = 0; L = 12$

15. (a) $y' = 0.2(1.5 - y)$ (b) $f(0) = 0; L = 1.5$

17. (a) $y' = 0.1(15.5 - y)$ (b) $f(0) = 0; L = 15.5$

19. (a) $y' = \dfrac{1}{30}y(30 - y)$ **(b)** $y(0) = \dfrac{30}{1 + e^0} = 15;\ \ L = 30$

21. (a) $y' = \dfrac{1}{50}y(50 - y);\ \ L = 50$

    (b) $y(0) = \dfrac{50}{1 + 2e^0} = \dfrac{50}{3};\ \ L = 50$

23. (a) $y' = \dfrac{1}{75}y(150 - y)$

    (b) $y(0) = \dfrac{150}{1 + e^{-2(0)}} = 75;\ \ L = 150$

25. (a) $y' = \dfrac{3}{400}y(400 - y)$

    (b) $y(0) = \dfrac{400}{1 + 2e^{-3(0)}} = \dfrac{400}{3};\ \ L = 400$

27. (a) $y' = 0.0009y(1000 - y)$

    (b) $y(0) = \dfrac{1000}{1 + 4.5e^{-0.9(0)}} = \dfrac{2000}{11} \approx 181.82;\ \ L = 1000$

29. (a) $P' = 0.05P,\ P(0) = 50{,}000$ **(b)** $P(t) = 50{,}000e^{0.05t}$

    (c) About 82,436 people

31. (a) $P' = 0.12P$ **(b)** $P = 5000e^{0.12t}$

    (c) $P(3) \approx 7167;\ P(12) \approx 21{,}103;\ P(24) \approx 89{,}071$

33. (a) $y = 3000e^{-(\ln 2/14.2)t}$ **(b)** About 8.31 days

35. (a) $P' = -0.01822P;\ P(0) = 920{,}000$

    (b) $P(t) = 920{,}000e^{-0.01822t}$ **(c)** About 662,760

37. (a) $y(2) \approx 1353.57;\ y'(2) \approx 494;$ Two weeks after installation about 1354 employees have learned the system and about 494 employees are learning it per week.

    (b) About 6 weeks

39. (a) $L = 300$ **(b)** $k = \ln\dfrac{15}{13}$

    (c)

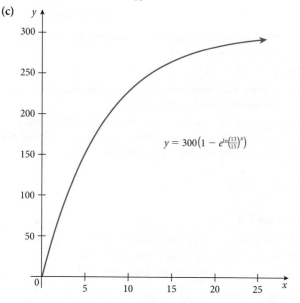

$$y = 300\left(1 - e^{\ln\left(\frac{13}{15}\right)x}\right)$$

    (d) No, it will take almost 4.78 weeks.

41. (a) About 86.51%

    (b) 11.75; In 1996, the percentage of households that owned video disc recorders was increasing at a rate of 11.75% per year.

    (c) 90% of households

43. (a) $L = 7{,}500$

    (b) $k = \dfrac{0.589}{7500} \approx 7.85 \times 10^{-5}$

    (c) $y = \dfrac{7500}{1 + 6.5e^{-0.589}}$

    (d) 3,750 catfish

45. (a) $\dfrac{dy}{dt} = 0.28y$

    (b) $f(t) = Ce^{0.28t}$

    (c) $f(t) = 1.28e^{0.28t}$

    (d) $f(14) \approx 64.51;$ If this growth rate continues, then Amazon's projected net revenue is expected to hit around 64.51 billion dollars in 2014.

47. (a) $f(t) = 160.95(1 - e^{-0.01t})$

    (b) $t = \dfrac{\ln\left(\dfrac{67}{87}\right)}{-0.01} \approx 26;$ In 2016, the annual cost of medical and hospital care for veterans is expected to reach around 37 billion dollars.

    (c) 160.95; As time continues without bound, the annual cost of medical and hospital care for veterans is expected to approach 160.95 billion dollars.

49. (a) 859.03; For the period of time from 1990 to 2000, the annual total in accounts in mutual funds was about 859.03 billion dollars.

    (b) In 2008, the total amount in mutual fund accounts will be 3,000 billion dollars.

51. The difference is that for a logistic growth, there is an initial period of unlimited growth, which makes the graph concave up. Then, it switches and remains at a limit growth model, which makes the graph concave down.

Chapter 7 Review Exercises

Section 7.1 Review Exercises

1. 13   **3.** 102

5. (a) About 11.8624

    (b) About 1.4828

7. (a) $4,900

    (b) $AC(x) = \dfrac{1300}{x} + 18;\ AC(400) = \$21.25$

    (c) Part (a) is the average cost of producing a total of 0 to 400 phones, while in part (b), we found the average cost of producing each phone if 400 phones are produced.

9. (a) About $2,457.29   **(b)** About $20.89

11. About $25,258   **13.** About $18,776

Section 7.2 Review Exercises

15. $\int_{\frac{-1-\sqrt{17}}{2}}^{\frac{-1+\sqrt{17}}{2}} (4 - x^2 - x)\,dx$ **17.** 36 **19.** $\dfrac{125}{6}$ **21.** $\dfrac{256}{3}$

23. About 211.43 **25.** $\dfrac{125}{6}$ **27.** $\dfrac{148}{3}$

29. About 2,706; If Lucky is used for the ad campaign, total sales in the first 15 months will increase by 2,706 tricycles.

31. About $795,766

33. (a) About 201 units **(b)** About 41 more units

Section 7.3 Review Exercises

35. $\int_{0}^{60} (-0.4x + 24)\,dx$

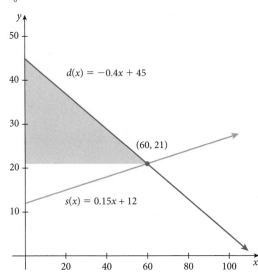

37. $1,260

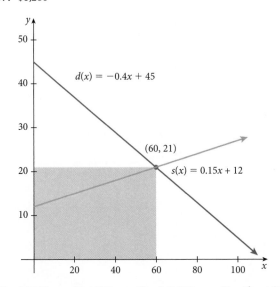

39. $4,000 **41.** $72 **43.** $40,000 **45.** About $1,667

47. (a) (380,136) **(b)** $21,660 **(c)** $14,440

49. (a) (200, 2,900) **(b)** About $213,333 **(c)** $160,000

51. (a) 16 **(b)** About $42.67

53. About $2,301 **55. (a)** (36, 6) **(b)** $155.52 **(c)** $72

Section 7.4 Review Exercises

57. $e^{2x}(2x - 1) + C$ **59.** $\dfrac{x^5}{25}(5\ln x - 1) + C$

61. $-40e^{-0.2t}(t + 5) + C$

63. $-e^{-x}(x^2 + 2x + 2) + C$

65. $x(\ln x)^4 - 4x(\ln x)^3 + 12x(\ln x)^2 - 24x\ln x + 24x + C$

67. u-substitution; $u = -x^3$

68. By parts; $u = \ln x$; $dv = 2x\,dx$

69. By parts; $u = x$; $dv = e^{5x}\,dx$

71. (a) $\int_{0}^{12} 8te^{-0.17t}\,dt$

(b) About 167.395 thousand barrels

73. About $598.717 thousand

75. 1.42; From 2008 to 2012, gas prices are, on average, 1.42 Euros per liter.

Section 7.5 Review Exercises

77. -154 **79.** About -0.65625

81. $n = 10$; ≈ 74.79658962

$n = 50$; ≈ 74.7981479

$n = 100$; ≈ 74.7981974

83. $n = 10$; ≈ 1.679481689

$n = 50$; ≈ 1.677478259

$n = 100$; ≈ 1.677420313

85. $12,007.50 **87.** About $153.50 thousand

Section 7.6 Review Exercises

89. Diverges **91.** Diverges **93.** Diverges

95. -2.5 **97.** Diverges **99.** About $266,667

101. 1,300 tons

103. (b) About 0.1818 **(c)** About 0.4493

Section 7.7 Review Exercises

105. $y = \dfrac{x}{1 + Cx}$

107. $y = Ce^{4x}$

109. $y = Ce^x + 2$

111. $y = \dfrac{-1}{4x + 1}$

113. $y = 2x^2 - \dfrac{x^3}{3} + C$

115. $y = x^2 - \dfrac{x^3}{3} + C$

117. (a) $f(x) = -0.22x + C$

(b) $f(x) = -0.22x + 17.8$

(c) 9.44 thousand traffic accidents

Section 7.8 Review Exercises

119. $f(t) = 8.2e^{1.37t}$ **121.** $f(t) = 384e^{-0.87t}$

123. (a) $P' = 0.08P$ **(b)** $P(t) = 300e^{0.08t}$

(c)

Number of hours, t	Number of bacteria, P
0	300
2	352.053
44	413.138
10	667.662
20	1485.91

(d)

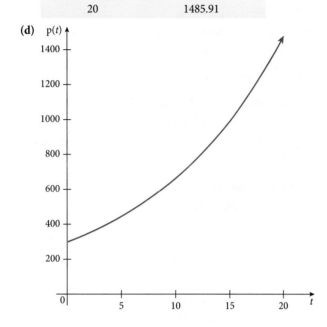

31.

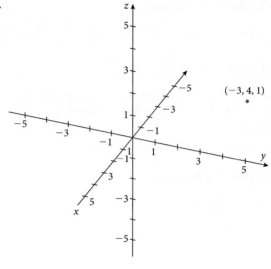

33.

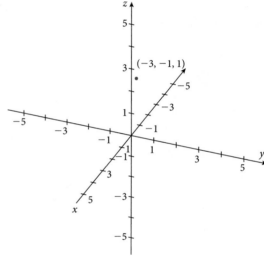

Chapter 8

Section 8.1

1. Independent   **3.** Triple   **5.** Cobb-Douglas   **7.** 19

9. 3   **11.** $\dfrac{5}{2}$ **13.** $-\dfrac{13}{5}$ **15.** 0   **17.** Undefined

19. $2 + e^2$ **21.** $2e$ **23.** $\{(x, y) | x \in R, y \in R\}$

25. $\{(x, y) | x \neq -y\}$ **27.** $\{(x, y) | x \neq 4y\}$

29.

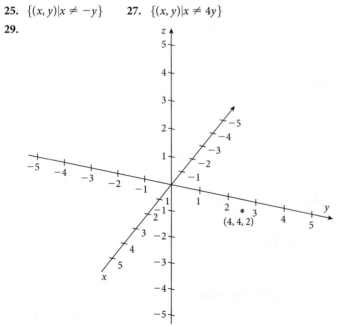

35.

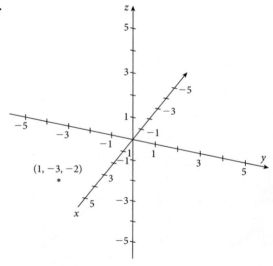

37.

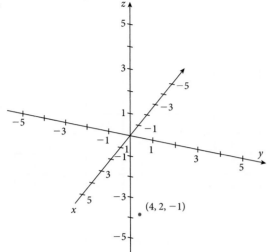

39.

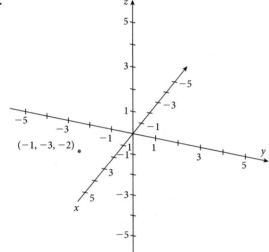

41.

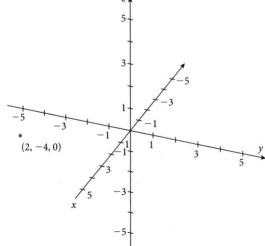

43. The point $(2, 2, 3)$ is above the xy-plane since the z value is positive.

45. The point $(-3, -2, 5)$ is above the xy-plane since the z value is positive.

47. The point $(4, -2, -1)$ is below the xy-plane; the z value is negative.

49. The point $(0, 3, -0.5)$ is below the xy-plane; the z value is negative.

51.

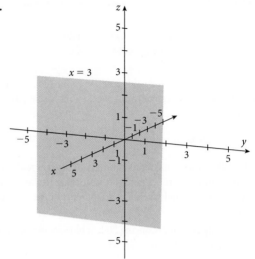

53.

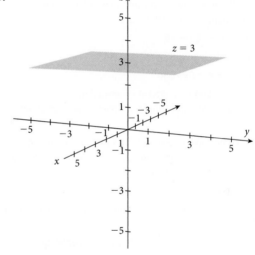

55.

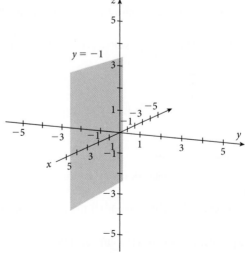

57. (a) 53; When Seamount spends $5,000 on newspaper ads and $3,000 on radio advertising each week, weekly sales are $530,000.

(b) 23; When Seamount spends $3,000 on newspaper ads and $5,000 on radio advertising each week, weekly sales are $230,000.

59. (a) 1.9; When Tube Town spends $1,000 on radio advertising and $500 on TV advertising each week, weekly sales are $19,000.

(b) 3.575; When Tube Town spends $500 on radio advertising and $1000 on TV advertising each week, weekly sales are $35,750.

61. $R(36,1) = 36K$; $R(36,2) = \dfrac{9}{4}K$

63. About -11; When the actual temperature is 10 degrees and the velocity of the wind is 25 miles per hour, according to the wind chill index model, it actually feels like it is about 11 degrees below zero.

65. 2795; The monthly cost of producing 50 leaf blowers per month and 30 attachments per month is $2795.

67. (a) $P(x, y) = 85x - 0.8x^2 - 0.25xy + 50.8y + 0.015y^2 - 1000$

(b) 2412.5; The monthly profit from sales of 50 leaf blowers and 30 attachments per month is $2,412.50.

69. (a) $C(x, y) = 5x + 4y + 50$

(b) $R(x, y) = 12x + 9.5y$

(c) $P(x, y) = 7x + 5.5y - 50$

71. About 6787.15; $2,000 invested at an annual interest rate of 6.125% ,compounded monthly for 20 years, will yield about $6,787.15.

73. About 12,251.49; Two thousand dollars invested at an annual interest rate of 7.25%, compounded continuously for 25 years, yields approximately $12,251.49.

75. About 7241 units

77. (a) About 133.33; A 9-year-old person with a 12-year-old mental age has an IQ of 133.33.

(b) 75; 12-year-old person with a 9-year-old mental age has an IQ of 75.

(c) 100; A 12-year-old person with a 12-year-old mental age has an IQ of 100.

(d) IQ = 100; An IQ of 100 represents a person with mental age equal to actual age.

79. (a)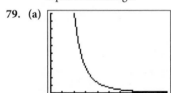

(b) About 199 units of capital are needed if 70 units of labor are used.

(c) About 68 units of capital are needed if 100 units of labor are used.

(d) Three possible answers are (119, 40), (81, 129), (40, 1067).

81. Various answers. Graphically, instead of lines and curves, we now have planes and surfaces. Also, we go from an ordered pair to an ordered triple. Analytically, finding the domain and evaluating a function with two independent variables is very similar to having just one variable.

83. $x = \pm 4$

85.

	$x = -3$	$x = -2$	$x = -1$	$x = 0$	$x = 1$	$x = 2$	$x = 3$
$y = -3$	0	3	6	9	12	15	18
$y = -2$	-5	-2	1	4	7	10	13
$y = -1$	-8	-5	-2	1	4	7	10
$y = 0$	-9	-6	-3	0	3	6	9
$y = 1$	-8	-5	-2	1	4	7	10
$y = 2$	-5	-2	1	4	7	10	13
$y = 3$	0	3	6	9	12	15	18

Section 8.2

1. Cross-section   **3.** Level curves   **5.** Isoquant

7. $y = -x - 1 - c$

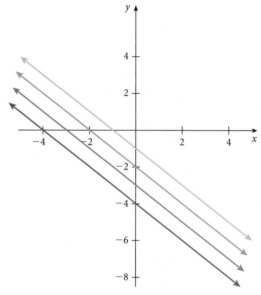

9. $y = \dfrac{-2x - 9 + c}{3}$

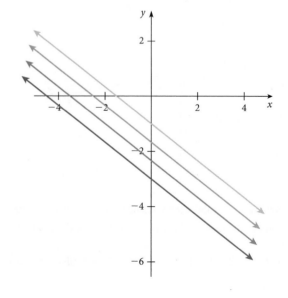

11. $y = \pm\sqrt{16 - x^2 - c^2}$

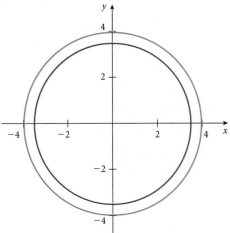

13. $y = c - e^x$

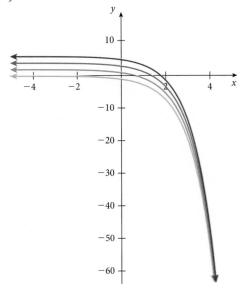

15. $y = \dfrac{c^2}{24^2 x}$

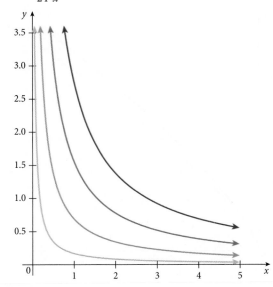

17. $y = \dfrac{c^{100/41}}{5.6^{100/41} x^{65/41}}$

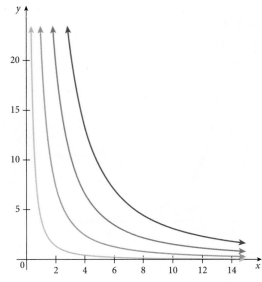

19. $y = \left(\dfrac{c}{180}\right)^{10/3} x^{-7/3}$

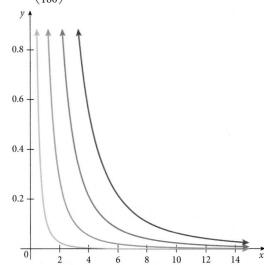

21. $y = c^2 x^{-1}$

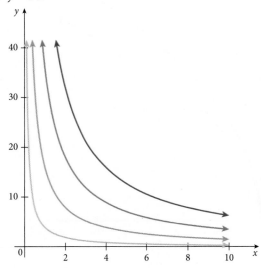

23. $y = c^4 x^{-3}$

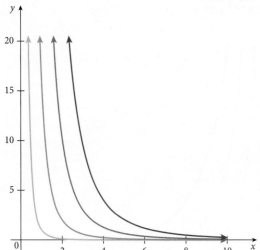

25. $y = c^{20/9} x^{-11/9}$

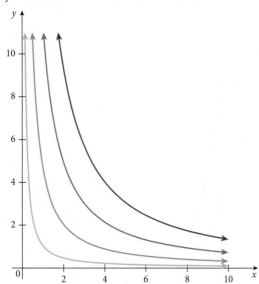

27.

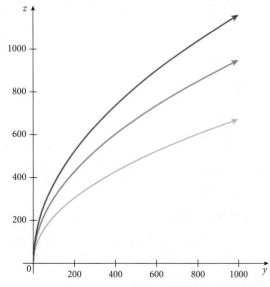

29.

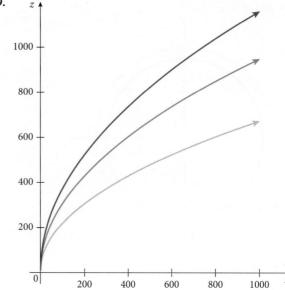

31.

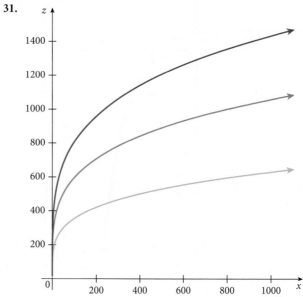

33.

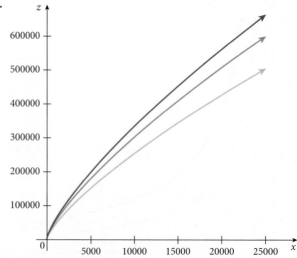

35. $y = \left(\dfrac{Q}{1.2}\right)^{10/3} x^{-7/3}$

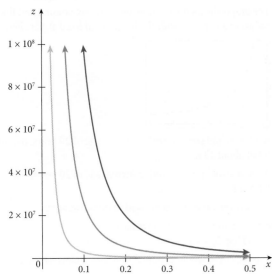

37. $y = c^3 x^{-2}$

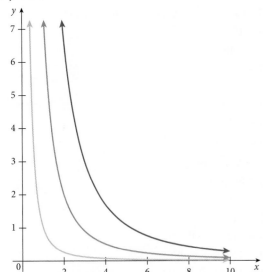

39. $y = \dfrac{100x}{c}$

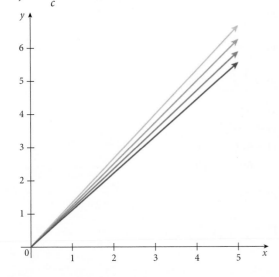

41. $y = \left(\dfrac{Q}{1.64}\right)^{5/2} x^{-3/2}$

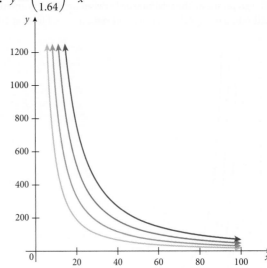

43. (a) Vertical

(b)

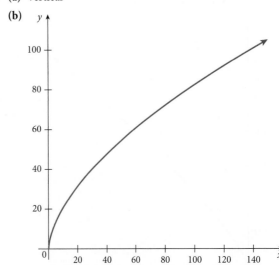

(c) About 137 thousand

(d) The graph shows the relationship between production and the number of labor hours when capital is held fixed at $18 million.

45. (a) Vertical

(b)

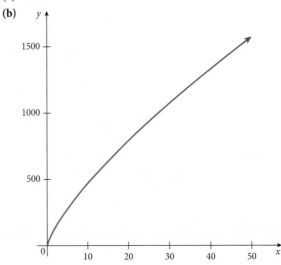

(c) About 47 units

(d) The graph shows the relationship between production and the utilization of capital when labor utilization is held fixed at 100.

47.

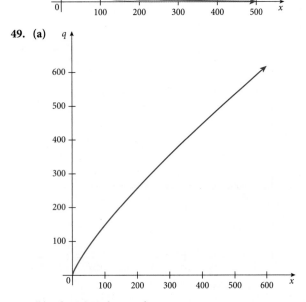

49. (a)

(b) About 528 thousand

51. (a)

(b) About 3 inches

(c) About 452 cubic inches

(d) The graph shows the relationship between volume and the radius of a cylinder with the height held fixed at 4 inches.

53. (a)

(b) A 15-year-old person, with a mental age of 20 years, has an IQ of about 133.

(c) A 16-year-old person, with a mental age of 20 years, has an IQ of 125.

55. Various answers. It can also be used in oceanography.

57. Since $c = 5$ and $x = -3$, then we have
$$5 = (-3)^2 + 2y \Rightarrow y = -2.$$

59. A parabolic curve would produce a mound-shaped surface.

Section 8.3

1. y 3. Labor   5. $f_{xx}(x, y)$ 6. $f_{yx}(x, y)$

7. $f_x(x, y) = 10x$
   $f_y(x, y) = -18y^2$

9. $f_x(x, y) = 2y$
   $f_y(x, y) = 2x - 2y$

11. $f_x(x, y) = 3x^2y^2$
    $f_y(x, y) = 2x^3y$

13. $f_x(x, y) = 2x + 6xy^3 - y$
    $f_y(x, y) = 9x^2y^2 - 6y^2 - x$

15. $f_x(x, y) = 6x^2 + 2$
    $f_y(x, y) = -2y$

17. $f_x(x, y) = 6(2x + 3y)^2$
    $f_y(x, y) = 9(2x + 3y)^2$

19. $f_x(x, y) = 5.5815x^{-0.85}y^{0.87}$
    $f_y(x, y) = 32.3727x^{0.15}y^{-0.13}$

21. $f_x(x, y) = e^x \ln y$
    $f_y(x, y) = \dfrac{e^x}{y}$

23. $f_x(x, y) = -\dfrac{y}{x^2}$
    $f_y(x, y) = \dfrac{1}{x}$

25. $f_x(x, y) = 6x - 6x^2y^4$
    $f_y(x, y) = -8x^3y^3$

27. $f_x(x, y) = 4x(x^2 + y^3)$
    $f_y(x, y) = 6y^2(x^2 + y^3)$

29. $f_x(x, y) = \dfrac{2x}{x^2 + y^3}$
    $f_y(x, y) = \dfrac{3y^2}{x^2 + y^3}$

31. $f_x(x, y) = \dfrac{2xy(y + x) - x^2 y}{(y + x)^2}$

$f_y(x, y) = \dfrac{x^2(y + x) - x^2 y}{(y + x)^2}$

33. $f_x(x, y) = 6xy^3 + e^{x+y}$

$f_y(x, y) = 9x^2 y^2 + e^{x+y}$

35. $f_x(x, y) = ye^{xy}$

$f_y(x, y) = xe^{xy} + \dfrac{1}{y}$

37. $f_x(x, y) = -10xy; f_x(1, 2) = -20$; If we are at the point $(1, 2)$ and move along the surface in the x-direction, parallel to the x-axis, the function is decreasing at a rate of $20 \frac{\text{units of } f}{\text{units of } x}$.

39. $f_y(x, y) = 2x^2 y - 6y - 1; f_y(2, 3) = 5$; If we are at the point $(2, 3)$ and move along the surface in the y-direction, parallel to the y-axis, the function is increasing at a rate of $5 \frac{\text{units of } f}{\text{units of } y}$.

41. $f_x(x, y) = \dfrac{-1}{y^2 - x}; f_x(2, 5) = \frac{-1}{23}$; If we are at the point $(2, 5)$ and move along the surface in the x-direction, parallel to the x-axis, the function is decreasing at a rate of $\frac{1}{23} \frac{\text{units of } f}{\text{units of } x}$.

43. $f_x(x, y) = 5.5815x^{-0.85} y^{0.87}; f_x(5, 2) \approx 2.5973$; If we are at the point $(5, 2)$ and move along the surface in the x-direction, parallel to the x-axis, the function is increasing at a rate of approximately $2.6 \frac{\text{units of } f}{\text{units of } y}$.

45. $f_x(x, y) = -6x^2 y - e^{x-y}; f_x(2,4) \approx -96.136$; If we are at the point $(2,4)$ and move along the surface in the x-direction, parallel to the x-axis, the function is decreasing at a rate of approximately $96.136 \frac{\text{units of } f}{\text{units of } y}$.

47. $f_x(x, y) = -10xy$

$f_{xx}(x, y) = -10y$

$f_{xy}(x, y) = -10x$

$f_y(x, y) = 12y^2 - 5x^2$

$f_{yy}(x, y) = 24y$

49. $f_x(x, y) = 6x - 6x^2 y^2 + 6x^2$

$f_{xx}(x, y) = 6 - 12xy^2 + 12x$

$f_{xy}(x, y) = -12x^2 y$

$f_y(x, y) = -4x^3 y$

$f_{yy}(x, y) = -4x^3$

51. $f_x(x, y) = 5y - 18x^2$

$f_{xx}(x, y) = -36x$

$f_{xy}(x, y) = 5$

$f_y(x, y) = 5x + 7$

$f_{yy}(x, y) = 0$

53. $f_x(x, y) = -6x^2 y^2 + 14x$

$f_{xx}(x, y) = -12xy^2 + 14$

$f_{xy}(x, y) = -12x^2 y$

$f_y(x, y) = 5y^4 - 4x^3 y$

$f_{yy}(x, y) = 20y^3 - 4x^3$

55. $f_x(x, y) = y^3 e^{xy} + \dfrac{1}{x}$

$f_{xx}(x, y) = y^4 e^{xy} - x^{-2}$

$f_{xy}(x, y) = 3y^2 e^{xy} + xy^3 e^{xy}$

$f_y(x, y) = 2ye^{xy} + xy^2 e^{xy}$

$f_{yy}(x, y) = 2e^{xy} + 4xye^{xy} + x^2 y^2 e^{xy}$

57. $f_x(x, y) = 3.38x^{-0.35} y^{0.4}$

$f_{xx}(x, y) = -1.18x^{-1.35} y^{0.4}$

$f_{xy}(x, y) = 1.352x^{-0.35} y^{-0.6}$

$f_y(x, y) = 2.08x^{0.65} y^{-0.6}$

$f_{yy}(x, y) = -1.248x^{0.65} y^{-1.6}$

59. $f_x(x, y) = -\dfrac{2y}{x^2}$

$f_{xx}(x, y) = \dfrac{4y}{x^3}$

$f_{xy}(x, y) = -\dfrac{2}{x^2}$

$f_y(x, y) = \dfrac{2}{x}$

$f_{yy}(x, y) = 0$

61. $f_x(x, y) = y^2 e^{xy}$

$f_{xx}(x, y) = y^3 e^{xy}$

$f_{xy}(x, y) = 2ye^{xy} + xy^2 e^{xy}$

$f_y(x, y) = e^{xy} + xye^{xy}$

$f_{yy}(x, y) = 2xe^{xy} + x^2 ye^{xy}$

63. **(a)** $S_x(x, y) = 4x; S_y(x, y) = 1$

(b) $S_x(3, 5) = 12$; If \$3,000 is spent on newspaper advertising and \$5,000 is spent on radio advertising each week, and the amount spent on radio advertising is kept fixed at \$5,000 per week, then sales will be increasing at a rate of \$120,000 (sales per week)/(thousands of dollars spent on newspaper advertising per week.)

$s_y(3, 5) = 1$; If \$3,000 is spent on newspaper advertising and \$5,000 is spent on radio advertising each week, and the amount spent on newspaper advertising is kept fixed at \$3,000 per week, then sales will be increasing at a rate of \$10,000 (sales per week)/(thousands of dollars spent on newspaper advertising per week.)

65. **(a)** $TS_x(x, y) = 3x; TS_y(x, y) = 6.4y$

(b) $TS_x(1, 0.5) = 3$; If \$1,000 is spent on radio advertising and \$500 is spent on television advertising each week, and the amount spent on television advertising is kept fixed at \$500 per week, then sales will be increasing at a rate of \$30,000 (sales per week)/(thousands of dollars spent on radio advertising per week.)

$TS_y(1, 0.5) = 3.2$; If \$1,000 is spent on radio advertising and \$500 is spent on television advertising each week, and the amount spent on television advertising is kept fixed at \$1,000 per week, then sales will be increasing at a rate of \$32,000 (sales per week)/(thousands of dollars spent on television advertising per week.)

67. $f_P(P, t) = e^{0.06t}$; The rate of change of the total amount accumulated when the point in time is held fixed and the amount of the initial investment changes.

$f_t(P, t) = 0.06Pe^{0.06t}$; The rate of change of the total amount accumulated when the amount of the initial investment is held fixed and time changes.

69. **(a)** $V_r(r, h) = \dfrac{2}{3}\pi hr; V_h(r, h) = \dfrac{1}{3}\pi r^2$

(b) $V(3, 5) = 15\pi$; The volume of a cone with a radius of 3 and a height of 5 is 15π.

$V_r(3, 5) = 10\pi$; For a cone with a radius of 3 units and a height of 5 units, if the radius increases by 1 unit while the height remains constant at 5 units, the volume will increase by approximately 10π cubic units.

$V_h(3, 5) = 3\pi$; For a cone with a radius of 3 units and a height of 5 units, if the height increases by 1 unit while the radius remains constant at 3 units, the volume will increase by approximately 3π cubic units.

71. $C_W(6, 8.2) \approx 12.2$; For a head with a width of 6 inches and a length of 8.2 inches, if the width of the head increases by 1 inch while the length remains constant at 8.2 inches, the cephalic index will increase by approximately 12.2 units.

$C_L(6, 8.2) \approx -8.92$; For a head with width 6 inches and a length of 8.2 inches, if the length increases by 1 inch while the width remains constant at 6 inches, the cephalic index will decrease by about 8.92 units.

73. (a) $R(x, y) = x(350 - 4x + y) + y(450 + 2x - 3y)$
$= -4x^2 + 350x + 3xy + 450y - 3y^2$

(b) $R_x(x, y) = -8x + 350 + 3y$; The marginal revenue from an increase in sales of racing bicycles.

$R_y(x, y) = -6y + 450 + 3x$; The marginal revenue from an increase in sales on mountain bicycles.

(c) $R(15, 20) = 13,050$; With a weekly demand of 15 racing bicycles and 20 mountain bicycles, the revenue is $13,050.

$R_x(15, 20) = 290$; With a weekly demand of 15 racing bicycles and 20 mountain bicycles, the marginal revenue from an increase in sales of racing bicycles, when the demand for mountain bicycles is held fixed at 20 is 290 dollars per bike.

75. (a) $R_x(x, y) = 70 + 0.5y - 0.08x$

(b) $R_y(x, y) = 95 + 0.5x - 0.08y$

77. (a) $P(x, y) = -0.5x^2 + 37x - 0.8xy + 35.5y - y^2$

(b) $P_x(x, y) = 37 - x - 0.8y$; The marginal profit function from an increase in sales of two-stroke engines, when the sales of four-stroke engines is held fixed.

$P_y(x, y) = 35.5 - 0.8x - 2y$; The marginal profit function from an increase in sales of four-stroke engines, when sales of two-stroke engines is held fixed.

(c) $P(18, 10) = 615$; The profit from sales of 18 thousand two-stroke engines and 10 thousand four-stroke engines is $615,000.

$P_x(18, 10) = 11$; With sales of 18 thousand two-stroke engines and 10 thousand four-stroke engines, the marginal profit from an increase in sales of two-stroke engines, when sales of four-stroke engines is held fixed at 10 thousand is 11 (thousand dollars)/(thousand of engines).

(d) $P_y(18, 10) = 1.1$; With sales of 18 thousand two-stroke engines and 10 thousand four-stroke engines, the marginal profit from an increase in sales of four-stroke engines, when sale of two-stroke engines is held fixed at 18 thousand is 1.1 (thousand dollars)/(thousand engines).

79. (a) $f_x(x, y) = 128.8x^{-0.2}y^{0.25}$; $f_y(x, y) = 40.25x^{0.8}y^{-0.75}$

(b) $f_x(111, 25) = 112.29$; If the company is now using 111 units of labor and 25 units of capital and keeps capital fixed at 25 units, production is increasing at a rate of 112.29 (calculators)/(thousand of hours of labor).

$f_y(111, 25) = 155.80$; If the company is now using 111 units of labor and 25 units of capital and keeps labor fixed at 111 units, production is increasing at a rate of 155.80 (calculators)/(capital equipment in millions).

81. (a) $f_x(x, y) = 860.598x^{-0.4}y^{0.45}$; $f_y(x, y) = 645.4485x^{0.6}y^{-0.55}$

(b) $f_x(47, 8) = 470.27$; If the company is now using 47 units of labor and 8 units of capital and keeps capital fixed at 8 units, production is increasing at a rate of 470.27 (engines)/(thousand of hours of labor).

$f_y(47, 8) = 2072.1$; If the company is now using 47 units of labor and 8 units of capital and keeps labor fixed at 47 units, production is increasing at a rate of 2072.1 (engines)/(capital equipment in millions).

83. (a) $f_x(x, y) = 33x^{-0.45}y^{0.5}$; $f_y(x, y) = 30x^{0.55}y^{-0.5}$

(b) $f_x(220, 140) = 34.474$; If the company is now using 220 units of labor and 140 units of capital and keeps capital fixed at 140 units, production is increasing at a rate of 34.474 (motorcycles)/(labor in millions of dollars).

$f_y(220, 140) = 49.248$; If the company is now using 220 units of labor and 140 units of capital and keeps labor fixed at 220 units, production is increasing at a rate of 49.248 (motorcycles)/(capital in millions of dollars).

(c) Production would increase more with increased spending on capital since the change in productivity with respect to capital is greater than the change in productivity with respect to labor (as shown in part b).

85. We are finding the slope of the tangent line to a cross section of $f(x, y)$ that is created by cutting the surface vertically at a given x-value and parallel to the y-axis.

87. $x = 4; y = -1$ **89.** $f(x, y) = 2x^2 + 3xy - 4y^2$

Section 8.4

1. Minimum   **3.** Second-ordered   **5.** Critical

7. $(2, -3)$ **9.** $(-1, 2)$ **11.** $(2, -1)$ **13.** $(1, 2)$

15. $(1, -4)$ **17.** $(3, 2, -10)$ is a relative minimum.

19. $(3, -2, -11)$ is a relative minimum.

21. $(-2, -\frac{2}{3}, \frac{13}{3})$ is a relative maximum.

23. $(-2, 4, -11)$ is a relative minimum.

25. $(3, 3, -7)$ is a relative minimum.

27. $(0, 1, 0)$ is a relative maximum. $(2, 1, -\frac{16}{3})$ is a saddle point.

29. $(1, -1, 20)$ is a saddle point. $(1, 3, -44)$ is a relative minimum.

31. $(0, 0, 0)$ is a saddle point. $(1, 1, -1)$ is a relative minimum.

33. $(0, 0, 1)$ is a relative maximum.

35. $(-2, \frac{1}{2}, -18)$ is a relative maximum.

37. (a) $3 million on labor; $2 million on robotics equipment

(b) $30 million

39. (a) 1500 regular size woks; 100 jumbo woks

(b) $14,999,800

41. (a) $P(x, y) = -x^2 - x + xy + 11y - 2y^2 - 2$

(b) 1 million bags of sour cream and chives; 3 million bags of barbecue

(c) $14 million

43. **(a)** $R(x, y) = -4x^2 + 349x + 3xy + 466y - 3y^2$

    (b) 88 touring bicycles; 118 mountain bicycles

    (c) About $41,744

45. **(a)** $R(x, y) = 228x - 8x^2 - 2xy + 31y - 0.15y^2$

    (b) 8 leaf blowers; 50 attachments

49. D must be positive.

51. $(3.5, 2)$

Section 8.5

1. Lagrange   **3.** Subject   **5.** Marginal productivity

7. $\dfrac{3}{4}$ **9.** 18   **11.** -16 **13.** $\dfrac{128}{9}$

15. 50   **17.** About 108.86   **19.** 5

21. 4   **23.** e^4

25. **(a)** 120 regular models and 60 joggers' models each day.

    (b) $2,160

27. **(a)** 80 pepper spray devices and 40 siren devices each day

    (b) $960

29. **(a)** 39 units by the old line; 65 units by the new line

    (b) $4,982

31. **(a)** About 857.1 labor units; 102.9 capital units

    (b) About 8268 golf carts per week

33. **(a)** About 60.7 labor units; 182.1 capital units

    (b) For each additional dollar value available to the budget, production increases by about 0.000075 units

    (c) About 15.1 units

35. **(a)** 245.3 labor units; 206.3 capital units

    (b) For each additional dollar available to the budget, production increases by about 0.002358 units.

    (c) About 118 units

37. **(a)** About 360 units of labor; 160 units of capital

    (b) For each additional dollar value available to the budget, production increases by about 0.4338 units.

    (c) 4338 units

39. 100 in $\times$ 50 in $\times$ 50 in

41. A constraint equation is a limitation placed on the problem used to help us find a maximum or minimum under this given limitation.

43. 156.25 ft $\times$ 312.5 ft

45. Looking at the constraint equation, if one variable could be easily written in terms of the other, then the Method of Lagrange is not needed.

Section 8.6

1. Partial   **3.** Double

5. Average   **7.** $26x^2$

9. $26y^2$ **11.** $4x + 6$ **13.** $4 + 6y$

15. $\dfrac{7x^3}{3} - 3x$ **17.** $\dfrac{15y^2}{4} - 3y$

19. **(a)**

    (b) $\displaystyle\int_0^2 \int_0^3 2xy\,dy\,dx = 18$ **(c)** $\displaystyle\int_0^3 \int_0^2 2xy\,dx\,dy = 18$

21. **(a)**

    (b) $\displaystyle\int_0^1 \int_0^2 (3x + y)\,dy\,dx = 5$ **(c)** $\displaystyle\int_0^2 \int_0^1 (3xy + y)\,dx\,dy = 5$

23. **(a)**

    (b) $\displaystyle\int_1^{16} \int_1^4 \sqrt{xy}\,dy\,dx = 196$ **(c)** $\displaystyle\int_1^4 \int_1^{16} \sqrt{xy}\,dx\,dy = 196$

25. (a)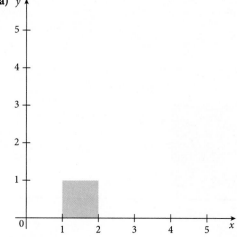

(b) $\int_1^2\int_0^1 x^2 y^2\,dy\,dx = \dfrac{7}{9}$ **(c)** $\int_0^1\int_1^2 x^2 y^2\,dx\,dy = \dfrac{7}{9}$

27. (a)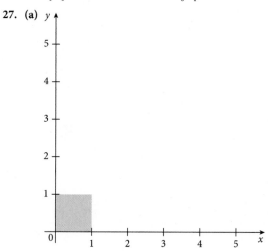

(b) $\int_0^1\int_0^1\left(2 - \dfrac{1}{2}x^2 + y^2\right)dy\,dx = \dfrac{13}{6}$

(c) $\int_0^1\int_0^1\left(2 - \dfrac{1}{2}x^2 + y^2\right)dx\,dy = \dfrac{13}{6}$

29. 1

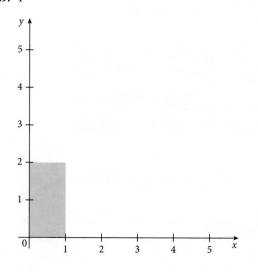

31. 16

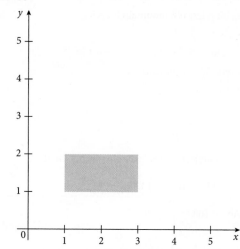

33. $\dfrac{8e}{3} - \dfrac{8}{3e}$

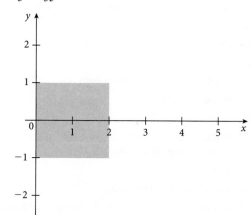

35. $\dfrac{2}{3}$ **37.** $\dfrac{25}{3}$ **39.** 2   **41.** $\dfrac{1}{2}\left(\dfrac{8e}{3} - \dfrac{8}{3}\right)$

43. $\dfrac{44}{3}$ un^3 **45.** 6 un^3 **47.** 3 un^3 **49.** $\dfrac{8e - 8}{3}$

51. About 13,919 solar cells   **53.** About 11,370 calculators

55. 36   **57.** About 5957.6 people per square mile

59. About 148,989 people   **61.** About 108.33 parts per million

63. About 126

65. By integrating from a to b, we would find the area under the curve in one direction. Then by integrating from c to d, we move perpendicularly along the surface $f(x, y)$ while we stretch out the 2D region into a solid.

69. $k(b - a)(c - d)$

Chapter 8 Review Exercises

Section 8.1 Review Exercises

1. $f(3, 5) = \dfrac{-6}{11}$

3. $f(10, 3) = 103$

5. $domf = \{(x, y)|x + y \geq 0\}$

7.

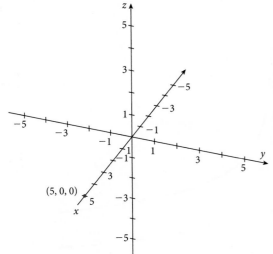

$(5, 0, 0)$

9.

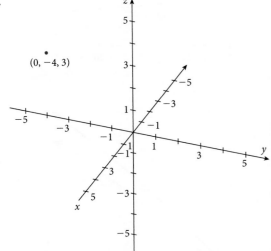

$(0, -4, 3)$

11. $(-3, -4, 2)$ is above the xy-plane; the z variable is positive.

13. $(6, -2, -1)$ is below the xy-plane; the z variable is negative.

15.

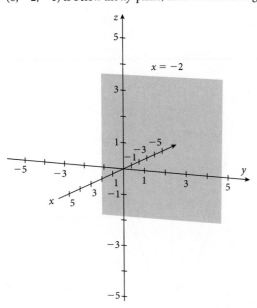

$x = -2$

17. $P(4000, 3000) = 800$; With sales of 4000 cups of coffee and 3000 cups of cappuccino per month, the monthly profits are \$800.

19. (a) $C(x, y) = 7x + 13y + 5000$

(b) $R(x, y) = 12x + 21y$

(c) $P(x, y) = 5x + 8y - 5000$

21. $f(4500, 0.045) \approx 7057.40$; If \$4,000 is invested at an annual rate of 4.5%, compounded continuously for 10 years, the total amount accumulated is approximately \$7057.40.

Section 8.2 Review Exercises

23. $y = 2x + 3 - c$

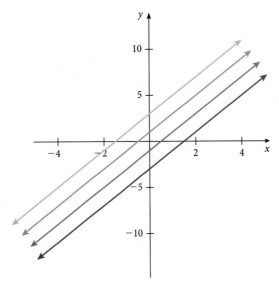

25. $y = \pm\sqrt{16 - x^2 - c}$

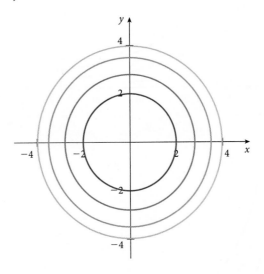

27. $y = \left(\dfrac{c}{10}\right)^2 x^{-1}$

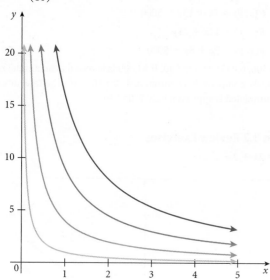

29. $y = \left(\dfrac{c}{8.9}\right)^{50/21} x^{-31/21}$

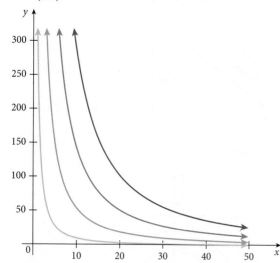

31. $y = c^{2/3} x^{-1/2}$

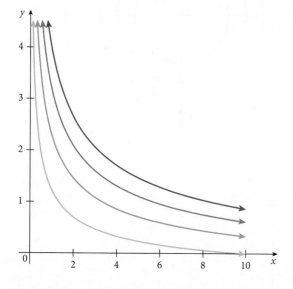

33. $y = c^{20/17} x^{-3/17}$

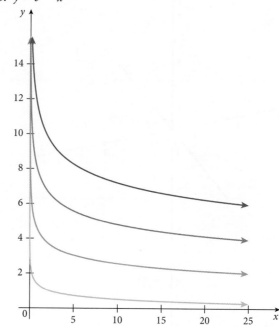

35.

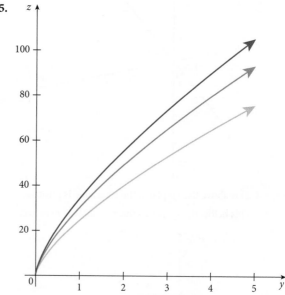

37.

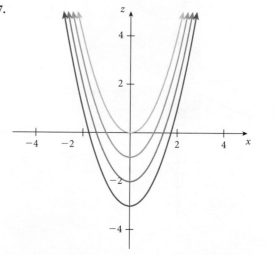

39.

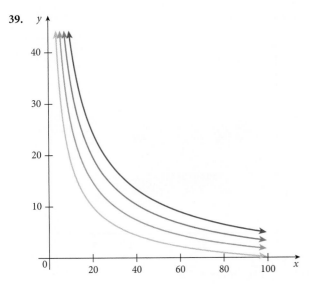

41. (a) Vertical

(b)

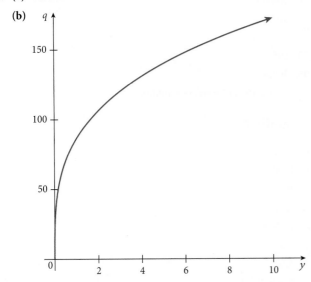

(c) About $6.2 million

Section 8.3 Review Exercises

43. $f_x(x, y) = 5$

$f_y(x, y) = -6y$

45. $f_x(x, y) = 5(8y^2 - 3x)^4(-3) = -15(8y^2 - 3x)^4$

$f_y(x, y) = 5(8y^2 - 3x)^4(16y) = 80y(8y^2 - 3x)^4$

47. $f_x(x, y) = \dfrac{3x - 3(x + y)}{(3x)^2} = \dfrac{-y}{3x^2}$

$f_y(x, y) = \dfrac{3x(1) - (x + y) \cdot 0}{(3x)^2} = \dfrac{1}{3x}$

49. $f_x(x, y) = e^{x-y} + 15x^2y$

$f_y(x, y) = -e^{x-y} + 5x^3$

51. $f_y(x, y) = 12xy^3 - 10y^4; f_y(3, 5) = -1750$; If we are at the point $(3, 5)$ and move along the surface in the y-direction, parallel to the y-axis, the function is decreasing at a rate of 1750 $\frac{\text{units of } f}{\text{units of } y}$.

53. $f_y(x, y) = 7.967x^{0.38}y^{-0.38}; f_y(3, 2) \approx 9.29$; If we are at the point $(-2, 5)$ and move along the surface in the y-direction, parallel to the y-axis, the function is increasing at a rate of approximately $9.29 \frac{\text{units of } f}{\text{units of } x}$.

55. $f_x(x, y) = -12x^2 + 10y + y^5$

$f_y(x, y) = 10x + 5xy^4$

$f_{xx}(x, y) = -24x$

$f_{xy}(x, y) = 10 + 5y^4$

$f_{yy}(x, y) = 20xy^3$

57. $f_x(x, y) = 2(x + 2y) \cdot 1 = 2x + 4y$

$f_y(x, y) = 2(x + 2y) \cdot 2 = 4x + 8y$

$f_{xx}(x, y) = 2$

$f_{xy}(x, y) = 4$

$f_{yy}(x, y) = 8$

59. $f_x(x, y) = 12x - 4y + 1$

$f_y(x, y) = -4x + 10y - 3$

$12x - 4y + 1 = 0 \Rightarrow y = 3x + \dfrac{1}{4}$

$-4x + 10y - 3 = 0$

$-4x - 4\left(3x + \dfrac{1}{4}\right) + 1 = 0 \Rightarrow x = \dfrac{1}{52}$

$y = 3\left(\dfrac{1}{52}\right) + \dfrac{1}{4} = \dfrac{4}{13}$

61. (a) $R(x, y) = -0.6x^2 + 480x + 0.5xy + 720y - 0.7y^2$

(b) $R_x(x, y) = -1.2x + 480 + 0.5y$

Marginal revenue; this gives us the amount that each additional 15-inch monitor sold will add to the total revenue.

$R_y(x, y) = 0.5x + 720 - 1.4y$

Marginal revenue; this gives us the amount that each additional 17-inch monitor sold will add to the total revenue.

(c) $R_x(640, 850) = 137$

If 640 15-inch monitors and 850 17-inch monitors are sold weekly, the company receives an additional $137 for the 641st 15-inch monitor sold.

$R_y(640, 850) = -150$

If 640 15-inch monitors and 850 17-inch monitors are sold weekly, the company loses $150 profit for the 851st 17-inch monitor sold.

63. (a) $P_x(x, y) = -3.5x^2 + 110x - 18.5xy + 130y + 1.2y^2 - 600$

(b) $P_x(x, y) = -7x + 110 - 18.5y$

This function gives us the marginal profit from an increase in sales of analog watches in hundreds of dollars per hundred watches.

$P_y(x, y) = -18.5x + 2.4y + 130$

This function gives us the marginal profit from an increase in sales of digital watches in hundreds of dollars per hundred watches.

(c) $P_x(8, 11) = -149.5$

With sales of 800 analog watches and 1100 digital watches, the marginal profit from an increase in sales of analog watches is $-14,950$ dollars per hundred watches.

$P_y(8, 11) = 8.4$

With sales of 800 analog watches and 1100 digital watches, the marginal profit from an increase in sales of digital watches is 840 dollars per hundred of watches.

(d) $P(8, 11) = 3.2$

The profit from sales of 800 analog watches and 1100 digital watches is $320.

Section 8.4 Review Exercises

65. $(-4, 3)$ **67.** $\left(-\dfrac{4}{3}, \dfrac{2}{3}\right)$

69. $(1, -4, -24)$ is a relative minimum.

71. $(-1, -1, 0)$ is a saddle point.

73. $(0, 0, 0)$ is a saddle point and $(2, -4, 8)$ is a relative minimum.

75. $(0, 2, e^4)$ is a saddle point.

77. **(a)** $R(x, y) = -0.3x^2 + 12x + 0.2xy + 20y - 0.2y^2$

(b) 44 basic haircuts; 72 deluxe haircuts daily

(c) $984

79. **(a)** 280,000 fine-tip pens; 440,000 large-tip pens

(b) $375,600

Section 8.5 Review Exercises

81. 125   **83.** -5

85. About 22.888

87. **(a)** About 4.8 units of labor; 2 units of capital

(b) About 94,692 telephones

89. 10 ft × 30 ft

Section 8.6 Review Exercises

91. $258y^3$ **93.** $16x^2 - 1280$

95. **(a)** y

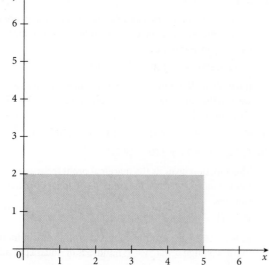

(b) $\displaystyle\int_0^5 \int_0^2 (3x - 4xy)\,dy\,dx = -25$

(c) $\displaystyle\int_0^2 \int_0^5 (3x - 4xy)\,dx\,dy = -25$

97. **(a)** y

(b) $\displaystyle\int_0^4 \int_1^3 (3xy^2)\,dy\,dx = 208$

99. 200 un^3

101. $\dfrac{20}{3}$

103. 195

105. 392.8

107. About 104.667 parts per million

Appendix A

1. $-4x^{-6}$

3. $1.6x^{1/2}$

5. $8x^{-2/3}$

7. $\sqrt[7]{x^3}$

9. $6.3\sqrt[5]{x^4}$

11. $\dfrac{2}{\sqrt[3]{x^2}}$

13. $x = 4; y = 6$

15. $x = 5; y = \dfrac{3}{2}$

17. $x = \dfrac{3}{4}; y = \dfrac{1}{2}; z = \dfrac{1}{3}$

19. $x = -2, y = 5$

21. $\log_b 8 + 3\log_b x - \log_b 125 - 3\log_b y$

23. $\log x + 2 \log y - \log (y + 4)$

25. $\dfrac{1}{2}\ln 5 + \dfrac{3}{2}\ln x - \dfrac{9}{2}\ln y$

27. $x = \dfrac{26}{3}$

29. $x = 4$

31. $x = \dfrac{\ln 6}{\ln 3} \approx 1.631$

33. $x = \dfrac{1}{5}$

Index

Applications Index